Viral Vectors:

Basic Science and Gene Therapy

Other BioTechniques® Books Titles

Gene Cloning and Analysis by RT-PCR
P.D. Siebert and J.W. Larrick (Eds.)

Apoptosis Detection and Assay Methods
L. Zhu and J. Chun (Eds.)

Protein Staining and Identification Techniques
R.C. Allen and B. Budowle

Immunological Reagents and Solutions: A Laboratory Handbook
B.B. Damaj

Affinity and Immunoaffinity Purification Techniques
T.M. Phillips and B.F. Dickens

Yeast Hybrid Technologies
L. Zhu and G. Hannon (Eds.)

Ribozyme Biochemistry and Biotechnology
G. Krupp and R. Gaur (Eds.)

Antigen Retrieval Techniques: Immunohistochemistry and Molecular Morphology
S.-R. Shi, J. Gu, and C.R. Taylor (Eds.)

Bioinformatics: A Biologist's Guide to Biocomputing and the Internet
S. Brown

Gene Transfer Methods: Introducing DNA into Living Cells and Organisms
P.A. Norton and L.F. Steel (Eds.)

SNP and Microsatellite Genotyping: Markers for Genetic Analysis
A. Hajeer, J. Worthington, and S. John (Eds.)

Viral Vectors:

Basic Science and Gene Therapy

Edited by

Angel Cid-Arregui
Institute of Applied Tumor Virology
German Cancer Research Center
Heidelberg, Germany

Alejandro García-Carrancá
Institute of Biomedical Research
National Autonomous University of Mexico
Mexico City, Mexico

A BioTechniques® Books Publication
Eaton Publishing

Angel Cid-Arregui, MD, PhD
Institute of Applied Tumor Virology
German Cancer Research Center
Im Neuenheimer Feld 242
69120-Heidelberg, Germany

Alejandro García-Carrancá, PhD
Institute of Biomedical Research
National Autonomous University of Mexico
04510 Mexico City, Mexico

Library of Congress Cataloging-in-Publication Data

Viral vectors : basic science and gene therapy / edited by Angel Cid-Arregui, Alejandro Manual García-Carrancá.
p. cm.
Includes bibliographical references and index.
ISBN 1-881299-35-X (hardcover)
1. Genetic vectors. 2. Viruses. 3. Gene therapy. 4. Molecular virology. I. Cid-Arregui, Angel, 1959- II. García-Carrancá, Alejandro, 1955-

QH442.2 .V573 2000
579.2--dc21

00-059287

ISBN 1-881299-35-X

Printed in the United States of America

9 8 7 6 5 4 3 2 1

Eaton Publishing/BioTechniques Books
154 E. Central Street
Natick, MA 01760
www.BioTechniques.com

Francis W. Eaton: *Publisher and President*
Stephen Weaver: *Director and Editor-in-Chief*
Christine McAndrews: *Managing Editor*
Sandy Lamont: *Production Manager*
Ken Strom: *Cover Designer*

Cover graphic depicts double immunofluorescence image of Vero cells infected (bottom) and noninfected (center) with HSV-1 amplicons expressing mouse Ephrin B1 (green). Both cells entered S-phase 24 h after addition of the virus, as shown by BrdU incorporation into the nucleus (red fluorescence, which appears yellow in the Ephrin B1 expressing cell). At the top, schematic representation of the HSV-1 virion (A. Cid-Arregui, unpublished observation).

Contributors

Manju Agarwal
SyStemix, Inc.
Palo Alto, CA, USA

Andrea Amalfitano
Department of Pediatrics
Division of Medical Genetics
Duke University Medical Center
Durham, NC, USA
(amalf001@mc.duke.edu)

Collette apRhys
Departments of Pharmacology and Molecular Sciences
The Johns Hopkins University School of Medicine
Baltimore, MD, USA

Tohru Arai
Department of Gene Regulation
Institute of Medical Science
University of Tokyo, Tokyo
and Tsukuba Research Laboratories
Eisai Co., Ltd.
Tsukuba, Japan

Ginette Balboni
Institute of Applied Tumor Virology
German Cancer Research Center
Heidelberg, Germany

Jordi Barquinero
Institute of Oncologic Research
Department of Criobiology and Cellular Therapy
Barcelona, Spain

Alison L. Bawden
Department of Molecular Genetics and Microbiology
University of Florida College of Medicine
Gainesville, FL, USA

Antonio Bernad
CNB-CIEMAT Asssociated Unit and
Department of Immunology and Oncology
National Biotechnology Center
Autonomous University
Madrid, Spain
(abernad@cnb.uam.es)

Frederick M. Boyce
Massachusetts General Hospital East
Charlestown, MA, USA
(boyce@helix.mgh.harvard.edu)

Juan A. Bueren
CNB-CIEMAT Asssociated Unit and
Unit of Molecular and Cellular Biology
CIEMAT
Madrid, Spain

Keith Bupp
Department of Biochemistry
Robert Wood Johnson Medical School
University of Medicine and Dentistry of New Jersey
Piscataway, NJ, USA

Sherry Bursztajn
Biomedical Research Institute
Louisiana State University Medical Center
Shreveport, LA, USA

Barry J. Byrne
Department of Molecular Genetics and Microbiology
Gene Therapy Center
and Department of Pediatrics
University of Florida
Gainesville, FL, USA
(byrne@college.med.ufl.edu)

Ling Chen
Merck Research Laboratories
Department of Human Genetics
Gene Therapy
West Point, PA, USA

Kye Chesnut
Department of Molecular Genetics and Microbiology
Gene Therapy Center
University of Florida
Gainesville, FL, USA

Marielle Christ
Phoenix International
L'Arbresle, France

Angel Cid-Arregui
Institute of Applied Tumor Virology
German Cancer Research Center
Heidelberg, Germany
(cid@dkfz-heidelberg.de)

Cesar Cobaleda
Department of Cellular Differentiation and Proliferation
Institute of Biochemical Microbiology
CSIC–University of Salamanca
Salamanca, Spain

James E. Conway
Departments of Pharmacology and Molecular Sciences
The Johns Hopkins University School of Medicine
Baltimore, MD, USA

Jan J. Cornelis
Institute of Applied Tumor Virology
German Cancer Research Center
Heidelberg, Germany

Alexandra Y. Dege
Institute of Applied Tumor Virology
German Cancer Research Center
Heidelberg, Germany

Christiane Dinsart
Institute of Applied Tumor Virology
German Cancer Research Center
Heidelberg, Germany

S. Monroe Duboise
Department of Applied Medical Sciences
University of Southern Maine
Portland, ME, USA

Alberto L. Epstein
Center for Molecular and Cellular Genetics
UMR 5534-CNRS
University Claude Bernard Lyon I
Villeurbanne, France
(epstein@biomserv.univ-lyon1.fr)

Cornel Fraefel
Institute of Virology
University of Zurich
Zurich, Switzerland
(cornelf@rzu-mailhost.unizh.ch)

Elizabeth A. Franco
Massachusetts General Hospital East
Charlestown, MA, USA

Alejandro García-Carrancá
Department of Molecular Biology
Institute of Biomedical Research and
National Institute of Cancerology
Mexico City, Mexico

Maria José García-Iglesias
Department of Animal Pathology
Faculty of Veterinary Science
University of León
León, Spain

Adolfo García-Sastre
Department of Microbiology
Mount Sinai School of Medicine
New York, NY, USA
(agarcia@smtplink.mssm.edu)

Claude P. Genain
Department of Neurology
San Francisco, CA, USA

Kathryn J. Glassberg
Department of Molecular Genetics and Microbiology
University of Florida College of Medicine
Gainesville, FL, USA

Linda Grave
Transgène S.A.
Strasbourg, France

Dirk Grimm
Institute of Applied Tumor Virology
German Center Research Center
Heidelberg, Germany

Guillermo Güenechea
CNB-CIEMAT Asssociated Unit and
Unit of Molecular and Cellular Biology
CIEMAT
Madrid, Spain

Andreas Haag
Institute of Applied Tumor Virology
German Cancer Research Center
Heidelberg, Germany

Wulf Haubensak
Max Planck Institute of Molecular Cell Biology and Genetics
Dresden, Germany

Martin Heinkelein
Institute of Virology and Immunobiology
University of Würzburg
Würzburg, Germany

Claire L. Hill
Department of G.U. Medicine and Communicable Diseases
Jefferiss Research Laboratories
Imperial College School of Medicine at St. Mary's
London, England, UK

Wieland B. Huttner
Max Planck Institute of Molecular Cell Biology and Genetics
Dresden, Germany
and Department of Neurobiology
University of Heidelberg
Heidelberg, Germany

Hideo Iba
Department of Gene Regulation
Institute of Medical Science
University of Tokyo
Tokyo, Japan
(iba@ims.u-tokyo.ac.jp)

Michael J. Imperiale
Department of Microbiology and Immunology
University of Michigan Medical School
Comprehensive Cancer Center and
Center for Gene Therapy
Ann Arbor, MI, USA
(imperial@umich.edu)

Kuan-Teh Jeang
Molecular Virology Section
National Institute of Allergy and Infectious Diseases
National Institutes of Health
Bethesda, MD,USA
(kjeang@atlas.niaid.gov)

Victoria Juárez
Laboratory of Animal Resources
European Molecular Biology Laboratory
Heidelberg, Germany

Jae U. Jung
Department of Microbiology and Molecular Genetics
New England Regional Primate Research Center
Harvard Medical School
Southborough, MA, USA
(jjung@warren.med.harvard.edu)

Jürgen A. Kleinschmidt
Institute of Applied Tumor Virology
German Cancer Research Center
Heidelberg, Germany
(j.kleinschmidt@dkfz-heidelberg.de)

Jay K. Kolls
Gene Therapy Program and
Departments of Medicine and Pediatrics
Louisiana State University Medical Center of New Orleans
New Orleans, LA, USA

Christopher Kornfeld
Department of Virology
Institute Pasteur
Paris, France

Pedro A. Lazo
Cancer Research Center
CSIC–University of Salamanca and
Carlos III Institute of Health
Majadahonda, Madrid, Spain
(plazozbi@isciii.es)

Yi Li
Department of Molecular Genetics and Microbiology
University of Florida College of Medicine
Gainesville, FL, USA

Filip Lim
Molecular Biology Center
CSIC
Autonomous University
Madrid, Spain

Dirk Lindemann
Institute of Virology and Immunology
University of Würzburg
Würzburg, Germany
(lindemann@mail.uni-wuerzburg.de)

Monika Lusky
Transgène S.A.
Strasbourg, France

William C. Manning
Chiron Corporation
Emeryville, CA, USA
(william.manning@cc.chiron.com)

Osam Mazda
Department of Microbiology
Kyoto Prefectural University of Medicine
Kamikyo, Kyoto, Japan
(mazda@basic.kpu-m.ac.jp)

Myra O. McClure
Department of G.U. Medicine and Communicable Diseases
Jeffriss Research Laboratories
Imperiale College School of Medicine at St. Mary's
London, England, UK
(mom32@mailhost.sm.med.ic.ac.uk)

Donna L. McPhie
Department of Psychiatry and Program in Neuroscience
Harvard Medical School
McLean Hospital
Belmont, MA, USA

Majid Mehtali
Introgene BV
Leiden, The Netherlands

Anne-Isabelle Michou
Transgène S.A.
Strasbourg, France

Nestor Morales-Peza
Department of Molecular Biology
Institute of Biomedical Research
UNAM
Mexico City, Mexico

Manal A. Morsy
Merck Research Laboratories
Department of Human Genetics
Gene Therapy
West Point, PA, USA
(morsy@merck.com)

Richard W. Moyer
Department of Molecular Genetics and Microbiology
University of Florida College of Medicine
Gainesville, FL, USA
(rmoyer@medmicro.med.ufl.edu)

Steffen Mueller
Department of Molecular Genetics and Microbiology
School of Medicine
State University of New York at Stony Brook
Stony Brook, NY, USA
(mueller@asterix.bio.sunysb.edu)

Nick Muzyczka
Department of Molecular Genetics and Microbiology
Gene Therapy Center
University of Florida
Gainesville, FL, USA

Bernd Neeb
Institute of Biochemistry
University of Mainz
Mainz, Germany

Rachael L. Neve
Department of Psychiatry and Program in Neuroscience
Harvard Medical School
McLean Hospital
Belmont, MA, USA
(neve@helix.mgh.harvard.edu)

Björn Oback
Reproductive Technologies
AgResearch
Ruakura Research Centre
Hamilton, New Zealand

Xavier Paliard
Chiron Corporation
Emeryville, CA, USA

Ivan Plavec
SyStemix, Inc.
Palo Alto, CA, USA
(ivan.plavec@pharma.novartis.com)

Colin Porter
Chester Beatty Laboratories
Institute of Cancer Research
London, England, UK
(colin@icr.ac.uk)

Manuel Ramírez
CNB-CIEMAT Asssociated Unit and
Unit of Molecular and Cellular Biology
CIEMAT
Madrid, Spain

Axel Rethwilm
Institute of Virology
Carl Gustav Carus Faculty of Medicine
Technical University of Dresden
Dresden, Germany

Jean Rommelaere
Institute of Applied Tumor Virology
German Cancer Research Center
Heidelberg, Germany
(e.steinmann@dkfz-heidelberg.de)

Ronald Rooke
Transgène S.A.
Strasbourg, France

Monica Roth
Department of Biochemistry
Robert Wood Johnson Medical School
University of Medicine and Dentistry of New Jersey
Piscataway, NJ, USA
(roth@waksman.rutgers.edu)

Yoshinaga Saeki
Department of Neurosurgery
Massachusetts General Hospital
Harvard Medical School
Charlestown, MA, USA and
Division of Gene Therapy Science
Osaka University Medical School
Osaka, Japan

Isidro Sánchez-García
Department of Cellular Differentiation and Proliferation
Institute of Biochemical Microbiology
CSIC–Universidad de Salamanca
Salamanca, Spain
(isg@gugu.usal.es)

Paul Schwarzenberger
Gene Therapy Program and
Department of Medicine
Louisiana State University Medical Center of New Orleans
New Orleans, LA, USA
(pschwa1@lsumc.edu)

José C. Segovia
CNB-CIEMAT Asssociated Unit and
Unit of Molecular and Cellular Biology
CIEMAT
Madrid, Spain

Stephen M. Smith
St. Michael's Medical Center
and The New Jersey Medical School–UMDNJ
Newark, NJ, USA

Arun Srivastava
Departments of Microbiology and Immunology, and Medicine
Walther Oncology Center
Indiana University School of Medicine
Walther Cancer Institute
Indianapolis, IN, USA
(asrivast@iupui.edu)

Tom A. Stavropoulos
Gene Therapy and Molecular Virology Group
The John P. Robarts Research Institute
and Department of Microbiology and Immunology
University of Western Ontario
London, ON, Canada

Craig A. Strathdee
Gene Therapy and Molecular Virology Group
The John P. Robarts Research Institute
and Department of Microbiology and Immunology
University of Western Ontario
London, ON, Canada
(cas@rri.on.ca)

Richard E. Sutton
Departments of Molecular Virology and Microbiology, and Medicine
and the Center for Cell and Gene Therapy
Baylor College of Medicine
Houston, TX, USA
(rsutton@bcm.tmc.edu)

Ernest F. Terwilliger
Division of Experimental Medicine
Harvard Institutes of Medicine
Boston, MA, USA
(eterwiligr@aol.com)

Motoyasu Ui
Department of Gene Regulation
Institute of Medical Science
University of Tokyo
Tokyo, Japan

Florencio Varas
CNB-CIEMAT Asssociated Unit and
Unit of Molecular and Cellular Biology
CIEMAT
Madrid, Spain

Eckard Wimmer
Department of Molecular Genetics and Microbiology
School of Medicine
State University of New York at Stony Brook
Stony Brook, NY, USA

Mónica Yunta
Cancer Research Center
CSIC–University of Salamanca and
Carlos III Institute of Health
Majadahonda, Madrid, Spain

Cécile Zaupa
Center for Molecular and Cellular Genetics
UMR 5534-CNRS
University Claude Bernard Lyon I
Villeurbanne, France

Irene Zolotukhin
Department of Molecular Genetics and Microbiology
Gene Therapy Center
and Department of Pediatrics
University of Florida
Gainesville, FL, USA

Sergei Zolotukhin
Department of Molecular Genetics and Microbiology
Gene Therapy Center
University of Florida
Gainesville, FL, USA

Foreword

From Foe to Friend

Virology celebrated its first century a decade ago. It was a century of discovery virtually unprecedented in the history of biological sciences. It took over a quarter of this century, to the time of the discovery of bacteriophages and the isolation of the first tumor virus by Payton Rous, to fully appreciate the unique features of viruses as biologic entities. It took more than half of the century until the advent of the age of antibiotics for the widespread use of cultured cells for identification and study of the biology of viruses. The end result was a rich harvest of viruses illustrative of the evolutionary power of selection of organisms that are highly divergent in size, structure, and composition, that at times occupy very narrow, unique niches in our bodies but nevertheless efficiently combat our defenses and feed on our cells. The last quarter of the century marks a turnabout, with the relentless dissection of their genomes and an analysis of the functions that permit them to target specific cells, multiply, and either drastically modify or destroy cells. The success of this endeavor is apparent from the fact that in recent years much of the emphasis has shifted from mere uncoding of viral functions to converting viral pathogens from foes to friends.

In the early eighties, methods became available to make specific insertions or deletions in the genomes of large DNA viruses such as the herpesviruses and poxviruses. Since that time, genetic engineering techniques have been extended to a variety of large and small DNA and RNA viruses. Today, members of virtually every virus family can be converted into vectors to deliver genes to prevent the multiplication of the same or other viruses, to replace cellular genes that are missing or damaged, or to destroy cancer cells. Many of these vectors have been in clinical trials to relieve genetic diseases, to kill cancer cells, or to immunize against disease, and more trials are yet to come. The beginning is encouraging and gene therapy with viral vectors remains an exciting, mesmerizing objective. There is in fact nothing better, but we are not there yet.

Viruses have a huge advantage over any other techniques to deliver genetic information or proteins to cells because this is what they are designed to do. Many viruses are target cell-specific; they normally deliver their own genomes to targets such as neurons, lymphoid cells, and the liver. The large DNA viruses in particular have considerable room for insertion of relatively large chunks of genetic material. Nevertheless many problems remain. The requirement for space to insert large amounts of genetic material is not always met by viruses with the desirable host range. Viruses seldom bother to regulate their gene expression in a fine fashion. Overproduction of a host gene product may be just as harmful as its absence. Elements that control gene expression in higher eukaryotes take up a lot of genetic space. Control of gene expression that is optimal to restore missing functions and yet not take up a huge chunk of available space even in the largest viral genomes is still in infancy. Targeting and control of expression are the objectives of the next decade. There is a safe bet, however, that in the second century of virology, viruses will emerge as the magic bullets humankind has always believed to exist. The remarkable collection of articles contained in this book is the signpost to this objective.

Bernard Roizman
The University of Chicago
May 2000

Preface

"Everything existing in the universe is the fruit of chance and of necessity."

Democritus (470 BC)

This quotation from the Greek philosopher Democritus was used by Jacques Monod in his essay on Natural Philosophy of Modern Biology (*Chance and Necessity*, Knopf, New York, 1971) and later by Christian de Duve (*Blueprint for a Cell. The Nature and Origin of Life*, Neil Patterson Publishers, Burlington, NC, 1991) to summarize the essence of the problem of how life originated. Although the nature of life is now better understood, the concepts of chance and necessity as decisive rules in the origin and evolution of cellular life and of its inevitable viral companion remain valid. However, life seems to possess an inherent drive that makes it less dominated by chance than by necessity. It could be seen, in thermodynamic terms, as governed by intrinsic rules leading toward levels of progressively higher entropy. Thus, evolution has reached a point where life has started to be engineered. Higher animals can be genetically modified and cloned. At a lower scale, viruses are being used as vectors to express recombinant genes into their target cells. A step further should result in the generation of new, chimeric viruses devoid of pathogenic capability while owning advantageous properties borrowed from different, unrelated viral and cellular functions to meet the needs of gene therapy. If so, these new beings will not be the fruit of chance, but of necessity.

Viruses were discovered, about a century ago, as elusive pathogens that parasitize and damage cells, causing disease and death. Since then, much effort has been put into fighting pathogenic viruses by developing vaccines to endow the organism with resistance to infection, sometimes for life. In some cases, as with smallpox, eradication of the virus worldwide has been achieved. Nevertheless, viruses have been and will continue to be extremely useful both as models and as tools for the study of the molecular basis of a long list of cellular processes. An overview of the great variety of mechanisms used by viruses to proliferate and spread or remain latent illustrates that they cannot merely be regarded as pathogens, but also as subcellular elements that have been actively involved in modeling cellular life during evolution. It follows that appropriate artificial manipulations of viruses should have predictable effects on specific target tissues and the whole organism.

As the molecular pathogenesis of a number of diseases has been dissected with the help of molecular biology techniques, it has become possible to design gene therapy approaches to treat them. The success of gene therapy, however, relies entirely on the development of more efficient methods for gene delivery in vivo and ex vivo. A variety of gene transfer techniques have been devised, some of which are currently being used in gene therapy studies that have reached the clinical trial phase. These studies include physical methods such as DNA-lipid complexes and formulations in which DNA is adhered to microscopic gold particles, as well as biological methods based mostly on viral vectors. Indeed, it is believed that future successes in the field of gene therapy will be closely linked to the development of new generations of viral vectors (M.B.A. Oldstone and A.J. Levine. 2000. Virology in the next millennium. Cell *100*:139-142).

Some features make viruses highly interesting as vehicles for gene transfer: *(i)* They offer a great variety of mechanisms of replication and gene expression. *(ii)* They target cells selectively by virtue of their tissue/cell-specificity. *(iii)* They are rapidly internalized by receptor-mediated endocytosis, which causes minimal perturbation in their target cells. *(iv)* Most of them can be easily purified in large amounts. Conveniently modified to eliminate adverse effects on cells and tissues and to provide room for insertion of recombinant DNA or RNA, any virus can be used as a vector to drive expression of heterologous protein. Nevertheless, no virus vector presently meets all the desired characteristics for each particular gene transfer protocol. Further work should solve problems such as side effects due to viral protein, neutralization by host immune response, latency and uncontrolled expression of recombinant protein. In addition, identification of new specific cellular receptors for viruses and a better understanding of virus-receptor interaction and internalization should make it possible to finely tune the cell-type specificity of the vectors. Meanwhile, researchers will need to discriminate among the available vectors and select those whose properties adjust better to their needs.

This volume provides the reader with comprehensive information on the molecular biology of the most relevant viral vectors developed to date, as well as detailed protocols for vector construction, purification and titration, and illustrative applications. It is intended for virologist and nonvirologist investigators and physicians involved in gene therapy studies as well as those interested in the application of virus vectors to fundamental research. Care has been taken to make the contents accessible to students and specialized technicians by supplying extensive background information and summaries together with abundant illustrations. *Viral Vectors* has been organized into two main sections, based on the nucleic acid composition of viruses. Part I deals with vectors derived from seven DNA virus genera, namely adeno-associated virus, autonomous parvovirus, adenovirus, herpesvirus, entomopoxvirus, and baculovirus. Part II focuses on vectors derived from RNA viruses, mainly retrovirus, but also influenza, poliovirus, and alphavirus vectors. Each part is preceded by an introduction that summarizes useful information on genome organization, mechanism of replication, and physical properties of the respective viruses. Two additional chapters close the contents with some considerations on the design of gene therapy studies and biosafety recommendations for handling virus vectors.

Collectively the articles presented outline the current state of the art and the major directions in which the viral vector field is expected to progress in the near future.

Angel Cid-Arregui
Alejandro García-Carrancá
June 2000

Acknowledgments

We would like to gratefully acknowledge all the authors for their enthusiasm and willingness to collaborate in this project by contributing valuable information on their field of expertise. We thank our colleagues at the Institute of Biomedical Research (UNAM, Mexico City), National Institute of Cancerology (INCAN, Mexico City), Institute of Neurobiology (University of Heidelberg), Max Plank Institute for Molecular Cell Biology and Genetics (Heidelberg-Dresden) and the German Cancer Research Center (Heidelberg) for helpful suggestions. Our special thanks to Professor B. Roizman for honoring this monograph with his Foreword. A. C.-A. is indebted to Professor H. zur Hausen for continuous encouragement, support, and valuable advice and is grateful to Professor W. B. Huttner for stimulating discussions and support. We are very grateful to S. Weaver and C. McAndrews for their patience and assistance during the compilation of this edition and their highly professional work thereafter. Our work has been supported by a joint grant (MXI 6 G0A 5A) from the German Federal Ministry for Education and Research (BMBF) and the Mexican National Council for Science and Technology (CONACYT).

Contents

Part I:
DNA Virus Vectors

Introduction to Part I: DNA Virus Vectors

Angel Cid-Arregui

This section (Chapters 1–22) focuses on the development and applications of vectors based on DNA viruses belonging to very different taxonomic families: adeno-associated virus (Chapters 1–5), autonomous parvovirus (Chapter 6), adenovirus (Chapters 7–11), herpes simplex virus (Chapters 12–18), rhadinovirus (Chapter 19), Epstein-Barr virus (Chapter 20), entomopoxvirus (Chapter 21), and baculovirus (Chapter 22). The following is a taxonomic description of the parental viruses from which those expression vectors have been derived. It is intended to be informative, but not exhaustive, on the most relevant aspects of the virions, genomes, proteins, viral replication, and physical characteristics of the respective viruses. A catalog of electron microscopy images of these viruses can be seen on the web sites recommended at the end of this section.

Family: Parvoviridae
Genus: Dependovirus
(Adeno-Associated Viruses [AAV])
Genus: Autonomous Parvoviruses

Virion: The virion particle is nonenveloped, has icosahedral symmetry (32 capsomeres in a T=3 arrangement), and is 18–26 nm in diameter. Virion Mr: 5.5–6.2 $\times 10^6$. Buoyant density in CsCl: 1.39–1.42 g/mL. Virions do not contain lipids or carbohydrates.

Physicochemical properties: Virions are sensitive to UV- and γ-irradiation, oxidizing agents, formaldehyde, hydroxylamine and β-propiolactone. They are resistant to heat, exposure to pH 3–9, detergents, and organic solvents.

Genome: The genome is a molecule of either positive or negative single-stranded DNA of about 5000 nucleotides in size. Viruses of the Parvovirus genus encapsidate mostly DNA^-, whereas Dependoviruses (AAV) encapsidate either DNA^+ or DNA^- in various proportions, which form a double strand upon extraction. Viral DNA replication is complex and proceeds via a self-priming mechanism involving base pairing of palindromic sequences to form a hairpin structure, chain extension by the host DNA polymerase(s), to generate a double-stranded molecule, and endonuclease cleavage.

Proteins: The viral proteins are produced from two major viral gene products: *(i)* REP or NS (for replication or nonstructural), which encodes the proteins required for viral transcription and replication, and *(ii)* CAP or S (for capsid or structural), which encodes the coat proteins (VP 1, 2, and 3).

Replication: Replication and virion assembly take place in the nucleus and

Viral Vectors: Basic Science and Gene Therapy
Edited by A. Cid-Arregui and A. García-Carrancá

involve host cell functions of the S phase of the cell cycle, suggesting a tight relationship between viral and host DNA synthesis. For productive infection, AAVS require co-infection with other (helper) viruses, such as adenoviruses or herpesviruses, or chemical or physical agents that affect replication functions of the host cell. Parvoviruses do not require helper viruses.

Family: Adenoviridae
Genus: Mastadenovirus (Mammalian Adenoviruses)

Virion: The virion particle is a nonenveloped icosahedron, 80–110 nm in diameter, containing 240 nonvertex capsomeres (hexons) 8–10 nm in diameter, and 12 vertex capsomeres (pentons) that project fibers of 10–35 nm from the virion surface. Virion Mr: $150–180 \times 10^6$. Buoyant density in CsCl: 1.32–1.35 g/mL. Virions contain at least 12 structural proteins and a number of nonstructural proteins. Virions do not contain lipids; fiber proteins and some nonstructural proteins are glycosylated.

Physicochemical properties: The viruses are insensitive to organic solvents and moderately resistant to low pH exposure and 70% ethanol. They are sensitive to heat (56°C for at least 30 minutes), 1% sodium hypochlorite, 2% glutaraldehyde, and formaldehyde.

Genome: The genome is a single molecule of linear, double-stranded DNA, 36–38 kb in size. It has inverted terminal repeats of 50–200 bp, which are involved in DNA replication. The genome contains two identical origins of DNA replication, one in each terminal repeat, and a *cis*-acting packaging sequence that must be located within a few hundred base pairs of a DNA end to be able to facilitate its encapsidation. A viral protein (terminal protein) is covalently bound to the 5′ end of each strand.

Proteins: The viral genome carries five early open reading frames (E1a, E1b, E2, E3, and E4), two delayed genes (IX and IVa2), and a late gene that is transcribed into five mRNAs (L1 through L5). Each of the viral genes is transcribed by the RNA polymerase II into multiple mRNA species, which may use different poly(A) signals and undergo alternative splicing. Gene expression proceeds in a cascade manner, by which early gene products activate expression of late genes leading to assembly of infectious viral particles.

Replication: Replication and virion assembly take place in the nucleus. Viral replication is divided into two phases, early and late, separated by the initiation of viral DNA replication. The early phase lasts a few hours and comprises viral adsorption, penetration, uncoating and expression of the viral early genes, whose products mediate a variety of functions, such as the control of expression and replication of the viral genome, induction of cell-cycle progression and inhibition of host antiviral functions. The late phase starts with the expression of the viral late genes, which drive synthesis of large amounts of hexon and penton capsomeres in the cytoplasm. These accumulate in the nucleus, where virion assembly occurs, often generating paracrystalline arrays of particles. Progeny virus is released by cell lysis.

Family: Herpesviridae
Genus: Alphasimplexvirus (Herpes Simplex Virus)

Subfamily: Gammaherpesvirinae
Genus: Rhadinovirus (Herpes Virus Saimiri [HVS]; Marek's Disease Virus)
Genus: Lymphocryptovirus (Epstein-Barr Virus [EBV])

Virion: The virion particle is 150–200 nm in diameter. The HSV-1 virion contains about 40 polypeptides, lipids derived from the host-cell nuclear membrane and carbohydrates of the complex high mannose type associated with the envelope glycoproteins. Host proteins have never been detected in purified virion preparations. Virions have four different components: *(i)* A core made of a linear double-stranded DNA molecule (150 kb). *(ii)* An icosahedral-shaped capsid (162 capsomeres, 100 nm in diameter). It is made of 4 proteins, which form 12 pentameric and 150 hexameric capsomeres. In addition, there are 4 proteins that are required for capsid assembly and encapsidation of viral DNA. *(iii)* A layer of viral protein (named tegument), which contains the rest of the virion proteins (VPs): a transactivation factor (also named α-TIF, VP16, or ICP25); a virion host shut-off protein (VHS); and VP1-2, associated to the complex binding the terminal "a" sequence (packaging sequence), among others. *(iv)* An envelope consisting of a lipid bilayer with spikes on its outer surface, which contain 12 glycoproteins (named gB through gM). In addition, it has two nonglycosylated integral membrane proteins.

Virion buoyant density in CsCl: 1.20–1.29 g/mL.

Physicochemical properties: Virions are sensitive to heat, UV- and γ-irradiation, detergents, acids, and organic solvents.

Genome: The HSV genome is a linear, double-stranded molecule of approximately 150 kb. It is organized in two unique segments: Unique Long (UL) and Unique Short (US), flanked by inverted repeats: *ab* (9 kb) and its inverted homologue *b'a'* flanking UL, and *a'c'* (6.5 kb) and *c'a'* flanking US. The structure of the repeats is highly conserved, although the number of repeat elements varies from strain to strain. The UL and US components can have either orientation relative to each other. Therefore, 4 isomers are possible. The viral DNA replicates by a rolling circle mechanism generating head-to-tail concatamers. During the process, the UL and US segments may undergo homologous recombination between their flanking repeat elements, thus generating the four isomers.

Proteins: The HSV genome contains 89 open reading frames: 65 in UL, 14 in US, and 10 in the repeats. The viral genes are expressed sequentially in three groups: *(i)* Immediate early genes (IE or α). Their expression does not require de novo viral protein synthesis. There are 5 IE genes, which are responsible for activating and coordinating viral gene expression and ultimately viral replication. ICP4 (infected cell polypeptide 4) and ICP 27 control expression of E and L genes. ICP0 is also an activator (although it is not a DNA binding protein). The IE genes map near the termini of the L and S components. *(ii)* Early (β) genes. Responsible for viral DNA synthesis. They are activated by IE genes. The b genes encoding the DNA polymerase and DNA binding proteins form a cluster close to the L origin of DNA synthesis. *(iii)* Late (γ) genes. They are expressed only after viral DNA synthesis has occurred. Encode structural proteins of the capsid, tegument, and envelope.

Replication: Replication and virion assembly take place in the nucleus. In nature, herpes simplex viruses infect exclusively humans, whereas under tissue culture conditions they infect and replicate in all tested mammalian cell lines. During its life cycle, HSV produces a variety of infections involving epithelial cells of mucocutaneous surfaces and neurons of the central nervous system. The following is a summary of events during HSV replication in fully permissive cultured cells. The lytic cycle is completed in 18–20 hours.

1. **Attachment**. Heparan sulfate on the cell surface binds HSV and facilitates virus entry (enzymatic breakdown of heparan sulfate from cells reduces infection by 80%–90%). In addition, one or more cell surface proteins (the immunoglobulin-like receptors nectin 1 and nectin 2) bind to HSV envelope

proteins, apparently at coated pits.

2. **Virus entry**. Penetration of virus into the cytoplasm requires fusion of the HSV envelope and the plasma membrane. Several HSV glycoproteins are involved in membrane fusion, although the precise mechanism has not been determined. Fusion leads to the release of at least two proteins from the viral tegument into the cytoplasm: VHS (virus host shut-off protein), which degrades mRNA molecules in polyribosomes; and α-TIF (or VP16, *trans* activator of α-genes), which is translocated into the nucleus.
3. **Transport of capsids to the vicinity of the nuclear envelope mediated by microtubules.** Proteins of the viral tegument are involved in both capsid movement and import of viral DNA through the nuclear pore. The released viral genome circularizes.
4. **Viral gene expression**. Transcription of viral genes proceeds in a precise temporal sequence: *(i)* Expression of the IE(α) genes is activated by α-TIF through Oct-1. The 5 α mRNAs are translated in the cytoplasm and the respective proteins translocated to the nucleus where they activate β gene expression. *(ii)* Expression of the E(β) genes takes place. The β proteins (e.g., thymidine kinase [TK], DNA polymerase) are translocated to the nucleus and start to produce nucleotide pools and synthesize viral DNA, generating head-to-tail concatamers of the viral genome. At the same time, the chromatin undergoes partial degradation and is displaced underneath the nuclear envelope, and the nucleoli disaggregate. *(iii)* L(γ) gene expression is activated and the γ proteins are synthesized and imported into the nucleus. They are mostly structural proteins involved in different functions: capsid proteins: at least five different proteins assemble to form empty capsids; envelope glycoproteins: 10 membrane proteins that accumulate in the inner nuclear membrane; and tegument proteins: surround the capsids as they enter the nucleus.
5. **Packaging of viral DNA.** Viral proteins bound to the internal *a* sequence in the viral DNA enter into the capsids pulling the DNA with them until the terminal *a* sequences are reached (a unit length viral genome).
6. **Nucleocapsids with tegument proteins attach to patches of the inner nuclear membrane enriched in viral glycoproteins, start budding, and finally pinch off into the endoplasmic reticulum.**
7. **Enveloped particles are transported to the cell surface and released into the extracellular space.**

Family: Poxviridae
Subfamily: Entomopoxvirinae (Insect poxviruses)
Genus: Entomopoxvirus B (Amsacta moorei)

Virion: The virion is a large enveloped particle, brick-shaped (vaccinia virus) or ovoid (entomopoxvirus B), 220–450 nm long × 150–260 nm wide × 150–260 nm thick. It has a complex structure composed of several layers: *(i)* external lipid envelope derived from the cell and contains viral proteins; *(ii)* a coat of regular spiral filaments (10–20 nm in diameter); and *(iii)* a cylindrical core containing the viral genome as a nucleoprotein complex and all enzymes required for mRNA synthesis. Virion buoyant density in CsCl: 1.25 g/mL and in sucrose: 1.16 g/mL.

Virions contain about 100 virus-encoded proteins (membrane associated, enzymes and factors), carbohydrates, in the form of glycoproteins, and lipids (predominantly cholesterol and phospholipids, but also glycolipids) mostly derived from cellular membranes and some synthesized de novo during infection.

Physicochemical properties: These viruses are sensitive to common detergents, oxidizing agents, and formaldehyde.

Genome: The genome consists of a single molecule of linear, double-stranded DNA, 130–375 kb (entomopoxvirus B, approximately 225 kb; vaccinia, approximately 200 kb), which contains inverted terminal repetitions of variable length (700 bp–12 kb), but identical and oriented oppositely at both ends. The two strands of viral DNA are connected by hairpin loops to form a covalently closed polynucleotide chain. A highly conserved sequence of nearly 100 bp, adjacent to the hairpin loops, is required for DNA replication. The viral open reading frames are under the control of promoters whose transcriptional activity is temporally regulated to specify three classes of genes: early, intermediate, and late.

Proteins: The viral genome contains about 185 open reading frames (vaccinia, Copenhagen strain). The viral proteins are expressed in three consecutive waves: *(i)* Early genes are expressed by a transcription system packaged within the core of the infectious particle. The early proteins are involved in viral DNA replication, intermediate gene expression and interaction with host functions. *(ii)* Intermediate gene expression is preceded by the onset of viral DNA replication. The proteins encoded are mainly transactivators of late gene expression. *(iii)* Expression of late genes follows that of intermediate genes. The products encoded are mostly virion structural proteins.

Replication: Replication and virion assembly take place in delimited, electrodense regions of the cytoplasm ("viral factories"). Replication of the viral genome is mainly directed by viral enzymes. The mechanism of replication is not completely understood but seems to proceed by a self-priming strand displacement (unidirectional) mechanism using concatamerized replicative intermediates as templates for the DNA synthesis. Virion morphogenesis is complex and proceeds via association of viral nucleoprotein with growing structures consisting of membrane coated with a brushlike border of spicules, which are later wrapped by membranes derived from the cisternae of the intermediate compartment between the endoplasmic reticulum and the *cis*-Golgi. Virions are released by budding (extracellular enveloped virions, containing an additional lipoprotein envelope) or following cell lysis (release of intracellular mature virions).

Family: Baculoviridae
Genus: Nucleopolyhedrovirus (NPV)
(*Autographa californica* NPV[AcNPV])

Virion: The virion particle is composed of a nucleoprotein core and a rod-shaped (bacilliform) proteinaceous sheath known as the capsid, which is wrapped by a membrane envelope. During their life cycle, baculoviruses form occlusion bodies (up to 5 μm in diameter) that are composed of numerous virions embedded in a crystalline matrix made of a single protein (polyhedrin). There are two types of NPV: multiple and single (MNPV and SNPV, respectively); the MNPV contain multiple nucleocapsids within a lipid envelope, whereas in SNPV, nucleocapsids are wrapped individually. Capsids are 250–300 nm in length and 30–60 nm in diameter. They are made of ringlike subunits. Virion contains proteins, lipids, and carbohydrates in the form of glycosylated protein. Virion Mr: 60–110 × 10^6. Buoyant density in CsCl: 1.20–1.30 g/mL.

Physicochemical properties: They are sensitive to heat, detergents, and formaldehyde.

Genome: The genome is a single molecule of circular, supercoiled, double-stranded DNA of 80–230 kb (AcNPV genome has 131 kb). Three to four percent of the genome consists of small repeats (60-bp motifs repeated two to eight times, known as homologous regions) intercalated within the viral DNA, which have been assigned functions as origins of DNA replication or as enhancers for early gene transcription. The genome can encode about 150 genes. Open reading frames are efficiently compacted on either strand of the DNA, most of them separated 2–200 bp, although there are some overlapping genes. Early, late, and very late genes are dispersed throughout the genome, rather than clustered.

Proteins: In the first hours of infection, the virus produces proteins required for viral DNA synthesis and proteins involved in regulation of early and late gene transcription. Early gene expression seems to be driven by host RNA polymerase II. Transcription of late genes is carried out by a viral polymerase. These transcripts specify the structural proteins of the virus and the stromal components. The very late phase (24–72 hours postinfection) is characterized by an enormous increase in the transcription of *polh*, the gene encoding polyhedrin. Since the *polh* gene is not essential for virus production, its promoter has been used for expression of heterologous genes.

Replication: Replication and virion assembly take place in the nucleus. Natural infection occurs when the insect ingests bodies of occluded virus; the matrix is dissolved in the midgut lumen releasing the virions, which fuse with the membrane of the microvilli. In cell culture, however, budded baculovirus are used that enter the cells by receptor-mediated endocytosis. Fusion to the endosomal membrane releases nucleocapsids in the cytoplasm, which move to the nuclear pore, enter the nucleus, and uncoat. DNA replication is somehow coupled to late gene expression (6–24 hours postinfection). Nucleocapsids assemble during the late and very late phases at the edge of the viral stroma, a nucleoprotein structure. In the late phase, nucleocapsids exit the nucleus through the nuclear pore and bud from the plasma membrane. During the very late phase, nucleocapsids are enveloped within the nucleus and later on virions are embedded in a polyhedrin matrix to form occlusion bodies.

RECOMMENDED READING

Botchan, M., T. Grodzicker, and P.A. Sharp (Eds.). 1986. DNA Tumor Viruses. Control of Gene Expression and Replication. CSH Laboratory Press, Cold Spring Harbor, New York.

Fields, B.N., D.M. Knipe, P.M. Howley, R.M. Chanock, J.L. Melnick, T.P. Monath, B. Roizman, and S.E. Straus (Eds.). 1996. Fundamental Virology, 3rd. ed. Lippincott-Raven, New York.

Muzyczka, N. 1992. Use of adeno-associated virus as a general transduction vector for mammalian cells. Curr. Top. Microbiol. Immunol. *158*: 97-129.

Webster, R.G. and A. Granoff (Eds.). 1994. Encyclopedia of Virology. Academic Press, San Diego.

RECOMMENDED URLs

www.tulane.edu/~dmsander/garryfavweb.html (Provides abundant information on virology. It contains listings of virology-related web sites of interest).

www.virology.net/Big_Virology/BVHomePage.html (The big picture book of viruses. It also serves as a resource for virus taxonomy).

FAMILY: Parvoviridae
GENUS: Dependovirus (Adeno-Associated Viruses [AAV])

1 Gene Transfer with Adeno-Associated Virus Vectors

Arun Srivastava
Department of Microbiology and Immunology, and Department of Medicine, Walther Oncology Center, Indiana University School of Medicine, Walther Cancer Institute, Indianapolis, IN, USA

INTRODUCTION

The idea of treating human diseases by introducing normal genes into appropriate target cells has become a clinical reality (4). Although a number of physical and chemical methods are available, in general, viruses are much more efficient in transducing cells. Indeed, retroviral and adenoviral vectors have been used in a number of clinical trials (23,94,130). Initial results with retroviral vectors have been encouraging (12,13, 38), but their use in nonhuman primate studies has been reported to lead to T-cell lymphoma (25). Similarly, questions have been raised with regard to the efficacy of first-generation adenoviral vectors (49). The fact that retroviruses and all other DNA-containing viruses are the etiologic agents of, or are intimately associated with, malignancy (116,123), has prompted the search for an alternative viral vector (43, 70). Interestingly, parvoviruses, which are among the smallest of the DNA-containing viruses that infect a wide variety of vertebrates (22,98), remain the only group of viruses that have thus far not been associated with malignant disorders (11,80). In fact, parvoviruses have been shown to possess antitumor properties (24,66,76). One of the members of the Parvoviridae family, the adeno-associated virus type 2 (AAV-2), has gained significant attention primarily because AAV-2 is a nonpathogenic human parvovirus (7–10) and because the wild-type (wt) AAV-2 genome establishes a latent infection in human cells where the viral genome integrates into the chromosomal DNA in a site-specific manner (52–55,97). Thus, the nonpathogenic nature of AAV-2, coupled with its remarkable site-specificity of integration, provided sufficient impetus towards the development of recombinant AAV-2 vectors (40,74).

RECOMBINANT AAV VECTORS

The overall strategy to construct recombinant AAV-2 vectors has previously been described (74,75,96). Briefly, a gene of interest is molecularly cloned in between the two inverted terminal repeats (ITRs), and this plasmid is co-transfected into ade-

Viral Vectors: Basic Science and Gene Therapy
Edited by A. Cid-Arregui and A. García-Carrancá

novirus-infected human cells with a helper plasmid that contains the entire AAV coding sequences (*rep* and *cap* genes) but lacks the AAV-ITRs (107). Recently, a number of investigators have devised alternative strategies that include the use of AAV-packaging cell lines, triple transfections involving adenovirus plasmid DNA, inclusion of AAV *rep* and *cap* genes in the adenovirus plasmid, and the use of an adeno-AAV hybrid virus (16,21,35,37,57,65,113,128). Other investigators have suggested alternative strategies to minimize or eliminate the wt AAV-like particles that are generated during the production of recombinant AAV vectors (3,121). More recently, affinity column chromatographic techniques have been utilized to significantly reduce the physical particle-to-infectious particle ratios of recombinant AAV vectors (19,133). Interestingly, recombinant AAV vectors based on serotypes other than AAV-2 have also been developed and are listed in Table 1 (5,17, 18,72,95,125). Thus, it is clear that at least two potential problems with regard to the generation of high-titer, clinical grade recombinant AAV vectors and the high degree of seropositivity against AAV-2, a potential limitation for repeat administrations, once deemed insurmountable, are now being overcome. A number of reviews on recombinant AAV-2 vectors have been published (9,15,34,44,51,63,74,102-104, 126). This article will attempt to highlight additional salient features of recombinant AAV vectors that provide further support to the suggestion that their use in potential gene therapy of human diseases is imminent and perhaps well-justified.

AAV-MEDIATED GENE TRANSFER IN VITRO

Although recombinant AAV vectors have been used to successfully transduce a variety of genes, the efficiency of transduction is cell type-specific. For example, there are at least two major obstacles that AAV vectors must overcome prior to transduction of cells (84,88). The first relates to the extent of expression of the cellular receptor for AAV, and the second concerns the rate-limiting step of the viral second-strand DNA synthesis (29,30). With reference to the first obstacle, although the ubiquitously expressed cell surface heparan sulfate proteoglycan (HSPG) has been reported to be a receptor for AAV (112), HSPG alone is insufficient for AAV infection. At least two different co-receptors, fibroblast growth factor receptor 1 (FGFR1) (91) and αVβ5 integrin (111), have been identified to mediate successful viral entry into the host cell. With reference to the second obstacle, a cellular protein, designated the single-stranded D-sequence-binding protein (ssD-BP) and phosphorylated at tyrosine residues, has been identified (90,92). It plays a crucial role in viral second-strand DNA synthesis. The ssD-BP is phosphorylated by the protein tyrosine kinase activity of the human epidermal growth factor receptor (EGFR) (64). Thus, at least two distinct cellular growth factor receptors play a crucial role in AAV-mediated transduction (63).

Table 1. Recombinant AAV Vectors

Serotype	References
AAV-1	125
AAV-2	74
AAV-3A	72
AAV-3B	95
AAV-4	18
AAV-5	5,17
AAV-6	95

In the absence of co-infection with a helper virus, the wt AAV genome establishes a latent infection and integrates into the host chromosomal DNA in a site-specific manner in human chromosome 19q13.3-qter with very high frequency in established human cell lines (52-55,97). Previous studies with the wt AAV have also revealed that integrated proviral genome is almost always present in linear, tandem repeats (8,10,74). Furthermore, a number of studies have suggested that recombinant AAV vector sequences persist in episomal forms for extended time periods (1,31,33,68). Recent studies suggest the existence of circular DNA intermediates as well (26,118). Using murine bone marrow primary as well as secondary transplant model systems, sustained presence of as well as expression from the recombinant proviral genomes in progeny hematopoietic cells approximately one year post-transplantation have been documented (87). These studies strongly suggest that successful transduction of the pluripotent hematopoietic stem cell, and by inference, stable integration of the proviral genome into host chromosomal DNA, does indeed occur (16,114). Again, although further studies are clearly warranted, these observations underscore the need to critically evaluate the interaction of recombinant AAV with each specific cell type being transduced.

Although the wt AAV is nonpathogenic, whether the recombinant AAV vectors are also nonpathogenic as well as capable of site-specific delivery of therapeutic genes has not been rigorously examined. Ponnazhagan et al. (83) have examined the underlying molecular basis of the site-specific integration of the AAV genome in human cells. In these studies, the integration patterns of the recombinant AAV genomes lacking one or both of the viral coding sequences were evaluated. The integration pattern of each of the recombinant AAV genomes in individual clonal isolates of human cells was noted to be distinctly different. In addition, in none of the clones examined was the proviral genome covalently linked to the previously described AAV right-junction human chromosomal DNA fragment (52), the putative specific site of integration for the wt AAV genome. Similar observations have been made using recombinant AAV genomes in established as well as primary human hematopoietic cells (32). Fluorescence in situ hybridization (FISH) analyses further revealed the lack of integration of the recombinant AAV into human chromosome 19, even in the presence a functional *rep* gene. These data are consistent with the conclusion that recombinant AAV genomes lack the site specificity of integration into human chromosome 19 (47,83) and suggest that in the absence of optimal levels of expression of the Rep proteins, the recombinant AAV genomes integrate at sites that are different from that characterized for the wt AAV genome. However, it is important to note that Kube et al. (56) have documented that AAV Rep proteins are encapsidated in highly purified mature progeny virions. Whether the AAV-encapsidated Rep proteins at high multiplicity of infection (moi) play a role *in trans* in site-specific integration of the viral genome on human chromosome 19, remains to be investigated.

The remarkable tissue tropism of a yet another human parvovirus, the parvovirus B19 for human hematopoietic cells in the erythroid lineage (77,78,105,106), has been exploited by Srivastava et al. (109) to construct a recombinant AAV-B19 genome, in which the B19-ITRs were replaced by the AAV-ITRs. This hybrid virus was biologically indistinguishable from the wt B19, thereby demonstrating autonomous replication in and cytotoxicity to human erythroid progenitor cells (109). Because the remarkable features of these two parvoviruses were combined into the AAV-B19 hybrid virus, investigators speculated that such a hybrid vector may prove useful for high-efficiency transduction of primary human hematopoietic progenitor cells since expression from parvovirus B19 promoter at map unit 6 (B19p6) is restricted to cells in the erythroid lineage (87,122).

The remarkable tissue tropism of parvovirus B19 for erythroid cells has also been attributed to the virus utilizing the erythrocyte P antigen as a cellular receptor for the selective infection of these progenitor cells (14). This suggestion has also prompted the development of recombinant B19 vectors in which a recombinant AAV genome is encapsidated into the B19 capsids, and these vectors are capable of erythroid cell-specific delivery of transgenes (89).

With the potential to target the erythroid lineage by these vectors, development of gene therapy approaches for human hemoglobinopathies has been contemplated (108,119). For example, sickle cell anemia and β-thalassemia are among the most well-characterized human hemoglobinopathies (110), and because the molecular defects in these diseases are known and can potentially be corrected by introduction of normal alleles of the human γ-globin gene, gene therapy of these diseases appears possible. A number of investigators have described the construction of recombinant AAV-globin vectors containing an upstream DNase I-hypersensitive site 2 (HS-2) from the locus control region (LCR) (117) of the human β-globin gene cluster and documented erythroid cell-specific expression of the transduced β-globin gene in vitro (28,69,120,132). These and additional studies indicate that it is feasible to obtain AAV-mediated transduction as well as erythroid cell-specific expression of human globin genes.

In addition to the potential benefit to patients with sickle cell anemia, transfer and expression of a normal human β-globin gene is also likely to ameliorate clinical symptoms of β-thalassemia, often a fatal disease. Although in β-thalassemia, which is characterized by the defective β-globin production, a mechanism to deliver a functional β-globin gene would be a preferred form of treatment. A second line of therapeutic measure that could ameliorate the severity of this disease would be to suppress the accumulation of excessive levels of the α-globin chains in red blood cells. Therefore, the potential usefulness of the AAV-based vectors to suppress the excess production of the human α-globin gene has also been investigated since, in the homozygous condition, β-thalassemia is also characterized by accumulation of excessive free α-globin chains, which results in premature death of erythroid progenitor cells before they reach the reticulocyte stage. This imbalanced production of globin chains results in red blood cell damage, thereby shortening the red blood cell life span. A mechanism that could control the excess production of free α-globin chains would be expected to ameliorate the clinical severity of this disease. For this purpose, recombinant AAV vectors have been generated that contain the human α-globin gene sequences in antisense orientation; these have been shown to achieve this objective, at least in vitro (86).

AAV-MEDIATED GENE TRANSFER

Table 2 lists the various organ systems that have been targeted for potentially therapeutic purposes by recombinant AAV vectors. Hematopoietic cells are particularly attractive targets for genetic intervention since well-established culture conditions are available and a number of diseases have been cured by bone marrow transplantation. Recombinant AAV-mediated β-globin gene transfer and expression in established human and murine erythroleukemia cell lines in vitro has been reported by a number of investigators (28,69,120,132). However, successful AAV-mediated globin gene transduction of hematopoietic stem cells and long-term expression in vivo in progeny cells has not been documented. Here, I summarize our experience with the bone marrow in a murine transplant model system.

Table 2. AAV-Mediated Gene Transfer In Vivo

Organ	References
Bone marrow	16,34,57,69,87,108,114,115,131
Brain	2,6,45,67
Ear	58–60
Eye	36,93
Heart	46,81
Joint	62,79
Liver	50,68,85,99,100,124
Lung	1,33,39
Muscle	1,20,31,41,42,48,71,73,101,127
Stomach	27

The AAV vector used in these studies, vHS2-βp-$^{A}\gamma$-globin, schematically depicted in Figure 1, which contains the genomic copy of the human $^{A}\gamma$-globin gene sequences under the control of human β-globin promoter and the HS-2 enhancer, was generated as previously described (87). This chimeric gene was useful for the unambiguous detection of the transduced sequences in murine cells by using the human βp- and the $^{A}\gamma$-globin-specific oligonucleotide primer pair. The vHS2-βp-$^{A}\gamma$-globin virus was used to infect murine low-density bone marrow cells obtained from B6.Hbb^{d}/Hbb^{d}-Gpi-1^{a}/Gpi-1^{a} mice at an moi of 1 and 10 ex vivo. The cells were transplanted into six lethally irradiated C57BL/6J mice (three for 1 moi and three for 10 moi). Six months post-transplantation, mice from both the groups along with the mock-transduced controls were sacrificed, and cells from bone marrow, spleen, and thymus tissues were analyzed by polymerase chain reaction (PCR) amplification of DNA sequences specific for the transduced human globin gene. DNA products were analyzed on Southern blots using a human $^{A}\gamma$-globin-specific DNA probe. These results are shown in Figure 2. Animals that were transplanted with the vHS2-βp-$^{A}\gamma$-globin-transduced bone marrow cells (two animals for 1 moi and three animals for 10 moi) showed the presence of the transduced human globin sequences in the bone marrow, and except for one animal for the 1 moi group, the remaining animals that were transplanted with the recombinant AAV-transduced cells, also showed the presence of the human globin sequences in the spleen. However, thymus tissues from none of these animals showed the presence of the

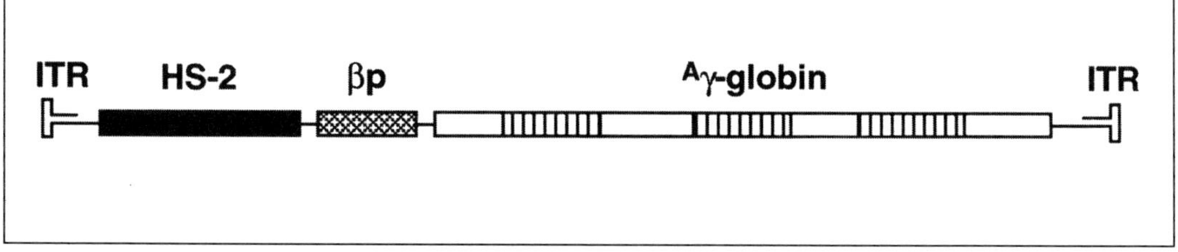

Figure 1. Schematic representation of the recombinant vHS2-βp-$^{A}\gamma$-globin vector. Various structural components of the recombinant AAV-globin vector are indicated. High-titer, purified viral stocks were generated as previously described (87).

transduced gene. The transplanted animals with mock-transduced cells did not show the presence of the human globin gene, as expected, confirming the specificity of detection of the transduced sequences. The results from semiquantitative PCR analyses, shown in Figure 3, indicated that the copy number of the transduced human $^{A}\gamma$-globin gene ranged from approximately 7 (± 0.4)% for 1 moi to approximately 8 (± 1)% for 10 moi as compared with that of the mouse endogenous β-actin gene.

The transcription potential of transduced human γ-globin gene sequences in bone marrow and spleen cells in primary recipient mice was also determined. Total RNA was isolated from bone marrow, spleen, and thymus tissues and subjected to reverse transcription (RT)-PCR analyses. Primers for PCR, following RT, were designed to eliminate amplification of the mouse endogenous globin mRNA as well as to distinguish amplification products from the viral sequences. The RT reactions were performed using random primers to facilitate subsequent amplification with primers specific for the mouse β-actin gene as well as to ensure the quality and structural integrity of RNA preparations. Figure 4 shows the results of expression analysis by RT-PCR. Figure 4A indicates the presence of the human $^{A}\gamma$-globin transcripts only in bone marrow cells of each animal transplanted with both 1 and 10 moi of the recombinant AAV, whereas no expression was detected in cells from spleen and thymus tissues from these animals. However, the quality of RNA from each of the cell types from all animals was confirmed to be good using the mouse β-actin gene transcript-specific primer pair (Figure 4B). No RT-PCR product was generated in the absence of RT (Figure 4, A and B), as expected. Semiquantitative RT-PCR assays were performed to determine the level of expression of the transduced human $^{A}\gamma$-globin gene relative to that of the endogenous mouse β^{major}-globin gene. These results are shown in Figure 5. The level of expression was approximately 4 (± 0.8)% for 1 moi and approximately 6 (± 0.9)% for 10 moi, compared with that of the endogenous mouse β^{major}-globin gene.

The potential of AAV-mediated stable

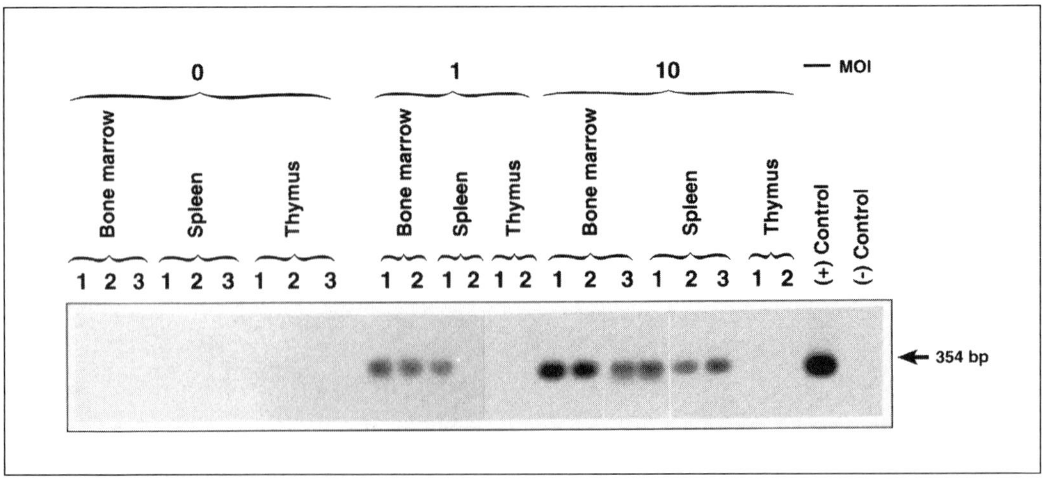

Figure 2. Southern blot analysis for the presence of the transduced human globin gene-specific sequences in hematopoietic cells from primary recipient mice. Bone marrow, spleen, and thymus tissues obtained from 3 mice transplanted with mock-infected (moi = 0), 2 mice transplanted with vHS2-βp-$^{A}\gamma$-globin virus-infected (moi = 1), and 3 mice transplanted with vHS2-βp-$^{A}\gamma$-globin virus-infected (moi = 10). Low-density bone marrow cells were obtained 6 months posttransplantation, subjected to PCR amplification, and analyzed on Southern blots. Positive and negative controls were the same as those described previously (87).

transduction and long-term expression of the human γ-globin gene in murine hematopoietic cells was determined by transplanting a portion of the bone marrow cells from primary recipients into secondary recipients. Approximately 3×10^6 low-density bone marrow cells, obtained six months post-transplantation from primary recipients, were transplanted into six lethally irradiated congenic mice along with appropriate controls. Cells obtained from bone marrow, spleen, and thymus tissues from the secondary recipients were also analyzed three months post-transplantation by PCR and RT-PCR analyses. Of the six animals transplanted with cells obtained from transduced and transplanted primary recipient, one animal showed the presence of the transduced human γ-globin sequences in both bone marrow and spleen cells (Figure 6A). The results of the semiquantitative PCR analysis indicated that the transduced gene copy number was approximately 0.5% compared to the endogenous mouse β-actin gene (Figure 6B). The expression of the transduced human globin gene was detected in the secondary recipient and was restricted to bone marrow cells as in prima-

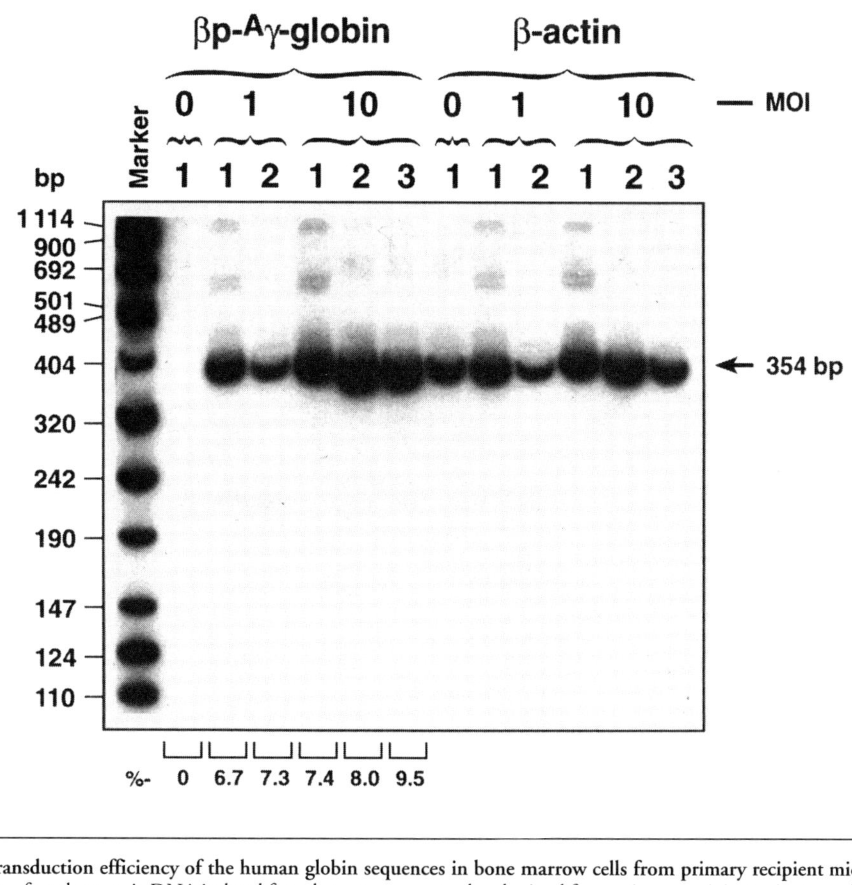

Figure 3. Transduction efficiency of the human globin sequences in bone marrow cells from primary recipient mice. Equivalent amounts of total genomic DNA isolated from bone marrow samples obtained from primary recipient mice transplanted with mock-infected (moi = 0), and vHS2-βp-^{A}γ-globin virus-infected (moi = 1 and 10) cells were used to amplify DNA products specific for the human βp-^{A}γ-globin gene and the mouse β-actin gene sequences, respectively, and analyzed on Southern blots. The transduction efficiency in bone marrow cells from each animal is indicated at the bottom of the blot and was determined as previously described (87).

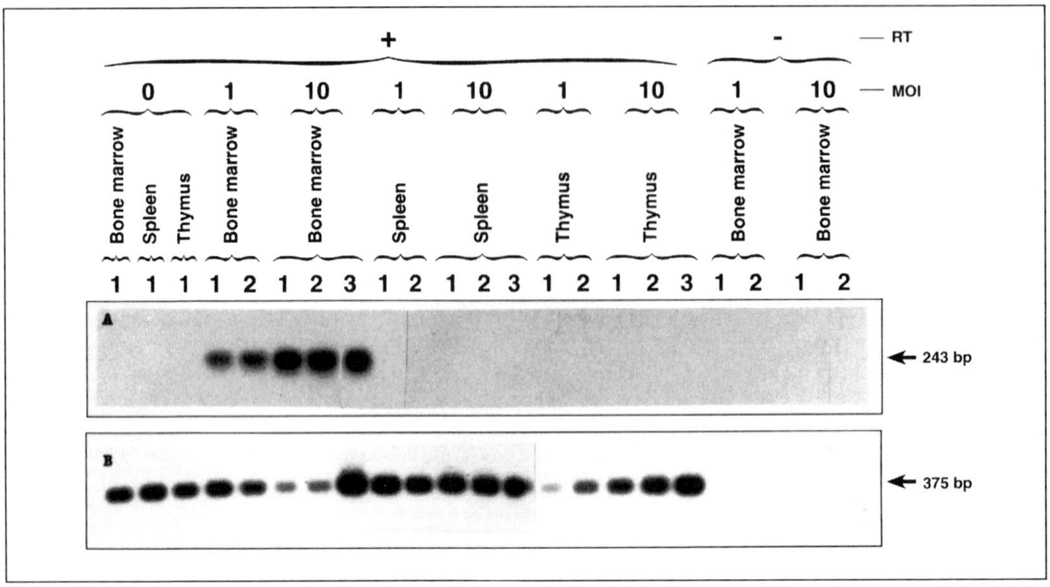

Figure 4. Expression of the transduced human globin gene in hematopoietic cells from primary recipient mice. Equivalent amounts of total RNA isolated from bone marrow samples obtained from primary recipient mice transplanted with mock-infected (moi = 0), and vHS2-βp-$^{A}\gamma$-globin virus-infected (moi = 1 and 10) cells were used to amplify DNA products specific for the human βp-$^{A}\gamma$-globin gene (A) and the mouse β-actin gene (B) transcripts, respectively, in the presence (+) or absence (-) of reverse transcriptase, and analyzed on Southern blots as previously described (87).

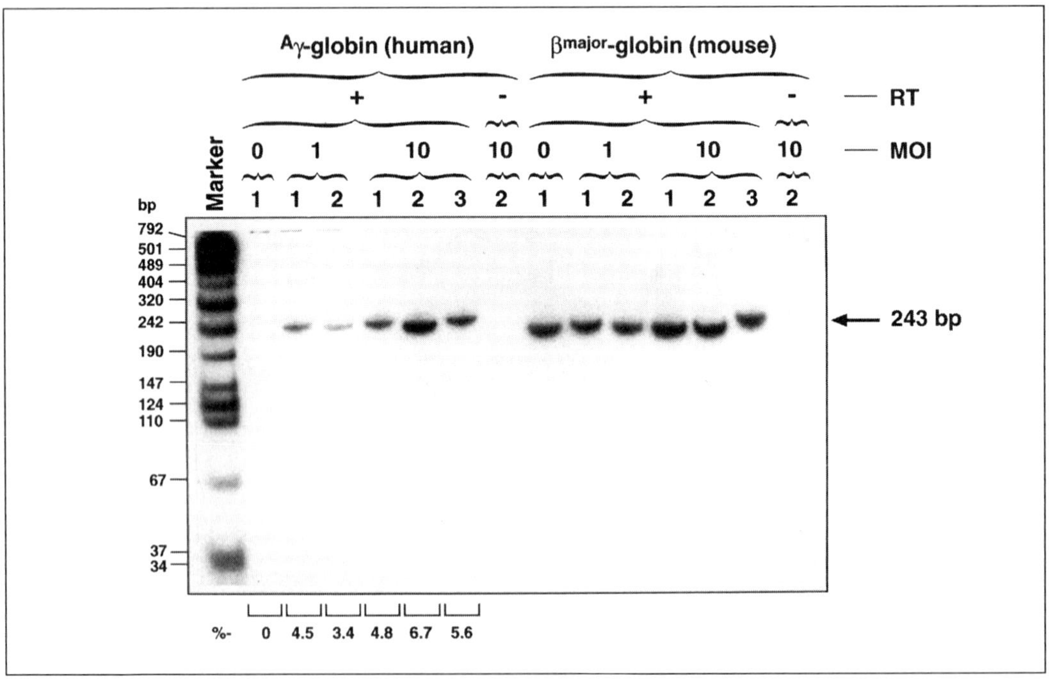

Figure 5. Level of expression of the transduced human globin gene in bone marrow cells from primary recipient mice. Equivalent amounts of total RNA isolated from bone marrow samples obtained from primary recipient mice transplanted with mock-infected (moi = 0), and vHS2-βp-$^{A}\gamma$-globin virus-infected (moi = 1 and 10) cells were used to amplify DNA products specific for the human βp-$^{A}\gamma$-globin gene and the mouse endogenous β^{major}-globin gene transcripts, respectively, in the presence (+) or absence (-) of reverse transcriptase, and analyzed on Southern blots. The level of expression of the transduced human γ-globin gene in bone marrow cells from each animal is indicated at the bottom of the blot and was determined as previously described (87).

ry transplant recipients (Figure 7A). No RT-PCR product was generated in the absence of RT (Figure 7, A and B). The level of expression of the transduced human $^{A}\gamma$-globin gene was determined by semiquantitative RT-PCR assays to be approximately 0.4% compared with that of the endogenous mouse β^{major}-globin gene (Figure 7C). In subsequent studies, the use of enriched hematopoietic stem cell populations and high-titer, purified recombinant AAV led to high-efficiency transduction (84), and more recent studies with a β-thalassemic murine bone marrow transplant model document that expression of the transduced human globin gene is restricted to the erythroid cell lineage in vivo (114). Thus, recombinant AAV vectors promise to prove useful in gene therapy of human hemoglobinopathies in general and β-thalassemia and sickle-cell anemia in particular. Current efforts are focused to evaluate the safety and the efficacy of AAV-globin vectors in nonhuman primates prior to their potential use in human gene therapy.

ADVANTAGES AND DISADVANTAGES OF AAV VECTORS

The nonpathogenic nature of AAV and stable integration of the viral genome make these extremely attractive vectors for human gene therapy. In contrast to the clinical application of retrovirus vectors, which are

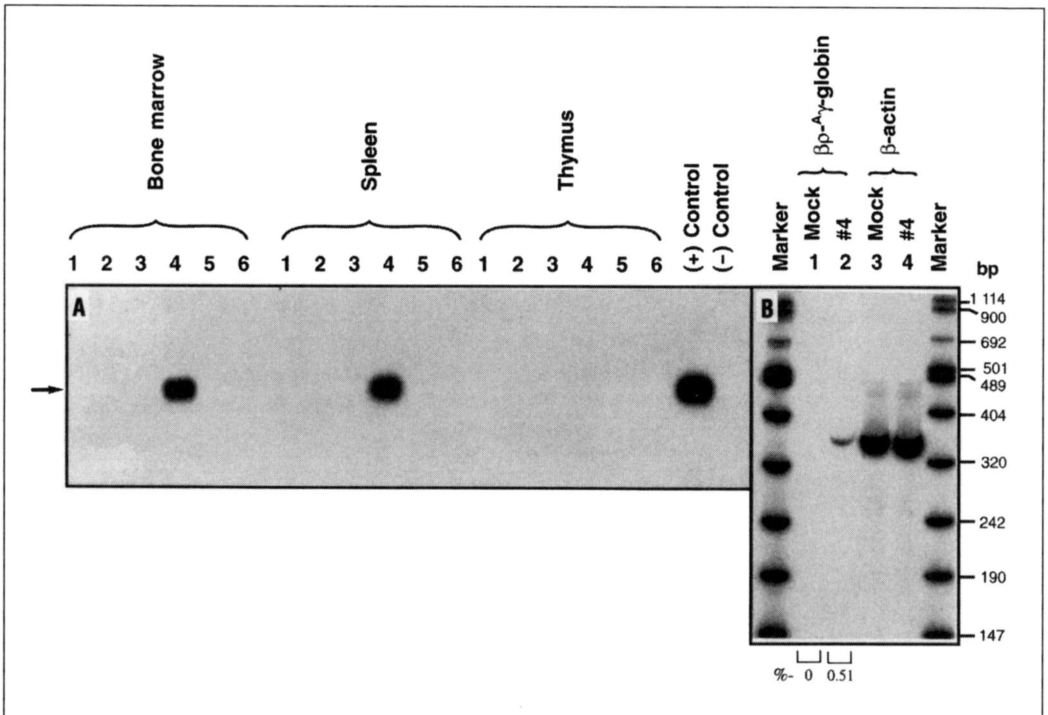

Figure 6. Transduction efficiency of the human globin gene sequences in hematopoietic cells from secondary recipient mice. Equivalent amounts of total genomic DNA isolated from bone marrow samples obtained from secondary recipient mice transplanted with bone marrow cells obtained from vHS2-βp-$^{A}\gamma$-globin virus-infected (moi = 10) cells were used to amplify DNA products specific for the human βp-$^{A}\gamma$-globin gene (A). Semiquantitative PCR analysis of DNA products specific for the human βp-$^{A}\gamma$-globin gene and the mouse β-actin gene sequences, respectively (B), was also performed as described in the legend to Figure 3. The transduction efficiency of the human γ-globin gene in bone marrow cells from the positive animal is indicated at the bottom of the blot and was determined as previously described (87).

of murine origin, AAV is a human virus and therefore may conceivably be physiologically more relevant for gene therapy in humans. The stability of AAV between pH 3.0 to 9.0, its resistance to lipid solvents, and its resistance to heat inactivation bode well for the eventual global application of gene therapy with these vectors. The wide host range and the ability of AAV to transduce nondividing cells (82,131), coupled with the potential for long-term transgene expression, are also desirable features. Vector mobilization is a finite possibility, yet it is unlikely in view of the required simultaneous infections by wt AAV and adenovirus. It is intriguing that little adverse immune response to these vectors occurs in vivo (31,85,127), and although a great majority of the human population is seropositive for AAV-2, it is clear that the availability of at least four additional serotype AAV vectors should permit repeat administrations. It has also become increasingly clear that the problem of large-scale,

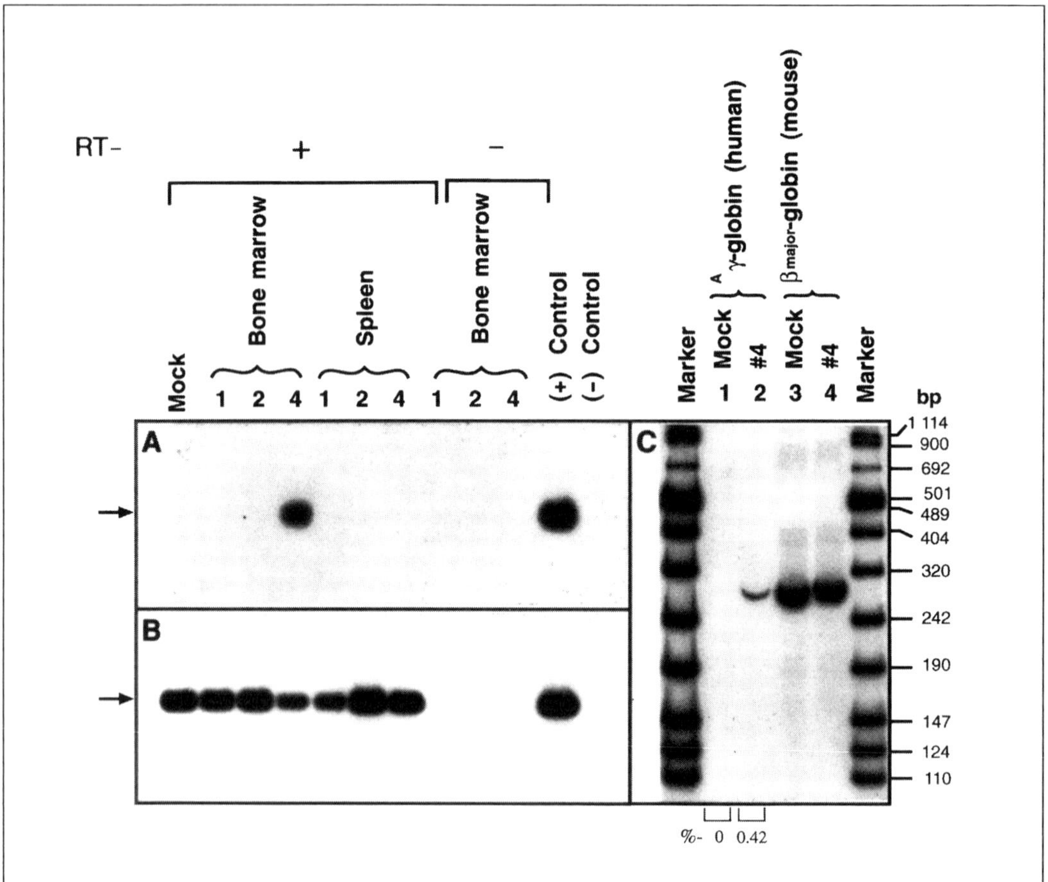

Figure 7. Level of expression of the transduced human globin gene in bone marrow cells from secondary recipient mice. Equivalent amounts of total RNA samples isolated from bone marrow samples obtained from secondary recipient mice transplanted with bone marrow cells from mock-infected (moi = 0), and vHS2-βp-$^{A}\gamma$-globin virus-infected (moi = 10) cells from primary recipient mice were used to amplify DNA products specific for the human βp-$^{A}\gamma$-globin gene (A) and the mouse endogenous β-actin gene (B) transcripts, respectively, in the presence (+) or absence (-) of reverse transcriptase, and analyzed on Southern blots. Semiquantitative RT-PCR analysis of DNA products specific for the human βp-Aγ-globin gene and the mouse endogenous β^{major}-globin gene transcripts, respectively (C), was also performed as described in the legend to Figure 5. The relative level of expression of the transduced human γ-globin gene in bone marrow cells from the secondary transplant recipient animal is indicated at the bottom of the blot and was determined as previously described (87).

commercial production of clinical grade recombinant AAV vectors is not as insurmountable as was initially perceived. Since superinfection immunity by AAV has not been observed in vitro or in vivo (61,129), one of the main limitations of these vectors—the size limitation of a DNA sequence that can be packaged into mature virions—has also been overcome. Thus, it is clear that the foregoing discussion attests to the fact that the advantages far outnumber the disadvantages of AAV vectors.

SUMMARY AND CONCLUSIONS

It has become increasingly clear that the AAV-based vector system is a safe and effective alternative to the more commonly used retrovirus- and adenovirus-based vectors in human gene therapy. Although a number of issues related to the basic molecular biology of AAV vectors still needs to be resolved, such as the precise underlying mechanism of viral entry, the role of cellular proteins in viral trafficking, uncoating, second-strand DNA synthesis, transgene expression, and viral DNA integration, extensive studies on these aspects are currently under way and will undoubtedly yield new and useful information that will be applicable in the further development of AAV vectors. A better understanding of the host immune response will also be critical in realizing the full potential of this vector system for human gene therapy.

FUTURE PROSPECTS

Phase II clinical trials with recombinant AAV vectors for cystic fibrosis are already under way, and additional clinical trials for hemophilia B and muscular dystrophy are on the horizon. After all safety and efficacy issues have been dealt with, these vectors will attract greater attention and widespread application in gene therapy of a variety of aquired and genetic clinical disorders in humans.

ACKNOWLEDGMENTS

I thank Drs. Kenneth I. Berns, Hal E. Broxmeyer, and Mervin C. Yoder for their encouragement and support. I also thank Dr. George Stamatoyannopoulos and Tim M. Townes for their kind gifts of recombinant plasmids, and the past and present members of my laboratory for helpful discussions.

The research in the author's laboratory was supported in part by grants from the National Institutes of Health (Public Health Service Grant Nos. HL-26323, HL-48342, HL-53586, HL-58881, and DK49218), Centers of Excellence in Molecular Hematology, Phi Beta Psi Sorority, and an Established Investigator Award from the American Heart Association.

REFERENCES

1. **Afione, S.A., C.K. Conrad, W.G. Kearns, S. Chunduni, R. Adams, T.C. Reynolds, W.G. Guggino, G.R. Cutting, B.J. Carter, and T.R. Flotte.** 1996. In vivo model of adeno-associated virus vector persistence and rescue. J. Virol. *70*:3235-3241.
2. **Alexander, I.A., D.W. Russell, A.M. Spence, and A.D. Miller.** 1996. Effects of gamma irradiation on the transduction of dividing and nondividing cells in brain and muscle of rats by adeno-associated virus vectors. Hum. Gene Ther. *7*:841-850.
3. **Allen, J.M., D.J. Debelak, T.C. Reynolds, and A.D. Miller.** 1997. Identification and elimination of replication-competent adeno-associated virus (AAV) that can arise by non-homologous recombination during AAV vector production. J. Virol. *71*:6816-6822.
4. **Anderson, W.F.** 1998. Human gene therapy. Nature *392*:25-30.
5. **Bantel-Schaal, U., H. Delius, R. Schmidt, and H.Z. Hausen.** 1999. Human adeno-associated virus type 5 is only distantly related to other known primate helper-dependent parvoviruses. J. Virol. *73*:939-947.
6. **Bartlett, J.S., R.J. Samulski, and T.J. McCown.** 1998. Selective and rapid uptake of adeno-associated virus type 2 in brain. Hum. Gene Ther. *9*:1181-1186.
7. **Berns, K.I.** 1990. Parvovirus replication. Microbiol. Rev. *54*:316-329.

8. **Berns, K.I. and R.A. Bohenzky**. 1987. Adeno-associated viruses: an update. Adv. Virus Res. *32*:243-306.
9. **Berns, K.I. and C. Giraud**. 1996. Adeno-associated virus (AAV) vectors in gene therapy. Curr. Top. Microbiol. Immunol. *218*:1-170.
10. **Berns, K.I. and R.M. Linden**. 1995. The cryptic life style of adeno-associated virus. BioEssays *17*:237-245.
11. **Blacklow, N.R.** 1988. *In* Parvoviruses and Human Disease. CRC Press, Boca Raton.
12. **Blaese, R.M., K.W. Culver, A.D. Miller, C.S. Carter, T. Fleisher, M. Clerici, G. Shearer, L. Chang, et al.** 1995. T lymphocyte-directed gene therapy for ADA-SCID: initial trial results after 4 years. Science *270*:475-480.
13. **Bordignon, C., L.G. Notarangelo, N. Nobili, G. Ferrari, G. Casorati, P. Panina, E. Mazzolari, D. Maggioni, et al.** 1995. Gene therapy in peripheral blood lymphocytes and bone marrow for ADA-immunodeficient patients. Science *270*:470-475.
14. **Brown, K.E., S.M. Anderson, and N.S. Young**. 1993. Erythrocyte P antigen: cellular receptor for Bl9 parvovirus. Science *262*:114-117.
15. **Carter, B.J.** 1993. Adeno-associated virus vectors. Curr. Opin. Biotechnol. *3*:533-539.
16. **Chatterjee, S., D. Lu, G. Podsakoff, and K.K. Wong, Jr.** 1995. Strategies for efficient gene transfer into hematopoietic cells: the use of adeno-associated virus vectors in gene therapy. Ann. NY Acad. Sci. *770*:79-90.
17. **Chiorini, J.A., F. Kim, L. Yang, and R.M. Kotin**. 1999. Cloning and characterization of adeno-associated virus type 5. J. Virol. *73*:1309-1319.
18. **Chiorini, J.A., L. Yang, Y. Liu, B. Safer, and R.M. Kotin**. 1997. Cloning of adeno-associated virus type 4 (AAV4) and generation of recombinant AAV4 particles. J. Virol. *71*:6823-6833.
19. **Clark, K.R., X. Liu, J.P. McGrath, and P.R. Johnson**. 1999. Highly purified recombinant adeno-associated virus vectors are biologically active and free of detectable helper and wild type viruses. Hum. Gene Ther. *10*:1031-1039.
20. **Clark, K.R., T.J. Sferra, and P.R. Johnson**. 1997. Recombinant adeno-associated viral vectors mediate long-term transgene expression in muscle. Hum. Gene Ther. *8*:659-669.
21. **Clark, K.R., F. Voulgaropoulou, D.M. Fraley, and P.R. Johnson**. 1995. Cell lines for the production of recombinant adeno-associated virus. Hum. Gene Ther. *6*:1329-1341.
22. **Cotmore, S.F. and P.J. Tattersall**. 1984. Characterization and molecular cloning of a human parvovirus genome. Science *226*:1161-1165.
23. **Crystal, R.G., N.G. McElvaney, M.A. Rosenfeld, C.S. Chu, A. Mastrangeli, J.G. Hay, S.L. Brody, H.A. Jaffe, N.T. Eissa, and C. Danel**. 1994. Administration of an adenovirus containing the human CFTR cDNA to the respiratory tract of individuals with cystic fibrosis. Nat. Genet. *270*:404-410.
24. **Cukor, G., N.R. Blacklow, S. Kibrick, and I.C. Swan**. 1975. Effect of adeno-associated virus on cancer expression by herpesvirus-transformed hamster cells. J. Natl. Cancer Inst. *55*:957-959.
25. **Donahue, R.E., S.W. Kessler, D.M. Bodine, K. McDonagh, C. Dunbar, S. Goodman, B. Agricola, E. Byrne, et al.** 1992. Helper virus induced T cell lymphoma in non-human primates after retroviral mediated gene transfer. J. Exp. Med. *176*:1125-1135.
26. **Duan, D., P. Sharma, J. Yang, Y. Yue, L. Dudus, Y. Zhang, K.J. Fisher, and J.F. Engelhardt**. 1998. Circular intermediates of recombinant adeno-associated virus have defined structural characteristics responsible for long-term episomal persistence in muscle tissue. J. Virol. *72*:8568-8577.
27. **During, M.J., R. Xu, D. Young, M.G. Kaplitt, R.S. Sherwin, and P. Leone**. 1998. Peroral gene therapy of lactose intolerance using an adeno-associated virus vector. Nat. Med. *4*:1131-1135.
28. **Einerhand, M.P.W., M. Antoniou, S. Zolotukhin, N. Muzyczka, K.I. Berns, F. Grosveld, and D. Valerio**. 1995. Regulated high-level human beta-globin gene expression in erythroid cells following recombinant adeno-associated virus-mediated gene transfer. Gene Ther. *2*:336-343.
29. **Ferrari, F.K., T. Samulski, T. Shenk, and R.J. Samulski**. 1996. Second-strand synthesis is a rate-limiting step for efficient transduction by recombinant adeno-associated virus vectors. J. Virol. *70*:3227-3234.
30. **Fisher, K.J., G.-P. Gao, M.D. Weitzman, R. DeMatteo, J.F. Burda, and J.M. Wilson**. 1996. Transduction with recombinant adeno-associated virus for gene therapy is limited by leading-strand synthesis. J. Virol. *70*:520-532.
31. **Fisher, K.J., K. Jooss, J. Alston, Y. Yang, S.E. Haecker, K. High, R. Pathak, S.E. Raper, and J.M. Wilson**. 1997. Recombinant adeno-associated virus for muscle directed gene therapy. Nat. Med. *3*:306-312.
32. **Fisher-Adams, G., K.K. Wong, G. Podsakoff, S.J. Forman, and S. Chatterjee**. 1996. Integration of adeno-associated virus vectors in CD34+ human hematopoietic progenitor cells after transduction. Blood *88*:492-504.
33. **Flotte, T.R., S.A. Afione, C. Conrad, S.A. McGrath, R. Solow, H. Oka, P.L. Zeitlin, W.G. Guggino, and B.J. Carter**. 1993. Stable in vivo expression of the cystic fibrosis transmembrane conductance regulator with an adeno-associated virus vector. Proc. Natl. Acad. Sci. USA *90*:10613-10617.
34. **Flotte, T.R. and B.J. Carter**. 1995. Adeno-associated virus vectors for gene therapy. Gene Ther. *2*:357-362.
35. **Gao, G.-P., G. Qu, L.Z. Faust, R.K. Engdahl, W. Xiao, J.V. Hughes, P.W. Zoltic, and J.M. Wilson**. High-titer adeno-associated viral vectors from a Rep/Cap cell line and hybrid shuttle vector. Hum. Gene Ther. *9*:2353-2362.
36. **Grant, C.A., S. Ponnazhagan, X.-S. Wang, A. Srivastava, and T. Li**. 1997. Evaluation of recombinant adeno-associated virus as a gene transfer vector for retina. Curr. Eye Res. *16*:949-956.
37. **Grimm, D., A. Kern, K. Rittner, and J. Kleinschmidt**. 1998. Novel tools for production and purification of recombinant adeno-associated virus vectors. Hum. Gene Ther. *9*:2745-2760.
38. **Grossman, M., S.E. Raper, K. Kozarsky, E.A. Stein, J.F. Engelhardt, D. Muller, P.J. Lupien, and J.M. Wilson**. 1994. Successful ex vivo gene therapy directed to liver in a patient with familial hypercholestero-

laemia. Nat. Genet. *6*:335-341.

39. **Halbert, C.L., T.A. Standaert, C.B. Wilson, and A.D. Miller.** 1998. Successful readminstration of adeno-associated virus vectors to the mouse lung requires transient immunosuppression during the initial exposure. J. Virol. *72*:9795-9805.
40. **Hermonat, P.L. and N. Muzyczka.** 1984. Use of adeno-associated virus as a mammalian DNA cloning vector: transduction of neomycin resistance into mammalian tissue culture cells. Proc. Natl. Acad. Sci. USA *81*:6466-6470.
41. **Herzog, R.W., J.N. Hagstrom, S.H. Kung, S.J. Tai, J. M. Wilson, K.J. Fisher, and K.A. High.** 1997. Stable gene transfer and expression of human blood coagulation factor IX after intramuscular injection of recombinant adeno-associated virus. Proc. Natl. Acad. Sci. USA *94*:5804-5809.
42. **Herzog, R.W., E.Y. Yang, L.B. Couto, J.N. Hagstrom, D. Elwell, P.A. Fields, M. Burton, D.A. Bellinger, et al.** 1999. Long-term correction of canine hemophilia B by gene transfer of blood coagulation factor IX mediated by adeno-associated viral vector. Nat. Med. *5*:56-63.
43. **Hodgson, C.P.** 1995. The vector void in gene therapy. Biotechnology *13*:222-225.
44. **Jindal, R.M., R.A. Sidner, M.R. Bochan, and A. Srivastava.** 1998. Adeno-associated virus vectors: potential for gene therapy. Graft *1*:146-152.
45. **Kaplitt, M.G., Leone, R.J. Samulski, X. Xiao, D.W. Pfaff, K.L. O'Malley, and M.J. During.** 1994. Long-term gene expression and phenotypic correction using adeno-associated virus vectors in the mammalian brain. Nat. Genet. *8*:148-153.
46. **Kaplitt, M.G., X. Xiao, R.J. Samulski, J. Li, K. Ojamaa, I.L. Klein, H. Makimura, M.J. Kaplitt, R.K. Strumpf, and E.B. Diethrich.** 1996. Long-term gene transfer in porcine myocardium after coronary infusion of an adeno-associated virus vector. Ann. Thor. Surg. *62*:1669-1676.
47. **Kearns, W.G., S.A. Afione, S.B. Fulmer, M.G. Pang, D. Erikson, M. Egan, M. Landrum, T.R. Flotte, and G.R. Cutting.** 1996. Recombinant adeno-associated virus (AAV-CFTR) vectors do not integrate in a site-specific fashion in an immortalized epithelial cell line. Gene Ther. *3*:748-755.
48. **Kessler, P.D., G.M. Podsakoff, X. Chen, S.A. McQuiston, P.C. Colosi, L.A. Matelis, G.J. Kurtzman, and B.J. Byrne.** 1996. Gene delivery to skeletal muscle results in sustained and expression and systemic delivery of a therapeutic protein. Proc. Natl. Acad. Sci. USA *93*:14082-14087.
49. **Knowles, M.R., K.W. Hohneker, Z. Zhou, J.C. Olsen, T.L. Noah, P.-C. Hu, M.W. Leigh, J.F. Engelhardt et al.** 1995. A controlled study of adenoviral vector-mediated gene transfer in the nasal epithelium of patients with cystic fibrosis. N. Engl. J. Med. *333*:823-831.
50. **Koeberl, D.D., I.E. Alexander, C.L. Halbert, D.W. Russell, and A.D. Miller.** 1997. Persistent expression of human clotting factor IX from mouse liver after intravenous injection of adeno-associated virus vectors. Proc. Natl. Acad. Sci. USA *94*:1426-1431.
51. **Kotin, R.M.** 1994. Prospects for the use of adeno-associated virus as a vector for human gene therapy. Hum. Gene Ther. *5*:793-801.
52. **Kotin, R.M. and K.I. Berns.** 1989. Organization of adeno-associated virus DNA in latently infected Detroit 6 cells. Virology *170*:460-467.
53. **Kotin, R.M., R.M. Linden, and K.I. Berns.** 1992. Characterization of a preferred site on human chromosome 19q for integration of adeno-associated virus DNA by non-homologous recombination. EMBO J. *11*:5071-5078.
54. **Kotin, R.M., J.C. Menninger, D.C. Ward, and K.I. Berns.** 1991. Mapping and direct visualization of a region-specific viral DNA integration site on chromosome 19q13-qter. Genomics *10*:831-834.
55. **Kotin, R.M., M. Siniscalco, R.J. Samulski, X. Zhu, L. Hunter, C.A. Laughlin, S. McLaughlin, N. Muzyczka, M. Rocchi, and K.I. Berns.** 1990. Site-specific integration by adeno-associated virus. Proc. Natl. Acad. Sci. USA *87*:2211-2215.
56. **Kube, D.M., S. Ponnazhagan, and A. Srivastava.** 1997. Encapsidation of adeno-associated virus type 2 Rep proteins in progeny virions: Rep-mediated growth inhibition of primary human cells. J. Virol. *71*:7361-7371.
57. **Luhovy, M., S. McCune, J.-Y. Dong, J.F. Prchal, T.M. Townes, and J.T. Prchal.** 1996. Stable transduction of adeno-associated virus into hematopoietic stem cells from normal and sickle-cell patients. Biol. Blood Marrow Transpl. *2*:24-30.
58. **Lalwani, A., B. Walsh, G.J. Carvalho, N. Muzyczka, and A. Mhatre.** 1998. Expression of adeno-associated virus integrated transgene within the mammalian vestibular organs. Am. J. Otol. *3*:390-395.
59. **Lalwani, A., B. Walsh, P. Reilly, G. Carvalho, S. Zolotukhin, N. Muzyczka, and A. Mhatre.** 1998. Long-term in vivo cochlear transgene expression mediated by recombinant adeno-associated virus. Gene Ther. *5*:277-281.
60. **Lalwani, A., B. Walsh, P. Reilly, N. Muzyczka, and A. Mhatre.** 1996. Development of in vivo gene therapy for hearing disorders: introduction of adeno-associated virus into the cochlea of the guinea pig. Gene Ther. *3*:588-592.
61. **Lebkowski, J.S., M.M. McNally, T.B. Okarma, and L.B. Lerch.** 1988. Adeno-associated virus: a vector system for efficient introduction of DNA into a variety of mammalian cell types. Mol. Cell. Biol. *8*:3988-3996.
62. **Li, J., D. Dressman, Y.P. Tsao, A. Sakamoto, E.P. Hoffman, and X. Xiao.** 1999. Recombinant AAV vector-mediated sarcogylcan gene transfer in a hamster model for limb girdle muscular dystrophy. Gene Ther. *6*:74-82.
63. **Mah, C., K.Y. Qing, J. Hansen, B. Khuntirat, M.C. Yoder, and A. Srivastava.** 1998. Gene transfer with adeno-associated virus vectors: the growth factor receptor connection. Gene Ther. Mol. Biol. *3*:57-65.
64. **Mah, C., K.Y. Qing, B. Khuntirat, S. Ponnazhagan, X.-S. Wang, D.M. Kube, M.C. Yoder, and A. Srivastava.** 1998. Adeno-associated virus type 2-mediated gene transfer: role of epidermal growth factor receptor protein tyrosine kinase in transgene expression. J. Virol. *72*:9835-9843.

65. **Mamounas, M., M. Leavitt, M. Yu, and F. Wong-Staal.** 1995. Increased titer of recombinant AAV vectors by gene transfer with adenovirus coupled to DNA-polylysine complexes. Gene Ther. *2*:429-432.

66. **Mayor, H.D., G.S. Houlditch, and D.M. Mumford.** 1973. Influence of adeno-associated satellite virus on adenovirus-induced tumors in hamsters. Nat. New Biol. *241*:44-46.

67. **McCown, T.J., X. Xiao, J. Li, G.R. Breese, and R.J. Samulski.** 1996. Differential and persistent expression patterns of CNS gene transfer by an adeno-associated virus (AAV) vector. Brain Res. *713*:99-107.

68. **Miao, C.H., R.O. Snyder, D.B. Schowalter, G.A. Patijn, B. Donahue, B. Winther, and M.A. Kay.** 1998. The kinetics of rAAV integration in the liver. Nat. Genet. *19*:13-15.

69. **Miller, J.L., R.E. Donahue, S.E. Sellers, R.J. Samulski, N.S. Young, and A.W. Nienhuis.** 1994. Recombinant adeno-associated virus (rAAV)-mediated expression of human γ-globin gene in human progenitor-derived erythroid cells. Proc. Natl. Acad. Sci. USA *91*:110183-1018.

70. **Miller, N., and R. Vile.** 1995. Targeted vectors for gene therapy. FASEB J. *9*:190-199.

71. **Monahan, P.E., R.J. Samulski, J. Tazelaar, X. Xiao, T.C. Nichols, D.A. Bellinger, M.S. Read, and C.E. Walsh.** 1998. Direct intramuscular injection with recombinant AAV vectors results in sustained expression in a dog model of hemophilia. Gene Ther. *5*:40-49.

72. **Muramatsu, S.I., H. Mizukami, N.S. Young, and K.E. Brown.** 1996. Nucleotide sequencing and generation of an infectious clone of adeno-associated virus 3. Virology 22:208-217.

73. **Murphy, J.E., S.Z. Zhou, K. Giese, L.T. Williams, J.A. Escobedo, and V.J. Dwarki.** 1997. Long-term correction of obesity and diabetes in genetically obese mice by a single intramuscular injection of recombinant adeno-associated virus encoding mouse leptin. Proc. Natl. Acad. Sci. USA *94*:13921-13926.

74. **Muzyczka, N.** 1992. Use of adeno-associated virus as a general transduction vector for mammalian cells. Curr. Top. Microbiol. Immunol. *158*:97-129.

75. **Nahreini, P., M.J. Woody, S.Z. Zhou, and Srivastava.** 1993. Versatile adeno-associated virus 2-based vectors for constructing recombinant virions. Gene *124*:257-262.

76. **Ostrove, J.M., D.H. Duckworth, and K.I. Berns.** 1981. Inhibition of adenovirus-transformed cell oncogenicity by adeno-associated virus. Virology *113*:521-533.

77. **Ozawa, K., G.J. Kurtzman, and N.S. Young.** 1986. Replication of the B19 parvovirus in human bone marrow cultures. Science *233*:883-886.

78. **Ozawa, K., G.J. Kurtzman, and N.S. Young.** 1987. Productive infection by B19 parvovirus of human erythroid bone marrow cells in vitro. Blood *70*:384-391.

79. **Pan, R., X. Xiao, S. Chen, J. Li, L. Lin, H. Wang, and Y. Tsao.** 1999. Disease-inducible transgene expression from a recombinant adeno-associated virus vector in a rat arthritis model. J. Virol. *73*:3410-3417.

80. **Pattison, J.R.** 1988. *In* Parvovirus and Human Disease. CRC Press, Boca Raton.

81. **Ping, P., Q. Yang, and H.K. Hammond.** 1996. Altered beta-adrenergic receptor signaling in heart failure, in vivo gene transfer via adeno and adeno-associated virus. Microcirculation *3*:225-228.

82. **Podsakoff, G., K.K. Wong, Jr., and S. Chatterjee.** 1994. Stable and efficient gene transfer into non-dividing cells by adeno-associated virus (AAV)-based vectors. J. Virol. *68*:5656-5666.

83. **Ponnazhagan, S., D. Erikson, W.G. Kearns, S.Z. Zhou, P. Nahreini, X.-S. Wang, and A. Srivastava.** 1997. Lack of site-specific integration of the recombinant adeno-associated virus 2 genomes in human cells. Hum. Gene Ther. *8*:275-284.

84. **Ponnazhagan, S., P. Mukherjee, X.-S. Wang, K.Y. Qing, D.M. Kube, C. Mah, C. Kurpad, M.C. Yoder, E.F. Srour, and A. Srivastava.** 1997. Adeno-associated virus type 2-mediated transduction of primary human bone marrow-derived CD34+ hematopoietic progenitor cells: donor variation and correlation of transgene expression with cellular differentiation. J. Virol. *71*:8262-8267.

85. **Ponnazhagan, S., P. Mukherjee, M.C. Yoder, X.-S. Wang, S.Z. Zhou, J. Kaplan, S. Wadsworth, and A. Srivastava.** 1997. Adeno-associated virus 2-mediated gene transfer in vivo: organ-tropism and expression of transduced sequences in mice. Gene *190*:203-210.

86. **Ponnazhagan, S., M.L. Nallari, and A. Srivastava.** 1994. Suppression of human a-globin gene expression mediated by the recombinant adeno-associated virus 2-based anti-sense vectors. J. Exp. Med. *179*:733-738.

87. **Ponnazhagan, S., M.C. Yoder, and A. Srivastava.** 1997. Adeno-associated virus type 2-mediated transduction of murine hematopoietic cells with long-term repopulating ability and sustained expression of a human globin gene in vivo. J. Virol. *71*:3098-3104.

88. **Ponnazhagan, S., X.-S. Wang, M.J. Woody, F. Luo, L.Y. Kang, M.L. Nallari, N.C. Munshi, S.Z. Zhou, and A. Srivastava.** 1996. Differential expression in human cells from the p6 promoter of human parvovirus B19 following plasmid transfection and recombinant adeno-associated virus 2 (AAV) infection: human megakaryocytic leukaemia cells are non-permissive for AAV infection. J. Gen. Virol. *77*:1111-1122.

89. **Ponnazhagan, S., K.A. Weigel, S.P. Raikwar, P. Mukherjee, M.C. Yoder, and A. Srivastava.** 1998. Recombinant human parvovirus B19 vectors: erythroid cell-specific delivery and expression of transduced genes. J. Virol. *72*:5224-5230.

90. **Qing, K.Y., B. Khuntirat, C. Mah, D.M. Kube, X.-S. Wang, S. Ponnazhagan, S.Z. Zhou, V.J. Dwarki, M.C. Yoder, and A. Srivastava.** 1998. Adeno-associated virus type 2-mediated gene transfer: correlation of tyrosine phosphorylation of the cellular single-stranded D sequence-binding protein with transgene expression in human cells in vitro and murine tissues in vivo. J. Virol. *72*:1593-1599.

91. **Qing, K.Y., C. Mah, J. Hansen, S.Z. Zhou, V.J. Dwarki, and A. Srivastava.** 1999. Human fibroblast growth factor receptor 1 is a co-receptor for infection by adeno-associated virus 2. Nat. Med. *5*:71-77.

92. **Qing, K.Y., X.-S. Wang, D.M. Kube, S. Ponnazhagan, A. Bajpai, and A. Srivastava.** 1997. Role of tyro-

sine phosphorylation of a cellular protein in adeno-associated virus 2-mediated transgene expression. Proc. Natl. Acad. Sci. USA *94*:10879-10884.

93. **Rolling, F., W. Shen, H. Tabarias, I. Constable, Y. Kanagasingam, C.J. Barry, and P.E. Rakoczy**. 1999. Evaluation of adeno-associated virus-mediated gene transfer into the rat retina by clinical fluorescence photography. Hum. Gene Ther. *10*:641-648.

94. **Rosenberg, S.A., P. Abersold, K. Cornetta, A. Kasid, R.A. Morgan, R. Moen, E.M. Karson, M.T. Lotze, et al.** 1990. Gene transfer into humans—immunotherapy of patients with advanced melanoma, using tumor-infiltrating lymphocytes modified by retroviral gene transduction. N. Engl. J. Med. *323*:570-577.

95. **Rutledge, E.A., C.L. Halbert, and D.W. Russell**. 1998. Infectious clones and vectors derived from adeno-associated virus (AAV) serotypes other than AAV type 2. J. Virol. *72*:309-319.

96. **Samulski, R.J., L.-S. Chang, and T. Shenk**. 1989. Helper-free stocks of recombinant adeno-associated viruses: normal integration does not require viral gene expression. J. Virol. *63*:3822-3828.

97. **Samulski, R.J., X. Zhu, X. Xiao, J. Brook, D.E. Houseman, N. Epstein, and L.A. Hunter**. 1991. Targeted integration of adeno-associated virus (AAV) into human chromosome 19. EMBO J. *10*:3941-3950.

98. **Siegl, G., R.C. Bates, K.I. Berns, B.J. Carter, D.C. Kelly, E. Kurstak, and P.J. Tattersall**. 1985. Characteristics and taxonomy of parvoviridae. Intervirology *23*:61-73.

99. **Snyder, R.O., C. Miao, L. Meuse, J. Tubb, B.A. Donahue, H.-F. Lin, W. Stafford, S. Patel, et al.** 1999. Correction of hemophilia B in canine and murine models using recombinant adeno-associated viral vectors. Nat. Med. *5*:64-70.

100. **Snyder, R.O., C.H. Miao, G.A. Patijn, S.K. Spratt, O. Danos, D. Nagy, A.M. Gown, B. Winther, et al.** 1997. Persistent and therapeutic concentrations of human factor IX in mice after hepatic gene transfer of recombinant AAV vectors. Nat. Genet. *16*:270-276.

101. **Snyder, R.O., S.K. Spratt, C. Lagarde, D. Bohl, B. Kaspar, B. Sloan, L.K. Cohen, and O. Danos**. 1997. Efficient and stable adeno-associated virus-mediated gene transduction in the skeletal muscle of adult immunocompetent mice. Hum. Gene Ther. *8*:1891-1900.

102. **Srivastava, A.** 1994. Parvovirus-based vectors for human gene therapy. Blood Cells *20*:531-538.

103. **Srivastava, A.** 1998. Delivery systems for gene therapy: adeno-associated virus 2, p. 257-288. *In* G. Stein, P. Quesenberry, B. Forget, and S. Weissman (Eds.), Stem Cell Biology and Gene Therapy. Wiley-Liss, New York.

104. **Srivastava, A.** 1999. Parvovirus vectors for human hematopoietic gene therapy, p. 89-122. *In* L.J. Fairbairn, and N. Testa (Eds.), Blood Cell Biochemistry: Hematopoiesis and Gene Therapy. Plenum Press, New York.

105. **Srivastava, A., E. Bruno, R. Briddell, R. Cooper, C. Srivastava, K. van Besien, and R. Hoffman**. 1990. Parvovirus B19-induced perturbation of human megakaryocytopoiesis in vitro. Blood *76*:1997-2000.

106. **Srivastava, A. and L. Lu**. 1988. Replication of B19 parvovirus in highly enriched hematopoietic progenitor cells from normal human bone marrow. J. Virol. *62*:3059-3063.

107. **Srivastava, A., E.W. Lusby, and K.I. Berns**. 1983. Nucleotide sequence and organization of the adeno-associated virus 2 genome. J. Virol. *45*:555-564.

108. **Srivastava, A., X.-S. Wang, S. Ponnazhagan, S.Z. Zhou, and M.C. Yoder**. 1996. Adeno-associated virus 2-mediated transduction and erythroid lineage-specific expression in human hematopoietic progenitor cells. Curr. Top. Microbiol. Immunol. *218*:93-117.

109. **Srivastava, C.H., R.J. Samulski, L. Lu, S.H. Larsen, and A. Srivastava**. 1989. Construction of a recombinant human parvovirus B19: adeno-associated virus 2 (AAV) DNA inverted terminal repeats are functional in an AAV-B19 hybrid virus. Proc. Natl. Acad. Sci. USA *86*:8078-8082.

110. **Stamatoyannopoulos, G. and A.W. Nienhuis**. 1994. The molecular basis of blood diseases, p. 107-155. *In* G. Stamatoyannopoulos, A. Nienhuis, P. Majerus, and H.E. Varmus (Eds.), Hemoglobin Switching. W.B. Saunders, Philadelphia.

111. **Summerford, C., J.S. Bartlett, and R.J. Samulski**. 1999. aVb5 integrin: a co-receptor for adeno-associated virus 2 infection. Nat. Med. *5*:78-81.

112. **Summerford, C. and R.J. Samulski**. 1998. Membrane-associated heparan sulfate proteoglycan is a receptor for adeno-associated virus 2 virions. J. Virol. *72*:1438-1445.

113. **Tamayose, K., Y. Hirai, and T. Shimada**. 1996. A new strategy for large-scale preparation of high-titer recombinant adeno-associated virus vectors by using sulfonated cellulose column chromatography. Hum. Gene Ther. *7*:507-513.

114. **Tan, M.-Q., K.Y. Qing, M.C. Yoder, and A. Srivastava**. 1999. Adeno-associated virus 2-mediated transduction of primary murine hematopoietic progenitor cells in vitro and in vivo. Blood *94*:365a-366a.

115. **Thrasher, A.J., M. de Alwis, C.M. Casimir, C. Kinnon, K. Page, J. Lebkowski, A.W. Segal, and R.J. Levinsky**. 1995. Functional reconstitution of the NADPH oxidase by adeno-associated virus gene transfer. Blood *86*:761-765.

116. **Tooze, J.** 1981. *In* DNA Tumor Viruses. CSH Laboratory Press, Cold Spring Harbor, New York.

117. **Tuan, D.Y.H., W.B. Solomon, I.M. London, and D.P. Lee**. 1989. An erythroid specific, development stage-independent enhancer far upstream of the human "β-like globin" genes. Proc. Natl. Acad. Sci. USA *86*:2554-2559.

118. **Vincent-Lacaze, N., R.O. Snyder, R. Gluzman, D. Bohl, C. Lagarde, and O. Danos**. 1999. Structure of adeno-associated virus vector DNA following transduction of the skeletal muscle. J. Virol. *73*:1949-1955.

119. **Walsh, C.E., J.M. Liu., J.L. Miller, A.W. Nienhuis, and R.J. Samulski**. 1993. Gene therapy for human hemoglobinopathies. Proc. Soc. Exp. Biol. Med. *204*:289-300.

120. **Walsh, C.E., J.M. Liu, X. Xiao, N.S. Young, A.W. Nienhuis, and R.J. Samulski**. 1992. Regulated high level expression of a human γ-globin gene introduced into erythroid cells by an adeno-associated virus vec-

tor. Proc. Natl. Acad. Sci. USA *89*:7257-7261.

121.**Wang, X.-S., B. Khuntirat, K.Y. Qing, S. Ponnazhagan, D.M. Kube, S.Z. Zhou, V.J. Dwarki, and A. Srivastava**. 1998. Characterization of wild-type adeno-associated virus type 2-like particles generated during recombinant viral vector production and strategies for their elimination. J. Virol. *72*:5472-5480.

122.**Wang, X.-S., M.C. Yoder, S.Z. Zhou, and A. Srivastava**. 1995. Parvovirus B19 promoter at map unit 6 confers replication competence and erythroid specificity to adeno-associated virus 2 in primary human hematopoietic progenitor cells. Proc. Natl. Acad. Sci. USA *92*:12416-12420.

123.**Weiss, R.N., N. Teich, H.E. Varmus, and J.M. Coffin**. 1984. *In* RNA Tumor Viruses. CSH Laboratory Press, Cold Spring Harbor, New York.

124.**Xiao, W., S.C. Berta, M.M. Lu, A.D. Moscioni, J. Tazelaar, and J.M. Wilson**. 1998. Adeno-associated virus as a vector for liver-directed gene therapy. J. Virol. *72*:10222-10226.

125.**Xiao, W., N. Chirmule, S.C. Berta, B. McCullough, G. Gao, and J.M. Wilson**. 1999. Gene therapy vectors based on adeno-associated virus type 1. J. Virol. *73*:3994-4003.

126.**Xiao, X., W. deVlaminck, and J. Monahan**. 1993. Adeno-associated virus (AAV) vectors for gene transfer. Adv. Drug Del. Rev. *12*:201-215.

127.**Xiao, X., J. Li, and R.J. Samulski**. 1996. Efficient long-term gene transfer into muscle tissue of immunocompetent mice by adeno-associated virus vector. J. Virol. *70*:8098-8108.

128.**Xiao, X., J. Li, and R.J. Samulski**. 1998. Production of high-titer recombinant adeno-associated virus vectors in the absence of helper adenovirus. J. Virol. *72*:2224-2232.

129.**Ye, X., V.M. Rivera, P. Zoltick, F. Cerasoli, Jr., M.A. Schnell, G.-P. Gao, J.V. Hughes, M. Gilman, and J.M. Wilson**. 1999. Regulated delivery of therapeutic proteins after in vivo somatic cell gene transfer. Science *283*:88-91.

130.**Zabner, J., L.A. Couture, R.J. Gregory, S.M. Graham, A.E. Smith, and M.J. Welsh**. 1993. Adenovirus-mediated gene transfer transiently corrects the chloride transport defect in nasal epithelia of patients with cystic fibrosis. Cell *75*:207-216.

131.**Zhou, S.Z., S. Cooper, L.Y. Kang, L. Ruggieri, S. Heimfeld, A. Srivastava, and H.E. Broxmeyer**. 1994. Adeno-associated virus 2-mediated high efficiency gene transfer into immature and mature subsets of hematopoietic progenitor cells in human umbilical cord blood. J. Exp. Med. *179*:1867-1875.

132.**Zhou, S.Z., Q. Li, G. Stamatoyannopoulos, and A. Srivastava**. 1996. Adeno-associated virus 2-mediated transduction and erythroid cell-specific expression of a human b-globin gene. Gene Ther. *3*:223-229.

133.**Zolotukhin, S., B.J. Byrne, E. Mason, I. Zolotukhin, M. Potter, K. Chesnut, C. Summerford, R.J. Samulski, and N. Muzyczka**. 1999. Recombinant adeno-associated virus purification using novel methods improves infectious titer and yield. Gene Ther. *6*:973-985.

2 Strategies for Production of Adeno-Associated Virus Vectors

Dirk Grimm and Jürgen A. Kleinschmidt
Institute of Applied Tumor Virology, German Cancer Research Center, Heidelberg, Germany

INTRODUCTION

Owing to a unique combination of attractive features, the adeno-associated virus type 2 (AAV-2), a human parvovirus, is one of the most promising viruses to develop vectors for human gene therapy. AAV-2 is classified as naturally defective, requiring co-infection by a helper virus (e.g., adenovirus [Ad] or herpes simplex virus [HSV]) to complete a full life cycle (16). In the absence of a helper virus, AAV-2 is able to stably integrate into a defined region on the long arm of human chromosome 19 (22,80,85,141). For this integration to occur efficiently, expression of one of the large nonstructural AAV-2 proteins, Rep78 or Rep68, is required (6,153). Common AAV-2 vectors, which are devoid of all viral genes (see Production of Wild-Type and Recombinant AAV-2), integrate at a lower efficiency and are no longer site-specific (77,95,112,120,134,164,176). On the other hand, recombinant AAV-2 (rAAV) genomes can transform into stable concatemers or interlocked circles and persist in an episomal form (40,163), thus providing the very appealing potential of AAV-2 vectors of mediating long-term gene expression in vivo. Indeed, permanent expression of a variety of genes transduced by AAV-2 vectors has already been observed for up to 18 months after rAAV administration in immunocompetent small and large animal models (32,47,57,67,78, 93,128,148,177,183).

Due to the absence of viral genes, common AAV-2 vectors were found to cause only a mild immune response, thus making readministration feasible, albeit with varying degrees of success (47,57,58,93, 97). In addition, no human disease has ever been shown to be associated with an AAV-2 infection, indeed approximately 80% of adults in the United States are seropositive (100). However, it should be mentioned that a relationship between AAV-2 infection and increased frequency of spontaneous abortion in mice and humans has been suggested although not convincingly proven (18,157). Accordingly, AAV-2 has established one of the highest biosafety ratings among all viral vector systems currently available.

Viral Vectors: Basic Science and Gene Therapy
Edited by A. Cid-Arregui and A. García-Carrancá

Another intriguing feature for human gene therapy applications is that AAV-2 vectors can transduce both mitotic and postmitotic cells (3,76,104,132) of a variety of cell types in different organs, such as muscle, lung, central nervous system, retina, and liver cells, albeit not equally efficiently (49,50,78,115,148). In general, the efficiency of rAAV-mediated gene transfer may vary with the particular mitotic state of the target cell (51,119,132). Several obstacles have recently been identified that can limit the expression of transduced genes, such as impaired intracellular trafficking of the virus (59) or alterations in the cellular redox state (142). Moreover, external stimuli, e.g. irradiation of infected cells (1,3), and in particular several distinct cellular or helper virus proteins (45,46,94,123,125) have been shown to influence the efficiency of AAV-2 vector-mediated gene transduction. The precise nature and role of some of these factors is reviewed by Srivastava in Chapter 1.

Despite AAV-2 vectors' appealing features, progress in the evaluation of these vectors for human gene therapy has long been hindered by the lack of methods for the production of high-titer vector stocks. The respective basic protocol (see Production of Wild-Type and Recombinant AAV-2) is quite cumbersome and yields only low titers of recombinant AAV-2 particles. In addition, AAV-2 vector stocks are contaminated with Ad and wild-type (wt) AAV-2 as by-products and, thus do not suit the requirements for clinical gene therapy trials. Consequently, a large number of modifications to the basic protocol have been proposed, which aim to improve AAV-2 vector production at different levels. This review is devoted to a critical description of the most recent progress in the production of AAV-2-derived vectors. Two more sections will also focus on advances in the technology of AAV-2 vector purification and titration.

AAV-2 STRUCTURE AND LIFE CYCLE

AAV-2 is a member of the family Parvoviridae (144). In common with the other members of this family, AAV-2 is a small virus with a nonenveloped, icosahedral capsid of approximately 18 to 24 nm in diameter. It carries a linear, single-stranded DNA genome of 4681 nucleotides, with both plus and minus strands being encapsidated with equal efficiency. Together with five other serotypes (AAV-1 and AAV-3 to AAV-6; see References 9,11,23,27,108, and 133), AAV-2 belongs to the genus *Dependovirus*, a name that implies their requirement of a helper virus (usually Ad or HSV) to undergo a productive infection. Indeed, the AAV-2 life cycle is characterized by two distinct phases, with the helper virus acting as a switch between them (Figure 1) (16). Either phase starts with AAV-2 virions entering the cell and the genome being uncoated. The process of virus entry has recently begun to be better understood. Heparan sulfate proteoglycans, which are present on a large variety of cells as an integral part of the plasma membrane, have been identified as the main receptors that bind AAV-2 virions and trigger their internalization (152). Moreover, the fibroblast growth factor (FGF) receptor as well as integrin $\alpha v\beta 5$ have been proposed to be second receptors that can enhance AAV-2 binding and uptake (124,151). However, the precise role of integrin $\alpha v\beta 5$ for AAV-2 infection has been questioned and is still a matter of controversy (126,127). Likewise, little is known about the fate of the virus after binding to its receptor and the uncoating of the viral genome within the cell. Two recent reports gave first evidence that AAV-2 internalizes predominantly via receptor-mediated endocytosis from clathrin-coated pits (12), a process that is particularly dependent on dynamin function (39).

After release from the endosome, AAV-2 then moves rapidly to the cell nucleus where it accumulates within 30 minutes after cell entry. Since later in the infection viral particles can be detected within the cell nucleus, it is currently assumed that AAV enters the nucleus prior to uncoating and subsequently releases its genome (12). Once there, the regulatory proteins of AAV-2 are expressed in limited amounts, and these subsequently repress further viral gene expression as well as replication of the AAV-2 genome (for review see Reference 17). At the same time, these proteins enhance integration of the viral genome into the host chromosomes, thereby establishing a latent infection. As a unique feature of AAV-2, its integration does not occur randomly as in the case of other viruses, such as retroviruses, but preferentially targets a defined region on the long arm of human chromosome 19 (80,141). This latent AAV-2 infection induces only subtle effects on the cell, such as an increased ability to respond to stress factors (8,79,165) or a perturbation of the cell cycle (61,172).

In the alternate phase (i.e., when the cells are superinfected by a helper virus) AAV-2 will undergo a productive infection. Helper virus infection can precede AAV-2 infection or, conversely, cells that already contain an integrated AAV-2 provirus can be superinfected. In both cases, AAV-2 gene expression is stimulated, the AAV-2 genome replicates (after excision from the host genome), and is packaged into newly formed AAV-2 capsids. The helper virus finally triggers cell lysis, and the AAV-2 as well as helper virus particles are released, thereby enabling a new round of infection. For Ad and HSV, the most common helper viruses, the gene products that are required

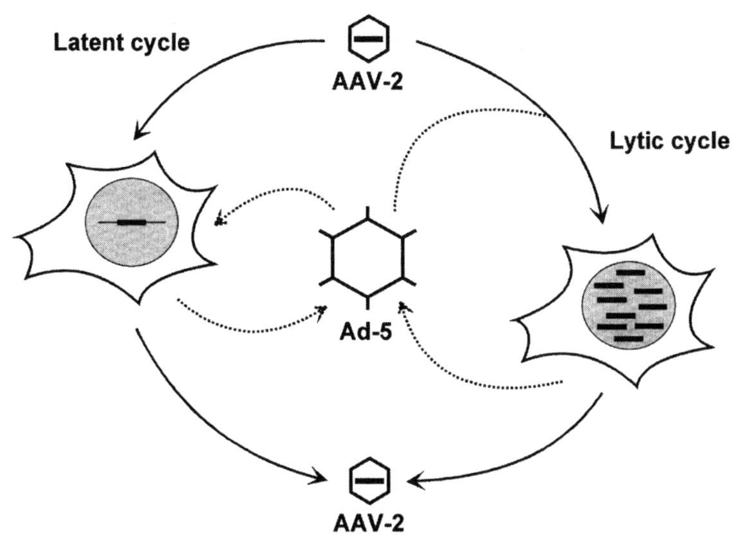

Figure 1. AAV-2 life cycle. Right half (lytic cycle). Under permissive conditions, i.e., in the presence of a helper virus (e.g., Ad-5), AAV-2 replicates in infected cells, thereby resulting in a viral stock containing a mixture of AAV-2 and helper virus particles. Left half (latent cycle). If cells are infected with AAV-2 in the absence of a helper virus (nonpermissive conditions), AAV-2 stably integrates into the q arm of human chromosome 19, where it remains until the cells are challenged with a helper virus. This leads to rescue of the integrated provirus and induction of the lytic AAV-2 life cycle.

for supporting AAV-2 production have been identified. Interestingly, while the Ad helper gene products are mainly involved in regulating gene expression (reviewed in Reference 99), HSV provide proteins involved in DNA replication (169). This difference may highlight the general role of the helper viruses in altering the cellular milieu, allowing AAV-2 gene expression, replication, and packaging to occur, while the specific mechanisms involved appear to be largely different.

Because of its apparent strict dependence upon a second virus, AAV-2 was long considered to be a defective virus. However, later studies showed that treatment of cells with genotoxic agents, such as UV irradiation or carcinogens, can render them permissive for a productive AAV-2 infection in the absence of a helper virus (179–181), although the yield of virus was somewhat lower in these cases. Nevertheless, it is now believed that AAV-2 is rather oriented to undergoing latent infection than being strictly dependent on a helper virus.

The AAV-2 genome has been cloned and completely sequenced (130,149). The linear single-stranded DNA genome comprises only two large open reading frames (ORFs), encoding the replication (rep ORF) and the capsid (cap ORF) proteins, respectively (Figure 2). Four closely related proteins with partially overlapping amino acid sequences are encoded by the rep ORF in the left half of the genome. On the basis of their apparent molecular weights, these proteins were designated Rep78, Rep68, Rep52, and Rep40. The two larger Rep proteins are translated from unspliced and spliced transcripts originating from a promoter at map unit 5 (p5 promoter), while Rep52 and Rep40 are expressed from identically spliced mRNAs initiating from a second promoter, p19. The *cap* gene, located in the right half of the AAV-2 genome, encodes the capsid proteins VP1, VP2, and VP3, which are translated from differentially spliced mRNAs originating at the third AAV-2 promoter, p40. Present at each end of the genome is a 145-nucleotide, inverted terminal repeat (ITR) sequence. Since the first 125 nucleotides constitute an overall palindrome, they can fold into a T-shaped stem loop structure when in the single-stranded form, which is important for replication of the AAV-2 genome. In addition, the presence of the ITR sequences *in cis* is sufficient for AAV-2 genome integration and rescue, as well as DNA encapsidation. Several lines of evidence suggest that additional, but as yet undefined *cis*-acting sequences within the AAV-2 genome may enhance the rate of DNA replication or encapsidation (55,63,105,160).

The Rep proteins have been well-characterized and shown to be directly involved in several aspects of AAV DNA replication and gene expression. The two large Rep proteins (Rep78 and Rep68) can homo-oligomerize and exhibit site-specific DNA binding, DNA helicase, and ATP-dependent DNA nicking activities, all of which are important for launching a new round of AAV-2 genome replication during a productive infection (for reviews see References 15 and 16). Moreover, these pleiotropic proteins play a major role in the positive and negative regulation of the AAV-2 promoters p5 and p19, which involves binding of Rep78 or Rep68 to sequences within the p5 promoter, as well as to distinct ITR sequences (14,81–83, 101–103,116,117,159,168). Expression of the VP proteins from the p40 promoter in the absence of ITRs, in contrast, is mainly stimulated by the helper virus in a manner independent of the Rep proteins (168). AAV-2 gene expression thus appears to be a highly complex process, having at least two different levels of control, depending on the presence or absence of a helper virus (116,168). Furthermore, a number of recent reports have demonstrated in vitro and in vivo interactions of Rep78 and

Rep68 with a variety of cellular proteins that might be involved in regulation of basic processes such as DNA replication and transcription (28,37,64–66,107,117, 167) Finally, the large Rep proteins are required for site-specific integration of the AAV-2 genome, probably by mediating recombination between the AAV-2 ITRs and homologous sequences within the so-called AAVS1 site of human chromosome 19 (6,26,42,43,84,90,91,114,118, 143,153,170). Interestingly, none of these processes seems to depend on the presence of the two smaller Rep proteins, Rep52 and Rep40. On the other hand, mutants that are deficient for Rep52/Rep40 are characterized by a decreased accumulation of infectious single-stranded viral genomes (21), suggesting that these Rep proteins play a role in stabilization or encapsidation of AAV-2 genomes during a productive infection. These processes probably also involve the recently described DNA helicase activity of Rep52 (146; J. King, unpublished observations).

By assembling in a 1:1:10 stoichiometry, the three structural VP proteins constitute the viral capsid, with VP3 being the most abundant protein. Capsid assembly is restricted to the nucleus, with initiation taking place in the nucleoli of infected cells (173) and the assembly efficiency being strongly dependent on the intracellular concentration of VP proteins (150,173). The assembly process does not require Rep proteins, but co-expression of Rep78 or Rep68 was shown to influence the intracellular distribution of VP proteins (173) and in fact, nuclear transport of the capsid proteins is an essential prerequisite for AAV capsid formation (70). Several lines of evidence suggest that direct or indirect interactions of Rep and VP proteins might be involved in formation of infectious AAV-2 particles (71,122,173,174). As discovered recently, this association of AAV-2 proteins is not

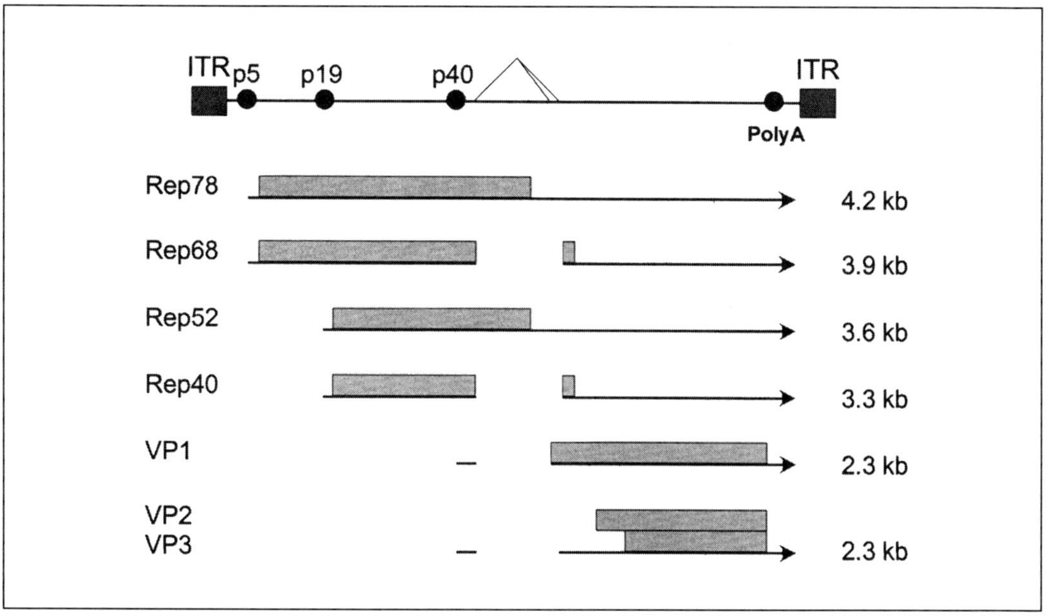

Figure 2. Map of the AAV-2 genome. Shown on top is a genetic map of AAV-2 including the three promoters (p5, p19, and p40), the single polyadenylation site (PolyA) and the two ITR at the ends of the genome. The position of the AAV-2 intron is depicted by a caret; the two sides on the right indicate that AAV-2 has two closely located splice acceptor sites. The arrows below the genome represent AAV-2 transcripts in the spliced (3.9, 3.3, and 2.3 kb) and unspliced (4.2 and 3.6 kb) forms, while the solid boxes above these lines indicate ORFs within these RNAs. Translation of these ORFs results in the proteins listed on the left.

dependent on the presence of packageable virus genomes, suggesting that Rep and VP proteins could form pre-encapsidation complexes which represent intermediates in the DNA packaging pathway (41). In fact, data from pulse-chase labeling experiments support a model in which empty capsids are preformed in the cells and then loaded with replicated viral genomes (110).

PRODUCTION OF WILD-TYPE AND RECOMBINANT AAV-2

As outlined above, wt AAV-2 can be easily propagated through infection of human culture cells with AAV-2 and a helper virus (Figure 1 and Figure 3, left panel), with typical yields of up to 10^4 infectious AAV-2 particles per cell (19). Information on an additional mode for producing wt AAV-2 arose from the discovery that prokaryotic clones of AAV-2 are infectious. So far, three bacterial plasmids carrying an entire AAV-2 genome have been successfully isolated, namely pSM620 (136), pAV1 (86) and psub201 (137). When either of these clones is transfected into Ad-infected cells, the AAV-2 genome is excised from the plasmid backbone, probably by mechanisms analogous to those for rescue of an integrated AAV-2 provirus from chromosomes. The AAV-2 genome is then replicated and encapsidated to yield infectious

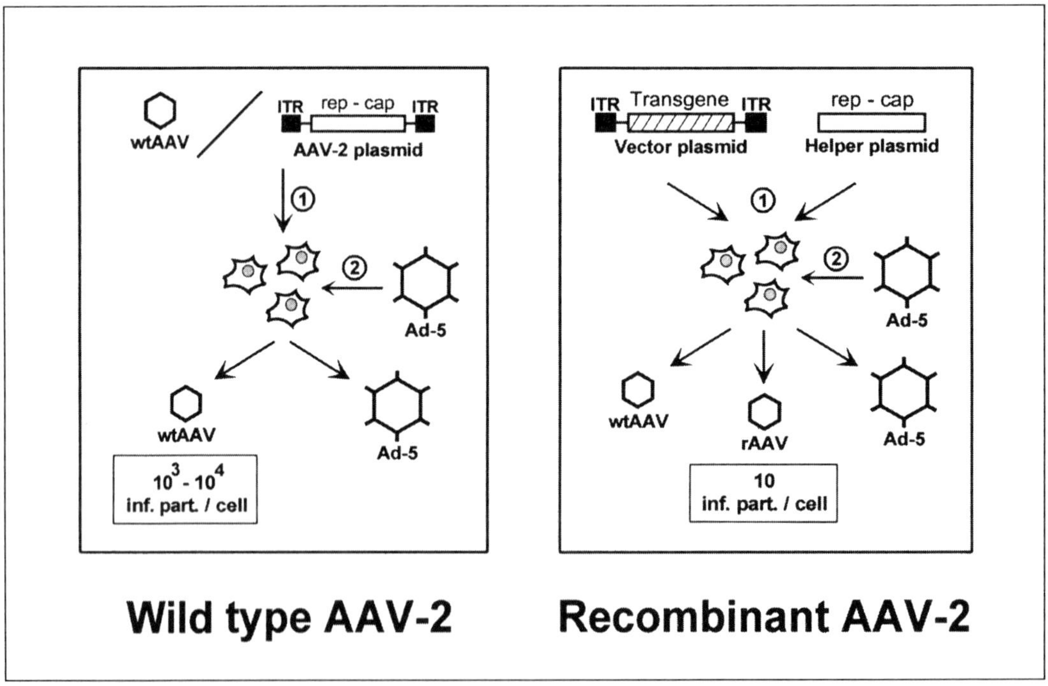

Figure 3. Production of AAV-2. Left panel, Generation of wt AAV-2. The wt AAV-2 genome is introduced into cells either via infection with AAV-2 virus or via transfection with a plasmid carrying a full-length AAV-2 genome, ①. In either case, the cells are additionally infected with a helper virus, ②, e.g., Ad-5. Within an incubation period of 72 hours, each cell that has received both the AAV-2 and Ad-5 genome produces 10^3 to 10^4 infectious AAV-2 particles. Right panel, Generation of recombinant AAV-2. Recombinant AAV-2 particles are produced through co-transfection of cells (e.g., 293 cells) with a rAAV vector plasmid, carrying the transgene to be packaged being flanked by the AAV-2 ITRs and a helper plasmid encoding the AAV-2 Rep and VP proteins, but lacking the ITRs to avoid its encapsidation, ①. Similar to wt AAV-2 production, the cells have to be infected with a helper virus, ②. Within 72 hours, each cell that was successfully co-transfected and infected generates a maximum of 10 infectious rAAV particles. Due to the initial helper virus infection, the cells also produce helper virus particles that contaminate the rAAV stocks. Moreover, this protocol often results in the unwanted generation of wt AAV-2, probably by means of nonhomologous recombination between vector and helper plasmid.

AAV-2 virions. This growth cycle is thus comparable to a normal productive infection, except that the viral genome is delivered via transfection rather than infection with AAV-2 (Figure 3, left panel). However, typical virus titers are somewhat lower, reaching only up to 10^3 infectious wt AAV-2 particles per cell (i.e., 10% of the yields that can be obtained from co-infection of cells with AAV-2 and Ad) (55,86).

The isolation of infectious AAV-2 plasmids provided a basis for the generation of AAV-2 vectors. As first reported by Hermonat and Muzyczka (63), co-transfection of helper virus-infected cells with a vector plasmid (i.e. a wild-type clone [pSM620] that had the AAV-2 *cap* gene replaced by heterologous genes) and a helper plasmid that carried an entire, but packaging-deficient, AAV-2 genome, resulted in production of recombinant AAV-2 vectors. A similar approach was also reported from Tratschin et al. (158). Unfortunately, these initial efforts were rather unsuccessful; recombinant particle titers reached only approximately 10^5/mL, and the vector preparations had wild-type contamination of up to 50%, which was derived from homologous recombination of the vector and helper plasmids at overlapping sequences.

To overcome these drawbacks, McLaughlin et al. (105) and Samulski et al. (138) introduced novel approaches for AAV-2 vector production (Figure 3, right panel). By deleting both the *rep* and *cap* genes from AAV-2 wild-type plasmids, minimal vector constructs were obtained that carried only the AAV-2 ITRs plus short adjacent stretches of noncoding sequences. The helper plasmids used to package these vector DNAs contained *rep* and *cap* genes, but lacked the ITRs, thus avoiding sequence identity to the vector constructs. As with the earlier systems, co-transfection of helper virus-infected cells with the two plasmids resulted in the production of recombinant AAV-2. In addition, this novel system allowed three major improvements of the AAV-2 vector production process. Firstly, the proof that the AAV-2 ITR sequences alone are sufficient for rescue, replication, and encapsidation of enclosed vector DNA sequences significantly increased the space for heterologous genes, thereby allowing efficient encapsidation of up to 5.2 kb DNA into AAV-2 virions (38). Secondly, the level of wt AAV-2 contamination was dramatically reduced because the helper and vector plasmids no longer shared any AAV-2 sequence identity. However, most AAV-2 vector preparations made using this protocol are still contaminated with 0.01% up to 10% replication-competent wt AAV-2, which probably results from nonhomologous recombination events between the two plasmids (4,56,135,166). In fact, contamination of vector stocks with wt AAV-2 has long remained a major inherent problem of rAAV production. Only in the last two years have there been described novel helper plasmids and packaging strategies resulting in rAAV stocks that are free of detectable wt AAV-2 (References 4,54,56, and 72; see also the following chapters for details). Thirdly, the titers of rAAV vectors were increased to up to 10^6 infectious particles per mL, resulting in production of up to 0.1 infectious particles per cell.

This improved packaging system has long been the method of choice for rAAV production, and we refer to it herein as the "basic protocol." Nevertheless, it soon became clear that its inherent limitations were hindering the further development of AAV-2 as a vector for human gene therapy. Based on the results of a variety of preclinical studies, it has been estimated that systemic treatment of a human patient could require more than 10^{12} infectious rAAV particles (e.g., 72). Unfortunately, with its low yield of about 0.1 recombinant particles per cell, the basic protocol does not allow such bulk virus production. Surprisingly, as opposed to rAAV production, transfection of wt AAV-2 plasmids into helper virus-infected cells is much more efficient, yielding titers of AAV-2 particles per cell that are at least 100-fold higher. Three reasons may account for this discrepancy. Firstly, since a wt AAV-2 genome contains the ITR sequences, it is amplified in Ad-infected cells to yield increased numbers of template for Rep and VP transcription. Moreover, by transactivating the AAV-2 promoters, in particular p40, the ITRs can further increase AAV-2 protein levels within the cells (168). In contrast, helper plasmids devoid of ITRs do not replicate, and their promoters are not activated, thereby resulting in a reduced AAV-2 protein expression and, subsequently, less efficient production of recombinant particles (44). Secondly, there is indirect evidence that wt AAV-2 genomes replicate or encapsidate more efficiently than rAAV genomes (55,63,105). Thirdly, efficient generation of AAV-2 vectors following the basic protocol is dependent on high transfection efficiencies. Only cells that acquire the helper plasmid will allow replication and package of the recombinant genomes. In contrast, cells that had initially received the wild-type plasmid will produce virus capable of spreading in cells infected by the helper virus.

In summary, the different efficiencies of wild-type and recombinant AAV-2 production can be explained by differences in the amount of particles generated per cell and in the overall number of cells capable of producing virus. Consequently, improvements in these two critical aspects of the basic protocol for rAAV generation should result in a significant increase in AAV-2 vector titers. In the following section, we describe the major strategies currently being pursued to achieve this goal. In order to compare the level of improvement achieved by the different protocols, the maximum number of infectious rAAV particles produced per cell are given along with each method (see also Table 1 and Titration of Recombinant AAV-2 Particles).

IMPROVED STRATEGIES FOR AAV-2 VECTOR PRODUCTION

The most relevant approaches to improve the basic protocol for rAAV production are summarized in Figure 4. These include the use of modified helper plasmids that express varying levels of Rep and VP proteins, as well as the generation of hybrid viruses to deliver the AAV-2 vector sequences to the cells. Alternative strategies are the establishment of stable producer cell lines or the attempts to generate recombinant AAV-2 particles in vitro. Finally, transfection protocols involving plasmids that contain essential parts of the adenoviral genome were developed to allow the helper virus-free production of AAV-2 vectors. All these strategies are discussed in detail below and are summarized in Table 1.

Modified (Transfection/Infection) Basic Protocols

A well-known observation in cells infected with replicating AAV-2 is that the levels of Rep and VP proteins are markedly up-regulated after superinfection with a helper virus, suggesting that strong AAV-2 protein expression is required for efficient AAV-2 production. In an effort to mimic this, Flotte et al. (52) constructed a novel helper plasmid in which the p5 promoter was replaced by the stronger long terminal repeat (LTR) promoter from human immunodeficiency virus (HIV) and that, when transfected into Ad-infected cells, should lead to increased expression levels of the large Rep proteins. Indeed, the use of this construct to package AAV-2 vectors was reported to provide better helper functions, resulting in up to 10-fold increases in

Table 1. Protocols for Production of Recombinant AAV-2

Protocol	Particular Approach	Reference	Maximum Particle Yield Per Cell[a]		Contamination	
			Transducing (t) or Infectious (i)	DNA-Containing	wt AAV-2	Helper Virus
Basic protocol	Transfection of Ad-infected cells with vector and helper plasmid	138	0.1 (t)	10	nd	+
Modified basic protocol	Exchange of p5 promoter for HIV-LTR/Stable vector cell line	52	500 (t)	10^4	nd	+
	Addition of SV40 replication origin to helper plasmid	25	0.5 (t)	10^3	0.1%	+
	Exchange of p40 promoter for CMV promoter	162	27 (t)	(3×10^3)	nd	+
	Mutation of ATG start codon for Rep78 and Rep68 to ACG	89	220 (t)	ni	ni	+
	Coupling of vector and helper plasmid to Ad virions	96	0.1 (t)	ni	ni	+
	Coupling of helper plasmid to Ad with integrated vector DNA	48	ni	(2×10^4)	10^{-4}%	+
	Transfection of vector cells with HSV amplicon carrying rep/cap	34	ni	320	nd	+
	Infection of vector cells with HSV vector carrying rep/cap	33	480(t)	ni	ni	+
Stable cell lines	Producer cell line (integrated vector and helper DNA)	29	36 (t)	(400)	nd	+
	Producer cell line	154	1 (t)	(500)	nd	+
	Producer or packaging (only rep/cap integrated) cell lines	72	276 (t)	2.9×10^4	nd	+
	Packaging cell lines	54	80 (t)	5.9×10^5	nd	+
In vitro packaging	Vector plasmid DNA as substrate	36	7×10^{-5}	ni	ni	ni
	Replicative vector DNA as substrate	185	0.01	(1)	ni	ni
Helper virus-free packaging	Triple transfection (plasmids: Ad, AAV-2 vector, and rep/cap)	135	10^3 (i)	10^4	10^{-3}%	+
	Triple transfection (plasmids: Ad, AAV-2 vector, and rep/cap)	178	1.1×10^3 (t)	9.4×10^5	ni	nd
	Double transfection (Ad plasmid and AAV-2 vector/rep/cap)	98	300 (t)	4×10^4	ni	nd
	Double transfection (Ad/rep/cap plasmid and AAV-2 vector)	56	300 (i)	ni	nd	nd

[a] In cases where maximum rAAV yields per cell were not provided by the genuine report, these values were calculated based on the respective rAAV particle numbers per mL, the final volume of the rAAV stocks, and the number of cells initially used to generate the recombinant particles. Yields of DNA-containing particles that are enclosed in brackets have been deduced from yields of functional (transducing or infectious) particles and the particle-to-infectivity ratios as reported by the respective authors. For details on rAAV titer nomenclature or methods for rAAV titration see Titration of Recombinant AAV-2 Particles. nd, not detected; ni, not investigated.

rAAV titers compared to a standard helper plasmid (52). However, when similar helper constructs were tested later in which stronger heterologous promoters (e.g., simian virus 40 [SV40], Rous sarcoma virus [RSV], or cytomegalovirus [CMV]) replaced the p5 promoter, it was observed that high-level Rep expression rather led to lower rAAV production (56,89,162).

An alternative approach was taken by Chiorini et al. (25), who engineered a helper plasmid to carry a SV40 origin of replication. When packaging was performed in COS cells, which contain and express the SV40 large T antigen, this helper plasmid was expected to replicate and thus yield increased template numbers for Rep and VP expression. Indeed, the novel approach yielded 60-fold more rAAV particles than protocols using a nonreplicating helper plasmid, thereby supporting the idea that the copy number of the *rep* and *cap* genes might be an important limiting parameter for rAAV production (25,44). Unfortunately, the rAAV stocks produced using this novel construct were mostly contaminated with wt AAV-2, probably because it contained sequences overlapping to vector plasmids, thus facilitating homologous recombination. Furthermore, this approach could not be adapted to 293T cells, which also express the SV40 large T antigen required for amplification of the helper plasmid, but are much easier to transfect than COS cells (D. Grimm and J.A. Kleinschmidt, unpublished data). This was probably because 293T cells contain stably integrated Ad E1 genes and thus support a high-level expression of AAV-2 Rep proteins, which are known to interfere with

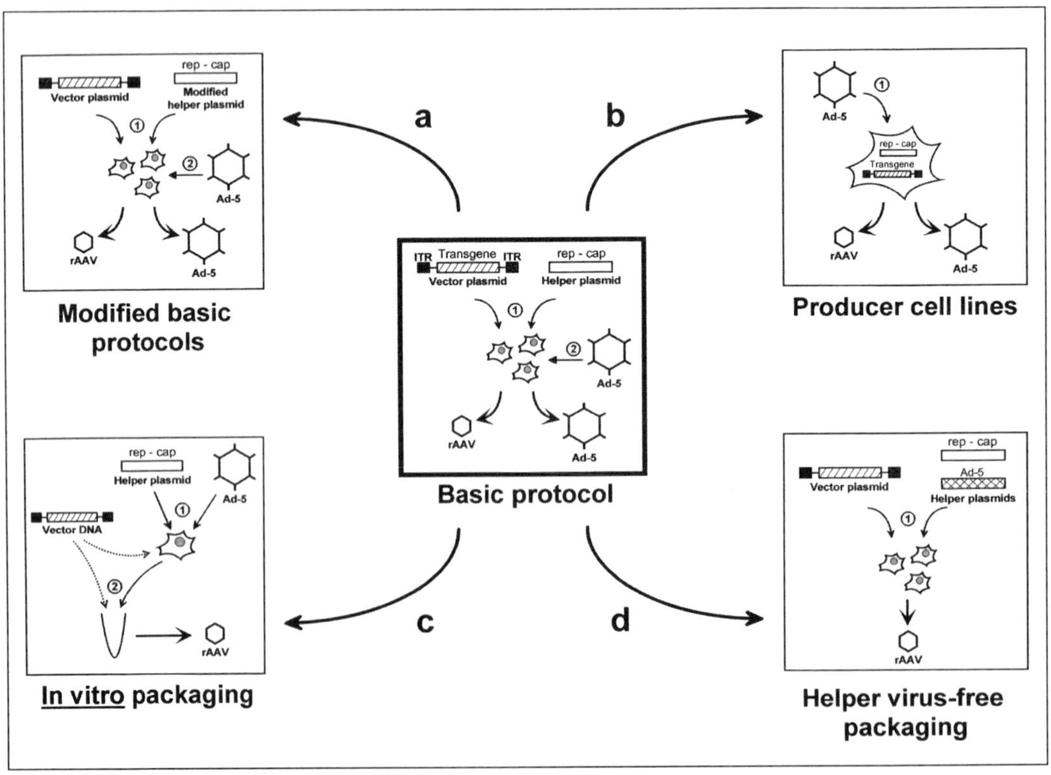

Figure 4. Improved strategies for rAAV production. Shown in the middle is the basic protocol for rAAV production, surrounded by the four most relevant approaches (a–d) that aim to simplify this method and/or increase its efficiency. For details on each protocol, see respective discussions in text.

SV40-induced DNA amplification (60).

A number of recent studies have begun to focus on *cap* gene expression. Vincent et al. (162) have shown that replacement of the p40 promoter, which drives *cap* gene expression from a wt AAV-2 genome, with the much stronger CMV promoter resulted in a 10-fold increase in rAAV yields. Surprisingly, when Rep expression was increased concurrently, by replacing the p5 promoter with the stronger RSV promoter, VP expression as well as AAV-2 vector titers were only twofold higher than those obtained using a standard helper plasmid. This provided the first evidence that, at least under certain circumstances, expression of Rep and VP proteins might correlate inversely, and that VP rather than Rep levels might be limiting for rAAV production (162). Very similar observations were reported in 1997 by Li et al. (90), who described a helper plasmid in which the start codon for Rep78 and Rep68 was mutated from ATG to an inefficient ACG. This mutation not only led to decreased Rep expression, but, in agreement with the findings of Vincent et al. (162), also yielded an increase in VP expression as well as rAAV titers. Finally, our own results also fit into this model. We constructed a helper plasmid, pDG (see Helper Virus-Free Packaging Systems), which drives high-level expression of VP in transfected cells, and yields up to 50-fold higher rAAV titers as compared to standard helper plasmids. In line with the findings of Li et al. (90), strong VP expression from pDG is linked to down-regulation of Rep expression. However, in the case of pDG, the decreased Rep levels result from the fact that the p5 promoter was replaced by the LTR of the mouse mammary tumor virus (MMTV), which requires an inducing agent, such as dexamethasone (dex), to gain full activity. Thus, when dex was added to pDG-transfected cells, Rep protein levels were elevated, while VP protein expression and AAV-2 vector titers decreased back to the levels obtained using a standard helper plasmid possessing the p5 promoter (56).

The exact mechanisms underlying these phenomena are not yet fully understood, but can be speculated. Wistuba et al. (173) and Weger et al. (168) showed that AAV-2 capsid assembly is dependent on the expression of high levels of VP. Therefore, low-level VP expression, as observed from standard helper plasmids, probably results in a limiting number of capsid substrates for encapsidation of the recombinant genomes. And vice versa, the stronger the VP proteins are expressed, the higher is the rate of capsid assembly, and consequently, the titers of rAAV. The observations described above on the use of improved helper constructs driving stronger VP expression (through heterologous promoters or replication of the *cap* gene) would fit exactly into this model. One possible explanation for the inverse correlation between Rep and VP expression is that high levels of Rep may inhibit the Ad helper functions required to activate all AAV-2 promoters, in particular the p40 promoter. Thus, in the reverse case of down-regulated Rep expression, the helper virus could stimulate more easily VP expression from the p40 promoter.

In summary, using the novel helper plasmids to package vector genomes, it is now possible to generate infectious rAAV at particle numbers per cell that are only slightly lower than those of wt AAV-2 (resulting from transfection of Ad-infected cells with a wild-type plasmid). The design of these modified helper plasmids that allow amplification of the *rep* and *cap* genes or, alternatively, lead to reduced Rep, but increased VP levels, thus represents a major improvement in the field of rAAV production. However, despite this success, one must remain aware of two inherent limitations of all packaging systems described above, i.e., they rely on transfection of cells (Figure 4a, ①) and require a helper virus (Figure 4a, ②; except when using the pDG plasmid, see Helper Virus-Free Packaging Systems). While the latter problem can be overcome by the use of adenoviral helper plasmids (see Helper Virus-Free Packaging Systems), a consequence to the dependence on transfection is that the scaling up of these protocols would be very expensive and labor-intensive. On the other hand, there have already been a variety of proposals that might help overcome this limitation. For example, Mamounas et al. (96) have reported the coupling of helper and vector plasmids to the surface of Ad capsids via transferrin-poly-L-lysine. With these plasmid-helper virus complexes, it is feasible to deliver into cells all *cis* and *trans* elements that are required for rAAV production via a simple and efficient infection step. In a similar approach, Fisher et al. (48) generated a recombinant Ad with an integrated rAAV vector and a helper plasmid coupled to the Ad capsid. Production of AAV-2 vectors was achieved via infection of cells with the recombinant Ad-plasmid complex. Similar recombinant Ad containing an AAV-2 vector DNA were also reported by Thrasher et al. (156) as well as Liu et al. (92). Unfortunately, so far all efforts to integrate the AAV-2 *rep* gene into a recombinant Ad have failed. This is most likely due to the strong inhibitory effects of even small amounts of Rep proteins on Ad amplification (56). Yet another strategy was pursued by Conway et al. (34) as well as Zhang et al. (184), who independently constructed HSV-1 amplicons containing the rep and cap genes. In the first report, transfection of cells carrying a stably integrated rAAV vector with the AAV-2/HSV-1 amplicon and concurrent infection of the cells with a disabled helper herpes virus (which stimulates replication of the HSV amplicon) resulted in rAAV titers approximately 10-fold higher than those from the standard protocol. An even 50-fold improvement in rAAV production was observed by the second group, who used an additional amplicon or, alternatively, a recombinant herpes virus to deliver AAV-2 vector sequences to the cells (184). However, this system is limited by the requirement for a particular cell line (baby hamster kidney cells), which is difficult to transfect, and moreover, contaminating wt AAV-2 was present in all rAAV preparations at levels of up to 1% (184). Recently, Conway et al. (33) reported a significant improvement to their former approach. A recombinant herpesvirus was described that carries the AAV-2 *rep* and *cap* genes and that leads to a strong expression of Rep and VP proteins following infection of 293 cells. Using this novel AAV-2/HSV-1 hybrid virus, rAAV production was achieved via infection of cells transfected with an AAV-2 vector plasmid or, alternatively, cells already containing an integrated vector genome. With maximum yields of up to 480 infectious particles per cell, this approach proved to be very efficient. Interestingly, the recombinant HSV could also be used to amplify rAAV vectors by a factor of 200 with each passage through coinfection of 293 cells with both viruses (33). However, with all these approaches, general questions such as stability and the ease with which the recombinant Ad or HSV-1 vectors and amplicons, respectively, can be prepared and propagated certainly require further investigation. Finally, it was recently demonstrated that 293 cells, which are commonly used for rAAV production, can be kept in suspension and efficiently transfected (75). This could allow the scaling-up of the transfection procedure for rAAV production and should result in further increased rAAV yields.

Packaging/Producer Cell Lines

In the case of adenoviral or retroviral vectors, the use of packaging cell lines represents the most efficient means for recombinant virus production (145). In general, such cell lines express constitutively viral proteins that are absent in the vector genome, thus helping to amplify and package the deficient recombinant genomes. In view of the rather cumbersome basic protocol for production of AAV-2 vectors, the development of efficient packaging cell lines would be highly desirable.

Unfortunately, the generation of cell lines that constitutively express AAV-2 proteins at high levels has turned out to be very difficult since AAV-2 and in particular the Rep proteins were found to exert antiproliferative effects on cells (7,10,61,172, 182). In a first effort to develop an AAV-2 packaging cell line, Vincent et al. (161) transfected HeLa cells with a construct carrying the AAV-2 *rep* and *cap* genes, and subsequently managed to isolate several clones with low copy numbers of integrated AAV-2 genes. However, to produce recombinant particles these cells had to be transfected with a vector plasmid and infected with Ad, thus providing no real improvement on the basic protocol. Moreover, the rAAV titers produced were very low, with maximum yields per cell being at least tenfold below those obtained using the basic packaging protocol, probably due to low Rep and VP expression in these cells.

In an effort to establish cell lines with higher expression of Rep proteins, two other laboratories expressed the *rep* gene under the control of an inducible promoter, namely the mouse methallothionein promoter (182) or the MMTV-LTR promoter (68,69). Under conditions in which these promoters were silent, the constructs were stably introduced into 293 or HeLa cells, respectively. After induction of the heterologous promoters, the cells expressed high levels of functional Rep proteins and were capable of supporting replication of rAAV genomes lacking the *rep* gene. Nevertheless, these cell lines could not be used for AAV-2 vector production, since they naturally failed to complement common *cap*-negative vector genomes.

A major improvement in the field came with the first reports of real producer cell lines by Clark et al. (29) and Tamayose et al. (154). In both cases, HeLa cell lines were established that expressed the *rep* and *cap* genes and an AAV-2 vector DNA. In contrast to the earlier systems, these lines no longer required transfection of exogenous DNA. Instead, AAV-2 vector production was simply accomplished upon infection with Ad, thereby leading the authors to propose that large-scale vector production could be easily achieved by adapting these cells to grow in suspension. However, this intriguing idea was not really explored. Unfortunately, with a maximum of 1 to 36 infectious rAAV particles per cell, the efficiency of AAV-2 vector production from these two cell lines was still rather low and did not exceed that of the basic protocol. This was clearly not due to saturation of rAAV production in the cells, since wt AAV-2 can be propagated at up to 10^4 infectious particles per cell. A more likely explanation for the low efficiency would be the limited copy number per cell of viral genes available as templates for Rep and VP expression, and/or of vector copies ready for encapsidation. In fact, in cells co-infected with AAV-2 and Ad, replication of the AAV-2 genome yields up to 10^5 copies of the *rep* and *cap* genes per cell (129), while the above-described cell lines carried only 10 to 30 copies which, due to the absence of the ITRs, could not amplify further.

In a recent effort to overcome this limitation, two novel rAAV-packaging/producer cell lines have been developed, which harbor amplifiable helper and vector constructs. Inoue and Russell (72) presented a HeLa cell line that had been first stably transfected with a plasmid expressing the SV40 large T antigen under the control of a doxycycline-responsive promoter. In a second step, a helper construct carrying the *rep* and *cap* genes and an SV40 replication origin was added to generate a packaging cell line. Finally, the stable introduction of a vector plasmid, also bearing an SV40 replication origin, led to the generation of a producer cell line. Infection of these producer cells with helper Ad and induction of SV40 large T antigen expression by addition of doxycycline had two effects: (*i*) *rep* and *cap* gene expression were stimulated by the helper virus, and (*ii*) the copy numbers of the integrated AAV-2 helper and vector sequences were raised 4- to 10-fold. As a result of this combination of effects, rAAV vector production was boosted 5- to 20-fold as compared with the basic packaging protocol, with maximum titers of approximately 280 infectious particles per cell. Moreover, infection of the doxycycline-induced packaging cell line with rAAV vectors and helper virus allowed a tenfold amplification of the recombinant virus, which is a very intriguing feature of these cells. On the other hand, this was not achieved with all types of vector plasmids analyzed, and the growth rates of the producer cells seemed to decline with increasing production of recombinant AAV-2 (72).

An alternative approach involving amplification of at least the vector sequences was pursued in 1998 by Gao et al. (54), who stably transfected HeLa cells with a *rep*/*cap*-containing plasmid utilizing endogenous AAV-2 promoters. Out of a total of 708 clones analyzed, they selected one packaging cell line (B50) that carries five copies of integrated *rep* and *cap* genes. Production of rAAV was then achieved by a two-step procedure: (*i*) the cells were infected with an E2B-mutant Ad to induce Rep and VP expression, and (*ii*) exactly 24 hours later, the cells were infected with a second (recombinant) Ad carrying an rAAV vector genome. This Ad was then allowed to replicate for 48 hours, thereby amplifying the AAV-2 vector sequences. During this period, the Rep and VP proteins concurrently mediated rescue, further replication, and finally encapsidation of the recombinant AAV-2 genomes, thereby resulting in high efficiency rAAV production. In fact, maximum titers of at least four different vectors reached 80 infectious particles per cell, which represents a 20- to 100-fold increase over the yield of the basic protocol. Consequently, a large scale preparation of AAV-2 vectors using 10^9 B50 cells resulted in the production of 7.8×10^{10} infectious vector particles. However, to obtain these optimal yields it was critical to maintain a precise sequence and timing of the two infection steps (54).

A very similar strategy based on HeLa cells carrying *rep* and *cap* genes and a recombinant Ad containing AAV-2 vector sequences was also published by Liu et al. (92). This group made the additional interesting observation that following Ad infection, the number of five to ten copies of *rep* and *cap* genes in their cell line increased by approximately 60-fold, which represents the first report of ITR-independent AAV-2 gene amplification (92). This phenomenon was restricted to stably integrated AAV-2 genes and was not observed following transient *rep* and *cap* gene transfection. However, with a maximum production of only 10 infectious particles per cell, this particular system was less efficient than the previously reported cell line-based methods.

In summary, the newly generated producer cell lines offer a highly efficient means to produce a defined AAV-2 vector, as they do not require transfection and yield rather high rAAV titers (Figure 4b). In addition, the resulting vector preparations were found in all cases to be free of replication-competent wt AAV-2 (29,72, 154). It is currently being analyzed whether these cell lines can be adapted for growth in suspension, which is an important prerequisite for establishing large-scale vector manufacture. However, a limitation for the use of producer cell lines is that they are restricted to producing only a defined type of recombinant AAV-2, since the vector genome is an integral part of the cells. Thus, novel producer cells have to be established for each new vector, which questions the use of this approach for small- and mid-scale vector preparations (e.g., for preclinical studies).

For this purpose, packaging cell lines, which contain only the *rep* and *cap* genes, but no vector DNA, might provide a more flexible means for rAAV production. However, these cell lines require introduction of the vector genome, either via plasmid transfection or via infection with a recombinant Ad—both of which make this particular approach more cumbersome than the use of truly producer cell lines. On the other hand, packaging cell lines might prove very useful for the simple amplification of already existing vector preparations through repeated infection with the recombinant particles and a helper virus.

Before packaging or producer cell lines can become a standard means to generate or amplify rAAV vectors, two more issues require further investigation. First, data concerning the long-term stability of these novel cell lines will have to be collected, especially in the case of producer cell lines that carry both integrated copies of helper and vector plasmids. Second, all cell lines described so far require infection with Ad, and consequently, lead to rAAV preparations that are contaminated with helper virus and have to be purified. In view of this limitation, it is being investigated whether, in addition to the AAV-2 helper and vector sequences, the Ad helper genes could also be incorporated into the cellular genome. If so, the resulting cells might continuously generate AAV-2 vectors that are free of contaminating helper virus. Unfortunately, most of the Ad genes that are responsible for helper functions have been found to be cytotoxic. In addition, as soon as helper virus functions are expressed in *rep/cap*-expressing cell lines, expression of Rep and VP will be activated, which might disturb the stability of the cell line. The establishment of such permanent AAV-2 vector producer cell lines will, therefore, require some intricate strategies.

In Vitro Packaging Systems

The first studies aimed to establish an in vitro packaging system for AAV-2 have only recently been described (Figure 4c) (36,185). According to these reports, the cell-free system requires all the components that are also needed for the in vivo vector assembly (i.e., vector DNA, Rep and VP proteins, Ad helper gene products, and yet-to-be-defined cellular proteins and metabolites). In both reports, a similar strategy was pursued to supply these factors; HeLa cells were infected with Ad and transfected with a *rep/cap*-expressing plasmid (Figure 4c, ①), and 24 to 48 hours later, crude cell extracts were prepared and subsequently incubated with vector DNA for up to 24 hours. However, there was a difference in the source of the recombinant AAV-2 DNA (Figure 4c, ②). While Ding et al. (36) were successful in adding an intact AAV-2 vector plasmid to the cell extract, Zhou and Muzyczka (185) found only poor packaging of vector substrates that contained even short plasmid sequences attached to the ends of the ITRs. Instead, replicative single- or double-stranded vector DNA prepared from Ad-infected cells that had been transfected with a vector and a *rep/cap*-expressing plasmid was packaged eightfold more efficiently. Both groups, however, agreed on the fact that intact ITR sequences were required to achieve packaging of the vector DNA. In addition, in 1998 Zhou and Muzyczka (185) thoroughly investigated the exact cofactor requirements for in vitro packaging. According to this study, the reaction requires Rep proteins and active DNA synthesis and is strongly dependent on the presence of ATP and in particular Mg^{2+} in the reaction mixture. A minor dependence was also noticed on the presence of the three other ribonucleoside triphosphates: UTP, CTP, and GTP.

Unfortunately, the output observed with the two in vitro packaging systems was rather low, with rAAV yields of less than 10^4 (36) or 10^5 (185) infectious particles per mL of reaction mixture. This may be due to inadequate amounts or functioning of Rep and VP proteins in the cell extracts, or to the different reaction volumes in vitro and in vivo (i.e., in the cell), which might have caused a dilution effect. In the approach pursued in 1997 by Ding et al. (36), inefficient in vitro replication of the vector plasmid might also have limited the rAAV packaging reaction.

In summary, it is obvious that a more extensive knowledge of the components involved in AAV-2 packaging is needed before this process can be efficiently mimicked in vitro. Moreover, a significant advancement would be achieved if at least some of the proteins involved in the reaction could be supplied as recombinant substrates, with the aim to eventually replace the crude cell extracts that have been used so far. A variety of earlier studies suggests that this goal might indeed be achievable. For example, Ruffing et al. (131) developed recombinant baculoviruses expressing large amounts of VP and showed that these proteins can self-assemble to form virus-like particles. Steinbach et al. (150) have shown the production of empty AAV-2 particles in cells transfected with a plasmid that expresses all three VP proteins. Alternatively, empty assembled AAV-2 capsids can be produced by infection of cells with a recombinant Ad that expresses the AAV-2 VP proteins (55). Finally, a variety of groups have reported the expression of Rep proteins in bacteria and have purified these proteins by different procedures (13,24, 102,113,175,186).

Helper Virus-Free Packaging Systems

The helper virus plays a pivotal role in the productive AAV-2 life cycle (see Figure 1). The gene products from Ad required for helping AAV-2 replication have been identified, but their particular roles are understood to varying degrees (see Reference 109 for a review). In brief, the E1A transactivator proteins bind to the AAV-2 p5 promoter and interact with the host repressor YY1, thereby derepressing the p5 and p19 promoters (88). Consequently, Ad infection results in an increase of Rep expression. Moreover, expression from the p40 promoter is also markedly stimulated, resulting in strongly elevated VP protein levels after Ad infection (73,74,111,168). The E1A proteins also exert an indirect helper function by transactivating other adenoviral promoters. Through interaction with E4ORF6, the p55-E1B protein of Ad stimulates efficient transport of viral transcripts to the cytoplasm, while at the same time inhibits export of cellular RNA (35,87,139). The 72-kDa single- and double-stranded DNA-binding protein (DBP) encoded by the E2A gene has multiple helper functions. Production of AAV-2 in the absence of functional DBP results in reduced *cap* gene expression, less accumulation of single- and double-stranded forms of the AAV-2 genome and lower levels of spliced Rep proteins (19). Finally, the adenoviral virus-associated (VA) transcripts can enhance translation from Ad and AAV-2 messenger RNAs (155,171).

Although helper functions are essential for rAAV production, infection with a helper virus poses two problems. First, an optimal generation of AAV-2 vectors requires a delicate balance between AAV-2 and Ad functions. If the multiplicity of infection (moi) of Ad is too high, AAV-2 gene expression will be higher. However, Ad-mediated cell lysis will proceed too fast, thus reducing the time available for rAAV production. Conversely, if the moi of Ad is too low, the AAV-2 proteins will prevent amplification of the helper virus, thereby resulting in a reduced number of cells infected by it. Consequently, rAAV vector yields strongly depend on the input of exact amounts of helper virus and AAV-2 helper plasmid when infecting or transfecting the cells, respectively. A second inherent problem when using a helper virus is the risk of contamination of rAAV stocks by the helper virus or its proteins. In fact, even the smallest amounts of residual Ad proteins can result in cytotoxic host immune responses (106). Unfortunately, complete inactivation and/or removal of adenoviral contaminants from rAAV preparations presents significant difficulty (see Purification of AAV-2 Particles). To overcome this problem, some groups replaced wild-type Ad helper virus with mutants that were rendered deficient for replication (e.g., through deletion of the Ad preterminal protein) (99). However, these approaches reduced, but failed to prevent helper virus contamination completely. Furthermore, these strategies still required the propagation of a (mutant) helper virus and an infection step to produce rAAV.

Efforts have been made to completely remove the Ad infection step from the rAAV production scheme. As a major progress in this direction, Ferrari et al. (45) showed that purified Ad DNA can supply helper functions to rAAV production. In this report, AAV-2 vectors were generated through triple transfection of 293 cells with an AAV-2 vector and helper plasmids as well as genomic Ad DNA isolated from an E3-mutant Ad. To prevent formation of infectious Ad, the origin and packaging signals were removed from the adenoviral genome prior to transfection. However, although successful in principle, this protocol suffered from two drawbacks: (*i*) it required laborious purification of Ad DNA as starting material, and (*ii*) the risk of wild-type Ad production still remained because of an overlap between transfected and endogenous adenoviral sequences within the 293 cells.

An improvement to this system came with the cloning of all parts of the adenoviral genome essential for rAAV production into plasmids, thus making the isolation of Ad DNA superfluous. In 1998 Salvetti et al. (135) showed that triple transfection of 293 cells with an AAV-2 vector and helper plasmids as well as a construct harboring the complete Ad type 5 genome leads to production of recombinant AAV-2. Importantly, the substitution of adenoviral virions by this construct did not affect the rAAV yields, which in either case reached up to 10^3 infectious particles per cell. However, since the adenoviral plasmid was infectious, greater than 20% of the resulting rAAV stocks were contaminated with Ad. Surprisingly, even after deletion of the replication origin and packaging signal from the adenoviral plasmid, such contamination could not be suppressed. Moreover, all rAAV preparations were contaminated with wt AAV-2 at an average extent of 0.001% of the infectious rAAV particles.

A significant further development was achieved by Xiao et al. in 1998 (178), who reported the first system for rAAV production really free of infectious Ad. Similar to Salvetti et al. in 1998 (135), this system relied on a triple transfection of cells with an AAV-2 vector as well as AAV-2 and Ad helper plasmids. However, the Ad plasmid used by Xiao et al. (178) represented a defective miniconstruct (i.e., a plasmid carrying an adenoviral genome from which large parts of the structural and replication genes have been deleted). Therefore, rAAV preparations made using this particular Ad miniplasmid were not only free of infectious Ad, but also of unwanted Ad structural proteins, such as the cytotoxic fiber protein. Surprisingly, the yield of rAAV obtained with this plasmid was about threefold higher than those obtained with a conventional Ad infection. This increase may be due to the fact that the Ad miniplasmid does neither replicate nor package and thus, in contrast to an Ad infection, competition for Ad gene products was eliminated. A further 15-fold increase in rAAV titers was achieved when the Ad miniconstruct was used in combination with an AAV-2 helper plasmid in which the ATG translation start codon for Rep78 and Rep68 was mutated to ACG (89), and which carried an extra copy of the p5 promoter. In sum, giving maximum yields of up to 1100 infectious rAAV particles per cell, these novel AAV-2 and Ad helper plasmids allowed for a 40-fold increase in rAAV titers as compared to the conventional protocol. Despite this intriguing advancement, an inherent disadvantage of this protocol is that it requires transfection of cells with three independent plasmids. Although it was argued that three plasmids can be efficiently introduced into the same target cell, suggesting that triple transfection is not rate limiting (178), the costs and work required to prepare three plasmids in large amounts remain a major concern of this approach.

The above problem was eventually resolved when Matsushita et al. in 1998 (98) presented a simplified, double-transfection approach involving an adenoviral helper construct and a plasmid that encoded both an AAV-2 vector and the *rep* and *cap* genes. For generation of the Ad helper construct, those regions that, in addition to the E1A/B genes, were thought to be required for support of AAV-2 propagation (i.e., E2A, E4, and VA RNA) were assembled from adenovirus type 2 or type 5 (Ad-2 or Ad-5) DNA. Using this construct for helping rAAV production in 293 cells proved to be even slightly more efficient than Ad infection, resulting in up to 300 infectious helper virus-free rAAV particles per cell. Unfortunately, no data were provided concerning possible contamination of these rAAV preparations with wt AAV-2. Interestingly, when only a subset of the above-mentioned Ad helper regions was cloned and co-transfected along with the AAV-2 vector/helper plasmid, some level of rAAV production also occurred, indicating that none of the various Ad gene products is essential for this process. However, the E2A region contributes much more to rAAV production than do the E4 and VA RNA since, in the absence of the E2A gene, rAAV titers dropped by approximately six orders of magnitude. In contrast, when E4 and/or VA RNA were eliminated from the transfection, titers were only 9- to 16-fold lower. It was also reported that the E4ORF6 34 kDa or the E2A 72-kDa proteins could functionally replace the E4 and E2A regions, suggesting that these proteins alone may account for the helper activity of the respective genomic Ad regions.

A further advancement in helper virus-free rAAV packaging came from our studies on AAV-2 vector production. A novel helper plasmid (pDG) was developed that provides all AAV-2 and Ad helper functions (i.e., the AAV-2 *rep* and *cap* genes as well as the adenoviral E2A, VA, and E4 regions) (56). Similar to the approach of Matsushita et al. (98), rAAV production is achieved through co-transfection of 293 cells with an AAV-2 vector plasmid and the pDG construct, thereby making the protocol easier, less expensive, and probably more reproducible than the triple transfection procedures described above. The rAAV titers obtained using pDG are comparable to those achievable with the other helper virus-free systems, reaching up to 300 infectious rAAV particles per cell and thus being 10 to 30-fold higher than those resulting from a standard packaging protocol. The reason for this increase, namely a strong expression of VP proteins and subsequent rise in formation of assembled capsids in pDG-transfected cells, has already been outlined [see Modified (Transfection/Infection) Basic Protocols]. However, as opposed to the other helper virus-free approaches reported so far, the pDG construct offers the particular advantage that the Ad and AAV-2 helper functions are provided by the same plasmid. Thus, while variations in transfection efficiencies or in determination of plasmid concentrations are prone to yield varying rAAV titers when using separate AAV-2 and Ad helper plasmids, this risk has been eliminated with the AAV-2/Ad hybrid construct pDG. Importantly, rAAV stocks prepared using the pDG helper plasmid are not only completely free of contaminating Ad, which is due to the absence of most Ad replication and structural genes in pDG, but also free of detectable replication-competent wt AAV-2.

In summary, an appealing advancement in AAV-2 vector technology, both in terms of rAAV quality and quantity, has been achieved with the establishment of helper virus-free packaging systems. The use of either a combination of Ad helper plasmids and modified AAV-2 helper constructs or the hybrid pDG plasmid allows production of rAAV at titers of up to 500 to 1100 vector particles per cell, while concurrently eliminating a number of safety and purity concerns associated with the conventional protocol. With these obvious benefits, the novel helper virus-free rAAV packaging protocols are probably excellent methods for generating sufficient and safe AAV-2 vectors for preclinical and phase I clinical studies. Future work, however, must be aimed at improving the only common limiting step in all these novel protocols, i.e., the requirement for transient transfection (see Figure 4d, ①). A number of promising approaches in this direction have already been outlined [see Modified (Transfection/Infection) Basic Protocols].

PURIFICATION OF AAV-2 PARTICLES

Until the recent development of helper virus-free technology for rAAV production, existing packaging protocols relied on the presence of infectious helper Ad. Since Ad replicated along with recombinant AAV-2, the resulting crude rAAV preparations were inevitably contaminated with helper virus particles and proteins. Because several Ad proteins are known to cause cellular host immune responses (106), these Ad-contaminated rAAV stocks required purification. Nevertheless, even vector preparations produced using a helper virus-free protocol have to be purified from contaminating cellular debris, medium ingredients, and adenoviral proteins that are expressed from the Ad helper plasmids. Besides, purification of rAAV preparations serves to prevent a phenomenon known as pseudotransduction (i.e. the transfer of proteins to the recipient cells, which were expressed from the vector DNA during rAAV production), thus simulating a higher transduction rate than actually achieved (2).

To date, these issues have been typically addressed by heat inactivating the Ad present in rAAV stocks at 56°C, followed by up to three consecutive CsCl density gradient centrifugations, resulting in purified and concentrated AAV-2 vectors (147). Importantly, this protocol allows an enrichment in full particles, which band at a density of 1.41 g/cm^3 (20). On the other hand, this method conveys a number of significant drawbacks. In addition to being time-consuming, it leads to dramatic losses of infectious vector particles of more than 50%. Furthermore, it does not allow the complete removal of structural Ad proteins (e.g., the cytotoxic fiber protein), thereby maintaining the risk of causing host immune response after AAV-2 vector application.

As an alternative approach to rAAV purification, a chromatography protocol

Table 2. Summary of Current Protocols for Titration of rAAV Particles

Type of particle	Titration method[a]	Measurement	References[b]
Transducing	Histochemical staining (*lacZ*), FACS analyses (*gfp*), selection of resistant cell colonies (neo^R)	Number of infectious and transgene expressing units	72,177,188
Infectious	Replication dot blot assay, replication center assay	Number of infectious and replication-competent units	5,56,135
DNA-containing	Extraction and quantification of viral DNA, QC-PCR	Number of units containing detectable viral DNA	30,34,147
Assembled virion	AAV-2 capsid ELISA	Number of assembled (full and empty) particles	55

[a]Denoted in brackets are the transgenes carried by the respective rAAV vectors
[b]Given are only some illustrative references reporting the use of the listed titration methods

involving the use of sulfonated cellulose has been reported that allows a quick and easy concentration of AAV-2 vectors, with a particle recovery of nearly 100% (154). However, this column cannot be used to separate AAV-2 from contaminating Ad virions. Furthermore, the column eluates contain significant levels of contaminants derived from the tissue culture supernatant, and with certain types of cell extract, purification using this column was not possible (56).

In an effort to combine high-throughput technology and high specificity, we have recently established a more favorable single-step chromatography method (56). The unique property of the monoclonal antibody A20 to specifically recognize assembled AAV-2 capsids, but not free VP proteins, was exploited to develop an affinity chromatography procedure. By coupling the A20 antibody to *N*-hydroxysuccinamide (NHS)-activated Sepharose® (Amersham Pharmacia Biotech, Piscataway NJ, USA), an affinity column was generated that was subsequently used to purify both wt AAV-2 and AAV-2 vector particles. The affinity-bound particles were eluted using 2.5 mol/L $MgCl_2$, a solution in which AAV-2 virions were found to be very stable. With a particle recovery of at least 65%, a minimal concentration factor of four, a purity of the eluted particles higher than 80%, and a maximum time span of less than one day to complete purification, this method is far more efficient than the traditional CsCl centrifugation. Moreover, in contrast to the previously described chromatography protocol involving sulfonated cellulose (see above), the A20 column allows total removal of contaminating Ad from AAV-2 vector preparations.

Comparable results have recently been reported from two other groups, who independently have developed single-step rAAV purification schemes involving the use of Iodixanol-based density gradients and/or heparin affinity chromatography. Based on the observation that heparan sulfate proteoglycans can serve as a cellular coreceptor for AAV-2 (152), Clark et al. (31) as well as Zolotukhin et al. (187) tested the feasibility of purifying AAV-2 vector particles via heparin-based affinity chromatography. Starting with crude lysates from rAAV-producing cells and using low-pressure or HPLC heparin columns, an average particle recovery of more than 70% was achieved in one working day. The resulting vector particles were biologically active and virtually free of Ad and cellular contaminants. Moreover, Zolotukhin et al. (187) as well as Hermens et al. (62) investigated the use of Iodixanol (an X-ray contrast solution) as an alternative to CsCl as a density gradient medium. Running either preformed gradients (ranging from 15%–60% Iodixanol) or continuous gradients for 1 to 3 hours, both groups achieved at least a 100-fold purification of crude cell lysates and a recovery of 70% to 80% of the starting infectious rAAV particles. The bulk of concentrated and pure rAAV virus was reproducibly found within the fractions containing 40% to 60% Iodixanol.

In summary, the feasibility of rAAV particle purification via affinity chromatography and improved density gradient methods provides a powerful means of generating preclinical grade AAV-2 vectors. A number of parameters, however, remain to be tested and optimized prior to scaling up these methods. For example, factors that are currently being investigated include the stability of the antibody column, as well as general aspects such as the maximum capacities for particle load and the possibility of reusing the antibody or heparin columns. As soon as these issues are resolved, the improved methodology should make a broad application of AAV-2 vectors feasible in the near future.

TITRATION OF RECOMBINANT AAV-2 PARTICLES

In the context of applying AAV-2 vectors as potential therapeutic agents, preclinical and clinical studies require the use of good manufacturing practice in rAAV preparation. In particular, titration of recombinant particles must be performed with sufficient accuracy and precision to ensure comparability of data obtained in such studies. However, none of the currently available protocols, which allow for the quantification of physical virions and of functional rAAV particles, is being used as a standard method for rAAV titration. Determination of rAAV titers is rather performed via a number of varying protocols and conditions, resulting in differing values for rAAV concentrations and confusing nomenclatures to express rAAV titers.

The methods used currently for rAAV titration can be divided roughly into biological and physical assays (Table 2). Biological assays rely on detection of infected cells in culture by evaluating events that depend on the functionality of the rAAV vector. This can be either replication of the recombinant genomes in the presence of AAV-2 and adenoviral helper functions, or expression of an heterologous gene packaged in the rAAV virions. In the first case, amplified rAAV genomes are detected by hybridizing blots with probes specific for the recombinant DNA sequences and calculating from them the titers of infectious or replicating particles. Typical examples of such methods are replication center assays (135,147) or replication dot blot assays (5,56). The second kind of biological assay allows the quantitation of transducing particles via measurement of transgene expression in rAAV-infected cells. These particular assays are, however, restricted to cases in which easily detectable reporter genes, e.g., *lacZ* (178) or *gfp* (188), were packaged, as quantitation of the products of most genes of therapeutic interest is usually difficult. In general, titers of transducing and infectious particles have been found to be very close, if not identical (55,135).

Physical methods serve to determine particle concentrations by assays that are independent of the biological function of the recombinant viruses. Typically, viral DNA is extracted from the recombinant particles by enzymatic digestion of the capsids, blotted onto a membrane, and quantified using serially diluted plasmid DNA including the transgene as standard (53,147). Alternatively, the extracted DNA can be quantified by means of quantitative-competitive polymerase chain reaction (QC-PCR)(34). The number of DNA-containing particles is then calculated assuming that each virion contains one single-stranded DNA molecule with a defined size. Generally, titers of DNA-containing rAAV particles were mostly found to be two to three orders of magnitude higher than titers of infectious or transducing particles (55,147).

The ratios of measured physical to functional particles are usually given as particle-to-infectivity ratios and are considered an important index for the quality of rAAV preparations. Most researchers determined a several hundredfold excess of physical to infectious particles (138,154), as has been described for wt AAV-2 (105), although ratios ranging from 1.4:1 up to larger than 10^4:1 have also been reported (25, 29,135,162). Likewise, the ratios of infectious to transducing particles were also a matter of controversy, ranging from nearly 1:1 to approximately 500:1 (29,135).

A problem of all titration methods described so far is that they are highly dependent on the assay conditions. For example, a crucial parameter when determining infectious rAAV titers is the delivery of the AAV-2 Rep proteins, which are required to allow replication of the rAAV genomes in the cells. Rep proteins have yet been supplied either by over-infection of rAAV-infected cells with wt AAV-2 (147) or by using various cell lines with integrated *rep* genes for rAAV titration (30,135). The quantification of transducing particles also depends on a number of variables, such as the cell line chosen, sensitivity of the assays used, and the rAAV incubation times (30,135). A more crucial parameter of these assays is the helper Ad, which can influence the efficiency of rAAV-mediated gene transduction and consequently, the calculated rAAV titers by three orders of

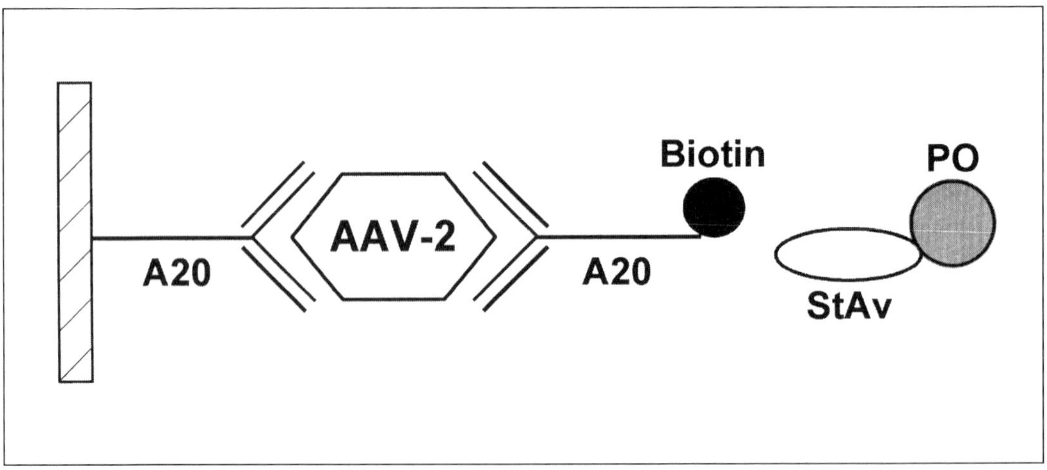

Figure 5. Principle of the AAV-2 capsid ELISA. Stocks of wild-type or recombinant AAV-2 are serially diluted and added to 96-well cell culture plates previously coated with purified monoclonal antibody A20. Assembled AAV-2 particles bind to the A20 antibody and are detected through incubation of the plate with biotinylated (Biotin) A20 antibody, followed by streptavidin-coupled peroxidase (StAv/PO). After measurement of absorption values at 450 nm, the titer of assembled capsids in the sample is determined by comparison with a standard curve that is obtained in parallel using an AAV-2 preparation of known capsid titer.

magnitude (30,45,46). Likewise, measurement of DNA-containing rAAV particles involves various steps and assumptions that might lead to errors that must not be overlooked. First, defective interfering genomes that are only partially packaged and thus not able to direct expression of the transgenes give an inflated signal that does not accurately represent the number of full-length encapsidated genomes. Second, it is likely that an unknown number of recombinant genomes are attached to the outside of the capsid in a state resistant to DNase digestion, instead of being encapsidated, thereby further falsifying the outcome of the measurement (185). Third, the reverse might also occur (i.e., physical particle titers might be underestimated, since some of the encapsidated genomes most certainly get lost in the course of the DNA extraction procedure). Therefore, titers of DNA-containing particles are probably not as reliable as often assumed.

As an alternative approach to quantify physical rAAV particles, we have recently established an AAV-2 capsid enzyme-linked immunosorbent assay (ELISA), which allows specific titration of assembled wild-type or recombinant AAV-2 particles (55). The principle of the ELISA is outlined in Figure 5. By design, this novel titration tool is very user-friendly. Making use of a simple and consistent protocol, nonradioactive measurement of particles in up to 22 different samples per microplate can be accomplished in less than 4 hours. In obvious contrast to the aforementioned titration method for quantification of DNA-containing particles, ELISA yields actual titers of physical particles, which are defined here as the sum of assembled virions, irrespective of being loaded with viral genomes or not. In view of the fact that application of AAV-2 vectors can result in a humoral immune response directed against capsid proteins, it will now be very interesting to investigate the correlation between numbers of intact rAAV virions applied and strength and extent of host immune reaction.

Using ELISA for titration of various preparations of wild-type and recombinant AAV-2, we discovered a previously unknown, but probably very important phenomenon, namely that deficient packaging of recombinant AAV-2 genomes may be responsible for the limited recovery of AAV-2 vectors. In fact, crude rAAV preparations are usually characterized by a preponderance of empty capsids (i.e., virions that are devoid of detectable viral DNA) with ratios of empty to full particles of up to 50:1. In contrast, these ratios were reduced to a maximum of 4:1 in the case of wt AAV-2, thus clearly depicting a difference in the efficiency of genome packaging (55). A possible explanation for this would be that as yet unidentified sequences within the AAV-2 genome, which are present within the wild-type genome, but absent in recombinant ones, are required for efficient encapsidation. Alternatively or in addition, such sequences might also play a role in replication of the viral genomes, thus providing an amplification advantage for wild-type over recombinant AAV-2 DNA—a hypothesis that is actually supported by earlier findings (109). The investigation of the mechanisms and factors that direct AAV-2 DNA replication and encapsidation thus will certainly have a major impact on the further development of AAV-2 vectors for human gene therapy.

In summary, there are a variety of different methods for titration of functional or physical rAAV particles, which, however, are not optimized with respect to versatility, ease, and reliability. In view of this, the recently developed AAV-2 capsid ELISA, which allows the rapid and dependable quantification of assembled (empty and full) AAV-2 particles, represents an additional useful tool for characterization of AAV-2 vectors.

SUMMARY AND CONCLUSIONS

Within the last 10 years, AAV-2 has turned into a promising viral vector system available for the treatment of several human diseases. Having started with cumbersome and inefficient methods for generation and purification of AAV-2 vectors, there have been a number of improvements that will help push the clinical use of this interesting virus. In particular, with the recent development of noninfectious adenoviral helper plasmids, the generation of high-titer, helper virus-free recombinant AAV-2 has now become feasible. The only limitation to scaling up the process is the requirement of a transient transfection step. In view of this, the ongoing establishment of new and improved stable packaging or producer cell lines, which might even include the adenoviral helper functions, could provide a means to solve the problem of scaling up. Thus, production of AAV-2 vectors in sufficient quantities for clinical studies is an achievable goal in the near future.

Likewise, the field of rAAV purification is currently advancing rapidly. Novel and more favorable chromatography methods are being developed that will eventually replace the traditional time-consuming CsCl centrifugation protocol. Finally, with the help of new and more reliable techniques for rAAV quantification, a standard terminology for vector titers may be established. This will hopefully provide a common basis ensuring comparability of data obtained in different studies using AAV-2 vectors.

REFERENCES

1. **Alexander, I.E., D.W. Russell, and A.D. Miller.** 1994. DNA-damaging agents greatly increase the transduction of nondividing cells by adeno-associated virus vectors. J. Virol. *68*:8282-8287.

2. **Alexander, I.E., D.W. Russell, and A.D. Miller.** 1997. Transfer of contaminants in adeno-associated virus vector stocks can mimic transduction and lead to artifactual results. Hum. Gene Ther. *8*:1911-1920.

3. **Alexander, I.E., D.W. Russell, A.M. Spence, and A.D. Miller.** 1996. Effects of gamma irradiation on the transduction of dividing and nondividing cells in brain and muscle of rats by adeno-associated virus vectors. Hum. Gene Ther. *7*:841-850.

4. **Allen, J.M., D.J. Debelak, T.C. Reynolds, and A.D. Miller.** 1997. Identification and elimination of replication-competent adeno-associated virus (AAV) that can arise by nonhomologous recombination during AAV vector production. J. Virol. *71*:6816-6822.

5. **Atkinson, E.M., D.J. Debelak, L.A. Hart, and T.C. Reynolds.** 1998. A high-throughput hybridization method for titer determination of viruses and gene therapy vectors. Nucleic Acids Res. *26*:2821-2823.

6. **Balague, C., M. Kalla, and W.W. Zhang.** 1997. Adeno-associated virus Rep78 protein and terminal repeats enhance integration of DNA sequences into the cellular genome. J. Virol. *71*:3299-3306.

7. **Bantel-Schaal, U.** 1990. Adeno-associated parvoviruses inhibit growth of cells derived from malignant human tumors. Int. J. Cancer *45*:190-194.

8. **Bantel-Schaal, U.** 1991. Infection with adeno-associated parvovirus leads to increased sensitivity of mammalian cells to stress. Virology *182*:260-268.

9. **Bantel-Schaal, U., H. Delius, R. Schmidt, and H. zur Hausen.** 1999. Human adeno-associated virus type 5 is only distantly related to other known primate helper-dependent parvoviruses. J. Virol. *73*:939-947.

10. **Bantel-Schaal, U. and M. Stöhr.** 1992. Influence of adeno-associated virus on adherence and growth properties of normal cells. J. Virol. *66*:773-779.

11. **Bantel-Schaal, U. and H. zur Hausen.** 1984. Characterization of the DNA of a defective human parvovirus isolated from a genital site. Virology *134*:52-63.

12. **Bartlett, J.S., R. Wilcher, and R.J. Samulski.** 2000. Infectious entry pathway of adeno-associated virus and adeno-associated virus vectors. J. Virol. *74*:2777-2785.

13. **Batchu, R.B., D.A. Miles, T.M. Rechtin, R.R. Drake, and P.L. Hermonat.** 1995. Cloning, expression and purification of full length Rep78 of adeno-associated virus as a fusion protein with maltose binding protein in *Escherichia coli*. Biochem. Biophys. Res. Commun. *208*:714-720.

14. **Beaton, A., P. Palumbo, and K.I. Berns.** 1989. Expression from the adeno-associated virus p5 and p19 promoters is negatively regulated in trans by the rep protein. J. Virol. *63*:4450-4454.

15. **Berns, K.I. and R.A. Bohenzky.** 1987. Adeno-associated viruses: an update. Adv. Virus Res. *32*:243-306.

16. **Berns, K.I. and C. Giraud.** 1996. Biology of adeno-associated virus. Curr. Top. Microbiol. Immunol. *218*:1-23.

17. **Berns, K.I. and R.M. Linden.** 1995. The cryptic life style of adeno-associated virus. Bioessays *17*:237-245.

18. **Botquin, V., A. Cid-Arregui, and J.R. Schlehofer.** 1994. Adeno-associated virus type 2 interferes with early development of mouse embryos. J. Gen. Virol. *75*:2655-2662.

19. **Carter, B.J.** 1992. Adeno-associated virus vectors.

Curr. Opin. Biotechnol. *3*:533-539.

20.**Carter, B.J., C.A. Laughlin, L.M. de la Maza, and M. Myers.** 1979. Adeno-associated virus autointerference. Virology *92*:449-462.

21.**Chejanovsky, N. and B.J. Carter.** 1989. Mutagenesis of an AUG codon in the adeno-associated virus rep gene: effects on viral DNA replication. Virology *173*:120-128.

22.**Cheung, A.K., M.D. Hoggan, W.W. Hauswirth, and K.I. Berns.** 1980. Integration of the adeno-associated virus genome into cellular DNA in latently infected human Detroit 6 cells. J. Virol. *33*:739-748.

23.**Chiorini, J.A., F. Kim, L. Yang, and R.M. Kotin.** 1999. Cloning and characterization of adeno-associated virus type 5. J. Virol. *73*:1309-1319.

24.**Chiorini, J.A., M.D. Weitzman, R.A. Owens, E. Urcelay, B. Safer, and R.M. Kotin.** 1994. Biologically active Rep proteins of adeno-associated virus type 2 produced as fusion proteins in *Escherichia coli.* J. Virol. *68*:797-804.

25.**Chiorini, J.A., C.M. Wendtner, E. Urcelay, B. Safer, M. Hallek, and R.M. Kotin.** 1995. High-efficiency transfer of the T cell co-stimulatory molecule B7-2 to lymphoid cells using high-titer recombinant adeno-associated virus vectors. Hum. Gene Ther. *6*:1531-1541.

26.**Chiorini, J.A., S.M. Wiener, L. Yang, R.H. Smith, B. Safer, N.P. Kilcoin, Y. Liu, E. Urcelay, and R.M. Kotin.** 1996. The roles of AAV Rep proteins in gene expression and targeted integration. Curr. Top. Microbiol. Immunol. *218*:25-33.

27.**Chiorini, J.A., L. Yang, Y. Liu, B. Safer, and R.M. Kotin.** 1997. Cloning of adeno-associated virus type 4 (AAV4) and generation of recombinant AAV4 particles. J. Virol. *71*:6823-6833.

28.**Chiorini, J.A., B. Zimmermann, L. Yang, R.H. Smith, A. Ahearn, F. Herberg, and R.M. Kotin.** 1998. Inhibition of PrKX, a novel protein kinase, and the cyclic AMP-dependent protein kinase PKA by the regulatory proteins of adeno-associated virus type 2. Mol. Cell. Biol. *18*:5921-5929.

29.**Clark, K.R., F. Voulgaropoulou, D.M. Fraley, and P.R. Johnson.** 1995. Cell lines for the production of recombinant adeno-associated virus. Hum. Gene Ther. *6*:1329-1341.

30.**Clark, K.R., F. Voulgaropoulou, and P.R. Johnson.** 1996. A stable cell line carrying adenovirus-inducible rep and cap genes allows for infectivity titration of adeno-associated virus vectors. Gene Ther. *3*:1124-1132.

31.**Clark, K.R., L. Xinglou, J.P. McGrath, and P.R. Johnson.** 1999. Highly purified recombinant adeno-associated virus vectors are biologically active and free of detectable helper and wild-type viruses. Hum. Gene Ther. *10*:1031-1039.

32.**Conrad, C.K., S.S. Allen, S.A. Afione, T.C. Reynolds, S.E. Beck, M. Fee-Maki, X. Barrazza-Ortiz, R. Adams, et al.** 1996. Safety of single-dose administration of an adeno-associated virus (AAV)-CFTR vector in the primate lung. Gene Ther. *3*:658-668.

33.**Conway, J.E., C.M.J. ap Rhys, I. Zolotukhin, S. Zolotukhin, N. Muzyczka, G.S. Hayward, and B.J. Byrne.** 1999. High-titer recombinant adeno-associated virus production utilizing a recombinant herpes simplex virus type I vector expressing AAV-2 Rep and Cap. Gene Ther. *6*:986-993.

34.**Conway, J.E., S. Zolotukhin, N. Muzyczka, G.S. Hayward, and B.J. Byrne.** 1997. Recombinant adeno-associated virus type 2 replication and packaging is entirely supported by a herpes simplex virus type 1 amplicon expressing Rep and Cap. J. Virol. *71*:8780-8789.

35.**Cutt, J.R., T. Shenk, and P. Hearing.** 1987. Analysis of adenovirus early region 4-encoded polypeptides synthesized in productively infected cells. J. Virol. *61*:543-552.

36.**Ding, L., S. Lu, and N.C. Munshi.** 1997. In vitro packaging of an infectious recombinant adeno-associated virus 2. Gene Ther. *4*:1167-1172.

37.**Di Pasquale, G. and S.N. Stacey.** 1998. Adeno-associated virus Rep78 protein interacts with protein kinase A and its homolog PRKX and inhibits CREB-dependent transcriptional activation. J. Virol. *72*:7916-7925.

38.**Dong, J.Y., P.D. Fan, and R.A. Frizzell.** 1996. Quantitative analysis of the packaging capacity of recombinant adeno-associated virus. Hum. Gene Ther. *7*:2101-2112.

39.**Duan, D., Q. Li, A.W. Kao, Y. Yue, J.E. Pessin, and J.F. Engelhardt.** 1999. Dynamin is required for recombinant adeno-associated virus type 2 infection. J. Virol. *73*:10371-10376.

40.**Duan, D., P. Sharma, J. Yang, Y. Yue, L. Dudus, Y. Zhang, K.J. Fisher, and J.F. Engelhardt.** 1998. Circular intermediates of recombinant adeno-associated virus have defined structural characteristics responsible for long-term episomal persistence in muscle tissue. J. Virol. *72*:8568-8577.

41.**Dubielzig, R., J.A. King, S. Weger, A. Kern, and J.A. Kleinschmidt.** 1999. Adeno-associated virus type 2 protein interactions: formation of pre-encapsidation complexes. J. Virol. *73*:8989-8998.

42.**Dutheil, N., O. Malhomme, N. Provost, P. Becquart, T. Burguette, J.R. Schlehofer, and T. Dupressoir.** 1997. Presence of integrated DNA sequences of adeno-associated virus type 2 in four cell lines of human embryonic origin. J. Gen. Virol. *78*:3039-3043.

43.**Dyall, J. and K.I. Berns.** 1998. Site-specific integration of adeno-associated virus into an episome with the target locus via a deletion-substitution mechanism. J. Virol. *72*:6195-6198.

44.**Fan, P.D. and J.Y. Dong.** 1997. Replication of rep-cap genes is essential for the high-efficiency production of recombinant AAV. Hum. Gene Ther. *8*:87-98.

45.**Ferrari, F.K., T. Samulski, T. Shenk, and R.J. Samulski.** 1996. Second-strand synthesis is a rate-limiting step for efficient transduction by recombinant adeno-associated virus vectors. J. Virol. *70*:3227-3234.

46.**Fisher, K.J., G.P. Gao, M.D. Weitzman, R.DeMatteo, J.F. Burda, and J.M. Wilson.** 1996. Transduction with recombinant adeno-associated virus for gene therapy is limited by leading-strand synthesis. J. Virol. *70*:520-532.

47.**Fisher, K.J., K. Jooss, J. Alston, Y. Yang, S.E. Haeck-

er, K. High, R. Pathak, S.E. Raper, and J.M. Wilson. 1997. Recombinant adeno-associated virus for muscle directed gene therapy. Nat. Med. *3*:306-312.

48.**Fisher, K.J., W.M. Kelley, J.F. Burda, and J.M. Wilson.** 1996. A novel adenovirus-adeno-associated virus hybrid vector that displays efficient rescue and delivery of the AAV genome. Hum. Gene Ther. *17*:2079-2087.

49.**Flannery, J.G., S. Zolotukhin, M.I. Vaquero, M.M. LaVail, N. Muzyczka, and W.W. Hauswirth.** 1997. Efficient photoreceptor-targeted gene expression in vivo by recombinant adeno-associated virus. Proc. Natl. Acad. Sci. USA *94*:6916-6921.

50.**Flotte, T.R., S.A. Afione, C. Conrad, S.A. McGrath, R. Solow, H. Oka, P.L. Zeitlin, W.B. Guggino, and B.J. Carter.** 1993. Stable in vivo expression of the cystic fibrosis transmembrane conductance regulator with an adeno-associated virus vector. Proc. Natl. Acad. Sci. USA *90*:10613-10617.

51.**Flotte, T.R., S.A. Afione, and P.L. Zeitlin.** 1994. Adeno-associated virus vector gene expression occurs in nondividing cells in the absence of vector DNA integration. Am. J. Respir. Cell Mol. Biol. *11*:517-521.

52.**Flotte, T.R., O.X. Barraza, R. Solow, S.A. Afione, B.J. Carter, and W.B. Guggino.** 1995. An improved system for packaging recombinant adeno-associated virus vectors capable of in vivo transduction. Gene Ther. *2*:29-37.

53.**Flotte, T.R., R. Solow, R.A. Owens, S. Afione, P.L. Zeitlin, and B.J. Carter.** 1992. Gene expression from adeno-associated virus vectors in airway epithelial cells. Am. J. Respir. Cell Mol. Biol. *7*:349-356.

54.**Gao, G.P., G. Qu, L.Z. Faust, R.K. Engdahl, W. Xiao, J.V. Hughes, P.W. Zoltick, and J.M. Wilson.** 1998. High-titer adeno-associated viral vectors from a Rep/Cap cell line and hybrid shuttle virus. Hum. Gene Ther. *9*:2353-2362.

55.**Grimm, D., A. Kern, M. Pawlita, F.K. Ferrari, R.J. Samulski, and J.A. Kleinschmidt.** 1999. Titration of AAV-2 particles via a novel capsid Elisa: packaging of genomes can limit production of recombinant AAV-2. Gene. Ther. *6*:1322-1330.

56.**Grimm, D., A. Kern, K. Rittner, and J.A. Kleinschmidt.** 1998. Novel tools for production and purification of recombinant adenoassociated virus vectors. Hum. Gene Ther. *9*:2745-2760.

57.**Halbert, C.L., T.A. Standaert, M.L. Aitken, I.E. Alexander, D.W. Russell, and A.D. Miller.** 1997. Transduction by adeno-associated virus vectors in the rabbit airway: efficiency, persistence, and readministration. J. Virol. *71*:5932-5941.

58.**Halbert, C.L., T.A. Standaert, C.B. Wilson, and A.D. Miller.** 1998. Successful readministration of adeno-associated virus vectors to the mouse lung requires transient immunosuppression during the initial exposure. J. Virol. *72*:9795-9805.

59.**Hansen, J., K. Qing, H.J. Kwon, C. Mah, and A. Srivastava.** 2000. Impaired intracellular trafficking of adeno-associated virus type 2 vectors limits efficient transduction of murine fibroblasts. J. Virol. *74*:992-996.

60.**Heilbronn, R., A. Bürkle, S. Stephan, and H. zur Hausen.** 1990. The adeno-associated virus rep gene suppresses herpes simplex virus-induced DNA amplification. J. Virol. *64*:3012-3018.

61.**Hermanns, J., A. Schulze, P. Jansen-Dürr, J.A. Kleinschmidt, R. Schmidt, and H. zur Hausen.** 1997. Infection of primary cells by adeno-associated virus type 2 results in a modulation of cell cycle-regulating proteins. J. Virol. *71*:6020-6027.

62.**Hermans, W.T.J.M.C., O. ter Brake, P.A. Dijkhuizen, M.A.F. Sonnemans, D. Grimm, J.A. Kleinschmidt, and J. Verhagen.** 1999. Purification of recombinant adeno-associated virus by iodixanol gradient ultracentrifugation allows rapid and reproducible preparation of vector stocks for gene transfer in the nervous system. Hum. Gene Ther. *10*:1885-1891.

63.**Hermonat, P.L. and N. Muzyczka.** 1984. Use of adeno-associated virus as a mammalian DNA cloning vector: transduction of neomycin resistance into mammalian tissue culture cells. Proc. Natl. Acad. Sci. USA *81*:6466-6470.

64.**Hermonat, P.L., A.D. Santin, and R.B. Batchu.** 1996. The adeno-associated virus Rep78 major regulatory/transformation suppressor protein binds cellular Sp1 in vitro and evidence of a biological effect. Cancer Res. *56*:5299-5304.

65.**Hermonat, P.L., A.D. Santin, R.B. Batchu, and D. Zhan.** 1998. The adeno-associated virus Rep78 major regulatory protein binds the cellular TATA-binding protein in vitro and in vivo. Virology *25*:120-127.

66.**Hermonat, P.L., A.D. Santin, C.A. Carter, G.P. Parhama, and J.G. Quirk.** 1997. Multiple cellular proteins are recognized by the adeno-associated virus Rep78 major regulatory protein and the amino-half of Rep78 is required for many of these interactions. Biochem Mol. Biol. Int. *43*:409-420.

67.**Herzog, R.W., J.N. Hagstrom, S.H. Kung, S.J. Tai, J.M. Wilson, K.J. Fisher, and K.A. High.** 1997. Stable gene transfer and expression of human blood coagulation factor IX after intramuscular injection of recombinant adeno-associated virus. Proc. Natl. Acad. Sci. USA *94*:5804-5809.

68.**Hölscher, C., M. Hörer, J.A. Kleinschmidt, H. Zentgraf, A. Bürkle, and R. Heilbronn.** 1994. Cell lines inducibly expressing the adeno-associated virus (AAV) rep gene: requirements for productive replication of rep-negative AAV mutants. J. Virol. *68*:7169-7177.

69.**Hölscher, C., J.A. Kleinschmidt, and A. Bürkle.** 1995. High-level expression of adeno-associated virus (AAV) Rep78 or Rep68 protein is sufficient for infectious-particle formation by a rep-negative AAV mutant. J. Virol. *69*:6880-6885.

70.**Hoque, M., K. Ishizu, A. Matsumoto, S.I. Han, F. Arisaka, M. Takayama, K. Suzuki, K. Kato, T. Kanda, H. Watanabe, and H. Handa.** 1999. Nuclear transport of the major capsid protein is essential for adeno-associated virus capsid formation. J. Virol. *73*:7912-7915.

71.**Hunter, L.A. and R.J. Samulski.** 1992. Colocalization of adeno-associated virus Rep and capsid proteins in the nuclei of infected cells. J. Virol. *66*:317-324.

72.**Inoue, N. and D.W. Russell.** 1998. Packaging cells based on inducible gene amplification for the production of adeno-associated virus vectors. J. Virol.

72:7024-7031.

73.**Janik, J.E., M.M. Huston, K. Cho, and J.A. Rose.** 1989. Efficient synthesis of adeno-associated virus structural proteins requires both adenovirus DNA binding protein and VA I RNA. Virology *168*:320-329.

74.**Jay, F.T., C.A. Laughlin, and B.J. Carter.** 1981. Eukaryotic translational control: adeno-associated virus protein synthesis is affected by a mutation in the adenovirus DNA-binding protein. Proc. Natl. Acad. Sci. USA *78*:2927-2931.

75.**Jordan, M., C. Köhne, and F.M. Wurm.** 1998. Calcium-phosphate mediated DNA transfer into HEK-293 cells in suspension: control of physicochemical parameters allows transfection in stirred media. Cytotechnology *26*:39-47.

76.**Kaplitt, M.G., P. Leone, R.J. Samulski, X. Xiao, D.W. Pfaff, K.L. O'Malley, and M.J. During.** 1994. Long-term gene expression and phenotypic correction using adeno-associated virus vectors in the mammalian brain. Nat. Genet. *8*:148-154.

77.**Kearns, W.G., S.A. Afione, S.B. Fulmer, M.C. Pang, D. Erikson, M. Egan, M.J. Landrum, T.R. Flotte, and G.R. Cutting.** 1996. Recombinant adeno-associated virus (AAV-CFTR) vectors do not integrate in a site-specific fashion in an immortalized epithelial cell line. Gene Ther. *3*:748-755.

78.**Kessler, P.D., G.M. Podsakoff, X. Chen, S.A. McQuiston, P.C. Colosi, L.A. Matelis, G.J. Kurtzman, and B.J. Byrne.** 1996. Gene delivery to skeletal muscle results in sustained expression and systemic delivery of a therapeutic protein. Proc. Natl. Acad. Sci. USA *93*:14082-14087.

79.**Klein-Bauernschmitt, P., H. zur Hausen, and J.R. Schlehofer.** 1992. Induction of differentiation-associated changes in established human cells by infection with adeno-associated virus type 2. J. Virol. *66*:4191-4200.

80.**Kotin, R.M., M. Siniscalco, R.J. Samulski, X.D. Zhu, L. Hunter, C.A. Laughlin, S. McLaughlin, N. Muzyczka, M. Rocchi, and K.I. Berns.** 1990. Site-specific integration by adeno-associated virus. Proc. Natl. Acad. Sci. USA *87*:2211-2215.

81.**Kyöstiö, S.R., R.A. Owens, M.D. Weitzman, B.A. Antoni, N. Chejanovsky, and B.J. Carter.** 1994. Analysis of adeno-associated virus (AAV) wild-type and mutant Rep proteins for their abilities to negatively regulate AAV p5 and p19 mRNA levels. J. Virol. *68*:2947-2957.

82.**Kyöstiö, S.R., R.S. Wonderling, and R.A. Owens.** 1995. Negative regulation of the adeno-associated virus (AAV) p5 promoter involves both the p5 rep binding site and the consensus ATP-binding motif of the AAV Rep68 protein. J. Virol. *69*:6787-6796.

83.**Labow, M.A., P.L. Hermonat, and K.I. Berns.** 1986. Positive and negative autoregulation of the adeno-associated virus type 2 genome. J. Virol. *60*:251-258.

84.**Lamartina, S., G. Roscilli, D. Rinaudo, P. Delmastro, and C. Toniatti.** 1998. Lipofection of purified adeno-associated virus Rep68 protein: toward a chromosome-targeting nonviral particle. J. Virol. *72*:7653-7658.

85.**Laughlin, C.A., C.B. Cardellichio, and H.C. Coon.** 1986. Latent infection of KB cells with adeno-associated virus type 2. J. Virol. *60*:515-524.

86.**Laughlin, C.A., J.D. Tratschin, H. Coon, and B.J. Carter.** 1983. Cloning of infectious adeno-associated virus genomes in bacterial plasmids. Gene *23*:65-73.

87.**Leppard, K.N. and T. Shenk.** 1989. The adenovirus E1B 55 kd protein influences mRNA transport via an intranuclear effect on RNA metabolism. EMBO J. *8*:2329-2336.

88.**Lewis, B.A., G. Tullis, E. Seto, N. Horikoshi, R. Weinmann, and T. Shenk.** 1995. Adenovirus E1A proteins interact with the cellular YY1 transcription factor. J. Virol. *69*:1628-1636.

89.**Li, J., R.J. Samulski, and X. Xiao.** 1997. Role for highly regulated rep gene expression in adeno-associated virus vector production. J. Virol. *71*:5236-5243.

90.**Linden, R.M., P. Ward, C. Giraud, E. Winocour, and K.I. Berns.** 1996. Site-specific integration by adeno-associated virus. Proc. Natl. Acad. Sci. USA *93*:11288-11294.

91.**Linden, R.M., E. Winocour, and K.I. Berns.** 1996. The recombination signals for adeno-associated virus site-specific integration. Proc. Natl. Acad. Sci. USA *93*:7966-7972.

92.**Liu, X.L., K.R. Clark, and P.R. Johnson.** 1999. Production of recombinant adeno-associated virus vectors using a packaging cell line and a hybrid recombinant adenovirus. Gene Ther. *6*:293-299.

93.**Lo, W.D., G. Qu, T.J. Sferra, R. Clark, R. Chen, and P.R. Johnson.** 1999. Adeno-associated virus-mediated gene transfer to the brain: duration and modulation of expression. Hum. Gene Ther. *10*:201-213.

94.**Mah, C., K. Qing, B. Khuntirat, S. Ponnazhagan, X.S. Wang, D.M. Kube, M.C. Yoder, and A. Srivastava.** 1998. Adeno-associated virus type 2-mediated gene transfer: role of epidermal growth factor receptor protein tyrosine kinase in transgene expression. J. Virol. *72*:9835-9843.

95.**Malik, P., S.A. McQuiston, X.J. Yu, K.A. Pepper, W.J. Krall, G.M. Podsakoff, G.J. Kurtzman, and D.B. Kohn.** 1997. Recombinant adeno-associated virus mediates a high level of gene transfer but less efficient integration in the K562 human hematopoietic cell line. J. Virol. *71*:1776-1783.

96.**Mamounas, M., M. Levitt, M. Yu, and F. Wong-Staal.** 1995. Increased titer of recombinant AAV vectors by gene transfer with adenovirus coupled to DNA-polylysine complexes. Gene Ther. *6*:429-432.

97.**Manning, W.C., S. Zhou, M.P. Bland, J.A. Escobedo, and V. Dwarki.** 1998. Transient immunosuppression allows transgene expression following readministration of adeno-associated viral vectors. Hum. Gene Ther. *9*:477-485.

98.**Matsushita, T., S. Elliger, C. Elliger, G. Podsakoff, L. Villarreal, G.J. Kurtzman, Y. Iwaki, and P. Colosi.** 1998. Adeno-associated virus vectors can be efficiently produced without helper virus. Gene Ther. *5*:938-945.

99.**Maxwell, I.H., F. Maxwell, and J. Schaack.** 1998. An adenovirus type 5 mutant with the preterminal protein gene deleted efficiently provides helper functions for the production of recombinant adeno-associated virus. J. Virol. *72*:8371-8373.

100.**Mayor, H.D., S. Drake, J. Stahmann, and D.M. Mumford.** 1976. Antibodies to adeno-associated satellite virus and herpes simplex in sera from cancer patients and normal adults. Am. J. Obstet. Gynecol. *126*:100-104.

101.**McCarty, D.M., M. Christensen, and N. Muzyczka.** 1991. Sequences required for coordinate induction of adeno-associated virus p19 and p40 promoters by Rep protein. J. Virol. *65*:2936-2945.

102.**McCarty, D.M., D.J. Pereira, I. Zolotukhin, X. Zhou, J.H. Ryan, and N. Muzyczka.** 1994. Identification of linear DNA sequences that specifically bind the adeno-associated virus Rep protein. J. Virol. *68*:4988-4997.

103.**McCarty, D.M., J.H. Ryan, S. Zolotukhin, X. Zhou, and N. Muzyczka.** 1994. Interaction of the adeno-associated virus Rep protein with a sequence within the A palindrome of the viral terminal repeat. J. Virol. *68*:4998-5006.

104.**McCown, T.J., X. Xiao, J. Li, G.R. Breese, and R.J. Samulski.** 1996. Differential and persistent expression patterns of CNS gene transfer by an adeno-associated virus (AAV) vector. Brain Res. *713*:99-107.

105.**McLaughlin, S.K., P. Collis, P.L. Hermonat, and N. Muzyczka.** 1988. Adeno-associated virus general transduction vectors: analysis of proviral structures. J. Virol. *62*:1963-1973.

106.**Monahan, P.E., R.J. Samulski, J. Tazelaar, X. Xiao, T.C. Nichols, D.A. Bellinger, M.S. Read, and C.E. Walsh.** 1998. Direct intramuscular injection with recombinant AAV vectors results in sustained expression in a dog model of hemophilia. Gene Ther. *5*:40-49.

107.**Muramatsu, S., A. Handa, S. Kajigaya, and K.E. Brown.** 1998. Transcription-positive cofactor 4 enhances rescue of adeno-associated virus genome from an infectious clone. J. Gen. Virol. *79*:2157-2161.

108.**Muramatsu, S., H. Mizukami, N.S. Young, and K.E. Brown.** 1996. Nucleotide sequencing and generation of an infectious clone of adeno-associated virus 3. Virology *221*:208-217.

109.**Muzyczka, N.** 1992. Use of adeno-associated virus as a general transduction vector for mammalian cells. Curr. Top. Microbiol. Immunol. *158*:97-129.

110.**Myers, M.W. and B.J. Carter.** 1980. Assembly of adeno-associated virus. Virology *102*:71-82.

111.**Myers, M.W., C.A. Laughlin, F.T. Jay, and B.J. Carter.** 1980. Adenovirus helper function for growth of adeno-associated virus: effect of temperature-sensitive mutations in adenovirus early gene region 2. J. Virol. *35*:65-75.

112.**Nahreini, P. and M.B. Mathews.** 1997. Transduction of the human immunodeficiency virus type 1 promoter into human chromosomal DNA by adeno-associated virus: effects on promoter activity. Virology *234*:42-50.

113.**Ni, T.H., X. Zhou, D.M. McCarty, I. Zolotukhin, and N. Muzyczka.** 1994. In vitro replication of adeno-associated virus DNA. J. Virol. *68*:1128-1138.

114.**Palombo, F., A. Monciotti, A. Recchia, R. Cortese, G. Ciliberto, and M.N. La Monica.** 1998. Site-specific integration in mammalian cells mediated by a new hybrid baculovirus-adeno-associated virus vector. J. Virol. *72*:5025-5034.

115.**Peel, A.L., S. Zolotukhin, G.W. Schrimsher, N. Muzyczka, and P.J. Reier.** 1997. Efficient transduction of green fluorescent protein in spinal cord neurons using adeno-associated virus vectors containing cell type-specific promoters. Gene Ther. *4*:16-24.

116.**Pereira, D.J., D.M. McCarty, and N. Muzyczka.** 1997. The adeno-associated virus (AAV) Rep protein acts as both a repressor and an activator to regulate AAV transcription during a productive infection. J. Virol. *71*:1079-1088.

117.**Pereira, D.J. and N. Muzyczka.** 1997. The cellular transcription factor SP1 and an unknown cellular protein are required to mediate Rep protein activation of the adeno-associated virus p19 promoter. J. Virol. *71*:1747-1756.

118.**Pieroni, L., C. Fipaldini, A. Monciotti, D. Cimini, A. Sgura, E. Fattori, O. Epifano, R. Cortese, F. Palombo, and N. La Monica.** 1998. Targeted integration of adeno-associated virus-derived plasmids in transfected human cells. Virology *249*:249-259.

119.**Podsakoff, G., K.J. Wong, and S. Chatterjee.** 1994. Efficient gene transfer into nondividing cells by adeno-associated virus-based vectors. J. Virol. *68*:5656-5666.

120.**Ponnazhagan, S., D. Erikson, W.G. Kearns, S.Z. Zhou, P. Nahreini, X.S. Wang, and A. Srivastava.** 1997. Lack of site-specific integration of the recombinant adeno-associated virus 2 genomes in human cells. Hum. Gene Ther. *10*:275-284.

121.**Ponnazhagan, S., M.C. Yoder, and A. Srivastava.** 1997. Adeno-associated virus type 2-mediated transduction of murine hematopoietic cells with long-term repopulating ability and sustained expression of a human globin gene in vivo. J. Virol. *71*:3098-3104.

122.**Prasad, K.M. and J.P. Trempe.** 1995. The adeno-associated virus Rep78 protein is covalently linked to viral DNA in a preformed virion. Virology *214*:360-370.

123.**Qing, K., B. Khuntirat, C. Mah, D.M. Kube, X.S. Wang, S. Ponnazhagan, S. Zhou, V. J. Dwarki, M.C. Yoder, and A. Srivastava.** 1998. Adeno-associated virus type 2-mediated gene transfer: correlation of tyrosine phosphorylation of the cellular single-stranded D sequence-binding protein with transgene expression in human cells in vitro and murine tissues in vivo. J. Virol. *72*:1593-1599.

124.**Qing, K., C. Mah, J. Hansen, S. Zhou, V. Dwarki, and A. Srivastava.** 1999. Human fibroblast growth factor receptor 1 is a co-receptor for infection by adeno-associated virus 2. Nat. Med. *5*:71-77.

125.**Qing, K., X.S. Wang, D.M. Kube, S. Ponnazhagan, A. Bajpaian, and A. Srivastava.** 1997. Role of tyrosine phosphorylation of a cellular protein in adeno-associated virus 2-mediated transgene expression. Proc. Natl. Acad. Sci. USA *94*:10879-10884.

126.**Qiu, J., H. Mizukami, and K.E. Brown.** 1999. Adeno-associated virus 2 co-receptors? Nat. Med. *5*:467-468.

127.**Qiu, J. and K.E. Brown.** 1999. Integrin alphaVbeta5 is not involved in adeno-associated virus type 2 (AAV2) infection. Virol. *264*:436-440.

128.**Rendahl, K.G., S.E. Leff, G.R. Otten, S.K. Spratt, D. Bohl, M.V. Roey, B.A. Donahue, L.K. Cohen, et al.** 1998. Regulation of gene expression in vivo following transduction by two separate rAAV vectors. Nat. Biotech. *16*:757-761.

129.**Rose, J.A. and F. Koczot.** 1972. Adenovirus-associated virus multiplication. VII. Helper requirement for viral deoxyribonucleic acid and ribonucleic acid synthesis. J. Virol. *10*:1-8.

130.**Ruffing, M., H. Heid, and J.A. Kleinschmidt.** 1994. Mutations in the carboxy terminus of adeno-associated virus 2 capsid proteins affect viral infectivity: lack of an RGD integrin-binding motif. J. Gen. Virol. *75*:3385-3392.

131.**Ruffing, M., H. Zentgraf, and J.A. Kleinschmidt.** 1992. Assembly of viruslike particles by recombinant structural proteins of adeno-associated virus type 2 in insect cells. J. Virol. *66*:6922-6930.

132.**Russell, D.W., A.D. Miller, and I.E. Alexander.** 1994. Adeno-associated virus vectors preferentially transduce cells in S phase. Proc. Natl. Acad. Sci. USA *91*:8915-8919.

133.**Rutledge, E.A., C.L. Halbert, and D.W. Russell.** 1998. Infectious clones and vectors derived from adeno-associated virus (AAV) serotypes other than AAV type 2. J. Virol. *72*:309-319.

134.**Rutledge, E.A. and D.W. Russell.** 1997. Adeno-associated virus vector integration junctions. J. Virol. *71*:8429-8436.

135.**Salvetti, A., S. Orève, G. Chadeuf, D. Favre, Y. Cherel, P. Champion-Arnaud, J. David-Ameline, and P. Moullier.** 1998. Factors influencing recombinant adeno-associated virus production. Hum. Gene Ther. *9*:695-706.

136.**Samulski, R.J., K.I. Berns, M. Tan, and N. Muzyczka.** 1982. Cloning of adeno-associated virus into pBR322 : Rescue of intact virus from the recombinant plasmid in human cells. Proc. Natl. Acad. Sci. USA *79*:2077-2081.

137.**Samulski, R.J., L.-S. Chang, and T. Shenk.** 1987. A recombinant plasmid from which an infectious adeno-associated virus genome can be excised in vitro and its use to study viral replication. J. Virol. *61*:3096-3101.

138.**Samulski, R.J., L.S. Chang, and T. Shenk.** 1989. Helper-free stocks of recombinant adeno-associated viruses : normal integration does not require viral gene expression. J. Virol. *63*:3822-3828.

139.**Samulski, R.J. and T. Shenk.** 1988. Adenovirus E1B 55-M_r polypeptide facilitates timely cytoplasmic accumulation of adeno-associated virus mRNAs. J. Virol. *62*:206-210

140.**Samulski, R.J., A. Srivastava, K.I. Berns, and N. Muzyczka.** 1983. Rescue of adeno-associated virus from recombinant plasmids: gene correction within the terminal repeats of AAV. Cell *33*:135-143.

141.**Samulski, R.J., X. Zhu, X. Xiao, J.D. Brook, D.E. Housman, N. Epstein, and L.A. Hunter.** 1991. Targeted integration of adeno-associated virus (AAV) into human chromosome 19. EMBO J. *10*:3941-3950.

142.**Sanlioglu, S., and J.F. Engelhardt.** 1999. Cellular redox state alters recombinant adeno-asociated virus transduction through tyrosine phosphatase pathways. Gene Ther. *6*:1427-1437.

143.**Shelling, A.N. and M.G. Smith.** 1994. Targeted integration of transfected and infected adeno-associated virus vectors containing the neomycin resistance gene. Gene Ther. *1*:165-169.

144.**Siegl, G., R.C. Bates, K.I. Berns, B.J. Carter, D.C. Kelly, E. Kurstak, and P. Tattersall.** 1985. Characteristics and taxonomy of parvoviridae. Intervirology *23*:61-73.

145.**Smith, A.E.** 1995. Viral vectors in gene therapy. Annu. Rev. Microbiol. *49*:807-838.

146.**Smith, R.H. and R.M. Kotin.** 1998. The rep52 gene product of adeno-associated virus is a DNA helicase with 3′-to-5′ polarity. J. Virol. *72*:4874-4881.

147.**Snyder, R.O., X. Xiao, and R.J. Samulski.** 1996. Production of recombinant adeno-associated viral vectors, p. 12.1.1-12.1.24. *In* N. Dracopoli, J. Haines, B. Krof, D. Moir, C. Morton, C. Seidman, J. Seidman, and D. Smith (Eds.), Current Protocols in Human Genetics. John Wiley & Sons, New York.

148.**Snyder, R.O., S.K. Spratt, C. Lagarde, D. Bohl, B. Kaspar, B. Sloan, L.K. Cohen, and O. Danos.** 1997. Efficient and stable adeno-associated virus-mediated transduction in the skeletal muscle of adult immunocompetent mice. Hum. Gene Ther. *8*:1891-1900.

149.**Srivastava, A., E.W. Lusby, and K.I. Berns.** 1983. Nucleotide sequence and organization of the adeno-associated virus 2 genome. J. Virol. *45*:555-564.

150.**Steinbach, S., A. Wistuba, T. Bock, and J.A. Kleinschmidt.** 1997. Assembly of adeno-associated virus type 2 capsids in vitro. J. Gen. Virol. *78*:1453-1462.

151.**Summerford, C., J.S. Bartlett, and R.J. Samulski.** 1999. AlphaVbeta5 integrin: a co-receptor for adeno-associated virus type 2 infection. Nat. Med. *5*:78-82.

152.**Summerford, C. and R.J. Samulski.** 1998. Membrane-associated heparan sulfate proteoglycan is a receptor for adeno-associated virus type 2 virions. J. Virol. *72*:1438-1445.

153.**Surosky, R.T., M. Urabe, S.G. Godwin, S.A. McQuiston, G.J. Kurtzman, K. Ozawa, and G. Natsoulis.** 1997. Adeno-associated virus Rep proteins target DNA sequences to a unique locus in the human genome. J. Virol. *71*:7951-7959.

154.**Tamayose, K., Y. Hirai, and T. Shimada.** 1996. A new strategy for large-scale preparation of high-titer recombinant adeno-associated virus vectors by using packaging cell lines and sulfonated cellulose column chromatography. Hum. Gene Ther. *7*:507-513.

155.**Thimmapaya, B., C. Weinberger, R.J. Schneider, and T. Shenk.** 1982. Adenovirus VA I RNA is required for efficient translation of viral mRNAs at late times after infection. Cell *31*:543-551.

156.**Thrasher, A.J., M. de Alwis, C.M. Casimir, C. Kinnon, K. Page, J. Lebkowski, A.W. Segal, and R.J. Levinsky.** 1995. Generation of recombinant adeno-associated virus (rAAV) from an adenoviral vector and functional reconstitution of the NADPH-oxidase. Gene Ther. *7*:481-485.

157.**Tobiasch, E., M. Rabreau, K. Geletneky, C.S. Larue, F. Severin, N. Becker, and J.R. Schlehofer.** 1994. Detection of adeno-associated virus DNA in human genital tissue and in material from spontaneous abortion. J. Med. Virol. *44*:215-222.

158.**Tratschin, J.D., I.L. Miller, M.G. Smith, and B.J. Carter.** 1985. Adeno-associated virus vector for high-frequency integration, expression, and rescue of genes in mammalian cells. Mol. Cell. Biol. *5*:3251-3260.

159.**Tratschin, J.D., J. Tal, and B.J. Carter.** 1986. Negative and positive regulation in trans of gene expression from adeno-associated virus vectors in mammalian cells by a viral rep gene product. Mol. Cell. Biol. *6*:2884-2894.

160.**Trempe, J.P.** 1996. Packaging systems for adeno-associated virus vectors. Curr. Top. Microbiol. Immunol. *218*:35-50.

161.**Vincent, K.A., G.K. Moore, and N.L. Haigwood.** 1990. Replication and packaging of HIV enevelope genes in a novel adeno-associated virus vector system. Vaccine *90*:353-359.

162.**Vincent, K.A., S.T. Piraino, and S.C. Wadsworth.** 1997. Analysis of recombinant adeno-associated virus packaging and requirements for rep and cap gene product. J. Virol. *71*:1897-1905.

163.**Vincent-Lacaze, N., R.O. Snyder, R. Gluzman, D. Bohl, C. Lagarde, and O. Danos.** 1999. Structure of adeno-associated virus vector DNA following transduction of the skeletal muscle. J. Virol. *73*:1949-1955.

164.**Walsh, C.E., J.M. Liu, X. Xiao, N.S. Young, A.W. Nienhuis, and R.J. Samulski.** 1992. Regulated high level expression of a human gamma-globin gene introduced into erythroid cells by an adeno-associated virus vector. Proc. Natl. Acad. Sci. USA *89*:7257-7261.

165.**Walz, C., J.R. Schlehofer, M. Flentje, V. Rudat, and H. zur Hausen.** 1992. Adeno-associated virus sensitizes HeLa cell tumors to gamma rays. J. Virol. *66*:5651-5657.

166.**Wang, X.S., B. Khuntirat, K. Qing, S. Ponnazhagan, D.M. Kube, S. Zhou, V.J. Dwarki, and A. Srivastava.** 1998. Characterization of wild-type adeno-associated virus type 2-like particles generated during recombinant viral vector production and strategies for their elimination. J. Virol. *72*:5472-5480.

167.**Weger, S., M. Wendland, J.A. Kleinschmidt, and R. Heilbronn.** 1999. The adeno-associated virus type 2 regulatory proteins rep78 and rep68 interact with the transcriptional coactivator PC4. J. Virol. *73*:260-269.

168.**Weger, S., A. Wistuba, D. Grimm, and J.A. Kleinschmidt.** 1997. Control of adeno-associated virus type 2 cap gene expression: relative influence of helper virus, terminal repeats, and Rep proteins. J. Virol. *71*:8437-8447.

169.**Weindler, F.W. and R. Heilbronn.** 1991. A subset of herpes simplex virus replication genes provides helper functions for productive adeno-associated virus replication. J. Virol. *65*:2476-2783.

170.**Weitzman, M.D., S.R. Kyöstiö, R.M. Kotin, and R.A. Owens.** 1994. Adeno-associated virus (AAV) Rep proteins mediate complex formation between AAV DNA and its integration site in human DNA. Proc. Natl. Acad. Sci. USA *91*:5808-5812.

171.**West, M.H., J.P. Trempe, J.D. Tratschin, and B.J. Carter.** 1987. Gene expression in adeno-associated virus vectors: the effects of chimeric mRNA structure, helper virus, and adenovirus VA1 RNA. Virology *160*:38-47.

172.**Winocour, E., M.F. Callaham, and E. Huberman.** 1988. Perturbation of the cell cycle by adeno-associated virus. Virol. *167*:393-399.

173.**Wistuba, A., A. Kern, S. Weger, D. Grimm, and J.A. Kleinschmidt.** 1997. Subcellular compartmentalization of adeno-associated virus type 2 assembly. J. Virol. *71*:1341-1352.

174.**Wistuba, A., S. Weger, A. Kern, and J.A. Kleinschmidt.** 1995. Intermediates of adeno-associated virus type 2 assembly: identification of soluble complexes containing rep and cap proteins. J. Virol. *69*:5311-5319.

175.**Wonderling, R.S., S.R. Kyöstiö, and R.A. Owens.** 1995. A maltose-binding protein/adeno-associated virus Rep68 fusion protein has DNA-RNA helicase and ATPase activities. J. Virol. *69*:3542-3548.

176.**Wu, P., M.I. Phillips, J. Bui, and E.F. Terwilliger.** 1998. Adeno-associated virus vector-mediated transgene integration into neurons and other nondividing cell targets. J. Virol. *72*:5919-5926.

177.**Xiao, X., J. Li, and R.J. Samulski.** 1996. Efficient long-term gene transfer into muscle tissue of immunocompetent mice by adeno-associated virus vector. J. Virol. *70*:8098-8108.

178.**Xiao, X., J. Li, and R.J. Samulski.** 1998. Production of high-titer recombinant adeno-associated virus vectors in the absence of helper adenovirus. J. Virol. *72*:2224-2232.

179.**Yakinoglu, A.O., R. Heilbronn, A. Bürkle, J.R. Schlehofer, and H. zur Hausen.** 1988. DNA amplification of adeno-associated virus as a response to cellular genotoxic stress. Cancer Res. *48*:3123-3129.

180.**Yakobson, B., T.A. Hrynko, M.J. Peak, and E. Winocour.** 1989. Replication of adeno-associated virus in cells irradiated with UV light at 254 nm. J. Virol. *63*:1023-1030.

181.**Yakobson, B., T. Koch, and E. Winocour.** 1987. Replication of adeno-associated virus in synchronized cells without the addition of helper virus. J. Virol. *61*:972-981.

182.**Yang, Q., F. Chen, and J.P. Trempe.** 1994. Characterization of cell lines that inducibly express the adeno-associated virus Rep proteins. J. Virol. *68*:4847-4856.

183.**Ye, X., V.M. Rivera, P. Zoltick, F. Cerasoli Jr., M.A. Schnell, G.-P. Gao, J.V. Hughes, M. Gilman, and J.M. Wilson.** 1999. Regulated delivery of therapeutic proteins after in vivo somatic cell gene transfer. Science *283*:88-91.

184.**Zhang, X., M. de Alwis, S.L. Hart, F.W. Fitzke, S.C. Inglis, M.E.G. Boursnell, R.L. Levinsky, C. Kinnon, R.R. Ali, and A.J. Thrasher.** 1999. High-titer recombinant adeno-associated virus production from replicating amplicons and herpes vectors deleted for glycoprotein H. Hum. Gene Ther. *10*:2527-2537.

185.**Zhou, X. and N. Muzyczka.** 1998. In vitro packaging of adeno-associated virus DNA. J. Virol. *72*:3241-3247.

186.**Zhou, X., I. Zolotukhin, D.S. Im, and N. Muzyczka.** 1999. Biochemical characterization of adeno-associated virus rep68 DNA helicase and ATPase activities. J. Virol. *73*:1580-1590.

187. **Zolotukhin, S., B.J. Byrne, E. Mason, I. Zolotukhin, M. Potter, K. Chesnut, C. Summerford, R.J. Samulski, and N. Muzyczka.** 1999. Recombinant adeno-associated virus purification using novel methods improves infectious titer and yield. Gene Ther. *6*:973-985.

188. **Zolotukhin, S., M. Potter, W.W. Hauswirth, J. Guy, and N. Muzyczka.** 1996. A "humanized" green fluorescent protein cDNA adapted for high-level expression in mammalian cells. J. Virol. *70*:4646-4654.

3 Production and Purification Systems for Recombinant Adeno-Associated Virus Type-2 Vectors

James E. Conway[1], Kye Chesnut[2], Irene Zolotukhin[2,3], Collette apRhys[1], Nick Muzyczka[2], Sergei Zolotukhin[2], and Barry J. Byrne[2,3]
[1]Departments of Pharmacology and Molecular Sciences, The Johns Hopkins University School of Medicine, Baltimore, MD;
[2]Department of Molecular Genetics and Microbiology, Gene Therapy Center, and [3]Department of Pediatrics, University of Florida, Gainesville, FL, USA

INTRODUCTION

Recombinant adeno-associated virus type 2 vectors (rAAV) are recombinant single-stranded DNA viruses that have been uniquely successful gene transfer vectors (12,28). One important goal of gene therapy—the long-term, high-level, in vivo expression of transduced genes—has been demonstrated after rAAV transduction in several different systems (3,7,8,10,17,18, 22,30,38,43). Erythropoietin has been produced from the skeletal muscle of mice after gene transduction with a rAAV vector encoding the human *epo* gene (17). Tyrosine hydroxylase and aromatic amino acid decarboxylase have been successfully expressed in the central nervous system in animal models of Parkinson's disease, including nonhuman primates (16). Factor IX has been produced from both the skeletal muscle and liver in animal models of hemophilia (13,20,27,29). A rAAV vector encoding the *cftr* gene has been administered in a human clinical trial (9,11,40). Importantly, the in vivo control of gene expression after rAAV transduction has also been demonstrated using the tetracycline transactivator and operon system (31).

While preclinical success has been attained with rAAV vectors, the successful clinical application of these vectors has, in part, been limited by rAAV vector production technology. For certain genetic diseases, the dose of rAAV required to achieve clinical therapeutic efficacy is orders of magnitude larger than doses previously administered and has been difficult to produce. In Pompe's disease, the administration of at least 10^{14} particles of a rAAV vector will be necessary to achieve therapeutic effect. For hemophilia B, greater than 5×10^{12} particles/kg of a Factor IX rAAV expression vector will have to be administered because this dose was not able to normalize blood clotting times in a canine model of hemophilia B (13,27). Traditional rAAV production techniques

Viral Vectors: Basic Science and Gene Therapy
Edited by A. Cid-Arregui and A. García-Carrancá

are not able to generate rAAV on the scale necessary for the successful clinical application of these vectors.

ADENOVIRUS-BASED rAAV PRODUCTION TECHNOLOGIES

Traditional protocols to produce rAAV vectors have generally been based on a three-component system. One component of this system is a proviral plasmid encoding the recombinant DNA to be packaged as rAAV. This recombinant DNA is located between 145-bp AAV type 2 (AAV-2) inverted terminal repeats (ITRs) that are the minimal *cis*-acting AAV-2 sequences that direct replication and packaging of the vector (12,36). A second component of the system is a plasmid encoding the AAV-2 genes, *rep* and *cap*. The AAV-2 *rep* gene encodes four Rep proteins (Rep 78, 68, 52, and 40) that act in *trans* to replicate the rAAV genome, resolve replicative intermediates, and then package single-stranded rAAV genomes. The AAV-2 *cap* gene encodes the three structural proteins (VP1, VP2, and VP3) that comprise the virus capsid. Because AAV-2 does not proficiently replicate on its own, the third component of a rAAV packaging system is a set of helper functions from another DNA virus. These helper functions create a cellular environment in which rAAV replication and packaging can efficiently occur. The helper functions provided by adenovirus (Ad) have almost exclusively been used to produce rAAV and are encoded by the genes E1A, E1B, E2A, E4ORF6, and VA (2,14,35,44). While the first two components of the system are generally introduced into cells in which replication and packaging is to occur by transfection, Ad helper functions are introduced by superinfection with wild-type (wt) Ad virus.

The traditional rAAV production techniques are limited in their abilities to produce large quantities of vector because of inherent inefficiencies in transfection and the difficulty of increasing the scale of transfection. The requirement for wt Ad may also reduce the amount of rAAV produced since Ad may compete for cellular and viral substrates that are required for viral replication, but are present only in limited amounts. In addition, superinfection with Ad requires effective procedures for purification of Ad from rAAV. While these purification processes are generally successful at eliminating Ad contamination of rAAV preparations, they also reduce rAAV titers. Stringent assays for Ad contamination of rAAV are also necessary.

Because standard rAAV production protocols are not capable of producing large quantities of high-titer rAAV vector and have limited development of this gene transfer agent, intensive research has focused on developing more efficient rAAV production technology. Some advancements have involved the construction of *rep*- and *cap*-inducible cell lines, adding translational control of Rep production and augmenting Cap production by driving *cap* expression with a strong heterologous promoter (4,23,39). The most significant advancement has been the creation of Ad-free methods of rAAV production (24,44). Ad-free methods introduce the helper functions required for rAAV production by transfecting a plasmid encoding these genes, thereby eliminating the requirement for Ad superinfection and the problems associated with it.

rAAV PRODUCTION SYSTEMS BASED ON HERPES SIMPLEX VIRUS TYPE I

Most efforts to increase the efficiency and capacity of rAAV production technologies have focused primarily upon improving Ad-based methods of rAAV production.

Significant restrictions to large-scale rAAV production still exist even with these modified Ad protocols, however. Altering transcriptional and translational control of Rep and Cap does not improve the efficiency of rAAV production more than tenfold (23, 39). The Ad-free methods require transfection on a large scale that is difficult to attain. The inducible cell lines require Ad superinfection, and the serious purification problems associated with Ad infection and replication must still be addressed.

In order to make a rAAV production system that does not possess the restrictions to large-scale rAAV production inherent to Ad systems, we have investigated the use of another DNA helper virus of AAV, herpes simplex virus type I (HSV-1), for the production of rAAV (34). Like Ad, HSV-1 is able to fully support AAV replication and packaging (1,15,26,34,41). While little is known about the nature of the interactions between HSV-1 and AAV, the minimal set of HSV-1 genes required to replicate and package AAV has been determined to be UL5, UL8, UL52, and UL29 (41). These genes encode components of the HSV-1 core replication machinery and by themselves are able to form nuclear prereplication centers that develop into mature replication foci (19).

Two different HSV-1 systems for the production of rAAV have been developed (5,6). In both of the systems, the goal has been the creation of a single infectious agent to deliver all AAV-2 genes and HSV-1 helper virus functions required to produce rAAV. Therefore, use of this infectious agent along with a rAAV proviral cell line would eliminate restrictions to large-scale rAAV production inherent in transfection protocols. In addition, use of a mutant HSV-1 that was replication-defective and not complemented by the rAAV proviral cell line would eliminate the HSV-1 helper virus from the rAAV preparation. Thus, the two significant problems associated with the Ad-based systems would not be encountered.

One of the HSV-1 systems developed for the production of rAAV was an HSV-1 amplicon (6). In this two-component system, the AAV-2 *rep* and *cap* genes are expressed from the HSV-1 amplicon. The HSV-1 functions required for rAAV production are expressed from the helper virus required to replicate and package the amplicon. The other HSV-1 system developed was a one-component, recombinant HSV-1 vector. In the recombinant HSV-1, the AAV-2 *rep* and *cap* genes are recombined into the viral genome. In this system, all AAV-2 and helper virus elements required for rAAV production are present in one infectious agent. While both of these HSV-1 systems developed for the production of rAAV are scalable, they have distinct advantages and disadvantages (5,6,42).

HSV-1 AMPLICON THAT EXPRESSES AAV-2 *rep* AND *cap* CAN PRODUCE rAAV

An HSV-1 amplicon system is composed of two elements. One component is the HSV-1 amplicon. The *cis*-acting sequences required for replication and packaging of an HSV-1 genome, the 190-bp origin of replication, (ori)S', and the 230-bp *a* sequence, respectively, are present on this vector (33,42). Also located on this vector is the recombinant DNA that is to be packaged into an HSV-1 capsid. The factors that are required for the replication and packaging of this amplicon into an HSV-1 viral capsid are supplied *in trans* by the second component of an HSV-1 amplicon system: the HSV-1 helper virus. Cotransfecting the amplicon vector along with HSV-1 DNA or transfecting the amplicon vector and superinfecting with HSV-1 virus results in the rolling circle replication of the amplicon vector. The serially repeated concatemers of replicated amplicon genomes are then processed into linear, double-stranded 150-kb genomes (the HSV-1 genome size) and packaged into HSV-1 capsids. Concurrent with amplicon replication and packaging are the replication and packaging of the helper virus genome. Thus, two different species of HSV-1 virus are produced—the recombinant virus and the helper virus.

Unique to the application of an HSV-1 amplicon for the production of rAAV is the requirement for helper virus in the final preparation. Most gene transduction applications utilizing HSV-1 amplicons require reduction or elimination of the helper virus. For the production of rAAV, however, the helper virus is essential since it, not the amplicon genome, provides the viral functions essential for rAAV production.

The first HSV-1 amplicon system constructed, HSV-RC/*wt*, consisted of the AAV-2 *rep* and *cap* genes cloned into an amplicon vector to create the plasmid pHSV-RC (Figure 1) (6). In this vector, the *rep* and *cap* genes were expressed under control of their native promoters. The p5 and p19 promoters of AAV-2 controlled expression of Rep 78 and 68, and Rep 52 and 40, respectively. The p40 promoter controlled expression of VP1, VP2, and VP3. The HSV-1 virus used to provide the *trans* factors for amplicon replication and packaging and rAAV production was wt HSV-1 (KOS strain).

The amplicon system HSV-RC/*wt* was capable of efficient rescue and replication of rAAV genomes from a rAAV proviral cell line, from proviral plasmids, or from infected rAAV genomes in both HeLa and 293 cells (6). Infectious particle production was limited, however, probably due to the cytotoxic effects of the wt HSV-1 helper virus. Expression of HSV-1 immediate early genes is known to be cytotoxic to cells and induces apoptosis. In addition, the complete lytic replication of the wt HSV-1 is clearly injurious to the cells in which rAAV packaging is also taking place. Thus, while the helper virus in HSV-RC/*wt* was needed for production of rAAV, it also caused significant cytotoxicity and rapidly induced cell death, thereby limiting rAAV production.

A second HSV-1 amplicon system was constructed with pHSV-RC using a mutant HSV-1 to reduce the cytotoxicity of the helper virus (5). This system, HSV-RC/*d27.1*, used the ICP27 mutant virus *d27.1* as the helper virus (32). The virus *d27.1* is replication-defective and should be less cytotoxic in a nonpermissive cell line than wt HSV-1. The virus *d27.1* still, however, expresses the early genes known to be required for rAAV production and thus should still provide efficient helper function for rAAV production (32,41). Using an ICP27 mutant as the helper virus for rAAV production has additional advantages. Host cell splicing of messenger RNA is inhibited by ICP27 (37). ICP27 probably also effects the appropriate splicing of the AAV-2 *rep* and *cap* messages. The splicing of the *rep* and *cap* messages should be less disrupted in an amplicon system using

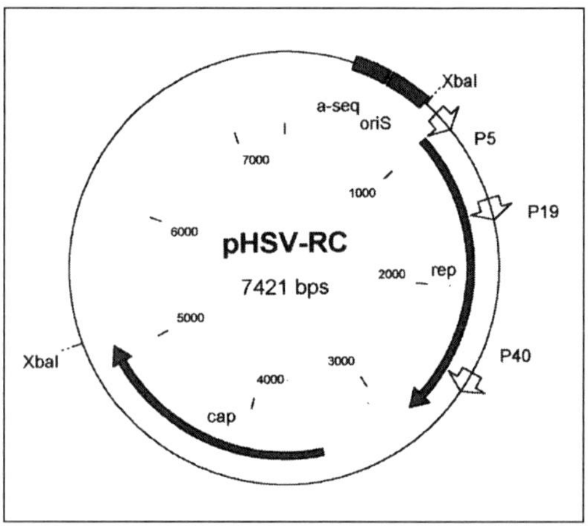

Figure 1. The HSV-1 amplicon designed to express the AAV-2 *rep* and *cap* genes. Located on the HSV-1 amplicon are the *cis*-acting sequences required for replication and packaging of the amplicon genome into an HSV-1 viral capsid, the HSV-1 ori*S*, and the HSV-1 *a* sequence, respectively. The AAV-2 *rep* and *cap* genes are also located on this vector. The helper virus provides all factors required for the replication and the packaging of the replicated amplicon genome into an HSV-1 viral capsid.

d27.1 as the helper virus. In addition, *d27.1* overexpresses ICP8, the single-stranded DNA binding protein that is the product of UL29, one of the HSV-1 genes essential for AAV replication and packaging (25,32,41). Increased expression of ICP8 should augment rAAV production.

The amplicon system, HSV-RC/*d27.1*, was able to rescue and replicate rAAV genomes (6). Most important, infectious rAAV particles could now be efficiently produced from the cell line green fluorescent protein (GFP)-92 (Figure 2). This 293 cell line has a rAAV provirus integrated in its genome that encodes the GFP. The amplicon system HSV-RC/*d27.1* was capable of producing up to 300 particles of rAAV per cell from the GFP-92 cell line. This efficiency of AAV-GFP production by HSV-RC/*d27.1* was comparable to the standard transfection-based methods used at the time.

Advantages of the amplicon system over the standard methods of rAAV production include the scalability of infection, which is not restricted by limitations found in transfection-based procedures. Another advantage is that wt AAV is not generated by the amplicon system in contrast to traditional rAAV methods where it is commonly produced. Purification of residual amplicon and helper virus from the rAAV produced is not difficult, either.

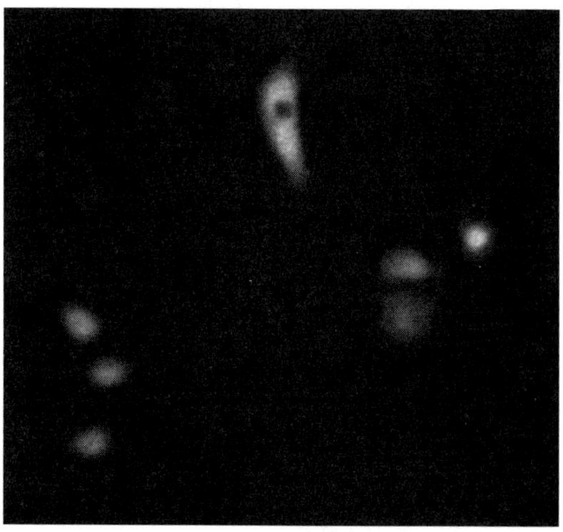

Figure 2. AAV-GFP produced by the amplicon system HSV-RC/*d27.1* is infectious. CsCl gradient-purified AAV-GFP was used to transduce HeLa cells in the presence of Ad (24 hours after transduction, AAV-GFP: moi of 0.1 eu; Ad: moi of 20 eu).

Other distinct advantages of using an HSV-1 amplicon system for producing rAAV exist. In general, it is not difficult to construct an HSV-1 amplicon. Amplicon vectors with altered transcriptional and translational control of AAV-2 Rep and Cap production could readily be constructed and tested for efficacy of rAAV production. Effects of changing the helper virus on the efficiency of rAAV production could also be readily determined. Amplicon systems using helper viruses that have other mutations or have multiple mutations can be quickly generated and tested. In addition, the proviral rAAV genome could be incorporated in the amplicon so that all components required for the production of rAAV are found in one agent, thereby avoiding the need to construct proviral rAAV cell lines.

The disadvantages of using an amplicon system for the production of rAAV are significant, however. The ratio of amplicon genomes to helper virus genomes in each preparation varies with serial passage. Since the ratio of the amplicon virus to the helper virus effects the efficiency of rAAV production, optimizing this system for rAAV production is difficult. There is also no selective pressure on the amplicon to retain the AAV-2 *rep* and *cap* genes. Serial passage of the amplicon eventually results in recombination and deletion of the recombinant AAV-2 genes, which adversely affects rAAV production.

A RECOMBINANT HSV-1 THAT EXPRESSES AAV-2 *rep* AND *cap* SUPPORTS EFFICIENT rAAV PRODUCTION

A recombinant HSV-1 vector that

expresses the AAV-2 *rep* and *cap* genes would eliminate disadvantages of the amplicon system while still providing all factors required for producing rAAV in an infectious agent. In a stable recombinant vector, the ratio of *rep* and *cap* to the helper virus genes would not vary with passage. Deletions of the recombinant genome would not occur as readily. A recombinant vector would be a simpler, one-component system that would facilitate optimizing large-scale production of rAAV.

The recombinant virus that expresses *rep* and *cap*, *d27.1-rc* was constructed by homologous recombination techniques (Figure 3) (5). The HSV-1 background used to make the recombinant vector that expresses the AAV-2 *rep* and *cap* was *d27.1.* This mutant vector was chosen because it effectively supported rAAV production as the helper virus in the amplicon system. The wt AAV-2 promoters controlled expression of the AAV-2 genes as in the amplicon.

The recombinant virus, *d27.1-rc*, was constructed in a two-step process. First, a β-galactosidase-expressing recombinant, *d27.1-lacz*, was generated by homologous recombination of a *lacz* expression cassette into the *tk* gene of *d27.1.* This integration disrupted the *tk* gene. This virus could not produce thymidine kinase and was resistant to 5-bromo-deoxycytidine. The *lacz* gene in *d27.1-lacz* was then knocked out by homologous recombination of the *rep* and *cap* genes into this locus to generate *d27.1-rc*. Southern blot analysis of infected cell DNA confirmed integration of the *rep* and *cap* genes. Loss of β-galactosidase production in the recombinant was also verified. The virus retained resistance to 5-bromo-deoxycytidine, indicating that the *tk* gene still did not produce functional thymidine kinase.

Western blot analysis and immunofluorescence assays (IFA) were used to characterize Rep production by *d27.1-rc*. Western blot analysis was used to determine the level of Rep production from *d27.1-rc* in different cell lines at different multiplicity of infection (moi). Western blot analysis

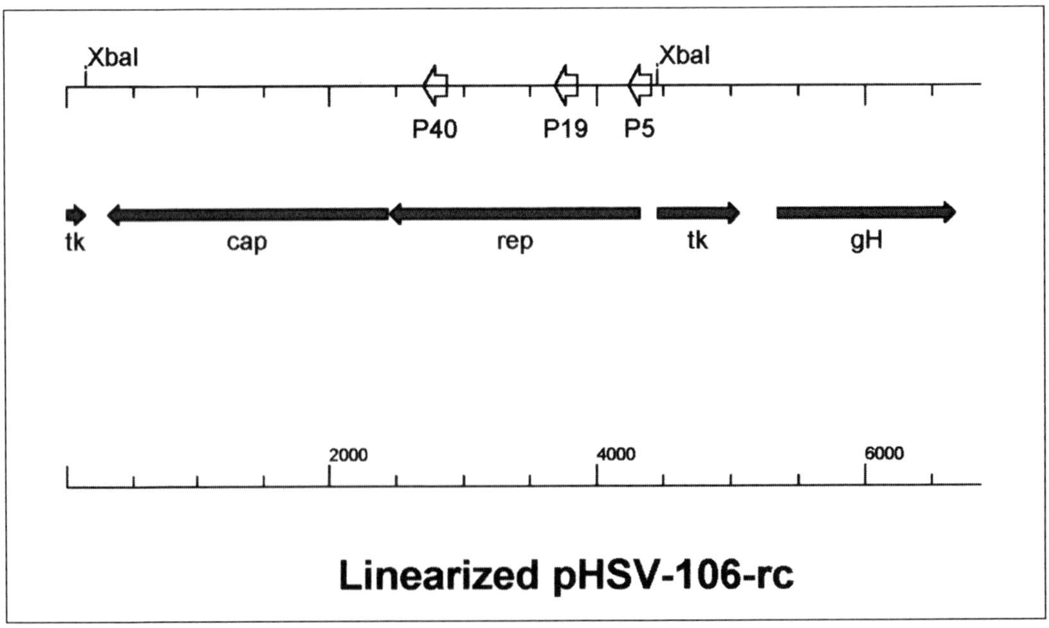

Figure 3. Map of a recombinant HSV-1 virus that was created to produce rAAV. The AAV-2 *rep* and *cap* genes, under control of the p5, p19, and p40 promoters, were cloned into the *tk* gene of the mutant HSV-1 virus *d27.1* to create *27.1-rc*.

indicated that both cell type and the moi independently affected Rep production (Figure 4). All forms of Rep were produced by *d27.1-rc*. Higher moi generally resulted in greater Rep. Different levels of Rep expression were also seen in different cell lines. The highest level of Rep expression was observed in 293 cells, possibly indicating that the Ad E1a present in these cells is up-regulating p5-driven *rep* expression from the viral genome. Interestingly, Rep expression was not observed in HeLa cells after infection by *d27.1-rc*. Because of the large differences in Rep production in different cell types, choice of cell line will be important for efficient rAAV production.

The distribution of Rep in 293 cells infected with *d27.1-rc* was determined using an IFA (Figure 5). The Rep proteins formed a punctate pattern in the infected cell's nucleus at 12 hours after infection. A nuclear distribution of the Rep proteins is required if *d27.1-rc* is to effectively produce rAAV.

Interestingly, the presence of the *rep* gene did not affect the replication or growth of *d27.1-rc* in the complementing cell line, V27. Differences were not observed in the formation of viral replication centers between *d27.1-rc* or *d27.1-lacz* in infected V27 cells. Mature HSV-1 viral replication centers were observed in *d27.1-rc*-infected V27 cells 12 hours after infection, indicating that the *rep* gene was not affecting viral replication. The titers *d27.1-rc* produced were not different from that of the parent virus either. Approximately 25 plaque-forming units (pfu) were produced per infected V27 cell.

The ability of the recombinant virus, *d27.1-rc*, to produce rAAV was analyzed (5). The amount of AAV-GFP produced after 293 cells were transfected with a proviral plasmid and superinfected with *d27.1-rc* was determined. The efficient rescue of rAAV from the plasmid pTR-UF5, a proviral rAAV plasmid that encodes the GFP, was observed. Up to 380 expression units (eu) of AAV-GFP were produced per cell after superinfection with *d27.1-rc* at a moi of ten.

Recombinant *d27.1-rc* was even more effective at producing rAAV from the

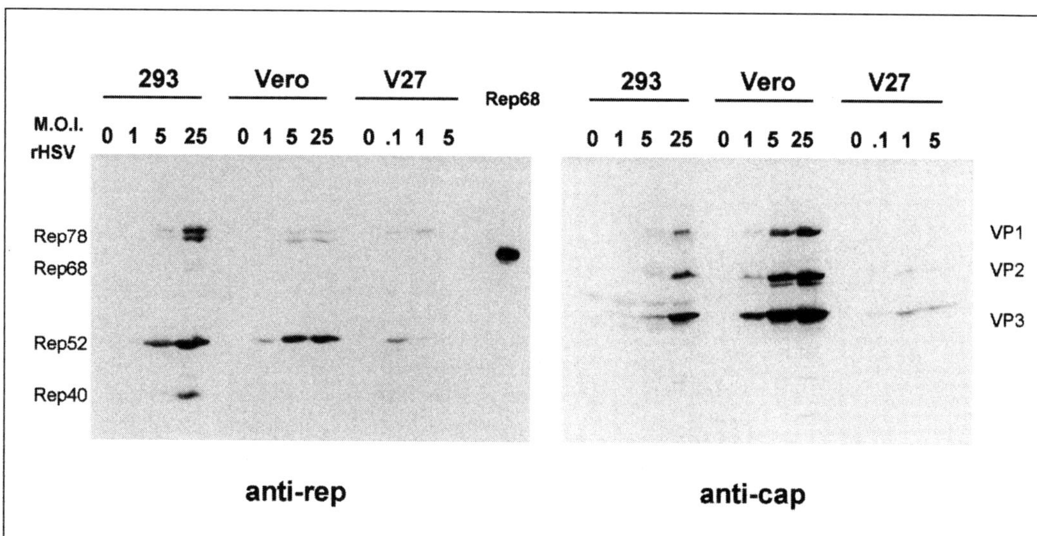

Figure 4. Western blot analysis of the AAV Rep and capsid proteins from *d27.1-rc* after infection of different cell types. Rep expression was detected by probing the membrane with an anti-Rep monoclonal antibody that recognizes all four Rep proteins (clone 1F11.8, 1:5000 dilution). Cap expression was detected by probing the membrane with B1 monoclonal anti-VP1, VP2, VP3 antibody at 1:1000 dilution. Secondary antibody is goat antimouse used at 1:5000 dilution.

rAAV proviral cell line GFP-92 (Figure 6). At a moi of ten, 480 eu of AAV-GFP per cell could be produced from this cell line. The amount of AAV-GFP produced from the cell line was observed to be a function of the moi of *d27.1-rc*. Increasing the moi up to ten resulted in an increase in rAAV production. At moi higher than ten, significantly less rAAV was produced.

Amplification of rAAV by co-infection with *d27.1-rc* was also attempted (5). Co-infection of 293 cells with both AAV-GFP and *d27.1-rc* resulted in the 200-fold amplification of the rAAV vector. While this amplification of AAV-GFP was not as efficient as the production of AAV-GFP from either cells transfected with a proviral plasmid and superinfected with *d27.1-rc* or a proviral cell line superinfected with *d27.1-rc*, significant amplification of rAAV by serial co-infection was possible. Amplification of the original titer of rAAV by 10^6 was accomplished by three serial co-infections of rAAV with *d27.1-rc*.

The efficient production of rAAV using *d27.1-rc* was maintained when the scale of infection was increased 100-fold (5). When 10^9 GFP-92 cells were infected with *d27.1-rc*, over 350 eu of AAV-GFP were

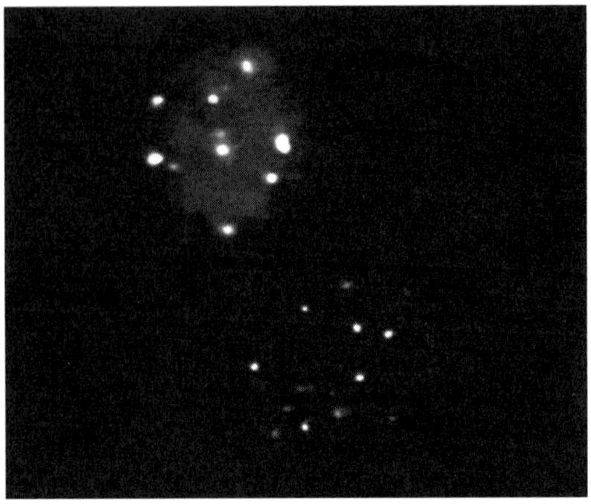

Figure 5. IFA of Rep expression. 293 cells were infected with *d27.1-rc* (moi of 10 eu) and assayed for Rep production at 12 hours postinfection. A monoclonal anti-Rep antibody that recognizes all four Rep proteins was utilized (clone 226.7, 1:1 dilution; American Research Products, Belmont, MA, USA). Magnification was 630×.

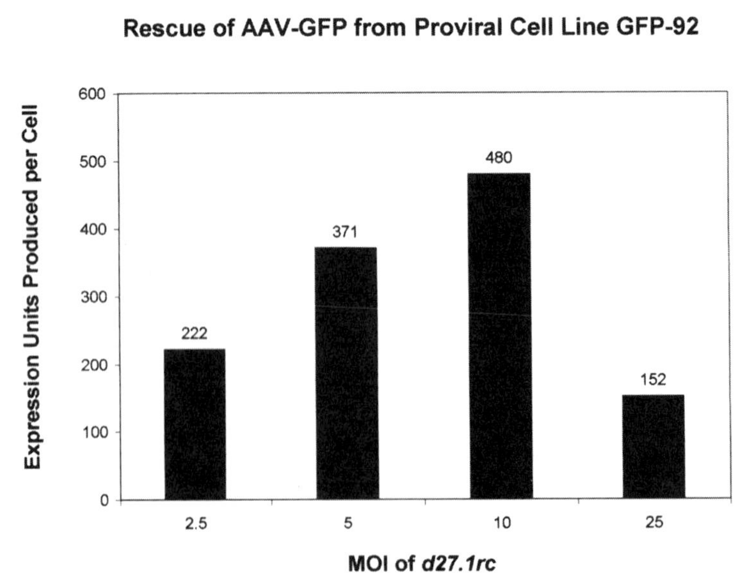

Figure 6. AAV-GFP production from the proviral 293 cell line GFP-92. GFP-92 cells were infected at different moi with *d27.1-rc*. A cell lysate was made, was heat-inactivated, and then titered on C12 cells that were co-infected with Ad (moi of 20 eu). The quantity of AAV-GFP produced per infected cell was determined.

produced per cell. Importantly, replication-competent AAV was not observed in any large preparation of AAV-GFP that was made using *d27.1-rc*.

CONCLUSION

One of the most important goals of gene therapy is the long-term, high-level, in vivo expression of transduced genes. This goal has been elusive and not attained by most gene transduction technologies. rAAVs have been the only vectors to consistently achieve this objective in a variety of different systems (17,20,21,43). Unfortunately, the development of rAAV vectors has been limited because of difficulty in producing the quantity of vector required for their successful in vivo development.

Traditional rAAV production protocols have been based upon transfection methods and Ad helper functions. Difficulties encountered in increasing the scale of these protocols have led to improvements. The new methods still possess significant restrictions to their implementation on the scale required for successful in vivo development of rAAV vectors, however.

All previous methods to generate rAAV have relied upon the Ad helper functions for AAV replication and packaging. In an attempt to make a scalable system for producing rAAV, we have investigated the use of HSV-1 to provide the helper functions for rAAV production. The purpose of this research was to develop a rAAV production system that delivered all AAV-2 and helper virus functions required for rAAV production in an infectious agent. Use of an infectious agent to make rAAV would permit the production of rAAV on the scale required for its successful application.

Two different HSV-1-based systems have been developed to produce rAAV. The first system developed was an HSV-1 amplicon system (6). In this system, the AAV-2 *rep* and *cap* genes were expressed from the amplicon virus and the helper functions required for rAAV production from a separate genome. An effective amplicon system for making rAAV was constructed using the ICP27 mutant HSV-1, *d27.1*, as the helper virus in the system. While this amplicon system was able to deliver all factors required for rAAV production through scalable infection, it possessed significant limitations to further development.

The second HSV-1 system was constructed to retain the advantages of delivering all helper functions required for rAAV production by infection while eliminating problems with the amplicon system. A recombinant *d27.1* virus was constructed, *d27.1-rc*, by recombining the AAV-2 *rep* and *cap* genes into the viral genome of *d27.1* (5). This recombinant virus, *d27.1-rc*, was observed to be 100 times more efficient than the amplicon system HSV-RC/*d27.1* at producing rAAV. This efficiency of rAAV production using *d27.1-rc* was maintained even when the scale of infection was increased 100-fold.

The use of *d27.1-rc* and a rAAV proviral cell line to make rAAV is probably the easiest and most efficient technology yet published for the production of large quantities of high-titer rAAV. In this two-component system, rAAV proviral cell lines could be grown in large capacity spinner cultures or cartridge systems. Scalable infection of these cells with *d27.1-rc* would then deliver all factors required *in trans* for the replication and packaging of rAAV. The need for transfection at any step in the procedure is thus eliminated. Only trace quantities of helper virus will have to be removed from the rAAV preparations since the proviral cell line will generally not complement the mutant *d27.1-rc*.

It is possible that rAAV vectors will be the first viral vectors to achieve clinically relevant, long-term, in vivo gene transduction. Unfortunately, progress with in vivo preclinical and clinical development of these vectors has been limited by inability to produce vector. Different methods of rAAV vector production will have to be considered. One possible alternative is the HSV-1-based system, *d27.1-rc*. This novel system could be used to achieve the efficient, large-scale production of rAAV that is essential for clinical success.

PROTOCOLS

Protocol for rAAV Virus Preparation

The following protocol is based on production of rAAV by a dual plasmid transfection method that utilizes 293 cells grown in a cell factory (Nalge Nunc International, Rochester, NY, USA) with 10^9 cells per production run.

1. Seeding Cell Factory with 293 Cells
   a. Split eight confluent T225s to seed one cell factory (10 plates; Nalge Nunc International).
   b. Collect cells in a 250-mL conical.
   c. Add approximately 1090 mL of media to aspirator bottle and add cells to aspirator bottle.
   d. Load cell factory and incubate at 37°C until ready for transfection.
2. Transfection (for One Cell Factory)
   a. Check cells for confluency; 75% is optimal.
   b. Thaw 2× HEPES-buffered saline (HBS) and keep at 37°C until ready to use.
   c. Prewarm media and fetal bovine serum (FBS) (1 L of DMEM, 90 mL FBS).
   d. Calculate amount of DNA and water to add:
      - Use 1867.5 µg of pDG (a gift from J.A. Kleinschmidt; Human Gene Therapy *10*:2445-2450, 1999), and 622.5 µg of rAAV per factory.
   e. Add 5.2 mL of 2.5-mol/L $CaCl_2$ to the tube.
   f. Add 52 mL of 2× HBS and mix well.
   g. Immediately add the transfection mixture to the media in an aspirator bottle and mix well.
   h. Discard old media from the cell factory and load transfection mixture.
   i. Incubate at 37°C for 48 to 60 hours.
3. Harvesting
   a. Discard media.
   b. Add 500 mL of phosphate-buffered saline (PBS) to aspirator bottle and wash the cell factory; discard wash.
   c. Add 500 mL PBS/0.5 mmol/L EDTA to cell factory; rock to dislodge cells from plastic.
   d. Pour cells into two 250-mL conical tubes.
   e. Add an additional 500 mL of PBS to the cell factory to recover remaining cells.
   f. Pour cells into two 250-mL conical tubes.
   g. Centrifuge at 5000× *g* at 4°C for 10 to 15 minutes to pellet cells.
   h. Discard supernatant.
   i. Store cell pellets at -20°C.

Protocol for rAAV Virus Purification

1. Freeze/Thaws:
   a. Thaw cells from harvest, as above (4× 250-mL conicals).
   b. Resuspend in 9 mL lysis buffer (150 mmol/L NaCl, 50 mmol/L Tris, pH 8.3).
   c. Pool suspension in one 250-mL conical and bring to 60 mL with lysis buffer.
   d. Divide cell suspension into four 50-mL conicals and vortex mix.
   e. Freeze for 10 minutes in dry ice/EtOH; thaw for 15 minutes at 37°C and vortex mix.
   f. Repeat freeze/thaw two more times.

2. Benzonase
   a. To each tube of lysed cell suspension, add 3 µL of 4.82 mol/L $MgCl_2$ and vortex mix.
   b. Add 3 µL of benzonase (pure grade, 250 U/mL; NYCOMED AS Pharma, Westbury, NY, USA) and vortex mix.
   c. Incubate at 37°C for 30 minutes.
   d. Centrifuge for 20 minutes at 3700 rpm in a Model J6-HC centrifuge (Beckman Instruments, Fullerton, CA, USA).
   e. Pipette lysate (supernatant) into two 50-mL conicals and store at -80°C until ready to load gradient.
3. Iodixanol (OptiPrep™; NYCOMED AS Pharma) Density Gradient
   a. In a Quick-Seal 25 × 89 mm ultracentrifuge tube (Beckman Instruments), underlay 15 mL of the clarified lysate with (in order):
      • 9 mL of 15% iodixanol
      • 6 mL of 25% iodixanol
      • 5 mL of 40% iodixanol
      • 5 mL of 60% iodixanol
   b. Cap and seal tubes with heat sealer.
   c. Centrifuge at 69 000 rpm at 18°C for 1 hour in a 70Ti rotor (Beckman Instruments).
   d. Puncture tube at the 60% to 40% interface and remove approximately 4 to 5 mL of the 40% and pool in 50-mL conical.

Do NOT take any of the interface between the 40% and 25% bands.

4. Heparin Column Chromatography

Several types of heparin columns can be used. We have routinely used prepacked Heparin Agarose Type I (Sigma, St. Louis, MO, USA) and also packed our own columns using the same medium. Alternatively, Heparin HiTrap (Amersham Pharmacia Biotech, Piscataway, NJ, USA) and POROS heparin columns have also been used with a FPLC format.

   a. Wash Heparin Agarose Type I column with:
      • 10 mL of 1× TD
      • 10 mL of 1× TD/1 mol/L NaCl
      • 20 mL of 1× TD
   b. Load virus on column (20 mL iodixanol pool).
   c. Wash column with 20 mL of 1× TD.
   d. Elute virus in 15 mL of 1× TD/1 mol/L NaCl.

(See Figure 7 for a silver-stained polyacrylamide gel of purified rAAV.)

5. Concentrate Virus
   a. Pretreat a 15-mL Biomax® 100K concentrator (Millipore, Bedford, MA, USA) with 2% serum in PBS overnight.
   b. Discard 2% serum from concentrator and rinse concentrator several times with lactated Ringer's solution.
   c. Add 5 mL lactated Ringer's solution to the concentrator and centrifuge at

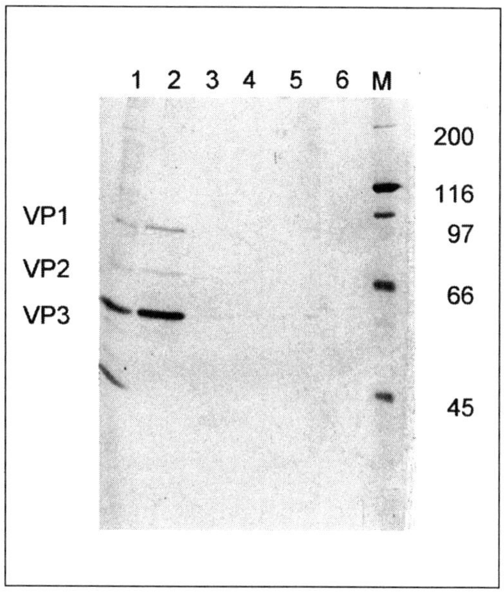

Figure 7. Silver-stained sodium dodecyl sulfate-polyacrylamide gel of the fractions of phenyl-sepharose® chromatography (Amersham Pharmacia Biotech). Lane 1, rAAV purified on iodixanol gradient and heparin-sepharose; lane 2, rAAV unbound to the phenyl-sepharose; lane 3, phenyl-sepharose washed with loading buffer; lanes 4, 5, and 6, phenyl-sepharose elution fractions; and lane M, molecular weight standard. VP1, VP2, and VP3 are rAAV capsid proteins.

2000× *g* to rinse concentrator.

d. Load virus and recentrifuge to bring volume down to 1 mL.

e. Add 9 mL of lactated Ringer's solution to wash virus (1:10 wash).

f. Again, centrifuge at 2000× *g* to bring volume down to 1 mL.

g. Repeat twice; bring volume down to 500 µL in final wash.

h. Transfer final virus stock to 1.5-mL microcentrifuge tube; remove aliquots as necessary for assay.

ASSAYS FOR VIRAL TITER

Slot Blot Assay and Quantitative Competitive Polymerase Chain Reaction (QC-PCR)

Viral DNA prepared as follows is used for the slot blot assay, and QC-PCR

1. To 4 µL of virus add:
   - 20 µL of 10× buffer
   - 2 µL of DNase (Roche Molecular Biochemicals, Indianapolis, IN, USA)
   - 174 µL of distilled water

   Total volume = 200 µL
2. Incubate at 37°C for 1 hour.
3. To the 200 µL DNase sample add:
   - 22 µL proteinase buffer
   - 2 µL of proteinase (Roche Molecular Biochemicals)

   Total volume = 224 µL
4. Incubate at 37°C for 1 hour.
5. To sample, add an equal volume (224 µL) of phenol/chloroform.
6. Vortex mix for 5 minutes and centrifuge for 5 minutes at 14 000 rpm in an Eppendorf microfuge.
7. Repeat phenol/chloroform extraction and follow with a chloroform extraction.
8. Vortex mix for 5 minutes and centrifuge for 5 minutes at 14 000 rpm.
9. Precipitate DNA in the presence of glycogen (1 µL of glycogen, 20 µg/µL; Roche Molecular Biochemicals) with 1/10 volume (22.4 µL) of 3 mol/L sodium acetate and 2.5 volumes of EtOH overnight at -20°C.
10. Centrifuge for 20 minutes at 14 000 rpm.
11. Wash pellet with 75% EtOH and air dry for 5 minutes.
12. Resuspend DNA in 40 µL of distilled water.

QC-PCR

1. Make up PCR master mixture:

	Final
distilled water	39.6 µL
10× buffer	5
dNTP	0.4
Primer 1 (100 ng/µL)	1
Primer 2 (100 ng/µL)	1
Taq DNA polymerase	1

2. Aliquot 336 µL of master mixture into a separate microcentrifuge tube.
3. Make a 1:100 dilution of the viral DNA prepared as above (already 1:10; 2 µL DNA 18 µL distilled water).
4. Add 7 µL of diluted viral DNA to master mixture aliquot.
5. Aliquot 49 µL of master mixture with viral DNA into six PCR tubes (five are standards; the sixth is virus only).
6. Starting with 100 fg standard, add 1 µL of each standard to the corresponding PCR tube (100 fg in first tube, 1 pg in second tube, 10 pg in third tube, 100 pg in fourth tube, and 1 ng in fifth tube).

   Negative (water only) and positive

controls (original and deleted size) are also run. PCR cycling is as follows: 25 cycles of

- 94°C 2 minutes
- 60°C 30 seconds
- 72°C 30 seconds; end with 75°C for 5 minutes

Samples are run on a 2% agarose gel.

Dot Blot Assay

1. Viral DNA prepared above is resuspended in 0.4 N NaOH/10 mmol/L EDTA.
2. Serially dilute a hybridization standard (rAAV plasmid DNA) twofold and resuspend in the same buffer.
3. Apply the DNA samples to the wells of a dot-blot manifold apparatus set with a prewetted nylon filter.
4. Rinse the wells with 400 µL of 0.4 N NaOH/10 mmol/L EDTA.
5. Disassemble the apparatus; rinse the membrane in 2× standard saline citrate (SSC) and air-dry.
6. Hybridize filters with a rAAV-specific ^{32}P-probe.
7. Compare viral DNA points to the standard curve and determine the amount of DNA (prism densitometry software by GraphPad Software, San Diego, CA, USA) and calculate rAAV genomes.
8. Compare the number of genomes, corresponding to the number of physical particles, to the infectious particle titer as determined below.

Infectious Center Assay

We currently use a C12 cell line (gift from P. Johnson, Children's Hospital, Department of Pediatrics, The Ohio State University, Columbus, OH, USA) to assay recombinant virus. To assay for rcAAV we use the 293 cell line.

Day 1: Seed a 96-well dish to be 75% confluent on day 2.

Day 2: Infect cells.

1. Set up 10× serial dilutions of virus in a separate 96-well dish as follows:
   a. Add 250 µL of media to the first wells, add 225 µL of media to the following wells.
   b. Add 2.5 µL of virus to the first well and serially dilute 10× by transferring 25 µL per dilution.
2. Transfer dilutions to the cells and Ad infect (moi approximately 5; good cytopathic effect (CPE) in 40 hours).

Day 4 (40 hours after infection): harvest cells.

1. Set up the ICA manifold (12 wells, 48-mm diameter, manifold custom constructed for this assay).
2. First, wet two pieces of Whatman® paper (Whatman, Clifton, NJ, USA) with PBS and apply it to the ICA manifold.
3. Wet the nylon transfer membrane with PBS and place on top of the Whatman paper.
4. Tighten the assembled ICA manifold.
5. Fill each well of the ICA manifold with 5 mL of PBS.
6. Without removing the media, detach cells from the 96-well dish by pipetting.
7. Apply to a well in the ICA manifold.
8. After all of the wells are complete, wait 5 minutes and apply vacuum.
9. Disassemble and allow the nylon membrane to air-dry for 5 minutes on Whatman paper.

10. Denature for 5 minutes (0.5 mol/L NaOH, 1.5 mol/L NaCl). Blot on Whatman paper.
11. Neutralize for 5 minutes (1.5 mol/L NaCl, 0.5 mol/L Tris-HCl, pH 7.8). Blot on Whatman paper.
12. Rinse in 4× SSC for 30 seconds.
13. Air-dry for 10 minutes.
14. Microwave nylon membrane on top of a piece of Whatman paper for 4 minutes to fix the DNA.
15. Prehybridize (7% sodium dodecyl sulfate [SDS], 0.25 mol/L NaHPO, pH 7.2, 1 mmol/L EDTA, pH 8.0) at 65°C for at least 2 hours before adding the ^{32}P-probe specific for the transgene or wt AAV.
16. Hybridize overnight, wash filters, and expose.
17. Count individual dots and calculate titers based on dilution.

REFERENCES

1. **Buller, R.M.L.** 1981. Herpes simplex virus types 1 and 2 completely help adnovirus-associated virus replication. J. Virol. *40*:241-247.
2. **Chejanovsky, N. and B.J. Carter.** 1990. Mutation of a consensus purine nucleotide binding site in the adeno- associated virus rep gene generates a dominant negative phenotype for DNA replication. J. Virol. *64*:1764-1770.
3. **Clark, K.R., T.J. Sferra, and P.R. Johnson.** 1997. Recombinant adeno-associated viral vectors mediate long-term transgene expression in muscle. Hum. Gene Ther. *8*:659-669.
4. **Clark, K.R., F. Voulgaropoulou, D.M. Fraley, and P.R. Johnson.** 1995. Cell lines for the production of recombinant adeno-associated virus. Hum. Gene Ther. *6*:1329-1341.
5. **Conway, J.E., C.M.J. ap Rhys, I. Zolotukhin, S. Zolotukhin, N. Muzyczka, G.S. Hayward, et al.** 1999. High-titer recombinant adeno-associated virus production utilizing a recombinant herpes simplex virus type I vector expressing AAV-2 Rep and Cap. Gene Ther. *6*:973-985.
6. **Conway, J.E., S. Zolotukhin, N. Muzyczka, G.S. Hayward, and B.J. Byrne.** 1997. Recombinant adeno-associated virus type 2 replication and packaging is entirely supported by a herpes simplex virus type 1 amplicon expressing Rep and Cap. J. Virol. *71*:8780-8789.
7. **Fisher, K.J., K. Jooss, J. Alston, Y. Yang, S.E. Haecker, K. High, et al.** 1997. Recombinant adeno-associated virus for muscle directed gene therapy. Nat. Med. 3:306-312.
8. **Flannery, J.G., S. Zolotukhin, M.I. Vaquero, M.M. LaVail, N. Muzyczka, and W.W. Hauswirth.** 1997. Efficient photoreceptor-targeted gene expression in vivo by recombinant adeno-associated virus. Proc. Natl. Acad. Sci. USA *94*:6916-6921.
9. **Flotte, T., B. Carter, C. Conrad, W. Guggino, T. Reynolds, B. Rosenstein, et al.** 1996. A phase I study of an adeno-associated virus-CFTR gene vector in adult CF patients with mild lung disease. Hum. Gene Ther. *7*:1145-1159.
10. **Flotte, T.R., S.A. Afione, C. Conrad, S.A. McGrath, R. Solow, H. Oka, et al.** 1993. Stable in vivo expression of the cystic fibrosis transmembrane conductance regulator with an adeno-associated virus vector. Proc. Natl. Acad. Sci. USA *90*:10613-10617.
11. **Flotte, T.R. and B.J. Carter.** 1998. Adeno-associated virus vectors for gene therapy of cystic fibrosis. Methods Enzymol. *292*:717-732.
12. **Hermonat, P.L. and N. Muzyczka.** 1984. Use of adeno-associated virus as a mammalian DNA cloning vector: transduction of neomycin resistance into mammalian tissue culture cells. Proc. Natl. Acad. Sci. USA *81*:6466-6470.
13. **Herzog, R.W., J.N. Hagstrom, S.H. Kung, S.J. Tai, J.M. Wilson, K.J. Fisher, et al.** 1997. Stable gene transfer and expression of human blood coagulation factor IX after intramuscular injection of recombinant adeno-associated virus. Proc. Natl. Acad. Sci. USA *94*:5804-5809.
14. **Huang, M.M. and P. Hearing.** 1989. Adenovirus early region 4 encodes two gene products with redundant effects in lytic infection. J. Virol. *63*:2605-2615.
15. **Johnston, K.M., D. Jacoby, P.A. Pechan, C. Fraefel, P. Borghesani, D. Schuback, et al.** 1997. HSV/AAV hybrid amplicon vectors extend transgene expression in human glioma cells. Hum. Gene Ther. *8*:359-370.
16. **Kaplitt, M.G., P. Leone, R.J. Samulski, X. Xiao, D.W. Pfaff, K.L. O'Malley, and M.J. During.** 1994. Long-term gene expression and phenotypic correction using adeno-associated virus vectors in the mammalian brain. Nat. Genet. *8*:148-154.
17. **Kessler, P.D., G.M. Podsakoff, X. Chen, S.A. McQuiston, P.C. Colosi, L.A. Matelis, et al.** 1996. Gene delivery to skeletal muscle results in sustained expression and systemic delivery of a therapeutic protein. Proc. Natl. Acad. Sci. USA *93*:14082-14087.
18. **Klein, R.L., E.M. Meyer, A.L. Peel, S. Zolotukhin, C. Meyers, N. Muzyczka, et al.** 1998. Neuron-specific transduction in the rat septohippocampal or nigrostriatal pathway by recombinant adeno-associated virus vectors. Exp. Neurol. *150*:183-194.
19. **Knipe, D.M.** 1989. The role of viral and cellular nuclear proteins in herpes simplex virus replication. (Review) (255 refs). Adv. Virus Res. *37*:85-123.
20. **Koeberl, D.D., I.E. Alexander, C.L. Halbert, D.W. Russell, and A.D. Miller.** 1997. Persistent expression of human clotting factor IX from mouse liver after

intravenous injection of adeno-associated virus vectors. Proc. Natl. Acad. Sci. USA *94*:1426-1431.

21.**Kotin, R.M.** 1994. Prospects for the use of adeno-associated virus as a vector for human gene therapy. Hum. Gene Ther. *5*:793-801.

22.**Lewin, A.S., K.A. Drenser, W.W. Hauswirth, S. Nishikawa, D. Yasumura, J.G. Flannery, et al.** 1998. Ribozyme rescue of photoreceptor cells in a transgenic rat model of autosomal dominant retinitis pigmentosa. Nat. Med. *4*:967-971.

23.**Li, J., R.J. Samulski, and X. Xiao.** 1997. Role for highly regulated rep gene expression in adeno-associated virus vector production. J. Virol. *71*:5236-5243.

24.**Matsushita, T., S. Elliger, C. Elliger, G. Podsakoff, L. Vilarreal, G.J. Kurtzman, Y. Iwaki, and P. Colosi.** 1998. Adeno-associated virus vectors can be efficiently produced without helper virus. Gene Ther. *5*:938-945.

25.**McCarthy, A.M., L. McMahan, and P.A. Schaffer.** 1989. Herpes simplex virus type 1 ICP27 deletion mutants exhibit altered patterns of transcription and are DNA deficient. J. Virol. *63*:18-27.

26.**Mishra, L. and J.A. Rose.** 1990. Adeno-associated virus DNA replication is induced by genes that are essential for HSV-1 DNA synthesis. Virology *179*:632-639.

27.**Monahan, P.E., R.J. Samulski, J. Tazelaar, X. Xiao, T.C. Nichols, D.A. Bellinger, et al.** 1998. Direct intramuscular injection with recombinant AAV vectors results in sustained expression in a dog model of hemophilia. Gene Ther. *5*:40-49.

28.**Muzyczka, N.** 1992. Use of adeno-associated virus as a general transduction vector for mammalian cells. Curr. Top. Microbiol. Immunol. *158*:97-129.

29.**Nakai, H., R.W. Herzog, J.N. Hagstrom, J. Walter, S.H. Kung, E.Y. Yang, et al.** 1998. Adeno-associated viral vector-mediated gene transfer of human blood coagulation factor IX into mouse liver. Blood *91*:4600-4607.

30.**Peel, A.L., S. Zolotukhin, G.W. Schrimsher, N. Muzyczka, and P.J. Reier.** 1997. Efficient transduction of green fluorescent protein in spinal cord neurons using adeno-associated virus vectors containing cell type-specific promoters. Gene Ther. *4*:16-24.

31.**Rendahl, K.G., S.E. Leff, G.R. Otten, S.K. Spratt, D. Bohl, M. Van Roey, et al.** 1998. Regulation of gene expression in vivo following transduction by two separate rAAV vectors. Nat. Biotechnol. *16*:757-761.

32.**Rice, S.A. and D.M. Knipe.** 1990. Genetic evidence for two distinct transactivation functions of the herpes simplex virus alpha protein ICP27. J. Virol. *64*:1704-1715.

33.**Roizman, B., A.E. Sears, B.N. Fields, D.M. Knipe, P. Howley, R.M. Chanock, M.S. Hirsch, J.L. Melnick, et al.** (Eds.). 1996. Fields Virology. 3rd ed., p. 2231-2295. Lippincott-Raven, Philadelphia.

34.**Rose, J.A. and F. Koczot.** 1972. Adenovirus-associated virus multiplication VII. Helper requirement for viral deoxyribonucleic acid and ribonucleic acid synthesis. J. Virol. *10*:1-8.

35.**Samulski, R.J. and T. Shenk.** 1988. Adenovirus E1B 55-Mr polypeptide facilitates timely cytoplasmic accumulation of adeno-associated virus mRNAs. J. Virol. *62*:206-210.

36.**Samulski, R.J., A. Srivastava, K.I. Berns, and N. Muzyczka.** 1983. Rescue of adeno-associated virus from recombinant plasmids: gene correction within the terminal repeats of AAV. Cell *33*:135-143.

37.**Sandri-Goldin, R.M. and G.E. Mendoza.** 1992. A herpesvirus regulatory protein appears to act post-transcriptionally by affecting mRNA processing. Genes Dev. *6*:848-863.

38.**Snyder, R.O., S.K. Spratt, C. Lagarde, D. Bohl, B. Kaspar, B. Sloan, et al.** 1997. Efficient and stable adeno-associated virus-mediated transduction in the skeletal muscle of adult immunocompetent mice. Hum. Gene Ther. *8*:1891-1900.

39.**Vincent, K.A., S.T. Piraino, and S.C. Wadsworth.** 1997. Analysis of recombinant adeno-associated virus packaging and requirements for rep and cap gene products. J. Virol. *71*:1897-1905.

40.**Wagner, J.A., T. Reynolds, M.L. Moran, R.B. Moss, J.J. Wine, T.R. Flotte, et al.** 1998. Efficient and persistent gene transfer of AAV-CFTR in maxillary sinus. Lancet *351*:1702-1703.

41.**Weindler, F.W. and R. Heilbronn.** 1991. A subset of herpes simplex virus replication genes provides helper functions for productive adeno-associated virus replication. J. Virol. *65*:2476-2483.

42.**Weller, S.K.** 1991. Genetic analysis of HSV-1 gene required for genome replication, p. 105-136. *In* E.K. Wagner (Ed.), Herpes Virus Transcription and Its Regulation. CRC Press, Boca Raton.

43.**Xiao, X., J. Li, and R.J. Samulski.** 1996. Efficient long-term gene transfer into muscle tissue of immunocompetent mice by adeno-associated virus vector. J. Virol. *70*:8098-8108.

44.**Xiao, X., J. Li, and R.J. Samulski.** 1998. Production of high-titer recombinant adeno-associated virus vectors in the absence of helper adenovirus. J. Virol. *72*:2224-2232.

4 Utility of rAAV for Gene Transfer into Neurons and Other Highly Differentiated Cells

Ernest F. Terwilliger
Division of Experimental Medicine, Harvard Institutes of Medicine, Boston, MA, USA

INTRODUCTION

A great deal of excitement, as well as controversy, has been generated in recent years regarding the promise and potential of gene therapy for treating various human illnesses. Much of the publicity in the general media has focused on diseases of the hematopoietic system, but the therapeutic manipulation of bone marrow stem cells poses especially formidable obstacles. The search for more suitable gene delivery vehicles for this complex and vexing target, as well as improved methods for culturing stem cells without compromising their pluripotentiality, are ongoing.

However, in addition to the hematopoietic system, many other potential targets for therapeutic gene delivery exist, and there seems little reason to doubt that specialized applications of gene transfer technologies will soon prove of great value for the treatment of a diverse group of conditions. Unlike with stem cells, the majority of these applications will require stable gene delivery into lineages that are already differentiated, many of which are in a permanent postmitotic state. For example, certain neurogenetic diseases and other chronic neurological disorders offer very attractive candidates for gene therapy strategies, either by the replacement of native neurotransmitters or provision of supplemental neurotropic factors to the central nervous system (CNS). Gene therapy treatments for Parkinson's and Alzheimer's diseases, as well as stroke, among others, are all under active investigation. Other novel strategies that are being developed include the direct injection of genes into muscle or other depot sites to achieve chronic systemic delivery of secreted factors. These should further broaden the therapeutic applications of gene therapy technology for CNS as well as other disorders.

Gene transfer into differentiated, postmitotic cells, which are resistant either to DNA plasmid transfection or to transduction by traditional Moloney murine leukemia virus-based retroviral vectors (RVs), imposes special requirements on the gene vehicle. For example, the distinctive nature of the CNS—with its populations

Viral Vectors: Basic Science and Gene Therapy
Edited by A. Cid-Arregui and A. García-Carrancá

of extremely long-lived neurons and other nondividing cells, their highly active metabolisms, and the complex cell-to-cell contacts and interactions—imposes stringent limitations upon gene delivery systems, rendering most viral vectors unsuitable for this purpose (14). RVs, with their general requirement for a round of cell division as a prerequisite for efficient integration and stable expression of transfected genes, are poor choices for targeting cells such as neurons or astrocytes. Significant concerns also remain regarding induction of oncogenesis by traditional RVs and contamination by harmful wild-type (wt) retroviruses reconstituted as a result of homologous recombination events (33,48).

Alternative recombinant viral vectors derived from adenovirus (Ad) (23) and herpes simplex virus (HSV) (13) are under active investigation for possible utilization in the CNS. Unfortunate drawbacks of these vectors include their cytotoxicity, immunogenicity (particularly Ad), and instability of expression from the transfected genes. For example, although Ad-based vectors have been used successfully to deliver neurotropic factors in vivo, the subject animals were newborn mice with immature immune systems (15). Human adults rapidly mount strong immune responses against vectors of this type.

For applications to the CNS and other refractory target sites, vectors derived from a human parvovirus, adeno-associated virus (AAV), are increasingly recognized as a promising alternative strategy. As will be discussed in more detail below, vectors derived from AAV offer uniquely valuable delivery vehicles, particularly for gene transfer into nondividing cells, because of their nonpathogenicity, low immunogenicity, small size, and ability to integrate into the host chromosome (22,45), among other features. These characteristics give recombinant AAV (rAAV) distinct advantages for diverse applications, including targeting within the CNS, without some of the key shortcomings associated with other classes of viral vector.

It was not that many years ago that vectors derived from this small DNA virus represented a novel and poorly characterized system used by few investigators. Pioneering work by a handful of laboratories did much to highlight the unique features of rAAV and define the capabilities of the system. Today, vectors derived from AAV are in increasingly widespread use, and the AAV field is advancing rapidly, regularly yielding new applications. One prominent factor hindering more widespread acceptance of the system has been the inconvenience of packaging the vectors due to the complex life cycle of the virus (4,6), with traditional methods requiring multiple transacting factors provided on combinations of one or more plasmids, together with a helper virus that must then be removed or inactivated; another factor has been the relatively low titers of the preparations most groups have been able to achieve (i.e., on the order of 10^9 to 10^{10} functional units, translating into roughly 10^{12} to 10^{13} particles per mL). New, entirely plasmid-based systems that promise higher titers have recently begun to appear; while not yet fully characterized, these should encourage more aggressive research into new applications of rAAV. Nevertheless, most of the achievements to date with this system, including the work of our own laboratory, have been accomplished using traditional protocols for packaging the vectors.

KEY FEATURES OF rAAV

While several distinct AAV viruses have been identified, and some intriguing recombinant vectors have been derived using related parvoviruses, the great majority of the vector development has been conducted using derivatives of AAV type 2 (AAV-2). Therefore, this discussion will be confined to this virus. As a nonpathogenic, replication-defective virus, AAV-2 is one of the smallest human DNA viruses (20–30 nm). Its genome consists of a linear, single-stranded DNA molecule 4680 nucleotides in length (12,34), containing two major open reading frames that encode the viral regulatory (Rep) and structural (Cap) proteins. Although a detailed understanding of how AAV infects the host cell is still being worked out, rAAVs exhibit a broad host tropism and are not restricted to human lineages, also efficiently transducing cells from other primates and rodents, among other species. This feature is important for modeling potential gene therapy treatments in animals. An additional key feature of wt AAV is that its DNA exhibits a strong tendency to integrate site-specifically at a single locus in the human genome (see additional details about this below).

Typically, all of the native viral coding sequences are deleted from rAAV, leaving only the inverted terminal repeats (ITRs), which flank the viral genes and contain the *cis* sequences required for packaging the DNA in the virus particle, as well as for its integration in the host cell. Unlike RVs, no promoter elements are contained within the ITRs, thereby making it relatively easy to design packaging systems for rAAV that do not permit significant levels of homologous recombination. Such rAAVs are capable of delivering a transgene cassette of up to approximately 4.9 kb. Once delivered into the target cell, the transgenes persist as a mixture of genuine integrants and stable episomes. Both forms are actively transcribed by the host (11), which again distinguishes this system from RVs that require integration as a prerequisite for expression. The episomal forms also exhibit a high degree of stability, with a half-life that can extend for months or even years, depending upon the cell type. Together with their ability to integrate in nondividing cells, these features give rAAV the capacity to transduce genes into many differentiated lineages with high efficiency. Also unlike RV, no block to superinfection is induced by exposure to rAAV, so that the same cells can be transduced with multiple vectors, or be susceptible to repeated challenges with either the same or different rAAV.

Because the native viral coding sequences have been removed from the vector, no viral products are expressed in cells transduced with rAAV, unlike the case with vectors based on Ad or HSV. This likely contributes to an additional key feature of these vectors—that they exhibit low immunogenicity compared with vectors derived from other viruses. As mentioned above, Ad as well as HSV vectors typically elicit strong cellular immune responses that can rapidly eliminate the transduced cell populations (5). As is clear from many examples in which persistent transgene expression in vivo has been achieved, this is generally not the case when using rAAV. Typically no cellular immune response is triggered against the viral capsid, and in some instances no cytotoxic T lymphocyte response is activated even against a foreign transgene product when delivered via rAAV (10). The latter phenomenon may be due to the fact that the simple shell of the AAV particle is composed of only a single major protein (4,6) and therefore acts as a poorer adjuvant than more complex particles such as Ad.

Exposure to rAAV does elicit a humoral immune response, with production of neutralizing antibodies against the capsid protein (10,27,52). One group (10) has reported that these antibodies did not interfere with successive secondary transductions into murine muscle using the same rAAV. However, two other laboratories (27,52) found that these capsid antibodies did block retransduction of muscle, unless the block was eliminated by prior injection of the animals with antibodies against the T-cell marker CD4 to transiently immunosuppress them. The differences between these findings may be the result of different doses of the vectors used, or the timing of the repeated treatments. However, even taking the most conservative view of the evidence, rAAVs are clearly much less immunogenic agents than available viral alternatives.

INTEGRATION OF rAAV

Efficient gene transfer and stable transgene expression are two key features that will be required for effective human gene therapy for congenital as well as for many acquired disorders. A unique feature of wt AAV is its ability to integrate site-specifically into the human host genome, at a particular locus (dubbed AAVS1) in chromosome 19 (22,30,45). This specific integration feature of AAV is mediated through its ITR sequences and the viral Rep proteins (7,12,39,50). Site-specific integration is a very attractive feature for the development of gene therapy vectors, and Rep-positive AAV vectors still retain this capability (2,46). Unfortunately, the *rep* gene cassette is undesirable for engineering rAAV because of strong inhibitory effects of its products on cell growth and other gene expression (4).

A few years ago, multiple groups (9,20,41,44) demonstrated that Rep-minus AAV vectors could still integrate into the chromosomes of the host, at least in cell lines. It quickly became clear, however, that Rep-minus vectors did not retain the preference of the wild-type for the AAVS1 site even though they possess the viral ITRs. More recent work, in which larger numbers of vector-cellular junctions were characterized, has led to the suggestion that AAV integration is mediated by distinct cellular pathways in the presence or absence of the Rep proteins.

Until recently, there was little fully convincing direct evidence for the integration of Rep-minus rAAV in nondividing cells. Clear determinations were made more difficult by the fact that both the integrated and episomal forms of the vector DNA express the transgene product; thus, there was no direct link between gene expression and integration. Not surprisingly, the status of rAAV-delivered transgenes in primary cells, and particularly in postmitotic cells, is more dependent on lineage than was the case with cell lines.

In support of rAAV integration in mouse muscle cells, one group (10) reported evidence that included polymerase chain reaction (PCR) detection of persistent head-to-tail vector concatemers, such as frequently occur in genuine integrants. However, because it was not certain whether episomal derivatives of rAAV might exist with this same structure, the possibility of extrachromosomal contributions to these signals could not be eliminated. Even Southern blot analysis data, also reported as evidence of AAV-mediated integration in mouse muscle cells (52), could not rule out contributions from high molecular weight concatemers of the extra chromosomal vectors.

In seeking methods to analyze the state of AAV-delivered transgenes in neural and other primary tissues where sample size was also a limiting factor, our group recently relied upon an Alu-PCR approach (35,51). The 300-bp Alu sequences, comprising a close family of dimeric retropseudogenes (47,49), represent the major class of short interspersed repeat DNAs (SINEs) in human chromosomes. The high frequency of Alu sequences—averaging one every 4 kb, or approximately 9×10^5 per genome—in combination with PCR techniques that amplify long DNA fragments, in theory allows the detection of any integrated exogenous DNA. In actual practice, this is not the case because the distribution of Alu sites is too uneven and also because the sequences of some Alu sites will be too divergent to be recognized by a consensus PCR primer. The technique, however, still allows detection of a significant subset of any series of random integration events.

One key advantage of Alu-PCR for detecting integrated rAAV is its ability to discriminate absolutely against episomal forms of the vector, because one primer specifically targets a consensus Alu sequence. Using this technique (for details see Reference 51), we were able to detect integrated rAAV in human neurons and primary macrophages, among other cell types, following in vitro exposure to an rAAV. The identity of the PCR amplicon product as vector-cellular junctions could be confirmed by sequencing, which also enabled us to examine rearrangements that had occurred during integration. This information provided the first direct evidence for the integration of Rep-minus rAAV in these clinically important postmitotic cell types. It could not be inferred from these findings that significant levels of vector integration can occur in any postmitotic target lineage, because the cell types chosen were notable for their strong susceptibility to rAAV. Nevertheless, this technique should provide useful information for further analysis of the mechanisms underlying AAV-mediated integration, including issues of frequency and site preference.

Because Alu-equivalent sequences, designated "B1", are present at a high frequency in the rat genome, this technique can also be extended into animal model systems. We have taken a first step in this direction ourselves, to detect integrated rAAV in rat brain tissue following intraventricular delivery of the vector (51). Stainings of rat brain sections in several previous experiments had detected transgene expression primarily, although not exclusively, in the hypothalamus following intraventricular injection of an rAAV, perhaps due to the relatively high exposure of this region to the vector following this type of injection (see also Some Therapeutic Applications of rAAV). To pursue this further, two weeks after intraventricular injection of the same rAAV used successfully in Alu PCR screens of vector integration sites in human cells, homogenates were prepared from the hypothalamus and several other regions of the brain from a treated rat. The samples were then subjected to PCR using primer sets designed to recognize a B1, rather than Alu, consensus sequence. Specific signals were only detected in the hypothalamic tissue, in good agreement with the earlier histochemical findings. Relatively few vector-specific bands were detected compared with the complex patterns observed in human neurons exposed in vitro to rAAV. We do not believe this indicated a restrictive site specificity of the vector in the tissue, but rather that the technique was near the limit of its present level of sensitivity, so that only a few preferred sites were successfully amplified. Nevertheless, to our knowledge, these findings represented the first conclusive demonstration of integration of an rAAV in the rodent brain. The question of whether the rAAV in the brain was integrating into neurons, or into their supporting networks of glial cells, or both, was not assessed in this in vivo experiment. However, by inference from the studies on cultured human neurons exposed to rAAV in vitro, some level of integration into brain neurons appears likely.

SOME THERAPEUTIC APPLICATIONS OF rAAV

AAV vectors are now being utilized for a broad spectrum of applications with therapeutic potential, as well as for basic research. The transduction efficiency as well as persistence of rAAV can vary widely with the target tissue, but findings from many groups make clear that these vectors can be used with considerable success to introduce genes into several clinically important targets, including muscle cells, which are typically refractory to transduction by RV, without the inflammation attendant upon use of Ad vectors (52). Long-term maintenance of expression after rAAV-mediated delivery of transgenes into either skeletal muscle or myocardium has been documented (20,21). These successes should speed the further development of depot gene strategies alluded to earlier in the Introduction.

Fortuitously, rAAV also appear to have a particular facility for gene transfer into neurons, in either the brain or spinal cord. High transduction efficiencies have been achieved using rAAV in populations of neurons both in vitro (8) and in vivo (19,24,31), making them attractive delivery vehicles for targeting the CNS. For example, one group (40) recently demonstrated long-term expression of green fluorescent protein under the control of either a neuron-specific enolase or platelet-derived growth factor promoter in the spines of rats receiving intrathecal injections of an rAAV. Our findings (References 5 and 8 and unpublished observations) are in good agreement with these reports. Experiments performed on cultures of freshly isolated rat brain neurons or astrocytes, using an rAAV bearing a cytomegalovirus immediate early-driver *lacZ* cassette, indicated that rAAV are able to transduce both cell types successfully, at least in vitro. More dramatic expression was detected in neurons than in astrocytes, but this may have reflected the choice of promoter element rather than an innate preference of the vector.

In vivo experiments also yielded strongly encouraging results. Individual cell morphologies in rat brain sections examined subsequent to intraventricular delivery of a different rAAV vector were more ambiguous, but again suggested at least some of the vector-positive cells were neurons (51). All of these findings indicate that control of rAAV vector tropism for particular cell types in the brain may need to be fine-tuned at the level of transcription of the gene cassette, rather than entry of the vector per se. Because both neuron-specific and myeloid-specific genetic elements are now available, this should not pose major obstacles.

There is a critical need for improved gene delivery strategies that will enhance global access of viral vectors within the brain, as opposed to focal delivery to a discrete site, in order to effectively treat brain tumors and other disorders. For example, gliomas are typically widely disseminated, while viral vectors generally do not disperse far from their site of injection, hence the deployment of antitumor gene delivery strategies that inject propagating, albeit attenuated, strains of live HSV virus, or even RV-producing cell lines, into the brain tissue to maximize exposure of a diffuse target to the gene therapy agent (1,28,29,38,42).

A crucial consideration for achieving effective dispersal of any vector in the brain is the permeability of the blood microvessels to the passage of nanoparticles such as viruses. For example, the relative hyperpermeability of tumor-associated microvessels compared with the normal blood brain barrier (BBB) has been widely reported (18). However, recent studies have demonstrated that even in this special case, the effective pore cutoff size is quite small for tumors in the cranial microenvironment (typically 100 nm or less) (16,53,54). By contrast, the sizes of the viral vectors commonly used in the brain at present are on the order of 100 to 300 nm, with HSV at the high end of this scale. This low threshold, not surprisingly, severely limits the transvascular transport of these viral vectors even within the comparatively leaky tumor vasculature.

Furthermore, delivery is not the only obstacle to effective gene therapy in the brain. Some enhanced penetration of even large viral vectors across the BBB, or across the brain tumor barrier, has been achieved using drugs that temporarily permeabilize blood vessel endothelium. However, even late reexposure in the periphery to an Ad vector following primary delivery to the brain has been shown to provoke inflammation and demyelination of the brain area given the original treatment (5).

The small size of rAAV, together with its low immunogenicity, offers a powerfully attractive combination, affording this vector system distinct advantages over other classes of vector for use in the brain. To date, rAAVs bearing the HSV thimidine kinase (*TK)* gene, either alone or in combination with the human IL-2 sequence, have been used successfully to regress human gliomas in nude mice and prolong the life of the animals (32,37). Also of relevance in this regard are several recent reports demonstrating effective gene transfer into vascular endothelium mediated by rAAV (25,26,43). However, a great deal of work remains to be done to maximize the potential of rAAV for these types of applications (36).

In addition to therapeutic applications, rAAVs are also finding increasing use in basic research. For example, a significant effort of our laboratory has been devoted to exploiting rAAV-mediated gene delivery into macrophages and other primary cells susceptible to targets of the human immunodeficiency virus (HIV) (17), to better assess the effects of HIV infection upon myeloid lineages. Use of rAAV has enabled us to examine directly the impact of individual HIV proteins, or antiviral gene strategies, in mature primary macrophages, as opposed to relying upon inferences from monocytic cell lines.

THE FUTURE

It is clear that no single vector system will be suitable for all gene therapy strategies. Each vector brings its own special set of advantages, as well as its own particular limitations, and doubtless each will find a niche for certain types of applications. Attempts to develop novel gene vectors from still other viruses, including even HIV, are being actively pursued. However, the unique features of AAV virtually guarantee that derivatives of this virus will continue to play important roles in the development of therapeutic approaches to gene delivery for the foreseeable future.

At present, several groups are attempting to combine the advantages afforded by AAV with other viral or nonviral delivery systems to develop next-generation gene vectors with improved capabilities and fewer limitations (3). Research is also being conducted aimed at retaining or manipulating the site-specific integration feature of wt AAV, without the deleterious effects produced by chronic expression of its Rep proteins. Recent preliminary yet tantalizing work further suggests that AAV may offer investigators a means by which transgenes could be directed to undergo efficient homologous recombination with their native cellular counterparts. Other novel approaches to controlling the targeting of either the viral particle itself or the insertion of its DNA are also under active investigation.

In addition to improvements in vector delivery, equally important advances are being made in controlling the expression of the delivered transgenes. For example, ARIAD (Cambridge, MA, USA) recently reported the successful regulation of an *Epo* transgene delivered by rAAV into rat muscle. Together with continuing advances in the cultivation and expansion of human cells outside the body, progress in these diverse avenues of investigation should lead in the near future to the fulfillment of the promise that gene therapy has held since its inception.

REFERENCES

1. **Andreansky, S.S., B. He, G.Y. Gillespie, L. Soroceanu, J. Markert, J. Chou, B. Roizman, and R.J. Whitley.** 1996. The application of genetically engineered herpes simplex viruses to the treatment of experimental brain tumors. Proc. Natl. Acad. Sci. USA *93*:11313-11318.
2. **Balague, C., M. Kalla, and W.W. Zhang.** 1997. Adeno-associated virus Rep78 protein and terminal repeats enhance integration of DNA sequences into the cellular genome. J. Virol. *71*:3299-3306.
3. **Baudard, M., T.R. Flotte, J.M. Aran, A.R. Thierry, I. Pastan, M.G. Pang, W.G. Kearns, and M.M. Gottesman.** 1996. Expression of the human multidrug resistance and glucocerebrosidase cDNAs from adeno-associated vectors: efficient promoter activity of AAV sequences and in vivo delivery via liposomes. Hum. Gene Ther. *7*:1309-1322.
4. **Berns, K.I.** 1990. Parvovirus replication. Microbiol. Rev. *54*:316-329.
5. **Byrnes, A.P., R.E. MacLaren, and H.M. Charlton.** 1996. Immunological instability of persistent adenovirus vectors in the brain: peripheral exposure to vector leads to renewed inflammation, reduced gene expression, and demyelination. J. Neurosci. *16*:3045-3055.
6. **Carter, B.J.** 1990. The growth cycle of adeno-associated virus, p. 155-168. *In* P. Tijssen (Ed.), Handbook of Parvoviruses. CRC Press, Boca Raton.
7. **Chiorini, J.A., S.M. Wiener, R.A. Owens, S.R. Kyostio, R.M. Kotin, and B. Safer.** 1994. Sequence requirements for stable binding and function of Rep68 on the adeno-associated virus type 2 inverted terminal repeats. J. Virol. *68*:7448-7457.
8. **Du, B., P. Wu, D.M. Boldt-Houle, and E.F. Terwilliger.** 1996. Efficient transduction of human neurons with an adeno-associated virus vector. Gene Ther. *3*:254-261.
9. **Duan, D., K.J. Fisher, J.F. Burda, and J.F. Engelhardt.** 1997. Structural and functional heterogeneity of integrated recombinant AAV genomes. Virus Res. *48*:41-56.
10. **Fisher, K.J., K. Jooss, J. Alston, Y. Yang, S.E. Haecker, K. High, R. Pathak, S.E. Raper, and J.M. Wilson.** 1997. Recombinant adeno-associated virus for muscle directed gene therapy. Nat. Med. *3*:306-312.
11. **Flotte, T.R., S.A. Afione, and P.L. Zeitlin.** 1994. Adeno-associated virus vector gene expression occurs in nondividing cells in the absence of vector DNA integration. Am. J. Respir. Cell Mol. Biol. *11*:517-521.
12. **Flotte, T.R. and B.J. Carter.** 1995. Adeno-associated virus vectors for gene therapy. Gene Ther. *2*:357-362.
13. **Freese, A., A.I. Geller, and R. Neve.** 1990. HSV-1 vector mediated neuronal gene delivery. Strategies for molecular neuroscience and neurology. Biochem. Pharmacol. *40*:2189-2199.
14. **Friedmann, T.** 1994. Gene therapy for neurological disorders. Trends Genet. *10*:210-214.
15. **Haase, G., P. Kennel, B. Pettmann, E. Vigne, S. Akli, F. Revah, H. Schmalbruch, and A. Kahn.** 1997. Gene therapy of murine motor neuron disease using adenoviral vectors for neurotrophic factors. (see comments). Nat. Med. *3*:429-436.
16. **Hobbs, S.K., W.L. Monsky, F. Yuan, W.G. Roberts, L. Griffith, V.P. Torchilin, and R.K. Jain.** 1998. Regulation of transport pathways in tumor vessels: role of tumor type and microenvironment. Proc. Natl. Acad. Sci. USA *95*:4607-4612.
17. **Inouye, R.T., B. Du, D. Boldt-Houle, A. Ferrante, I.W. Park, S.M. Hammer, L. Duan, J.E. Groopman, R.J. Pomerantz, and E.F. Terwilliger.** 1997. Potent inhibition of human immunodeficiency virus type 1 in primary T cells and alveolar macrophages by a combination anti-Rev strategy delivered in an adeno-associated virus vector. [published erratum appears in J. Virol. 1998. *72*(4):3506]. J. Virol. *71*:4071-4078.
18. **Jain, R.K.** 1996. Delivery of molecular medicine to solid tumors. Science *271*:1079-1080.
19. **Kaplitt, M.G., P. Leone, R.J. Samulski, X. Xiao, D.W. Pfaff, K.L. O'Malley, and M.J. During.** 1994. Long-term gene expression and phenotypic correction using adeno-associated virus vectors in the mammalian brain. Nat. Genet. *8*:148-154.
20. **Kaplitt, M.G., X. Xiao, R.J. Samulski, J. Li, K. Ojamaa, I.L. Klein, H. Makimura, M.J. Kaplitt, R.K. Strumpf, and E.B. Diethrich.** 1996. Long-term gene transfer in porcine myocardium after coronary infusion of an adeno-associated virus vector. Ann. Thorac. Surg. *62*:1669-1676.
21. **Kessler, P.D., G.M. Podsakoff, X. Chen, S.A. McQuiston, P.C. Colosi, L.A. Matelis, G.J. Kurtzman, and B.J. Byrne.** 1996. Gene delivery to skeletal muscle results in sustained expression and systemic delivery of a therapeutic protein. Proc. Natl. Acad. Sci. USA *93*:14082-14087.
22. **Kotin, R.M., R.M. Linden, and K.I. Berns.** 1992. Characterization of a preferred site on human chromosome 19q for integration of adeno-associated virus DNA by non-homologous recombination. EMBO J. *11*:5071-5078.
23. **Le Gal La Salle, G., J.J. Robert, S. Berrard, V. Ridoux, L.D. Stratford-Perricaudet, M. Perricaudet, and J. Mallet.** 1993. An adenovirus vector for gene transfer into neurons and glia in the brain. Science *259*:988-990.
24. **Lo, W.D., R. Chen, G. Qu, T. Sferra, R. Clark, and P. Johnson.** 1996. Society for Neuroscience: Abstracts *22*:955.
25. **Lynch, C.M., P.S. Hara, J.C. Leonard, J.K. Williams, R.H. Dean, and R.L. Geary.** 1997. Adeno-associated virus vectors for vascular gene delivery. Circ. Res. *80*:497-505.
26. **Maeda, Y., U. Ikeda, Y. Ogasawara, M. Urabe, T. Takizawa, T. Saito, P. Colosi, G. Kurtzman, K. Shimada, and K. Ozawa.** 1997. Gene transfer into vascular cells using adeno-associated virus (AAV) vectors. Cardiovasc. Res. *35*:514-521.
27. **Manning, W.C., S. Zhou, M.P. Bland, J.A. Escobedo, and V. Dwarki.** 1998. Transient immunosuppres-

sion allows transgene expression following readministration of adeno-associated viral vectors. Hum. Gene Ther. *9*:477-485.

28.**Markert, J.M., A. Malick, D.M. Coen, and R.L. Martuza.** 1993. Reduction and elimination of encephalitis in an experimental glioma therapy model with attenuated herpes simplex mutants that retain susceptibility to acyclovir. Neurosurgery *32*:597-603.

29.**Martuza, R.L., A. Malick, J.M. Markert, K.L. Ruffner, and D.M. Coen.** 1991. Experimental therapy of human glioma by means of a genetically engineered virus mutant. Science *252*:854-856.

30.**McCarty, D.M., J.H. Ryan, S. Zolotukhin, X. Zhou, and N. Muzyczka.** 1994. Interaction of the adeno-associated virus Rep protein with a sequence within the A palindrome of the viral terminal repeat. J. Virol. *68*:4998-5006.

31.**McCown, T.J., X. Xiao, J. Li, G.R. Breese, and R.J. Samulski.** 1996. Differential and persistent expression patterns of CNS gene transfer by an adeno-associated virus (AAV) vector. Brain Res. *713*:99-107.

32.**Mizuno, M., J. Yoshida, P. Colosi, and G. Kurtzman.** 1998. Adeno-associated virus vector containing the herpes simplex virus thymidine kinase gene causes complete regression of intracerebrally implanted human gliomas in mice, in conjunction with ganciclovir administration. Jpn. J. Cancer Res. *89*:76-80.

33.**Morgan, R.A. and W.F. Anderson.** 1993. Human gene therapy. Annu. Rev. Biochem. *62*:191-217.

34.**Muzyczka, N.** 1992. Use of adeno-associated virus as a general transduction vector for mammalian cells. Curr. Top. Microbiol. Immunol. *158*:97-129.

35.**Nelson, D.L., S.A. Ledbetter, L. Corbo, M.F. Victoria, R. Ramirez-Solis, T.D. Webster, D.H. Ledbetter, and C.T. Caskey.** 1989. Alu polymerase chain reaction: a method for rapid isolation of human-specific sequences from complex DNA sources. Proc. Natl. Acad. Sci. USA *86*:6686-6690.

36.**Nguyen, J.T., P. Wu, M.E. Clouse, L. Hlatky, and E.F. Terwilliger.** 1998. Adeno-associated virus-mediated delivery of antiangiogenic factors as an antitumor strategy. Cancer Res. *58*:5673-5677.

37.**Okada, H., K. Miyamura, T. Itoh, M. Hagiwara, T. Wakabayashi, M. Mizuno, P. Colosi, G. Kurtzman, and J. Yoshida.** 1996. Gene therapy against an experimental glioma using adeno-associated virus vectors. Gene Ther. *3*:957-964.

38.**Oldfield, E.H., Z. Ram, K.W. Culver, R.M. Blaese, H.L. DeVroom, and W.F. Anderson.** 1993. Gene therapy for the treatment of brain tumors using intratumoral transduction with the thymidine kinase gene and intravenous ganciclovir. Hum. Gene Ther. *4*:39-69.

39.**Owens, R.A., M.D. Weitzman, S.R. Kyostio, and B.J. Carter.** 1993. Identification of a DNA-binding domain in the amino terminus of adeno-associated virus Rep proteins. J. Virol. *67*:997-1005.

40.**Peel, A.L., S. Zolotukhin, G.W. Schrimsher, N. Muzyczka, and P.J. Reier.** 1997. Efficient transduction of green fluorescent protein in spinal cord neurons using adeno-associated virus vectors containing cell type-specific promoters. Gene Ther. *4*:16-24.

41.**Ponnazhagan, S., D. Erikson, W.G. Kearns, S.Z. Zhou, P. Nahreini, X.S. Wang, and A. Srivastava.** 1997. Lack of site-specific integration of the recombinant adeno-associated virus 2 genomes in human cells. Hum. Gene Ther. *8*:275-284.

42.**Ram, Z., K.W. Culver, S. Walbridge, R.M. Blaese, and E.H. Oldfield.** 1993. In situ retroviral-mediated gene transfer for the treatment of brain tumors in rats. (see comments). Cancer Res. *53*:83-88.

43.**Rolling, F., Z. Nong, S. Pisvin, and D. Collen.** 1997. Adeno-associated virus-mediated gene transfer into rat carotid arteries. Gene Ther. *4*:757-761.

44.**Rutledge, E.A. and D.W. Russell.** 1997. Adeno-associated virus vector integration junctions. J. Virol. *71*:8429-8436.

45.**Samulski, R.J., X. Zhu, X. Xiao, J.D. Brook, D.E. Housman, N. Epstein, and L.A. Hunter.** 1991. Targeted integration of adeno-associated virus (AAV) into human chromosome 19 [published erratum appears in EMBO J. 1992 *11*(3):1228]. EMBO J. *10*:3941-3950.

46.**Shelling, A.N. and M.G. Smith.** 1994. Targeted integration of transfected and infected adeno-associated virus vectors containing the neomycin resistance gene. Gene Ther. *1*:165-169.

47.**Singer, M.F.** 1982. Highly repeated sequences in mammalian genomes. Int. Rev. Cytol. *76*:67-112.

48.**Walsh, C.E., J.M. Liu, J.L. Miller, A.W. Nienhuis, and R.J. Samulski.** 1993. Gene therapy for human hemoglobinopathies. Proc. Soc. Exp. Biol. Med. *204*:289-300.

49.**Weiner, A.M., P.L. Deininger, and A. Efstratiadis.** 1986. Nonviral retroposons: genes, pseudogenes, and transposable elements generated by the reverse flow of genetic information. Annu. Rev. Biochem. *55*:631-661.

50.**Weitzman, M.D., S.R. Kyostio, R.M. Kotin, and R.A. Owens.** 1994. Adeno-associated virus (AAV) Rep proteins mediate complex formation between AAV DNA and its integration site in human DNA. Proc. Natl. Acad. Sci. USA *91*:5808-5812.

51.**Wu, P., M.I. Phillips, J. Bui, and E.F. Terwilliger.** 1998. Adeno-associated virus vector-mediated transgene integration into neurons and other nondividing cell targets. J. Virol. *72*:5919-5926.

52.**Xiao, X., J. Li, and R.J. Samulski.** 1996. Efficient long-term gene transfer into muscle tissue of immunocompetent mice by adeno-associated virus vector. J. Virol. *70*:8098-8108.

53.**Yuan, F., Y. Chen, M. Dellian, N. Safabakhsh, N. Ferrara, and R.K. Jain.** 1996. Time-dependent vascular regression and permeability changes in established human tumor xenografts induced by an anti-vascular endothelial growth factor/vascular permeability factor antibody. Proc. Natl. Acad. Sci. USA *93*:14765-14770.

54.**Yuan, F., M. Dellian, D. Fukumura, M. Leunig, D.A. Berk, V.P. Torchilin, and R.K. Jain.** 1995. Vascular permeability in a human tumor xenograft: molecular size dependence and cutoff size. Cancer Res. *55*:3752-3756.

5 Adeno-Associated Virus Vectors for Genetic Immunization

William C. Manning and Xavier Paliard
Chiron Corporation, Emeryville, CA, USA

INTRODUCTION

In recent years, adeno-associated virus (AAV) vectors have received much attention in the field of gene therapy for their ability to establish long-lasting transgene expression in a variety of tissues. Recombinant AAV (rAAV) vectors have been used to successfully transduce skeletal muscle (4,9,12,20,32,35), cardiac muscle (19), vascular endothelial cells (23), retina (10), brain (18,26), and liver (21,31). Some of the most impressive data have come from skeletal muscle in the mouse where transgene expression after a single injection persists for over one year (35). The efficacy of rAAV-mediated muscle delivery is not limited to small animal models. Successful skeletal muscle transduction by rAAV vectors has also been reported in dogs (13,27) and nonhuman primates (37).

Most genetic immunization protocols involve the direct injection of "naked" plasmid DNA into skeletal muscle (34). While injection of very little plasmid DNA is needed to prime immune responses in rodents, in large animals, multiple injections of relatively large quantities of plasmid are often required (6). We hypothesized that the robust and persistent transgene expression observed in rAAV-transduced muscle might afford genetic immunization with a single injection even in larger animals.

Other viral vectors have been used for genetic immunization. Among the most studied are adenovirus (Ad) (15), vaccinia (28), and canary pox vectors (5). Unlike these viral vectors that express viral proteins in addition to the transgene, rAAV vectors do not encode any AAV proteins. The only viral sequences in an rAAV vector are the 145 nucleotide inverted terminal repeats. Another potential benefit is safety, since AAV is a nonpathogenic human parvovirus.

There is precedence in the literature for the immunogenicity of rAAV vectors. Several studies using rAAV for transduction of skeletal muscle have reported the development of antibodies to the transgene (4,13,27,32).

In this chapter, the ability of intramuscular injection of rAAV to prime both cellular and humoral immune responses in mice and nonhuman primates is reviewed,

Viral Vectors: Basic Science and Gene Therapy
Edited by A. Cid-Arregui and A. García-Carrancá

and the potential of rAAV vaccination strategies is discussed.

MATERIALS AND METHODS

Construction and Preparation of rAAV

The construction of rAAV-gB (24) and rAAV-LacZ (25) has been described. To construct rAAV-NS3-4a-E2, the following were cloned in order into our standard cytomegalovirus (CMV) promoter-driven AAV vector: *(i)* the 25 amino acid human tissue plasminogen activator leader peptide, *(ii)* hepatitis C virus (HCV) NS3 (amino acids 1027–1657), *(iii)* HCV NS4a (amino acids 1658–1713), *(iv)* two alanine residues, *(v)* HCV E2 (amino acids 704–750), and *(vi)* a TAG stop codon. rAAV vector particles were produced by a modified transfection/infection protocol (17,36). Briefly, human embryonic kidney 293 cells, grown to 60% confluence in 15-cm dishes, were co-transfected with 12.5 µg of helper plasmid pKSrep/cap (25) and 12.5 µg of vector plasmid using the calcium phosphate co-precipitation method. Eight hours after co-transfection, medium was replaced with Iscore's modified Dulbecco's medium (IMDM) plus 10% fetal bovine serum (FBS) containing Ad type 5 dl312 at a multiplicity of infection (moi) of 2. Seventy-two hours after infection, cells were harvested and resuspended in HEPES buffer (2.5 mL per dish) and lysed by three cycles of freezing and thawing. Cell debris was removed by centrifugation at 12 000× *g* for 20 minutes. Packaged rAAV was purified from Ad by two rounds of CsCl equilibrium density gradient centrifugation. Residual Ad contamination was inactivated by heating at 56°C for 45 minutes. To estimate total number of rAAV particles, the virus stock was treated with DNase I, and encapsidated DNA was extracted with phenol-chloroform and precipitated with ethanol. DNA dot blot analysis against a known standard was used to determine titer (33). To assay for Ad contamination, 293 cells were infected with 10 µL of purified rAAV stock and followed for any signs of cytopathic effect. All stocks were negative for Ad contamination (level of detection greater than or equal to 100 plaque forming units [pfu]/mL). To assay for wild-type (wt) AAV, 293 cells were co-infected with serial dilutions of rAAV stocks and Ad dl312 at a moi of 2. Three days later the cells were harvested, lyzed by three cycles of freezing and thawing, and centrifuged to remove cell debris. The supernatant was heat inactivated (56°C for 10 minutes) and fresh plates of 293 cells (6×10^6) were infected in the presence of Ad dl312 at a moi of 2. Forty-eight hours after infection, low molecular weight DNA was isolated (14), subjected to agarose gel electrophoresis, and transferred to a nylon membrane. The membrane was hybridized with a biotinylated oligonucleotide probe specific for the AAV capsid region. The wt AAV titer was defined as the highest dilution of rAAV vector stock that demonstrated a positive hybridization signal. Our rAAV preparations never contained more than one wt AAV genome per 10^6 rAAV genomes.

Immunoassays

Protocols for assaying gB-specific cytotoxic T lymphocyte (CTL), gB-specific lymphoproliferation, and gB-specific antibody have been described (7,24). To perform the HCV NS3 enzyme-linked immunosorbent assay (ELISA), microplates were coated with recombinant HCV NS3 made as a superoxide dismutase (SOD) fusion in *Escherichia coli* (125 ng/well). Dilutions of serum samples were added to the wells and incubated for 1 hour at 37°C. Plates were washed and incubated with goat antimouse peroxidase conjugate (diluted 1:5000) or goat anti-monkey peroxidase conjugate (diluted 1:160 000; Organon Teknika, West Chester, PA, USA) for 1 hour at 37°C. O-phenylenediamine substrate was used to develop the plates. The plates were read at 492 nm with a cutoff of 1.0 absorbance used to determine titer.

To perform the AAV capsid ELISA, microplates were coated overnight at 4°C with 10^9 rAAV-LacZ particles/well in phosphate-buffered saline (PBS). The following day the plates were washed and then blocked for 30 minutes at 37°C with PBS containing 1% goat serum and 0.3% Tween® 20. Serial threefold dilutions of sample and control sera were loaded onto the plate starting at 1:75 (control sera were from mice that had received multiple injections of AAV). Microplates were then incubated for 1 hour at 37°C. Plates were washed and incubated at 37°C for 30 minutes with goat antimouse Ig-horseradish peroxidase (HRP) at 1:2000 (Dako, Carpenteria, CA, USA). O-phenylenediamine substrate was used to develop the plates. Plates were read at 492 nm with a cutoff of 0.2 OD.

To perform the AAV neutralizing antibody assay, 293 cells were plated at 3×10^4 cells/well in a 96-well microplate. The following day, prebleed, positive control, and sample sera were inactivated at 56°C for 30 minutes. Threefold serial dilutions of sera in IMDM without fetal calf serum (FCS) (BioWhittaker, Walkersville, MD, USA) were then incubated with 10^8 particles of rAAV-Luc at 37°C for 1 hour. The medium was removed from the 293 cells, and diluted sera plus virus was added and incubated for 1 hour at 37°C. After this incubation, the plates were washed and fresh IMDM containing 10% FCS was added. Twenty-four hours later, the cells were rinsed with PBS and lysed in Reporter Lysis Buffer (Promega, Madison, WI, USA). Cells were then harvested and assayed for luciferase activity. The AAV neutralizing antibody titer was defined as the dilution of serum required to obtain 50% of the luciferase activity in 293 cells infected with rAAV-Luc preincubated with negative control serum.

Animals

Six-week-old female C57BL/6 and CB6F1 mice were purchased from Charles River Laboratories (Wilmington, MA, USA). Prior to injection, mice were anesthetized with a mixture of ketamine and xylazine. The rAAV was diluted in 0.9% saline, and a final volume of 50 µL was injected into the tibialis anterior (TA) muscle.

Four 2-year-old baboons, which weighed between 14 and 20 kg, were used in the study. Animals were sedated with 10 mg/kg ketamine, and each animal received four 0.5-mL injections of rAAV—two injections in the right TA and two injections in the left TA. Animals were housed at Southwest Foundation for Biomedical Research (San Antonio, TX, USA). The study was performed under the National Institutes of Health *Guidelines for Care and Use of Laboratory Animals.*

RESULTS

Cellular and Humoral Immune Responses to Herpes Simplex Virus gB After Immunization with rAAV-gB

To investigate the ability of rAAV immunization to elicit both humoral and cellular immune responses, C57Bl/6 (H-2^b) mice were injected intramuscularly with doses of rAAV-gB ranging from 1×10^6 to 5×10^{10}. At six weeks after immunization, mice were assayed for major histocompatibility complex (MHC) class I-restricted, gB-specific CTL in a standard ^{51}Cr release assay. As shown in Table 1, gB-

Table 1. Generation of MHC Class I-Restricted, gB-Specific CTL after Intramuscular Injection of rAAV-gB

No. of Particles Injected	E:T Ratio	Percent Specific Lysis MC57/Ø ($H-2^b$)	MC57/gB ($H-2^b$)	SvBalb/gB ($H-2^d$)
1×10^8	50:1	9	21	10
	12:1	6	17	15
	3:1	6	4	1
1×10^{10}	50:1	7	21	18
	12:1	7	17	19
	3:1	4	15	16
1×10^{10}	50:1	7	66	19
	12:1	0	19	2
	3:1	5	14	1
5×10^{10}	50:1	10	45	17
	12:1	12	25	9
	3:1	9	16	8

specific, MHC class I-restricted CTL were detected in the groups that had received at least 1×10^{10} particles of vector. The data for the mice that received 10^6 and 10^7 particles is not shown since all animals were negative for gB-specific CTL. To examine the ability of rAAV to prime a helper T cell response, lymphoproliferation assays (LPA) were also performed on splenocytes from the immunized mice. gB-specific lymphoproliferation was also observed at the same doses of virus that primed CTL (Table 2). Serum was collected from the mice at the time of sacrifice and assayed for gB-specific antibody by ELISA. Similarly, no seroconversion to the transgene was seen until injection of at least 10^{10} rAAV-gB particles (Table 2).

Immunization of Mice and Baboons with an rAAV Vector Expressing HCV NS3-NS4a-E2

The immunogenicity of a rAAV expressing HCV 1a NS3-NS4a-E2 was also studied. In this study, groups of four CB6F1 mice were immunized with rAAV-NS3-NS4a-E2 by intramuscular injection. As shown in Figure 1, the mice developed NS3-specific antibody that persisted for the duration of the study.

To examine the immunogenicity of an rAAV vector in a larger animal, four baboons were injected intramuscularly with 3×10^{12} particles of rAAV-NS3-NS4a-E2. As shown in Figure 2, three of the four baboons developed antibodies to NS3 by 16 weeks postimmunization. To look for priming of a cellular immune response to NS3, peripheral blood leukocytes were collected at 0, 10, and 16 weeks postimmunization and assayed for NS3-specific CTL. NS3-specific CTL were not detected at any of these time points.

Notably, one of the four baboons failed to develop antibodies to NS3. One explanation for the lack of a NS3-specific antibody response in this baboon is the presence of AAV neutralizing antibody in the serum of that baboon at the time of immunization. Although AAV type 2 (AAV-2), the serotype used in this study, is a human virus, it is possible that this baboon could have contracted AAV-2 infection from one

Table 2. Effect of rAAV Dose on Priming of gB-Specific Lymphoproliferation and gB-Specific Antibody

rAAV-gB Dose	gB-Specific LPA (stimulation index)	Reciprocal gB Antibody Titer
1×10^6	1.0	<25
1×10^7	0.9	<25
1×10^8	1.6	<25
1×10^9	0.9	<25
1×10^{10}	4.9	5675
5×10^{10}	3.0	15577

of its human handlers. Prevaccination serum samples from all four baboons were screened for the presence of AAV-specific antibody by both an AAV capsid ELISA and an AAV neutralization assay. As shown in Table 3, the prevaccination serum from baboon 1, the animal that did not seroconvert to NS3, contained AAV-specific antibodies as measured by both ELISA and viral neutralization. This result indicated that the existence of neutralizing antibody to the vector blocked vector transduction and thus the development of an immune response to the transgene.

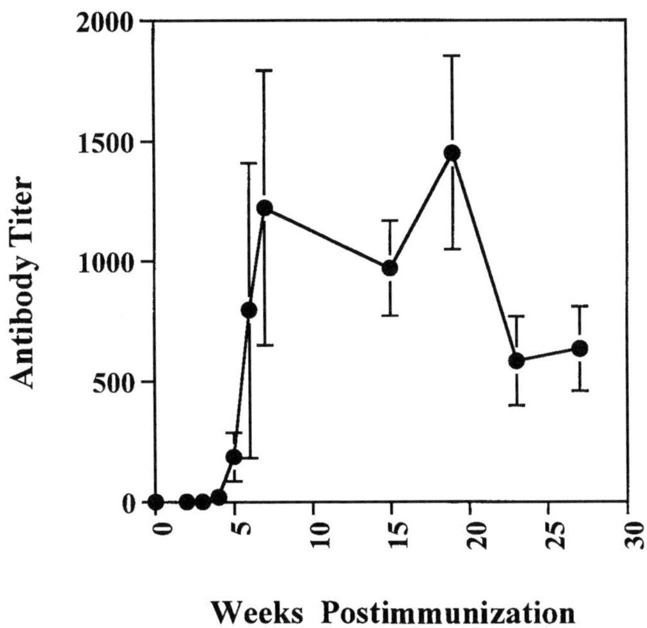

Figure 1. Intramuscular immunization of mice with rAAV-NS3-NS4a-E2 results in the production of NS3-specific antibody. Four CB6F1 mice were immunized intramuscularly with 6×10^9 particles of rAAV-NS3-NS4a-E2. Titers shown are the mean ± standard error of the mean.

Table 3. AAV Capsid and Neutralizing Antibody in Baboon Prebleeds

Baboon	AAV Capsid Titer	AAV Neutralizing Titer
1	4993	400
2	480	<25
3	524	<25
4	113	<25

Does Primary rAAV Vaccination Block Subsequent rAAV-Mediated Vaccination?

The results from the previous baboon vaccination experiment indicated that pre-existing neutralizing antibodies to the AAV vector particle could block subsequent use of a rAAV vector for genetic immunization. To study this observation in more detail, we examined whether prior immunization with one rAAV vector precluded subsequent immunization with another rAAV vector. In this study, rAAV vectors that expressed either lacZ or gB were used. C57Bl/6 mice were immunized with 5×10^{10} particles of rAAV-gB, rAAV-LacZ, or no virus at week 0 in the right TA muscle. At week 7, mice were immunized a second time in the left TA muscle with the same dose of the virus they did not receive at week 0. The various combinations of immunizations are shown in Table 4. At week 11, the animals were sacrificed and assayed for gB-specific CTL, gB antibody,

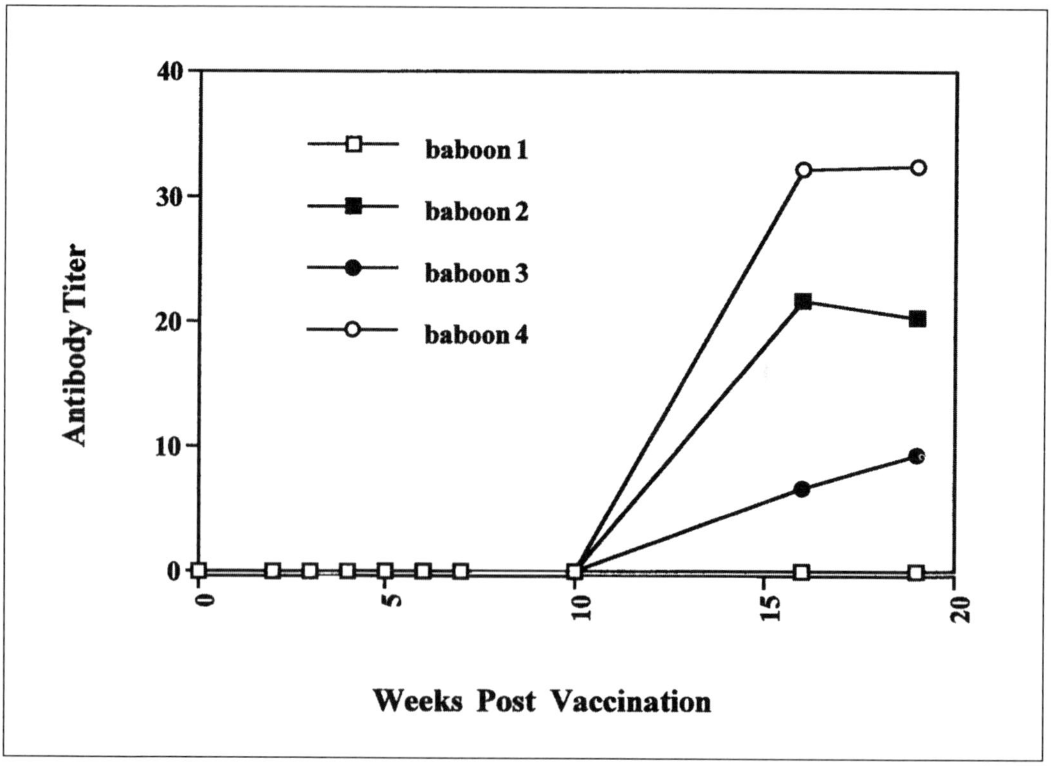

Figure 2. Intramuscular immunization of baboons with rAAV-NS3-NS4a-E2 results in the production of NS3-specific antibody. Four baboons were immunized intramuscularly with 3×10^{12} particles of rAAV-NS3-NS4a-E2. Individual baboon titers are shown.

Table 4. Prior Immunization with an rAAV Vector Precludes Subsequent Immunization

Virus Injected (wk 0/wk 7)	gB-Specific CTL (wk 11)	Reciprocal Antibody Titers		
		gB (wk 11)	LacZ (wk 11)	AAV Neutralizing (wk 7)
gB/0	yes	20 000	NA	7250
0/gB	yes	10 500	NA	NA
gB/LacZ	yes	15 000	<25	6700
LacZ/gB	no	<25	7500	6000

NA, not applicable.

and lacZ antibody. As shown in Table 4, mice that received rAAV-LacZ first demonstrated antibody to lacZ, but failed to mount an immune response to gB. Similarly, mice that received rAAV-gB first developed an immune response to gB, but failed to mount an immune response to lacZ. Serum samples were taken from all groups at week 7 prior to the second injection. In all cases in which the second injection failed to elicit an immune response, AAV neutralizing antibodies were present in the serum sample.

DISCUSSION

The immunogenicity of rAAV vectors has been controversial. Some studies have suggested that intramuscular injection of rAAV vectors does not elicit a humoral or cellular immune responses to the transgene (9,16). Other studies, however, have documented the development of antibodies to the transgene after intramuscular injection of rAAV (4,13,24,27,32). Many of these studies, however, did not address whether cellular immune responses to the transgene were primed. Our studies were the first to clearly demonstrate the priming of both humoral and cellular immune responses to the transgene after rAAV injection (24). Our results in mice and baboons have also demonstrated the potential utility of rAAV for genetic immunization. In addition to herpes simplex virus (HSV) gB, HCV NS3, and lacZ, we have demonstrated transgene-specific antibody and CTL in mice injected with an rAAV vector expressing human immunodeficiency virus type 1 $(HIV\text{-}1)_{SF2}$ p55gag (W.C. Manning and X. Paliard, unpublished data). p55gag-specific antibodies were detected in baboons immunized with this vector (W.C. Manning and X. Paliard, unpublished data). As was the case for HCV NS3, we have not been able to detect p55gag-specific CTL in the immunized baboons. One possible explanation for our inability to detect transgene-specific CTL in baboons primed by rAAV vector immunization is that the frequency of antigen-specific CTL in the peripheral blood was too low to detect. It might be possible, for example, to detect CTL in the draining lymph nodes. Alternatively, it might also be possible to increase the frequency of antigen-specific CTL by boosting the baboons. Because of the difficulty in readministering rAAV, it would be interesting to look at the ability of “naked” plasmid DNA or protein to boost the immune response after rAAV vector-mediated priming. Both Ad (22) and canary pox (5) vectors have been used to prime the immune response in such prime/boost vaccination strategies.

In addition to an immune response to the transgene, intramuscular injection of rAAV also resulted in the development of antibodies to the capsid of the vector particle. Previous studies (4,11,25,35) have shown that these antibodies are neutralizing and can block readministration of rAAV vectors. We have not looked for AAV capsid-specific CTL, but our readministration studies in MHC class I and class II deficient mice indicated that it was the host humoral immune response to the particle that was responsible for the block in readministration (25).

The ability of neutralizing antibody to block rAAV-mediated transduction raises an important issue to consider before using rAAV vectors for either genetic immunization or gene therapy. It has been reported that 50% to 80% of the adult population contains neutralizing antibodies to AAV-2 in their serum (2). It is not known, however, if the neutralizing titer present is sufficient to block transduction by vector. Since the majority of the population is infected with AAV during early childhood, by adulthood, neutralizing titers may not be high enough to block vector transduction. If AAV-2 neutralizing antibodies prove to be a problem, it may be necessary to use a different AAV serotype as a vector. To date, six AAV serotypes have been identified—AAV type 1 (AAV-1) through AAV type 6 (AAV-6) (1,29,30). One of these, AAV type 4 (AAV-4), is believed to be of monkey origin (8,29), and thus, neutralizing antibodies are rare in humans. There is some evidence that AAV-2 and AAV-4 may use different receptors (3). Therefore, before AAV-4-derived vectors can be used for vaccination, it will be important to verify the ability of AAV-4 vectors to transduce muscle as efficiently as AAV-2 vectors.

PERSPECTIVES

Many of the features that make rAAV vectors attractive for gene therapy also make them promising vectors for genetic immunization. Though our work has clearly demonstrated the immunogenicity of rAAV vectors, there are still several important issues to address before AAV vectors can be considered for genetic immunization in humans. Since rAAV vectors can probably be used only once, it will be appropriate to use rAAV vectors for serious diseases without effective vaccines, such as HIV or HCV. Because boosting with AAV may be problematic, the use of DNA, recombinant protein, or even different AAV serotypes, to boost should be examined in more detail. If effective means of boosting the immune response can be found, AAV may turn out to be most useful in priming immune responses. Finally, work needs to be done to determine whether the level of AAV-2 neutralizing antibody in the population's serum is sufficient to preclude the use of AAV-2-based vectors in humans. If this is indeed the case, then other serotypes of AAV, in particular AAV-4, may need to be used for vectors.

ACKNOWLEDGMENTS

The authors would like to acknowledge Christopher Walker, Michael Houghton, Varavani Dwarki, and Jaime Escobedo for their valuable contributions to this project.

REFERENCES

1. **Arella, M., S. Garzon, J. Bergeron, and P. Tijssen.** 1990. Physicochemical properties, production, and purification of parvoviruses, p. 11-30. *In* P. Tijssen (Ed.), Handbook of Parvoviruses, Vol. 1. CRC Press, Boca Raton.
2. **Blacklow, N.R., M.D. Hoggan, and W.P. Rowe.** 1968. Serologic evidence for human infection with adeno-associated viruses. J. Natl. Cancer Inst. *40*:319-327.
3. **Chiorini, J.A., L. Yang, Y. Liu, B. Safer, and R.M. Kotin.** 1997. Cloning of adeno-associated virus type 4 (AAV4) and generation of recombinant AAV4 particles. J. Virol. *71*:6823-6833.
4. **Clark, K.R., T.J. Sferra, and P.R. Johnson.** 1997.

Recombinant adeno-associated viral vectors mediate long-term transgene expression in muscle. Hum. Gene Ther. *8*:659-669.

5. **Clements-Mann, M.L., K. Weinhold, T.J. Matthews, B.S. Graham, G.J. Gorse, M.C. Keefer, M.J. McElrath, R.H. Hsieh et al., and the NIAID AIDS Vaccine Evaluation Group.** 1998. Immune responses to human immunodeficiency virus (HIV) type 1 induced by canarypox expressing HIV-1_{MN} gp120, HIV-1_{SF2} recombinant gp 120, or both vaccines in seronegative adults. J. Infect. Dis. *177*:1230-1246.
6. **Davis, H.L.** 1998. DNA-based immunization against hepatitis B: experience with animal models, p. 57-68. *In* H. Koprowski and D.B. Weiner (Eds.), Current Topics in Microbiology and Immunology: DNA Vaccination/Genetic Vaccination. Springer-Verlag, Berlin.
7. **Doe, B., M. Selby, S. Barnett, J. Baenziger, and C.M. Walker.** 1996. Induction of cytotoxic T lymphocytes by intramuscular immunization with plasmid DNA is facilitated by bone marrow-derived cells. Proc. Natl. Acad. Sci. USA *93*:8578-8583.
8. **Dreizin, S., T.F. Zhuravel, A.B. Tarasova, S.G. Sobolev, G. Kozlov, and V.P. Grachev.** 1981. Experimental infection of green monkeys by adeno-associated virus. Vorosy Virusol. *1*:82-89.
9. **Fisher, K.J., K. Jooss, J. Alston, Y. Yang, S.E. Haecker, K. High, R. Pathak, S.E. Raper, and J.M. Wilson.** 1997. Recombinant adeno-associated virus for muscle directed gene therapy. Nat. Med. *3*:306-312.
10. **Flannery, J.G., S. Zolotukhin, M.I. Vaquero, M.M. Lavail, N. Muzyczka, and W.W. Hauswirth.** 1997. Efficient photoreceptor-targeted gene expression *in vivo* by recombinant adeno-associated virus. Proc. Natl. Acad. Sci. USA *94*:6916-6921.
11. **Halbert, C., T.A. Standaert, C.B. Wilson, and A.D. Miller.** 1998. Successful readministration of adeno-associated virus vectors to the mouse lung requires transient immunosuppression during the initial exposure. J. Virol. *72*:9795-9805.
12. **Herzog, R.W., J.N. Hagstrom, S.-H. Kung, S.J. Tai, J.M. Wilson, K.J. Fisher, and K.A. High.** 1997. Stable gene transfer and expression of human blood coagulation factor IX after intramuscular injection of recombinant adeno-associated virus. Proc. Natl. Acad. Sci. USA *94*:5804-5809.
13. **Herzog, R.W., E.Y. Yang, L.B. Couto, N. Hagstrom, D. Elwell, P.A. Fields, M. Burton, D.A. Bellinger et al.** 1999. Long-term correction of canine hemophilia B by gene transfer of blood coagulation factor IX mediated by adeno-associated viral vector. Nat. Med. *5*:56-63.
14. **Hirt, B.** 1967. Selective extraction of polyoma DNA from infected mouse cultures. J. Mol. Biol. *26*:365-369.
15. **Imler, J.L.** 1995. Adenovirus vectors as recombinant viral vaccines. Vaccine *13*:1143-1151.
16. **Joos, K., Y. Yang, K.J. Fisher, and J.M. Wilson.** 1998. Transduction of dentritic cells by DNA viral vectors directs the immune response to transgene products in muscle fibers. J. Virol. *72*:4212-4223.
17. **Jordan, M., A. Schallhorn, and F.M. Wurm.** 1996. Transfecting mammalian cells: optimization of critical parameters affecting calcium-phosphate precipitate formation. Nucleic Acids Res. *24*:596-601.
18. **Kaplitt, M.G., P. Leone, R.J. Samulski, X. Xiao, D.W. Pfaff, K.L. O'Malley, and M.J. During.** 1994. Long-term gene expression and phenotypic correction using adeno-associated virus vectors in the mammalian brain. Nat. Genet. *8*:148-154.
19. **Kaplitt, M.G., X. Xiao, R.J. Samulski, J. Li, K. Ojamaa, I.L. Klein, H. Makimura, M.J. Kaplitt, R.K. Strumpf, and E.B. Diethrich.** 1996. Long-term gene transfer in porcine myocardium after coronary infusion of an adeno-associated virus vector. Ann. Thorac. Surg. *62*:1669-1676.
20. **Kessler, P.D., G.M. Podsakoff, X. Chen, S.A. Mcquiston, P.C. Colosi, L.A. Matelis, G.J. Kurtzman, and B.J. Byrne.** 1996. Gene delivery to skeletal muscle results in sustained expression and systemic delivery of a therapeutic protein. Proc. Natl. Acad. Sci. USA *93*:14082-14087.
21. **Koeberl, D.D., I.E. Alexander, C.L. Halbert, D.W. Russell, and A.D. Miller.** 1997. Persistent expression of human clotting factor IX from mouse liver after intravenous injection of adeno-associated virus vectors. Proc. Natl. Acad. Sci. USA *94*:1426-1431.
22. **Lubeck, M.D., R. Natuk, M. Myagkikh, N. Kalyan, K. Aldrich, F. Sinangil, S. Alipanah, S.C.S. Murthy et al.** 1997. Long-term protection of chimpanzees against high-dose HIV-1 challenge induced my immunization. Nat. Med. *3*:651-658.
23. **Lynch, C.M., P.S. Hara, J.C. Leonard, J.K. Williams, R.H. Dean, and R.L. Geary.** 1997. Adeno-associated virus vectors for vascular gene delivery. Circ. Res. *80*:497-505.
24. **Manning, W.C., X. Paliard, S. Zhou, M.P. Bland, A.Y. Lee, K.H. Hong, C.M. Walker, J.A. Escobedo, and V. Dwarki.** 1997. Genetic immunization with adeno-associated virus vectors expressing herpes simplex virus type 2 glycoproteins B and D. J. Virol. *71*:7960-7962.
25. **Manning, W.C., S. Zhou, M.P. Bland, J.A. Escobedo, and V. Dwarki.** 1998. Transient immunosuppression allows transgene expression following readministration of adeno-associated viral vectors. Hum. Gene Ther. *9*:477-485.
26. **McCown, T.J., X. Xiao, J. Li, G.R. Breese, and R.J. Samulski.** 1996. Differential and persistent expression patterns of CNS gene transfer by an adeno-associated virus (AAV) vector. Brain Res. *713*:99-107.
27. **Monahan, P.E., R.J. Samulski, T. Tazelaar, X. Xiao, T.C. Nichols, D.A. Bellinger, M.S. Read, and C.E. Walsh.** 1998. Direct intramuscular injection with recombinant AAV vectors results in sustained expression in a dog model of hemophilia. Gene Ther. *5*:40-49.
28. **Moss, B.** 1994. Replicating and host-restricted non-replicating vaccinia virus vectors for vaccine development. Dev. Biol. Stand. *82*:55-63.
29. **Parks, W.P., D.W. Boucher, J.L. Melnick, L.H. Taber, and M.D. Yow.** 1970. Seroepidemiological and ecological studies of the adeno-associated satellite viruses. Infect. Immun. *2*:716-722.
30. **Rutledge, E.A., C.L. Halbert, and D.W. Russell.** 1998. Infectious clones and vectors derived from adeno-associated virus (AAV) serotypes other than

AAV type 2. J. Virol. *72*:309-319.

31. **Snyder, R.O., C.H. Miao, G.A. Patijn, S.K. Spratt, O. Danos, D. Nagy, A.M. Gown, B. Winther et al.** 1997. Persistent and therapeutic concentrations of human factor IX in mice after hepatic gene transfer of recombinant AAV vectors. Nat. Genet. *16*:270-276.
32. **Song, S., M. Morgan, T. Ellis, A. Poirier, K. Chesnut, J. Wang, M. Brantly, N. Muzyczka et al.** 1998. Sustained secretion of human alpha-1-antitrypsin from murine muscle transduced with adeno-associated virus vectors. Proc. Natl. Acad. Sci. USA *95*:14384-14388.
33. **Srivastava, A., E. Bruno, R. Briddell, R. Cooper, C. Srivastava, K.V. Besien, and R. Hoffman.** 1990. Parvovirus B19-induced perturbation of human megakaryocytopoiesis in vitro. Blood *76*:1997-2000.
34. **Ulmer, J.B., J.J. Donnelly, S.E. Parker, G.H. Rhodes, P.L. Felgner, V.J. Dwarki, S.H. Gromkowski, R.R. Deck et al.** 1992. Heterologous protection against influenza by injection of DNA encoding a viral protein. Science *259*:1745-1749.
35. **Xiao, X., J. Li, and R.J. Samulski.** 1996. Efficient long-term gene transfer into muscle tissue of immunocompetent mice by adeno-associated virus vector. J. Virol. *70*:8098-8108.
36. **Zhou, S., S. Cooper, L.Y. Kang, L. Ruggieri, S. Heimfeld, A. Srivastava, and H.E. Broxmeyer.** 1994. Adeno-associated virus 2 mediated high efficiency gene transfer into immature and mature subsets of hematopoietic progenitor cells in human umbilical cord blood. J. Exp. Med. *179*:1867-1875.
37. **Zhou, S., J.E. Murphy, J.A. Escobedo, and V.J. Dwarki.** 1998. Adeno-associated virus-mediated delivery of erythropoietin leads to sustained elevation of hematocrit in nonhuman primates. Gene Ther. *5*:665-670.

FAMILY: Parvoviridae
GENUS: Autonomous Parvoviruses

6 Autonomous Parvovirus Vectors

Jan J. Cornelis[1], Andreas Haag[1], Christopher Kornfeld[1,2], Ginette Balboni[1], Alexandra Y. Dege[1], Bernd Neeb[1,3], Christiane Dinsart[1], and Jean Rommelaere[1]

[1]Institute of Applied Tumor Virology, German Cancer Research Center, Heidelberg, Germany; [2]Department of Virology, Institute Pasteur, Paris, France; [3]Institute of Biochemistry, University of Mainz, Mainz, Germany

BIOLOGY OF AUTONOMOUS PARVOVIRUSES

Viruses belonging to the Parvoviridae family are small (20 nm in diameter), nonenveloped particles containing a linear single-stranded DNA genome flanked by short inverted repeat sequences (3). A great number of parvoviruses have been isolated from tumor-bearing animals, tumor cell filtrates, or transplanted tumor cells (35). These observations initially suggested that the viruses play a role in the etiology of cancer. However, it was soon recognized that the special relationship between these viruses and cancer is an opportunistic one. In vitro experiments to be discussed in the following sections have confirmed this relationship in more detail for some rodent parvoviruses and suggest that many tumor cells apparently provide a physiological milieu that is especially suitable for virus replication. These parvoviruses are now grouped in the genus *Parvovirus* and are—together with members of the genus *Erythrovirus*—referred to as the autonomous parvoviruses. In contrast, the replication of other vertebrate parvoviruses was found to be tightly associated with adenovirus (Ad) infections. These viruses are known as adeno-associated viruses (AAV) and belong to the genus *Dependovirus*, as they are defective and typically require a co-infecting helper virus (such as Ad or herpes virus) for their efficient replication (3).

In this chapter, we review the development and possible applications of recombinant autonomous parvoviruses. As a way of introduction, the biology, genome organization, and replication of autonomous parvoviruses is briefly presented in so far as these features are relevant for the construction and use of recombinant derivatives. Readers should also refer to recent reviews concerning specific aspects of parvovirus research such as oncotropism and oncosuppression (65), replication (21), cell and tissue tropism (80), vectors (15,69), and in vivo biology and pathogenicity (35,87).

Life Cycle of *Parvoviruses*

Members of the genus *Parvovirus* bind to as yet uncharacterized cell surface recep-

Viral Vectors: Basic Science and Gene Therapy
Edited by A. Cid-Arregui and A. García-Carrancá

tor(s) that are known to be *N*-acetylneuraminic acid (sialyl)-containing on the cell surface (19). The receptor for the erythrovirus B19 has been identified as the erythrocyte P antigen (globoside) (87). Internalization, nuclear transport, and subsequent uncoating of the virions remain ill-defined processes. Canine parvovirus uses the endocytic pathway, a route that may be necessary for their ability to achieve a productive infection (83). Replication of the viral genome takes place in the nucleus and is restricted to cells in the S-phase of the mitotic cycle. The input single-stranded DNA is converted into a duplex replication form in an S-phase-dependent fashion (86), followed by the expression and amplification of this replicative intermediate (19,21). Furthermore, it has recently been shown that the cellular transcription factor E2F mediates a strong induction of the parvoviral promoter P4 at the G1/S transition, thereby leading to production of the viral replicative, large nonstructural (NS) protein NS1 in amounts high enough for the initiation of viral DNA amplification (23,24). S-phase cells provide all the other factors required to initiate and carry out this process (19,21). In contrast to AAV, autonomous parvovirus DNA has not so far been found to integrate into the host genome (63). A productive parvovirus infection is usually associated with cell death, for which the viral protein NS1 has been made (at least in part) responsible (81,87). The expression of NS proteins has been correlated with the induction of apoptosis in human erythrocyte progenitors, leukemic promonocytic cells, and in hepatocellular carcinoma cells (References 50,61,87, and M. Moehler et al., unpublished data). Parvovirus H-1 was also reported to induce apoptosis in rat glioblastoma cells and cerebellum (54). These observations do not rule out, however, that necrotic cell death occurs concomitantly or prevails in other cell types (60).

Genome Organization and Replication

Autonomous parvoviruses contain a 4.6 to 5.6 kb single-stranded linear DNA genome. Both ends of the genome consist of palindromic sequences that can form hairpin structures and serve as self-priming origins of DNA replication (3,21). It is worth noting that the genus *Erythrovirus* is similar to *Dependovirus* in that both ends of the genome have the same palindromic sequence, while the genus *Parvovirus* (with the exception of goose parvovirus) is characterized by the uniqueness of each telomere (3,6,19,88). Genetic analyses have shown that the genomes of autonomous parvoviruses contain two large open reading frames (ORFs) and either one (genus *Erythrovirus*) or two (genus *Parvovirus*) promoters (see Figure 1 and Reference 3). The left- and right-hand ORFs code for the NS polypeptides and viral capsid proteins (VP), respectively. The internal promoter among members of the genus *Parvovirus* directs the transcription unit encoding the structural proteins. The capsid proteins VP1 and VP2 are translated using two distinct initiation codons as a result of differential splicing of the precursor messenger RNA. In the genus *Parvovirus*, all transcripts terminate close to the right end of the genome, while in B19 (*Erythrovirus*), an additional termination site is present in the middle of the genome (87). To form the parvoviral capsids, two types of structural proteins associate in a defined stoichiometry (19). As a result of proteolytic cleavage of virion VP2, full but not empty capsids may contain copies of a third capsid protein (VP3). NS1 also plays a role in the production of the capsid proteins. Indeed, besides its viral DNA replication functions, NS1 is a transcription factor by virtue of its ability to transactivate the internal P38 promoter, which directs expression of the genes encoding the capsid proteins (11,42).

The single-stranded genome of members of the genus *Parvovirus*, of which the minute virus of mice (MVM) is the prototype, replicates through a continuous elongation mechanism resulting in the formation of double-stranded (duplex) monomeric and multimeric intermediates (19, 21). As stated above, the *Erythrovirus* and goose parvovirus genomes share structural features with AAV DNA and may be replicated according to the AAV mode, which does not require the further processing of duplex monomeric intermediates into multimers (3).

NS1 is the major viral effector of DNA replication (21). It displays helicase, DNA nicking, ATPase, and sequence-specific DNA binding activities. These functions are involved in the initiation of strand displacement synthesis from terminal and

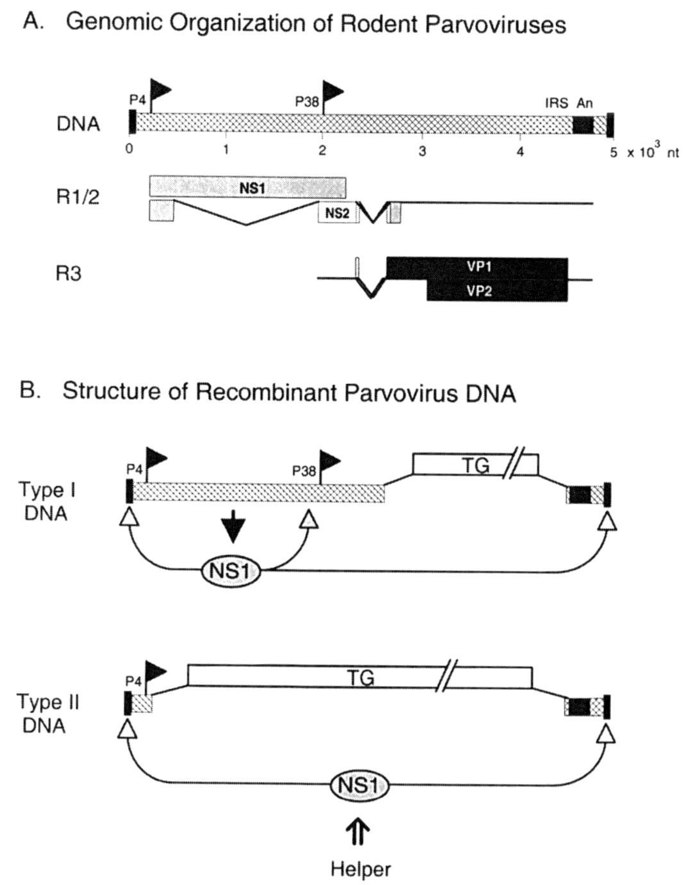

Figure 1. Schematic representations of the genomic organization of rodent autonomous parvoviruses (A) and derived recombinant viruses (B). DNA (double-hatched boxes) represent viral genomes, flanked by palindromic telomeres (filled rectangles). The positions of promoters (P4 and P38) transcription start sites (flags), polyadenylation signal (An), and IRS are indicated along the genome. (A) Transcripts (R) are depicted with their splicing pattern and open reading frames (gray [3], open [2], and black [1] boxes). (B) Transgenes (TG) are placed under control of the parvoviral promoters P38 (type I recombinants) or P4 (type II recombinants). In some of the latter constructs, a heterologous promoter is substituted for P4. NS1 proteins expressed from the recombinant (type I) or a helper plasmid (type II) interact with promoter and telomere sequences, as schematized by open arrows, leading more particularly to P38 transactivation and initiation of DNA replication.

internal origin of replication (ori) located in monomer and multimer replicative forms, respectively (3,12,19,22). NS1 becomes covalently attached to the DNA 5′ end following its site-specific, single-strand nicking activity. Cellular factors required in addition to NS1 for the initiation of MVM DNA synthesis from origins located either at the junction of the two genomes in the dimer-length replicative intermediate or at the 5′ end have been identified (12,22). The production of single-stranded progeny genomes of parvovirus MVM from duplex replicative intermediates has been shown to result from the displacement synthesis reaction and to require viral capsids (64). Thus, packaging and replication are tightly coupled processes.

As discussed above, NS1 is also endowed with a cell-killing activity. While the *Erythrovirus* B19 appears to produce only one NS polypeptide (NS1) (87), MVM and related parvoviruses express a second class of NS polypeptides that are smaller in size (around 22 kDa) and referred to as NS2. NS2 and NS1 proteins overlap in their N-terminal regions. The inactivation of NS2 severely affects several steps of the viral life cycle, particularly in cells originating from the natural host (18,40,53). Viruses that express mutated forms of NS2 are handicapped in their ability to replicate duplex DNA, accumulate viral transcripts, and produce single-strand genomes and progeny viruses. The latter defect could recently be assigned to the requirement of NS2 for progeny particle assembly (18). As a consequence, the expression of NS2 is required for parvoviruses to exert pathological effects upon their natural host (8).

Cell Differentiation and Virus Host Range

Although necessary, S-phase functions only partially fulfill the cellular requirements of parvoviruses for replication. There are tissues in the adult animal and in vitro cell cultures that contain actively dividing cells, but are unable to support virus replication. In a few cases, the restriction appears to be due to the absence of specific receptors for the virus on the cell surface. For instance, mouse B cells are defective for adsorption of parvovirus MVM (19). In most cells refractory to infection, however, the virus can enter, but becomes blocked at an intracellular step in its life cycle. Thus, the fibrotropic strain "p" of parvovirus MVM is taken up by T lymphocytes, but is unable to initiate its transcriptional program in these cells due to an early block in the virus life cycle possibly during uncoating (33,59), while both pretranscriptional and posttranscriptional barriers to gene expression have recently been identified for the *Erythrovirus* B19 in some nonpermissive cells (55,84).

The cellular factors responsible for these S-independent restrictions have not yet been characterized, but appear to be associated with the differentiation state of the host cell (19,80), as illustrated by the following observations. *(i)* Under in vitro conditions, MVM does not replicate in undifferentiated mouse teratoma stem cells and most of its differentiated derivatives, with the exception of a fibroblastlike cell (74). *(ii)* Although mouse embryonal carcinoma cells are resistant to MVM infection, certain differentiated derivatives are sensitive (49). *(iii)* Rare H-1 virus-resisting cell clones have been isolated from the susceptible human promonocytic leukemia cell line U937. These variants were found to display distinct differentiation markers and an altered c-MYC turnover compared to the original U937 cells (41,61). *(iv)* The erythrotropic human virus B19 replicates only efficiently in a narrow range of committed erythroid precursor cells (72).

Based on pathology findings in infected

organisms, the tissue tropism of mammalian parvoviruses can be expected to include—depending on the virus considered—some epithelial and mesenchymal cells, endothelial cells, and hematopoetic cells (35,80). In several instances, this tropism could be confirmed by in situ detection of virus components and replication centers (7). The tropism for endothelial cells may be an advantage for antitumor strategies (see below), since it may contribute to the inhibition of tumor neoangiogenesis and thus slow down cancer progression. It should also be stated that the above-mentioned target tissues may serve as viral reservoirs, thereby allowing the persistence of parvovirus during the chronic infections that typify these agents (35). Yet, in most instances, adult animals contain persisting parvoviruses without showing any clinical signs (35). Thus, under normal conditions, maintenance of autonomous parvoviruses often occurs in an asymptomatic fashion. Accordingly, the only acute deleterious effects of these viruses in humans are caused by the *Erythrovirus* B19 and only then under exceptional conditions in which sites of highly proliferating target cells are provided, notably in patients suffering from hemolytic anemias (aplastic crises) and in pregnant women undergoing a primary infection followed by placental virus transmission (fetal losses) (87).

The host range varies between different parvoviruses and therefore depends on viral determinants as well as cellular receptors. This can be exemplified by closely related parvovirus strains differing in their pathogenic potential and infectivity in vitro. For instance, the parvovirus MVM exists as two closely related variants designated "p" and "i", which can be distinguished by the ability of the former to replicate in fibroblasts but not in T cells, while the latter has the converse host range (80). This variation could be traced back to a few substitutions within the NS-encoding and more particularly in the VP-encoding regions (2, 14, and references therein). A region of the capsid genes (mu 54–73) also appears to determine the tissue and species tropism of canine, porcine, feline, and mink parvoviruses (77,80). It remains to be seen how the viral and cellular determinants interact to define the host ranges of a given parvoviral strain.

Oncotropic and Oncolytic Activities of Certain Autonomous Parvoviruses

The isolation of a number of autonomous parvoviruses from tumor cells originally raised questions concerning the oncogenic potential of these viruses. This possibility was not supported, however, by subsequent experimental parvovirus infections of laboratory animals, which were in fact found to be associated with a lower incidence of spontaneous and induced tumors (for reviews see References 3, 35, and 65). Moreover, the development of human tumor implants in immunocompromised mice could be delayed or suppressed by parvovirus H-1 inoculation. In a few cases, preestablished engrafted tumors even regressed after the intratumoral or systemic injection of H-1 virus (29–31).

The malignant transformation of human and murine cells was reported to modify the outcome of their interaction with rodent parvoviruses in various systems. PVR cells, mouse fibroblasts that resist MVM infection by virtue of an intracellular blockage in virus replication, were found to be susceptible to virus-induced killing after their transformation by simian virus 40 (SV40) (51). From these data, it was proposed that the oncogenic state of the host cell might create a physiological environment that favors parvovirus replication and eventually killing of the infected cells. This assumption was verified by comparing a number of transformed versus parental cells for their permissiveness to parvovirus infection (Reference 63 and F. Dupont et al., unpublished data). In short, oncogenic transformation of human and murine fibroblasts or epithelial cells by various oncogenes of cellular (c-Ha-*ras*) or viral (SV40, v-*ras*, v-*src*, v-*myc*) origin correlated with an increase in parvovirus H-1 and MVM replication, NS accumulation, and cell killing. This targeting of parvovirus gene expression to transformed cells (oncotropism) is consistent with the above-mentioned accumulation of certain parvoviruses in tumor material under in vivo conditions. The oncotropism of certain autonomous parvoviruses further argues for the oncolysis hypothesis, that is, the assumption that their oncosuppressive activity could be assigned, at least in part, to oncolysis, which is a preferential killing of neoplastic cells (51,65,78).

In all these examples, virus was taken up to a similar extent by susceptible and resistant cells, while intracellular steps of the virus life cycle—notably viral DNA replication and gene expression—were stimulated. Thus, some specific barriers to virus replication appear to have been overcome as a result of host cell oncogenic transformation. It should be stressed that interesting exceptions to this rule were detected, which may shed light on oncoprotein-mediated signalling pathways to which parvovirus replication is responsive. Oncoproteins can contribute to transcriptional stimulation from the pivotal P4 promoter, as shown for *ras*-transformation of rat fibroblasts (33 and references therein). In addition, oncogenic transformation appears to stimulate specific step(s) of viral DNA replication (39,63). Conversely, tumor suppressor proteins may inhibit parvovirus replication in normal cells. In this respect, it is noteworthy that the resistance of certain human erythroleukemia cell clones to H-1 virus-induced killing correlated with the reactivation of p53 tumor suppressor expression (76). Thus, the parvoviruses H-1 and MVM exhibit oncotropic and oncolytic properties under various cell culture conditions. Although the intratumoral accumulation of infectious particles (see above) and the destruction of virus-infected tumor cells (30) have also been demonstrated in vivo, the contribution of direct oncolysis to the oncosuppressive activity of these viruses in animals remains to be demonstrated.

PARVOVIRAL VECTORS

Potential of Autonomous Parvovirus-Based Vectors

The above-mentioned S-phase dependence of autonomous parvoviruses, together with the oncotropism of at least some of these agents, makes them attractive vectors for the gene therapy of cancer. As discussed in the previous section, natural parvoviruses, which lack additional transgenes, are already interesting in this respect since they encode the toxic NS proteins that have proven to be oncogene-responsive in both their production and cytopathogenicity. Yet, the mere fact that a number of parvoviruses have been isolated from growing tumors indicates that wild-type virus replication and expression are not always sufficient to prevent neoplastic cells from multiplying. This led to the idea of producing recombinant viruses that retain the cytotoxic NS genes and are further supplemented with an additional anticancer transgene. Through the insertion of various transgenes into the parvoviral expression system, it can be expected that they will be overexpressed in neoplastic cells to yield a more efficient oncosuppressive effect compared to the wild-type viruses. Besides possible tumor-killing applications, parvovirus vectors retaining the cytotoxic NS genes may conceivably be used to destroy other undesirable cells, in particular, cells infected with pathogenic viruses.

It should be stated that in order to accommodate most transgenes, recombinant parvoviral vectors retaining the NS genes need to be relieved of at least part of the structural VP genes, given the low tolerance of the capsid for an increase in genome size (37) and the compactness of parvovirus genomic organization (see Figure 1). These type I $NS^{+}VP^{-}$ vectors achieve a one-hit type of transduction without being able to further propagate in infected organisms. In addition, transduced cells are not expected to retain a long-term expression of the viral and foreign genes harbored by NS^{+} vectors. On the other hand, the capacity for NS protein production should allow the DNA of NS^{+} vectors to become amplified in permissive cells. Therefore, NS^{+} parvoviral vectors appear to be especially suitable for the transient, acute expression of factors that jeopardize the survival of proliferating target cells. The transience of transgene expression may be an advantage where cytotoxic or immunomodulatory products are concerned, as in the case of the above-mentioned applications. In contrast, a long-term transduction is desirable for gene replacement therapies. In spite of the cytotoxic considerations, use of parvovirus vectors to this end should not be dismissed.

To avoid the killing of transduced cells, NS-deficient parvovirus vectors can be produced (Figure 1B). However, the persistence of NS^{-} vectors is questionable because, in addition to the fact that autonomous parvoviruses do not integrate their DNA into the host genome, they are also unable to replicate their DNA in the absence of the NS proteins. As previously stated, additional drawbacks are the inability of the parvoviral capsid to accomodate enlarged genomes (limiting the size of inserts in $NS^{-}VP^{-}$ vectors to approximately 4000 nucleotides [nt]) and the probable requirement of cell proliferation for transduction to be efficient (66). Should these limitations be overcome, an intriguing alternative could involve the introduction of parvovirus elements into other vectors, thereby allowing their maintenance in transduced cells. Parvoviral components of interest for the elaboration of such chimeric vectors include promoter elements contributing to the strength and tissue (notably onco-) tropism of gene expression.

A major concern associated with the use of parvoviruses as vectors relates to the safety of such vectors. In this respect, it is worth mentioning that at least one autonomous parvovirus, the rat H-1 virus, has proven to be infectious in humans (28,79), albeit without being associated with any significant harmful effects that would limit a therapeutic application.

Production and Titration of Recombinant Vector Virus

Infectious molecular clones have been engineered from several parvoviruses. They consist of full-length viral genomes cloned into bacterial plasmids that are capable of producing infectious virus upon transfection of permissive cells. The partial or complete deletion of sequences coding for the structural proteins from these infectious molecular clones, and their replacement with foreign genes, result in type I $NS^{+}VP^{-}$ recombinant genomes (Figure 1B). Type II recombinants are characterized by the removal of all parvoviral genes $NS1^{-}$ VP^{-}. From these recombinant molecular clones, viruses can be prepared (see below) by providing the missing gene products (NS or NS + VP) *in trans* from a helper plasmid or by using a packaging cell line.

Titers of initial type I and II recombinant parvoviruses were low (5,15,44,45), until it was realized that existing full-length parvovirus DNA clones terminate on the left-hand side in an incomplete resolution sequence (37). Resolution refers to cleavage and processing of the left telomere, which normally takes place during viral DNA replication (after the telomere becomes an internal bridge linking the head-to-head arranged subunits of dimeric intermediates) but is also most likely involved in the excision of parvovirus DNA from recombinant plasmids (48,62). Therefore, parvovirus molecular clones may not be efficiently excised from the plasmid backbone. Indeed, engineering of these clones to introduce a minimal origin of replication at the left end fulfilled the *cis*-requirement for resolution and thus led to a marked increase in parvovirus DNA replication and progeny particle formation in transfected cells (37). In some systems, these second-generation molecular clones yielded virus titers that were up to 1000-fold higher than those obtained with standard infectious clones (15,27,44,68).

Several strategies have been used to produce parvovirus-based vector particles. The most conventional technique—co-transfection of the recombinant molecular clone and a helper plasmid that expresses the missing viral genes—is still the method of choice (37,56). Supplying the H-1 viral capsid proteins VP1 and VP2 by means of a recombinant Ad has also been attempted, with no improvement of the packaging efficiency as compared with the co-transfection approach (J.J. Cornelis, unpublished data). A packaging cell line expressing the MVM capsid genes has also been developed; however, its capacity for producing recombinant viruses is low (5).

Because the capsids are known to drive packaging of single-stranded genomes (20,64), and a threshold level of intracellular capsid proteins is required for the assembly of particles (85), a high expression of the VP genes is desirable in the producer cells. Helper shuttle plasmids were constructed that contain an SV40 ori aimed at enabling the generation of multiple plasmid copies (including VP genes) in SV40 T-antigen-expressing primate cells. When VP gene expression from the helper plasmid is driven by the genuine parvoviral promoter P38, the yields of VP proteins are low in cells transfected with the helper construct alone, due to the promoter's weak constitutive activity (Figure 2A, lanes 1–4). In contrast, a striking induction of VP protein production took place in cells co-transfected with the same helper plasmid and a type I (NS1-expressing) recombinant molecular clone due to the capacity of NS1 for P38 transactivation (Figure 2A, lanes 6–8). The titers of recombinant viruses produced under these conditions were not significantly different when capsid genes were driven by P38 or when driven by the constitutively strong cytomegalovirus (CMV) promoter (unpublished data). In order to determine whether replication of the helper construct contributed to the enhanced VP production, the SV40 ori in the helper plasmid was destroyed. Surprisingly, VP expression from ori$^+$ and ori$^-$ helper constructs was similar in 293T cells (Figure 2B). We assume that in cells such as 293T, which are highly transfectable, the large amount of incoming DNA saturates the capacity of the system for further amplification and overexpression of the helper construct. In cells that cannot easily be transfected, replication of the helper plasmid may enhance recombinant virus titers, as shown for the production of recombinant AAV (rAAV) (10).

Recombinant virus titers are routinely determined using an infected cell hybridization assay, a method based on the ability of the viral genome to replicate in the presence of NS1 (44,45,68). Type I recombinant viruses are viral DNA replication-proficient by virtue of their NS gene expression. Type II recombinant viruses are devoid of the NS coding region, yet their replication can be detected upon co-infection (or delayed infection) with the NS1-expressing wild-type virus. Titers determined in this way are expressed as replication units (RU). A standardized infected cell hybridization assay was developed for the MVMp and H-1 viruses, which yielded wild-type virus titers similar to those obtained from the conventional plaque assays (A. Haag, unpublished data). Whereas the titers achieved with first-generation type I recombinants were no higher than 5 $\times 10^3$ RU/mL crude cell lysate, use of the improved infectious molecular clones containing a minimal replication origin at the left end of the MVM or H-1 genome routinely allows titers of at least 10^7 RU/mL crude cell extract to be obtained (Table 1 and References 37 and 56). Titers of first-generation type II recombinants range from 10^3 to 10^5 RU/mL crude extract (15,44). A protocol for the purification and concentration of parvovirus particles has been established to allow virus from crude cell extracts to be concentrated up to 100-fold without significant losses (1).

A problem inherent to type I and to a lesser extent to type II recombinant virus production is the generation of wild-type virus through homologous recombination of regions of recombinant genome and the VP-expressing helper plasmid used. Initially, the homology between recombinant and helper construct was high, thereby resulting in the generation of large amounts (up to 80%) of wild-type virus in the recombinant virus stocks (5,27,68). Reduction of the homology between the helper plasmid and the recombinant genome to be packaged led to recombinant virus stocks that were contaminated with only 0.1% to 1% wild-type virions (13,37). To eliminate the generation of wild-type virions, type I recombinant MVM genomes were constructed in which partial VP sequences remaining in the vector were replaced with equivalent sequences from a related parvovirus (57). This strategy allowed the production of high-titer recombinant MVM stocks, without detectable wild-type viruses, using a helper plasmid expressing the MVM VPs. Another approach involved the

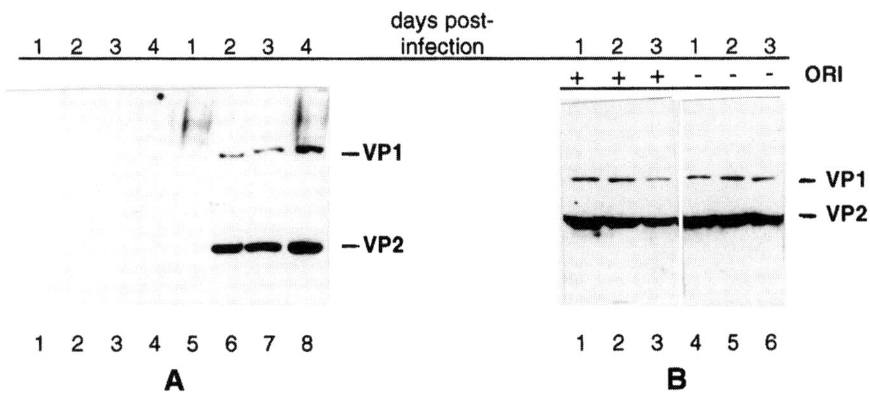

Figure 2. Production of the parvoviral capsid proteins VP1 and VP2 from various helper constructs. 293T cells were transfected with helper constructs alone or in combination with the empty type I recombinant DNA. 293T cells express the SV40 large T antigen. The production of capsid proteins was monitored from day 1 to day 3 or 4 posttransfection by immunoblotting using rabbit antisera directed against MVMp or H-1 VP proteins. The helper constructs are shuttle plasmids that express the MVMp or H-1 VP genes from the genuine parvoviral promoter P38 (pP38VP) or the constitutively strong human CMV immediate early promoter (pBK-CMVVP and pBKΔori-CMVVP), respectively. (A) Conditional expression of MVMp capsid proteins from the helper pP38VP. The P38 promoter directing VP genes is constitutively weak (lanes 1–4) but strongly transactivated by NS1 from the DNA clone pdBMVΔ800 (lanes 5–8), leading the NS1-expressing recombinant construct to turn on VP protein production from the helper. (B) Effect of helper plasmid replication on capsid protein production. The helper plasmids pBK-CMVVP and pBKΔori-CMVVP contain (+) or lack (-) a functional SV40 ori, respectively.

introduction of point mutations into the VP sequences of the helper plasmid, which strongly reduced the VP homology between the helper and recombinant genomes while conserving the amino acid sequence of the capsid proteins (36).

Requirement of *cis*-Acting Sequences for the Production of High-Titer Virus Stocks

Several limitations to the production of high-titer recombinant virus stocks have been identified. Progressive deletion of the VP gene from the P38 promoter end in recombinant genomes results in a steady reduction in recombinant virus titer (37). This is illustrated in Table 1. The recombinant hH1Hinluc is obtained with approximately 400-fold lower titers than the virus hH1HinMCP1-UTR, although both have a similar genome size. Whereas the hH1hinluc genome has almost the entire VP gene deleted and replaced by a transgene, the hH1HinMCP1-UTR virus still possesses a thousand nucleotides or so of VP coding sequence (37). Since vector DNA amplification is not impaired by VP gene deletion, it was hypothesized that a DNA sequence(s) present towards the end of the VP gene is/are able to promote higher levels of single-strand genome production and packaging (37). Packaging signals are thought to be located in the viral palindromic regions, as it is these regions that are retained by small defective viruses (31). In addition, reiterated motifs are found in a more or less degenerated form at many sites along the genomes of MVM and H-1 virus (4). These sequences were found to constitute potential NS1 binding sites, upon which it was proposed that the association of NS1 with these motifs may improve the packaging of single-stranded progeny genome into preformed capsids (11,17). The finding that sequences in the VP gene affect the packaging efficiency is in agreement with this assumption. It implies that the size of a transgene that can be inserted under the control of the P38 promoter for the production of high-titer virus stocks is limited. Yet, foreign genes up to 800 bp in length can be substituted for VP sequences without great losses in virus yield compared

Table 1. Titers of Recombinant and Wild-Type H-1 Viruses Harvested from Producer Cells Transfected with Corresponding DNA Clones

Molecular Clone	Virus Titer (RU/mL) 24 h	48 h
phH1	1.7×10^7	5.4×10^7
phH1Hinluc	7.3×10^3	3.0×10^4
phH1HinMCP1-UTR	3.1×10^6	1.0×10^7

293T cultures (3×10^6 cells) were transfected with 3 µg of wild-type (phH1) or recombinant (phH1Hinluc and phH1HinMCP1-UTR) type 1 DNA clone and 6 µg of helper plasmid providing the H-1 virus capsid proteins (37). At 24 and 48 hours posttransfection, the medium was discarded, and 1 mL of 50 mmol/L Tris-HCl, pH. 8.7, 0.5 mmol/L EDTA were added to each culture. The cultures were frozen and thawed, and virus titers (expressed as RU/mL of crude extract) were determined by infected cell hybridization assays. The size of the wild-type H-1 virus DNA is 5176 nucleotides (nt). In the hH1Hinluc clone (5098 nt long), the firefy luciferase gene is substituted for the *Hin*dIII-*Sma*I (nt 2653–4424) restriction fragment of H-1 virus DNA, leaving only 150 nt from the VP coding sequence in the vector. hH1HinMCP1-UTR (4955 nt in size) contains the 718 nt long cDNA corresponding to the human MCP-1 mRNA with its 3′ untranslated region, in replacement of the nt 2653 to 3597 fragment of the H-1 virus VP coding region, which leaves in the vector a residual sequence from the VP coding region of 976 nt long.

to wild-type virus produced from the infectious molecular clone (Table 1).

Type I recombinant viruses retain the origins of replication and express the replicative protein NS1, and are therefore competent for genome amplification. The NS and VP coding regions are not required *in cis* for DNA replication, as is apparent from the amplification proficiency of genomes lacking these sequences (37,62, 73). On the other hand, an element termed the internal replication sequence (IRS) has been mapped to the extreme right end of the MVMp genome close to the terminal palindrome and shown to drastically affect the replication of MVM minigenomes in certain cells (9,73). In keeping with this observation, we found that the replication of recombinant H-1 virus DNA was drastically reduced in 293T cells when most of the homologous putative IRS sequence was removed. This correlated with extremely poor yields of IRS-deleted recombinant H-1 virus production in the presence of helper plasmid, so that retention of the IRS element may be an essential *cis*-requirement to obtain high-titer vector stocks. Further studies are required, however, to better define the role of the IRS region and its interaction with adjacent elements during genome replication (9).

A further limitation to recombinant parvovirus production has been identified at the level of viral genome size. Small increases in genome length, of as little as 5% of the standard parvovirus genome, were found to correlate with significant reductions in the titers of recombinant H-1 viruses (37). This restriction is likely to take place at the level of packaging since the amplification of enlarged genomes is not impaired. These observations are in agreement with reports concerning the related helper-dependent virus AAV-2, which cannot efficiently accommodate a 10% oversized genome (25,52).

In summary, the efficient production of recombinant parvoviruses requires the expression of NS and capsid proteins *in trans*, the presence of intact terminal palindromes, and as much of the region upstream from the right-hand palindrome as possible *in cis*, taking care that the total length of the recombinant genome does not exceed the size of the wild-type genome. When all these conditions are taken into account, the titers of recombinant H-1 viruses approach those of the wild-type virus produced from the infectious molecular clone. Depending on the particular preparation, between 10^7 and 10^8 RU of recombinant H-1 virus can routinely be obtained from 3×10^6 293T or COS cells transfected with the second-generation molecular clones (Table 1 and Reference 37). This represents a 10 000-fold increase in virus titers over the yields initially reported for similarly designed type I MVM recombinants under the same transfection conditions (1,5,27,68).

APPLIED PROSPECTS OF PARVOVIRUS-BASED VECTORS

In Vitro and In Vivo Gene Transduction by Recombinant Parvoviruses

Type I recombinant MVM and H-1 viruses were produced to transduce various reporter genes encoding the firefly luciferase, bacterial chloramphenicol transferase, the humanized green fluorescent protein, or effector cDNAs (encoding the human interleukin-2 [IL-2], IL-4, or human monocyte chemotactic protein 1 [MCP-1]) (References 27,34,37,56,68, and F. Dupont et al., unpublished data). In these constructs, the transgene was placed under the control of the late viral promoter P38 and is therefore transcriptionally induced by the vector-encoded NS1 protein. In addition, NS1 drives vector DNA amplification in infected cells. In keeping with their retention of the *cis* (telomeres and promoters) and *trans* (NS proteins) determinants of oncotropism, these recombinants support a much higher expression of the transgene in various in vitro transformed and tumor-derived cells compared to the parental cells (27,68). This makes using parvoviral vectors for the targeting of neoplastic cells a most attractive possibility. Other features of parvovirus-mediated transgene expression are its transience (27, 68), which is likely due to the cytotoxic activity of NS proteins, and its saturation at relatively low multiplicities of infection (moi) (34), which is in agreement with the competence of type I vectors for DNA replication. It is also worth noting that type I recombinant parvoviruses are especially efficient in transgene expression. Thus, HeLa cells (34) and transformed human fibroblasts (Figure 3, upper panel) secrete approximately 4 μg of human MCP-1 per 10^6 cells infected with recombinant H-1 virus at a moi as low as 1 RU/cell. Around seven days postinfection, cytokine production has ceased (data not shown), but owing to the high stability of the transgene product, its concentration in the culture medium did not decrease. Type I parvoviral vectors are thus able to induce large amounts of cytokines in transformed or tumor cells in vitro at low input multiplicities.

HeLa cells were infected in vitro with a recombinant type I virus harboring the cDNA of human MCP-1 at the moi of 1 RU/cell and then subcutaneously implanted into nude mice. Nanogram levels of human MCP-1 were detected in tumor material with a cytokine production kinetics similar to that observed in cell culture (Figure 3, lower panel). Thus, the parvovirus vector is able to efficiently transduce cytokines to animals and cell cultures.

Type II recombinants have been produced from parvovirus LuIII (16,43, 45–47,70). Since these vectors are DNA replication-deficient, they are expected to have a reduced oncotropism and appear to be less suitable for tumor cell targeting. As exemplified below, these vectors may be engineered to become directed towards other target tissues. Type II recombinants express transgenes in a transient way in vitro (45). Gene transduction mediated by these vectors is also dependent upon the moi used (43,45).

Targeting of Parvoviruses

It may be useful for certain clinical applications to modify parvoviral vectors in order to target specific cells, or to make the virus more appropriate for infections in

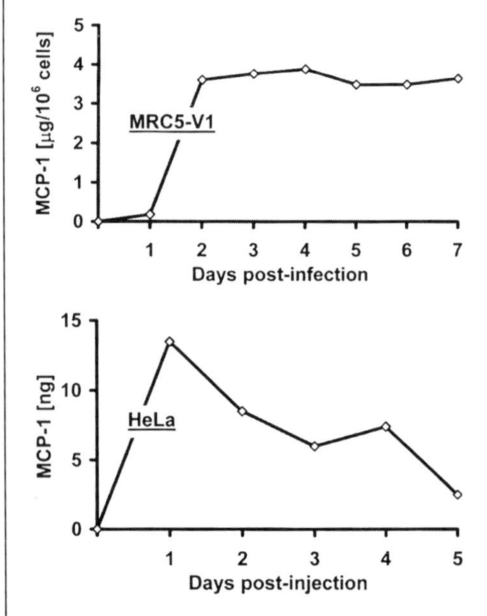

Figure 3. Production of human MCP-1 upon cell infection with recombinant H-1 virus. MRC-5V1 (SV40-transformed human fibroblasts) and HeLa (human cervical carcinoma) cell cultures were infected with the recombinant virus hH1Δ800MCP1 at a moi of 1 RU/cell. This type I recombinant carries the human MCP-1 cDNA deprived of the 3′ untranslated region (37). MRC5V1-infected cells were further cultivated in vitro (upper panel), or 2×10^6 infected HeLa cells were immediately injected into recipient nude mice to form tumor implants (lower panel). MCP-1 levels were determined in the culture medium (MRC5-V1) or in tumor extracts using a specific enzyme-linked immunosorbent assay (ELISA) (37).

humans. The following strategies can be used to this end. A modified MVMi virus has been constructed to contain the specific TATA and TAR elements from the human immunodeficiency virus type 1 (HIV-1) long terminal repeat in place of the MVMi P4 promoter's SP1 and TATAA motifs and part of the downstream 5′ untranslated region (75). Infectious parvovirus was obtained from the chimeric DNA in a cell line constitutively expressing the HIV-1 TAT transactivator protein. Further improvements of the modified virus may possibly lead to the development of a therapeutic parvoviral agent that selectively infects and kills HIV-1-infected human T cells in vivo.

Modified MVMp viruses were also produced for fundamental purposes to analyze the importance of specific cellular transcription factors in the regulation of virus gene expression. The viral P4 promoter contains a binding site for the transcription factor E2F rendering P4 activity dependent on the S-phase of the cell cycle (23). Mutations introduced into this site to prevent E2F from binding to promoter P4 have been shown to suppress the infectivity of an MVM molecular clone. E2F motif-mutant viruses can still be produced, provided that NS1 proteins are supplied *in trans.* This suggests that a P4 promoter deficient in E2F binding cannot achieve sufficiently high viral RNA levels to direct the synthesis of the NS proteins needed for the initiation of viral DNA replication in the S-phase (24). Thus, the E2F site should be retained in the P4 promoter of recombinant vectors to allow the targeting of S-phase cells. This study paves the way for other P4 modifications that may increase the value of parvoviral vectors (e.g., through enhancing the oncotropic character of these agents), a property that has been assigned, at least in part, to defined P4 elements (Reference 32 and references therein).

Another investigation was aimed at assessing whether a recombinant LuIII parvovirus could be engineered to introduce and specifically express a transgene in liver cells. To this end, a chimeric construct consisting of the minimal promoter elements of the herpes simplex thymidine kinase gene fused to the liver-specific enhancer of the alpha-1 protease inhibitor gene was placed in front of a reporter gene in place of the genuine P4 promoter of a type II virus (46). Recombinant parvoviruses constructed in this way can be produced that express higher levels in a liver tumor cell versus a reference cell line, although the chimeric promoter is extremely weak.

As discussed above, the virus host range is determined in a number of instances by the capsid proteins through their effect on early (pre- and postentry) steps of the viral life cycle. Pseudotyping is feasible between members of the rodent and feline groups of autonomous parvoviruses (43,47,70). This provides an additional level at which targeting of parvoviral vectors can be achieved. For example, virions containing chimeric B19/AAV genomes that are pseudotyped with B19 capsids are capable of erythroid cell-specific gene delivery and expression, owing to the erythropism of the B19 P6 promoter and the specific binding of the capsids to erythroid cells. Such recombinants may therefore be useful as therapeutic agents for hemoglobinopathies, in particular sickle cell anemia and β-thalassemia (71). Similarly, expression of the luciferase gene in the context of a type II LuIII vector can be directed to murine lymphocytes or fibroblasts, depending on whether the recombinant genome is packaged in the capsids of the lymphotropic or fibrotropic MVM strain, respectively (43).

Another promising development is the production of parvoviruses containing chimeric genomes to combine beneficial properties of partners not belonging to the same genus. An example of this is a chimeric genome that has been engineered from

AAV and the autonomously growing human erythropic parvovirus B19, whereby the B19 P6 promoter is substituted for the AAV-2 P5 promoter. It was reported that B19 P6 promoter makes AAV-2 competent for gene expression and autonomous replication competence in primary human hematopoietic progenitor cells (84).

Parvovirus-Based Vectors for the Gene Therapy of Cancer

Although some natural autonomous parvoviruses exert an oncosuppressive activity in animals, there is a need to reinforce this activity, since in many instances virus infection alone is not sufficient for suppression of tumorigenesis. The oncotropism of certain parvoviruses raises the possibility of using recombinant derivatives to achieve a more efficient antineoplastic effect. As is evident from the previous sections, type I recombinants have a certain potential for tumor-targeted gene therapy, as they combine parvoviral (NS) and heterologous genes of therapeutic value in a genetic setting allowing their preferential amplification and expression in oncogene-transformed cells. Parvoviruses of particular interest in this respect are MVM and the closely related H-1, KRV, and LuIII viruses, which can grow in a variety of human tumor cell lines, nonestablished tumor cell cultures, and explants (References 65 and 82, and F. Dupont et al. unpublished data). It should be stated, however, that these cellular systems are artificial and that it remains to be established to what extent human tumors are targets for parvoviral infections under natural conditions. Yet, it is noteworthy that the H-1 virus can induce viremia in humans without pathological consequences (28,79).

It is unlikely that any type of viral vector will be able to hit all cells of the target tumor, even if high virus doses are used. For this reason, cytokine-encoding transgenes may be suitable candidates for parvovirus-mediated gene therapy, as their products are secreted. The released cytokines would be expected to stimulate host defense mechanisms locally, not only against the cells expressing the transduced gene but also against noninfected neighboring tumor cells (67). In addition, the size of cytokine cDNAs is often small and readily accomodated by parvoviral vectors. Certain interleukines and chemokines are especially interesting in this respect, as they have been shown to jeopardize the in vivo survival of tumor cells engineered to release these immunomodulators (58). This antineoplastic effect is thought to be due to paracrine, cytokine-induced stimulation of host effector cells resident within the tumor, or attracted to it as a result of the expression of specific chemokines.

A first step in assessing whether recombinant parvoviruses carrying a cytokine transgene have a stronger oncosuppressive capacity than the natural viruses, human cervical tumor cells (HeLa) were infected ex vivo with an H-1 virus-based vector containing (or not) the human IL-2 cDNA. Infected cells were then implanted subcutaneously in nude mice and tested for their ability to generate tumors (34). IL-2 was chosen as a model cytokine to be transduced by the recombinant H-1 virus because of its strong antitumor activity in several mouse tumor systems (58). The H-1/IL-2 virus achieved a potential (90%) tumor suppression under infection conditions at a low input moi (1 RU/cell), in which the empty vector and wild-type virus gave no significant protection. It is also worth noting that the residual tumors, arising from cells infected at moi of 0.1 and 1 RU/cell, grow at much slower rates than the empty vector-treated tumors, which are themselves delayed relative to the uninfected controls (34). The effects of the empty vector are similar to those of the wild-type virus in this system. In conclusion, parvovirus H-1 has a significant but limited intrinsic antineoplastic effect that can be potentiated by supplementing the viral vector with a cytokine gene, at least in the present experimental model.

Recombinant parvoviruses may also be used to transduce tumors with genes coding for conditional toxins. Such suicide vectors typically express enzymes converting prodrugs to cytotoxic products. One of the most widely used suicide genes encodes the herpes simplex virus type 1 (HSV-1) thymidine kinase (HSV-tk), which converts relatively innocuous nucleoside analogues, such as acyclovir, to toxic DNA precursors. The main problem with existing chemotherapy protocols is their lack of selectivity. If parvoviruses could transfer and preferentially express drug-activating genes in tumors, then the administration of the appropriate prodrugs should lead to the selective obliteration of neoplastic tissues. Results from a reconstituted system mimicking infections with MVM/HSV-tk recombinants (38) and from in vitro and in vivo infections with recombinants (26,27) suggest that the strategy outlined above is worth exploring.

CONCLUSION

Recombinant parvovirus stocks, virtually free of contaminating wild-type virus, can be obtained with titers of up to 5×10^7 RU/mL cell lysate. Such stocks can be concentrated by equilibrium CsCl centrifugation and sucrose gradient sedimentation to yield virus suspensions containing more than 10^{10} RU/mL. Although these preparations allow an initial evaluation of recombinant parvoviruses in cell cultures and animal models, the production procedure should be further improved, in particular through the development of packaging cell lines sustaining the propagation of recombinants while minimizing the risk of wild-type virus reconstitution.

Several types of recombinant parvoviruses have proved useful in the elucidation of the virus life cycle. As well as being of fundamental interest, such modified parvoviruses are useful tools. One promising development is the alteration of the cell tropism of infectious cytolytic parvoviruses, allowing these agents to obliterate specific cells whose existence is harmful to the organism. Present evidence also supports the possibility of using parvovirus components (capsids or DNA elements) to provide other viruses with novel properties (e.g., a specific cell tropism) through the formation of chimeric viruses. Furthermore, the data reviewed in this paper indicate that therapeutic genes can be efficiently transduced and expressed in (pre)neoplastic cell cultures and animals using recombinant parvoviruses, while the corresponding normal cells are poor recipients. Success in using recombinant parvoviruses as anticancer agents will depend on how selectively they target tumor cells, which remains to be determined under in vivo conditions. However, further investigation into the possibility of exploiting the parvoviral oncotropism for the further development/refinement of tumor-targeting vectors is certainly justified, based on the experimental data collected to date.

ACKNOWLEDGMENTS

This work was supported by grants from the European Union (Biotechnology program), Ligue Nationale contre le Cancer (France), German-Israeli Foundation for Scientific Research and Development, and Bundesministerium für Bildung, Wissenschaft, Forschung und Technologie (Germany). We are grateful to Dr. J. King for his critical reading of the manuscript and acknowledge the help of A. Domeyer for part of the artwork.

REFERENCES

1. **Avalosse, B., F. Dupont, P. Spegelaere, N. Mine, and A. Burny.** 1996. Method for concentrating and purifying recombinant autonomous parvovirus vectors designed for tumour-cell-targeted gene therapy. J. Virol. Methods *62*:179-183.
2. **Ball-Goodrich, L.J. and P. Tattersall.** 1992. Two amino acid substitutions within the capsid are coordinately required for acquisition of fibrotropism by the lymphotropic strain of minute virus of mice. J. Virol. *66*:3415-3423.
3. **Berns, K.I.** 1996. *Parvoviridae*: the viruses and their replication, p. 2173-2197. *In* B.N. Fields, D.M. Knipe, P.M. Howley, D.M. Chanock, J.L. Melnick, T.P. Monath, B. Roizman, and S.E. Straus (Eds.), Fields Virology. Lipincott-Raven, Philadelphia.
4. **Bodnar, J.W.** 1988. Sequence organization in regulatory regions of DNA of minute virus of mice. Virus Genes *2*:167-182.
5. **Brandenburger, A. and S.J. Russell.** 1996. A novel packaging system for the generation of helper-free oncolytic MVM vector stocks. Gene Ther. *3*:927-931.
6. **Brown, K.E., S.W. Green, and N.S. Young.** 1995. Goose parvovirus—an autonomous member of the dependovirus genus? Virology *210*:283-291.
7. **Brownstein, D.G., A.L. Smith, R.O. Jacoby, E.A. Johnson, G. Hansen, and P. Tattersall.** 1991. Pathogenesis of infection with a virulent allotropic variant of minute virus of mice and regulation by host genome. Lab. Invest. *65*:357-364.
8. **Brownstein, D.G., A.L. Smith, E.A. Johnson, D.J. Pintel, L.K. Naeger, and P. Tattersall.** 1992. The pathogenesis of infections with minute virus of mice depends on expression of the small nonstructural protein NS2 and on the genotype of the allotropic determinants VP1 and VP2. J. Virol. *66*:3118-3124.
9. **Brunstein, J. and C.R. Astell.** 1997. Analysis of the internal replication sequence indicates that there are three elements required for efficient replication of minute virus of mice minigenomes. J. Virol. *71*: 9087-9095.
10. **Chiorini, J.A., C.M. Wendtner, E. Urcelay, B. Safer, M. Hallek, and R.M. Kotin.** 1995. High-efficiency transfer of the T cell co-stimulatory molecule B7-2 to lymphoid cells using high-titer recombinant adeno-associated virus vectors. Hum. Gene Ther. *6*:1531-1541.
11. **Christensen, J., S.F. Cotmore, and P. Tattersall.** 1995. The minute virus of mice transcriptional activator protein NS1 binds directly to the *trans*-activating region (tar) of the viral P38 promoter in a strictly ATP-dependent manner. J. Virol. *69*:5422-5430.
12. **Christensen, J., S.F. Cotmore, and P. Tattersall.** 1997. A novel cellular site-specific DNA binding protein cooperates with the viral NS1 polypeptide to initiate parvovirus DNA replication. J. Virol. *71*:1405-1416.
13. **Clement, N., T. Velu, and A. Brandenburger.** 1998. Elaboration of new recombinant parvoviruses for the production of wild-type free vector stocks. J. Gene Med. *1*:49.
14. **Colomar, M.C., B. Hirt, and P. Beard.** 1998. Two segments in the genome of the immunosuppressive minute virus of mice determine the host-cell specificity, control viral DNA replication and affect viral RNA metabolism. J. Gen. Virol. *79*:581-586.
15. **Corsini, J., B. Afenasiev, I.H. Maxwell, and J.O. Carlson.** 1996. Autonomous parvovirus and densovirus gene vectors. Adv. Virus Res. *47*:303-351.
16. **Corsini, J., J.O. Carlson, F. Maxwell, and I.H. Maxwell.** 1995. Symmetric-strand packaging of recombinant parvovirus LuIII genomes that retain only the terminal regions. J. Virol. *69*:2692-2696.
17. **Cotmore, S.F., J. Christensen, J.P.F. Nüesch, and P.Tattersall.** 1995. The NS1 polypeptide of the murine parvovirus minute virus of mice binds to DNA sequences containing the motif $[ACCA]_{2-3}$. J. Virol. *69*:1652-1660.
18. **Cotmore, S.F., A.M. D'Abramo, Jr., L.F. Carbonell, J. Bratton, and P. Tattersall.** 1997. The NS2 polypeptide of parvovirus MVM is required for capsid assembly in murine cells. Virology *231*:267-280.
19. **Cotmore, S.F. and P. Tattersall.** 1987. The autonomously replicating parvoviruses of vertebrates. Adv. Virus Res. *33*:91-174.
20. **Cotmore, S.F. and P. Tattersall.** 1989. A genome-linked copy of the NS-1 polypeptide is located on the outside of infectious parvovirus particles. J. Virol. *63*:3902-3911.
21. **Cotmore, S.F. and P. Tattersall.** 1995. DNA replication in the autonomous parvoviruses. Semin. Virol. *6*:271-281.
22. **Cotmore, S.F. and P. Tattersall.** 1998. High-mobility group proteins are essential for initiating rolling-circle-type DNA replication at a parvovirus hairpin origin. J. Virol. *72*:8477-8484.
23. **Deleu, L., F. Fuks, D. Spitkovsky, R. Hörlein, S. Faisst, and J. Rommelaere.** 1998. Opposite transcriptional effects of cyclic AMP-responsive elements in confluent or p27KIP-overexpressing cells versus serum-starved or growing cells. Mol. Cell. Biol. *18*:409-419.
24. **Deleu, L., A. Pujol, S. Faisst, and J. Rommelaere.** 1999. Activation of promoter P4 of the autonomous parvovirus minute virus of mice (MVM) at the early S phase is required for productive infection. J. Virol. *73*:3877-3885.
25. **Dong, J.-Y., P.-D. Fan, and R.A. Frizzell.** 1996. Quantitative analysis of the packaging capacity of recombinant adeno-associated virus. Hum. Gene Ther. *7*:2101-2112.
26. **Dupont, F., B. Avalosse, A. Karim, N. Mine, and A. Burny.** 1998. MVM-based vectors as oncoselective vehicles for tumor-cell-targeted gene therapy [abstr 388, p. 98A]. First Annual Meeting of American Society for Gene Therapy. May 28-31, Seatttle, WA.
27. **Dupont, F., L. Tenenbaum, L.P. Guo, P. Spegelaere, M. Zeicher, and J. Rommelaere.** 1994. Use of an autonomous parvovirus vector for selective transfer of a foreign gene into transformed human cells of different tissue origins and its expression therein. J. Virol. *68*:1397-1407.
28. **Dupressoir, T., A. Le Cesne, N. Janin, M. Spielman, T. Le Chevalier, H. Sancho-Garnier, C. Paoletti, D. Stehelin, T. Tursz, and J. Rommelaere.** 1993. Phase I

clinical study on the possible use of H-1 parvovirus in cancer treatment [abstr P2-6]. 5th Parvovirus Workshop. Crystal River, FL.

29. **Dupressoir, T., J.-M. Vanacker, J.J. Cornelis, N. Duponchel, and J. Rommelaere.** 1989. Inhibition by parvovirus H-1 of the formation of tumors in nude mice and colonies *in vitro* by transformed human mammary epithelial cells. Cancer Res. *49*:3203-3208.

30. **Faisst, S., D. Guittard, A. Benner, J.Y. Cesbron, J.R. Schlehofer, J. Rommelaere, and T. Dupressoir.** 1998. Dose-dependent regression of HeLa cell-derived tumours in SCID mice after parvovirus H-1 infection. Int. J. Cancer *75*:584-589.

31. **Faust, E. and D.C. Ward.** 1979. Incomplete genomes of the parvovirus minute virus of mice: conservation of genome termini, including the origin of DNA replication. J. Virol. *32*:276-292.

32. **Fuks, F., L. Deleu, C. Dinsart, J. Rommelaere, and S. Faisst.** 1996. *Ras* oncogene-dependent activation of the P4 promoter of minute virus of mice through a proximal P4 element interacting with the Ets family of transcription factors. J. Virol. *70*:1331-1339.

33. **Gardiner, E.M. and P. Tattersall.** 1988. Evidence that the developmentally regulated control of gene expression by a parvoviral allotropic determinant is particle mediated. J. Virol. *62*:1713-1722.

34. **Haag, A., P. Menten, J. Van Damme, C. Dinsart, J. Rommelaere, and J.J. Cornelis.** 2000. Highly efficient transduction and expression of cytokine genes in human tumor cells by means of autonomous parvovirus vectors; generation of antitumor responses in recipient mice. Hum. Gene Ther. *11*:597-609.

35. **Jacoby, R.O., L.J. Ball-Goodrich, D.G. Besselsen, M.D. McKisic, L.K. Riley, and A.L. Smith.** 1996. Rodent parvovirus infections. Lab. Animal Sci. *46*:370-380.

36. **Karim, A., B. Avalosse, A. Maron, N. Mine, A. Burny, and F. Dupont.** 1997. Development of an advanced MVMp-based vector system designed specifically to prevent the generation of replication competent helper virus via homologous recombination. 7th International Parvovirus Workshop. Heidelberg, Germany. Sept 3-7, Abstract 5.3.

37. **Kestler, J., B. Neeb, S. Struyf, J. van Damme, S.F. Cotmore, A. D'Abramo, P. Tattersall, J. Rommelaere, C. Dinsart, and J.J. Cornelis.** 1999. *Cis*-requirements for the efficient production of recombinant DNA vectors based on autonomous parvoviruses. Hum. Gene Ther. *10*:1619-1632.

38. **Koering, C.E., T. Dupressoir, S. Plaza, D. Stehelin, and J. Rommelaere.** 1994. Induced expression of the conditionally cytotoxic herper simplex virus *thymidine kinase* gene by means of a parvoviral regulatory circuit. Hum. Gene Ther. *5*:457-463.

39. **Kuntz-Simon, G., T. Bashir, J. Rommelaere, and K. Willwand.** 1999. Neoplastic transformation-associated stimulation of the *in vitro* resolution of concatemer junction fragments from minute virus of mice DNA. J. Virol. *73*:2552-2558.

40. **Li, X.L. and S.L. Rhode.** 1990. Nonstructural protein NS2 of parvovirus H-1 is required for efficient viral protein synthesis and virus production in rat cells *in vivo* and *in vitro*. Virology *184*:117-130.

41. **Lopez-Guerrero, J.-A., B. Rayet, J. Rommelaere, and C. Dinsart.** 1997. Constitutive activation of U937 promonocytic cell clones selected for their resistance to parvovirus H-1 infection. Blood *89*:1642-1653.

42. **Lorson, C., J. Pearson, L. Burger, and D.J. Pintel.** 1989. An SP1-binding site and TATA element are sufficient to support full transactivation by proximally bound NS1 protein of minute virus of mice. Virology *240*:326-337.

43. **Maxwell, I.H., C.J. Long, J.O. Carlson, S.L. Rhode III, and F. Maxwell.** 1993. Encapsidation of a recombinant LuIII parvovirus genome by H1 virus and the fibrotropic or lymphotropic strains of minute virus of mice. J. Gen. Virol. *74*:1175-1179.

44. **Maxwell, I.H. and F. Maxwell.** 1994. A modified plaque assay and infected cell hybridization assay for wild-type and recombinant LuIII autonomous parvovirus. BioTechniques *16*:877-881.

45. **Maxwell, I.H., F. Maxwell, S.L. Rhode III, J. Corsini, and J.O. Carlson.** 1993. Recombinant LuIII autonomous parvovirus as a transient transducing vector for human cells. Hum. Gene Ther. *4*:441-450.

46. **Maxwell, I.H., A.L. Spitzer, C.J. Long, and F. Maxwell.** 1996. Autonomous parvovirus transduction of a gene under control of tissue-specific or inducible promoters. Gene Ther. *3*:28-36.

47. **Maxwell, I.H., A.L. Spitzer, F. Maxwell, and D.J. Pintel.** 1995. The capsid determinant of fibrotropism of the MVMp strain of minute virus of mice functions via VP2 and not VP1. J. Virol. *69*:5829-5832.

48. **Merchlinsky, M.J., P.J. Tattersall, J.J. Leary, S.F. Cotmore, E.M. Gardiner, and D.C. Ward.** 1983. Construction of an infectious molecular clone of the autonomous parvovirus minute virus of mice. J. Virol. *47*:227-232.

49. **Miller, R.A., D.C. Ward, and F.H. Ruddle.** 1977. Embryonal carcinoma cells (and their somatic cell hybrids) are resistant to infection by the murine parvovirus MVM, which does infect other teratocarcinoma-derived cell lines. J. Cell. Physiol. *91*:393-402.

50. **Morey, A.L., D.J. Ferguson, and K.A. Fleming.** 1993. Ultrastructural features of fetal erythroid precursors infected with parvovirus B19 in vitro: evidence of cell death by apoptosis. J. Pathol. *169*:213-220.

51. **Mousset, S. and J. Rommelaere.** 1982. Minute virus of mice inhibits cell transformation by simian virus 40. Nature *300*:537-539.

52. **Muzyczka, N.** 1992. Use of adeno-associated virus as a general transduction vector for mammalian cells. Curr. Topics Microbiol. Immunol. *158*:97-129.

53. **Naeger, L.K., J. Cater, and D.J. Pintel.** 1990. The small nonstructural protein (NS2) of the parvovirus minute virus of mice is required for efficient DNA replication and infectious virus production in a cell-type specific manner. J. Virol. *64*:6166-6175.

54. **Ohshima, T., M. Iwama, Y. Ueno, F. Sugiyama, T. Nakajima, A. Fukamizu, and K.-I. Yagami.** 1998. Induction of apoptosis *in vitro* and *in vivo* by H-1 parvovirus infection. J. Gen. Virol. *79*:3067-3071.

55. **Pallier, C., A. Greco, J. Le Junter, A. Saib, I. Vassias, and F. Morinet.** 1997. The 3′ untranslated region of the B19 parvovirus capsid protein mRNAs inhibit its own mRNA translation in nonpermissive cells. J.

Virol. *71*:9482-9489.

56. **Palmer, G.A. and P. Tattersall.** 1997. Construction of a MVM-based gene therapy vector that suppresses replication-competent virus production. 7th International Parvovirus Workshop. Heidelberg, Germany, Sept 3-7, Abstract P79.

57. **Palmer, G.A. and P. Tattersall.** 1998. MVM-based gene therapy vectors that transduce genes into human and murine T cells. Am. Soc. Virol. 17th Ann. Meeting. Vancouver, Canada. July 11-15, Abstract W34-6, p119.

58. **Parmiani, G., M.P. Colombo, C. Melani, and F. Arienti.** 1997. Cytokine gene transduction in the immunotherapy of cancer. Adv. Pharmacol. *40*:259-307.

59. **Previsiani, N., S. Fontana, B. Hirt, and P. Beard.** 1997. Growth of the parvovirus minute virus of mice MVMp3 in EL4 lymphocytes is restricted after cell entry and before viral DNA amplification: cell-specific differences in virus uncoating *in vitro*. J. Virol. *71*:7769-7780.

60. **Ran, Z-H., B. Rayet, J. Rommelaere, and S. Faisst.** 1999. Parvovirus H1-induced cell death: influence of intracellular NAD consumption on the regulation of necrosis and apoptosis. Virus Res. *65*:161-174.

61. **Rayet, B., J.-A. Lopez-Guerrero, J. Rommelaere, and C. Dinsart.** 1998. Induction of programmed cell death by parvovirus H-1 in U937 cells: connection with the tumor necrosis factor alpha signalling pathway. J. Virol. *72*:8893-8903.

62. **Rhode, S.L.** 1989. Both excision and replication of cloned autonomous parvovirus DNA require the NS1 (rep) protein. J. Virol. *63*:4249-4256.

63. **Richards, R.G. and R.W. Armentrout.** 1979. Early events in parvovirus replication: lack of integration by minute virus of mice into host cell DNA. J. Virol. *30*:397-399.

64. **Richards, R.G., P. Linser, and R.W. Armentrout.** 1977. Kinetics of assembly of a parvovirus, minute virus of mice, in synchronized rat brain cells. J. Virol. *22*:778-793.

65. **Rommelaere, J. and J.J. Cornelis.** 1991. Antineoplastic activity of parvoviruses. J. Virol. Methods *33*:233-250.

66. **Russell, D.W., A.D. Miller, and I.E. Alexander.** 1994. Adeno-associated virus vectors preferentially transduce cells in S phase. Proc. Natl. Acad. Sci. USA *91*:8915-8919.

67. **Russell, S.J.** 1990. Lymphokine gene therapy for cancer. Immunol. Today *11*:196-200.

68. **Russell, S.J., A. Brandenburger, C.L. Flemming, M.K.L. Collins, and J. Rommelaere.** 1992. Transformation-dependent expression of interleukine genes delivered by a recombinant parvovirus. J. Virol. *66*:2821-2828.

69. **Shaughnessy, E., D. Lu, S. Chatterjee, and K.K. Wong.** 1996. Parvoviral vectors for the gene therapy of cancer. Sem. Oncol. *23*:159-171.

70. **Spitzer, A.L., F. Maxwell, J. Corsini, and I.H. Maxwell.** 1996. Species specificity for transduction of cultured cells by recombinant LuIII rodent parvovirus genome encapsidated by canin parvovirus or feline panleukopenia virus. J. Gen. Virol. *77*:1787-1792.

71. **Srivastava, A., S. Ponnazhagan, X.-S. Wang, K.Y. Qing, B. Khuntirat, D.M. Kube, C. Mah, and M.C. Yoder.** 1997. AAV and parvovirus B19 vectors for human gene therapy. 7th International Parvovirus Workshop. Heidelberg, Germany. Sept 3-7, Abstract 3.5.

72. **Takahashi, T., K. Ozawa, K. Takahashi, S. Asamo, and F. Takabu.** 1990. Susceptibility of human erythopoietic cells to B19 parvovirus in vitro increases with differentiation. Blood *75*:603-610.

73. **Tam, P. and C.R. Astell.** 1993. Replication of minute virus of mice minichromosomes: novel replication elements required for MVM DNA replication. Virology *193*:812-824.

74. **Tattersall, P.** 1978. Susceptibility to minute virus of mice as a function of host-cell differentiation, p. 131-149. *In* D.C. Ward and P. Tattersall (Eds.), Replication of Mammalian Parvoviruses. CSH Laboratory Press, Cold Spring Harbor, NY.

75. **Tattersall, P., A. D'Abramo, Jr., and S.F. Cotmore.** 1998. HIV TAT transactivator dependent, replication-competent chimeric MVM vectors. Am. Soc. Virol. 17th Ann. Meeting. Vancouver, Canada. July 11-15, Abstract W34-7, p. 120.

76. **Telerman, A., M. Tuynder, T. Dupressoir, B. Robaye, F. Sigaux, E. Shaulian, M. Oren, J. Rommelaere, and R. Amson.** 1993. A model of tumor suppression using H-1 parvovirus. Proc. Natl. Acad. Sci. USA *90*:8702-8706.

77. **Tijssen, P., J. Bergeron, R. Dubuc, and B. Hébert.** 1995. Minor genetic changes among porcine parvovirus group are responsible for major distinguishing biological properties. Sem. Virol. *6*:319-328.

78. **Toolan, H. and N. Ledinko.** 1965. Growth and cytopathogenicity of H-viruses in human and simian cell cultures. Nature *208*:812-813.

79. **Toolan, H.W., E.L. Saunders, C.M. Southam, A.E. Moore, and A.G. Levin.** 1995. H-1 virus viremia in the human. Proc. Soc. Exp. Biol. Med. *119*:711-715.

80. **Truyen, U. and C.R. Parrish.** 1995. The evolution and control of parvovirus host ranges. Sem. Virol. *6*:311-317.

81. **Vanacker, J.-M. and J. Rommelaere.** 1995. Non-structural proteins of autonomous parvoviruses: from cellular effects to molecular mechanisms. Semin. Virol. *6*:291-297.

82. **Van Pachterbeke, C., M. Tuynder, J.P. Cosyn, L. Lespagnard, D. Larsimont, and J. Rommelaere.** 1993. Parvovirus H-1 inhibits growth of short-term tumor-derived but not normal mammary tissue cultures. Int. J. Cancer *55*:672-677.

83. **Vihinen-Ranta, M., A. Kalela, P. Mäkinen, L. Kakkola, V. Marjomäki, and M. Vuento.** 1996. Intracellular route of canine parvovirus entry. J. Virol. *72*:802-806.

84. **Wang, X.-S., M.C. Yoder, S.Z. Zhou, and A. Srivastava.** 1995. Parvovirus B19 promoter confers autonomous replication competence and erythroid specificity to adeno-associated virus 2 in primary human hematopoietic progenitor cells. Proc. Natl. Acad. Sci. USA *92*:12416-12420.

85. **Wistuba, A., A. Kern, S. Weger, D. Grimm, and J. Kleinschmidt.** 1997. Subcellular compartmentaliza-

tion of adeno-associated virus type 2 assembly. J. Virol. *71*:1341-1352.

86. **Wolter, S., R. Richards, and R.W. Armentrout.** 1980. Cell cycle-dependent replication of the DNA of minute virus of mice, a parvovirus. Biochim. Biophys. Acta *607*:420-431.

87. **Young, N.S.** 1996. Parvoviruses, p. 2199-2220. *In* B.N. Fields, D.M. Knipe, P.M. Howley, et al. (Eds.), Fields Virology. Lipincott-Raven, Philadelphia.

88. **Zadori, Z., R. Stefancsik, T. Rauch, and J. Kisary.** 1995. Analysis of the complete nucleotide sequences of goose and muscovy duck parvoviruses indicates common ancestral origin with adeno-associated virus 2. Virology *212*:562-573.

Family: Adenoviridae
Genus: Mastadenovirus (Mammalian Adenoviruses)

7 Molecular Biology of Adenovirus Gene Therapy Vectors

Michael J. Imperiale
Department of Microbiology and Immunology, Comprehensive Cancer Center and Center for Gene Therapy, University of Michigan Medical School, Ann Arbor, MI, USA

INTRODUCTION

Over the past decade, adenovirus (Ad) has emerged as an attractive vector for gene therapy. The ability of recombinant viruses to be grown to high titers, to infect most cell types, and to accommodate reasonably large transgenes, combined with the extensive knowledge of viral genetics, have made Ad the vector of choice for a large number of preclinical studies and clinical trials. In this chapter, the replication cycle of the virus will be reviewed, and the features of the virus that affect its use as a gene therapy vector will be discussed. The biology of Ad has been presented recently in an excellent volume (17) to which the reader is referred for more detail.

VIRAL STRUCTURE

Ads are nonenveloped viruses that replicate in the nucleus of the cell. The viral capsid is an icosahedral structure, the major components of which are the hexon, penton, and fiber proteins. These virions contain a linear, double-stranded genome that is approximately 36 kb long. The 5′ end of each strand is covalently attached to a protein called terminal protein. Similarly, an inverted terminal repeat (ITR) is found at both ends of the genome (Figure 1). These two ITRs serve as the origins of replication. Genes are encoded on both strands of the DNA, and the virus has many overlapping transcription units. This property can sometimes be a hindrance in terms of constructing vectors containing deletions in specific genes. Almost all of the viral transcription units encode multiple mRNA molecules that are produced by alternative RNA processing of the primary transcript.

BINDING AND ENTRY

It is only recently that the details of the molecular interactions that are involved in Ad attachment to and entry into the host cell have begun to be understood. The initial attachment to the surface of the cell is mediated by the fiber protein, which is a trimeric structure extruding from the ver-

Viral Vectors: Basic Science and Gene Therapy
Edited by A. Cid-Arregui and A. García-Carrancá

tices of the icosahedron (16,49). Specifically, the knob domain of the fiber protein, which is located at the end of the protein farthest from the surface of the capsid, binds to a receptor called the coxsackievirus and adenovirus receptor (CAR) (5,66). After this attachment step, there is a subsequent interaction between the penton base and α_V integrins on the cell surface (74), leading to internalization of the virus through receptor-mediated endocytosis (9,68). The virus then escapes the endosome and translocates to the nuclear pore complex, where the viral DNA is released into the nucleus and the transcriptional program begins (28).

EARLY GENE EXPRESSION AND DNA REPLICATION

The first viral transcription unit to be expressed is E1A (Figure 1). This region encodes a number of transcripts, the most critical of which are referred to as the 13S and 12S mRNAs, based on their sedimentation profiles. The proteins encoded by these mRNAs transactivate transcription from the other viral early genes and also stimulate cells that are not cycling to begin entry into S phase (reviewed in Reference 35). Thus, in the absence of E1A proteins, replication of the virus in most cell types is severely impaired. The ability of the E1A proteins to immortalize primary cells in culture is a by-product of their ability to activate early gene expression. This is because one of the cellular transcription factors that E1A activates, E2F, is required not only for transcription of the viral E2 promoter but also stimulates a variety of cellular genes that are involved in S phase progression and DNA synthesis (19).

The other early viral transcription units include E1B, E2, E3, and E4. As alluded to above, each of these units encodes multiple mRNAs and therefore multiple proteins. The E1B proteins are necessary for transport of viral mRNAs from the nucleus to the cytoplasm (4,51,78) and play an important role in inhibiting apoptosis (50,53,62,70). The latter role is thought to be important in counteracting apoptotic induction by E1A and involves both p53-dependent and p53-independent pathways. Together, the E1A and E1B genes, referred to collectively as the E1 region, are the transforming genes of the virus, acting to stimulate cell growth and block apoptotic death. This is one reason why the E1

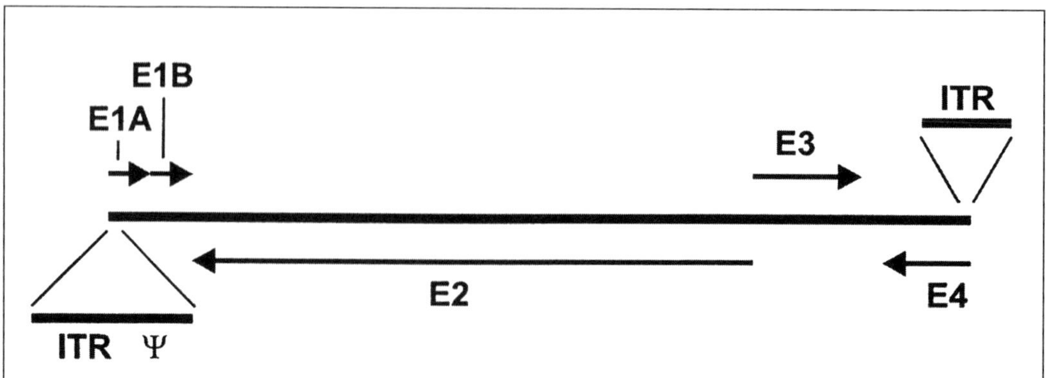

Figure 1. Map of the adenovirus genome and early transcription units. The dark line at the middle of the figure represents the viral genome. The positions of the left and right ITRs, the packaging sequences (ψ), and the early regions (E1A, E1B, E2, E3, and E4) are shown. Arrows indicate the direction of transcription.

region is commonly deleted from Ad vectors. The other reasons are to impair viral replication and to make room for the transgene. The ability to delete the E1 region is made possible by the existence of cell lines that provide these functions *in trans*. The classic cell line for this purpose is the 293 cell line, a human embryonic kidney-derived line that has been transformed by Ad (1). More recently, other cell lines such as 911 and PER.C6 have been used for this purpose (21,22). Ads that do not contain the E1 region grow just as well in these cells as do wild-type viruses.

The viral E2 region encodes the DNA replication genes of the virus: DNA polymerase, preterminal protein, and the 72 kDa single-stranded DNA binding protein. Even though Ad replicates in the nucleus, it requires its own enzymatic machinery due to its chromosomal structure. Because it does not have telomeres, the integrity of the DNA ends is ensured by the use of a viral protein, preterminal protein, as the primer for the viral polymerase. DNA replication is mediated by the polymerase, requires the 72-kDa protein, and occurs via a strand displacement mechanism. Until recently, it has been difficult to construct viruses deleted for these genes, as the toxicity of these E2 proteins has made the derivation of complementing cell lines tricky. Progress in this area is described in the chapter entitled Generation of Multiply Deleted Adenovirus Vectors that Are Helper Virus Independent.

The E3 transcription unit encodes a number of proteins, many of which allow the virus to blunt the host immune response. These effects include down-modulation of major histocompatibility complex (MHC) Class I molecules on the surface of the cell (2,8), resistance to killing of the infected cell mediated by the cytokine tumor necrosis factor (TNF)α or by Fas-ligand interactions (26,63), and other functions. Many of the early vectors that were used for gene therapy applications contained deletions in the E3 region, mainly for a practical reason. One of the methods used to construct recombinant viruses is the overlap recombination method, in which viral DNA restricted in the E1 region is co-transfected into cells along with a shuttle plasmid containing the left ITR, packaging signal, transgene of interest, and some residual E1 sequences (11). Recombination between the two DNA molecules places both ITRs on the same molecule, allowing for replication. To optimize the yield of recombinants, investigators used the Ad type 5 (Ad5) mutant *dl*309 or its derivatives, which contain two unique restriction sites, *Cla*I and *Xba*I, in the E1 region due to the partial deletion of E3. Thus, the likelihood of regeneration of the starting wild-type virus, which could arise due to either incomplete restriction digestion or religation of viral DNA in the cell, is minimized. Since the level of E3 protein expression in E1-deleted viruses is likely to be very low due to the absence of transactivation, the presence or absence of the E3 region in these vectors may be insignificant. Indeed, it has been reported that even artificial enhancement of E3 expression does not result in prolonged gene expression (56) although overexpression of just the gp19k product does (7,38). Therefore, while its functional role in Ad gene therapy vectors remains unclear, deletion of E3, which is dispensable for viral growth in culture, does afford the investigator more room for the transgene.

The E4 proteins also have numerous activities, including the induction and inhibition of apoptosis, cooperation with E1B to effect viral mRNA nucleocytoplasmic transport, and regulation of viral pre-mRNA splicing (see Reference 39 for recent review). A number of investigators have recently begun deleting various amounts of the E4 region from their vectors, with the hope that this will diminish replication and thereby permit longer persistence of the viral chromosome and, hence, transgene expression in the cell by reducing the immune response against viral-encoded proteins. The results of initial experiments along these lines have been mixed, however, and indicate that the exact effect of deleting E4 may depend upon the target tissue (3,6,12,15,24,42,69).

LATE GENE EXPRESSION AND VIRUS ASSEMBLY

The Ad late genes are, for the most part, expressed from a single promoter, the major late promoter (MLP) (Figure 2). The major late transcription unit encodes approximately 15 to 20 different mRNAs, all of which are derived from a single pre-mRNA by alternative splicing and 3′ end processing (see Reference 32 for review). By and large, these transcripts encode structural proteins of the virus and other proteins that are involved in virion assembly. After the onset of DNA replication, transcription from the MLP is induced to very high levels, thereby ensuring the production of adequate amounts of these structural proteins for the assembly of progeny virions (48,57). Assembly occurs in the nucleus in a multistep process. The DNA is packaged into preformed capsids in what appears to be a unidirectional process, beginning with packaging signals found just internal to the left ITR (27,29). Core proteins are also added, and the final maturation step is carried out by the viral protease. Progeny virions are released from the cell, which is killed by a combination of the toxicity of some viral proteins, apoptotic mechanisms activated by E4ORF4 (44), and the E3-encoded adenovirus death protein (ADP) (64,65).

ISSUES FACING THE GENE THERAPIST

For any gene therapy vector to attain utility in patients, two major issues must be addressed. First, the vector must be able to deliver genes to the intended target with limited toxicity to other tissues and organs. Second, the transgenes must be expressed for an appropriate length of time to achieve a therapeutic effect. In the case of Ad, much progress has been made along both these lines, but there are still some hurdles that must be crossed. Table 1 lists Ad vectors with various replication defects that have been developed to date; their biology is described in the following paragraphs.

Most of the trials to date using Ad as a vector have been aimed at controlling local-regional diseases, such as cystic fibro-

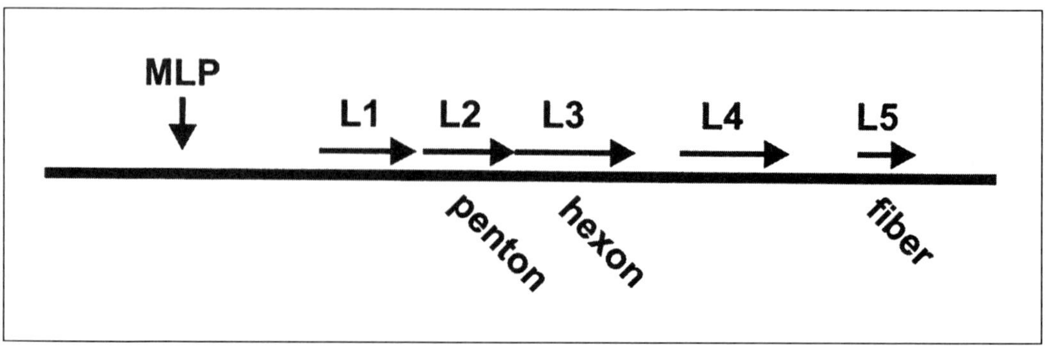

Figure 2. Map of the major late transcription unit. The MLP and the five late regions (L1–L5) are shown above the viral genome. The major capsid proteins are shown below their respective genes. Arrows indicate the direction of transcription.

Table 1. Genotypes of Common Adenovirus Vectors

Mutated Gene(s)	Properties
ΔE1 (E1A + E1B)	Replication defective in most cell types; so-called "first generation" vector.
ΔE1B	Replication defective in some cell types; molecular basis not understood.
ΔE1/ts E2A	Slightly more impaired than E1-deleted virus due to temperature-sensitive mutation in single-stranded DNA binding protein.
ΔE1/ΔE2B	More severely impaired than E1-deleted virus due to deletion of polymerase and preterminal protein genes.
ΔE1/ΔE4	Various deletions in E4 impair viral gene expression and replication.

sis or primary tumors. In these protocols, the virus is delivered directly to the affected tissue, and the spread of virus to surrounding nondiseased tissue is limited. This is an important issue, because it has been shown that systemic delivery of Ad can cause severe hepatotoxicity (80) due to targeting of a great majority of the inoculum to the liver (31,34,40,58). Both CAR and α_v integrins are expressed on a wide variety of cell types and so, at least from the point of view of binding and entry, there is nothing to prohibit the virus from infecting nontargeted cells. An active area of current investigation, then, is the retargeting of Ad to specific tissues. Such retargeting also has applications in cases in which the target cell does not express significant amounts of the natural viral receptors.

Two approaches have been taken towards redirecting Ad to different tissues. The first is a biochemical approach in which the normal interaction between the fiber and the CAR molecule is blocked and a different receptor specificity is imparted on the virus. This has been accomplished using Fab-ligand complexes, in which the Fab fragment of a monoclonal, anti-fiber antibody that neutralizes infection is covalently linked to a ligand for which there is a receptor on the cell to be infected. This molecule is mixed with the viral stock and applied to cells (18,54). In similar experiments, investigators have produced bispecific antibodies in which the neutralizing moiety is linked to one that recognizes the receptor of interest (72,73,76). The second approach is a genetic one, in which the fiber protein is mutated to alter its receptor specificity by changing the knob domain so that it no longer binds CAR efficiently, but instead binds an alternative receptor (23,36,47,60,75,77). Changes in the Arg-Gly-Asp (RGD) motif of the penton base, which contacts the integrin molecule, can also alter the specificity of infection (71).

Another issue relating to toxicity is that of viral replication and potential spread to either normal tissues and organs in the patient or to other persons such as family, friends, and health care workers. There are three causes for concern in this regard. First, as mentioned above, deletion of the E1 region does not fully disrupt the ability of the virus to replicate in all cells. This is because the other viral promoters are driven by cellular transcription factors whose activity is increased by the action of E1A. For example, in tumor cells containing mutations in the retinoblastoma susceptibility gene pathway, levels of E2F are likely to be elevated, allowing for E1A-independent E2 expression and subsequent viral DNA replication and late gene expression. Because it is known that E1A-independent gene expression is more robust at higher multiplicities of infection (25,33), this could be especially relevant at the very high doses used in most gene therapy protocols. Nevertheless, replication of E1-deleted viruses can be avoided by using recombinant vectors that cannot express one or more of the E2 functions. Viruses that contain E4 deletions are also severely impaired in their replication ability and demonstrate less toxicity. Second, one must assay recombinant viral stocks for contaminating replication-competent adenovirus (RCA) (41). RCA arises because of recombination between the E1 region sequences in the complementing cell line and the recombinant virus, giving rise to viral progeny that contain an intact E1 region and are therefore phenotypically wild-type. Attempts are underway to produce new helper cell lines in which the overlap between the E1 sequences in the cell and those commonly present on recombinant chromosomes are minimized, thereby reducing the likelihood of homologous recombination (21). In addition, the use of vectors that are deleted in more than one viral gene, such as E1/E4 or E1/E2B, greatly reduces the chances of recombination to form RCA. Finally, there is always the possibility in the patient setting that superinfection can occur with environmental strains of Ad, which would then act as helper viruses to allow replication of the recombinants. This last possibility is commonly dealt with by excluding patients with apparent respiratory infections from receiving Ad gene therapy.

Obtaining persistence of transgene expression may be a more difficult task. It has been known since the earliest attempts to use Ad as a vector that even E1-deleted viruses can stimulate a robust immune response (14,79-81). This is probably due in part to immunogenicity of the infecting virions and in part to leaky replication for the reasons discussed above. For example, inactivated viruses still induce an immune response (46), while building an E2 mutation into the vector dampens the response (20) and, as discussed above, E4 mutations also lessen the synthesis of immunogenic proteins. In addition, the transgene product itself can be immunogenic. Thus, clearance of infected cells can result in the loss of therapeutic gene expression. While this may not be a problem in the treatment of diseases in which one wishes to kill the cell, such as cancer, it is a problem in other instances when long-term expression is desired. The immune response could ultimately prevent repeated dosing of the virus if a high titer of neutralizing antibodies is induced, although anecdotal evidence indicates that repeated administrations are possible in some applications even in the presence of high antiviral titers. Investigators are currently examining ways of blunting the immune response using immunosuppressive drugs, or avoiding the possibility of neutralization by developing recombinant vectors with backbones derived from other Ad serotypes that do not cross react with Ad5 (43,45). Killing of infected cells by the immune system, however, is not the only contributor to loss of gene expression. In some instances, the infected cell is still present, but the transgene is not transcribed (3). There is some evidence that cytokines may contribute to turning off commonly used promoters (52), but the mechanisms by which transcription is shut down are not well understood. One tissue in which long-term expression seems attainable is muscle, where expression has been reported for up to one year (61,67).

The approach that holds perhaps the most promise for long-term gene expression in the absence of complicating effects due to the presence of viral genes is that of gutted, or helper-dependent, Ad vectors (10,13,37). In this strategy, all of the viral structural genes are deleted from the viral chromosome, leaving just the two ITRs and the packaging signal. Such a chromosome can accommodate at least 30 kb of transgene sequences. To propagate such a genome, however, the presence of a helper virus that provides the functions required for replication and assembly is necessary. A number of investigators have been pursuing this avenue of investigation, and the main problem to date is the inability to completely separate virions containing the gutted chromosome from those containing the helper virus genome. Strategies to reduce helper virus contamination have included the use of a helper virus carrying a mutated packaging signal and keeping the size of the gutted chromosome smaller than that of the helper virus chromosome with the hope that the two types of virions can be separated based on their slightly different densities. Even using these techniques, however, gutted virus preparations to date contain at least 0.1% contaminating helper virus (30,55,59). Ideally, a complementing cell line that could replace the helper virus would be preferred, but the need for very high levels of some virion components, together with the toxicity of some of these proteins to the cell, make this approach unlikely to be viable.

CONCLUDING REMARKS

It is likely that Ad will prove to be a useful vector in certain disease applications. A great deal is being learned about its potential and its limitations from the large number of ongoing preclinical and clinical studies. In addition, it is clear that further laboratory investigation is needed to fully understand the biology of the virus, both at the cellular and organismal levels. As this knowledge grows, the utility of Ad as a tool to treat human diseases will also grow.

REFERENCES

1. **Aiello, L., R. Guilfoyle, K. Huebner, and R. Weinmann.** 1979. Adenovirus 5 DNA sequences present and RNA sequences transcribed in transformed human embryo kidney cells (HEK-Ad-5 or 293). Virology *94*:460-469.
2. **Andersson, M., S. Paabo, T. Nilsson, and P.A. Peterson.** 1985. Impaired intracellular transport of class I MHC antigens as a possible means for adenoviruses to evade immune surveillance. Cell *43*:215-222.
3. **Armentano, D., J. Zabner, C. Sacks, C.C. Sookdeo, M.P. Smith, J.A. St. George, S.C. Wadsworth, A.E. Smith, and R.J. Gregory.** 1997. Effect of the E4 region on the persistence of transgene expression from adenovirus vectors. J. Virol. *71*:2408-2416.
4. **Babiss, L.E., H.S. Ginsberg, and J.E. Darnell, Jr.** 1985. Adenovirus E1B proteins are required for accumulation of late viral mRNA and for effects on cellular mRNA translation and transport. Mol. Cell. Biol. *5*:2552-2558.
5. **Bergelson, J.M., J.A. Cunningham, G. Droguett, E.A. Kurt-Jones, A. Krithivas, J.S. Hong, M.S. Horwitz, R.L. Crowell, and R.W. Finberg.** 1997. Isolation of a common receptor for Coxsackie B viruses and adenoviruses 2 and 5. Science *275*:1320-1323.
6. **Brough, D.E., C. Hsu, V.A. Kulesa, G.M. Lee, L.J. Cantolupo, A. Lizonova, and I. Kovesdi.** 1997. Activation of transgene expression by early region 4 is responsible for a high level of persistent transgene expression from adenovirus vectors in vivo. J. Virol. *71*:9206-9213.
7. **Bruder, J.T., T. Jie, D.L. McVey, and I. Kovesdi.** 1997. Expression of gp19K increases the persistence of transgene expression from an adenovirus vector in the mouse lung and liver. J. Virol. *71*:7623-7628.
8. **Burgert, H.G. and S. Kvist.** 1985. An adenovirus type 2 glycoprotein blocks cell surface expression of human histocompatibility class I antigens. Cell *41*:987-997.
9. **Chardonnet, Y. and S. Dales.** 1970. Early events in the interaction of adenoviruses with HeLa cells. I. Penetration of type 5 and intracellular release of the DNA genome. Virology *40*:462-477.
10. **Chen, H.H., L.M. Mack, R. Kelly, M. Ontell, S. Kochanek, and P.R. Clemens.** 1997. Persistence in muscle of an adenoviral vector that lacks all viral genes. Proc. Natl. Acad. Sci. USA *94*:1645-1650.
11. **Chinnadurai, G., S. Chinnadurai, and J. Brusca.** 1979. Physical mapping of a large-plaque mutation of adenovirus type 2. J. Virol. *32*:623-628.
12. **Chirmule, N., J.V. Hughes, G.P. Gao, S.E. Raper, and J.M. Wilson.** 1998. Role of E4 in eliciting CD4 T-cell and B-cell responses to adenovirus vectors delivered to murine and nonhuman primate lungs. J. Virol. *72*:6138-6145.
13. **Clemens, P.R., S. Kochanek, Y. Sunada, S. Chan, H.H. Chen, K.P. Campbell, and C.T. Caskey.** 1996. In vivo muscle gene transfer of full-length dystrophin with an adenoviral vector that lacks all viral genes. Gene Ther. *3*:965-972.
14. **Dai, Y., E.M. Schwarz, D. Gu, W.W. Zhang, N. Sarvetnick, and I.M. Verma.** 1995. Cellular and humoral immune responses to adenoviral vectors containing factor IX gene: tolerization of factor IX and vector antigens allows for long-term expression. Proc. Natl. Acad. Sci. USA *92*:1401-1405.
15. **Dedieu, J.F., E. Vigne, C. Torrent, C. Jullien, I. Mahfouz, J.M. Caillaud, N. Aubailly, C. Orsini, et al.** 1997. Long-term gene delivery into the livers of

immunocompetent mice with E1/E4-defective adenoviruses. J. Virol. *71*:4626-4637.

16.**Defer, C., M.T. Belin, M.L. Caillet-Boudin, and P. Boulanger.** 1990. Human adenovirus-host cell interactions: comparative study with members of subgroups B and C. J. Virol. *64*:3661-3673.

17.**Doerfler, W. and P. Böhm (Eds.).** 1995. The Molecular Repertoire of Adenoviruses. *In* Current Topics in Microbiology and Immunology, Vol. 199. Springer, New York.

18.**Douglas, J.T., B.E. Rogers, M.E. Rosenfeld, S.I. Michael, M. Feng, and D.T. Curiel.** 1996. Targeted gene delivery by tropism-modified adenoviral vectors. Nat. Biotechnol. *14*:1574-1578.

19.**Dyson, N.** 1998. The regulation of E2F by pRB-family proteins. Genes Dev. *12*:2245-2262.

20.**Engelhardt, J.F., X. Ye, B. Doranz, and J.M. Wilson.** 1994. Ablation of E2A in recombinant adenoviruses improves transgene persistence and decreases inflammatory response in mouse liver. Proc. Natl. Acad. Sci. USA *91*:6196-6200.

21.**Fallaux, F.J., A. Bout, I. van der Velde, D.J. van den Wollenberg, K.M. Hehir, J. Keegan, C. Auger, S.J. Cramer, H. van Ormondt, A.J. van der Eb, D. Valerio, and R.C. Hoeben.** 1998. New helper cells and matched early region 1-deleted adenovirus vectors prevent generation of replication-competent adenoviruses. Hum. Gene Ther. *9*:1909-1917.

22.**Fallaux, F.J., O. Kranenburg, S.J. Cramer, A. Houweling, H. van Ormondt, R.C. Hoeben, and A.J. van der Eb.** 1996. Characterization of 911: a new helper cell line for the titration and propagation of early region 1-deleted adenoviral vectors. Hum. Gene Ther. *7*:215-222.

23.**Gall, J., A. Kass-Eisler, L. Leinwand, and E. Falck-Pedersen.** 1996. Adenovirus type 5 and 7 capsid chimera: fiber replacement alters receptor tropism without affecting primary immune neutralization epitopes. J. Virol. *70*:2116-2123.

24.**Gao, G.P., Y. Yang, and J.M. Wilson.** 1996. Biology of adenovirus vectors with E1 and E4 deletions for liver-directed gene therapy. J. Virol. *70*:8934-8943.

25.**Gaynor, R.B. and A.J. Berk.** 1983. Cis-acting induction of adenovirus transcription. Cell *33*:683-693.

26.**Gooding, L.R., L.W. Elmore, A.E. Tollefson, H.A. Brady, and W.S. Wold.** 1988. A 14,700 MW protein from the E3 region of adenovirus inhibits cytolysis by tumor necrosis factor. Cell *53*:341-346.

27.**Grable, M. and P. Hearing.** 1990. Adenovirus type 5 packaging domain is composed of a repeated element that is functionally redundant. J. Virol. *64*:2047-2056.

28.**Greber, U.F., M. Willetts, P. Webster, and A. Helenius.** 1993. Stepwise dismantling of adenovirus 2 during entry into cells. Cell *75*:477-486.

29.**Hammarskjold, M.L. and G. Winberg.** 1980. Encapsidation of adenovirus 16 DNA is directed by a small DNA sequence at the left end of the genome. Cell *20*:787-795.

30.**Hartigan-O'Connor, D., A. Amalfitano, and J.S. Chamberlain.** 1999. Improved production of gutted adenovirus in cells expressing adenovirus preterminal protein and DNA polymerase. J. Virol. *73*:7835-7841.

31.**Huard, J., H. Lochmuller, G. Acsadi, A. Jani, B. Massie, and G. Karpati.** 1995. The route of administration is a major determinant of the transduction efficiency of rat tissues by adenoviral recombinants. Gene Ther. *2*:107-115.

32.**Imperiale, M.J., G. Akusjärvi, and K.N. Leppard.** 1995. Post-transcriptional control of adenovirus gene expression. Curr. Top. Microbiol. Immunol. *199* (Pt 2):139-171.

33.**Imperiale, M.J., H.T. Kao, L.T. Feldman, J.R. Nevins, and S. Strickland.** 1984. Common control of the heat shock gene and early adenovirus genes: evidence for a cellular E1A-like activity. Mol. Cell. Biol. *4*:867-874.

34.**Jaffe, H.A., C. Danel, G. Longenecker, M. Metzger, Y. Setoguchi, M.A. Rosenfeld, T.W. Gant, S.S. Thorgeirsson, et al.** 1992. Adenovirus-mediated in vivo gene transfer and expression in normal rat liver. Nat. Genet. *1*:372-378.

35.**Jones, N.** 1995. Transcriptional modulation by the adenovirus E1A gene. Curr. Top. Microbiol. Immunol. *199* (Pt 3):59-80.

36.**Krasnykh, V.N., G.V. Mikheeva, J.T. Douglas, and D.T. Curiel.** 1996. Generation of recombinant adenovirus vectors with modified fibers for altering viral tropism. J. Virol. *70*:6839-6846.

37.**Kumar-Singh, R. and J.S. Chamberlain.** 1996. Encapsidated adenovirus minichromosomes allow delivery and expression of a 14 kb dystrophin cDNA to muscle cells. Hum. Mol. Genet. *5*:913-921.

38.**Lee, M.G., M.A. Abina, H. Haddada, and M. Perricaudet.** 1995. The constitutive expression of the immunomodulatory gp19k protein in E1-, E3- adenoviral vectors strongly reduces the host cytotoxic T cell response against the vector. Gene Ther. *2*:256-262.

39.**Leppard, K.N.** 1997. E4 gene function in adenovirus, adenovirus vector and adeno-associated virus infections. J. Gen. Virol. *78*:2131-2138.

40.**Li, Q., M.A. Kay, M. Finegold, L.D. Stratford-Perricaudet, and S.L. Woo.** 1993. Assessment of recombinant adenoviral vectors for hepatic gene therapy. Hum. Gene Ther. *4*:403-409.

41.**Lochmuller, H., A. Jani, J. Huard, S. Prescott, M. Simoneau, B. Massie, G. Karpati, and G. Acsadi.** 1994. Emergence of early region 1-containing replication-competent adenovirus in stocks of replication-defective adenovirus recombinants (delta E1 + delta E3) during multiple passages in 293 cells. Hum. Gene Ther. *5*:1485-1491.

42.**Lusky, M., M. Christ, K. Rittner, A. Dieterle, D. Dreyer, B. Mourot, H. Schultz, F. Stoeckel, A. Pavirani, and M. Mehtali.** 1998. In vitro and in vivo biology of recombinant adenovirus vectors with E1, E1/E2A, or E1/E4 deleted. J. Virol. *72*:2022-2032.

43.**Mack, C.A., W.R. Song, H. Carpenter, T.J. Wickham, I. Kovesdi, B.G. Harvey, C.J. Magovern, O.W. Isom, et al.** 1997. Circumvention of anti-adenovirus neutralizing immunity by administration of an adenoviral vector of an alternate serotype. Hum. Gene Ther. *8*:99-109.

44.**Marcellus, R.C., J.N. Lavoie, D. Boivin, G.C. Shore, G. Ketner, and P.E. Branton.** 1998. The early region 4 orf4 protein of human adenovirus type 5 induces p53-independent cell death by apoptosis. J. Virol. *72*:7144-7153.

45. **Mastrangeli, A., B.G. Harvey, J. Yao, G. Wolff, I. Kovesdi, R.G. Crystal, and E. Falck-Pedersen.** 1996. "Sero-switch" adenovirus-mediated in vivo gene transfer: circumvention of anti-adenovirus humoral immune defenses against repeat adenovirus vector administration by changing the adenovirus serotype. Hum. Gene Ther. *7*:79-87.

46. **McCoy, R.D., B.L. Davidson, B.J. Roessler, G.B. Huffnagle, S.L. Janich, T.J. Laing, and R.H. Simon.** 1995. Pulmonary inflammation induced by incomplete or inactivated adenoviral particles. Hum. Gene Ther. *6*:1553-1560.

47. **Michael, S.I., J.S. Hong, D.T. Curiel, and J.A. Engler.** 1995. Addition of a short peptide ligand to the adenovirus fiber protein. Gene Ther. *2*:660-668.

48. **Nevins, J.R.** 1987. Regulation of early adenovirus gene expression. Microbiol. Rev. *51*:419-430.

49. **Philipson, L., K. Lonberg-Holm, and U. Pettersson.** 1968. Virus-receptor interaction in an adenovirus system. J. Virol. *2*:1064-1075.

50. **Pilder, S., J. Logan, and T. Shenk.** 1984. Deletion of the gene encoding the adenovirus 5 early region 1b 21,000-molecular-weight polypeptide leads to degradation of viral and host cell DNA. J. Virol. *52*:664-671.

51. **Pilder, S., M. Moore, J. Logan, and T. Shenk.** 1986. The adenovirus E1B-55K transforming polypeptide modulates transport or cytoplasmic stabilization of viral and host cell mRNAs. Mol. Cell. Biol. *6*:470-476.

52. **Qin, L., Y. Ding, D.R. Pahud, E. Chang, M.J. Imperiale, and J.S. Bromberg.** 1997. Promoter attenuation in gene therapy: interferon-gamma and tumor necrosis factor-alpha inhibit transgene expression. Hum. Gene Ther. *8*:2019-2029.

53. **Rao, L., M. Debbas, P. Sabbatini, D. Hockenbery, S. Korsmeyer, and E. White.** 1992. The adenovirus E1A proteins induce apoptosis, which is inhibited by the E1B 19-kDa and Bcl-2 proteins. Proc. Natl. Acad. Sci. USA *89*:7742-7746.

54. **Rogers, B.E., J.T. Douglas, C. Ahlem, D.J. Buchsbaum, J. Frincke, and D.T. Curiel.** 1997. Use of a novel cross-linking method to modify adenovirus tropism. Gene Ther. *4*:1387-1392.

55. **Sandig, V., R. Youil, A.J. Bett, L.L. Franlin, M. Oshima, D. Malone, F. Wang, M.L. Metzker, R. Savino, and C.T. Caskey.** 2000. Optimization of the helper-dependent adenovirus system for production and potency in vivo. Proc. Natl. Acad. Sci. USA *97*:1002-1007.

56. **Schowalter, D.B., J.C. Tubb, M. Liu, C.B. Wilson, and M.A. Kay.** 1997. Heterologous expression of adenovirus E3-gp19K in an E1a-deleted adenovirus vector inhibits MHC I expression in vitro, but does not prolong transgene expression in vivo. Gene Ther. *4*:351-360.

57. **Sharp, P.A.** 1984. Adenovirus transcription, p. 173-204. *In* H.S. Ginsberg (Ed.), The Adenoviruses. Plenum Press, New York.

58. **Smith, T.A., M.G. Mehaffey, D.B. Kayda, J.M. Saunders, S. Yei, B.C. Trapnell, A. McClelland, and M. Kaleko.** 1993. Adenovirus mediated expression of therapeutic plasma levels of human factor IX in mice. Nat. Genet. *5*:397-402.

59. **Steinwaerder, D.S., C.A. Carlson, and A. Lieber.** 1999. Generation of adenovirus vectors devoid of all viral genes by recombination between inverted repeats. J. Virol. *73*:9303-9313.

60. **Stevenson, S.C., M. Rollence, J. Marshall-Neff, and A. McClelland.** 1997. Selective targeting of human cells by a chimeric adenovirus vector containing a modified fiber protein. J. Virol. *71*:4782-4790.

61. **Svensson, E.C., H.B. Black, D.L. Dugger, S.K. Tripathy, E. Goldwasser, Z. Hao, L. Chu, and J.M. Leiden.** 1997. Long-term erythropoietin expression in rodents and non-human primates following intramuscular injection of a replication-defective adenoviral vector. Hum. Gene Ther. *8*:1797-1806.

62. **Takemori, N., C. Cladaras, B. Bhat, A.J. Conley, and W.S. Wold.** 1984. cyt gene of adenoviruses 2 and 5 is an oncogene for transforming function in early region E1B and encodes the E1B 19,000-molecular-weight polypeptide. J. Virol. *52*:793-805.

63. **Tollefson, A.E., T.W. Hermiston, D.L. Lichtenstein, C.F. Colle, R.A. Tripp, T. Dimitrov, K. Toth, C.E. Wells, P.C. Doherty, and W.S. Wold.** 1998. Forced degradation of Fas inhibits apoptosis in adenovirus-infected cells. Nature *392*:726-730.

64. **Tollefson, A.E., J.S. Ryerse, A. Scaria, T.W. Hermiston, and W.S. Wold.** 1996. The E3-11.6-kDa adenovirus death protein (ADP) is required for efficient cell death: characterization of cells infected with adp mutants. Virology *220*:152-162.

65. **Tollefson, A.E., A. Scaria, T.W. Hermiston, J.S. Ryerse, L.J. Wold, and W.S. Wold.** 1996. The adenovirus death protein (E3-11.6K) is required at very late stages of infection for efficient cell lysis and release of adenovirus from infected cells. J. Virol. *70*:2296-2306.

66. **Tomko, R.P., R. Xu, and L. Philipson.** 1997. HCAR and MCAR: the human and mouse cellular receptors for subgroup C adenoviruses and group B coxsackieviruses. Proc. Natl. Acad. Sci. USA *94*:3352-3356.

67. **Tripathy, S.K., H.B. Black, E. Goldwasser, and J.M. Leiden.** 1996. Immune responses to transgene-encoded proteins limit the stability of gene expression after injection of replication-defective adenovirus vectors. Nat. Med. *2*:545-550.

68. **Varga, M.J., C. Weibull, and E. Everitt.** 1991. Infectious entry pathway of adenovirus type 2. J. Virol. *65*:6061-6070.

69. **Wang, Q., G. Greenburg, D. Bunch, D. Farson, and M.H. Finer.** 1997. Persistent transgene expression in mouse liver following in vivo gene transfer with a delta E1/delta E4 adenovirus vector. Gene Ther. *4*:393-400.

70. **White, E., T. Grodzicker, and B.W. Stillman.** 1984. Mutations in the gene encoding the adenovirus early region 1B 19,000-molecular-weight tumor antigen cause the degradation of chromosomal DNA. J. Virol. *52*:410-419.

71. **Wickham, T.J., M.E. Carrion, and I. Kovesdi.** 1995. Targeting of adenovirus penton base to new receptors through replacement of its RGD motif with other

receptor-specific peptide motifs. Gene Ther. *2*:750-756.

72. **Wickham, T.J., D. Haskard, D. Segal, and I. Kovesdi.** 1997. Targeting endothelium for gene therapy via receptors up-regulated during angiogenesis and inflammation. Cancer Immunol. Immunother. *45*:149-151.

73. **Wickham, T.J., G.M. Lee, J.A. Titus, G. Sconocchia, T. Bakacs, I. Kovesdi, and D.M. Segal.** 1997. Targeted adenovirus-mediated gene delivery to T cells via CD3. J. Virol. *71*:7663-7669.

74. **Wickham, T.J., P. Mathias, D.A. Cheresh, and G.R. Nemerow.** 1993. Integrins alpha v beta 3 and alpha v beta 5 promote adenovirus internalization but not virus attachment. Cell *73*:309-319.

75. **Wickham, T.J., P.W. Roelvink, D.E. Brough, and I. Kovesdi.** 1996. Adenovirus targeted to heparan-containing receptors increases its gene delivery efficiency to multiple cell types. Nat. Biotechnol. *14*:1570-1573.

76. **Wickham, T.J., D.M. Segal, P.W. Roelvink, M.E. Carrion, A. Lizonova, G.M. Lee, and I. Kovesdi.** 1996. Targeted adenovirus gene transfer to endothelial and smooth muscle cells by using bispecific antibodies. J. Virol. *70*:6831-6838.

77. **Wickham, T.J., E. Tzeng, L.L. Shears, P.W. Roelvink, Y. Li, G.M. Lee, D.E. Brough, A. Lizonova, and I. Kovesdi.** 1997. Increased in vitro and in vivo gene transfer by adenovirus vectors containing chimeric fiber proteins. J. Virol. *71*:8221-8229.

78. **Williams, J., B.D. Karger, Y.S. Ho, C.L. Castiglia, T. Mann, and S.J. Flint.** 1986. The adenovirus E1B 495R protein plays a role in regulating the transport and stability of the viral late messages. Cancer Cells *4*:275-284.

79. **Yang, Y., Q. Li, H.C. Ertl, and J.M. Wilson.** 1995. Cellular and humoral immune responses to viral antigens create barriers to lung-directed gene therapy with recombinant adenoviruses. J. Virol. *69*:2004-2015.

80. **Yang, Y., F.A. Nunes, K. Berencsi, E.E. Furth, E. Gonczol, and J.M. Wilson.** 1994. Cellular immunity to viral antigens limits E1-deleted adenoviruses for gene therapy. Proc. Natl. Acad. Sci. USA *91*:4407-4411.

81. **Zsengeller, Z.K., S.E. Wert, W.M. Hull, X. Hu, S. Yei, B.C. Trapnell, and J.A. Whitsett.** 1995. Persistence of replication-deficient adenovirus-mediated gene transfer in lungs of immune-deficient (nu/nu) mice. Hum. Gene Ther. *6*:457-467.

8 Helper-Dependent Adenoviral Vectors: A New Generation of Gene Delivery Vehicles

Manal A. Morsy and Ling Chen
Merck Research Laboratories, Department of Human Genetics, Gene Therapy, West Point, PA, USA

INTRODUCTION

Adenoviruses (Ads) are characterized by many features that make them very attractive tools for gene delivery, and therefore, they have been a focus of interest in the field of gene therapy. The extensive data on the lack of morbid pathology associated with the live Ad virus in upper respiratory tract infections and a favorable safety history as vaccines contributed to the appeal of this vector system as a potential vehicle for gene/antigen delivery. A wealth of examples in the literature establish the potential usefulness of these vectors as carriers for heterologous antigens/proteins (16). Similarly, investigators interested in gene therapy approaches find the safety history and Ad's other attractive features promising. Ads are characterized with a well-studied genome (Figure 1A) that is tolerant of significant manipulation and can accommodate large insert capacities. Recombinant vectors can infect a wide range of dividing and nondividing cells, and their uptake is efficient both in vitro and in vivo. In addition, there are well-established techniques that allow growth of vector stocks to titers that exceed 10^{12} viral particles (14,15,19). Early generations of recombinant Ad vectors are currently being used in clinical trials to determine their potential efficacy in the treatment of cancer and genetic disorders. Modifications of this vector system continue to evolve, mainly to overcome some of the immunogenic setbacks of Ad vectors. The most advanced modification towards a less immunogenic and safer vehicle is the creation of Ad vectors that are completely devoid of viral coding sequences, namely the helper-dependent (HD) system.

HD VECTORS

The HD vectors are a new generation of Ad-based vectors in which the coding sequence of the Ad genome is completely deleted, retaining only the inverted terminal repeats (ITRs) and packaging signal sequence (Ψ), which are the only essential *cis*-elements required for viral replication and packaging (Figure 1). The deleted genome is replaced with a foreign fragment

Viral Vectors: Basic Science and Gene Therapy
Edited by A. Cid-Arregui and A. García-Carrancá

of DNA that encompasses the expression cassette of interest with or without additional "stuffer" DNA, which may be required to maintain the optimal genome size (approximately 30–35 kb) necessary for the stability of these pseudoadenoviruses (19,20,23). In order for this genome to be propagated and rescued as a recombinant virus, it needs to be provided with the proteins necessary for viral DNA replication and packaging. These proteins are supplied *in trans* by a helper virus, which is typically an E1-deleted (first generation [FG]), replication-deficient Ad virus that is co-cultured with the HD vector on E1-complementing cell lines. The helper FG Ad virus can be one that is crippled such that it is rendered unable to compete with the propagation and rescue of the HD virus and/or incapable of efficient packaging into viral capsids (17–19). We have successfully developed and used a *cre-lox*-based system for eliminating the helper Ad viral contamination from HD stocks. HD vector stocks can be routinely generated with insignificant amounts of helper contamination (less than 1 plaque-forming unit [pfu] per 10 000 optical density [OD]) particles of HD virus in 1 mL of (CsCl) banded stock (20). The *cre-lox* system is a site-specific recombination system based on the efficiency by which *cre*—a bacteriophage Pl-encoded recombinase (38 kDa)—recognizes and catalyzes recombination between DNA target sites called *loxP*. These target sites are 34-bp sequences that consist of two inverted 13-bp repeats and an asymmetric spacer sequence of 8 bp. The recombination between two *loxP* sites arranged in parallel orientation on the same DNA fragment results in excision of the intervening sequences, whereas if the *loxP* sites are arranged in opposite orientations, the recombination event results in the inversion of the intervening DNA segment. Our method of HD vector propagation and preparation uses a helper virus system consisting of a modified FG E1-deleted vector with *lox* sites flanking (floxed) the packaging signals (AdLC8clucl) and a 293 cell line derivative expressing *cre* recombinase (293-cre4) (22). Alternatively, the HD and helper Ad can be separated physically on CsCl density gradient, a less efficient way of isolating the HD vector that may result in lower yields and high contamination levels with the helper virus (17).

A BASIC APPROACH FOR GENERATING HD VECTORS USING THE *cre-lox* SYSTEM

Described below is a basic approach that can be adapted for generating HD vectors in the laboratory. The method is based on the use of the *cre-lox* helper Ad and cell line system (6,20,22).

Components of the *cre-lox* System

We will give a brief description of the components of the system and some suggestions regarding some basic properties that may be helpful in designing an HD system for laboratory research use. For propagation and titration protocols, as well as a detailed description of the helper virus AdLC8cluc, the HD backbone plasmid pSTK120, and the 293-cre4 cell line, refer to prior publications (14,15,20,22).

AdLC8luc

The helper virus is an E1-deleted virus with an expression cassette inserted in E3 (cytomegalovirus [CMV]-luciferase) and *lox* sites flanking the packaging signal sequences. A large-scale stock of this helper virus on 293 cells should be grown. For best yields a multiplicity of infection (moi) equivalent to approximately 5 pfu of helper virus to a 293 cell should be used. It should be noted though that this virus does not grow to very high titers. Titers of CsCl-banded stocks are typically approximately 10^8 to 10^{10} pfu. The enzyme restriction pattern of the AdLC8luc and the luciferase expression should be analyzed to confirm the stability and authenticity of the stock. AdLC8luc is supplied at an moi of 1/293-cre4 cell during rescue and propagation of the HD vector.

pSTK120 (HD backbone plasmid)

This is a pBluescript® IIKS (Stratagene, La Jolla, CA, USA)-based plasmid that carries the HD backbone. Inserted between two *Pme*I unique enzyme restriction sites are the following structures (in order with the intervening multiple cloning sites between junctions of the different fragments): (*i*) the adenovirus type 5 (Ad-5) ITR and the Ψ, 440 bp, (nucleotides [nt] 1–440); (*ii*) a 16 054 bp of hypoxanthine guanine phosphoribosyltransferase (HPRT) (nt 17 99–17 853 in gb: humhprtb); (*iii*) a *Hind*-III 9063-bp fragment of C346 cosmid (nt 12 421–21 484 in GenBank®: L31948); and (*iv*) the right end terminus of Ad-5, composed of the ITR sequence, 117 bp (nt

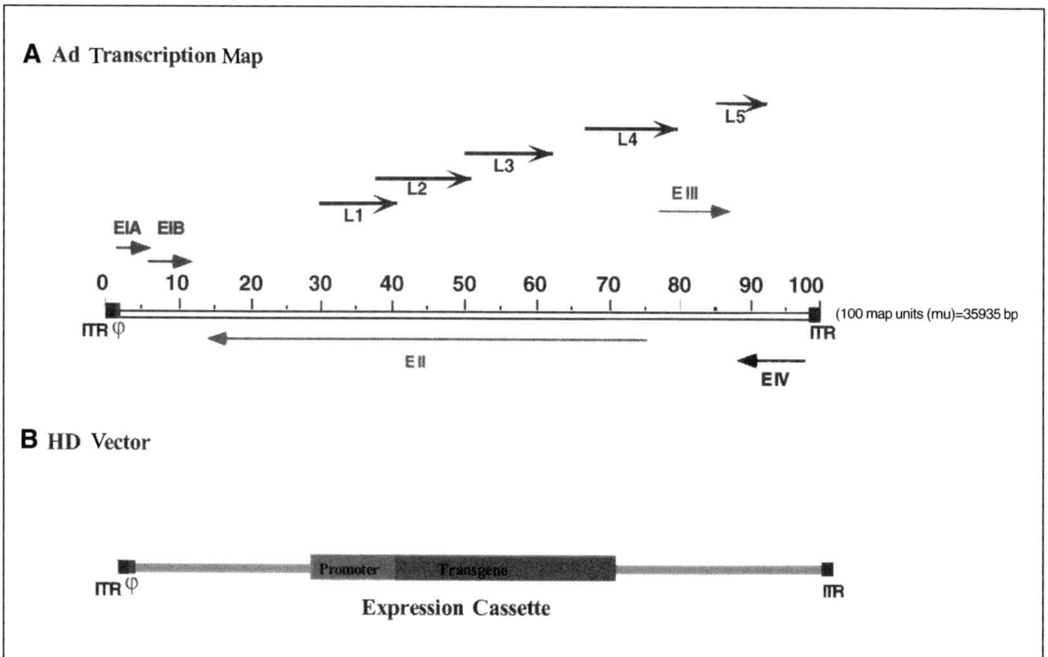

Figure 1. Ad and HD backbone structure. (A) A simplified schematic of the Ad genome modified from Chow et al. (11). The genome is flanked by ITRs which serve as replication origins. The Ψ mediates DNA-capsid recognition. The five early transcription units are E1A, E1B, E11, E111, and E1V. The two delayed early units are 1X and 1Va2 (not shown), and the late transcription unit (major late) is processed to generate five families of late mRNAs (L1–L5). By convention, the map is drawn with ElA on the left end. Both strands are transcribed with ElA, ElB, IX, major late, VA RNA (not shown), and Elll units encoded for by the rightward strand, and ElV, Ell, and lVa2 (not shown) units encoded for by the reverse strand. ElA produces proteins involved in transcription activation and inducing the host cell to enter the S phase of the cell cycle. ElB produces proteins, which in cooperation with ElA products induce cell growth. Ell (EllA and EllB) encodes three proteins that are involved in DNA replication and transcription modulation (i.e., terminal protein, DNA-binding protein involved in DNA elongation, and viral DNA polymerase protein). Elll encodes proteins that modulate the immune response to the Ad infection. ElV encodes proteins that are involved in a number of functions, including transcription regulation and transition from early to late gene expression, mRNA transport, shutoff of host cell gene expression, viral DNA replication, and assembly of the virion. The late family of mRNAs (L1–L5) is mainly involved in the production and assembly of capsid proteins. (B) This shows a simplified schematic of the HD genome. The sequences retained from the Ad genome are the ITR and Ψ sequences. The expression cassette along with stuffer DNA (if needed) are inserted between the ITR sequences. The total size can range between 30 and 35 kb.

35818–35935). The total size is approximately 30 kb including 2.9 kb of the pBluescript IIKS. The 2.9 kb of pBluescript IIKS is eliminated prior to HD vector rescue by linearizing the plasmid by digestion of the two unique *Pme*I flanking sites. There are a number of unique sites distributed within the stuffer DNA. A preferred site that we used in a number of our constructs is the *Swa*I site, which is imbedded within the HPRT fragment. Cassettes of any size can be used to either directly insert into *Swa*I or replace stretches of deletions in the stuffer DNA so that the final size of the recombinant HD (STK120-X) is between 28 and 35 kb. It is essential to extensively analyze the recombinant HD plasmid after insertion of the expression cassette either by extensive restriction mapping or by sequencing. The large plasmid size and the nature of the stuffer DNA used in pSTK120 render this specific plasmid unstable at times. If replacements of some of the stuffer DNA are attempted, careful selection of segments that have a reduced number of repeat elements and open reading frames is preferred. We typically linearize the HD plasmid by digesting pSTK120-X with *Pme*I prior to transfection into 293-cre4 cells, provided that the sites are unique. We did not find purification of the linear fragment necessary; the digest mixture can be used directly for transfection.

293-cre4

This derivative of 293 expresses *cre*. We routinely split cells at a 1:3 ratio; this accounts for a passage. Cells that were passaged over 45 passages should be avoided. It may be useful to generate an extensive cell bank that would allow for a long-term supply of such low passages. Alternatively, clonal derivation can be established, and a stable cell line for constant levels of *cre* expression can be selected in which passage is not an issue.

Culture Medium

The culture medium consists of minimum essential medium (MEM) alpha, 1% Geneticin (12561-056 and 10131-019, respectively; Life Technologies, Rockville, MD, USA), and 10% fetal bovine serum (FBS).

METHODS

Figures 2 and 3 show a flowchart and an illustration of the steps used to generate HD vectors. The process takes approximately three weeks from transfection to cesium chloride (CsCl) banding of P5.

The HD backbone (with expression cassette of interest inserted) should be designed so that there are two flanking unique restriction sites that can be used to linearize the HD backbone and liberate it from the plasmid DNA prior to transfection.

Vial Propagation 1: P1

1. 4 µg of HD DNA (pSTK120-X linearized) are transfected to 6-cm culture plates of 70% to 80% confluent 293-cre4 cells prepared the day before.
2. 24 hours after transfection, cultures are infected with helper virus AdLC-8cluc at an moi of 1.
3. When *cpe* (cytopathic effect) is detected (at approximately 72–96 hours), cells are scraped into media, followed by a 3× freeze/thaw step.
4. Cellular debris is spun down, and crude cell lysate is used for the next propagation step.

Serial Propagation: P2

1. 2 mL of P1 lysate are used to infect two or three, 70% to 80% confluent, 6-cm plates of 293-cre4 cells.
2. Steps 2 to 4 of P1 are followed, and an

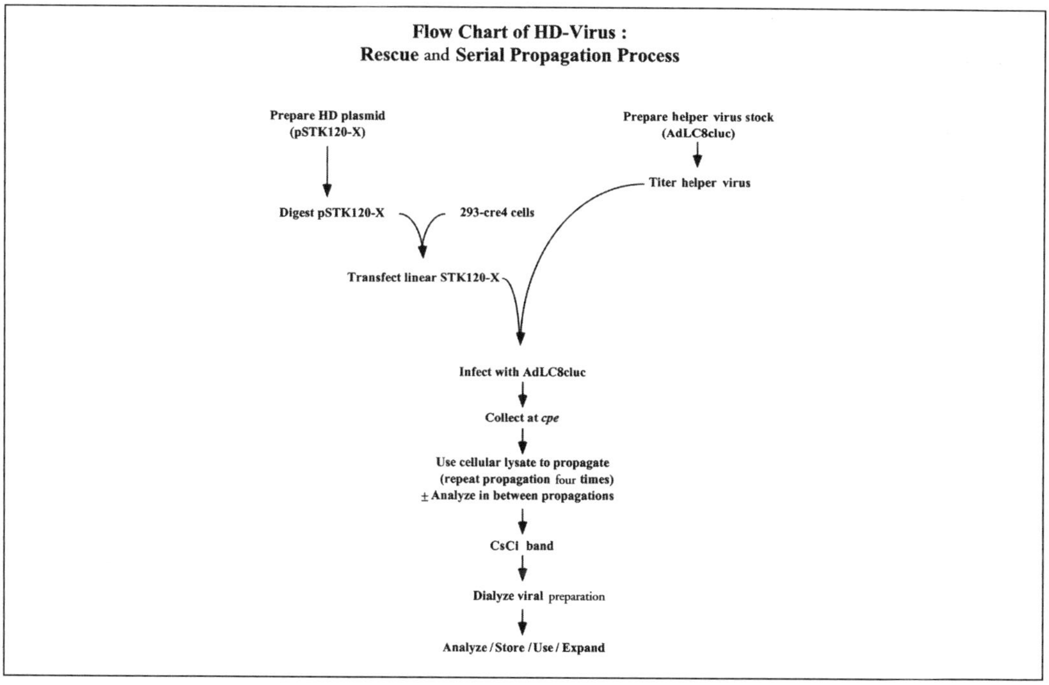

Figure 2. Schematic flowchart that shows the steps required to prepare an HD viral vector.

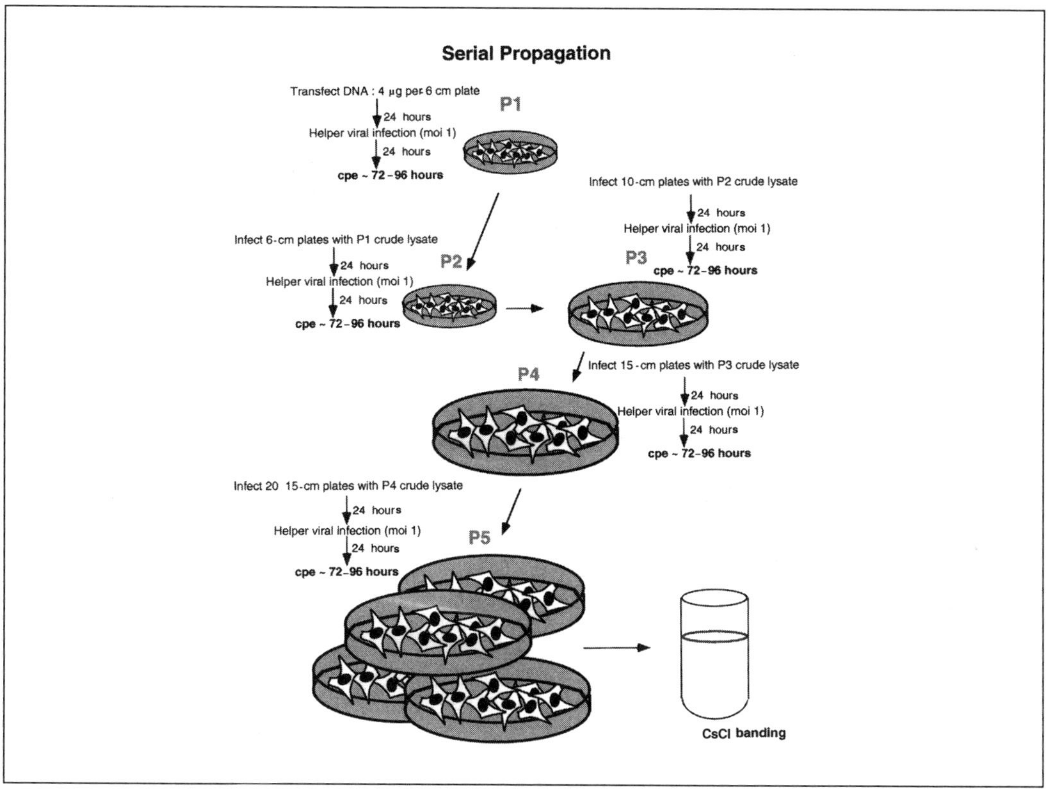

Figure 3. Diagram that illustrates the steps required for the serial propagation of the HD viral vectors P1 through P5.

Table 1. HD Ad Versus FG Ad for Cancer Gene Therapy

Therapeutic	HD Ad	FG Ad
Insert capacity	35 kb	8 kb
Leaky viral protein	No	Yes
Transgene expression	Long-term for isogenic products	Short-term
Host CTL to vector	Unlikely	Against de novo synthesized viral proteins
Allow regulatory gene expression	Yes	Difficult, limited by size
Multiple gene/gene cassette	More	Less, limited by size

aliquot of the collected cells or lysate is analyzed to detect propagation of transgene expression and/or HD viral DNA pattern.

Serial Propagation: P3

1. 1 mL of P2 lysate is used to infect two or three, 70% to 80% confluent, 10-cm plates of 293-cre4 cells.
2. Repeat step 2 of P2.

Serial Propagation: P4

1. 5 mL of P3 lysate are used to infect two or three, 70% to 80% confluent, 15-cm plates of 293-cre4 cells.
2. Repeat step 2 of P2.

Serial Propagation: P5

1. 1 mL of P4 lysate is used to infect twenty 70% to 80% confluent, 15-cm plates of 293-cre4 cells.
2. Repeat step 2 of P2 (save supernatant media for future propagation of HD virus).
3. P5 lysate is CsCl banded (14). CsCl banded stock is then analyzed for HD DNA pattern, transgene activity, OD value, and pfu value (to determine the helper virus contamination level, typically we find less than 1 pfu/1000 OD particle).

Expansion: P6

2 mL of P5 supernatant media are used to infect 60 to 80, 70% to 80% confluent, 15-cm plates of 293-cre4 cells.

Storage

All viral stocks and supernatant media should be stored at -80C.

HD PROOF OF PRINCIPLE

There is evidence that the generation of HD vectors has significantly reduced the immunogenic consequences associated with earlier generation of Ad vectors. Also, these vectors have been shown to persist in vivo for very long periods of time, far exceeding E1-deleted Ad vectors (20,27). Persistence of transgene expression was shown by Morsy et al. (20). Leptin expression in the context of an HD vector (HD-leptin) delivered to both leptin-deficient and normal C57Bl/6J mice persisted beyond what was seen with FG Ad vectors (Ad-leptin) expressing the same transgene in the same strain of mice. The HD stocks were highly purified with less than 0.1% (approximately 7.5×10^5)/dose of the estimated 1 to 2×10^9 infectious units (or $1–2 \times 10^{11}$ OD particles) of helper virus contamination (20). Longevity of expression in animals treated with HD-leptin was associated with significant enhancement in the safety/toxicity profile compared with that seen in animals treated with Ad-leptin vectors (20).

The HD system, however, has yet to overcome the immunogenicity of the capsid proteins. As would be expected, these vectors can induce a neutralizing antibody immune response, which can result in a reduced efficiency of repeat systemic administration/uptake. This may be a limitation for some applications in which repeated delivery is required. However, for a variety of other applications, especially where systemic delivery is not required or even desired, and local delivery and uptake are not affected to a significant extent with the presence or absence of circulating Ad-neutralizing antibodies, it is conceivable that such a system with its multiple advantages would be useful. One area that comes to mind is cancer gene therapy.

Application of HD Ad for Cancer

Although the initial assumption of gene therapy was to treat genetic disorders, the field has evolved so that now 72% of Food and Drug Administration-approved gene therapy clinical trials are for cancer. Cancer gene therapy can be defined as transferring genetic materials into specific cells to confer tumor killing and host protection. An increasing number of FG Ad vectors are being tested in clinical trials. In most cases, therapeutic genes are delivered into the tumor cells or cells such as peripheral blood cells or muscle cells. The HD Ad vector with deletion of all Ad-coding proteins have several advantages over other vector systems for cancer treatment (Table 1).

HD Vectors for Multiple Insert Delivery

Recent reports have demonstrated that the combination of multiple genes-based therapy can achieve more efficient tumor remission and less toxicity. For example, the combination of thymidine kinase/ganciclovir prodrug activation system with interleukin (IL)-2 gene therapy (8), the combination of intratumor delivery of B7 with IL-2 or IL-12 gene therapy (2,25), and the combination of p53 gene therapy with IL-2 gene therapy (24). The large insert capacity of HD vector is an ideal system to accommodate multiple genes under the control of either a constitutive or regulatable promoter.

HD Ad for Antiangiogenic Gene Therapy

Based on the concept that angiogenesis is an essential process for growth of tumors and metastases, there is an intense interest in the use of antiangiogenic agents to inhibit tumor growth through suppression of neovascularization (13). Gene transfer provides a strategy for local (i.e., intratumor or peritumor injection of the vector) or systemic (i.e., intramuscular injection of the vector) delivery of antiangiogenic factors, depending on the nature of the therapeutic gene and the disease state. A number of antiangiogenic factors, including platelet factor 4 (28), urokinase, angiostatin (5), and endostatin (3) have been exploited in animal models using various vector systems. Antiangiogenic gene therapy will most likely be effective if it can provide persistent, high concentrations of antiangiogenic factors in the tumor microenvironment. Since the HD vectors do not express viral coding proteins while retaining their high gene transfer efficiency, they are the vectors of choice for expression of isogenic proteins (20). On the other hand, it would be most desirable to avoid any potential adverse effects by being able to turn the expression off whenever necessary, such as during the wound healing processes. The large capacity of HD vector provides the advantage of being able to contain multiple expression cassettes in the same vector backbone. Recently, several ligand-regulated gene expression systems have been reported (4). The expression of therapeutic genes can be induced by progesterone analog (21) or rapamycin-related compounds (26).

HD Ad for Immunostimulatory Gene Therapy

A variety of strategies have been used to modulate the host immune response against tumor cells. Enhanced host immune surveillance has been observed to be part of successful antitumor response in people, as evidenced by the spontaneous tumor remission in certain patients, such as those with melanoma, renal cell carcinoma, and retinoblastoma. Recently, an increasing number of tumor antigens such as MAGE, MART-1, and HER2/erbB2 have been identified as targets for immune recognition, in particular cytotoxic T lymphocytes (CTLs). These tumor antigens are presented by major histocompatibility class I molecules (MHC I) and allow T cells to distinguish tumor cells from normal cells. Immunostimulatory cytokines play critical roles in mediating the differentiation, proliferation, and activation of the various immune effector cells. Recombinant IL-2, IL-12, and IFNa showed potent antitumor activities in animal tumor models. However, clinical trials demonstrated that these cytokine proteins appeared to have limited usefulness due to severe toxicity resulting from systemic administration. The recent advancements in gene transfer technologies have enabled transfer of cytokine genes directly into tumors. The release of cytokine locally allows maximal stimulation of infiltrated T lymphocytes and other immune effector cells in the tumor microenvironment (1,7). The HD vector provides a novel approach to deliver cytokines into tumors for a relatively sustained period until the tumor cells are eliminated either by immune response or by the combined anticancer treatments such as chemotherapy, radiation therapy, or surgery. In addition to cytokines, the HD vector can mediate codelivery of costimulatory molecules such as B7, which may allow further immune modulation.

HD Vectors for Suicide Gene Therapy

One mechanism designed to kill tumor cells has been devised to either induce tumor apoptosis by overexpression of wild-type p53 or to enable tumor cells to activate a nontoxic prodrug to the toxic compound. Three widely used prodrug activation systems are: (*i*) herpes simplex virus thymidine kinase, which activates ganciclovir (10), (*ii*) bacterial cytosine deaminase, which converts nontoxic 5-fluorocytosine to 5-fluorouracil (12), and *(iii)* P450 enzymes, which activate cyclophosphamide to phospharamide mustard (9). It is possible that the potential immunogenicity of nonhuman gene products in the context of FG Ad vectors may undermine their efficacy during repeated usage. Therefore, the use of a humanized prodrug activation enzyme or an isogenic gene product such as human P450 CYP2B6 in the context of HD vector should provide a more consistent efficacy of prodrug activation and tumor killing without the immune interference. Accordingly, HD vectors may be useful for repeated local delivery of p53 or other apoptotic factors through the patient's lifetime.

CONCLUSION

The HD system is remarkably enhanced in comparison to other existing viral vector delivery systems. It has a significantly safer profile than that of earlier generations of Ad vectors. Despite the extensive deletion of the viral genome, the vector retained its highly efficient in vitro and in vivo delivery characteristics. In addition, the significant increase in insert capacity adds tremendous value to this vector. With a capacity of up to 35 kb, cassettes of large sizes or multiple cassettes can be inserted, an advantage for a wide range of applications. At this juncture, it must be noted that the HD system is not a very simple system to work with, and careful evaluation of the constructs is necessary. No doubt there is a lot of room for modifications, but at this stage of development the system is complex. The need for two cell lines, two viruses, and maintaining the low helper contamination levels are all issues that have to be dealt with especially for large-scale production. Nonetheless, the proof of principle is very encouraging, and it will only be a matter of time before this system becomes simplified, it certainly has all the potentials.

REFERENCES

1. **Addison, C.L., T. Braciak, R. Ralston, W.J. Muller, J. Gauldie, and F.L. Graham.** 1995. Intratumoral injection of an adenovirus expressing interleukin 2 induces regression and immunity in a murine breast cancer model. Proc. Natl. Acad. Sci. USA *92*:8522-8526.
2. **Addison, C.L., J.L. Bramson, M.M. Hitt, W.J. Muller, J. Gauldie, and F.L. Graham.** 1998. Intratumoral coinjection of adenoviral vectors expressing IL-2 and IL-12 results in enhanced frequency of regression of injected and untreated distal tumors. Gene Ther. *5*:1400-1409.
3. **Blezinger, P., J. Wang, M. Gondo, A. Quezada, D. Mehrens, M. French, A. Singhal, S. Sullivan et al.** 1999. Systemic inhibition of tumor growth and tumor metastases by intramuscular administration of the endostatin gene. Nat. Biotechnol. *17*:343-348.
4. **Burcin, M.M., G. Schiedner, S. Kochanek, S.Y. Tsai, and B.W. O'Malley.** 1999. Adenovirus-mediated regulatable target gene expression in vivo. Proc. Natl. Acad. Sci. USA *96*:355-360.
5. **Cao, Y., M.S. O'Reilly, B. Marshall, E. Flynn, R.W. Ji, and J. Folkman.** 1998. Expression of angiostatin cDNA in a murine fibrosarcoma suppresses primary tumor growth and produces long-term dormancy of metastases. J. Clin. Invest. *101*:1055-1063.
6. **Chen, L., M. Anton, and F.L. Graham.** 1996. Production and characterization of human 293 cell lines expressing the site-specific recombinase Cre. Somat. Cell Mol. Genet. *22*:477-488.
7. **Chen, L., D. Chen, E. Block, M. O'Donnell, D.W. Kufe, and S.K. Clinton.** 1997. Eradication of murine bladder carcinoma by intratumor injection of a bicistronic adenoviral vector carrying cDNAs for the IL-12 heterodimer and its inhibition by the IL-12 p40 subunit homodimer. J. Immunol. *159*:351-359.
8. **Chen, L., D. Chen, Y. Manome, Y. Dong, H.A. Fine, and D.W. Kufe.** 1995. Breast cancer selective gene expression and therapy mediated by recombinant adenoviruses containing the DF3/MUC1 promoter. J. Clin. Invest. *96*:2775-2782.
9. **Chen, L., D.J. Waxman, D. Chen, and D.W. Kufe.** 1996. Sensitization of human breast cancer cells to cyclophosphamide and ifosfamide by transfer of a liver cytochrome P450 gene. Cancer Res. *56*:1331-1340.
10. **Chen, S.H., X.H. Chen, Y. Wang, K. Kosai, M.J. Finegold, S.S. Rich.** 1995. Combination gene therapy for liver metastasis of colon carcinoma in vivo. Proc. Natl. Acad. Sci. USA *92*:2577-2581.
11. **Chow, L.T., J.B. Lewis, and T.R. Broker.** 1980. RNA transcription and splicing at early and intermediate times after adenovirus-2 infection. Cold Spring Harbor Symp. Quant. Biol. *44* (Pt 1) :401-414.
12. **Dong, Y., P. Wen, Y. Manome, M. Parr, A. Hirshowitz, L. Chen, E.A. Hirshowitz, R. Crystal et al.** 1996. In vivo replication-deficient adenovirus vector-mediated transduction of the cytosine deaminase gene sensitizes glioma cells to 5-fluorocytosine. Hum. Gene Ther. *7*:713-720.
13. **Folkman, J.** 1998. Antiangiogenic gene therapy. Proc. Natl. Acad. Sci. USA *95*:9064-9066.
14. **Graham, F.L. and L. Prevec.** 1991. Gene transfer and expression protocols, p. 109-128. *In* E.J. Murray (Ed.), Methods in Molecular Biology. Humana Press, Clinton, NJ.
15. **Graham, F.L., J. Smiley, W.C. Russell, and R. Nairn.** 1977. Characteristics of a human cell line transformed by DNA from human adenovirus type 5. J. Gen. Virol. *36*:59-74.
16. **Imler, J.L.** 1995. Adenovirus vectors as recombinant viral vaccines. Vaccine *13*:1143-1151.
17. **Kochanek, S., P.R. Clemens, K. Mitani, H.H. Chen, S. Chan, and C.T. Caskey.** 1996. A new adenoviral vector: replacement of all viral coding sequences with 28 kb of DNA independently expressing both full-length dystrophin and beta-galactosidase. Proc. Natl. Acad. Sci. USA *93*:5731-5736.
18. **Mitani, K., F.L. Graham, C.T. Caskey, and S. Kochanek.** 1995. Rescue, propagation, and partial purification of a helper virus-dependent adenovirus vector. Proc. Natl. Acad. Sci. USA *92*:3854-3858.
19. **Morsy, M.A. and C.T. Caskey.** 1999. Expanded-capacity adenoviral vectors—the helper-dependent vectors. Mol. Med. Today *5*:18-24.
20. **Morsy, M.A., M. Gu, S. Motzel, J. Zhou, J. Lin, Q. Su, H. Allen, L. Franlin et al.** 1998. An adenoviral vector deleted for all viral coding sequences enhanced safety and extended expression of the leptin transgene. Proc. Natl. Acad. Sci. USA *95*:7866-7871.
21. **Morsy, M.A., D. Harvey, and C.T. Caskey.** 1999.

Helper-dependent (gutless) adenoviral vectors as gene delivery/gene regulation vehicles. Gene Ther. Mol. Biol. *3*:223-232.

22. **Parks, R.J., L. Chen, M. Anton, U. Sankar, M.A. Rudnicki, and F.L. Graham.** 1996. A helper-dependent adenovirus vector system: removal of helper virus by Cre-mediated excision of the viral packaging signal. Proc. Natl. Acad. Sci. USA *93*:13565-13570.
23. **Parks, R.J. and F.L. Graham.** 1997. A helper-dependent system for adenovirus vector production helps define a lower limit for efficient DNA packaging. J. Virol. *71*:3293-3298.
24. **Putzer, B.M., J.L. Bramson, C.L. Addison, M. Hitt, P.M. Siegal, W.J. Muller, and F.L. Graham.** 1998. Combination therapy with interleukin-2 and wild-type p53 expressed by adenoviral vectors potentiates tumor regression in a murine model of breast cancer. Hum. Gene Ther. *9*:707-718.
25. **Putzer, B.M., M. Hitt, W.J. Muller, P. Emtage, J. Gauldie, and F.L. Graham.** 1997. Interleukin 12 and B7-1 costimulatory molecule expressed by an adenovirus vector act synergistically to facilitate tumor regression. Proc. Natl. Acad. Sci. USA *94*:10889-10894.
26. **Rivera, V.M., T. Clackson, S. Natesan, R. Pollock, J.F. Amara, T. Keenan, S.R. Magari, T. Phillips et al.** 1996. A humanized system for pharmacologic control of gene expression. Nat. Med. *2*:1028-1032.
27. **Schiedner, G., N. Morral, R.J. Parks, Y. Wu, S.C. Koopmans, C. Langston, F.L. Graham, A.L. Beaudet, and S. Kochanek.** 1998. Genomic DNA transfer with a high-capacity adenovirus vector results in improved in vivo gene expression and decreased toxicity. Nat. Genet. *18*:180-183.
28. **Tanaka, T., Y. Manome, P. Wen, D.W. Kufe, and H.A. Fine.** 1997. Viral vector-mediated transduction of a modified platelet factor 4 cDNA inhibits angiogenesis and tumor growth. Nat. Med. *3*:437-442.

9 Generation of Multiply Deleted Adenovirus Vectors that are Helper Virus Independent

Andrea Amalfitano
Department of Pediatrics, Division of Medical Genetics, Duke University Medical Center, Durham, NC, USA

INTRODUCTION

The causes of genetic diseases are numerous and complex; yet, in many instances, conventional therapies are limited or nonexistent. Most gene therapy approaches attempt to subvert, correct, or prevent the pathogenic mechanisms of specific genetic disease processes by the introduction of corrective genetic information into appropriate target cells. There are a variety of systems currently being investigated for their potential use as gene transfer vectors in vivo. The use of nonviral vectors (i.e., liposomes, molecular conjugates, and naked DNA) can minimize host immune responses directed against the vector, but in many situations they cannot introduce genetic material in vivo with the efficiency of virus-based vectors. Virus-based vectors can transduce a significant number of cells in vivo, but they also have limitations. One of the more commonly used virus-based vectoring systems has been derived from the modification of the human adenovirus (Ad).

Ad-based vectors hold great promise for the genetic therapy of a number of human disorders. Ad vectors can infect a variety of both dividing and nondividing cells in vivo, such as those found in the liver, kidney, muscle (skeletal and cardiac), respiratory, and nervous systems (5,6,11,13,18,19,34, 38). Diseases that affect these cell types are therefore potentially amenable to Ad-mediated gene therapy. Conventional Ad vectors have a large cloning capacity (theoretically approaching 8 kb), and they can be concentrated to very high titers, more than 10^{12} particles/mL. The high titer preparations contribute enormously to the ability of Ad vectors to infect and transduce a significant number of target cells after a single administration in vivo. For example, more than 90% of all the liver cells in a mouse can be demonstrated to express an Ad-delivered transgene after a single tail vein administration—an ability that cannot be approximated by any other vectoring system currently under development (20,52). Ad vectors can express transgenes episomally and have never been demonstrated to integrate into the host genome. Therefore, the risk of inadvertently activating an endogenous oncogene is minimized with

Viral Vectors: Basic Science and Gene Therapy
Edited by A. Cid-Arregui and A. García-Carrancá

the use of Ad vectors. Despite these significant attributes, there are major shortcomings of first-generation (FG) [E1-] Ad vectors that must be addressed before their full clinical potential may be realized. The most serious problem with Ad vectors is their inability to persist in immune-competent animals. If one is to understand some of the limitations of Ad vectors, the author strongly encourages the reader to become intimate with the details of the normal Ad life cycle, described in Chapter 7.

Briefly, FG Ad vectors are constructed in a manner such that a transgene (flanked by an appropriate enhancer/promoter element) replaces the Ad E1 genes, thereby generating an [E1-] Ad vector. The recombinant Ad vector genomes can be replicated and subsequently packaged only in human cells that express high levels of the E1 proteins *in trans* (16). Several interesting observations have been noted when [E1-] Ad vectors are utilized in vivo. Many in vivo models of [E1-] Ad-mediated gene transfer demonstrated that vector persistence is limited to two to three weeks, at which time host immune responses eliminated the transduced cells. Furthermore, a potent neutralizing antibody response is generated after initial viral infection (a limitation common to all virus-based vectoring systems) that precludes the readministration of Ad vectors into immune-competent animals. In contrast, the infection of immune-deficient animals can allow for long term persistence of Ad vectors in vivo (42,54,55). [E1-] Ad infection of animals deficient solely for immunoglobulin production demonstrated that transgene expression was also transient, suggesting that the humoral arm of the immune response was not responsible for the loss of [E1-] Ad-transduced cells (50,51). Numerous studies have demonstrated that a CD8$^+$ cytotoxic T lymphocyte (CTL)-mediated immune response is likely responsible for the elimination of Ad vector-infected tissues in vivo (14,49,52).

The elimination of [E1-] Ad vector-infected cells by the immune system can in part be explained by the following observations. Theoretically, the deletion of the E1 gene should block the Ad life cycle just after viral entry into any host cell lacking E1 activity; see Chapter 7. However, it has been known that in tissue culture systems [E1-] Ad are not completely replication defective. For example, after [E1-] Ad infection of HeLa cells, the E2, E3, and E4 promoters are known to be active and allow viral replication to proceed, despite the deletion of the E1 gene in the viral backbone (32,33). When utilized in vivo, [E1-] Ad vectors have also been demonstrated to express the early genes, undergo genome replication, and express the L1→L5 structural proteins (49,52,53). More dramatically, human and nonhuman primate trials utilizing Ad vectors demonstrated that [E1-] Ad vectors can have associated toxicity, possibly because of their E1-independent abilities to express Ad genes (10). Thus, the sole deletion of E1 sequences in [E1-] Ad vectors is insufficient to prevent Ad vector replication and/or gene expression and likely contributes to the immune recognition of [E1-] Ad-infected cells in vivo.

Several studies have suggested that foreign protein expression derived from the transgene is the major determinant responsible for the immune-mediated elimination of [E1-] Ad-infected cells (27,29,44). However, several vectoring systems (such as adeno-associated virus [AAV] vectors, and direct DNA injection) can stably transduce foreign transgenes into immune-competent hosts (20,47,48).

Based upon these observations, a general hypothesis can be reached. Perhaps the isolation of Ad vectors that are unable to express critical Ad genes in addition to E1 will not trigger as potent an immune response in vivo, thereby resulting in persistent transgene expression. Two forms of Ad vector modification have evolved in response to the need to improve [E1-] Ad vectors. The first involves the isolation of "empty-Ad" vectors that are helper virus-dependent for growth, but devoid of most of the Ad genome (see Chapter 8). The second form of Ad vector modification involves the use of multiply deleted, helper virus-independent Ad vectors, a form of vector modification our group has focused upon.

WHY GENERATE MULTIPLY DELETED Ad VECTORS?

Our group has concentrated on the construction and isolation of multiply deleted, helper virus-independent Ad vectors. There are several theoretical benefits of such vectors, including increased carrying capacities, decreased immunogenicity, and a decreased propensity to generate replication-competent Ad (RCA) during their propagation. One of the important goals of our studies has been to achieve these advantages without forfeiting one of the key attributes of [E1-] Ad vectors, namely their ability to be grown to high levels in the absence of any contaminating helper viruses. This characteristic is critical when considering the large-scale production of vectors for potential clinical use. Helper virus-independent, multiply deleted Ad vectors are not dependent upon CsCl-based centrifugation methods for their isolation and/or purification; therefore, they are still amenable to all large-scale procedures (such as anion exchange chromatography) currently available to mass produce [E1-] Ad vectors.

In contrast, empty-Ad systems currently require some form of centrifugation method to separate the helper virus from the fully deleted vector because the only difference between the two viruses is their buoyant densities (8,12,17,23,24,26,28, 35). Although the absolute amounts of helper virus packaged during empty-Ad production can be limited by *cre/lox*-mediated excision of the helper virus packaging signals, all preparations to date are contaminated with significant amounts of helper virus (0.1%–1%) (17,30,35,41). As a further caveat, some of the purest helper virus-dependent Ad vector preparations have been found to be dependent upon low levels of helper virus contamination (and *trans*-acting functions) for the stability of their highly modified genomes (25,26). The impact of helper virus contamination upon host immune responses, genomic stability, and the feasibility of high level production must be considered by investigators utilizing empty-Ad vectors.

The isolation and production of multiply deleted (helper virus-independent) Ad vectors also has a number of limitations that must also be addressed. These vectors are critically dependent upon the isolation of packaging cell lines for their high level growth. Therefore, packaging cell lines, rather than a helper virus, supply each of the deleted gene functions *in trans*.

We have previously described the generation of a new series of Ad packaging cell lines co-expressing the Ad E1 and polymerase genes, or the E1, polymerase, and the preterminal protein (pTP) genes without the need for inducible expression of the respective genes (1,2). With these cell lines, the production of new Ad vectors deleted for each of these gene activities is now possible (3). The isolation of Ad packaging cell lines co-expressing other Ad genes, such as the E1 and E4 genes, the E1 and E2a (single-stranded DNA binding protein [ssDBP]) genes, the E1, E4, and protein IX genes, and the E1 and pTP genes have also been described (7,15,40,46,56). Our selection of the polymerase and pTP genes as targets for vector modification was predicted to significantly improve overall Ad vector performance in vivo because each of the respective gene activities are critically important for several aspects of the Ad life cycle described below.

Ad REPLICATION AND LATE GENE EXPRESSION

The Ad genome is a linear, double-stranded DNA molecule that is 35 995 bp in length. The genome is divided into 100 map units (mu), each mu being equal to approximately 360 bp (Figure 1). The three Ad replication proteins are encoded within the E2 region of the Ad genome. Each of these three proteins are absolutely required for replication of the linear viral genome (22,37,45). For example, in the presence of these three proteins, at least three host nuclear proteins, and the inverted terminal repeats (ITRs) at each end of the viral genome, Ad DNA replication can be demonstrated to proceed in vitro (9,22). The Ad E1 proteins normally act to up-regulate E2 promoter activity, thereby resulting in the high level transcription of a 23 kb primary RNA transcript. Two mRNAs are spliced from the E2 primary RNA; the E2a mRNA, which encodes the 72 kDa ssDBP; and the E2b mRNA, which is differentially spliced to yield the 87 kDa pTP; and the 140 kDa DNA polymerase proteins. With the expression of these three proteins, Ad genome replication can proceed.

Ad replication is the event that marks a switch in the viral life cycle from the early to the late phases, culminating in high-level transcription from the Ad major late promoter (MLP). Transcription from the MLP results in the translation of the capsid proteins and the subsequent packaging of the replicated Ad genomes (36). Significantly, if a cell that is in the late phase of primary Ad infection is superinfected with a second Ad, transcription from the MLP of the second Ad cannot be demonstrated until it first replicates its genome (43). Therefore, high-level MLP-derived transcription is in part a *cis*-dependent activation event—an event that is dependent upon Ad genome replication. This sophisticated form of temporally regulated gene expression is common to the biology of many DNA viruses, such as polyoma and simian virus 40 (SV40) (4,21,39).

We had surmised that the E2 region of genes would be a desirable target for modification in Ad vectors for a variety of reasons. Their deletion would be predicted to

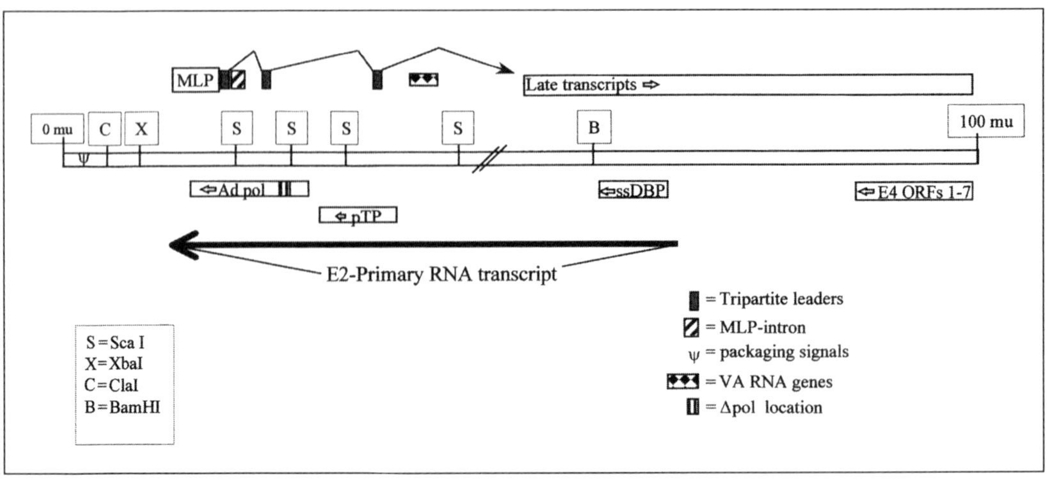

Figure 1. Map of the Ad genome.

result in new vectors incapable of replicating the Ad genome even in the presence of high levels of E1 activity. The replication blockade would significantly prevent the *cis*-mediated activation of the MLP and result in a dramatic lack of expression of the late genes without the need for their physical deletion. Therefore, a single gene deletion of the polymerase gene could potentially result in the lack of its expression, as well as that of major antigenic epitopes (the viral structural proteins) transcribed from the MLP.

The ability to grow an Ad vector with multiple gene deletions requires the construction and isolation of new Ad packaging cell lines capable of supplying the respective gene functions *in trans*. Therefore, a screening method was developed to successfully isolate new Ad packaging cell lines that co-expressed the E1, polymerase, and pTP genes.

ISOLATION OF PACKAGING CELL LINES CO-EXPRESSING THE Ad E1, POLYMERASE, AND PRETERMINAL PROTEINS

Transfection was undertaken into human 293 cells (293 cells constitutively express the E1 genes) of hygromycin resistance plasmids encoding the functional polymerase gene under the transcriptional control of the Rous sarcoma virus (RSV) enhancer/long terminal repeat (LTR) elements. Hygromycin-resistant cell lines were screened for the ability to support the growth and replication of Ad viruses with temperature-sensitive (*ts*) mutations within the Ad polymerase genes. These viruses were the only viable mutants available for these genes at the time. The *ts* mutants are defective for normal growth at the nonpermissive temperature of 38.5°C. Of 20 hygromycin-resistant cell clones, seven were found to complement the high-level growth of H5*ts*36 (H5*ts*36 contained a *ts* mutation within the Ad polymerase gene) at the nonpermissive temperature (1). This was the first report demonstrating that two Ad genes (E1 and polymerase) could be constitutively co-expressed within a cell without the need for the inducible control of their expression. Five of the cell lines that had received the pRSV-Adpol plasmid were also transfected with a pTP encoding plasmid. One clonal isolate derived from this transfection was capable of transcomplementing not only Ad polymerase mutants but also Ad-pTP mutants such as H5*sub*100 and H5*in*36. The latter represented the first Ad packaging cell line that constitutively expressed high levels of three Ad genes (E1, pTP, and polymerase) without apparent toxicity (2).

THE ISOLATION OF VECTORS DELETED FOR Ad POLYMERASE ACTIVITY

The main open reading frames (ORFs) of the Ad polymerase and pTP genes span approximately 5.5 kb of the viral genome and are transcribed (by convention) in a right-to-left orientation. A deletion within the polymerase ORF of 608 bp (Figure 1) was engineered into a new shuttling plasmid, pAdALΔpol (Figure 2). This plasmid allowed for the rapid and efficient isolation of Ad vectors deleted for the Ad E1, polymerase, and E3 genes. We co-transfected the pAdALΔpol/cytomegalovirus (CMV)/β-galactosidase (LacZ) shuttle plasmid (a bacterial β-galactosidase transgene expression cassette was subcloned into pAdALΔ–pol) with digested Ad virion DNA (isolated from an [E3-] Ad, *dl*7001) into the packaging cell lines (Figure 3). After co-transfection, the cells were separated. Those cells that showed evidence of viral cytopathic effect (CPE) were harvested, and the subcloned Ad vector was amplified. This procedure allowed us to isolate AdLacZΔpol, a new vector deleted for the E1, polymerase, and E3 gene activities. A restriction digest of the vector genome demonstrated that AdLacZΔpol retained all regions of the Ad genome except an approximately 2900-bp E1 deletion (replaced by the CMV-LacZ-pA transgene cassette), a 608-bp deletion within the polymerase gene, and the approximately 2700-bp E3 deletion (Figure 4).

The kinetics of AdLacZΔpol growth in either 293 cells or B-6 cells was determined. One-step burst assays of 2×10^6 B-6 cells infected at a multiplicity of infection (moi) of 5 with either AdLacZΔpol or Adsub360LacZ (an E1, E3- traditional, or FG Ad vector) demonstrated that the AdLacZΔpol vector could be produced to as high a titer as Adsub360LacZ (3). High titer growth was important not only for future clinical production issues but also because some previously described cell lines designed to allow the growth of modified Ad vectors can be inefficient because of the cumulative toxicity of the co-expressed Ad genes (57).

AdLacZΔpol IS BLOCKED IN REPLICATION

Infection of the [E1+, polymerase+] cells allowed high-level replication of the AdLacZΔpol genomes, unlike an identical infection of 293 cells (3). Furthermore, infection of 293 cells with an [E1+, polymerase-] Ad also prevented replication of the modified Ad (3). This series of results demonstrated that even in the presence of excessive levels of E1 activities, the polymerase deletion conveyed a severe replication blockade upon vectors containing this deletion.

In support of these observations, high titer infection of HeLa cells (lacking both E1 and polymerase activities) with AdLacZΔpol demonstrated that the vector

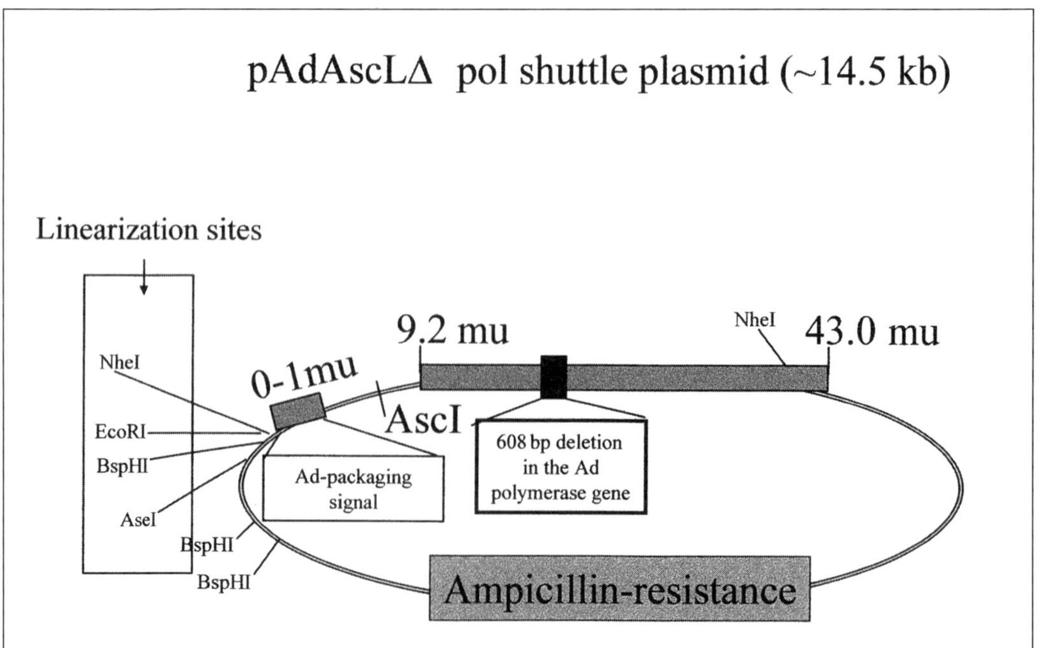

Figure 2. The pAdAscLΔpol shuttle plasmid. Transgenes may be inserted into the *Asc*I site for eventual [E1-, polymerase-] Ad vector production.

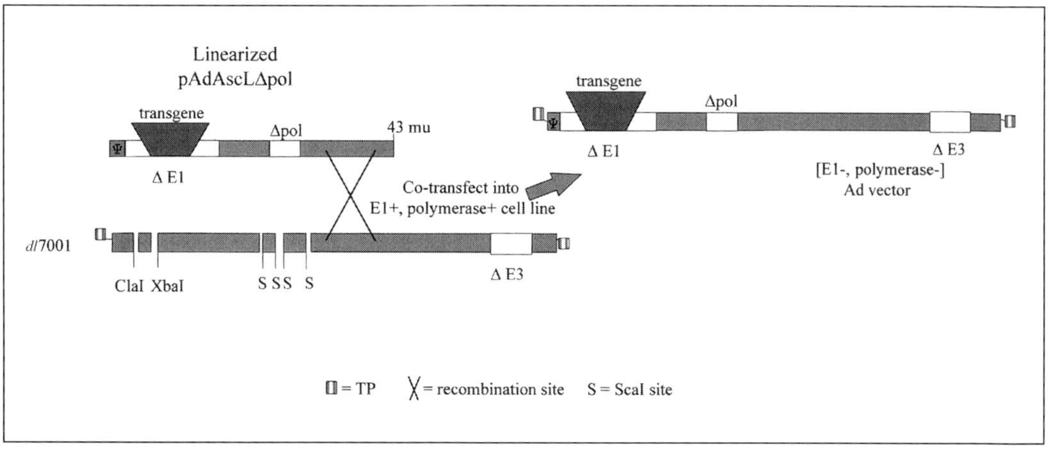

Figure 3. Co-transfection/recombination strategy to generate multiply deleted Ad vectors. The linearized shuttle plasmid containing the transgene of interest is co-transfected into the packaging cells along with restriction enzyme-digested virion DNA. Recombination between homologous portions of the shuttle plasmid and the virus derived right-end sequences results in the generation of the desired [E1-, polymerase-] Ad vector. Note that the *dl*7001 virus contains a deletion within the E3 genes; these gene activities are not required for growth of the vectors in tissue culture.

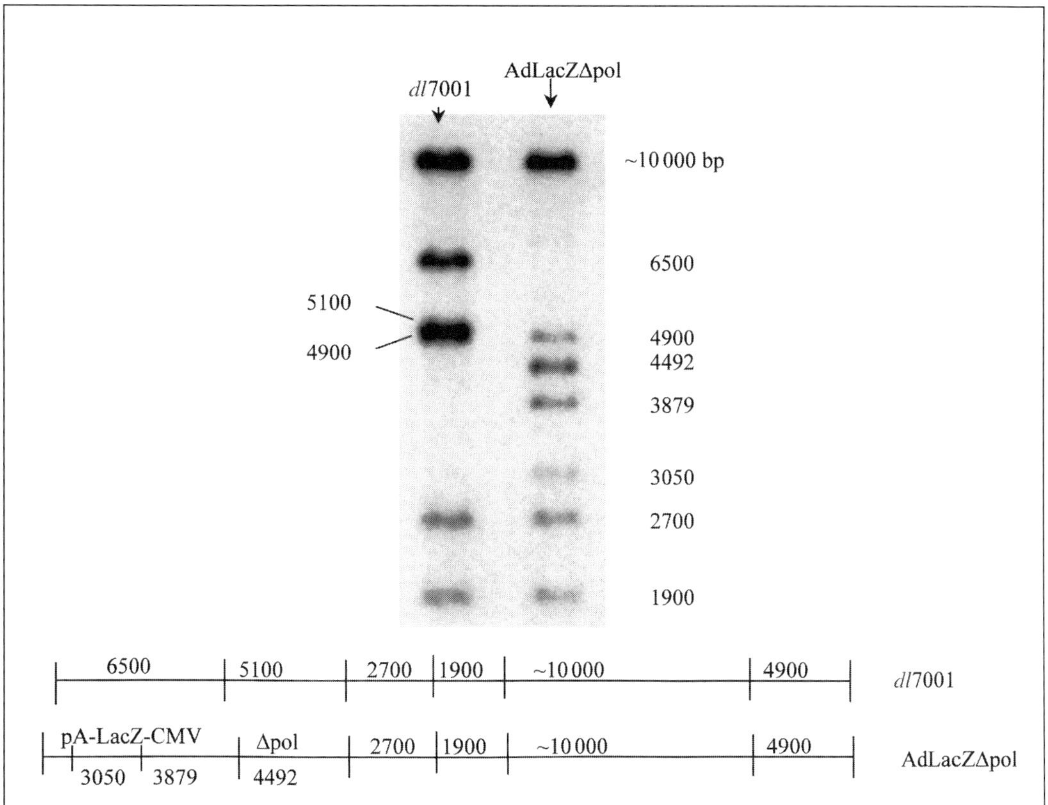

Figure 4. Restriction enzyme mapping of AdLacZΔpol DNA. [E1+, polymerase+] cells were infected at an moi of 5 with AdLacZΔpol; total DNA was extracted, digested with *Eco*RI and *Not*I, electrophoretically separated, blotted, and probed with an Ad genome-specific probe. The predicted structure is depicted below the image.

was blocked in its ability to replicate, in contrast to the [E1-] Adsub360LacZ vector (Figure 5). In summary, our replication studies demonstrated that [polymerase-] Ad viruses were severely blocked in their ability to replicate (even in the presence of excessive levels of E1 activity)—a blockade that did not simultaneously result in a rapid loss or destabilization of their genomes.

AdLacZΔpol LATE GENE EXPRESSION IS BLOCKED IN NONCOMPLEMENTING CELL LINES

The late gene expression blockade predicted to occur by the deletion of E1 and polymerase activities was demonstrated after HeLa cells were infected with either the Adsub360LacZ or AdLacZΔpol vectors (Figure 6). Fiber expression was readily detected after infection of HeLa cells with the Adsub360LacZ vector, due to post-replication MLP activation. In contrast, infection with the AdLacZΔpol vector (even at an moi of 500) did not result in detectable fiber expression in Hela cells. These results demonstrated that [E1-, polymerase-] Ad vectors have a significantly decreased potential to express viral late genes (at least 100-fold less late protein expressed than [El-] FG vectors), because of the severe replication blockade afforded by the presence of both the E1 and polymerase deletions (31).

IN VIVO TRANSDUCTION WITH THE AdLacZΔpol VECTOR

We were readily able to generate high titer stocks of the AdLacZΔpol vector (>1×10^{12} bacterial β-galactosidase-forming units [BFU]/mL). Therefore, 1×10^9 BFU of AdLacZΔpol were injected intramuscularly into outbred C3H immune-competent mice. Twenty-eight days after infection, β-galactosidase expression was still evident in AdLacZΔpol-infected mice, in contrast to a lack of activity in mice identically infected with the Adsub360LacZ (Figure 7). Therefore, the deletion of both the E1 and polymerase genes in the AdLacZΔpol vector allowed for persistent transgene expression in vivo. This result has also been confirmed in the livers of C57BL/6 mice that were nontolerant of the bacterial β-galactosidase transgene; in fact, the modified Ad vector was found to be capable of persisting for more than two months (experiment duration) in the latter situation (20). The demonstration that multiple vector modifications can overcome the immune respons-

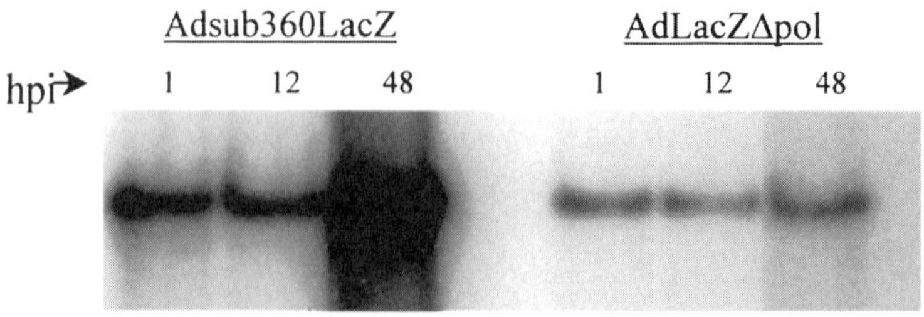

Figure 5. Multiply deleted Ad vectors cannot replicate in HeLa cells. HeLa cells were infected at an moi of 50 with the indicated vectors; total DNA was harvested at the indicated time points, resuspended in 200 µL of 10 mmol/L Tris-HCl, 1 mmol/L EDTA, and 20 µL of each sample were digested with *Eco*RI. The DNA samples were then electrophoretically separated and visualized with a probe specific to the approximately 25-kb subfragment of the Ad genome common to both vectors. Note that despite a complete lack of replication, the AdLacZΔpol vector genomes were present at near input levels even at the extended time points.

es previously noted to occur after the Ad-mediated expression of a foreign gene is critical. Our results suggest that the Ad-mediated transfer of a foreign transgene elicits two stimuli—one derived from the vector (*hit 1*) and one derived from the transgene (*hit 2*). The presence of both *hits* after Ad infection results in the immune-mediated recognition and elimination of the Ad-infected cells, a concept we have referred to as the two-*hit* hypothesis (20). The use of the multiply deleted Ad vector abrogated the *hit 1* stimulus and prevented the immune-mediated elimination of the infected cells, despite the transduction of a foreign transgene (*hit 2*). This hypothesis suggests that the complete analysis of modified Ad vector performance should always include evaluations of their ability to persist after the transfer of foreign genes into non-tolerant, immune-competent hosts.

SUMMARY

We have demonstrated that the modification of Ad vectors can improve their biological profiles. The evaluation of this class of vector has shed light upon several important aspects of Ad vector biology as well. For example, despite the profound replicational blockade afforded by the codeletion of E1 and polymerase gene activities, the [E1-, polymerase-] Ad vector genome is stable both in vitro and in vivo. Therefore, low-level replication of Ad vec-

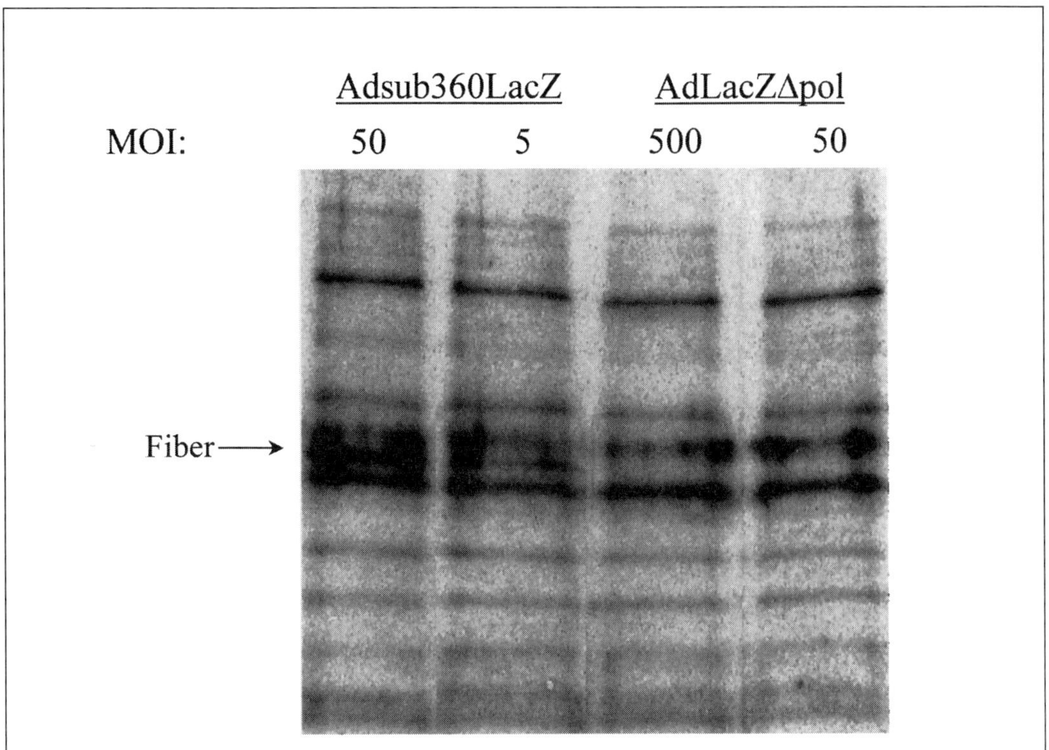

Figure 6. Blockade of late gene expression with the use of [E1-, polymerase-] Ad vectors. HeLa cells were infected at the indicated moi with either the FG [E1-] Ad vector or the multiply deleted [E1-, polymerase-] Ad vector. Twenty hours after infection, protein extracts of the cells were harvested; equivalent amounts of the extracts were electrophoretically separated, electroblotted onto a nylon membrane, and probed with a fiber-specific polyclonal antibody (kindly provided by R. Gerard, University of Michigan).

tor genomes is not required for Ad persistence in vivo. The replication blockade occurred even in the presence of excessive amounts of E1 activity, a useful attribute when considering gene transfer into tissues thought to have high levels of E1-like activity, such as hepatocytes. This class of vector will also have significantly increased carrying capacities theoretically approaching 11 kb with further deletion of the Ad polymerase ORF. [E1-, polymerase-] Ad vectors are also less likely to generate replication-competent Ads (RCAs) during large-scale production. This is because

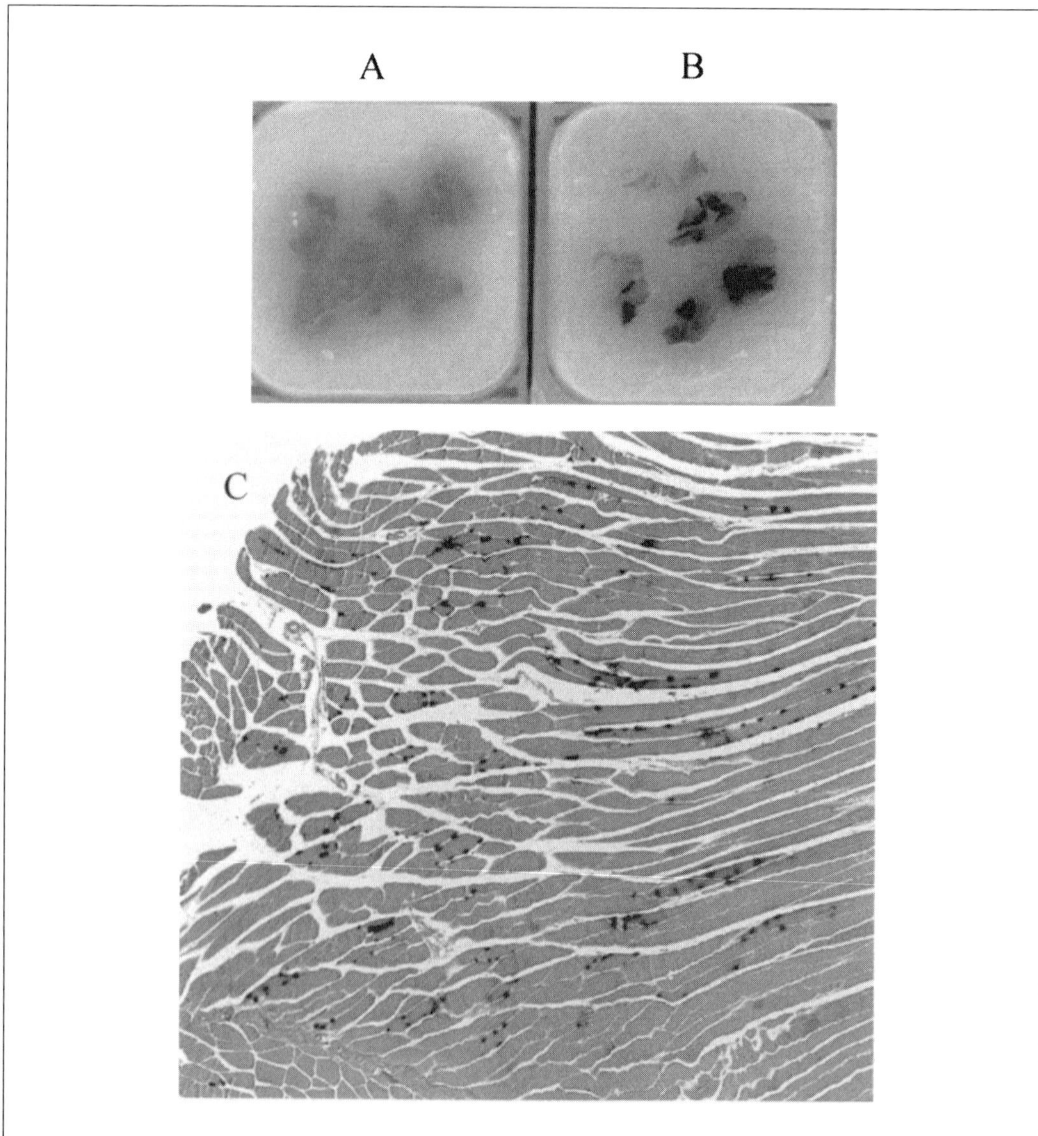

Figure 7. Persistence of foreign gene expression in immune-competent mice with the use of an [E1-, polymerase-] Ad vector. Outbred C3H mice were intramuscularly injected with 1×10^9 BFU of a conventional [E1-] LacZ Ad vector (A) or with the multiply deleted [E1-, polymerase-] LacZ Ad vector (B). Animals were sacrificed after 28 days, and the injected muscles were stained with 5-bromo-4 chloro-3 indolyl β-D-galactopyranoside (X-Gal) in situ. (C) A higher magnification view of panel B is shown, which demonstrates that LacZ staining was localized to the nuclei of the muscle cells.

multiple recombination events between the vector genome and the integrated polymerase and E1 sequences would theoretically need to occur before regeneration of an RCA could ensue. RCA contamination is an issue that limits all modes of high-level Ad vector production. Finally, multiply deleted Ad vectors can persist after the transfer of foreign transgenes, thereby overcoming one of the most critical limitations of these vectors in vivo (20).

Despite these benefits, multiply deleted Ad vectors still retain significant portions of the Ad genome. These sequences potentially encode noxious viral proteins that may limit the overall usefulness of these vectors in specific clinical situations. Because of this, the benefit-to-risk ratio of any vector must be ascertained under conditions that mimic the clinical reality of specific disease states.

The isolation of an Ad vector deleted for both the Ad polymerase and E1 activities has been proven to be achievable. The further analysis of this class of vector will improve our understanding of the biology and limitations of Ad vectors and may one day allow for the routine use of gene therapies in a variety of patient populations.

ACKNOWLEDGMENT

The author would like to acknowledge support from the NIH, ROI# DK-52925-01.

REFERENCES

1. **Amalfitano, A., C.R. Begy, and J.S. Chamberlain.** 1996. Improved adenovirus packaging cell lines to support the growth of replication-defective gene-delivery vectors. Proc. Natl. Acad. Sci. USA *93*:3352-3356.
2. **Amalfitano, A. and J.S. Chamberlain.** 1997. Isolation and characterization of packaging cell lines that coexpress the adenovirus E1, DNA polymerase, and preterminal proteins—implications for gene therapy. Gene Ther. *4*:258-263.
3. **Amalfitano, A., M.A. Hauser, H. Hu, D. Serra, C.R. Begy, and J.S. Chamberlain.** 1998. Production and characterization of improved Adenovirus vectors with the E1, E2b, and E3 genes deleted. J. Virol. *72*:926-933.
4. **Amalfitano, A., L.G. Martin, and M.M. Fluck.** 1992. Different roles for two enhancer domains in the organ- and age-specific pattern of polyomavirus replication in the mouse. Mol. Cell. Biol. *12*:3628-3635.
5. **Askari, F.K., Y. Hitomi, M. Mao, and J.M. Wilson.** 1996. Complete correction of hyperbilirubinemia in the Gunn rat model of Crigler-Najjar syndrome Type I following transient in vivo adenovirus-mediated expression of human bilirubin UDP-glucuronosyltransferase. Gene Ther. *3*:381-388.
6. **Barr, E., J. Carroll, A.M. Kalynych, S.K. Tripathy, K. Kozarsky, J.M. Wilson, and J.M. Leiden.** 1994. Efficient catheter-mediated gene transfer into the heart using replication-defective adenovirus. Gene Ther. *1*:51-58.
7. **Brough, D.E., A. Lizonova, C. Hsu, V.A. Kulesa, and I. Kovesdi.** 1996. A gene transfer vector-cell line system for complete functional complementation of adenovirus early regions E1 and E4. J. Virol. *70*:6497-6501.
8. **Clemens, P.R., S. Kochanek, Y. Sunada, S. Chan, H.H. Chen, K.P. Campbell, and C.T. Caskey.** 1996. In vivo muscle gene transfer of full-length dystrophin with an adenoviral vector that lacks all viral genes. Gene Ther. *3*:965-972.
9. **Coenjaerts, F.E. and P.C. Van Der Vliet.** 1995. Adenovirus DNA replication in a reconstituted system. Methods Enzymol. *262*:548-560.
10. **Crystal, R.G., A. Jaffe, S. Brody, A. Mastrangeli, N.G. McElvaney, M. Rosenfeld, C.S. Chu, C. Danel, J. Hay, and T. Eissa.** 1995. A phase 1 study, in cystic fibrosis patients, of the safety, toxicity, and biological efficacy of a single administration of a replication deficient, recombinant adenovirus carrying the cDNA of the normal cystic fibrosis transmembrane conductance regulator gene in the lung. Hum. Gene Ther. *6*:643-666.
11. **Engelhardt, J.F., R.H. Simon, Y. Yang, M. Zepeda, S. Weber-Pendleton, B. Doranz, M. Grossman, and J.M. Wilson.** 1993. Adenovirus-mediated transfer of the CFTR gene to lung of nonhuman primates: biological efficacy study. Hum. Gene Ther. *4*:759-769.
12. **Fisher, K.J., H. Choi, J. Burda, S.J. Chen, and J.M. Wilson.** 1996. Recombinant adenovirus deleted of all viral genes for gene therapy of cystic fibrosis. Virol. *217*:11-22.
13. **Ghadge, G.D., R.P. Roos, U.J. Kang, R. Wollmann, P.S. Fishman, A.M. Kalynych, E. Barr, and J.M. Leiden.** 1995. CNS gene delivery by retrograde transport of recombinant replication-defective adenoviruses. Gene Ther. *2*:132-137.
14. **Gilgenkrantz, H., D. Duboc, V. Juillard, D. Couton, A. Pavirani, J.G. Guillet, P. Briand, and A. Kahn.** 1995. Transient expression of genes transferred in vivo into heart using first-generation adenoviral vectors: role of the immune response. Hum. Gene Ther. *6*:1265-1274.
15. **Gorziglia, M.I., M.J. Kadan, S. Yei, J. Lim, G.M.**

Lee, R. Luthra, and B.C. Trapnell. 1996. Elimination of both E1 and E2a from adenovirus vectors further improves prospects for in vivo human gene therapy. J. Virol. *70*:4173-4178.

16. **Graham, F.L., J. Smiley, W.C. Russell, and R. Nairn.** 1977. Characteristics of a human cell line transformed by DNA from human adenovirus type 5. J. Gen. Virol. *36*:59-74.

17. **Hardy, S., M. Kitamura, T. Harrisstansil, Y.M. Dai, and M.L. Phipps.** 1997. Construction of adenovirus vectors through *cre-lox* recombination. J. Virol. *71*: 1842-1849.

18. **Heikkila, P., T. Parpala, O. Lukkarinen, M. Weber, and K. Tryggvason.** 1996. Adenovirus-mediated gene transfer into kidney glomeruli using an ex vivo and in vivo kidney perfusion system: first steps towards gene therapy of Alport syndrome. Gene Ther. *3*:21-27.

19. **Horellou, P., E. Vigne, M.N. Castel, P. Barneoud, P. Colin, M. Perricaudet, P. Delaere, and J. Mallet.** 1994. Direct intracerebral gene transfer of an adenoviral vector expressing tyrosine hydroxylase in a rat model of Parkinson's disease. Neuroreport *6*:49-53.

20. **Hu, H., D. Serra, and A. Amalfitano.** 1999. Persistence of an [E1-, polymerase-] adenovirus vector despite transduction of a neoantigen into immune-competent mice. Hum. Gene Ther. *10*:355-364.

21. **Jansen-Durr, P., G. Mondesert, and C. Kedinger.** 1989. Replication-dependent activation of the adenovirus major late promoter is mediated by the increased binding of a transcription factor to sequences in the first intron. J. Virol. *63*:5124-5132.

22. **Kelly, T.J.** 1984. Adenovirus DNA replication, p. 271-308. *In* H.S. Ginsberg (Ed.), The Adenoviruses. Plenum Press, New York.

23. **Kochanek, S., P.R. Clemens, K. Mitani, H.H. Chen, S. Chan, and C.T. Caskey.** 1996. A new adenoviral vector—replacement of all viral coding sequences with 28 kb of DNA independently expressing both full-length dystrophin and beta-galactosidase. Proc. Natl. Acad. Sci. USA *93*:5731-5736.

24. **Kumar-Singh, R. and J.S. Chamberlain.** 1996. Encapsidated adenovirus minichromosomes allow delivery and expression of a 14 kb dystrophin cDNA to muscle cells. Hum. Mol. Genet. *5*:913-921.

25. **Lieber, A., C.Y. He, and M.A. Kay.** 1997. Adenoviral preterminal protein stabilizes mini-adenoviral genomes *in vitro* and *in vivo*. Nat. Biotechnol. *15*:1383-1387.

26. **Lieber, A., C.Y. He, I. Kirillova, and M.A. Kay.** 1996. Recombinant adenoviruses with large deletions generated by *cre*-mediated excision exhibit different biological properties compared with first-generation vectors *in vitro* and *in vivo*. J. Virol. *70*:8944-8960.

27. **Lusky, M., M. Christ, K. Rittner, A. Dieterle, D. Dreyer, B. Mourot, H. Schultz, F. Stoeckel, A. Pavirani, and M. Mehtali.** 1998. In vitro and in vivo biology of recombinant adenovirus vectors with E1, E1/E2a, or E1/E4 deleted. J. Virol. *72*:2022-2032.

28. **Mitani, K., F.L. Graham, C.T. Caskey, and S. Kochanek.** 1995. Rescue, propagation, and partial purification of a helper virus-dependent adenovirus vector. Proc. Natl. Acad. Sci. USA *92*:3854-3858.

29. **Morral, N., W. Oneal, H.S. Zhou, C. Langston, and A. Beaudet.** 1997. Immune responses to reporter proteins and high viral dose limit duration of expression with adenoviral vectors—comparison of E2a wild type and E2a deleted vectors. Hum. Gene Ther. *8*:1275-1286.

30. **Morsy, M.A., M.C. Gu, S. Motzel, J. Zhao, Q. Su, H. Allen, L. Franlin, R.J. Parks et al.** 1998. An adenoviral vector deleted for all viral coding sequences results in enhanced safety and extended expression of a leptin transgene. Proc. Natl. Acad. Sci. USA *95*:7866-7871.

31. **Mullbacher, A., A.J. Bellett, and R.T. Hla.** 1989. The murine cellular immune response to adenovirus type 5. Immunol. Cell. Biol. *67*:31-39.

32. **Nevins, J.R.** 1981. Mechanism of activation of early viral transcription by the adenovirus E1a gene product. Cell *26*:213-220.

33. **Nevins, J.R., M.J. Imperiale, H.T. Kao, S. Strickland, and L.T. Feldman.** 1984. Detection of an adenovirus E1a-like activity in mammalian cells. Curr. Top. Microbiol. Immunol. *113*:15-19.

34. **Ohashi, T., K. Watabe, Y. Sato, I. Saito, J.A. Barranger, and Y. Eto.** 1995. Successful transduction of oligodendrocytes and restoration of arylsulfatase A deficiency in metachromatic leukodystrophy fibroblasts using an adenovirus vector. Gene Ther. *2*:443-449.

35. **Parks, R.J., L. Chen, M. Anton, U. Sankar, M.A. Rudnicki, and F.L. Graham.** 1996. A helper-dependent adenovirus vector system—removal of helper virus by cre-mediated excision of the viral packaging signal. Proc. Natl. Acad. Sci. USA *93*:13565-13570.

36. **Philipson, L.** 1984. Adenovirus assembly, p. 309-338. *In* H.S. Ginsberg (Ed.), The Adenoviruses. Plenum Press, New York.

37. **Pronk, R., M.H. Stuiver, and P.C. Van Der Vliet.** 1992. Adenovirus DNA replication: the function of the covalently bound terminal protein. Chromosoma *102*:S39-S45.

38. **Ragot, T., N. Vincent, P. Chafey, E. Vigne, H. Gilgenkrantz, D. Couton, J. Cartaud, P. Briand et al.** 1993. Efficient adenovirus-mediated transfer of a human minidystrophin gene to skeletal muscle of mdx mice. Nature *361*:647-650.

39. **Rio, D.C. and R. Tjian.** 1984. Multiple control elements involved in the initiation of SV40 late transcription. J. Mol. Appl. Genet. *2*:423-435.

40. **Schaack, J., X. Guo, W.Y. Ho, M. Karlok, C. Chen, and D. Ornelles.** 1995. Adenovirus type 5 precursor terminal protein-expressing 293 and HeLa cell lines. J. Virol. *69*:4079-4085.

41. **Schiedner, G., N. Morral, R.J. Parks, Y. Wu, S.C. Koopmans, C. Langston, F.L. Graham, A.L. Beaudet, and S. Kochanek.** 1998. Genomic DNA transfer with a high-capacity adenovirus vector results in improved in vivo gene expression and decreased toxicity. Nat. Genet. *18*:180-183.

42. **Stratford-Perricaudet, L.D., I. Makeh, M. Perricaudet, and P. Briand.** 1992. Widespread long-term gene transfer to mouse skeletal muscles and heart. J. Clin. Invest. *90*:626-630.

43. **Thomas, G.P. and M.B. Mathews.** 1980. DNA replication and the early to late transition in adenovirus infection. Cell *22*:523-533.

44. **Tripathy, S.K., H.B. Black, E. Goldwasser, and J.M. Leiden.** 1996. Immune responses to transgene-encoded proteins limit the stability of gene expression after injection of replication-defective adenovirus vectors. Nat. Med. *2*:545-550.
45. **Van Der Vliet, P.C.** 1995. Adenovirus DNA replication [review]. Curr. Top. Microbiol. Immunol. *199*:1-30.
46. **Wang, Q., X.C. Jia, and M.H. Finer.** 1995. A packaging cell line for propagation of recombinant adenovirus vectors containing two lethal gene-region deletion. Gene Ther. *2*:775-783.
47. **Wolff, J.A., J.J. Ludtke, G. Acsadi, P. Williams, and A. Jani.** 1992. Long-term persistence of plasmid DNA and foreign gene expression in mouse muscle. Hum. Mol. Genet. *1*:363-369.
48. **Xiao, X.A., J.A. Li, and R.J. Samulski.** 1996. Efficient long-term gene transfer into muscle tissue of immunocompetent mice by adeno-associated virus vector. J. Virol. *70*:8098-8108.
49. **Yang, Y., H.C. Ertl, and J.M. Wilson.** 1994. MHC class I-restricted cytotoxic T lymphocytes to viral antigens destroy hepatocytes in mice infected with E1-deleted recombinant adenoviruses. Immunity *1*:433-442.
50. **Yang, Y., K. Greenough, and J.M. Wilson.** 1996. Transient immune blockade prevents formation of neutralizing antibody to recombinant adenovirus and allows repeated gene transfer to mouse liver. Gene Ther. 3:412-420.
51. **Yang, Y., K.U. Jooss, Q. Su, H.C.J. Ertl, and J.M. Wilson.** 1996. Immune responses to viral antigens versus transgene product in the elimination of recombinant adenovirus-infected hepatocytes in vivo. Gene Ther. *3*:137-144.
52. **Yang, Y., F.A. Nunes, K. Berencsi, E.E. Furth, E. Gonczol, and J.M. Wilson.** 1994. Cellular immunity to viral antigens limits E1-deleted adenoviruses for gene therapy. Proc. Natl. Acad. Sci. USA *91*:4407-4411.
53. **Yang, Y., F.A. Nunes, K. Berencsi, E. Gonczol, J.F. Engelhardt, and J.M. Wilson.** 1994. Inactivation of E2a in recombinant adenoviruses improves the prospect for gene therapy in cystic fibrosis. Nat. Genet. *7*:362-369.
54. **Yang, Y. and J.M. Wilson.** 1995. Clearance of adenovirus-infected hepatocytes by MHC class I-restricted CD4+ CTLs in vivo. J. Immunol. *155*:2564-2570.
55. **Yang, Y.P., Z.Q. Xiang, H.C.J. Ertl, and J.M. Wilson.** 1995. Upregulation of class I major histocompatibility complex antigens by interferon gamma is necessary for T-cell-mediated elimination of recombinant adenovirus-infected hepatocytes in vivo. Proc. Natl. Acad. Sci. USA *92*:7257-7261.
56. **Yeh, P., J.F. Dedieu, C. Orsini, E. Vigne, P. Denefle, and M. Perricaudet.** 1996. Efficient dual transcomplementation of adenovirus E1 and E4 regions from a 293-derived cell line expressing a minimal E4 functional unit. J. Virol. *70*:559-565.
57. **Zhou, H.S., W. Oneal, N. Morral, and A.L. Beaudet.** 1996. Development of a complementing cell line and a system for construction of adenovirus vectors with E1 and E2a deleted. J. Virol. *70*:7030-7038.

10 Adenovirus-Mediated Gene Transfer: In Vivo Persistence of Transduced Cells and Control of Transgene Expression

Ronald Rooke[1], Linda Grave[1], Marielle Christ[2], Monika Lusky[1], Anne-Isabelle Michou[1], and Majid Mehtali[3]

[1]Transgène S.A., Strasbourg, France; [2]Phoenix International, L'Arbresle, France; [3]Introgene BV, Leiden, The Netherlands

INTRODUCTION

Adenoviruses (Ads) have been contemplated as tools for gene therapy for almost a decade. They offer numerous advantages, such as strong and transient expression of transgenes necessary in some applications such as cancer gene therapy. In such protocols, the immunogenicity of either the transgene, the Ad proteins, or both may enhance treatment efficiency by inducing or enhancing an antitumor-directed immune response. Alternatively, to correct inherited or cardiovascular diseases, long-term expression of a therapeutic gene is required. However, it became obvious early on that the major hurdle for these latter kinds of applications is the transient presence of the transgene product in vivo (4,17,18,43,55,61,86). Classically, in immunocompetent animals, transgene expression peaks 2 to 3 days postinjection and declines to undetectable levels over the following 3 to 4 weeks (reviewed in Reference 14). The fact that transgenes are generally expressed over much longer time periods in immunodeficient animals has incriminated the cellular immune response directed against the transgene product and/or the viral proteins as the major cause of elimination of transduced cells in vivo (14). Although the humoral response is thought not to play a direct role in the elimination of transduced cells, it is believed that the anti-Ad neutralizing antibodies prevent efficient vector readministration (39,102). In addition, other mechanisms associated with short-term transgene expression have been described, including genetic component(s) (4), cellular and humoral response to the transgene product, presence or absence of regulatory sequences in the expression cassette, degradation of vector DNA, and promoter shut off. With the obvious goal of improving vector DNA persistence and transgene expression, there has been a tremendous effort to dissect the mechanisms underlying these phenomena and the interplay between them. In parallel, the administration of immunosuppressive molecules at the time of vector infusion has also been used to obtain long-term expression of transgenes. This review describes the work responsible for our present view of Ad vectors in long-term therapeutic applications.

Viral Vectors: Basic Science and Gene Therapy
Edited by A. Cid-Arregui and A. García-Carrancá

ELIMINATION OF THE VECTOR DNA

Nonspecific Mechanisms

We and others have looked at the survival of vector DNA from very early to late time points in animals injected via various routes. This work has shown that 70% to 95% of the viral DNA is eliminated within the first 3 days postadministration (9,58,98). This rapid elimination of viral DNA happened in all organs looked at (liver, lung, and spleen) (9,14), irrespective of the route of administration (9,58,98) and is apparently independent of the initial vector dose (10). This elimination can be explained by the importance of the role played by nonspecific mechanisms in viral clearance (7,21,33). By pharmacologically eliminating macrophages, it was shown that it is this cell type that is essential for rapidly eliminating viral particles (87,98). Because this sharp reduction in viral copy numbers at early time points is seen in all the organs examined, it can be concluded that the reticular endothelial system plays a major role in eliminating the majority of vector particles. After this rapid elimination, the amount of the injected DNA remaining in the transduced organ stabilizes and is eliminated at a much slower rate (see the section on Specific Mechanisms) (Figure 1).

Since the administration of UV-inactivated virus particles also induces a strong inflammatory response, we can also conclude that this initial inflammation seen upon vector administration is, at least in part, caused by the high doses of vectors used, independent of viral replication and/or transgene expression (61). Nonspecific inflammatory processes have been shown important in preventing efficient gene transfer and their inhibition by dexamethasone to favor in vivo transduction

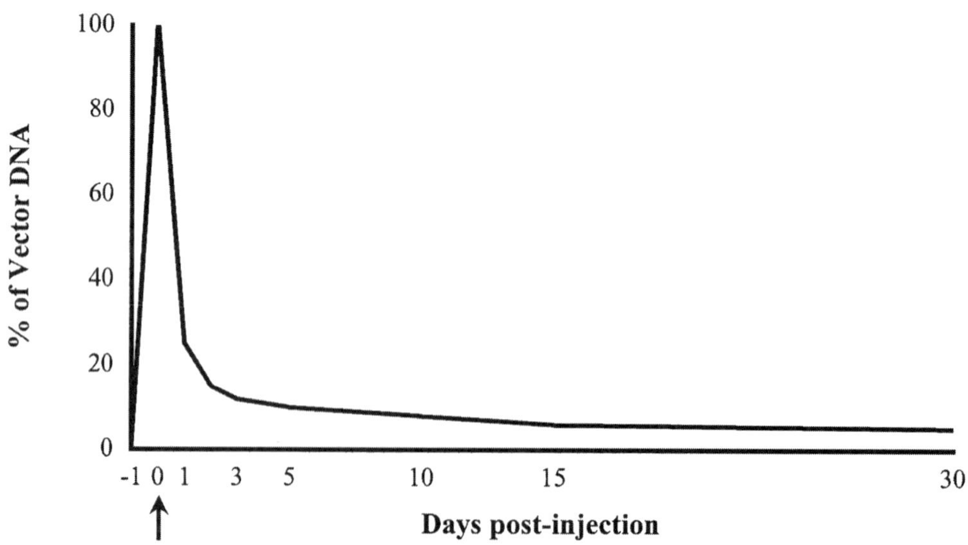

Figure 1. Kinetics of DNA persistence in injected mice. Typical profile of persistence of viral DNA in organs of transduced animals. Animals were injected on day 0 with a defined amount of an E1–E3-deleted vector. Target organs are sampled at indicated times, the total DNA is extracted, digested with relevant restriction enzymes, resolved on an agarose gel, blotted to a nylon membrane, and probed with a randomly primed Ad sequence. Autoradiograms of hybridized membranes are scanned, and amount of DNA is expressed as percent of DNA found in the same organ 2 hours postinjection.

of the lung (70). In turn, this may affect long-term transgene persistence.

Other mechanisms can be invoked to explain the nonspecific decrease in transduced DNA. They include natural cell turnover and intracellular degradation of episomal DNA. These are poorly characterized and documented, primarily because they are difficult to study. While their role in elimination of vector DNA probably represents only a fraction of all mechanisms, their importance may vary as a function of the transduced tissue. Some of these mechanisms can probably explain the fact that transgene expression was lost 3 weeks postinfection when a recombinant virus was injected in the thymus of immunodeficient animals (77). Because intrathymic delivery of proteins is a classic way to induce central tolerance and at no point were there any activated T cells in the transduced thymus, alternative mechanisms may explain the loss of transgene expression.

Specific Mechanisms

Cellular Immune Response

Extensive data demonstrate that in addition to the innate response, the specific CD8+ cytotoxic T lymphocyte (CTL)-mediated response is responsible for reducing long-term transgene expression (18,29, 45,102,104,109,114). A panoply of models were used to demonstrate this point, including adoptive transfers in major histocompatibility (MHC) molecule deficient animals, use of immunodeficient animals, isolation of MHC class I- and class II-restricted CTLs from transduced animals, and the use of immunosuppressive drugs or cytokines to achieve longer transgene expression. All in all, these data have shown that CTLs are responsible for the elimination of the transduced cells, which is reflected by a decrease in the amount of vector DNA in the target organs or of the transgene product in immunocompetent, unmanipulated animals (14). Early work on vector DNA persistence has suggested that the antiviral CTLs were stimulated by a residual de novo expression of viral structural proteins when E1-deleted vectors were administered at high multiplicity of infection (moi) (99,101,104). With the rationale of further preventing viral replication or protein expression, these results have prompted the development of so-called second generation vectors that contain at least one additional deletion (besides the E1 deletion) in regions that encode products involved in the regulation of the viral life cycle (Figures 2 and 3). Vectors with deletions in the E2A or E4 regions and cell lines that complement for the E1 products along with the product(s) encoded by the other mutated gene(s) (Figure 3) were generated and studied (10,19,23,27,32,58,65,74,95,96,108,112). The E2A region codes for a DNA binding protein (DBP) that is essential for viral DNA synthesis early during infection and for activation of the major late promoter (MLP) late in the infection cycle. The E4 region is complex and encodes for functions that are involved in DNA replication, late gene expression, host cell protein shut off, and host cell apoptosis (reviewed in Reference 54). Other vectors being developed are called "gutless" in that all virus coding regions have been removed (15,26,48,63). For all these tools, in vitro and in vivo studies have shown that further crippling the virus does indeed prevent viral DNA replication and expression of viral proteins (19,27,32,58). While numerous studies report that the genomes of E1–E2A-deleted, E1–E4-deleted, and gutless vectors have a longer life span in vivo than first generation (FG) viruses (19,23, 24,31,32,65–67,82,95,103) and that transgene expression is prolonged, other work has shown that viral DNA persistence is equivalent for all types of vectors examined (58,65) and even shortened for gutless vectors (56). While some of these discrepancies remain unexplained, some can be accounted for, at least in part, by the differences between the model systems used for these studies (i.e., presence or absence of a transgene, expression by the vector of a self or a foreign protein, animal model and mouse strain used).

More recent data has put in question the actual role played by the CTLs directed to Ad proteins in the elimination of transduced cells (27,62,81,90,94). This was brought about by the fact that long-term expression was achieved when animals were injected with vectors expressing self-proteins, proteins with low immunogenicity or proteins to which the animals were tolerized. Alternatively, a transient expression and rapid elimination of vector DNA is seen when animals are injected with similar vectors that express known immunogenes. In both instances, potent Ad-targeted CTL and humoral responses are seen in the host. Additionally, the in situ survival time of hepatocytes or pulmonary cells transduced with either E1-deleted or E1–E4-deleted vectors that do not contain a transgene is similar, irrespective of the immune status of the injected strain of mouse (58). Keeping in mind that E1–E4-deleted viruses do not express any structural proteins, these data show that the administration of replication-deficient Ad vectors provokes a CTL response directed towards Ad proteins. These data are in agreement with the development of an antiviral CTL response in animals that have been injected with UV-inactivated virus (40). Similarly, the expression by the vector of a transgene-encoded protein that is foreign to the host may also stimulate a CTL response directed against that protein. Taken together, these results suggest that the role of the antiviral CTL response

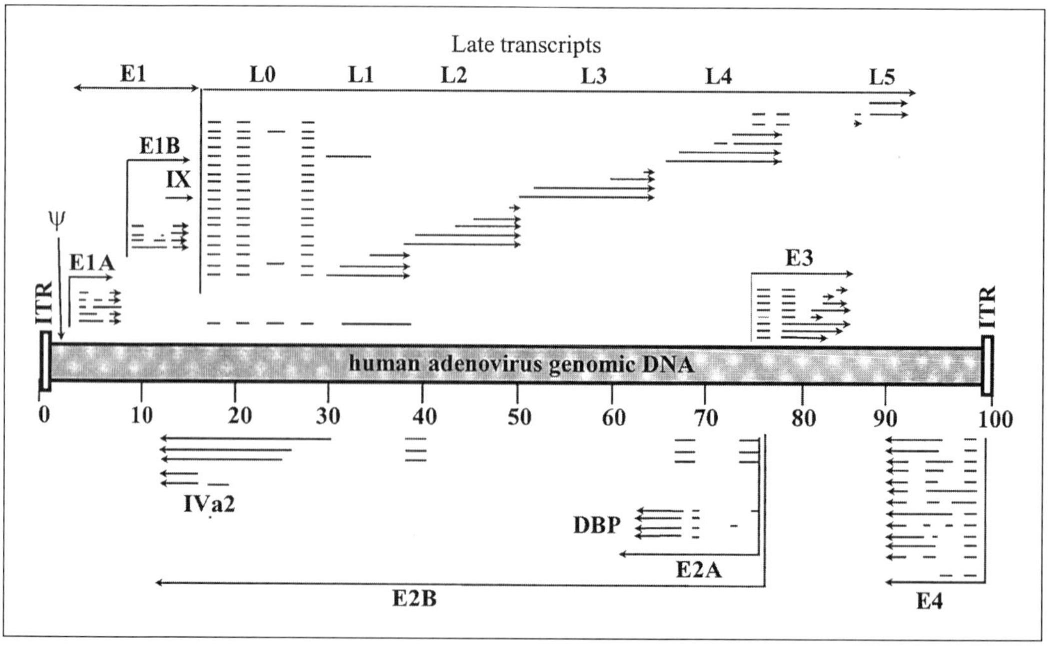

Figure 2. **Schematic representation of the genomic structure of the human Ad type 5.** The early genes encoding the regulatory proteins (E1–E4), the late genes encoding the structural proteins (L0–L5), the packaging signal (ψ), and the left and right inverted terminal repeats (ITRs) are indicated. The viral transcripts are represented by discontinuous thin lines with their 3′ extremities indicated by arrows (1 map unit [mu] = 360 bp).

in the elimination of transduced cells is only marginal, and that it is the latter response that causes the demise of transduced cells. This possibility is exemplified when persistence of vector DNA is compared between C57BL/6 and CBA mice that have been injected with the same dose of a human cystic fibrosis transmembrane conductance regulator (hCFTR)-expressing vector. Both strains of mice mount a detectable anti-Ad CTL response. However, CTLs specific for the hCFTR are only seen in CBA animals. Concomitantly, the persistence of transduced cells is much shorter in CBA animals (Reference 61 and M. Lusky, unpublished data). Little data is available on the possible synergy between both of these responses, although antigen spreading models and the existence of minor immunogenic epitopes in Ad structural proteins suggest that such mechanisms may exist (69,75).

One important aspect concerning Ad-based vectors is the role of preexisting immunity to Ad proteins in the establishment and the persistence of transgene

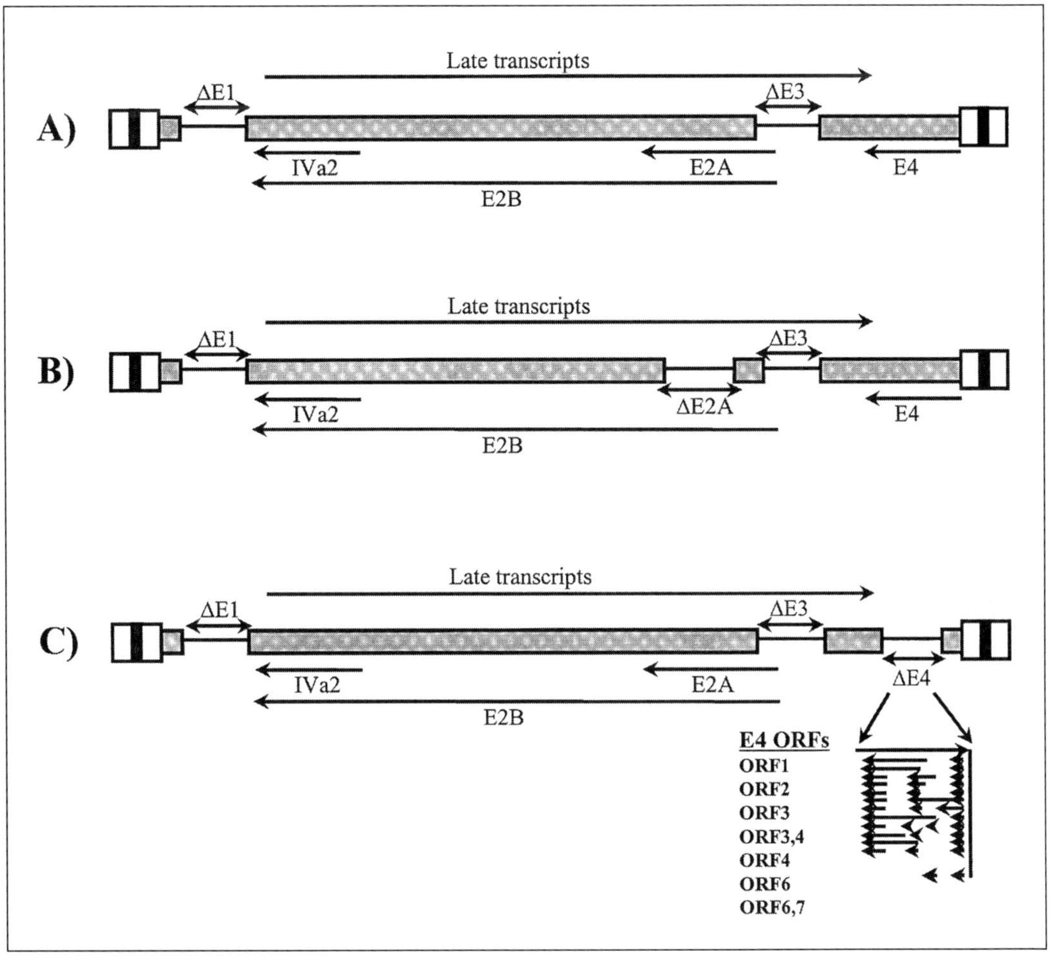

Figure 3. Schematic representation of first and second generation Ad vector genomes. (A) Transcription regions of a FG Ad vector in which the E1 and E3 regions are deleted. The E3 region deletion does not disrupt the late transcript because it is localized in an intron. (B and C) Transcription units of multiply deleted second generation vectors. The E1, E3, and E2A regions (B) or the E1, E3, and E4 regions (C) are deleted. The late region transcripts are not shown for simplicity, but their organization is not disrupted by the deletions of the early regulatory genes. Also shown are the ORFs of the E4 region (see text for detail).

expression. On the one hand, since most animals are naive for human Ad products, these issues cannot be addressed in animal models unless preimmunization is included in the protocols. On the other hand, a good proportion of individuals in the human population have been infected by Ad and have preexisting anti-Ad antibodies and CTLs (64). Data is sparse relative to efficiency of transduction and transgene persistence in the human population (64,71). These points are complex to address since transgene delivery and persistence must be evaluated as a function of the immune status of target populations and relative to the mode and route of administration. For example, the anti-Ad immunity in untreated hemophilia patients may be quite different than that seen in CF patients. Also, in the former case, the target organ is the liver, while it is the lung in the latter. The immunological constituents of these organs are quite different, and vector particles will be exposed to different types of effector immune mechanisms. Consequently, they will generate very different responses.

Relative to clinical applications in which long-term transgene expression is required, the advantages of using multiply deleted vector backbones remains unclear. Such vectors should still induce a detectable antiviral CTL response, but the impact CTLs elicit in human patients with a preexisting antiviral immunity on the persistence of the transduced cells is unknown. This comment also applies to the intrinsic toxic reaction induced by Ad proteins. This nonspecific toxic response of the host has an important role in mounting an efficient immune response to the vector. While early data have shown that the amount and/or the nature of the input Ad proteins are sufficient in inducing strong inflammatory and cellular immune responses (40,61), more detailed studies of second generation vectors have incriminated some Ad early products as playing an important role in the inflammatory processes (see section on Humoral Response).

Humoral Response

As mentioned previously, there is no evidence that the anti-Ad antibodies play a significant role in the elimination of transduced cells in vitro (39,100,107). However, we have shown that the expression of a transgene product may go unnoticed or seem short-term because of the antibody response to the transgene product (62). This result came about when the levels of human coagulation factor IX (hFIX) protein in the circulation of different immunocompetent and immunodeficient mouse strains were compared to the amount of DNA in the liver of injected animals. We found that in spite of strong CTL response to the Ad proteins in all immunocompetent mouse strains analyzed, transduced hepatocytes persisted for a relatively long time. However, this result did not correlate with the plasmatic levels of the transgene product, which dropped rapidly after injection. Surprisingly, circulating levels of hFIX in C57BL/6 mice were prolonged and found to be similar to the ones in immunocompromised mice (Figure 4). Further analysis of this finding correlated the disappearance of hFIX from the serum with the development of anti-hFIX antibodies in all mouse strains examined except in C57BL/6 mice (62). The antibodies may cause rapid clearance of the hFIX by forming immune complexes or mask the epitopes needed for its capture in the enzyme-linked immunosorbent assay (ELISA) used in its detection. The same type of FIX-directed humoral response has been noticed in a canine model of hemophilia (H. Homann, unpublished data). Taken together, these observations emphasize the minor role played by Ad-directed CTLs in the elimination of transduced cells. Second, they indicate that the animal model used and/or the animal strain used may influence the persistence of the transgene. For example, it has been shown that C57BL/6 mice are incapable of mounting strong humoral responses to a number of different antigens and have defective immune responses (34). Finally, the method utilized to detect transgene expression must be chosen carefully when looking at persistence of a transgene product. Ideally, many different parameters should be correlated.

While the anti-Ad antibodies have no evident impact on the persistence of the transduced cells, they play a crucial role in preventing the use of repeated vector administration as a way to palliate for short-term transgene expression. The first injection of vector is usually sufficient to induce a potent anti-Ad humoral response in which neutralizing anti-Ad antibodies can be found (51,86,102,106,109). The presence of neutralizing antibodies does not preclude gene transfer in vivo, but drastically reduces its efficiency (64,86). In addition, multiple vector injections in immune individuals increase the anti-Ad antibody titers and CTL activities. Besides the use of immunosuppressive agents (see section on Immunosuppression), other strategies being developed to circumvent this problem include reinjection of a vector of a different serotype or of a different species (47,60,79). In this situation, the antibodies developed in response to the vector used in the first injection are useless against the vector used for the second injection. This approach is, however, limited by the number of vector systems available for Ad-based gene therapy and by the industrial costs associated with the development of such a multiple vector system.

Immunosuppression

The importance of the host's immune

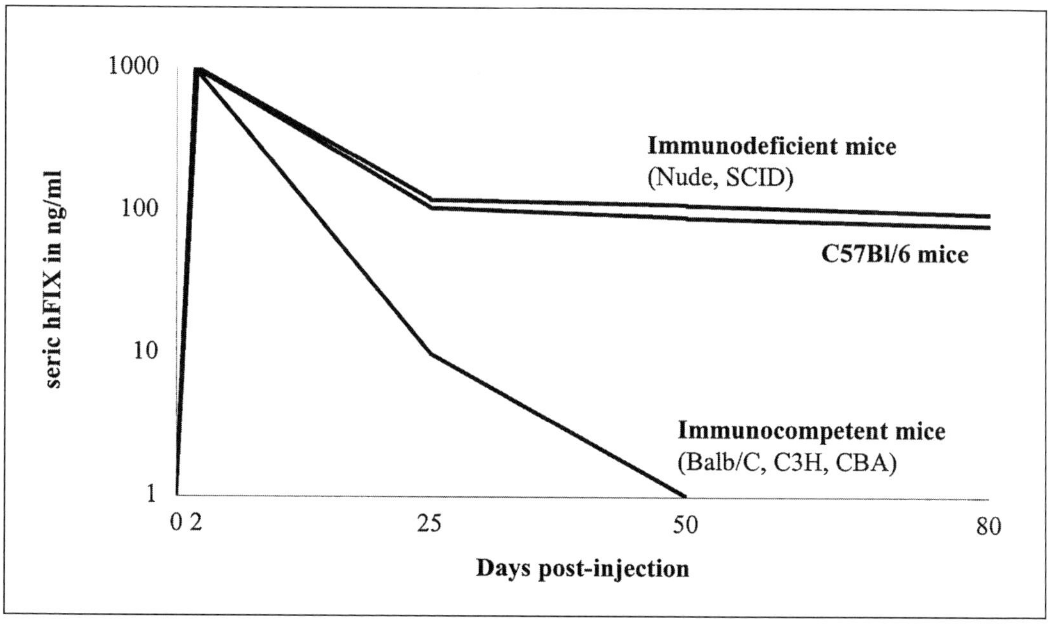

Figure 4. Influence of the mouse strain on the in vivo persistence of transgene expression. Immunocompetent (Balb/C, C3H, CBA, and C57Bl/6) and immunodeficient (SCID and nude) mice were injected intravenously with a defined number of infectious units of an E1–E3-deleted vector encoding hFIX. Mice were bled at the indicated time points, and the seric levels of hFIX were determined by an ELISA. Controls are sera from animals that had been injected with the same number of infectious units of a transgene-empty vector and were negative for hFIX throughout the experiment. Typical results of seric levels of hFIX are shown.

response in limiting the persistence of expression of immunogenic transgene products is highlighted by the numerous examples of better persistence when immunosuppressive protocols are included in the evaluation of vectors. An impressive array of methodologies has been used either to increase the survival of transduced cells, presumably by limiting or eliminating the antitransgene CTL response, or to prevent the development of neutralizing antibodies, which in turn allows vector readministration. One approach is the induction of central tolerance to the immunogenic viral and transgene proteins (41,44,100). Such tolerance has been achieved by exposing neonatal thymus to antigens, by oral administration of the antigen or by using animals that are transgenic for the transgene product (hence recognizing this protein as self). Others have used thymectomy and/or T-cell-depleting antibodies (anti-CD4 or T-cell receptor) to induce a similar state of tolerance and extend transgene expression (20,50,80, 113). The data generated from this work emphasize the central role played by CTL in the elimination of transduced cells. However, it is difficult to envisage how these methodologies can be adapted to clinical situations.

The importance of the cytokine network in the control of the immune response has prompted some groups to study how these molecules affect transgene persistence. Exposure of transduced animals to inhibitory cytokines such as interleukin (IL)-10 or IL-12 results in a decreased immune response to transduced products and/or prolonged expression of the transgene (73,100,106). These cytokines can be co-administered along with the vector, expressed by the vector itself, or expressed by a vector co-injected with the vector of interest. A similar approach is based on results showing that tumor necrosis factor α (TNFα)-deficient mice do not mount an efficient immune response to Ad vectors, that they express the transgene product for a long time, and that vector readministration is readily possible (7,21). In agreement with these data, Zhang et al. (111) have used soluble TNFα receptor to reduce the early inflammatory response and prolonged the expression of the highly immunogenic transgene-encoded β-galactosidase protein. However, this treatment altered neither the Ad-directed humoral response nor the induction of CTLs, thereby showing the important role of nonspecific inflammatory processes in the elimination of vector DNA in situ. Immunosuppressive drugs have also been used successfully to improve the duration of viral DNA persistence and transgene expression (8,37,42,92). These molecules (cyclophosphamide, cyclosporine, FK506, etoposide, and deoxyspergualin), when administered along with the vector, are reported to inhibit the antibody and CTL productions. Although the methodologies described above show clearly the importance of the immune response in the transient expression of immunogenic transgene products and in preventing repeated vector administrations, patients can in most situations only be treated for a finite time period with these drugs, and therefore, their clinical value is limited.

One interesting way by which tolerance may be induced is by blocking the costimulatory pathways of T-cell activation. Two signals are required for a T cell to be activated. One is mediated by the clonotypic T-cell receptor (TCR) that recognizes and binds a major histocompatibility antigen molecule loaded with the proper peptide. The other signal is provided by the interaction of costimulatory molecules on the surface of the T cell with their ligand on the surface of antigen presenting cells (APCs). T-cell activation can be inhibited, and a long-term anergic state induced by blocking the interaction between the CD28 or

CTLA4 costimulatory molecules and the B7 ligand molecule on the surface of the APCs. Linsley and coworkers (57) have shown that a soluble form of CTLA4, CTLA4Ig, can bind with high affinity to the B7 molecule and inhibit T-cell activation in vivo. This molecule has been used as an immunosuppressant in the context of Ad vector injection with success (39,45). Systemic co-injection of CTLA4Ig and vector resulted in a long-term transgene expression, but did not affect neutralizing antibody titers. Efficient reinjection of vector in CTLA4Ig-treated animals fails unless an antibody directed to the CD40-ligand molecule is injected at the same time as the vector (46). This latter antibody binds to the CD40-ligand molecule present on the surface of activated T cells and prevents B cells from getting their own costimulatory signal. More recently, Ad vectors expressing CTLA4Ig have been shown to persist and express at least 3 months in vivo, probably because of the sustained immunosuppression induced by this molecule (68,83). While long-term therapeutic expression of immunogenic transgene products may be considered by co-injection of a CTLA4Ig-expressing vector, the immune status induced by long-term presence of a potent immunomodulatory molecule remains to be determined in the transduced host.

GENETIC REGULATION OF TRANSGENE EXPRESSION

Transgene Expression Cassette

An important issue in improving long-term transgene expression is the design of the expression cassette. However, in the context of specific applications of gene therapy, our knowledge on optimal promoters and expression cassettes are limited. Some of the issues raised in the previous sections may be solved by introducing strong, tissue-specific expression cassettes in the vector backbone. For example, for secreted proteins, stronger expression of a self-transgene may allow one to inject a smaller amount of vector particles and hence of viral proteins. In turn, the inflammatory and immune responses to the virus may be decreased and vector persistence prolonged. Alternatively, lower levels of expression may allow a sequestered transgene product to go unnoticed by the host's immune system and to be expressed for longer periods. In addition, there is some precedence for the demonstration of a threshold effect of the transgene product in that the same efficacy of treatment was seen with two vectors expressing the same transgene from promoters of different strengths (22). This result argues that in some instances the use of a strong promoter is unnecessary and may cause unwanted toxicity or immunogenicity. Also, judicious selection of promoters allowing tissue specificity of transgene expression may resolve some targeting issues.

In one study, we have compared the strength and persistence of a number of different promoters (human cytomegalovirus [hCMV] immediate-early promoter, SRα, Rous sarcoma virus long terminal repeat [RSV-LTR], Ad MLP, and the mouse phosphoglycerate kinase promoter [PGK] (L. Grave, unpublished data). We have found a good correlation between the in vitro and in vivo expression data relative to the strength of the promoters as determined by the amount of hFIX found either in the culture supernatant of infected A549 cells (a human lung carcinoma cell line) or in the circulation of intravenously injected immunodeficient animals; the hCMV being the strongest, the RSV-LTR reaching intermediary levels, and the three other promoters yielding the lowest amounts. While promoter strength is of issue, chronicity of expression is also important. In the study described above, the serum levels of the hFIX in immunocompromised mice were followed over a 90-day period. We have found that even at 90 days postinjection, the concentration of seric hFIX in animals injected with vectors expressing the transgene from either the hCMV or RSV-LTR promoters are at acceptable therapeutic levels (>1 mg/mL) albeit the levels in the CMV–FIX-injected animals are at 1% the levels found in the same animal at early time points (Figure 5). The serum concentration of hFIX in animals receiving the RSV-LTR/hFIX vector remained constant throughout the experiment (Figure 5). These results are in agreement with results describing the hCMV promoter as being susceptible to down-modulation in vitro and in vivo (3,5,110). It is, therefore, important to differentiate between promoter strength and stability of expression. In a clinical setting, it might not be necessary to get the highest possible expression but rather the longest. Another important parameter in the choice of promoter is its tissue specificity. A major limitation is that when taken out of their natural context, viral promoters may show different specificities, and their behavior can be difficult to predict. For instance, while studying the CMV promoter in transgenic mice, Koedood et al. (49) suggested that this promoter works best in the tissue naturally infected by that virus. In partial contradiction, others have described that, at least in the context of an Ad vector, the murine CMV immediate-early promoter expresses higher levels of reporter genes in mouse cells than its human equivalent but that both promoters have comparable strengths in human cells (1).

Other transcription and/or translation regulatory sequences may be essential for the chronicity of transgene expression. We and others have also shown that the addition of intronic sequences usually significantly increases the amount of steady state levels of mRNAs of the transgene in transduced cells, both in vitro and in vivo (Ref-

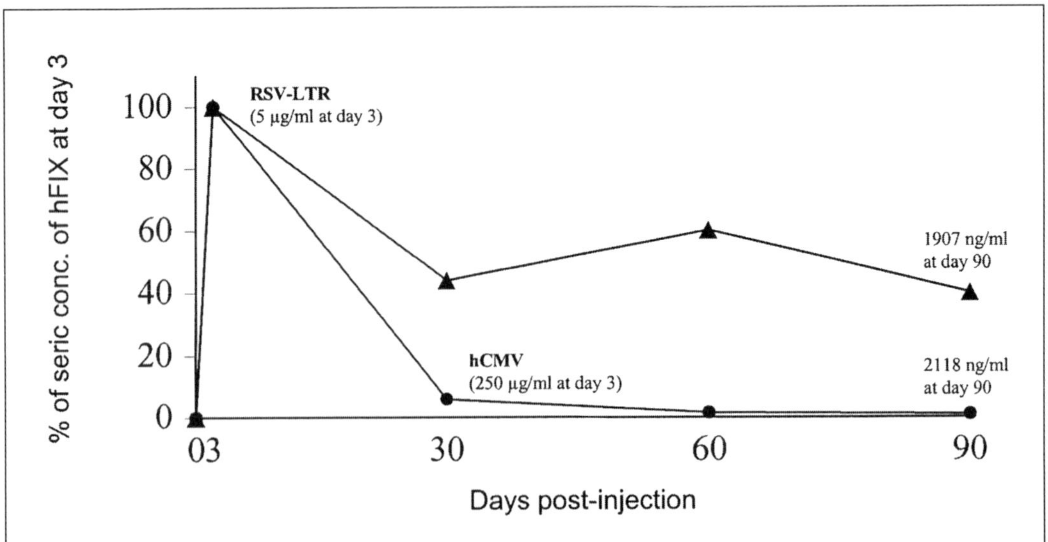

Figure 5. Influence of the promoter sequence on the in vivo strength and persistence of transgene expression. The same number of infectious units of E1–E3-deleted vectors expressing the hFIX from either the CMV promoter or the RSV-LTR promoter were injected in immunocompetent C57Bl/6 mice. Serum was sampled at the indicated time points, and the levels of hFIX were determined by a hFIX-specific ELISA. Results are expressed as the percent of hFIX found in the serum 3 days after vector injection. Note that although the CMV promoter shows a more drastic decrease in expression relative to the RSV-LTR, the levels of hFIX at day 90 remain higher in animals injected with the vector using the CMV promoter.

erence 16 and L. Grave, unpublished data). In spite of the limited amount of information available, it can be said that the polyadenylation sequences have very little or no influence on levels or persistence of transduced gene expression (Reference 16 and L. Graves, unpublished data).

Finally, the orientation of the expression cassette in the E1 region of the vector has been shown to play a role in the levels of transgene expression attained. By comparing the same expression cassette placed in either orientation in the E1 region of two different vectors, a sevenfold higher expression level was reached for the vector that contained the cassette in the sense orientation relative to the positive DNA viral strand (1).

Influence of Viral Regulatory Gene Products on Transgene Expression

E4 Region

For reasons evoked in the first section of this review (i.e., eliminating leaky expression of potentially immunogenic viral genes), Ad vectors containing multiple deletions of viral regulatory genes have been created. As mentioned above, the impact of deleting the E4 region on vector DNA persistence has been controversial. An additional explanation for the discrepancies is that E4-encoded products modulate transgene expression (References 2 and 19; M. Lusky, unpublished data; and L. Grave, unpublished data). To identify the product(s) required for efficient transgene expression, we have dissected the E4 region by creating a battery of E1–E4-deleted vectors that express single or combinations of different E4 open reading frames (ORFs) and compared the levels of expression reached by cassettes driving a transgene from the hCMV promoter or the RSV-LTR in vitro and in vivo (Reference 58; M. Lusky, unpublished data; and L. Grave, unpublished data) (Figure 3). In a model using immunocompetent mouse strains, no difference in the specific immune response to these vectors was noticed (15), but important variations in transgene expression were seen (59). The levels of transgene transcripts ranged from no expression, when looking in the liver after an intravenous injection of an E1–E4-deleted vector with or without E4ORF4, to expression levels higher than the ones obtained with a vector retaining a wild-type E4 region, when looking in the lung after an intratracheal injection of a vector retaining the E4ORFS3,6,7. On the whole, the data indicate that E4ORF3 alone in the liver or E4ORF3 with ORF4 in the lung are sufficient for long-term expression of the hCFTR gene. This also argues in favor of an important function played by a tissuespecific factor in transgene expression that is compensated for by the E4ORF4 product in the mouse lung. In addition, we have evaluated the toxicity and inflammation of these vectors and found that the E4 region products play an important role in hepatotoxicity (M. Christ, unpublished data). Some ORFs of the E4 region allow very high levels of expression of the transgene from the hCMV promoter, but their use also correlates with increased hepatotoxicity and inflammation. Interestingly, vectors containing E4ORF3 or E4ORF3 and ORF4 allow long-term transgene expression, but induce only minimal inflammation in injected animals, thereby indicating that these vectors are particularly attractive for clinical applications. To add to this complicated picture of tissue-specific effect of individual E4ORFs on transgene expression, we have also seen that the impact of deleting the E4 region on transgene expression varies with the promoter used and the nature of the encoded product (Reference 2 and L. Grave, unpublished data).

E3 Region

FG Ads have one deletion in the E1 region, which renders them replication-deficient, and one deletion in the E3 region (Figure 3). Vectors with the latter deletion are used because this region is nonessential for virus replication, and its absence allows a larger size transgene. In wild-type viruses, the E3 region produces six known products, most of which have immunomodulatory functions (Table 1) (reviewed in Reference 36). The idea of using all or some of the products encoded in this region to decrease the immune reaction to transduced cells has been tantalizing. The difficulty faced in achieving this goal has been the stringent requirement of E1 region-encoded transacting factors to realize efficient transcription from the E3 promoter (reviewed in Reference 97). Nevertheless, early work has demonstrated a role for the E3 region in the pathogenesis caused by Ads and has motivated the efforts to improve vector efficacy (28,76).

The E3gp19K protein is well described and is known to associate noncovalently with MHC class I molecules in the endoplasmic reticulum of an infected cell and prevent its egress to the cell surface (6,12). In this way, it is believed that infected cells may go unnoticed by Ad-specific CTLs. Reports on the efficacy of this protein in prolonging transgene expression or vector DNA persistence have been contradictory. Some have reported a beneficial effect of the E3gp19K on these parameters while others report no effect (11,53,78, 84). These discrepancies are difficult to reconcile, but could be explained in part by the inherent differences in the models used. For example, the haplotype of the animals used in these studies is important because the E3gp19K protein is known to interact with different affinities to different class I MHC molecules (6,88). Another point that has not been addressed is the sensitivity of cells that have down-modulated the class I MHC molecule from their surface to natural killer (NK) cell-induced cytolysis. These issues remain to be solved.

Table 1. Immunomodulatory Functions of the E3 Proteins

E3 Proteins	Functions
E3gp19K	Retention of class I MHC
14.7 K	Inhibits TNF-induced apoptosis
RIDαβ	Inhibits TNF-induced apoptosis
	Inhibits Fas-induced apoptosis
	Down-modulation of Fas from the cell surface

Other E3-encoded proteins can inhibit TNF- and/or Fas-induced apoptosis in vitro (Table 1). The receptor for internalization and degradation complex made of a 10.4 K and a 14.5 K protein, recently termed RIDαβ complex, or the 14.7 K protein have partly overlapping functions in that they both inhibit TNF-induced lysis, while only the RIDαβ complex can down-modulate the Fas molecule from the surface of infected cells and hence, prevent their lysis by Fas ligand (30,35,52,85,89). However, their effect on transgene persistence remains to be determined. The expression by a vector of all products encoded in the E3 region by introducing a constitutive promoter upstream of this region has been reported to prolong transgene expression or, when used in transgenic mouse lines, to protect animals from developing autoimmunity (25,38,72,93).

Gutless Vectors

Recently, vectors deleted of all structural and regulatory genes have been developed (26,48). Similar to the E1–E4-deleted vectors, they offer the advantage of low immunogenicity and toxicity (13,66,67, 82). In animal models, these new vectors allow transgene expression for extended periods, thereby implying that the transduced promoters are not submitted to the same regulation than when they are in the context of the viral genome (13,66,67). However, their clinical use is hampered by the limits in the methodologies available to separate the gutless vector from the helper virus required for their production.

CONCLUSION

The goal of Ad vectorology is to adapt a virus in such a way that it can perform tasks for which it was not originally meant. This has motivated a tremendous effort to dissect the in vivo biology of vectors. Many problems have arisen, and controversial results remain to be solved, especially in terms of making vectors that can express transgene for long time periods without readministration. Based on early data, it seemed that the development of second generation vectors deleted simultaneously from several viral regulatory genes was a promising avenue to achieve this goal. More recent information shows that these vectors do not outperform FG vectors in terms of immunogenicity and long-term transgene expression. However, their lower toxicity may render them interesting tools in the clinic. The specificity of transgene expression relative to different ORFs encoded in the E4 region of the Ad, along with the information available on expression cassettes, indicates that the best combination of parameters to get long-term expression of a therapeutic gene remains trial and error, as described by Verma and Somia (91). Although their use is limited by purification methodologies, gutless vectors are promising tools to obtain chronic expression of a transgene with minimal toxicity. The administration of immunosuppressive molecules at the time of vector delivery was also shown to have an important impact on transgene chronicity and reduction of the anti-Ad antibody titers. Nevertheless, the disparity in the different model systems used to study the biological properties of these vectors may explain some of the disparities seen in the data and motivate caution in their interpretation.

In summary, Ads constitute the most efficient gene transfer vector available for in vivo gene therapy. They are well adapted for therapeutic applications in which a strong and transient transgene expression is required, as in most cancer gene therapy protocols. However, their use for the treatment of inherited diseases, in which the transgene must be expressed during the whole life of the patient, is still hampered by the current inability to repeatedly administer the vector and by the absence of knowledge regarding the role of the preexisting antiviral immunity on the persistence of the transduced cells.

ACKNOWLEDGMENTS

We thank Drs. M. Courtney and A. Pavirani for the critical reading of the manuscript. This work was supported in part by the Association Française contre la Mucoviscidose (AFLM) and the Association Française contre les Myopathies (AFM).

REFERENCES

1. **Addison, C.L., M. Hitt, D. Kunsken, and F.L. Graham.** 1997. Comparison of the human versus murine cytomegalovirus immediate early gene promoters for transgene expression by adenoviral vectors. J. Gen. Virol. *78*:1653-1661.

2. **Armentano, D., J. Zabner, C. Sacks, C.C. Sookdeo, M.P. Smith, J.A. St. George, S.C. Wadsworth, A.E. Smith, and R.J. Gregory.** 1997. Effect of the E4 region on the persistence of transgene expression from adenovirus vectors. J. Virol. *71*:2408-2416.

3. **Ascadi, G., D. O'Hagan, H. Lochmuller, S. Prescott,**

N. Larochelle, J. Nalbantoglu, A. Jani, and G. Karpati. 1998. Interferons impair early transgene expression by adenovirus-mediated gene transfer in muscle cells. J. Mol. Med. *76*:442-450.

4. **Barr, D., J. Tubb, D. Ferguson, A. Scaria, A. Lieber, C. Wilson, J. Perkins, and M.A. Kay.** 1995. Strain related variations in adenovirally mediated transgene expression from mouse hepatocytes in vivo: comparisons between immunocompetent and immunodeficient inbred strains. Gene Ther. *2*:151-155.

5. **Baskar, J.F. P.P. Smith, G.S. Ciment, S. Hoffmann, C. Tucker, D.J. Tenney, A.M. Colberg-Poley, J.A. Nelson, and P. Ghazal.** 1996. Developmental analysis of the cytomegalovirus enhancer in transgenic animals. J. Virol. *70*:3215-3226.

6. **Beier, D.C., J.H. Cox, D.R. Vining, P. Creswell, and V.H. Engelhard.** 1994. Assocation of human class I MHC alleles with the adenovirus E3/19K protein. J. Immunol. *152*:3862-3872.

7. **Benihoud, K., I. Saggio, P. Opolon, B. Salone, F. Amiot, E. Connault, C. Chianale, F. Dautry, P. Yeh, and M. Perricaudet.** 1998. Efficient, repeated adenovirus-mediated gene transfer in mice lacking both tumor necrosis factor alpha and lymphotoxin α. J. Virol. *72*:9514-9525.

8. **Bouvet, M., B. Fang, S. Ekmekcioglu, L. Ji, C.D. Bucana, K. Hamada, E.A. Grimm, and J.A. Roth.** 1998. Suppression of the immune response to an adenovirus vector and the enhancement of intratumoral transgene expression by low-dose etoposide. Gene Ther. *5*:189-195.

9. **Brough, D.E., C. Hsu, V.A. Kulesa, G.M. Lee, L.J. Cantolupo, A. Lizonova, and I. Kovesdi.** 1997. Activation of transgene expression by early region 4 is responsible for a high level of persistent transgene expression from adenovirus vectors *in vivo.* J. Virol. *71*:9206-9213.

10. **Brough, D.E., A. Lizonova, C. Hsu, V.A. Kulesa, and I. Kovesdi.** 1996. A gene transfer vector cell-line system for complete functional complementation of adenovirus early regions E1 and E4. J. Virol. *70*:6197-6201.

11. **Bruder, J.T., T. Jie, D.L. McVey, and I. Kovesdi.** 1997. Expression of gp19K increases the persistence of transgene expression from an adenovirus vector in the mouse lung and liver. J. Virol. *71*:7623-7628.

12. **Burgert, H.G., J.L. Maryanski, and S. Kvist.** 1987. "E3/19K" protein of adenovirus type 2 inhibits lysis of cytolytic T lymphocytes by blocking cell-surface expression of histocompatibility class I antigens. Proc. Natl. Acad. Sci. USA *84*:1356-1360.

13. **Chen, H.H., L.M. Mack, R. Kelly, M. Ontell, S. Kochanek, and P.R. Clemens.** 1997. Persistence in muscle of an adenoviral vector that lacks all viral genes. Proc. Natl. Acad. Sci. USA *94*:1645-1650.

14. **Christ, M., M. Lusky, F. Stoeckel, D. Dreyer, A. Dieterle, A.I. Michou, A. Pavirani, and M. Mehtali.** 1997. Gene therapy with recombinant adenovirus vectors: evaluation of the host immune response. Immunol. Lett. *57*:19-25.

15. **Clemens, P.R., S. Kochanek, Y. Sunada, S. Chan, H.H. Chen, K.P. Campbell, and C.T. Caskey.** 1996. In vivo muscle gene transfer of full-length dystrophin with an adenoviral vector that lacks all viral genes. Gene Ther. *3*:965-972.

16. **Connelly, S., J.M. Gardner, A. McClelland, and M. Kaleko.** 1996. High-level tissue-specific expression of functional human factor VIII in mice. Hum. Gene Ther. *7*:183-195.

17. **Connelly, S., T.A. Smith, G. Dhir, J.M. Gardner, M.G. Mehaffey, K.S. Zaret, A. McClelland, and M. Kaleko.** 1995. In vivo gene delivery and expression of physiological levels of functional human factor VIII in mice. Hum. Gene Ther. *2*:185-193.

18. **Dai, Y., E.M. Schwarz, D. Gu, W.W. Zhang, N. Sarvetnick, and I.M. Verma.** 1995. Cellular and humoral immune responses to adenoviral vectors containing factor IX gene: tolerization of factor IX and vector antigens allows for long-term expression. Proc. Natl. Acad. Sci. USA *92*:1401-1405.

19. **Dedieu, J.F., E. Vigne, C. Torrent, C. Jullien, I. Mahfouz, J.M. Caillaud, N. Aubailly, C. Orisini et al.** 1997. Long-term gene delivery into the livers of immunocompetent mice with E1/E4-defective adenoviruses. J. Virol. *71*:4626-4637.

20. **DeMatteo, R.P., J.F. Markmann, K.F. Kozarsky, C.F. Baker, and S.E. Raper.** 1996. Prolongation of adenoviral transgene expression in mouse liver by T lymphocyte subset depletion. Gene Ther. *3*:4-12.

21. **Elkon, K.B, C.C. Liu, J.G. Gall, J. Trevejo, M.W. Marino, K.A. Abrahamsen, X. Song, J.L. Zhou, et al.** 1997. Tumor necrosis factor alpha plays a central role in immune-mediated clearance of adenoviral vectors. Proc. Natl. Acad. Sci. USA *94*:9814-9819.

22. **Elshami, A.A., J.W. Cook, K.M. Amin, H. Choi, J.Y. Park, L. Coonrod, J. Sun, K. Molnar-Kimber, et al.** 1997. The effect of promoter strength in adenoviral vectors containing herpes simplex virus thymidine kinase on cancer gene therapy *in vitro* and *in vivo.* Cancer Gene Ther. *4*:213-221.

23. **Engelhardt, J.F., L. Litzky, and J.M. Wilson.** 1994. Prolonged transgene expression in cotton rat lung with recombinant adenoviruses defective in E2a. Hum. Gene Ther. *5*:1217-1229.

24. **Engelhardt, J.F., X. Ye, B. Doranz, and J.M. Wilson.** 1994. Ablation of E2a in recombinant adenoviruses improves transgene persistence and decreases inflammatory response in mouse liver. Proc. Natl. Acad. Sci. USA *91*:6196-6200.

25. **Fejer, G., I. Gyory, J. Tufariello, and M.S. Horwitz.** 1994. Characterization of transgenic mice containing adenovirus early region 3 genomic DNA. J. Virol. *68*:5871-5881.

26. **Fisher, K.J., H. Choi, J. Burda, S.J. Chen, and J.M. Wilson.** 1996. Recombinant adenovirus deleted of all genes for gene therapy of cystic fibrosis. Virology *217*:11-22.

27. **Gao, G.P., Y. Yang, and J.M. Wilson.** 1996. Biology of adenovirus vectors with E1 and E4 deletions for liver-directed gene therapy. J. Virol. *70*:8934-8943.

28. **Ginsberg, H.S., U. Lundholm-Beauchamp, R.L. Horswood, B. Pernis, W.S. Wold, R.M. Chanock, and G.A. Prince.** 1989. Role of early region 3 (E3) in pathogenesis of adenovirus disease. Proc. Natl. Acad. Sci. USA *86*:3823-3827.

29. **Ginsberg, H.S. and G.A. Prince.** 1994. The molecu-

lar basis of adenovirus pathogenesis. Infect. Agents Dis. *1*:1-8.

30.**Gooding, L.R., T.S. Ranheim, A.E. Tollefson, L. Aquino, P. Duerksen-Hughes, T.M. Horton, and W.S.M. Wold.** 1991. The 10,400 and 14,500 dalton proteins encoded by region E3 of adenovirus function together to protect many but not all mouse cell lines against lysis by tumor necrosis factor. J. Virol. *65*:4114-4123.

31.**Goldman, M.J., L.A. Litzsy, J.F. Engelhardt, and J.M. Wilson.** 1995. Transfer of the CFTR gene to the lung of non-human primates with E1-deleted, E2A-defectinve recombinant adenovirus: a preclinical toxicology study. Hum. Gene Ther. *6*:839-851.

32.**Gorziglia, M.I., M.J. Kadan, S. Yei, J. Lim, G.M. Lee, R. Luthra, and B.C. Trapnell.** 1996. Elimination of both E1 and E2A from adenovirus vectors further improves the prospects for in vivo human gene therapy. J. Virol. *70*:4173-4178.

33.**Guidotti, L.G. and F.V. Chisari.** 1996. To kill or to cure: options in host defense against viral infection. Curr. Opin. Immunol. *4*:478-483.

34.**Gustavsson, S., S. Hjulstrom-Chomez, B.M. Lidstrom, N. Ahlborg, R. Andersson, and B. Heyman.** 1998. Impaired antibody responses in H-2Ab mice. J. Immunol. *161*:1765-1771.

35.**Horton, T.M., T.S. Ranheim, L. Aquino, D.I. Kusher, S.K. Saha, C.F. Ware, W.S.M. Wold, and L.R. Gooding.** 1991. Adenovirus 14.7K protein functions in the absence of other adenovirus proteins to protect transfected cells from tumor necrosis factor cytolysis. J. Virol. *65*:2629-2639.

36.**Horwitz, M.S., J. Tufariello, A. Grunhaus, and G. Fejer.** 1995. Model systems for studying the effects of the adenovirus E3 genes on virulence in vivo, p. 195-211. *In* W. Doefler and P. Bohm (Eds.), Current Topics in Microbiology and Immunology: The Molecular Repertoire of Adenoviruses III: Biology and Pathogenesis. Springer Publishing, Berlin.

37.**Howell, J.M., H. Lochmuller, A. O'Hara, S. Fletcher, B.A. Kakulas, B. Massie, J. Nalbantoglu, and G. Karpati.** 1998. High-level dystrophin expression after adenovirus-mediated dystrophin minigene transfer to skeletal muscle of dystrophic dogs: prolongation of expression with immunosuppression. Hum. Gene Ther. *9*:629-634.

38.**Ilan, Y., G. Droguett, N.R. Chowdury, Y. Li, K. Sengupta, N.R. Thummala, A. Davidson, J.R. Chowdury, and M.S. Horwitz.** 1997. Insertion of the adenoviral E3 region into a recombinant viral vector prevents antiviral humoral and cellular immune responses and permits long-term gene expression. Proc. Natl. Acad. Sci. USA *94*:2587-2592.

39.**Joos, K., L.A. Turka, and J.M. Wilson.** 1998. Blunting of immune responses to adenoviral vectors in mouse liver and lung with CTLA4Ig. Gene Ther. *5*:309-319.

40.**Kafri, T., D. Morgan, T. Krahl, N. Sarvetnick, L. Sherman, and I. Verma.** 1998. Cellular immune response to adenoviral vector infected cells does not require *de novo* viral gene expression: implications for gene therapy. Proc. Natl. Acad. Sci. USA *95*:11377-11382.

41.**Kagami, H., J.C. Atkinson, S.M. Michalek, B. Handelman, S. Yu, B.J. Baum, and B. O'Connel.** 1998. Repetitive adenovirus administration to the parotid gland: role of immunologic barriers and induction of oral tolerance. Hum. Gene Ther. *9*:305-313.

42.**Kaplan, J.M. and A.E. Smith.** 1997. Transient immunosuppression with deoxyspergualin improves longevity of transgene expression and the ability to readminister adenoviral vector to the mouse lung. Hum. Gene Ther. *8*:1095-1104.

43.**Kass-Eisler, A., E. Falck-Pedersen, D.H. Elfenbein, M. Alvira, P.M. Buttrick, and L.A. Leinwand.** 1994. The impact of developmental stage, route of administration and the immune system on adenovirus-mediated gene transfer. Gene Ther. *6*:395-402.

44.**Kass-Eisler, A., L. Leinwand, J. Gall, B. Bloom, and E. Falk-Pedersen.** 1996. Circumventing the immune response to adenovirus-mediated gene therapy. Gene Ther. *3*:154-162.

45.**Kay, M.A., A.X. Holterman, L. Meuse, A. Gown, H.D. Ochs, P.S. Linsley, and C.B. Wilson.** 1995. Long-term hepatic adenovirus-mediated expression in mice following CTLA4Ig administration. Nat. Genet. *11*:191-197.

46.**Kay, M.A., L. Meuse, A.M. Gown, P. Linsley, D. Hollenbaugh, A. Aruffo, H. Ochs, and C.B. Wilson.** 1997. Transient immunosuppression with CD-40 ligand antibody and CTLA4Ig enhances persistence and secondary adenovirus-mediated gene transfer into mouse liver. Proc. Natl. Acad. Sci. USA *94*:4686-4691.

47.**Klonjkowski, B., P. Gilardi-Hebenstreit, J. Hadchouel, V. Randrianarison, S. Boutin, P. Yeh, M. Perricaudet, and E.J. Kremer.** 1997. A recombinant E1-deleted canine adenoviral vector capable of transduction and expression of a transgene in human-derived cells and *in vivo*. Hum. Gene Ther. *8*:2103-2115.

48.**Kochanek, S., P.R. Clemens, K. Mitani, H.H. Chen, S. Chan, and C.T. Caskey.** 1996. A new adenoviral vector: replacement of all viral coding sequences with 28 kb of DNA independently expressing both full-length dystrophin and beta-galactosidase. Proc. Natl. Acad. Sci. USA *93*:5731-5736.

49.**Koedood, M., A. Fichtel, P. Meier, and P.J. Mitchell.** 1995. Human cytomegalovirus (HCMV) immediate-early enhancer/promoter specificity during embryogenesis defines target tissues of congenital HCMV infection. J. Virol. *69*:2194-2207.

50.**Kolls, J.K., D. Lei, G. Odom, S. Nelson, W.R. Summer, M.A. Gerber, and J.E. Shellito.** 1996. Use of transient CD4 lymphocyte depletion to prolong transgene expression of E1-deleted adenovirus vectors. Hum. Gene Ther. *7*:489-497.

51.**Kozarsky, K.F., D.R. McKinley, L.L. Austin, S.E. Raper, L.D. Stratford-Perricaudet, and J.M. Wilson.** 1994. *In vivo* correction of low density lipoprotein receptor deficiency in the Watanabe heritable hyperlipidemic rabbit with recombinant adenoviruses. J. Biol. Chem. *269*:13695-13702.

52.**Krajcsi, P., T. Dimitrov, T.W. Hermiston, A.E. Tollefson, T.S. Ranheim, S.B. Vande Pol, A.H. Stevenson, and W.S.M. Wold.** 1996. The adenovirus

E3-14.7K protein and the E3-10.4K/14.5K complex of proteins, which independently inhibit tumor necrosis factor (TNF)-induced apoptosis, also independently inhibit TNF-induced release of arachidonic acid. J. Virol. *70*:4904-4913.

53. **Lee, M.G., M.A. Abina, H. Haddada, and M. Perricaudet.** 1995. The constitutive expression of the immunomodulatory gp19K protein in E1-, E3- adenoviral vectors strongly reduces the host cytotoxic T cell response against the vector. Gene Ther. *2*:256-262.

54. **Leppard, K.N.** 1997. E4 gene function in adenovirus, adenovirus vector and adeno-associated virus infections. J. Gen. Virol. *78*:2131-2138.

55. **Li, Q., M.A. Kay, M. Finegold, L.D. Stratford-Perricaudet and S.L. Woo.** 1993. Assessment of recombinant adenoviral vectors for hepatic gene therapy. Hum. Gene Ther. *4*:403-409.

56. **Lieber, A., C.Y. He, I. Kirillova, and M.A. Kay.** 1996. Recombinant adenoviruses with large deletions generated by Cre-mediated excision exhibit different biological properties compared with first-generation vectors *in vitro* and *in vivo*. J. Virol. *70*:8944-8960.

57. **Linsley, P.S., P.M. Wallace, J. Johnson, M.G. Gibson, J.L. Greene, J.A. Ledbetter, J.A. Singh, and M.A. Tepper.** 1992. Immunosuppression *in vivo* by a soluble form of the CTLA-4 T cell activation molecule. Science *257*:792-795.

58. **Lusky, M., M. Christ, K. Rittner, A. Dieterle, D. Dreyer, B. Mourot, H. Schultz, F. Stoeckel, A. Pavirani, and M. Mehtali.** 1998. *In vitro* and *in vivo* biology of recombinant adenovirus vectors with E1, E1/E2A, or E1/E4 deleted. J. Virol. *72*:2022-2032.

59. **Lusky, M., L. Grave, A. Dieterlé, D. Dreyer, M. Christ, C. Ziller, P. Furstenberger, J. Kintz, D. Ali Hadji, A. Pavirani, and M. Mehtali.** 1999. Regulation of adenovirus-mediated transgene expression by viral E4 gene products: requirements for E4 ORF3. J. Virol. *73*:8308-8319.

60. **Mastrangeli, A., B.G. Harvey, J. Yao, G. Wolff, I. Kovesdi, R.G. Crystal, and E. Falck- Pedersen.** 1996. "Sero-switch" adenovirus-mediated *in vivo* gene transfer: circumvention of anti-adenovirus humoral immune defenses against repeat adenovirus vector administration by changing the adenovirus serotype. Hum. Gene Ther. *7*:79-87.

61. **McCoy, R.D., B.L. Davidson, B.J. Roessler, G.B. Huffnagle, S.L. Janich, T.J. Laing, and R.H. Simon.** 1995. Pulmonary inflammation induced by incomplete or inactivated adenoviral particles. Hum. Gene Ther. *12*:1553-1560.

62. **Michou, A.I., L. Santoro, M. Christ, V. Julliard, A. Pavirani, and M. Mehtali.** 1997. Adenovirus-mediated gene transfer: influence of transgene, mouse strain and type of immune response on persistence of transgene expression. Gene Ther. *5*:473-482.

63. **Mitani, K., F.L. Graham, C.T. Caskey, and S. Kochane.** 1995. Rescue, propagation, and partial purification of a helper virus-dependent adenovirus vector. Proc. Natl. Acad. Sci. USA *92*:3854-6858.

64. **Molnar-Kimber, K.L., D.H. Sterman, M. Chang, E.H. Kang, M. Elbash, M. Lanuti, A. Elshami, K. Gelfand, J.M. Wilson, L.P. Kaiser, and S.M. Albelda.** 1998. Impact of preexisting and induced humoral and cellular immune responses in an adenovirus-based gene therapy phase-I clinical trial for localized mesothelioma. Hum. Gene Ther. *9*:2121-2133.

65. **Morral, N., W. O'Neal, H. Zhou, C. Langstrom, and A. Beaudet.** 1997. Immune responses to reporter proteins and high viral dose limit duration of expression with adenovirus vectors: comparison of E2A wild-type and E2A-deleted vectors. Hum. Gene Ther. *8*:1275-1286.

66. **Morral, N., R.J. Parks, H. Zhou, C. Langston, G. Schiedner, J. Quinones, F.L. Graham, S. Kochanek, and A.L. Beaudet.** 1998. High doses of a helper-dependent adenoviral vector yield supraphysiological levels of alpha1-antitrypsin with negligible toxicity. Hum. Gene Ther. *9*:2709-2716.

67. **Morsy, M.A., M. Gu, S. Motzel, J. Zhao, J. Lin, Q. Su, H. Allen, L. Franlin, et al.** 1998. An adenoviral vector deleted for all viral coding sequences results in enhanced safety and extended expression of a leptin transgene. Proc. Natl. Acad. Sci. USA *95*:7866-7871.

68. **Nakagawa, I., M. Murakami, K. Ijima, S. Chikuma, I. Saito, Y. Kanegae, H. Ishikura, T. Yoshiki, et al.** 1998. Persistent and secondary adenovirus-mediated hepatic gene expression using adenovirus vector containing CTLA4IgG. Hum. Gene Ther. *9*:1739-1745.

69. **Nanda, N.K., K.K. Arzoo, H.M. Geysen, A. Sette, and E.E. Sercarz.** 1995. Recognition of multiple peptide cores by a single T cell receptor. J. Exp. Med. *182*:531-539.

70. **Otake, K., D.L. Ennist, K. Harrod, and B.C. Trapnell.** 1998. Nonspecific inflammation inhibits adenovirus-mediated pulmonary gene transfer and expression indendent of specific acquired immune reponses. Hum. Gene Ther. *9*:2207-2222.

71. **Piedra, P.A., G.A. Poveda, B. Ramsey, K. McCoy, and P.W. Hiatt.** 1998. Incidence and prevalence of neutralizing antibodies to the common adenoviruses in children with cystic fibrosis: implication for gene therapy with adenovirus vectors. Pediatrics *101*:1013-1019.

72. **Poller, W., S. Schneider-Rasp, U. Liebert, F. Merklein, P. Thalheimer, A. Haack, R. Schwaab, C. Schmitt, and H.H. Brackmann.** 1996. Stabilization of transgene expression by incorporation of E3 region genes into an adenovirus factor IX vector and by transient anti-CD4 treatment of the host. Gene Ther. *3*:521-530.

73. **Qin, L., Y. Ding, D.R. Pahud, N.D. Robson, A. Shaked, and J.S. Bromberg.** 1997. Adenovirus-mediated gene transfer of viral interleukin-10 inhibits the immune response to both alloantigen and adenoviral antigen. Hum. Gene Ther. *8*:1365-1374.

74. **Raper, S.E., Z.J. Haskal, X. Ye, C. Pugh, E.E. Furth, G.P. Gao, and J.M. Wilson.** 1998. Selective gene transfer into the liver of non-human primates with E1-deleted, E2A-defective, or E4 deleted recombinant adenoviruses. Hum. Gene Ther. *9*:671-679.

75. **Rawle, F.C., B.B. Knowles, R.P. Ricciardi, V. Brahmacheri, P. Duerksen-Hughes, W.S.M. Wold, and L.R. Gooding.** 1991. Specificity of the mouse cytotoxic T lymphocyte response to adenovirus 5. J. Immunol. *146*:3977-3984.

76. **Rawle, F.C., A.E. Tollefson, W.S. Wold, and L.R.**

Gooding. 1989. Mouse anti-adenovirus cytotoxic T lymphocytes: inhibition of lysis by E3 gp19K but not E3 14.7K. J. Immunol. *143*:2031-2037.

77. **Rooke, R., C. Waltzinger, C. Benoist, and D. Mathis.** 1997. Targeted complementation of MHC class II deficiency by intrathymic delivery of recombinant adenovirus. Immunity *7*:123-134.

78. **Routes, J.M. and J.L. Cook.** 1990. Resistance of human cells to the adenovirus E3 effect on class I MHC antigen expression: implications for antiviral immunity. J. Immunol. *144*:2763- 2770.

79. **Roy, S., P.S. Shirley, A. McClelland, and M. Kaleko.** 1998. Circumvention of immunity to the adenovirus major coat protein hexon. J. Virol. *72*:6875-6879.

80. **Sawchuk, S.J., G.P. Boivin, L.E. Duwel, W. Ball, K. Bove, B. Trapnell, and R. Hirsch.** 1996. Anti-T cells receptor monoclonal antibody prolongs transgene expression following adenovirus-mediated *in vivo* gene transfer to mouse synovium. Hum. Gene Ther. *7*:499-506.

81. **Scaria, A., J.A. St. George, C. Jiang, J.M. Kaplan, S.C. Wadsworth, and R.J. Gregory.** 1998. Adenovirus-mediated persistent cystic fibrosis transmembrane conductance regulator expression in mouse airway epithelium. J. Virol. *72*:7302-7309.

82. **Schiedner, G., N. Morral, R.J. Parks, Y. Wu, S.C. Koopman, C. Langston, F.L. Graham, A.L. Beaudet, and S. Kochanek.** 1998. Genomic DNA transfer with a high-capacity adenovirus vector results in improved *in vivo* gene expression and decreased toxicity. Nat. Genet. *18*:180-183.

83. **Schowalter, D.B., L. Meuse, C.B. Wilson, P.S. Linsley, and M.A. Kay.** 1997. Constitutive expression of murine CTLA4Ig from a recombinant adenovirus vector results in prolonged transgene expression. Gene Ther. *4*:853-860.

84. **Schowalter, D.B., J.C. Tubb, M. Liu, C.B. Wilson, and M.A. Kay.** 1997. Heterologous expression of adenovirus E3-gp19K in an E1a-deleted adenovirus vector inhibits MHC I expression *in vitro*, but does not prolong transgene expression *in vivo*. Gene Ther. *4*:351-360.

85. **Shisler, J., C. Yang, B. Walter, C.F. Ware, and L.R. Gooding.** 1997. The adenovirus E3-10.4K/14.5K complex mediates loss of cell surface Fas (CD95) and resistance to Fas-induced apoptosis. J. Virol. *71*:8299-8306.

86. **Smith, T.A., M.G. Mehaffey, D.B. Kayda, J.M. Saunders, S. Yei, B.C. Trapnell, A. McClelland, and M. Kaleko.** 1993. Adenovirus mediated expression of therapeutic plasma levels of human factor IX in mice. Nat. Genet. *4*:397-402.

87. **Stein, C.S., J.L. Pemberton, N. Van Rooijen, and B.L. Davidson.** 1998. Effects of macrophage depletion and anti-CD40 ligand on transgene expression and redosing with recombinant adenovirus. Gene Ther. *5*:431-439.

88. **Tanaka, Y. and S.S. Tevethia.** 1988. Differential effect of adenovirus E3/19K glycoprotein on the expression of H-$2K^b$ and H-$2D^b$ class I antigens and H-$2K^b$- and H-$2D^b$-restricted SV40-specific CTL-mediated lysis. Virology *165*:357-366.

89. **Tollefson, A.E., T.W. Hermiston, D.L. Lichtenstein, C.F. Colle, R.A. Tripp, T. Dimitrov, K. Toth, C.E. Wells, P.C. Doherty, and W.S.M. Wold.** 1998. Forced degragation of Fas inhibits apoptosis in adenovirus-infected cells. Nature *392*:726-730.

90. **Tripathy, S.K., H.B. Black, E. Goldwasser, and J.M. Leiden.** 1996. Immune responses to transgene-encoded proteins limit the stability of the gene expression after injection of replication-defective adenovirus vectors. Nat. Med. *2*:545-550.

91. **Verma, I.M. and N. Somia.** 1997. Gene therapy—promises, problems and prospects. Nature *389*:239-242.

92. **Vilquin, J.T., B. Guerette, I. Kinoshita, B. Roy, M. Goulet, C. Gravel, R. Roy, and J.P. Tremblay.** 1995. FK506 immunosuppression to control the immune reactions triggered by first-generation adenovirus-mediated gene transfer. Hum. Gene Ther. *6*:1391-1401.

93. **Von Herrath, M.G., S. Efrat, M.B.A. Oldstone, and M.S. Horwitz.** 1997. Expression of adenovirus E3 transgenes in β cells prevents autoimmune diabetes. Proc. Natl. Acad. Sci. USA *94*:9808-9813.

94. **Wadsworth, S.C., H. Zhou, A.E. Smith, and J.M. Kaplan.** 1997. Adenovirus vector-infected cells can escape adenovirus antigen-specific cytotoxic T-lymphocyte killing in vivo. J. Virol. *71*:5189-5196.

95. **Wang, Q., C. Greenburg, D. Bunch, D. Farson, and M.H. Finer.** 1997. Persistent transgene expression in mouse liver following *in vivo* gene transfer ΔE1/ΔE4 adenovirus vectors. Gene Ther. *4*:393-400.

96. **Wang, Q., X.C. Jia, and M.H. Finer.** 1995. A packaging cell line for the propagation of recombinant adenovirus vectors containing two lethal gene-region deletions. Gene Ther. *2*:775-783.

97. **Wold, W.S.M., A.E. Tolleffson, and T.W. Hermiston.** 1995. E3 transcription unit of adenovirus, p. 237-274. *In* W. Doefler and P. Bohm (Eds.), Current Topics in Microbiology and Immunology: The Molecular Repertoire of Adenoviruses I: Virion Structure and Infection. Springer Publishing, Berlin.

98. **Worgall, S., P.L. Leopold, G. Wolff, B. Ferris, N. Van Roijen, and R.G. Crystal.** 1997. Role of alveolar macrophages in rapid elimination of adenovirus vectors administered to the epithelial surface of the respiratory tract. Hum. Gene Ther. *8*:1675-1684.

99. **Yang, Y., H.C. Ertl, and J.M. Wilson.** 1994. MHC class I-restricted cytotoxic T lymphocytes to viral antigens destroy hepatocytes in mice infected with E1-deleted recombinant adenoviruses. Immunity *1*:433-442.

100. **Yang, Y., K. Greenough, and J.M. Wilson.** 1996. Transient immune blockade prevents formation of neutralizing antibody to recombinant adenovirus and allows repeated gene transfer to mouse liver. Gene Ther. *3*:412-420.

101. **Yang, Y., K.U. Jooss, Q. Su, H.C. Ertl, and J.M. Wilson.** 1996. Immune responses to viral antigens versus transgene product in the elimination of recombinant adenovirus-infected hepatocytes in vivo. Gene Ther. *3*:137-144.

102. **Yang, Y., Q. Li, H.C. Ertl, and J.M. Wilson.** 1995. Cellular and humoral immune responses to viral antigens create barriers to lung-directed gene therapy with

recombinant adenoviruses. J. Virol. *69*:2004-2015.

103. **Yang, Y., F.A. Nunes, K. Berencsi, E. Gonczol, J.F. Engelhardt, and J.M. Wilson.** 1994. Inactivation of E2a in recombinant adenoviruses improves the prospect for gene therapy in cystic fibrosis. Nat. Genet. *3*:362-369.

104. **Yang, Y., F.A. Nunes, K. Berencsi, E. Gonczol, E.E. Furth, and J.M. Wilson.** 1994. Cellular immunity to viral antigen limits E1-deleted adenovirus for gene therapy. Proc. Natl. Acad. Sci. USA *91*:4407-4411.

105. **Yang, Y., Q. Su, and J.M. Wilson.** 1996. Role of viral antigens in destructive cellular immune responses to adenovirus vector-transduced cells in mouse lungs. J. Virol. *70*:7209-7212.

106. **Yang, Y., G. Trinchieri, and J.M. Wilson.** 1995. Recombinant IL-12 prevents formation of blocking IgA antibodies to recombinant adenovirus and allows repeated gene therapy to mouse lung. Nat. Med. *1*:890-893.

107. **Yang, Y., Z. Xiang, H.C. Ertl, and J.M. Wilson.** 1995. Upregulation of class I major histocompatibility complex antigens by interferon gamma is necessary for T-cell-mediated elimination of recombinant adenovirus-infected hepatocytes *in vivo*. Proc. Natl. Acad. Sci. USA *92*:7257-7261.

108. **Yeh, P., J.F. Dedieu, C. Orsini, E. Vigne, P. Denefle, and M. Perricaudet.** 1996. Efficient dual transcomplementation of adenovirus E1 and E4 regions from a 293-derived cell ligne expressing a minimal E4 functional unit. J. Virol. *70*:559-565.

109. **Yei, S., N. Mittereder, K. Tang, C. O'Sullivan, and B.C. Trapnell.** 1994. Adenovirus-mediated gene transfer for cystic fibrosis: quantitative evaluation of repeated *in vivo* vector administration to the lung. Gene Ther. *3*:192-200.

110. **Zabner, J., S.C. Wadsworth, A.E. Smith, and M.J. Welsh.** 1996. Adenovirus-mediated generation of cAMP-stimulated Cl-transport in cystic fibrosis airway epithelia in vitro: effect of promoter and administration method. Gene Ther. *3*:458-465.

111. **Zhang, H.G., T. Zhou, P. Yang, C.K. Edwards III, D.T. Curiel, and J.D. Mountz.** 1998. Inhibition of tumor necrosis factor a decreases inflammation and prolongs adenovirus gene expression in the lung and liver. Hum. Gene Ther. *9*:1875-1884.

112. **Zhou, H., W. O'Neal, N. Morral, and A.L. Beaudet.** 1996. Development of a complementation cell line and a system for construction of adenovirus with E1 and E2A deleted. J. Virol. *70*:7030-7038.

113. **Zsengeller, Z.K., G.P. Boivin, S.S. Sawchuk, B.C. Trapnell, J.A. Whitsett, and R. Hirsch.** 1997. Anti-T cell receptor antibody prolongs transgene expression and reduces lung inflammation after adenovirus-mediated gene transfer. Hum. Gene Ther. *8*:935-941.

114. **Zsengeller, Z.K., S.E. Wert, W.M. Hull, X. Hu, S. Yei, B.C. Trapnell, and J.A. Whitsett.** 1995. Persistence of replication-deficient adenovirus-mediated gene transfer in lungs of immune-deficient (nu/nu) mice. Hum. Gene Ther. *6*:457-467.

11 Molecular Conjugate Vectors

Paul Schwarzenberger[1,2] and Jay K. Kolls[1,2,3]
[1]Gene Therapy Program, [2]Department of Medicine, [3]Department of Pediatrics, Louisiana State University Medical Center of New Orleans, New Orleans, LA, USA

BASIC CONSIDERATIONS FOR GENE TRANSFER VECTORS

The concept of transferring genetic information and efficiently expressing it in cells has tremendous potential in developing novel therapeutic strategies to treat human disease (18,45,57,60). Success of this methodology depends largely on the development of vehicles to transport the genetic information into its target destination, in vitro and in vivo. Requirements for this ideal vector may vary and depend on the desired goals for each individual application. For instance, integration into primitive hematopoietic and long-term gene expression, which preferably is physiologically regulated, would be sought for deficiency syndromes such as adenosine deaminase (ADA) deficiency and Gaucher's disease (45,60). On the other side, transient expression of a toxin or suicide gene may be sufficient for cancer therapy such as the purging of contaminated bone marrow from leukemia cells. This task, however, requires a high selectivity of the vector to target leukemia cells, either by exclusivity of vector uptake or transgene expression. The ideal vector would transduce all cancer cells and the toxin gene would be expressed, whereas normal cells would not be transfected or at least the toxin gene would remain transcriptionally silent (3,5,16,17,25,44,46,48,59,61,66,80). The higher the specificity, the better the therapeutic index is expected to be, which is a measurement of toxic versus beneficial effects of a treatment. Similar selectivity is critical for the reverse situation, in which hematopoietic stem cells are made more drug-resistant than cancer cells with a vector encoding the multidrug-resistant (MDR) 1 gene (8,67,69). Transferring the drug resistance gene to cancer cells would defeat the purpose by creating a more drug-resistant cancer. For in vivo applications, vector stability and correct and efficient delivery to the target tissue is a requirement for the ideal vector. Furthermore, the vector should remain unrecognized by the host's immune system, which may permit repeated administration (stealth vector) (21,50,52). None of the presently available vector systems is even close to being considered ideal for any category of the requirements for therapeutic purpose.

Viral Vectors: Basic Science and Gene Therapy
Edited by A. Cid-Arregui and A. García-Carrancá

PREVIOUS DEVELOPMENT AND PROPERTIES OF MOLECULAR CONJUGATE VECTORS

In vector development, scientists have taken advantage of naturally occurring vehicles—viruses that were genetically altered with recombinant DNA technology to encode reporter genes or other genes of interest (1,11,35,41). However, disadvantages with these viral vectors are the limited size of the transgene, safety concerns, restricted tropism and gene expression depending on the virus, and inherent host defense mechanisms against viruses. Since these may not be encountered with synthetic compounds, nonviral vector systems are an attractive alternative that may offer certain advantages. Nevertheless, viral vectors have certain properties, ensuring their survival in a foreign environment, that cannot be mimicked entirely with synthetic vectors (e.g., endosomal lysis escape mechanisms with adenovirus [Ad] and the ability to integrate with retrovirus) (19). Therefore, viral elements are likely to be a necessary complement to synthetic vectors.

Several nonviral delivery systems have been developed, of which the simplest form is free plasmid DNA as a mammalian expression vector. Intracellular uptake of unmodified plasmid DNA is low, and intracellular degradation of the small DNA amount entering is high, resulting in very low and short gene expression. Therefore, to increase gene transfer efficiency, either some other form of delivery vehicle or the modification of the target cells is required. Modification of target cells can be accomplished with electric current (electroporation) or calcium phosphate co-precipitation causing transient cell membrane damage (40). Encapsulation of DNA with different lipid formulations (liposomes) significantly enhances DNA uptake (30, 31). This procedure is also less toxic, since no additional physical or chemical cell membrane modification for DNA entry is needed (e.g., osmosis or electric current). Cationic molecules (lipids, proteins, and carbohydrates) bind negatively charged DNA and also facilitate cellular uptake. These constructs have been designated as molecular conjugate vectors (MCVs), although this term usually refers to cationic polyamines bound to DNA by electrostatic forces (17,62,63,84).

Further modifications of these MCVs by incorporating receptor ligands have led to the development of targeted MCVs or transferrin (T)-MCVs (12,13,19,56,74, 84). Figure 1 depicts the targetability of a MCV. The idea of utilizing cell surface receptors for gene transfer has been previously described in the literature (22). Wu et al. were the first group to successfully utilize this strategy using molecular conjugate vectors targeted to the asialoglycoprotein receptor of hepatocytes in vitro and in vivo (78,80,81). Such T-MCVs consisted of poly-L-lysine, chemically linked to asialoglycoprotein and DNA, which encoded the gene of interest. It has been demonstrated that electrostatic forces condense the components to form toroid-shaped structures with a diameter of 80 to 100 nm (34,62). Several modifications of the T-MCV have been described, although the principle remains the same; a polycation (polylysine [PL], protamin sulfate) is linked either chemically or via a biotin-streptavidin technique to an epitope binding moiety (usually a receptor ligand or antibody) and is condensed with negatively charged DNA (13,63,74,81,84). One significant advantage over recombinant virus systems is that there are theoretically no size limitations for transferred DNA. Cotton et al. showed efficient transfer of 48 kb genes (14).

Although gene delivery across the cell membrane with MCVs is highly efficient, only a small portion of vector initially entering the cell will reach the nucleus intact for transcription of the introduced gene. Most of the DNA entering the cells is degraded in the endosome-lysosome system. The half-life of DNA complexed to PL as delivered in MCVs is significantly prolonged over that of naked DNA (9). To further enhance escape of endosomal lysis, Curiel et al. (19) introduced replication-deficient Ad particles into the vector, which increased gene expression by 2000-fold (19,73). In this regard, despite the development of other agents—mostly virus-derived compounds capable of mediating endosomal lysis—Ad remains the most potent agent (14,19,64,72,73). For instance, in direct comparison, Ad showed an 800-fold and 1200-fold higher gene expression when used as endosomal lysis agent over synthetic influenza peptides and chloroquine, respectively (References 64 and 72, and P. Schwarzenberger and J.K. Kolls, unpublished data). The Ad protein(s) mediating endosomal lysis have not been identified yet. Physical or chemical manipulations (UV-psoralen) of the virus particle for safety purposes resulted in decreased endosomal lytic properties (14). Although gene transfer efficiency and transgene expression is higher with MCVs than with most other gene transfer systems in most cell lines and primary cells, the introduced DNA does not integrate and thus is distributed to daughter cells and diluted with each cell division. Therefore, gene expression is transient, although it is cell line-specific with the expression interval varying from 3 to 28 days (68).

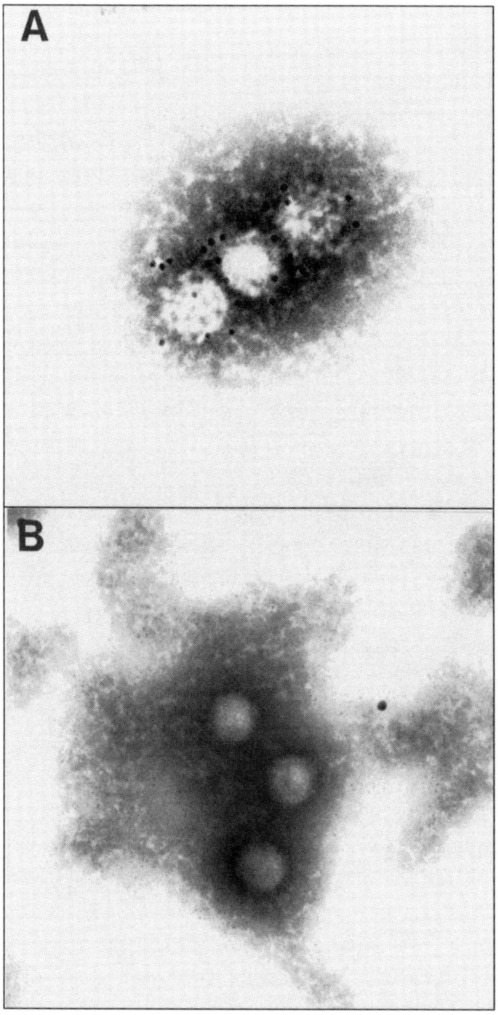

Figure 1. A mixture of recAd-PL (recombinant Ad linked to poly-L-lysine) and SA-PL (streptavidin linked to poly-L-lysine) was condensed with carrier (yeast tRNA). The complex was incubated with biotin, and after removal of excess biotin, incubated with streptavidin conjugated gold particles. Excess gold was washed off. Analysis by electron microscopy demonstrates gold-particle binding to the biotinylated complex (A), whereas complex not exposed to biotin did not retain streptavidin conjugated gold (B). This experiment visualizes in principle the targetability of recMCV with biotinylated ligands.

LIMITATIONS OF RECEPTOR TARGETING WITH MCV

The MCV system involves several factors regarding specificity for receptor-targeted gene transfer, which need to be considered. These are directly related to the components used for vector assembly.

Adenovirus

The MCV can enter the cell with its Ad component via the coxsackie-Ad receptor (CAR) (6). This receptor is found on most cells at variable density; therefore a high density as found in HeLa cells is expected to result in increased background (6,56).

Polylysine

PL is a highly efficient agent at coating cells. It binds to heparin receptors on the cell surface, but also nonspecifically through electrostatic forces. It was shown that MCVs can enter cells via this PL-mediated mechanism. We have found that a variety of cell lines that are resistant to Ad transfection, presumably due to low levels of CAR, could be successfully transfected with Ad–PL-based MCV with high efficiency (References 4 and 28, and P. Schwarzenberger and J.K. Kolls, unpublished data). Furthermore, Wickham et al. showed the incorporation of repeated lysine sequences into the penton base or the knob region greatly enhances Ad uptake in cells containing low levels of CAR (75).

DNA

The transfection efficiency greatly depends on the charges of the MCV. The combination of positively charged PL and negatively charged DNA with excess positive charges of the complex results in toroid formation of the MCV (34,63). Further charge neutralization with excess DNA leads to loss of the transfection ability (P. Schwarzenberger and J.K. Kolls, unpublished data).

The addition of a ligand (e.g., transferrin) to the MCV (T-MCV) preferentially directs vector uptake through its receptor. The process requires binding to the cell surface and internalization of the receptor-T-MCV complex. Receptor-mediated amplification of gene transfer is influenced by the different competing mechanisms involved (Figure 2). Therefore, this procedure may not be universally applicable to every cell line, even if the cell lines express the receptor at the same density. For instance, although gene transfer to K562 cells results in a 3- to 10-fold amplification with the T-MCV over the untargeted MCV vector, this effect is not seen in other leukemia cell lines. Thus, no difference in uptake is seen between T-MCV and control MCV vectors in TF1, KG1, MO7e, MBO2, U937, or THP1 cells. We could not correlate such discrepancy with differences in transferrin receptor internalization or receptor density on the surface of those cells (P. Schwarzenberger and J.K. Kolls, unpublished data). Since CAR expression in these cell lines is very low, we concluded that the overriding mechanism for gene transfer with the T-MCV is mediated through the PL component.

The uptake in HeLa cells of Ad–PL-based MCV tagged with transferrin was dramatically decreased when the vector was preincubated with anti-Ad antibodies. This indicated that in HeLa cells, which have a high density of CAR, PL is a less prominent contributor to receptor-independent vector uptake than the Ad component (56).

Different groups have reported successful receptor targeting in vivo to liver and lung with T-MCV (32,39,62,76,77). We have been unable to demonstrate receptor-mediated amplification of gene transfer in mouse liver after intravenous injection of transferrin-T-MCV (P. Schwarzenberger and J.K. Kolls, unpublished data).

RECOMBINANT MOLECULAR CONJUGATE VECTORS

The novel concept of recombinant molecular conjugate vectors (recMCVs) has been recently introduced. Essentially, recMCV is constructed identically to MCV, except that the DNA is a recombinant adenovirus (recAd) encoding the gene of interest (e.g., a reporter gene), whereas in MCV, the DNA is a plasmid. In recMCV, a recAd is chemically linked to PL, which behaves both as gene delivery vehicle and as endosomal lytic agent, whereas in MCV, Ad (e.g., *dl*312) is used for the sole purpose of endosomal lysis.

Similar to MCV, recMCV can be targeted by incorporating ligands to constitute T-recMCV (see Figure 2).

The recMCV vector system exhibited a significantly higher gene transfer efficiency over the conventional systems (MCV, T-MCV, and recAd), both in short- and long-term tests (10-fold increase) (66). Furthermore, durable reporter gene expression (>1 year without selection) was observed, which was associated with a high rate of genomic integration (30%–60%). To explain these findings, several possibilities can be considered. The location within the virus capsid further improves the protective environment for the delivered DNA (9). Therefore, intracellular survival time of the transgene is expected to be enhanced. Although Ad is generally believed to be a nonintegrating vector, infrequent integra-

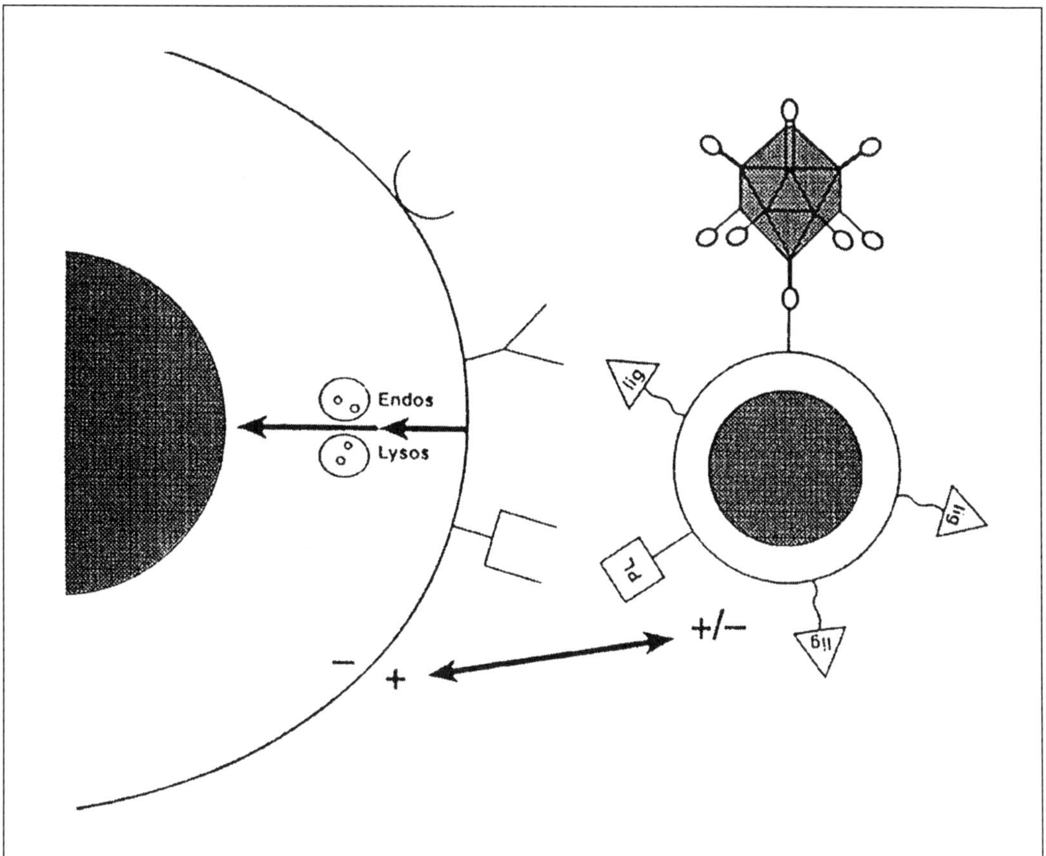

Figure 2. Targeted MCV and cellular interactions. The vector can enter the cell via different mechanisms: receptor-mediated internalization via cell surface receptor and corresponding ligand interactions (e.g., transferrin/transferrin-receptor); Ad and CAR receptor, with CAR being expressed at varying densities on most mammalian cells; or PL-mediated. This latter mechanism is charge-dependent. It is unclear whether it is in part receptor-mediated (heparan cell surface receptor). If the overriding mechanism of cell entry is via the ligand/receptor (transferrin), receptor targeting can be established. If entry via CAR or PL is significant, background is high, and receptor targeting cannot be established. Receptor-mediated endocytosed molecules are directed into the endosome/lysosome compartment. Fusion of the Ad with the membrane complex results in rupture and escape of the complex (endosomal lysis). The vector complexes remain mostly in the cytosol, although penetration of the nuclear membrane occurs at high frequency. The MCV nomenclature is determined by individual compounds utilized to assemble the vector: Ad, adenovirus (nonrecombinant, e.g., *dl*312); recAd, recombinant Ad encoding a gene of interest, e.g., luciferase; PL, poly-L-lysine; Ad-PL, Ad chemically linked to PL; SA, streptavidin; SA-PL, SA chemically linked to PL; MCV, molecular congugate vector (Ad-PL + DNA); T-MCV, receptor-targeted MCV (biotinylated ligand + SA-PL + Ad-PL + DNA); recMCV, MCV constructed with recAd instead of Ad; T-recMCV, receptor-targeted recMCV (biotinylated ligand + SA-PL + recAd-PL + DNA).

tion events have been reported, which depend on virus subtype and cell type (8, 24,29,38,49,70).

Uptake of recAd is limited by the cell surface density of CAR (6). We found that the uptake of Ad can be significantly increased if delivered as a PL conjugate (66). The PL component may also enhance nuclear import. It was concluded that a threshold dose is necessary for Ad integration to occur. Cellular uptake of unmodified Ad is limited by the number of CAR available on the cell surface. This limit can be overcome by the use of recMCV or T-recMCV. The PL component increases unspecific cellular uptake of recAd conglomerates, and retargeting through the transferrin receptor can further increase uptake of the complex. Moreover, the PL-Ad conglomerate appears to stabilize the complex. However, recombinant vector conglomerates are not restricted to Ad, but can be applied to other viral vectors as well (e.g., retrovirus) (P. Schwarzenberger and J.K. Kolls, unpublished data). Most of these studies were performed in K562 cells.

In addition, we have used a c-kit receptor to target recMCV in MO7e cells. The kinetics of reporter gene expression was similar to that obtained with T-MCV (up to 96 hours after transfection), although the absolute levels increased with T-recMCV and recMCV. However, in T-recMCV-transfected cells maintained in culture over three months, the transgene could be detected by polymerase chain reaction (PCR), whereas in cells infected with recAd or transfected with T-MCV could not (P. Schwarzenberger and J.K. Kolls, unpublished data). In contrast, T-MCV-transfected cells consistently became PCR-negative after two weeks in culture. These findings are suggestive of transgene integration. Our working hypothesis is that gene expression becomes silenced at a posttranscriptional level, mainly through promoter inactivation. Current experiments with hematopoietic cell lines using different promoters should determine whether long-term gene expression with recMCV can be accomplished.

POTENTIAL APPLICATIONS AND FUTURE DEVELOPMENTS

In vivo use of MCV or recMCV is subjected to similar concerns than those applying to recAd, particularly in regard to host immune response. Therefore, a single application of vector is possible in an immune-naïve host for a given Ad subtype, since subsequent administration would lead to accelerated inactivation of the vector and prevent expression of the transgene (50,52,82). Present data are insufficient to support a clinical trial utilizing T-MCV for in vivo targeting.

Considering the current limitations of this system, possible clinical applications can only derive from the biological characteristics of the T-MCV. Examples are its use to enhance host antitumor immune response by transferring cytokine genes or to express co-stimulatory molecules into tumor cells for cancer vaccination procedures (10,83). A phase I clinical trial using the IL-2 gene and reinfusion of transfected tumor cells in melanoma patients has recently been published (71). Another potential application is ex vivo targeting of tumor cells to purge contaminated bone marrow. Given the high gene transfer efficiency that can be obtained with vectors carrying PL, strategies for differential gene expression with specific promoter elements appear to be more promising than receptor targeting.

In vivo data with recMCV is not available, however, it may be useful mainly for ex vivo approaches or local administration (e.g., localized tumor). Unmodified Ad encoding the suicide gene herpes simplex virus thymidine kinase (HSV-tk) or p53 has been utilized in several cancer gene therapy trials (43,65). Success of this approach depends largely on efficient virus uptake into the tumor cells. Unlike many well-characterized cancer cell lines, most primary epithelial tumors express very low levels of CAR, resulting in poor Ad transduction efficiency (20,47). Therefore, modification of Ad to increase cellular uptake may be associated with increased antitumor efficacy. Miller et al. have demonstrated that targeting Ad to the epidermal growth factor receptor (EGFR) significantly enhances vector uptake to tumor cells (58). Replacement of the complete Ad 5 knob region with serotype Ad 3 knob has also been demonstrated previously to modify native Ad 5 tropism (51). Presently, one aspect of the development of novel Ad-based vectors focuses on generating receptor targeted, recombinantly engineered Ad constructs that encode a targeting moiety in the knob region. Such enhanced specificity would be associated with decreased transduction of normal tissue if the receptor is predominantly found on cancer cells as is the case with EGFR. Dmitriev et al. recently reported results obtained with a recAd encoding the 3 amino acid Arg-Gly-Asp (RGD) motif in the knob region, which conferred expanded tissue tropism to this vector (23). However, rescue of a virus encoding a receptor ligand such as EGF in the knob region has not been reported yet.

Replacement of unmodified Ad in clinical trials by targeted recombinant constructs or recMCV after appropriate preclinical evaluation warrants serious consideration. Gene transfer of unmodified Ad could be significantly enhanced to primary normal bronchial epithelium by calcium phosphate precipitation (27). It is conceivable that gene transfer in cystic fibrosis models could be enhanced using recMCV. To ameliorate immunogenicity of recAd, deletions have been performed resulting in "gutted" vectors (2,33). It is unclear if these offer advantages over conventional Ad and therefore require further investigation.

Other cationic polyamines presently under development for MCV are protaminesulfate, dioctadecylamidoglycylspermine (DOGS), and the cationic polymer polyethylenimine (PEI). PEI in particular appears to be a promising compound because its cytotoxicity appears to be low and is seen only at concentrations well above those required for optimal transfection (7). The formation of PEI-DNA-Ad complexes increased transfection efficiencies by 140- to 300-fold when compared with those obtained with PEI alone (55). The cationic polysaccharide Chitosan demonstrated favorable characteristics for gene transfer, although only under very defined and cell-specific conditions (26). Further research will evaluate its potential role for gene therapy.

Harbottle et al. synthesized a linear, bifunctional peptide that comprises an integrin-targeting domain and a DNA-binding moiety consisting of a short stretch of 16 lysine residues. The authors demonstrated targeted delivery of plasmid DNA with these constructs (42).

An elegant and novel MCV-related strategy is being pursued in which single chain antibodies targeted to a receptor (e.g., ErbB-2) are fused to the exotoxin A translocation domain and a DNA binding domain derived from the yeast GAL4 protein (36,37,54). The translocation domain of bacterial toxins (pseudomonas and diphtheria) facilitates endosome escape (36,37, 54). After condensation of these DNA-molecular fusion proteins with poly-L-lysine, an up to 150-fold increase in reporter gene expression in EGF receptor-expressing cells in comparison to poly-L-lysine-DNA complexes alone was observed.

SUMMARY

PL-based MCVs containing Ad as an endosomal lysis agent are high-efficiency gene transfer vehicles that result in transient high-level gene expression. Despite some limitations, which are mainly nonspecific gene transfer and limited targetability, transient gene expression, and cytotoxicity, T-MCV is one of the most effective gene transfer systems available. These biological properties could be applied for various ex vivo clinical applications such as transfection of tumor cells or lymphocytes with cytokine molecules or co-stimulatory molecules for immunotherapy protocols. They permit gene transfer of larger sized DNA than recombinant viral vectors. Although versatile in principle, T-MCVs are suitable constructs for receptor-targeted gene transfer only under very defined conditions that are determined by interactions of the cell with the individual components of the vector (cell membrane properties, ligand-receptor, PL, charges, and Ad). In different cell lines, one or the other component may become the overriding means for gene transfer.

The development of targeted, receptor-mediated gene transfer vectors remains a major challenge in the field. Work has been directed to reduce background gene expression, and progress has been made in addressing immune-mediated vector destruction, the results of which are applicable for T-MCV (50,52). Recent results support the concept of developing T-MCV as an integrating vector that would enable its utilization for a new spectrum of potential clinical applications, such as replacement therapy or the introduction of therapeutic genes into stem cells (66).

PROTOCOLS

Isolation of Ad Particles

1. Plate cells (911 or 293) into 40 dishes of 150 mm ø using Dulbecco's modified Eagle's medium (DMEM) with 5% fetal bovine serum (FBS) and 1% penicillin/streptomycin. The cells are examined for overall growth and signs of microbiological contamination.
2. Infect cells when they are approximately 95% confluent (approximately 10^7 cells/plate). An extra 150-mm plate should be harvested and counted so that a more accurate cell count can be performed.
3. Prewarm phosphate-buffered saline (PBS) (pH 7.4) and media in a 37°C water bath.
4. Remove media from the 40 plates, being careful not to disturb the cell monolayer.
5. The monolayer is carefully rinsed with PBS to wash away any residual media.
6. It is preferable to infect at a multiplicity of infection (moi) of 5 to 10. To calculate the volume of viral stock that is needed for an infection, multiply the following: (moi wanted) × (number of plates) × (number of cells/plate) = the number of plaque forming units (pfu)/mL needed. Divide this number by the concentration of virus in the stock (in pfus/mL) to obtain the volume needed. This amount will be diluted into 120 mL so that 3 mL can be put on each of 40 plates.
7. Add 3 mL of diluted virus to the middle of each 150-mm plate and allow them to sit for approximately 30 minutes at room temperature in the laminar flow hood. The medium with the virus should completely cover the monolayer.
8. After 30 minutes, carefully add 20 mL of medium (DMEM with 5% FBS and 1% penicillin/streptomycin). The plates are then placed back into the 37°C/5% CO_2 incubator.

9. Check the cells for cytopathic effect (CPE) after 24 to 48 hours. Cells will become more and more rounded and refractile, and it is advisable to incubate until they are just about to detach from the plate. This will depend on the virus and moi used, as well as the cell density in the plates at the time of infection. If the cells have not detached from the plate after 48 hours, the medium can be discarded. However, if cells are already detached, it is best to harvest both media and cells. In both cases, the cells must be scraped off with a sterile cell scraper and/or pipetting medium to wash the monolayer off the surface of the plate.
10. Pellet the cells in a 4°C tabletop centrifuge.
11. The tube with pellet and supernatant is frozen/thawed a minimum of three times using a dry ice/ethanol bath and a 37°C water bath.
12. Prepare the following CsCl solution in virus storage buffer (VSB) (150 mmol/L NaCl, 20 mmol/L HEPES, pH 7.8):
    a. 1.45 g/mL CsCl solution:
       73.11 g CsCl into 100 mL of VSB. Dissolve completely and filter-sterilize using a 0.22-µm filter or by steam sterilization.
    b. 1.25 g/mL CsCl solution:
       33.8 g CsCl into 100 mL of VSB. Dissolve completely and filter-sterilize using a 0.22-µm filter or sterilize by steam sterilization.
13. Prepare four SW41 tubes (Beckman Instruments, Fullerton, CA, USA) (one for balance) as follows:
    a. Place 4 mL of heavy CsCl solution onto the bottom of each centrifuge. Quick-freeze this layer by placing in a dry ice/ethanol bath.
    b. Next, onto the heavy CsCl solution, carefully layer 4 mL of the light CsCl solution. Quick-freeze this layer by placing in a dry ice/ethanol bath.
    c. Carefully add the viral supernatant on the top of the second layer. Allow gradient to thaw at room temperature before loading into the rotor buckets. Alternatively, the gradient can be made by carefully underlying the supernatant with the 1.25 g/mL CsCl solution and then underlying that with the heavier CsCl solution.
14. Load the tubes carefully into an SW41 rotor (Beckman Instruments) and spin for 2 hours at 30 000 rpm at 4°C.
15. After centrifugation is complete, remove the tubes from the rotor and transfer them to a rack in a laminar flow hood.
16. Collect the virus by carefully performing the following steps:
    a. Prepare an 18-gauge needle attached to a 5-cc syringe.
    b. Fix a centrifuge tube in a clamping device or secure rack and insert the needle directly below the viral layer. Be careful to only puncture one side of the tube (*Note*: be extremely careful not to puncture yourself). There should at least two visible layers, the lower corresponding to the Ad. Be careful not to pull the upper layer as this corresponds to the 911 chromosomal DNA. This can be removed by careful aspiration. The virus should be recovered in the minimum volume possible.
    c. Transfer the contents of the syringe to a 15-mL conical tube and place on ice.

17. Prepare four SW41 tubes (or less, depending on the volume of virus) as follows:
    a. Pipet 4 mL of 1.6 g/mL CsCl solution (70 g into 100 mL of virus storage solution) onto the bottom of each centrifuge tube.
    b. Freeze as above.
    c. Onto the heavy CsCl solution, carefully layer 4 mL of 1.4 g/mL CsCl (53.00 g/100 mL of virus storage solution).
    d. Freeze as above.
    e. Pipet viral band on top of the gradient. Allow gradient to thaw at room temperature before loading on rotor.
18. Spin at 30 000 rpm for a minimum of 18 hours at 4°C.
19. Pull bands as above. There may be three bands present; the top being that of 911 chromosomal DNA. The other two bands will be below this and will be very close together. The upper band corresponds to empty capsids, while the lower band corresponds to viable Ad particles. Be very careful to only pull the lower band.
20. CsCl is removed by loading onto an Econo-Pac® 10 DG disposable chromatography column (10 mL; Bio-Rad, Hercules, CA, USA). Follow instructions provided by the manufacturer. Never load more than 3 mL of virus onto a column, as this increases the amount of CsCl contamination in the product. Alternatively, use Sepharose® CL-6B spin columns (Amersham Pharmacia Biotech, Piscataway, NJ, USA). The sepharose is prepared by washing seven times in VSB and aliquoting in sterile 50-cc tubes. Spin columns are prepared by removing caps of Falcon® 2059 tubes (Becton Dickinson Labware, Bedford, MA, USA) and placing a 3-mL syringe without plunger into each tube. Sterilize hemostats or tweezers by dipping into 70% ethanol and use them to transfer a small amount of siliconized glass wool to the bottom of each syringe. Fill with freshly vortex mixed CL-6B and spin at 1500 rpm in a Sorvall Table Top centrifuge (Kendro Laboratory Products, Newton, CT, USA) for 1 minute. Decant the fluid off the tube and repeat. This should result in a column of approximately 2 cc of packed Sepharose. Place spin columns into a new Falcon tube.
21. Add 1.5 mL of virus suspension to each spin column and spin at 1500 rpm for 1 minute. The fluid should appear slightly cloudy. Transfer to a new spin column and repeat the spin.
22. Filter-sterilize the product by using a 0.22-μM syringe filter.
23. Add sterile glycerol to the virus so that it is 10% of the final volume. Alternatively, bovine serum albumin (BSA) can be used as a preservative. To do this, add 10 μL of 10 mg/mL BSA to each mL of virus.
24. Measure absorbance in a spectrophotometer reading at optical density (OD) 280 using VSB with 10% glycerol as the blank. This reading multiplied by the dilution factor and by 10^{12} gives the number of particles/mL. This number divided by 100 will give an approximate value of pfus/mL. However, a plaque assay should be performed to determine this value (see below). A replication competency assay (RCA) should also be performed to determine the amount of replication-competent virus.

Plaque Assay and Replication Compentency Assay

These assays can be set up at the same time so that serial dilutions of the virus can be used for both assays.

Plaque Assay

1. Plate 911 or 293 cells on 60-mm plates using DMEM with 5% FBS and 1% penicillin/streptomycin. Usually five plates per viral lot are plated, in addition to a negative control plate. Allow cells to grow until they are approximately 95% confluent.
2. Thaw an aliquot of virus on ice.
3. Pipet 2 µL of virus suspension on the bottom of a well in a 24-well plate. Add 2 mL of PBS and mix by pipetting back and forth. This well should be marked 10^{-3}.
4. Transfer 230 µL from this well into an empty adjacent well. Add 2070 µL of PBS to this well and mix. This will be the 10^{-4} dilution. Repeat transferring 230 µL from the previous well to a new one containing 2070 µL of PBS to obtain the following dilutions: 10^{-5}, 10^{-6}, 10^{-7}, 10^{-8}, 10^{-9}, and 10^{-10}.
5. Aspirate the medium from the plates. Gently, wash the monolayer with PBS to remove residual media. Discard PBS.
6. Carefully distribute 200 µL of the following dilutions over the monolayers: 10^{-6}, 10^{-7}, 10^{-8}, 10^{-9}, and 10^{-10}. For the negative control, add 200 µL of PBS to the monolayer. To get a more accurate range of pfu for a lot of virus, one can take the 280 OD reading and multiply by the dilution factor and 10^{12} to get the approximate number of particles/mL. This number divided by 100 will give the approximate pfu/mL.
7. Allow the virus to adsorb at room temperature for 30 minutes, swirling plates at the 15-minute time point. Also, at this time, place 2× DMEM medium containing 4% FBS and 4% penicillin/streptomycin in a water bath at 44°C. Melt 1.3% SeaKem® ME agarose (FMC BioProducts, Rockland, ME, USA) in a microwave and transfer to the 44°C water bath.
8. In the hood, mix equal volumes of agar and 2× minimum essential medium (MEM) in a sterile 50-mL conical tube and slowly overlay the monolayer with 5 mL of this mixture by tilting the plate while adding the mixture by its side and then slowly returning the plate back to a level position. Allow the agarose to sit flat for five minutes unstacked. Place plates back into the incubator. Plaques should become visible after 6 to 9 days, and can be counted by 10 to 14 days.
9. Count the plaques and calculate the titer of virus according to the dilution factor.

Replication Competency Assay

To detect replicating virus, infect A549 cells at a high moi as described below and check for cytopathic effects.

1. A549 cells are plated in DMEM/F12 (Kaign's) with 5% FBS and 1% penicillin/streptomycin. Cells are allowed to grow until they are approximately 95% confluent. Three plates are seeded per lot of virus, in addition to three plates for the negative control.
2. Removed medium from the plates.
3. The monolayer is carefully rinsed with PBS to remove residual media.
4. 200 µL of the 10^{-3} dilution from

above is gently added to the cells by dropping it onto the monolayer in the center of the plate. For the negative control, use 200 µL of PBS.

5. The virus is allowed to adsorb at room temperature for 30 minutes after which 5 mL of DMEM/F12 with 2% FBS and 1% penicillin/streptomycin is added to each plate.
6. Plates are allowed to incubate at 37°C with 5% CO_2 for one week.
7. Evidence of CPE in test plates is an indication that the virus can replicate. These plates should be compared to the negative control.

Chemical Linkage of Ad and PL

1. Freshly prepared virus after the second CsCl purification step is resuspended in 1.33 g/mL CsCl to achieve a volume of 2.5 mL. Cryopreserved virus stored in viral preservation medium (VPM) should not be used.
2. Load the solution onto a PD10 column (Amersham Pharmacia Biotech) that has been preequilibrated with Hank's balanced salt solution (HBSS), 5 mmol/L HEPES, 150 mmol/L NaCl, pH 7.8.
3. Elute with 2 mL HBSS. Resuspend with HBSS to a final volume of 3.6 mL.

Reagent Preparation

1. Dissolve 6 mg of poly-L-lysine (P-2636; Sigma, St. Louis, MO, USA) in 1 mL of HBSS. Aliquots can be stored at -20°C.
2. EDC (22980G; Pierce Chemical, Rockford, IL, USA) 1 g in 4.0 mL of distilled water. This solution must be prepared immediately before use and discarded thereafter. Once opened, we recommend using EDC powder only two more times.

Reaction

3.6 mL virus
0.4 mL poly-L-lysine
0.08 mL EDC
Incubate on ice for 4 hours.

Purification and Storage

1. Mix the reaction mixture with 8 mL 1.45 g/mL CsCl.
2. Centrifuge at 25 000 rpm for 18 hours in SW41 at 12°C.
3. Harvest the opalescent band. A final concentration of 1×10^{11} virus particles/mL should be achieved.
4. Determine concentration by measuring OD at 210 nm (1 mg of virus protein goes to 8.4×10^{12} particles). Poly-L-lysine does not interfere with measurement. Establish a PL standard curve to allow accurate determination of sample concentration.
5. Dilute with VPM and store in aliquots at -70°C (10% glycerol VPM, recipe in isolation of Ad protocol). One viral preparation (25 dishes) should provide 10 to 14 mL of Ad-PL.

Titering and Bioassay of the Ad-PL Complex

1. Expand HeLa cells in six-well plates to 80% confluency.
2. Prior to transfection, remove medium, carefully wash with PBS (37°C) and cover cells with 0.75 mL of starvation medium.

Preparation of Plasmid DNA

DNA has to be endotoxin-free to minimize cytotoxicity. This can be accomplished using the Triton phase extraction procedure described by Manthorpe et al. (53):

1. Gently mix DNA solution (dissolved in water or buffer) with Triton® X-114 at 1% (vol/vol).
2. Place tube on ice for 15 minutes (formation of single phase).
3. Warm in water bath (55°C) for 20 minutes (phase separation occurs at 22°C).
4. Centrifuge at 2000× *g* for 10 minutes in prewarmed tabletop centrifuge (40°C).
5. Remove the upper phase (DNA-containing). Discard lower phase.
6. Repeat steps 1 to 5 twice.
7. Precipitate DNA from final phase separation step with EtOH.
8. Resuspend DNA in buffer and determine concentration by OD.

Alternatively, use columns to remove endotoxin, which are available from Qiagen (Valencia, CA, USA). Endotoxin concentration of <1 U should be confirmed with the Limulus amoeba assay.

Determination of the Optimum Ad-PL-DNA Ratio

1. Mix aliquots of 100 µL of Ad-PL with 10 ng, 100 ng, 500 ng, and 1 µg plasmid DNA (1 µg/µL) (e.g., pluc4 plasmid encoding the luciferase reporter gene or another eukaryotic expression construct encoding a reporter gene).
2. Gently vortex mix a few seconds and incubate at room temperature for 30 minutes.
3. Pipet a fixed amount of Ad-PL (10 µL, may be scaled up or down) into each well in triplicate and disperse gently by rocking the plate. Return plates to incubator for 2 hours (37°C, 5% CO_2). Thereafter, add 2 cc complete medium/well.
4. Harvest cells after 48 hours using a rubber policeman. Centrifuge using a standard tabletop centrifuge (e.g., GS-GR; Beckman Instruments) at 1000 rpm at 40°C for 10 minutes. Discard supernatant. Analyze cell pellets for reporter gene expression as recommended by manufacturer (e.g., Promega [Madison, WI, USA] for luciferase).
5. The maximum gene expression determines the optimal ratio of Ad-PL (positive charges) to DNA (negative charges). The results can be further refined by repeating the experiment using different Ad-PL-DNA ratios as determined from this experiment. Results from a typical experiment are depicted in Figure 3.

Determination of the Optimum Cell Vector Ratio

1. Mix 500 µL of Ad-PL with the optimum amount of reporter DNA as determined above (the amount may be scaled up or down). Vortex mix gently for a few seconds and incubate at room temperature for 30 minutes.
2. Add (in duplicate) different amounts of the solution (10, 25, 50, 100, and 150 µL) to HeLa cells (six-well plates, 80% confluent, 0.75 mL starved 2% FCS medium). Disperse by gently rocking the plates a few times. Return plates to incubator for 2 hours (37°C, 5% CO_2). Thereafter, add 2 mL complete medium/well.
3. Harvest cells at 48 hours and analyze the pellets for reporter gene expression. The maximum gene expression determines the best ratio of gene transfer to toxicity of the vector that dominates at higher

doses. Results from a typical experiment are depicted in Figure 4.

4. Each individual lot of Ad-PL should be tested and optimized using the same cell type as planned for the experiments.

Synthesis of Streptavidin-PL

Reagents

- Neutravidin (Pierce Chemical): 2.5 mL (10 mg/mL in HBSS)
- PL: 0.4 mL (6 mg/mL in HBSS)
- EDC: 0.04 mL (250 mg/mL in distilled water)

1. Mix and briefly vortex mix, and incubate on ice for 4 hours.
2. Buffer exchange with molecular porous membrane tubing (25225-260; Spectrum Laboratories, Rancho Dominguez, CA, USA) in PBS, 1000 mL three times.
3. Store at -20°C.

A summary of quality control steps is depicted in Figure 5.

Receptor-Targeted Transfection Using Ad-PL and Streptavidin-Polylysine Molecular Conjugate Vectors

1. Mix freshly made Ad-PL with streptavidin polylysine (SA-PL) at a PL

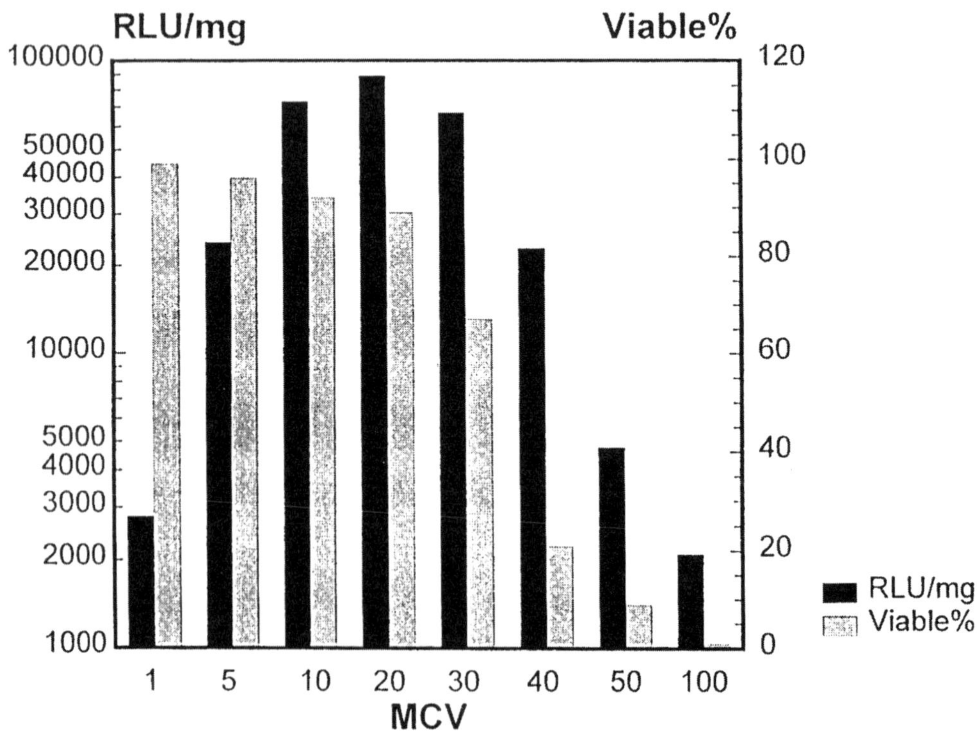

Figure 3. Optimization for charges of MCVs. 10 mL of Ad-PL was condensed with increasing doses of DNA (luciferase reporter gene *pluc4*). HeLa cells at 80% confluence were transfected with these vectors in six-well plates. Results reflect the median of triplicate experiments and are expressed as relative light units (RLU)/mg protein. Increasing amounts of DNA resulted in enhanced reporter gene expression up to a maximum of 10 ng/10 mL of Ad-PL. Gene expression declined with further increase of DNA because of decreased vector uptake.

ratio of 2:1 (determine PL concentration at 210 nm) and aliquot. Determine the optimum transfection conditions for the Ad-PL/SA-PL DNA MCV as determined above. Incubate for 30 minutes at room temperature.

2. Determine the optimum receptor-mediated gene transfer amplification with a dose response: add increasing amounts of biotinylated ligand (transferrin) (0.5–500 ng) to MCV; very gently vortex mix for a few seconds. The control series uses unbiotinylated ligand. Incubate for 30 minutes at room temperature. This will result in the T-MCV. Perform experiments in duplicate or triplicate.
3. Add the T-MCV to HeLa cells at 80% confluence in transfection medium. Gently rock several times during a 2-hour incubation period.
4. Analyze cells at 48 hours for reporter gene expression. Calculate protein amount by determining protein concentration using a standard assay. These experiments will determine the optimum concentration and ratio of MCV and ligand to cells.

Transfection of Suspension Cells

1. 10^6 K562 cells (harvested at log phase) are washed two times with PBS (37°C) and resuspended in starved medium (2% FCS RPMI)

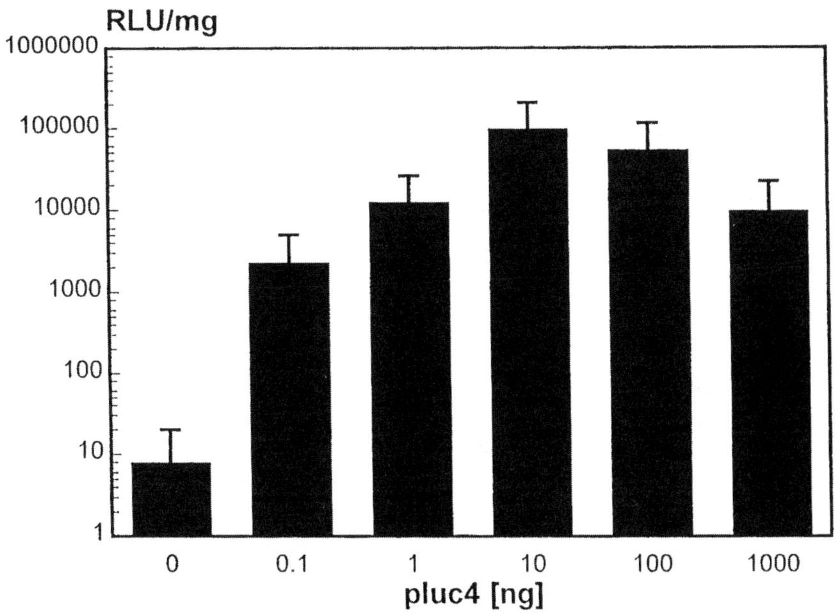

Figure 4. Optimization for cytotoxicity of MCVs. HeLa cells at 80% confluence were transfected in six-well plates with increasing amounts of MCV (the luciferase reporter gene encoding plasmid pluc4 complexed to Ad-PL). The optimum Ad-PL to DNA ratio was used that had been determined in a previous experiment (1 ng/1 µL Ad-PL; see Figure 3). Gene expression was measured and depicted as RLUs on the left axis (RLU/mg protein), cytotoxicity by trypan blue dye inclusion determined by light microscopy with no inclusion equaling 100% (right y-axis). Gene expression increased with vector concentration to a maximum of 20 µL MCV. Decreasing gene expression was observed with higher concentration, which was associated with an increase in cytotoxicity.

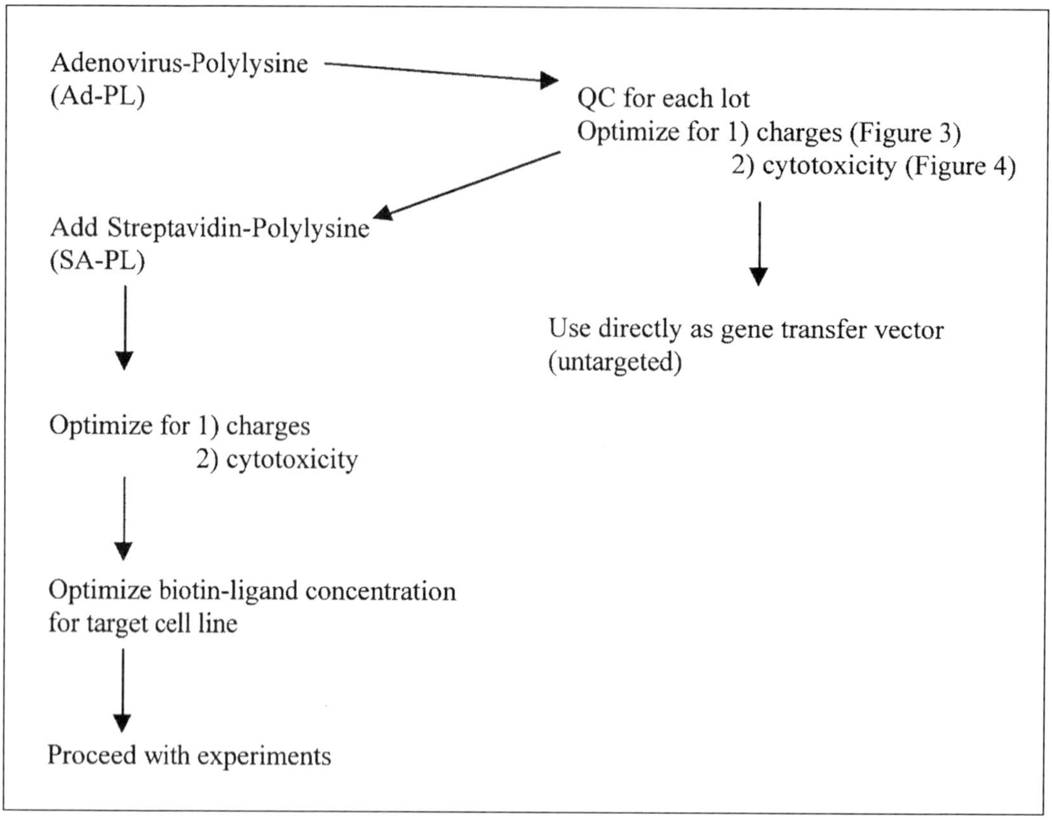

Figure 5. Schema of quality control (QC) and MCV lot reagent testing. Rigorous optimization of the vector for each given cell line and cell number is necessary for each newly generated vector lot prior to successful utilization.

in an Eppendorf (Eppendorf Scientific, Westbury, NY, USA) or cryopreservation tube. Phosphate contained in medium has been reported to interfere at least in some cells with transfection efficiency. Should this be of concern, substitution with HBSS, pH 7.3, is recommended.

2. Add vector (MCV or T-MCV). Secure cap.
3. Place vials into a rotating hybridization oven (37°C) for 2 hours.

ACKNOWLEDGMENTS

The authors thank Dr. David Curiel of the University of Alabama for establishing the initial adenovirus conjugate technology and for generously sharing it. The authors also thank Jill Schurr of the Louisiana State University Medical Center of New Orleans Gene Therapy Program for her expertise in improving and refining the adenovirus methodology employed in our laboratory.

REFERENCES

1. **Afione, S.A., C.K. Conrad, and T.R. Flotte.** 1995. Gene therapy vectors as drug delivery systems. Clin. Pharmacokinet. *28*:181-189.
2. **Amalfitano, A., M.A. Hauser, H. Hu, D. Serra, C.R. Begy, and J.S. Chamberlain.** 1998. Production and characterization of improved adenovirus vectors with the E1, E2b, and E3 genes deleted. J. Virol. *72*:926-933.
3. **Andersen, J.K., D.A. Garber, C.A. Meaney, and X.O. Breakfield.** 1992. Gene transfer into mammalian central nervous system using herpes virus vectors: extended expression of bacterial *lacZ* in neurons using the neuron-specific enolase promoter. Hum. Gene Ther.

3:487-499.

4.**Arcasoy, S.M., J.D. Latoche, M. Gondor, B.R. Pitt, and J.M. Pilewski.** 1997. Polycations increase the efficiency of adenovirus-mediated gene transfer to epithelial and endothelial cells in vitro. Gene Ther. *4*:32-38.

5.**Barry, M.A., W.J. Dower, and A. Johnston.** 1996. Toward cell-targeting gene therapy vectors: selection of cell-binding peptides from random peptide-presenting phage libraries. Nat. Med. *2*:299-305.

6.**Bergelson, J.M., J.A. Cunningham, G. Droguett, E.A. Kurt-Jones, A. Krithivas, J.S. Hong, M.S. Horwitz, R.L. Crowell, and R.W. Finberg.** 1997. Isolation of a common receptor for coxsackie B viruses and adenoviruses 2 and 5. Science *275*:1320-1323.

7.**Boussif, O., F. Lezoualch, M.D. Mergny, M.A. Zanta, D. Scherman, B. Demeneix, and J.P. Behr.** 1995. A versatile vector for gene and oligonucleotide transfer into cells in culture and in vivo: polyethylenimine. Proc. Natl. Acad. Sci. USA *92*:7297-7301.

8.**Brenner, M.K., J.M. Cunningham, B.P. Sorrentino, and H.E. Heslop.** 1995. Gene transfer into human hematopoietic progenitor cells. Br. Med. Bull. *51*:167-191.

9.**Chiou, H.C., M.V. Tangco, S.M. Levine, D. Robertson, K. Kormis, C.H. Wu, and G.Y. Wu.** 1994. Enhanced resistance to nuclease degradation of nucleic acids complexed to asialoglycoprotein-poly-lysine carriers. Nucleic Acids Res. *22*:5439-5446.

10.**Conry, R.M., A.F. LoBuglio, and D.T. Curiel.** 1996. Polynucleotide-mediated immunization therapy of cancer. Semin. Oncol. *23*:135-147.

11.**Corbeau, P., G. Kraus, and F. Wong-Staal.** 1996. Efficient gene transfer by a human immunodeficiency virus type 1 (HIV-1)-derived vector utilizing a stable HIV packaging cell line. Proc. Natl. Acad. Sci. USA *93*:14070-14075.

12.**Cotten, M., F. Langle-Rouault, H. Kirlappos, E. Wagner, K. Mechtler, M. Zenke, H. Beug, and M.L. Birnstiel.** 1990. Transferrin—polycation-mediated introduction of DNA into human leukemic cells: stimulation by agents that affect the survival of transfected DNA or modulate transferrin receptor levels. Proc. Natl. Acad. Sci. USA *87*:4033-4037.

13.**Cotten, M., E. Wagner, and M.L. Birnstiel.** 1993. Receptor-mediated transport of DNA into eukaryotic cells. Methods Enzymol. *217*:618-644.

14.**Cotton, M., E. Wagner, E. Zatloukal, S. Phillips, D.T. Curiel, and M. Birnstiel.** 1992. High efficiency receptor-mediated delivery of small and large (48 kilobase) gene constructs using the endosome-disruption activity of defective or chemically inactivated adenovirus particles. Proc. Natl. Acad. Sci. USA *89*:6094-6098.

15.**Cotten, M., E. Wagner, K. Zatloukal, and M.L. Birnstiel.** 1993. Chicken adenovirus (CELO virus) particles augment receptor-mediated DNA delivery to mammalian cells and yield exceptional levels of stable transformants. J. Virol. *67*:3777-3785.

16.**Couture, L.A., G.A. Mullen, and R.A. Morgan.** 1994. Retroviral vectors containing chimeric promoter/enhancer elements exhibit cell-type-specific gene expression. Hum. Gene Ther. *5*:667-677.

17.**Cristiano, R.J. and J.A. Roth.** 1995. Molecular conjugates: a targeted gene delivery vector for molecular medicine. J. Mol. Med. *73*:479-486.

18.**Culver, K.W. and M.R. Blaese.** 1994. Gene therapy for cancer. Trends Genet. *10*:174-178.

19.**Curiel, D.T., S. Agarwal, E. Wagner, and M. Cotten.** 1991. Adenovirus enhancement of transferrin-polylysine-mediated gene delivery. Proc. Natl. Acad. Sci. USA *88*:8850-8854.

20.**Curiel, D.T.** 1999. Strastegies to adapt adenoviral vectors for targeted delivery. Ann. N.Y. Acad. Sci. *886*:158-171.

21.**Dai, Y., E. Schwarz, D. Gu, W.W. Zhang, N. Sarvetnick, and I. Verma.** 1995. Cellular and humoral immune responses to adenoviral vectors containing factor IX gene: tolerization of factor IX and vector antigens allows for long-term expression. Proc. Natl. Acad. Sci. USA *92*:1401-1405.

22.**Dautry-Varsat, A., A. Ciechanover, and H.F. Lodish.** 1983. pH and the recycling of transferrin during receptor-mediated endocytosis. Proc. Natl. Acad. Sci. USA *80*:2258-2262.

23.**Dmitriev, I., V. Krasnykh, C.R. Miller, M. Wang, E. Kashentseva, G. Mikheeva, N. Belousova, and D.T. Curiel.** 1998. An adenovirus vector with genetically modified fibers demonstrates expanded tropism via utilization of a coxsackie virus and adenovirus receptor-independent cell entry mechanism. J. Virol. *72*:9706-9713.

24.**Doerfler, W.** 1993. Adenoviral DNA integration and changes in DNA methylation patterns: a different view of insertional mutagenesis. Prog. Nucleic Acids Res. Mol. Biol. *46*:1-36.

25.**Douglas, J.T., B.E. Rogers, M.E. Rosenfeld, S.I. Michael, M. Feng, and D.T. Curiel.** 1996. Targeted gene delivery by tropism-modified adenoviral vectors. Nat. Biotechnol. *14*:1574-1578.

26.**Erbacher, P., S. Zou, T. Bettinger, A.M. Steffan, and J.S. Remy.** 1998. Chitosan-based vector/DNA complexes for gene delivery: biophysical characteristics and transfection ability. Pharm. Res. *15*:1332-1339.

27.**Fasbender, A., J.H. Lee, R.W. Walters, T.O. Moninger, J. Zabner, and M.J. Welsh.** 1998. Incorporation of adenovirus in calcium phosphate precipitates enhances gene transfer to airway epithelia in vitro and in vivo. J. Clin. Invest. *102*:184-193.

28.**Fasbender, A., J. Zabner, M. Chillon, T.O. Moninger, A.P. Puga, B.L. Davidson, and M.J. Welsh.** 1997. Complexes of adenovirus with polycationic polymers and ctionic lipids increase the efficiency of gene transfer in vitro and in vivo. J. Biol. Chem. *272*:6479-6489.

29.**Fechteler, K., J. Tatzelt, S. Huppertz, P. Wilgenbus, and W. Doerfler.** 1995. The mechanism of adenovirus DNA integration: studies in a cell-free system. Curr. Top. Microbiol. Immunol. *199*:109-137.

30.**Felgner, P.L., T.R. Gadek, M. Holm, R. Roman, H.W. Chan, M. Wenz, J.P. Northrop, G.M. Ringold, and M. Danielsen.** 1987. Lipofection: a highly efficient, lipid-mediated DNA-transfection procedure. Proc. Natl. Acad. Sci. USA *84*:7413-7417.

31.**Felgner, P.L., Y.J. Tsai, L. Sukhu, C.J. Wheeler, M. Manthorpe, J. Marshall, and S.H. Cheng.** 1995. Improved cationic lipid formulations for in vivo gene therapy. Ann. NY Acad. Sci. *772*:126-139.

32.**Ferkol, T., J.C. Perales, E. Eckman, C. Kaetzel, R. Hanson, and P.B. Davis.** 1995. Gene transfer into the epithelium of animals by targeting the polymeric immunoglobulin receptor. J. Clin. Invest. *95*:493-502.

33.**Fisher, K.J., H. Choi, J. Burda, S. Chen, and J.M. Wilson.** 1996. Recombinant adenovirus deleted of all viral genes for gene therapy of cystic fibrosis. Virology *217*:11-22.

34.**Fisher, K.J. and J.M. Wilson.** 1994. Biochemical and functional analysis of an adenovirus-based ligand complex for gene transfer. Biochem. J. *299*:49-58.

35.**Flotte, T.R. and B.J. Carter.** 1995. Adeno associated virus vectors for gene therapy. Gene Ther. *2*:357-362.

36.**Fominaya, J., C. Uherek, and W. Wels.** 1998. A chimeric fusion protein containing transforming growth factor-alpha mediates gene transfer via binding to the EGF receptor. Gene Ther. *5*:521-530.

37.**Fominaya, J. and W. Wels.** 1996. Target cell-specific DNA transfer mediated by a chimeric multidomain protein. Novel non-viral gene delivery system. J. Biol. Chem. *271*:10560-10568.

38.**Gahlmann, R. and W. Doerfler.** 1983. Integration of viral DNA into the genome of the adenovirus type 2 hamster cell line HE5 without loss or alteration of cellular nucleotides. Nucleic Acids Res. *11*:7347-7359.

39.**Gao, L., M. Cotten, E. Wagner, S. Agarwal, C. Harris, M. Romer, L. Miller, P.-C. Hu, and D. Curiel.** 1993. Direct in vivo gene transfer to airway epithelium employing adenovirus-poly-lysine-DNA complexes. Hum. Gene Ther. *4*:17-24.

40.**Graham, F.L. and A.J. van der Eb.** 1973. Transformation of rat cells by DNA of human adenovirus 5. Virology *54*:536-539.

41.**Guenzburg, W.H., R.M. Saller, and B. Salmons.** 1995. Retroviral vectors directed to predefined cell types for gene therapy. Biologicals *23*:5-12.

42.**Harbottle, R.P., R.G. Cooper, S.L. Hart, A. Ladhoff, T. McKay, A.M. Knight, E. Wagner, A.D. Miller, and C. Coutelle.** 1998. An RGD-oligolysine peptide: a prototype construct for integrin-mediated gene delivery. Hum. Gene Ther. *9*:1037-1047.

43.**Hortobagyi, G.N., M.C. Hung, and G. Lopez-Berestein.** 1998. A phase I multicenter study of E1A gene therapy for patients with metastatic breast cancer and epithelial ovarian cancer that overexpresses HER-2/neu or epithelial ovarian cancer. Hum. Gene Ther. *9*:1775-1798.

44.**Kabat, D.** 1995. Targeting retroviral vectors to specific cells. Science *269*:417.

45.**Karlsson, S.** 1991. Treatment of genetic defects in hematopoietic cell function by gene transfer. Blood *78*:2481-2492.

46.**Kasahara, N., A.M. Dozy, and Y.W. Kan.** 1994. Tissue-specific targeting of retroviral vectors through ligand-receptor interactions. Science *266*:1373-1376.

47.**Kasono, K., J.L. Blackwell, J.T. Douglas, I. Dmitriev, T.V. Strong, P. Reynolds, D.A. Kropf, W.R. Carroll, G.E. Peters, and R.P. Bucy.** 1999. Selective gene delivery to head and neck cancer cells via an integrin targeted adenoviral vector. Clin. Cancer Res. *5*:2571-2579.

48.**Kim, M., M. Wright, J. Deshane, M.A. Accavitti, A. Tilden, M. Saleh, W.P. Vaughan, M.H. Carabasi, et al.** 1997. A novel gene therapy strategy for elimination of prostate carcinoma cells from human bone marrow. Hum. Gene Ther. *8*:157-170.

49.**Knoblauch, M., J. Schroer, B. Schmitz, and W. Doerfler.** 1996. The structure of adenovirus type 12 DNA integration sites in the hamster cell genome. J. Virol. *70*:3788-3796.

50.**Kolls, J.K., D. Lei, G. Odom, S. Nelson, W.R. Summer, M.A. Gerber, and J.E. Shellito.** 1996. Use of transient CD4 lymphocyte depletion to prolong transgene expression of E1-deleted adenoviral vectors. Hum. Gene Ther. *7*:489-497.

51.**Krasnykh, V.N., G.V. Mikheeva, J.T. Douglas, and D.T. Curiel.** 1996. Generation of recombinant adenovirus vectors with modified fibers for altering viral tropism. J. Virol. *70*:6839-6846.

52.**Lei, D., M. Lehmann, J.E. Shellito, S. Nelson, A. Siegling, H.D. Volk, and J.K. Kolls.** 1996. Nondepleting anti-CD4 antibody treatment prolongs lung-directed E1-deleted adenovirus-mediated gene expression in rats. Hum. Gene Ther. *7*:2273-2279.

53.**Manthorpe, M., F. Cornefert-Jensen, J. Hartikka, J. Felgner, and V. Dwarki** 1993. Gene therapy by intramuscular injection of plasmid DNA: studies on firefly luciferase gene expression in mice. Hum. Gene Ther. *4*:419-431.

54.**Maurer-Gebhard, M., M. Schmidt, M. Azemar, U. Altenschmidt, E. Stocklin, W. Wels, and B. Groner.** 1998. Systemic treatment with a recombinant erbB-2 receptor-specific tumor toxin efficiently reduces metastases in mice injected with genetically modified carcinoma cells. Cancer Res. *58*:2661-2666.

55.**Meunier-Durmort, C., H. Griamal, L.M. Sachs, B.A. Demeneix, and C. Forest.** 1997. Adenovirus enhancement of polyethyleneimine-mediated transfer of regulated genes in differentiated cells. Gene Ther. *4*:808-814.

56.**Michael, S.I., C.H. Huang, M.U. Romer, E. Wagner, P.C. Hu, and D.T. Curiel.** 1993. Binding-incompetent adenovirus faciltates molecular conjugate-mediate gene transfer by the receptor-mediated endocytosis pathway. J. Biol. Chem. *268*:6866-6869.

57.**Miller, A.D.** 1990. Progress toward human gene therapy. Blood *76*:271-278.

58.**Miller, C.R., D.J. Buchsbaum, P.N. Reynolds, J.T. Douglas, G.Y. Gillespie, M.S. Mayo, D. Raben, and D.T. Curiel.** 1998. Differential susceptibility of primary and established human glioma cells to adenovirus infection: targeting via the epidermal growth factor receptor achieves fiber receptor-independent gene transfer. Cancer Res. *58*:5738-5748.

59.**Miller, N. and R. Vile.** 1995. Targeted vectors for gene therapy. FASEB J. *9*:190-199.

60.**Nienhuis, A.W.** 1994. Gene transfer into hematopoietic stem cells. Blood Cells *20*:141-148.

61.**Parr, M.J., Y. Manome, T. Tanaka, P. Wen, D.W. Kufe, W.G. Kaelin, Jr., and H.A. Fine.** 1997. Tumor-selective transgene expression in vivo mediated by an E2F-responsive adenoviral vector. Nat. Med. *3*:1145-1149.

62.**Perales, J.C., T. Ferkol, H. Beegen, and O.D. Rat-**

noff. 1994. Gene transfer in vivo: sustained expression and regulation of genes introduced into the liver by receptor targeted uptake. Proc. Natl. Acad. Sci. USA *91*:4086-4090.

63. **Perales, J.C., T. Ferkol, M. Molas, and R.W. Hanson.** 1994. An evaluation of receptor-mediated gene transfer using synthetic DNA-ligand complexes. Eur. J. Biochem. *226*:255-266.

64. **Plank, C., B. Oberhauser, K. Mechtler, C. Koch, and E. Wagner.** 1994. The influence of endosome-disruptive peptides of gene transfer using synthetic virus-like gene transfer systems. J. Biol. Chem. *269*:12918-12924.

65. **Roth, J.A., S.G. Swisher, J.A. Merritt, D.D. Lawrence, B.L. Kemp, D.S. Schrump, D.M. Shin, and G.L. Walsh.** 1998. Gene therapy for non-small cell lung cancer: a preliminary report of a phase I trial of adenoviral p53 gene replacement. Semin. Oncol. *25*:33-37.

66. **Schwarzenberger, P., J.D. Hunt, E. Robert, C. Theodossiou, and J.K. Kolls.** 1997. Receptor-targeted recombinant adenovirus conglomerates: a novel molecular conjugate vector with improved expression characteristics. J. Virol. *71*:8563-8571.

67. **Schwarzenberger, P., N. Lohrey, T. Kmiecik, D.L. Longo, W.J. Murphy, F.W. Ruscetti, and J.R. Keller.** 1996. Gene transfer of multi drug resistace into a factor dependent human hematopoietic progenitor cell line. In vivo model for genetically transferred chemoprotection. Blood *87*:2723-2731.

68. **Schwarzenberger, P., S.E. Spence, J.M. Gooya, D.M. Michiel, F.W. Ruscetti, and J.R. Keller.** 1996. Targeted gene transfer to human hematopoietic progenitor cell lines through the c-kit receptor. Blood *87*:472-478.

69. **Sorrentino, B.P., S.J. Brandt, D. Bodine, M. Gottesman, I. Pastan, A. Cline, and A.W. Nienhuis.** 1992. Selection of drug-resistant bone marrow cells in vivo after retroviral transfer of human *MDR*1. Science *257*:99-103.

70. **Stabel, S., W. Doerfler, and R.R. Friss.** 1980. Integration sites of adenovirus type12 DNA in transformed hamster cells and hamster tumor cells. J. Virol. *36*:22-40.

71. **Stingl, G., R. Mertelsmann, K. Wolff, S. Schreiber, E. Kampgen, A. Schneeberger, U. Brennscheidt, H. Veelken, et al.** 1999. Phase I study to the immunotherapy of metastatic malignant melanoma by a cancer vaccine consisting of autologous cancer cells transfected with the human IL-2 gene. J. Mol. Med. *75*:297-299.

72. **Wagner, E., C. Plank, K. Zatloukal, M. Cotten, and M.L. Birnstiel.** 1992. Influenza virus hemagglutinin HA-2 N-terminal fusogenic peptides augment gene transfer by transferrin polysine—DNA complexes: toward a synthetic virus-like gene-transfer vehicle. Proc. Natl. Acad. Sci. USA *89*:7934-7938.

73. **Wagner, E., K. Zatloukal, M. Cotton, and M. Birnstiel.** 1992. Coupling of adenovirus to transferrin-polysine/DNA complexes greatly enhances receptor-mediated gene delivery and expression of transfected genes. Proc. Natl. Acad. Sci. USA *89*:6099-6103.

74. **Wagner, E., M. Zenke, M. Cotton, H. Beug, and M. Birnstiel.** 1990. Transferrin-polycation conjugates as carriers for DNA uptake into cells. Proc. Natl. Acad. Sci. USA *87*:3410-3414.

75. **Wickham, T.J., P.W. Roelvink, D.E. Brough, and I. Kovesdi.** 1996. Adenovirus targeted to heparin-containing receptors increases its gene delivery efficiency to multiple cell types. Nat. Biotechnol. *14*:1570-1573.

76. **Wilson, J.M., M. Grossman, C.H. Wu, N.R. Chowdhury, G.Y. Wu, and J.R. Chowdhury.** 1992. Hepatocyte directed gene transfer in vivo leads to transient improvement of hyperchlesterolemia in low density lipoprotein receptor deficient rabbits. J. Biol. Chem. *267*:963-967.

77. **Wu, G.Y., J.M. Wilson, F. Shalaby, M. Grossman, D.A. Shafritz, and C.H. Wu.** 1991. Receptor mediated gene delivery in vivo-partial correction of genetic analbuminemia in nagase rats. J. Biol. Chem. *266*:14338-14343.

78. **Wu, G.Y. and C.H. Wu.** 1986. Receptor mediated in vitro gene transformation by a soluble DNA carrier system. J. Biol. Chem. *262*:4429-4432.

79. **Wu, G.Y. and C.H. Wu.** 1986. Targeted inhibition of transferrin mediated iron uptake in hepG2 hepatoma cells. J. Biol. Chem. *261*:16384-16837.

80. **Wu, G.Y. and C.H. Wu.** 1988. Receptor mediated gene delivery and expression in vivo. J. Biol. Chem. *263*:14621-14626.

81. **Wu, G.Y. and C.H. Wu.** 1991. Delivery systems for gene therapy. Biotherapy *3*:87-95.

82. **Yang, Y., Q. Li, H.C. Ertl, and J.M. Wilson.** 1995. Cellular and humoral immune responses to viral antigens create barriers to lung-directed gene therapy with recombinant adenoviruses. J. Virol. *69*:2004-2015.

83.. **Zatloukal, K., W. Schmidt, M. Cotten, E. Wagner, G. Stingl, and M.L. Birnstiel.** 1993. Somatic gene therapy for cancer: the utility of transferrinfection in generating 'tumor vaccines'. Gene *135*:199-207.

84. **Zenke, M., P. Steinlein, E. Wagner, and M. Birnstiel.** 1990. Receptor mediated endocytosis of transferrin-polycation conjugates: an efficient way to introduce DNA into hematopoietic cells. Proc. Natl. Acad. Sci. USA *87*:3655-3659.

FAMILY: Herpesviridae
GENUS: Alphasimplexvirus (Herpes Simplex Virus)

12 Construction of Defective Herpes Simplex Recombinant Vectors and Helper-Dependent Amplicons

Angel Cid-Arregui[1,3] and Filip Lim[2]
[1]Max Planck Institute of Molecular Cell Biology and Genetics, and Institute of Neurobiology, University of Heidelberg, Heidelberg, Germany, [2]Molecular Biology Center, CSIC, Autonomous University, Madrid, Spain; [3]Institute of Applied Tumor Virology, German Cancer Research Center, Heidelberg, Germany

INTRODUCTION

Herpes simplex viruses (HSVs) are among the most intensively characterized of all viruses at the clinical, molecular, and structural levels (3,25,26). Yet the precise mechanisms underlying their life cycle, and in particular their ability to remain latent in their host, sometimes for life, are only partially understood. This fact, however, has not hampered the development of efficient expression vectors that take advantage of the biologic properties of HSV to target expression of recombinant proteins in a variety of cell types in vitro and in vivo (25). There are several features of HSV that make these viruses particularly attractive for gene transfer applications. They *(i)* can package large DNA fragments, *(ii)* infect a wide range of hosts and cell types, *(iii)* are suitable for transduction of nondividing cells, and *(iv)* do not integrate into the host genome, but remain episomal, thus avoiding positional effects on transgene expression and insertional mutagenesis.

HSV-1 is a complex virus (26). Its genome is a linear double-stranded DNA molecule of approximately 152 kb that contains 83 open reading frames (ORFs). Of these, 45 appear not to be essential for viral replication, at least in vitro and individually. In permissive cells, the viral proteins are expressed in a coordinated sequence to generate with their diverse functions a highly efficient machinery for viral replication (25), which leads to the lysis of the cell. Alternatively, the virus may establish a latent state (9,33) in which neither viral proteins nor particles are detectable. During latency, only a small region of the viral genome, which persists in the nucleus as an episome, is transcriptionally active and generating nonpolyadenylated, latency-associated transcripts of an as yet unknown function (4,31,34). Viral replication appears not to be required for the latent state to be established. Moreover, the mechanisms for latency and reactivation of the virus remain unknown.

Thus, there are two main problems to

Viral Vectors: Basic Science and Gene Therapy
Edited by A. Cid-Arregui and A. García-Carrancá

solve to convert HSV into a useful expression vector: *(i)* the cytopathogenicity caused by the viral proteins, and *(ii)* the tendency of HSV to establish a latent state that may switch off transgene expression. Therefore, HSV-based vectors should be necessarily defective viruses (i.e., viruses unable to transcribe the viral ORFs either because of mutations or, ideally, deletions of entire regions of the genome that are not essential for viral replication). In addition, they should retain their ability to replicate and package into infective virions.

These requirements have been approached differently in the two types of defective viruses that have been engineered to date. In the first one, the viral genome carries mutations that inactivate essential ORFs, usually one or more of the immediate early genes such as ICP4 or IE3, and ICP27 or IE2. These mutations are combined with deletions of nonessential genes to make room for the transgenes. Theoretically, at least 40 kb of the genome could be replaced (9,10,22,23,25,27). These mutants need to be propagated in complementing helper cell lines that express the corresponding viral genes. The second type of vector was devised upon the observation that minimal HSV genome subunits, called amplicons, that contain the terminal *a* sequence and an origin of viral DNA synthesis can be packaged into virions (29). Amplicons are propagated in vitro with the help of a replication-deficient HSV that provides the viral proteins needed for replication and packaging. The result is a mixture of amplicon and helper viruses. Alternatively, the viral proteins are expressed from cosmids (11) or from bacterial artificial chromosomes (32) (see also Chapters 13 and 14).

Both systems convey some burdens that need to be overcome in order to produce new generations of improved HSV vectors. First, there is a significant risk of replication-competent helper and/or wild-type virus rescue by homologous recombination, which may occur during vector production (see Construction of HSV Mutants Using "Knock-in" Rescue and Cre-Mediated Excision). Second, the titers of the vectors tend to be lower than those obtained with the wild-type virus, probably as a consequence of suboptimal expression of viral proteins. Third, mutant viruses present some residual cytotoxicity caused by leaky expression of some viral proteins such as the virion-associated virus host shut-off protein (VHS), which causes nonspecific degradation of mRNA. Fourth, mutant HSV-based vectors may enter a latent state leading to cessation of transgene expression.

Current efforts to solve these problems should result in highly improved vectors useful for gene and cancer therapies, vaccine development, and fundamental research. This chapter describes methods for the generation of mutant HSV-1 by a simple and efficient procedure using the *loxP*/Cre recombination system and for the production of high titers of helper-containing stocks of HSV amplicons. In addition, it contains some examples of reporter gene expression in PC12 cells and cultured hippocampal neurons infected with amplicons.

CONSTRUCTION OF HSV MUTANTS USING "KNOCK-IN" RESCUE AND CRE-MEDIATED EXCISION

In the past few years, a number of different techniques for manipulating large genomes have been adapted for making mutations in the 150-kb HSV-1 genome. These tools enable the construction of insertions and deletions much more rapidly than previously possible. The following example illustrates the use of "knock-in" marker rescue and site-specific, recombinase-mediated excision to construct a tagged HSV-1 mutant haploid for all genes by replacement of one of the viral diploid regions with a

LacZ expression cassette (see Figure 1).

(i) In the first step of the construction, the lacZ cassette (see below) was cloned into the plasmid plpRH between two arms that correspond to the HSV-1 sequences immediately flanking the right-hand copy (21) of the HSV-1 region containing *ICP4*, *a*, and *ICP0.* To select for recombination of the plasmid into the HSV-1 genome, the *ICP27* gene, which is essential for viral growth, was also cloned along with the insertion cassette between the arms.

(ii) The linearized plasmid plpRH was transfected into Vero cells that were subsequently infected with 5*dl*1.2, an HSV-1 mutant deleted in the ICP27 gene (19). Since 5*dl*1.2 is unable to grow on Vero cells, the only viable viruses from this culture were recombinants (PL; see Figure 1) that had acquired the *ICP27* gene from the plasmid plpRH. Since incorporation of the *ICP27* gene into the HSV-1 genome requires crossover events in both of the flanking arms, the *lacZ* cassette was also transferred into the virus during the recombination process, which also resulted in the deletion of one copy of *ICP4*, *a*, and *ICP0.*

(iii) The *lacZ* cassette in plpRH originates from the plasmid pSVpAX1 (2) and contains the simian virus 40 (SV40) early promoter, a polyadenylation signal flanked by Cre recombination targets (*loxP* sites),

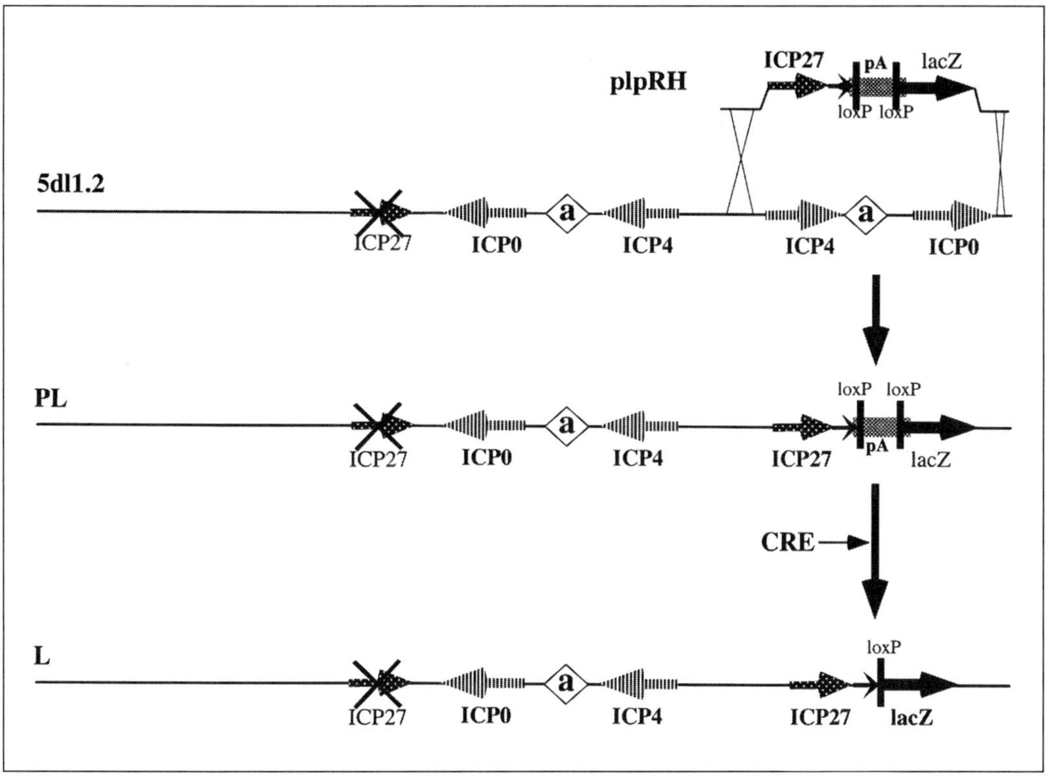

Figure 1. Construction of an HSV-1 mutant L, which is haploid for all genes and contains the *lacZ* reporter gene. The intermediate mutant PL contains a silent *lacZ* cassette that can be used as a reporter to detect Cre recombinase activity in vertebrate cells. A cassette from the plasmid pSVpAX1 containing the SV40 early promoter, a polyadenylation signal flanked by *loxP* sites and the *lacZ* gene was recombined into the HSV-1 mutant 5*dl*1.2 by marker rescue of *ICP27*, simultaneously deleting one copy each of *ICP4*, *a*, and *ICP0.* Transcription of the *lacZ* gene in the PL virus is blocked by the polyadenylation signal positioned immediately 5′ to the lacZ ORF. Excision of the polyadenylation signal via the flanking Cre recombination targets (*loxP* sites) allows the readthrough transcription of the *lacZ* gene. Since this HSV-1 mutant can be used to infect virtually any vertebrate cell type, this provides a simple method to screen for Cre transgenes in cell lines and animals.

Table 1. Summary of the Successive Generations of Amplicon Packaging Systems

Description	References
Wild-type HSV-1 as helper virus and first amplicon plasmid	29,30
TsK mutant of HSV-1 as helper virus	12
D30EBA deletion mutant and M64A complementing cell line	13
5*dl*1.2 deletion mutant and Vero 2-2 cell line	19
Helper-free packaging by transfection of HSV cosmids	11
Helper-free HSV-1 bacmid with *pac* elements deleted	32

and the *lacZ* gene. Transcription of the *lacZ* gene from the SV40 promoter is blocked by the polyadenylation signal positioned immediately 5′ to the *lacZ* ORF. Cre recombinase-mediated excision of the polyadenylation signal via the flanking *loxP* sites allows the readthrough transcription of the *lacZ* gene. This was accomplished by transfecting a Cre-expressing plasmid into Vero cells followed by infection of these cells with PL virus (from step *ii*). Viral supernatant from this culture was used to infect fresh Vero cells, which were then fixed and stained with 5-bromo-4-chloro-3-indolyl-β-D-galactopyranoside (X-gal). The presence of blue-staining plaques indicated the successful activation of the *lacZ* gene in Cre-excised mutants (L; see Figure 1).

CONSTRUCTION AND PURIFICATION OF HELPER-DEPENDENT HSV-1 AMPLICONS

Amplicons are eukaryotic expression plasmids that contain an HSV-1 packaging site (terminal *a* sequence) and an HSV-1 origin of DNA synthesis (29,30). When transfected into cells that express the viral proteins required for viral replication (packaging cell lines), these plasmids are incorporated into virions substituting the viral genome. Since a DNA molecule approximating the full-length HSV-1 genome size (approximately 152 kb) is required to be packaged (26), the multicopy plasmid concatemers being synthesized by the viral DNA polymerase are cleaved in genome length units, and these are incorporated into virions. Expression of the viral proteins essential for packaging is provided by a helper virus that has been rendered replication-defective by deletion of one of the essential α genes. The mutation of the helper virus is complemented by a packaging cell line, which derives from a stably transfected clone expressing constitutively the gene inactivated in the helper virus.

The different versions of amplicon-packaging systems used to date are listed in Table 1. In the helper-dependent (HD) systems, the amplicon plasmid is transfected transiently into the packaging cell line before it is infected with the helper virus. Viral replication takes place and results in a mixture of amplicon and helper viruses that are easily purified from the medium. Initial protocols used thermosensitive HSV mutants, replaced soon by mutants carrying deletions in one of the α genes, most commonly ICP4$^-$ (IE3$^-$) (5,12) and ICP27$^-$ (IE2$^-$) (19) mutants. More recently, the use of a group of cosmids or bacterial artificial chromosomes (BACs) carrying the HSV genome has made possible the purification of amplicon stocks free from helper virus (Table 1; see also Chapters 13 and 14). These approaches represent an important step toward the ideal amplicon system (i.e., devoid of viral contamination). However, in practice, the yields that can be obtained by these methods are one to two orders of magnitude lower than those using HD systems.

Amplicon vectors have been used to express cDNAs and genomic clones under the control of viral or cellular promoters, in vitro in cultured cells (12,35), ex vivo in brain slices (1), and in vivo (7,8,16,17).

Components of the Packaging System

The protocol for amplicon packaging described here uses the 5*dl*1.2 HSV-1 mutant, which carries a deletion in the *ICP27*$^-$ (*IE2*$^-$) ORF. This mutant has been shown to yield high-titer vector stocks with relatively low contamination by helper virus and undetectable levels of wild-type

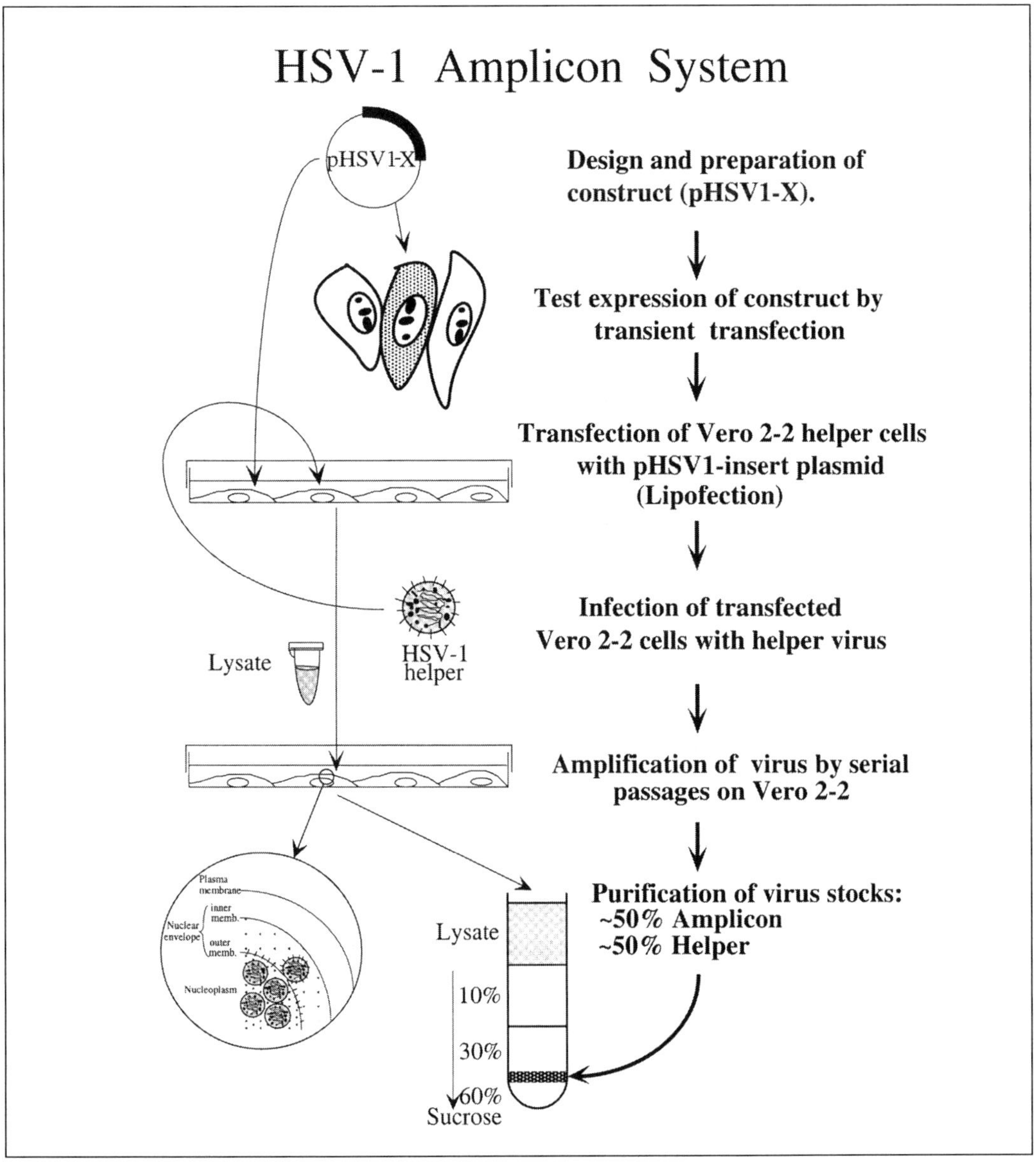

Figure 2. Schematic overview of the protocol for packaging HD HSV-1 amplicons.

virus (19). The packaging procedure is summarized in Figure 2. This system consists of three components.

(i) Plasmid pHSVpUC (13). This plasmid contains the HSV-1 oriS sequence followed by the immediate early gene IE4/5 promoter plus intron, 3′ of which a polylinker has been introduced where foreign cDNAs can be subcloned. Downstream from the polylinker there is a poly(A) signal from SV40, followed by the HSV-1 *a* packaging sequence (see Figure 3).

(ii) Helper virus 5*dl*1.2 (20). Mutant virus derived from HSV-1 strain KOS by deletion of most of the IE2 (or ICP27) gene.

(iii) Vero 2-2 cell line (28). Clone of Vero cells stably transfected to express constitutively the IE2 (*ICP27*) herpes simplex protein.

Figure 3 shows the structure of plasmids pHSVpUC and pHSVLacZ (12) and of various derivatives that we have constructed.

Materials

Virus and Cells

1. HSV-1 helper virus: 5*dl*1.2 helper virus derived from HSV-1 strain KOS by deletion of most of the IE2 (or ICP27) gene (20).
2. Hippocampal neurons: rat hippocampal neurons were explanted from E18 fetuses and maintained in culture as described elsewhere (14).
3. PC12 cell line (CRL1721; ATCC, Rockville, MD, USA): this cell line is derived from a transplantable rat adrenal pheochromocytoma (15). The morphology consists of clusters of small squared, phase-bright cells (Figure 4, G–J). The culture medium consists of RPMI, 2 mmol/L glutamine, 5% fetal bovine serum (FBS), and 10% horse serum. The cell line is cultured under 10% CO_2 at 37°C. Split the culture weekly (80%–90% confluent cultures) in a 1:3 to 1:4 ratio; dislodge the cells by pipetting medium directly onto them. Since the cells adhere poorly to plastic, use collagen or PL-coated dishes. These cells are useful for neurobiological studies and produce catecholamine and dopamine. When challenged with physiological levels of nerve growth factor (NGF), these cells cease dividing and acquire characteristics of mature sympathetic neurons.
4. Vero cell line (CCL81; ATCC): this cell line was established from the kidney of a normal adult African Green monkey and is composed of fibroblasts with a karyotype of 2n = 60; modal no. 58. The culture medium consists of Dulbecco's modified Eagle medium (DMEM), 2 mmol/L glutamine, and 10% FBS. Culture conditions should be 5% CO_2 at 37°C. Split confluent cultures 1:3 to 1:6 (i.e., seeding at 1 to 3×10^4 cells/cm^2 using 0.25% trypsin or trypsin/EDTA). This cell type is useful in virus replication studies because it is susceptible to a wide range of viruses, including polio, rubella, herpes viruses, arboviruses, and reoviruses.
5. Vero 2-2 cell line: this is derived from the Vero cell line by transfection with a plasmid expressing the immediate early IE2 gene (ICP27) of HSV-1 (28).

Equipment

- Laminar flow hood and 37°C humidified incubator with controlled CO_2 atmosphere.
- Photomicroscope (Axiophot [Carl Zeiss, Thornwood, NY, USA] or equivalent).
- Plastic dishes, cell scrapers, and screw cap vials for virus storage.

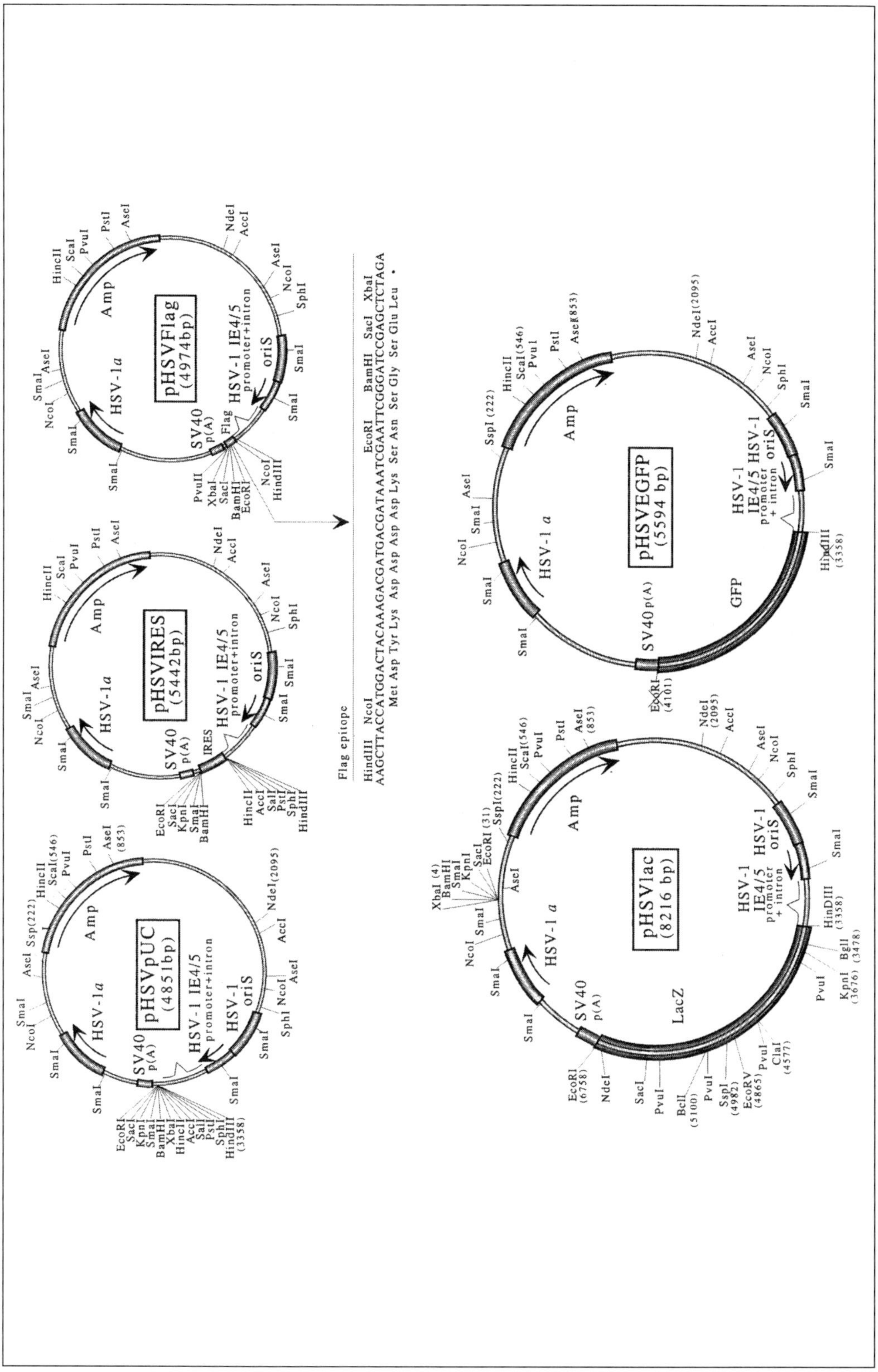

Figure 3. Structure of some available plasmids for production of amplicon vectors.

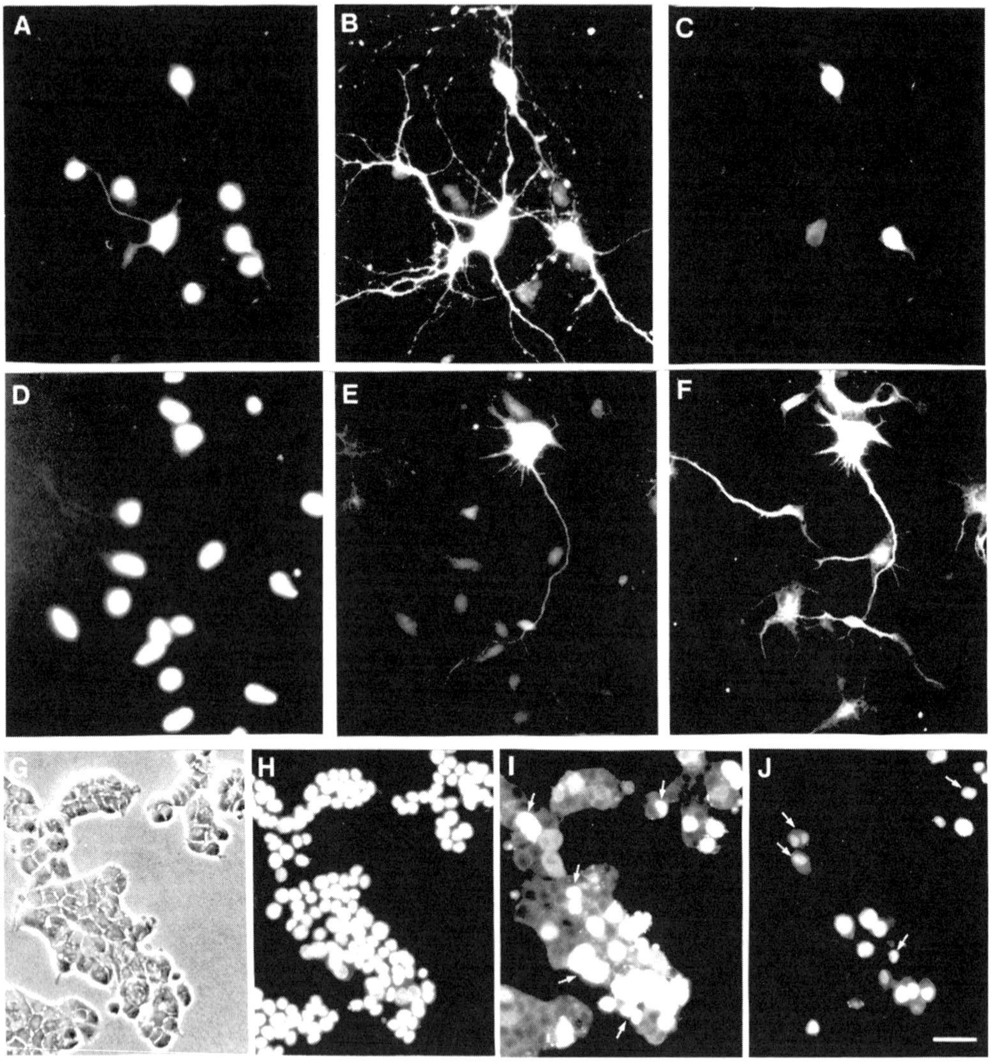

Figure 4. Gene transfer into rat hippocampal neurons in culture (A–F) and PC12 cells (G–J) with HSV-1 amplicons expressing a *lacZ* reporter gene. (A–C) Mature, fully polarized neurons (four days in culture) were infected at an moi of 5 with HSVLIL amplicon (A. Cid-Arregui, unpublished data), which expresses a β-galactosidase (β-gal) reporter carrying a nuclear localization signal (nls-β-gal). After incubating 1 hour in medium containing virus, the cells were washed and transferred to a dish with fresh medium. The cells were fixed with PFA 24 hours later and processed for immunofluorescence using anti-HSV-1 (B) and anti-β-gal (C) antibodies. Finally, the nuclei were counterstained with DAPI (A). Note that of the three cells infected in this field, two (the brighter in panel C) were infected by amplicon and helper virus, whereas the third only by helper virus. The faint fluorescence of the latter in panel C is likely due to the β-gal carried with the virion. The fluorescence in some processes visible in panel A is due to cross talk between both channels. (D–E) Young neurons (2 days in culture) were infected, cultured, and fixed as above. Immunofluorescence was performed using antibodies against HSV-1 (E) and microtubule-associated protein 2 (MAP-2) (F), and the nuclei were labeled with DAPI (D). (G–J) PC12 cells were infected with HSVLIL amplicon as above and cultured for 48 hours before being stained with anti-HSV-1 (I), anti-β-gal (J) antibodies and DAPI (H). Panel G, phase contrast. Arrows in panels I and J point to cells infected exclusively by helper virus or by amplicon, respectively. Bar: 16 μm (A–C), 20 μm (D–J).

- Sonicator (both cup- and bath-type sonicators can be used with little difference). *Note:* use maximum power settings and keep samples cool with ice water if necessary.
- Tissue culture microscope and hemocytometer.

Reagents and Solutions

- Alkaline phosphatase (AP) substrates from Sigma (St. Louis, MO, USA) (store at -80°C in aliquots).
- Antibodies: anti-β-galactosidase antibodies (Cat. No. 1083104; Roche Molecular Biochemicals, Indianapolis, IN, USA; Cat. No. 55976; ICN Biomedicals, Costa Mesa, CA, USA; Cat. No. G-8021, mAb; Sigma), anti-HSV-1, and rabbit polyclonal (Cat. No. B0114; DAKO, Carpinteria, CA, USA; recognizes all viral glycoproteins except gK).
- AP buffer (store at room temperature): mix 100 mmol/L Tris-HCl, pH 8.5, 150 mmol/L NaCl, and 5 mmol/L $MgCl_2$.
- AP substrate solution: add 330 μL nitroblue tetrazolium (NBT) stock to 10 mL AP buffer. Mix, then add 33 μL of 5-bromo-4-chloro-3-indolyl phosphate (BCIP) stock.
- BCIP (Cat. No. B-0274; Sigma). Stock solution: dissolve one tablet in 0.5 mL of dimethyl formamide. Store at -20°C.
- Blocking solution: mix 2% fetal calf serum (FCS), 2% bovine serum albumin (BSA) (Cat. No. A-6793; Sigma), and 0.2% fish gelatin (Cat. No. G-7765; Sigma) in phosphate-buffered saline (PBS). Store at -20°C. For blocking, use this solution undiluted. For diluting first and secondary antibodies, use 1:10 blocking solution in PBS.
- Crystal violet stain: prepare a solution containing 0.5% crystal violet and 0.2% sodium acetate, adjusting the pH to 3.6 with acetic acid before making it up to final volume. Filter through 3MM paper (Whatman, Clifton, NJ, USA). The stain can be reused several times.
- DAPI (4′-6′-diamidino-2-phenylindol; Cat. No. D-1306; Molecular Probes, Eugene, OR, USA), stock solution consists of 0.1 mg/mL in water. Store at -20°C. Use a working concentration of 1 μg/mL in PBS.
- Fe solution: mix 200 mL PBS, 0.332 g potassium ferricyanide, 0.424 g potassium ferrocyanide, 0.2 mL 1 mol/L $MgCl_2$, 0.2 mL 20% Nonidet® P-40, and 0.2 mL 10% sodium deoxycholate. Store at 4°C in a darkened container because this reagent is light-sensitive.
- Fixatives: in general, it is advisable to fix the cells with paraformaldehyde (PFA) as described below. However, for an optimal immunostaining, the most appropriate conditions should be established for each particular antibody. Prepare buffered 4% PFA/4% sucrose by dissolving 2 g of PFA in 25 mL of PBS by heating to 60°C while adding 10 mol/L NaOH (1 small drop/minute) until PFA is dissolved and the solution becomes clear (it takes 5–10 minutes). Let cool to room temperature. If necessary, adjust pH to 7.0 to 7.2 with 1 mol/L NaOH. To better preserve the morphology of the cells, add sucrose to a final concentration of 4% (e.g., 3.33 mL of a 60% stock of sucrose). Complete volume to 50 mL with deionized water. This fixative can be stored at -20°C for months or at 4°C for a few days. To fix, *(i)* wash the coverslips in PBS; *(ii)* incubate in PFA for 10 to 15 minutes at room temperature; *(iii)* rinse briefly in PBS; *(iv)* quench autofluorescence by incubating in 50 mmol/L NH_4Cl in PBS twice for 10 minutes each; *(v)* if

coverslips are to be processed for immunofluorescence, permeabilize the cells by incubating in 0.2% Triton® X-100 in PBS for 5 to 10 minutes at room temperature. Alternatively, the cells can be permeabilized by passing the coverslips through an ethanol series: 25%, 50%, 70%, 90%, 100%, 90%, 70%, 50%, and 25% ethanol, for 1 minute each. Finally, incubate in PBS for 1 minute to rehydrate and proceed for immunofluorescence.

- FuGENE™ 6 transfection reagent (Cat. No. 1814443; Roche Molecular Biochemicals).
- G418 (neomycin analog): mix 100 mg/mL stock solution in 100 mmol/L HEPES, pH 7.3. This can be stored at -20°C.
- HBCSS buffer: mix 86.5 mmol/L NaCl (17.3 mL 1 mmol/L NaCl); 5.4 mmol/L KCl (540 μL 2 mol/L KCl); 0.8 mmol/L $MgCl_2$ (160 μL 1 mol/L $MgCl_2$); 1.8 mmol/L $CaCl_2$ (360 μL 1 mol/L $CaCl_2$); 20 mmol/L glucose (3.96 mL 20% glucose); 20 mmol/L HEPES (20 mL 200 mmol/L HEPES); 26 mmol/L sodium bicarbonate (5.2 mL 1 mol/L $NaHCO_3$); 10 mmol/L NaOH (2 mL 1 mol/L NaOH). Make up to 200 mL with sterile deionized water.
- HBCSS/60% sucrose: sterilize by filtration through a 0.45-μm filter.
- 5% methanol, 10% acetic acid.
- Mowiol mounting medium is prepared as follows: *(i)* add 10 g of Mowiol 4-88 (Cat. No. 475904; Calbiochem-Novabiochem, San Diego, CA, USA) to 40 mL PBS; *(ii)* stir for 2 hours; *(iii)* add 20 mL glycerin and stir for an additional 10 minutes; *(iv)* centrifuge at 3000 rpm for 15 minutes; *(v)* make aliquots of 1 mL and store at -20°C. This solution is very dense and should not be vortex mixed, but mixed gently by inversion after thawing to avoid bubble formation. Do not refreeze aliquots, but keep them at 4°C.
- NBT (Cat. No. N-5514; Sigma). Prepare a stock solution by dissolving one tablet in 1 mL water.
- OPTI-MEM™ reduced serum medium (Life Technologies, Rockville, MD, USA).
- PBS (Dulbecco's phosphate-buffered saline, without Ca^{2+} and Mg^{2+}): use 8.0 g NaCl, 0.2 g KCl, 0.24 g KH_2PO_4, and 1.44 g Na_2HPO_4 anhydrous (alternatively 2.16 g $Na_2HPO_4.7\ H_2O$). Note: adjust the pH using HCl to 7.0 to 7.2 (for tissue culture) or 7.3 to 7.4 (for histochemistry). Make up to 1 L with deionized water. Autoclave.
- PC12 culture medium: mix RPMI, 2 mmol/L glutamine, with 5% FCS, 10% horse serum, and 1% penicillin/streptomycin. Filter the serum through a 0.45-μm filter. Note: do not heat-inactivate the sera.
- Plaque agar (DMEM + 5% FCS and 1% agarose). Place bottles of 2× DMEM and FCS into a 42°C water bath. Prepare a solution of hot molten 2% agarose (tissue culture grade) in water and place into the 42°C water bath. After temperature equilibration, mix the reagents together at a ratio of 20:20:1 (DMEM:agarose:FCS). Remove just before required and use when the temperature is below 40°C.
- Poly-D-lysine (PDL) (MW 70 000–150 000, Cat. No. P6407; Sigma). Prepare a 50× stock solution of 1 mg/mL PDL in water. Filter, sterilize, and store at -20°C. Dilute to 20 μg/mL in water immediately prior to use.
- TBS: mix 100 mmol/L Tris-HCl, pH 7.5, and 150 mmol/L NaCl.
- TBS-serum-triton (TST): use 1% goat serum and 0.1% Triton X-100 in Tris-buffered saline (TBS).
- Trypsin-EDTA: use 0.5 mg/mL

trypsin and 0.2 mg/mL EDTA in PBS. This is also commercially available (e.g., Cat. No. 45300-019; Life Technologies).

- Vero 2-2 culture medium, use DMEM containing 1% penicillin/streptomycin, 4 mmol/L glutamine (50× glutamine in water can be stored at -20°C), with either 5% or 10% FCS as indicated.
- X-gal (Cat. No. B-4252; Sigma) stock: mix 40 mg/mL in dimethyl-formamide or in dimethyl-sulphoxide. Store in a glass container (never polycarbonate or polystyrene) at -20°C in the dark.
- X-gal staining solution: mix 1 mg/mL X-gal (stock 40 mg/mL in dimethylformamide), 0.02% Nonidet P-40, 2 mmol/L $MgCl_2$, 5 mmol/L $K_3Fe(CN)_6$ (stock 50 mmol/L in PBS, pH 7.3), and 5 mmol/L $K_4Fe(CN)_6$ (stock 50 mmol/L in PBS, pH 7.3). For long incubations, add Tris, pH 7.3, to 50 mmol/L. Can be reused for months if stored at 4°C and protected from light. Filter through 0.22-µm membranes after each use. Add X-gal last and after warming up the premixture (in a 37°C water bath). Thaw out X-gal stock and warm it before pipetting into the premixture. If solutions are not warm, precipitate forms.

Packaging of Amplicons

Day 1: Transfection of Vero 2-2 Cells

1. Vero 2-2 cells are plated two days before transfection at a density of 3 × 10^5 cells per 6-cm ø dish in 5 mL of Vero medium.
2. To prepare cells:
   - Prewarm serum-free medium (DMEM or OPTI-MEM) for at least 20 minutes at 37°C.
   - In a sterile microcentrifuge tube, add 200 µL of serum-free medium. Add 10 µL of FuGENE 6 transfection reagent directly into the medium. It is not advisable to pipet it on the tube wall because adsorption of reagent to the plastic and also solubilization of plasticizers by the reagent may occur. Incubate for 5 minutes at room temperature.
   - Pipet 2 to 4 µg DNA (high quality; purified with Qiagen™ columns) into a second sterile microcentrifuge tube. The total volume of DNA solution should not exceed 12 µL.
   - Add the diluted FuGENE to the tube containing the DNA dropwise. Mix gently. Incubate for 15 minutes at room temperature for liposomes to form.
   - Add 800 µL of Vero medium to the DNA/FuGENE mixture.
   - Dropwise, add the mixture from the previous step to the cells growing on the 6-cm ø dish in 3 mL of Vero medium (5% FCS), distributing it around the well. Swirl the dish to ensure even dispersal of liposomes. Incubate overnight at 37°C with 5% CO_2.
   - Allow the cells to recover at least 20 hours after the transfection.

Day 2: Infection of Transfected Cells with 5dl*1.2 Helper Virus*

1. Infect transfected cells with helper virus:
   - Prewarm DMEM and 5% FCS for at least 20 minutes at 37°C.
   - Remove the medium from the plates and add 3 mL DMEM and 10% FCS.
   - Add 60 µL of 5*dl*1.2 helper virus (approximately 6 × 10^5 pfu).
   - Incubate at 37°C until cells start to detach and look spherical (about 40 hours).

2. Plate out Vero 2-2 cells (4×10^5/6-cm ø dish, in 5 mL Vero medium with 10% FCS) for amplification p1 (see day 4 below).

Day 3

1. Check that the cells infected on day 2 are not contaminated with bacteria or fungi.
2. Plate out Vero 2-2 cells (7×10^5/6-cm ø dish, in 5 mL Vero medium with 10% FCS) for amplification p1 (see day 5 below).

Day 4: Harvesting p0; Amplification p1

1. To harvest p0:
   - Harvest cells by pipetting the medium onto them until they have all detached from the plate.
   - Alternatively, harvest by scraping cells into the medium they grow in. Transfer into a 15-mL plastic tube.
   - Freeze/thaw three times using a dry ice/ethanol bath and a 37°C water bath. The cell extract can be stored at -80°C at this stage.
   - Sonicate for 2 minutes.
   - Centrifuge 5 minutes at low speed (3 000 rpm, approximately 800× *g*) to pellet cell debris, but not virus particles. Keep supernatant (3 mL).
   - The supernatant can be stored at -80°C.
2. Amplification p1:
   - Remove medium from the 6-cm ø dish and replace with 4 mL fresh medium (10% FCS).
   - Plate supernatant p0 (3 mL) on cells for p1.
   - These cells were plated two days before at a density of 4×10^5/6-cm ø dish or one day before at 7×10^5 per 6-cm ø dish. In both cases, they were plated out with 5 mL medium and 10% FCS.
3. Plate fresh Vero 2-2 cells (2×10^6/10-cm ø dish in 10 mL Vero medium with 10% FCS; two dishes per sample) for amplification p2 (see below, day 6).

Day 5: Harvesting p1; Amplification p2

1. Check Vero 2-2 cells infected with supernatant p0. Most of them should have rounded up after 24 hours of incubation.
2. To harvest p1:
   - Proceed exactly as on day 4. Pipet harvest into a 15-mL plastic tube.
   - After centrifugation, transfer supernatant (7 mL) into a sterile 15-mL plastic tube.
3. Amplification p2:
   - Remove medium from the two 10-cm ø dishes from day 5 and replace with 2 mL DMEM and 2% FCS.
   - Add p1 supernatant (approximately 3.5 mL 10-cm ø dish)
   - Incubate cells for at least 1.5 hours.
   - Replace medium with 2 mL HBCSS and 1% BSA.
4. Plate out Vero 2-2 cells in 6-cm ø dishes (2×10^6/6-cm ø dish, in 5 mL Vero medium with 10% FCS). Prepare one dish for each dilution of virus stock to be assayed.

Day 6: Harvesting and Purification of Virus; Plaque Assay

1. Check Vero 2-2 cells infected with supernatant p1. Most cells should have rounded up after 24 hours of incubation.
2. To harvest p2:
   - Proceed exactly as on day 4. Pipet harvest into a 50-mL plastic tube.
   - After centrifugation, transfer supernatant (4 mL) into a sterile 15-mL plastic tube.
3. To purify virus:

- Prepare a sucrose step gradient in 20-mL SW28 ultracentrifuge tubes (Beckman Instruments, Fullerton, CA, USA) (or equivalent). Layer the sucrose solutions into the tube in the following order: 7 mL 60% sucrose in PBS, 6 mL 30% sucrose in PBS, and 3 mL 10% sucrose in PBS.
- Gently load p2 supernatant (approximately 4 mL) onto the gradient.
- Centrifuge for 1 hour at 125 000× *g* (SW28 rotor [Beckman Instruments], 25 000 rpm). The virus bands at the interface between the 30% and 60% sucrose layers and can be best visualized against a dark background using side illumination of the tube with a lamp.
- Carefully remove the top layers of sucrose above the white band containing the virus, and then recover the virus in approximately 2 mL.
- Dilute the purified virus with PBS and 1% BSA to fill a 20-mL SW28 tube (or equivalent), mix gently, and centrifuge at 125 000× *g* for 1 hour.
- Carefully, pipet the supernatant, and resuspend the viral pellet in 200 µL 10% sucrose in HBCSS and 1% BSA. Dispense 30-µL aliquots into cryotubes. Freeze quickly in liquid nitrogen and store at -80°C.

4. To infect for plaque assay:
   - Infect the cells by removing 3 mL of medium from the plate and adding 100 µL virus dilution to the plate. Allow the virus to adsorb for at least 90 minutes, but less than 6 hours.
   - Remove the medium and add 3 mL plaque agar (see Appendix 1). Leave to solidify before returning to the 37°C incubator. The following day, add 2 mL DMEM and 5% FCS on top of the agarose. Repeat the next day.
   - Three days after infection, check the plates with a microscope for plaques. If they are too small to reveal clear holes in the cell monolayer, the plates may be left for another day before staining.

Notes

1. The efficiency of transfection of Vero 2-2 cells with the pHSVpUC derivatives is a critical parameter to obtain high vector titers. In our hands, lipofection with FuGENE results in at least threefold higher packaging efficiencies than with other lipofection reagents or using calcium phosphate transfection. In addition, lipofection with this reagent has some advantages: *(i)* the efficiency of transfection is the same in the presence or absence of serum and *(ii)* there is no need to remove the transfection medium containing the DNA/FuGENE reagent complex, as it has no adverse effects on the cells.
2. The packaging procedure has an optimum efficiency when the multiplicity of infection (moi) of the helper virus is between 0.2 and 0.6 (19). It is advised that a large amount of helper virus be kept aside and stored in aliquots specifically for packaging. An initial experiment to titrate for the optimum amount of this helper stock can then be set up by transfecting several identical plates with pHSVEGFP and/or by varying the volume of helper virus (to cover a range from 10^5–10^6 pfu) added in the infection step. The cells are harvested the next day, and virus stocks are prepared and assayed for vector by X-gal staining. The optimum volume of helper stock that results in the highest number of fluorescent cells can be recorded and used for subsequent packaging. Different helper viruses and the corre-

sponding complementary cell lines can be used [e.g., the IE3 deletion mutant d120 (5) grown on E5 cells (6) and the IE3 deletion mutant D30EBA (21) grown on RR1 cells) (18)]. However, packaging with either of these systems is usually approximately ten times less efficient than with the 5*dl*1.2/2-2 system.

3. Thorough characterization of viral vector stocks involves measuring infectious vector units per milliliter (ivu/mL) on PC12 cells, helper pfu/mL on Vero 2-2 cells, and wild-type HSV-1 pfu/mL on Vero cells.
4. As a control for the transfection procedure, include a dish transfected with pHSVEGFP and check for expression one day later. To monitor the packaging efficiency during the subsequent steps, transfect another dish with pHSVEGFP and monitor vector titers at each step.
5. Virus particles are thermolabile and should be kept on ice during the manipulations. Approximately 50% loss in titer can be expected after storage of unpurified virus at -80°C for one year. Concentrated virus in sucrose solutions has higher stability. After one week at 4°C, we observed 50% loss of vector titer. After one thawing of virus stored at -80°C, between 10% to 50% loss can be expected, depending on the size of the aliquot, the temperature, and the time the virus remains thawed.

COMPLEMENTARY PROTOCOLS

Preparation of Helper Stocks of HSV-1

1. Seed 60-mm tissue culture plates with 1×10^6 2-2 cells/plate in 5 mL DMEM and 10% FCS.
2. On the next day, remove 3 mL of the medium and add an amount of virus that will give well-separated plaques (about 50 pfu/plate). Incubate at 37°C for 1.5 to 6 hours.
3. Remove the medium and add 3 mL plaque agar (see Materials). Leave to solidify before returning to the 37°C incubator. The following day, add 2 mL DMEM and 5% FCS on top of the agarose. Repeat the next day.
4. Two to four days after infection, when plaques are visible, pick a single plaque by using a Pasteur pipet to stab through the agar and into the cells, and then transfer the plug plus the cells to a 60-mm plate of cells plated as in step 1. It is advisable to pick several different plaques to ensure that at least one isolate has the appropriate phenotype for the virus.
5. After one or two days (depending on the amount of virus obtained from the plaque), the cells should show cytopathic effects (CPEs). The cells should round up, but remain stuck to the plate. Harvest the cells and prepare the virus. These plaque-purified stocks should be stored at -80°C and used only as seed stocks for inoculation for large-scale virus preparations (steps 6–8).
6. Seed four 100-mm plates with 1×10^6 2-2 cells/plate in 10 mL DMEM and 10% FCS and grow for two days.
7. Change the medium to 10 mL DMEM and 5% FCS and add 50 µL of seed stock virus. The cells should be infected at an moi of 0.1 or lower. At high moi, spontaneously arising defective viruses are able to grow and interfere with virus replication, resulting in lower pfu/mL. The 2-2 cells should reach a density of 0.5 to 2×10^7 cells per plate by the second day. Seed stocks prepared as outlined in steps 1 to 5 usually have titers of 0.5 to 1×10^7 pfu/mL.

8. Harvest the cells when they have rounded up, but still remain stuck to the plates. Purify the virus as described in the previous section, and store in aliquots of 100 μL at -80°C. See notes on how to determine the optimum amount of helper virus stock to use for packaging.

Vector Assays

The assay described here has been adapted for PC12 cells; if a different cell type is used, plate the cells as required to obtain a monolayer.

1. To plate cells, start by coating 24-well plates with PDL to make the PC12 cells stick properly. Pipet 500 μL of a 20 μg/mL PDL solution into each well, and incubate for at least 5 minutes at room temperature. Just before use, aspirate PDL and wash once with PBS. Pass the PC12 cells through a 21 × 11/2 syringe gently to dissociate aggregates. Count cells with hemocytometer and plate 3×10^5 cells/well.
2. On the next day, change the medium and add virus at different dilutions in OPTI-MEM (sample volumes of 100 μL). Always include an uninfected well for negative control background staining. Incubate for 1 hour to let the virus attach and enter the cells. Remove the medium with the virus, wash once with PBS, and add fresh medium. Incubate for 20 to 24 hours.
3. Wash the cells with PBS. Fix with 0.5 mL 4% PFA in 0.1 mol/L phosphate, pH 7.0, for at least 15 minutes at room temperature. If the PFA is old or the fixation is not long enough, the cells will collapse and shrink, and background staining problems will occur.
4. Wash cells with PBS or TBS (500 μL) once. Note: phosphate inhibits AP. Therefore, PBS should be avoided if staining with an AP-conjugated antibody follows.
5. The staining procedures follow.

   X-gal staining procedure

   - Warm 500 μL Fe solution/well of cells to be stained in 37°C water bath.
   - Thaw out X-gal stock and add to warm Fe solution to a final concentration of 1 mg/mL (dilute 50 times). If solutions are not warm, precipitate may form.
   - Remove TST from the wells and add X-gal/Fe solution (500 μL/well). Incubate at 37°C for 30 minutes to overnight. Stop reaction by washing twice with PBS.

   Immunocytochemical staining procedure

   - Use first antibody diluted in TST (e.g., rabbit anti-HSV at 1:10 000). Leave overnight at 4°C.
   - Wash with TBS twice, and then leave for 10 minutes.
   - Use AP-antirabbit conjugate (or antimouse if appropriate) at 1:2000 in TST. Leave 1 hour at room temperature.
   - Wash with TBS twice, and then wash with AP buffer twice.
   - Add substrate solution and incubate in dimly lit area. Check with microscope for color development.
   - Stop color development by washing twice with PBS. Leave cells in PBS and 10 mmol/L EDTA to prevent any further darkening.

Plaque Assays

1. Plate Vero 2-2 (or Vero) cells in 60-mm tissue culture plates (1×10^6/plate) in 5 mL DMEM and 10% FCS and grow for one day. Prepare one plate for each dilution of virus stock to be titered (see step 2).

2. Infect the cells by removing 3 mL of medium from the plate and adding 100 µL virus dilution to the plate. Allow the virus to adsorb for at least 90 minutes, but less than 6 hours.
3. Remove the medium and add 3 mL plaque agar (see Materials). Leave to solidify before returning to the 37°C incubator. The following day, add 2 mL DMEM and 5% FCS on top of the agarose. Repeat the next day.
4. Three days after infection, check the plates with a microscope for plaques. If they are too small to reveal clear holes in the cell monolayer, the plates may be left for another day before staining.
5. Remove the medium and fix the cells with 3 mL 5% methanol, 10% acetic acid for at least 15 minutes. Remove the fixing solution and add 1 mL crystal violet stain. Remove the stain (which can be reused), wash with 1 mL water, and air-dry. Plaques should be visible to the naked eye as holes.
6. Count the number of plaques and, using the dilution factor of the virus stock tested, calculate the virus plaque-forming titer in pfu/mL.

GENE TRANSFER INTO CULTURED CELLS WITH HSV-1 AMPLICONS

With the plasmids described in the Figure 3, we have produced HSV-1 amplicons expressing the reporter proteins β-galactosidase and green fluorescent protein (GFP) in various ways and have used them to perform infections of different cell types and tissues. These include primary rat hippocampal neurons, PC12, HeLa, MDCK, Vero, BHK-21, CV-1, NIH3T3 cells, and embryonic neuroepithelial tissue (see Figure 4; see also Chapters 16 and 17). All these cell types were susceptible to infection, albeit with varying degrees of susceptibility (data not shown). For instance, differences were noticed among hippocampal neurons at different stages of differentiation. Thus, a higher number of infected cells were seen in cultures of fully polarized neurons than in cultures of younger neurons exposed to amplicon at an equivalent moi (Figure 4, compare A–C with D–F). This can be explained by the fact that mature neurons have longer and more branched processes, which offer a wider surface for the virus to attach to.

Since the amplicon stocks contained a mixture of amplicon and helper viruses, three subpopulations of infected cells were obtained, as illustrated in Figure 4: (*i*) cells infected by the helper virus (30%–35%), (*ii*) cells infected by the amplicon separately (30%–35%), and (*iii*) cells infected simultaneously by both (30%–40%). The proportion of each subpopulation was variable depending on the amplicon/helper ratio in the stock. The percents given were estimated in cultures of hippocampal neurons and PC12 cells infected with β-galactosidase expressing amplicon at an moi of 1 to 5, using stocks in which the amplicon/helper ratio was approximately 1:1.

Reporter expression was detectable by immunofluorescence as soon as 1 hour after infection and lasted for at least one week, although some oscillations in the amounts of reporter protein could be detected by Western blotting. It has been shown that the IE4/5 promoter, which drives foreign gene expression in these amplicons, is transactivated by factors encoded by the 5*dl*1.2 helper, most likely IE3 (ICP4) (19). Accordingly, we observed by immunofluorescence a stronger expression of reporter protein in cells infected simultaneously by amplicon and helper viruses than in those infected only by the amplicon (Figure 4J).

When the cells were infected with amplicons containing the IRES sequence of the EMC virus to express two different genes from the IE4/5 promoter, the expression of the gene downstream from the IRES sequence appeared relatively more efficient, as estimated by immunofluorescence (data not shown). In addition, a fraction of cells (10%–20%) expressed only one of the proteins. The reason for this is unknown and is currently under study. A possible explanation is that significant amounts of amplicon carry mutations or deletions, but this needs to be confirmed.

ACKNOWLEDGMENTS

We gratefully acknowledge C.G. Dotti and A. Clement for advice with the hippocampal cultures, R. Jelinek for technical assistance, and W.B. Huttner and the Max-Planck Society for their support.

REFERENCES

1. **Bergold, P., P. Casaccia-Bonnefil, Z. Xiu-Liu, and H. Federoff.** 1993. Transsynaptic neuronal loss induced in hippocampal slice cultures by a herpes simplex virus vector expressing the GluR6 subunit of the kainate receptor. Proc. Natl. Acad. Sci. USA *90*:6165-6169.
2. **Buchholz, F., P.-O. Angrand, and A.F. Stewart.** 1996. A simple assay to verify the functionality of CRE and FLP recombination targets in genomic manipulation constructs. Nucleic Acids Res. *24*:3118-3119.
3. **Corey, L.** 1998. Herpes simplex viruses, p. 692-697. *In* A.S. Fauci et al. (Ed.), Principles of Internal Medicine, 14th ed., Vol. 1. McGraw-Hill, Hamburg.
4. **Croen, K.D., J.M. Ostrove, L.J. Dragovic, J.E. Smialek, and S.E. Strauss.** 1987. Latent herpes simplex virus in human trigeminal ganglia. Detection of an immediate early gene "anti-sense" transcript by in situ hybridizaton. N. Engl. J. Med. *317*:1427-1432.
5. **DeLuca, N.A., A.M. McCarthy, and P.A. Schaffer.** 1985. Isolation and characterization of deletion mutants of herpes simplex virus type 1 in the gene encoding immediate-early regulatory protein ICP4. J. Virol. *56*:558-570.
6. **DeLuca, N.A. and P. Schaffer.** 1987. Activities of herpes simplex virus type 1 (HSV-1) ICP4 genes specifying nonsense peptides. Nucleic Acids Res. *15*:4491-4511.
7. **During, M.J., J.R. Naegele, K.L. O'Malley, and A.I. Geller.** 1994. Long-term behavioral recovery in parkinsonian rats by an HSV vector expressing tyrosine hydroxilase. Science *266*:1399-1403.
8. **Federoff, H.J., M.D. Geschwind, A.I. Geller, and J.A. Kessler.** 1992. Expression of nerve growth factor in vivo from a defective herpes simplex virus 1 vector prevents effects of axotomy on sympathetic ganglia. Proc. Natl. Acad. Sci. USA *89*:1636-1640.
9. **Fink, D.J., N.A. DeLuca, W.F. Goins, and J.C. Glorioso.** 1996. Gene transfer to neurons using herpes simplex virus-based vectors. Annu. Rev. Neurosci. *19*:265-287.
10. **Fink, D.J., L.R. Sternberg, P.C. Weber, M. Mata, W.F. Goins, and J.C. Glorioso.** 1992. In vivo expression of β-galactosidase in hippocampal neurons by HSV-mediated gene transfer. Hum. Gene Ther. *3*:11-19.
11. **Fraefel, C., S. Song, F. Lim, P. Lang, L. Yu, Y. Wang, P. Wild, and A.I. Geller.** 1996. Helper virus-free transfer of herpes simplex virus type 1 plasmid vectors into neural cells. J. Virol. *70*:7190-7197.
12. **Geller, A.I. and X.O. Breakfield.** 1988. A defective HSV-1 vector expresses *Escherichia coli* beta-galactosidase in cultured peripheral neurons. Science *241*:1667-1669.
13. **Geller, A.I., K. Keyomarsi, J. Bryan, and A.B. Pardee.** 1990. An efficient deletion mutant packaging system for defective herpes simplex virus vectors: potential applications to human gene therapy and neuronal physiology. Proc. Natl. Acad. Sci. USA *87*:8950-8954.
14. **Goslin, K. and G. Banker.** 1991. Rat hippocampal neurons in low density cultures, p. 251-281. *In* G. Banker and K. Goslin (Eds.), Culturing Nerve Cells. MIT Press, Cambridge, MA.
15. **Greene, L.A. and A.S. Tischler.** 1976. Establishment of a noradrenergic clonal line of rat adrenal pheochromocytoma cells which respond to nerve growth factor. Proc. Natl. Acad. Sci. USA *73*:2424-2428.
16. **Ho, D.Y., J.R. McLaughlin, and R.M. Sapolsky.** 1996. Inducible gene expression from defective herpes simplex virus vectors using the tetracycline-responsive promoter system. Brain Res. Mol. Brain Res. *41*:200-209.
17. **Ho, D.Y., E.S. Mocarski, and R.M. Sapolsky.** 1993. Altering central nervous system physiology with a defective herpes simplex virus vector expressing the glucose transporter gene. Proc. Natl. Acad. Sci. USA *90*:3655-3659.
18. **Johnson, P.A., A. Miyanohara, F. Levine, T. Cahill, and T. Friedmann.** 1992. Cytotoxicity of a replication-defective mutant herpes simplex virus type 1. J. Virol. *66*:2952-2965.
19. **Lim, F., D. Hartley, P. Starr, P. Lang, S. Song, L. Yu, Y. Wang, and A.I. Geller.** 1996. Generation of high-titer defective HSV-1 vectors using an IE 2 deletion mutant and quantitative study of expression in cultured cortical cells. BioTechniques *20*:460-469.
20. **McCarthy, A.M., L. McMahan, and P.A. Schaffer.** 1989. Herpes simplex virus type 1 ICP27 deletion mutants exhibit altered patterns of transcription and are DNA deficient. J. Virol. *63*:18-27.
21. **McGeoch, D.J., M.A. Dalrymple, A.J. Davison, A. Dolan, M.C. Frame, D. McNab, J. Perry, J.E. Scott, and P. Taylor.** 1988. The complete DNA sequence of the long unique region in the genome of herpes simplex virus type 1. J. Gen. Virol. *69*:1531-1574.

22.**Meignier, B., R. Longnecker, P. Mavromara-Nazos, A.E. Sears, and B. Roizman.** 1988. Virulence and establishment of latency by genetically engineered deletion mutants of herpes simplex virus type 1. Virology *162*:251-254.
23.**Palella, T.D., L.J. Silverman, C.T. Schroll, F.L. Homa, M. Levine, and V.M. Kelley.** 1988. Herpes simplex virus-mediated human hypoxanthine-guanine phosphoribosyl-transferase gene transfer into neuronal cells. Mol. Cell. Biol. *8*:457-460.
24.**Paterson, T. and R.D. Everett.** 1990. A prominent serine-rich region in Vmw175, the major transcriptional regulator protein of herpes simplex virus type 1, is not essential for virus growth in tissue culture. J. Gen. Virol. *71*:1775-1783.
25.**Roizman, B.** 1996. The function of herpes simplex virus genes: a primer for genetic engineering of novel vectors. Proc. Natl. Acad. Sci. USA *93*:11307-11312.
26.**Roizman, B. and A.E. Sears.** 1996. Herpes simplex viruses and their replication, p. 1043-1107. *In* B.N. Fields, D.M. Knipe, P.M. Howley, et al. (Eds.), Fundamental Virology, 3rd. ed. Lippincott-Raven, New York.
27.**Shih, M.-F., M. Arsenakis, P. Tiollais, and B. Roizman.** 1984. Expression of hepatitis B virus S gene by herpes simplex virus type 1 vectors carrying alpha- and beta-regulated gene chimeras. Proc. Natl. Acad. Sci. USA *81*:5867-5870.
28.**Smith, I.L., M.A. Hardwicke, and R.M. Sandri-Goldin.** 1992. Evidence that the herpes simplex virus immediate early protein ICP27 acts post-transcriptionally during infection to regulate gene expression. Virology *186*:74-86.
29.**Spaete, R.R. and N. Frenkel.** 1982. The herpes simplex virus amplicon: a new eucaryotic defective-virus cloning-amplifying vector. Cell *30*:295-304.
30.**Spaete, R.R. and N. Frenkel.** 1985. The herpes simplex virus amplicon: analyses of cis-acting replication functions. Proc. Natl. Acad. Sci. USA *82*:694-698.
31.**Spivack, J.G. and N.W. Fraser.** 1987. Detection of herpes simplex virus type 1 transcripts during latent infection in mice. J. Virol. *61*:3841-3847.
32.**Stavropoulos, T.A. and C.A. Strathdee.** 1998. An enhanced packaging system for helper-dependent herpes simplex virus vectors. J. Virol. *72*:7137-7143.
33.**Stevens, J.G.** 1989. Human herpes viruses: a consideration of the latent state. Microbiol. Rev. *53*:318-332.
34.**Stevens, J.G., E.K. Wagner, G.B. Devi-Rao, M.L. Cook, and L.T. Feldman.** 1987. RNA complementary to a herpes viruses gene mRNA is prominent in latency infected neurons. Science *255*:1056-1059.
35.**Xu, H., H. Federoff, J. Maragos, L.F. Parada, and J.A. Kessler.** 1994. Viral transduction of trkA into cultured nodose and spinal motor neurons conveys NGF responsiveness. Dev. Biol. *163*:152-161.

13 Gene Transfer Using Helper Virus-Free Herpes Simplex Virus Type 1 Amplicon and Hybrid Amplicon Vectors

Yoshinaga Saeki[1,2] and Cornel Fraefel[3]

[1]Department of Neurosurgery, Massachusetts General Hospital, Harvard Medical School, Charlestown, MA, USA; [2]Division of Gene Therapy Science, Osaka University Medical School, Osaka, Japan; [3]Institute of Virology, University of Zurich, Zurich, Switzerland

INTRODUCTION

The *Herpesviridae* are a family of viruses that have a large genome of linear, double-stranded DNA (>120 kb), a unique virion structure consisting of an electron-opaque core, an icosadeltahedral capsid surrounding the core, an amorphous tegument surrounding the capsid, and an outer envelope exhibiting spikes on its surface. To date, more than 100 different herpesviruses have been isolated and partially characterized; all of these share the ability to establish both lytic and latent infections.

Herpes simplex virus type 1 (HSV-1) is one of eight human herpesviruses and among the most extensively investigated of all viruses. To establish a primary infection, virion glycoproteins C and B (gC and gB) bind to glycosaminoglycans, preferentially heparan sulfate, on the cellular surface of skin or mucosa (9,31,43,53,74). At least four virion glycoproteins—gB, gD, and the gH-gL complex—mediate fusion of the virion envelope with the cell membrane (8,16,44). Two specific cell surface receptors that interact with gD have been identified and designated herpesvirus entry mediator (Hve) A, a member of the tumor necrosis factor receptor family (59,86), and HveC, which is the poliovirus receptor-related protein 1 (28). After entry into the cytoplasm, virus tegument proteins and capsids are transported along microtubules to nuclear pores, and the genome is released into the nucleoplasm, where it circularizes before the approximately 80 virus genes are transcribed. The HSV-1 genome is approximately 152 kb long. It consists of two long stretches of quasi-unique sequences, unique long *(U_L)* and unique short *(U_S)*, both of which are flanked by inverted repeats (Figure 1A). U_L is flanked by the approximately 9-kb terminal repeat of the long segment *(TR_L)* and its inversion, the internal repeat of the long segment *(IR_L)*; U_S is flanked by the approximately 6.5-kb internal repeat of the short segment *(IR_S)* and its inversion, the terminal repeat of the short segment *(TR_S)*. HSV-1 genes are transcribed in a tightly regulated fashion as a cascade with three temporal phases (67,68): *(i)* immediate early (*IE* or α) genes encode proteins that

Viral Vectors: Basic Science and Gene Therapy
Edited by A. Cid-Arregui and A. García-Carrancá

have regulatory functions and are required for the transcription of subsequent gene groups; *(ii)* early (E or β) genes encode some of the tegument proteins and envelope glycoproteins, as well as enzymes required for DNA metabolism and therefore initiate replication of the virus genome; *(iii)* virus genome replication enhances, or is required for, the expression of the late (L or γ) genes, which encode most of the structural virion components. The replication of the HSV-1 genome is initiated at the origins of DNA replication, ori_S and ori_L (49,82), and is followed by an early phase of uni- or bidirectional DNA synthesis in specific replication compartments (52). Recombination between the inverted repeats results in inversions of the long and short segments (5,54). Late-phase DNA replication has been thought to proceed by a rolling circle mechanism (29). However, recent work suggests a mechanism similar to that of T4 DNA or yeast 2-µm plasmids, which produces large, branched concatamers of head-to-tail linked genomes (73,97). The concatameric products of genome replication are cleaved into unit-length genomes at the DNA cleavage/packaging signals *(pac)* after filling the capsids that are preformed at the nuclear membrane (14,34,88). Several hypotheses about the source of the virus envelope and egress of the mature virions exist (80). In both models, the capsids and surrounding tegument are believed to bud through the nuclear membrane, thereby

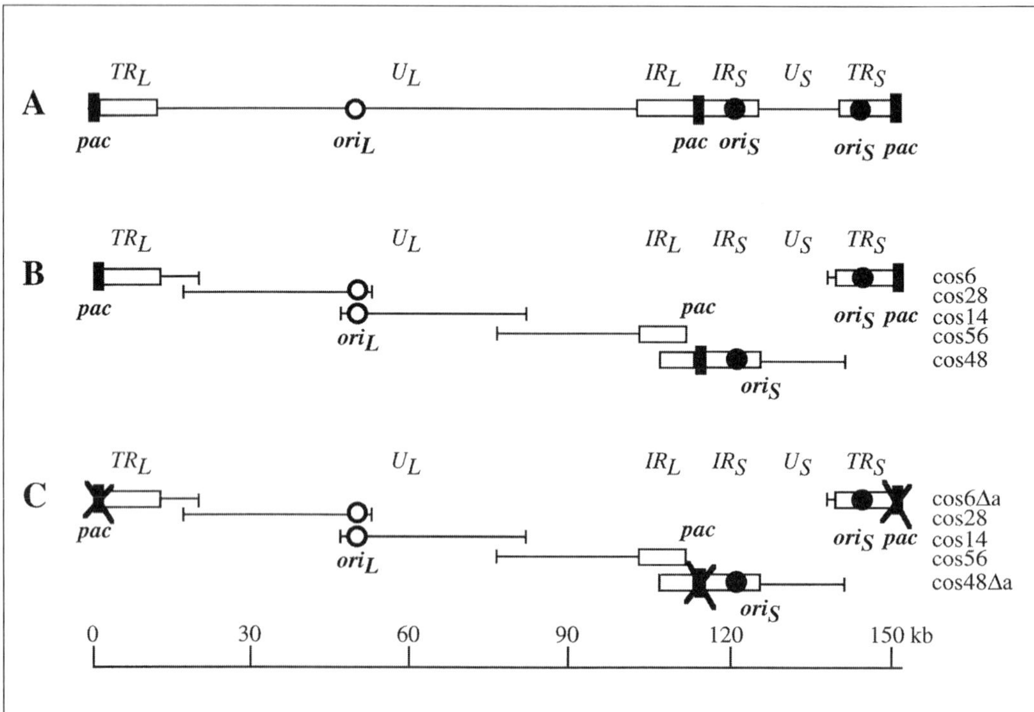

Figure 1. Schematic diagrams of the HSV-1 genome and HSV-1 cosmid sets. (A) The HSV-1 genome (approximately 152 kb) is composed of U_L and U_S segments (horizontal lines), which are flanked by inverted repeats (open rectangles): IR_S, internal repeat of the short segment; TR_S, terminal repeat of the short segment; IR_L, internal repeat of the long segment; and TR_L, terminal repeat of the long segment. The ori_S (solid circle), ori_L (open circle), and the *pac* (solid rectangles) are shown. (B) Cosmid set C represents the HSV-1 genome in five overlapping clones (cos 6, cos 14, cos 28, cos 48, and cos 56). (C) Cosmid set C6Δa48Δa was derived from cosmid set C by deleting the *pac* signals *(X)* and represents a replication-competent, packaging-defective HSV-1 genome in five overlapping clones (cos 6Δa, cos 14, cos 28, cos 48Δa, and cos 56).

acquiring an envelope. However, one theory claims that the virion retains the nuclear envelope, while the other states that the nuclear membrane is lost after fusion with the cytoplasmic membrane (e.g., Golgi apparatus) where a new envelope is acquired. There is also evidence supporting exocytosis and budding as release mechanisms from cells.

HSV-1 has many features that make it a promising gene transfer vehicle: *(i)* the genome has been fully sequenced (56,64); *(ii)* at least 30 kb of foreign DNA can be inserted by removing genes dispensable for virus replication in cell culture (50); *(iii)* HSV-1 can transduce most cell types, both dividing or nondividing; and *(iv)* HSV-1 can enter a state of latency in sensory neurons with very limited transcriptional activity (81).

Two types of distinct vectors, recombinant and amplicon, have been developed based on HSV-1. For recombinant HSV-1 vectors, one or more HSV-1 genes are replaced with the gene(s) of interest. These vectors can be further subdivided into two groups, replication-defective and replication-competent/-conditional. In contrast, amplicon vectors are based on bacterial plasmids that carry one or several genes of interest and two *cis*-acting HSV-1 signals, an *ori* and a *pac* signal, to support replication and packaging of amplicon DNA into HSV-1 virions in the presence of helper functions. Amplicon vectors have been derived from naturally occurring defective viruses more than two decades ago (24,77).

PACKAGING SYSTEMS FOR HSV-1-BASED AMPLICON VECTORS

Defective Viruses and Amplicons

It has been demonstrated that the serial passage of HSV-1 at high multiplicities of infection (moi) results in the formation of defective viruses (23,47). The DNA molecules isolated from these defective viruses were 150 kb long, similar to the wild-type (wt) HSV-1 genome. However, unlike the wt genome, the defective genomes consisted of uniform head-to-tail reiterations of short, specific HSV-1 sequences (repeat units). The repeat units ranged in size from 3 to 30 kb and contained at least a viral origin of DNA replication and a *pac* signal (23,24,47,48). Co-transfection of cells with wt HSV-1 DNA and bacterial plasmids containing single repeat units of defective genomes resulted in the generation of packaged defective genomes of approximately 150 kb and wt HSV-1 progeny. Because the defective genomes were composed of amplified head-to-tail linked concatamers of the seed plasmid, the chimeric plasmid was termed "amplicon" (77).

Because the HSV-1 *cis*-elements required for replication and packaging are smaller than 1 kb, the amplicon has the potential to accommodate large fragments of foreign DNA (theoretically up to 150 kb). This large transgene capacity, combined with the broad host range, promoted the HSV-1 amplicon to the status of an ideal candidate for gene delivery vehicle. This possibility was first demonstrated by Kwong and Frenkel (40), who reported on the HSV-1 amplicon-mediated expression of a truncated chicken ovalbumin protein in Hep-2 cells. However, because of the extreme cytopathic effects caused by coexisting lytic helper virus, the usefulness of HSV-1 amplicon vectors was at that time limited to transient overexpression of the genes of interest. Consequently, recent developments have been aimed at reducing the toxicity associated with the helper virus.

Deletion Mutant Helper Viruses

Initially, wt HSV-1 or an attenuated but replication-competent *UL41* (virion host shut-off [VHS] protein) deletion mutants have been used as a helper virus to package amplicons into HSV-1 virions (40,77). As a first step to reduce cytotoxicity, these helper viruses were replaced by a temperature-sensitive *(ts)* mutant of HSV-1, HSV-1 *tsK* (66). This mutant expresses a *ts* form of infected cell protein (ICP) 4 and therefore replicates at 31°C but not at 37°C. Amplicons packaged at 31°C in the presence of HSV-1 *tsK* were successfully used to transfer the *E. coli lacZ* gene into primary cultures of rat peripheral neurons (26). However, although replication of HSV-1 *tsK* was restricted at 37°C, gene expression from the helper virus genome caused cytopathic effects. Moreover, reversion to wt ICP4 phenotype occurred at a relatively high frequency. To reduce the risk of reversion, replication-defective deletion mutants of HSV-1 were introduced as helper viruses (Figure 2A) (27,45). Although these mutants carry deletions in genes essential for virus replication, such as the *IE3* (ICP4) and *IE2* (ICP27) genes, they can support amplicon packaging in cells that complement the deleted genes. An *IE2* deletion mutant packaging system, which used HSV-1 5*dl*1.2 as the helper virus and Vero 2-2 cells as the *IE2*-complementing cells, consistently produced stocks with relatively high titers of amplicon vector (1×10^6 to 1×10^7 transducing units [tu] per mL), an acceptable vector to helper virus ratio (up to 1), and low levels of revertants with wt phenotype ($<1 \times 10^{-7}$) (45,55,76). However, many problems associated with the inevitable coexistence of helper virus and amplicon vectors still remained. These included: *(i)* acute cytopathic effects and pronounced immune responses induced by gene expression from the helper virus, *(ii)* possible interactions between the helper virus and endogenous viruses, *(iii)* the risk of reversion of the helper virus to wt HSV-1 phenotype, and *(iv)* instability of transgene expression. Many of these problems have been reduced or eliminated by the recent development of a helper virus-free packaging system (22).

Helper Virus-Free Packaging of HSV-1 Amplicon Vectors

The development of a helper virus-free packaging system for HSV-1 amplicon vectors (Figure 2B) was based on two previous observations: *(i)* cosmid sets representing the entire HSV-1 genome in overlapping clones can form a complete, circular HSV-1 genome via homologous recombination and generate infectious progeny virus upon transfection into mammalian cells (Figure 1B) (12); and *(ii)* the HSV-1 *pac* signal is the essential *cis*-acting signal for cleavage and packaging of the replicating HSV-1 genome. Consequently, a HSV-1 cosmid set lacking the *pac* signals should not be packageable, but would still be able to provide all the helper functions required for the replication and the packaging of co-transfected amplicon DNA. Based on this hypothesis, a cosmid set with deleted *pac* signals (set C6Δa48Δa; Figure 1C) was constructed (20). Vector stocks produced following co-transfection of DNA from cosmid set C6Δa48Δa, and amplicon DNA contained undetectable amounts of helper virus and efficiently transduced many different cell types, including neural cells and hepatocytes both in vitro and in vivo, while causing minimal to no cytopathic effects (2,20,21,22,37). Moreover, readministration of stocks of helper virus-free amplicon vector was shown to be as effective as the first administration, even in immunocompetent mice (21).

Because of a small sequence homology between the cosmids and the amplicon (ori_S; 0.2–1 kb), the reconstitution of a replication-competent HSV-1 genome containing a *pac* signal is theoretically possible, although rare, since it requires at least six recombination events. Furthermore, such helper viruses would be sensitive to ganci-

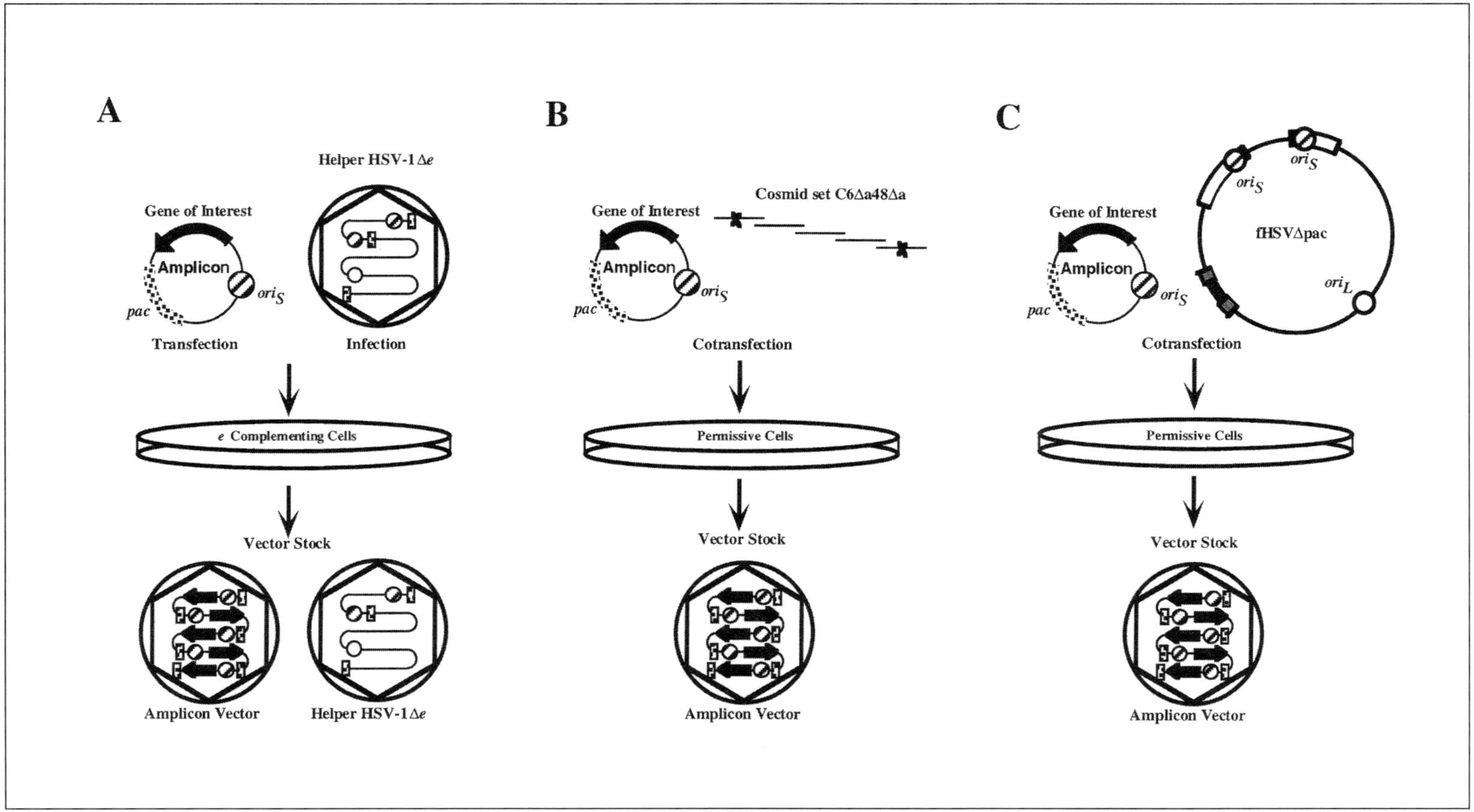

Figure 2. Packaging systems for HSV-1 amplicon vectors. (A) Helper virus-dependent packaging system using a replication-defective mutant of HSV-1 containing a deletion in an essential gene *(Δe)*. Cells that complement the deleted virus gene *(e)* are transfected with amplicon DNA and then superinfected with the deletion mutant helper virus. The resulting vector stocks contain both packaged amplicon vector and helper virus. (B) Cosmid-based, helper virus-free packaging system. Cells that are permissive for HSV-1 replication are co-transfected with amplicon DNA and DNA from cosmid set C6Δa48Δa. While helper HSV-1 genomes reconstituted from the cosmid clones are not packageable in the absence of *pac* signals, they still can provide all the *trans*-acting functions required for the replication and packaging of the co-transfected amplicon DNA. (C) F-plasmid-based helper virus-free packaging system. Permissive cells are co-transfected with amplicon DNA and fHSVΔpac plasmid DNA. The fHSVΔpac helper DNA is not packageable in the absence of *pac* signals, but it can still provide all the *trans*-acting functions required for the replication and packaging of co-transfected amplicon DNA.

clovir treatment because of a functional thymidine kinase gene and would be attenuated because both copies of the *γ-34.5 (RL1)* gene, which encodes a neurovirulence factor (10), were inactivated in the course of deleting the *pac* signals. Other limitations of the helper virus-free amplicon system include: *(i)* the genetic instability of the cosmid clones, *(ii)* difficulties in large-scale production of vector stocks, and *(iii)* the instability of amplicon vector-mediated transgene expression. The cloning of the entire 152-kb genome of HSV-1 in a bacterial plasmid has been a first step to resolve some of these problems (71,78).

Packaging of HSV-1 Amplicon Vectors Using a Bacterial Artificial Chromosome Containing a Replication-Competent, Packaging-Defective HSV-1 Genome

Bacterial artificial chromosomes (BACs) are based on the single-copy F-plasmid from *E. coli* and have been widely used in genome research (61,75). Because of their capacity to accommodate large inserts (>300 kb) and their genetic stability in bacteria, BACs are suitable for cloning large genomes of DNA viruses, as first demonstrated by Luckow et al. (51) with baculovirus. Recently, the genomes of several herpesviruses, including those of murine cytomegalovirus, Epstein-Barr virus (EBV), and HSV-1 have successfully been cloned in *E. coli* (15,57,70,78).

We developed the following strategy to clone the entire HSV-1 genome as a BAC in *E. coli* (Figure 3) (70). First, mini-F transfer plasmid fH98-102pac was constructed to contain: *(i)* a HSV-1 *pac* signal, flanked by recognition sites for *Pac*I, a restriction endonuclease that does not cut HSV-1 DNA, and *(ii)* an approximately 4-kb fragment of the HSV-1 genome between nucleotides 98 968 and 102 732 to facilitate homologous recombination (Figure 3C). Following co-transfection of cells with transfer plasmid fH98-102pac and DNA from cosmid set C6Δa48Δa, a replication- and packaging-competent virus progeny was generated, presumably via a single recombination event within nucleotides 98 968 to 102 732, which integrates the entire transfer plasmid, including the F-plasmid backbone and the *pac* signal, into cos56 (Figure 3, B–D). The progeny virus was amplified, and DNA isolated from purified virions was cleaved with *Pac*I to remove the *pac* signals (Figure 3D). A resulting DNA fragment of approximately 160 kb was ligated, electroporated into *E. coli*, and clones that contained the entire HSV-1 genome with *pac* signals deleted were identified and designated fHSVΔpac. These clones appeared to be very stable, as restriction enzyme patterns were entirely conserved even after serial passages of bacterial cultures (Figure 3E).

Although fHSVΔpac alone did not generate replication-competent virus progeny, it efficiently packaged co-transfected amplicon DNA, resulting in vector stocks with titers of up to 1×10^7 tu/mL (an approximately twofold increase compared to the cosmid-based packaging system; Figure 2C). When a *pac* signal was introduced into fHSVΔpac through ligation, the resulting clone, designated fHSVpac+, generated more than 400 times the number of plaques upon transfection than did unmodified HSV-1 cosmid set C (Figure 1B), which contains intact *pac* signals (70). The discrepancy observed between the efficiencies of amplicon packaging (twofold) and generation of virus progeny (400-fold) may suggest that while complete reconstitution of the HSV-1 genome from five cosmids is a prerequisite for production of infectious virus progeny, it may not entirely be necessary to support replication and packaging of amplicon DNA. Products from partial and/or aberrant recombinations among the five cosmid clones may serve as helper HSV-1 genomes as well.

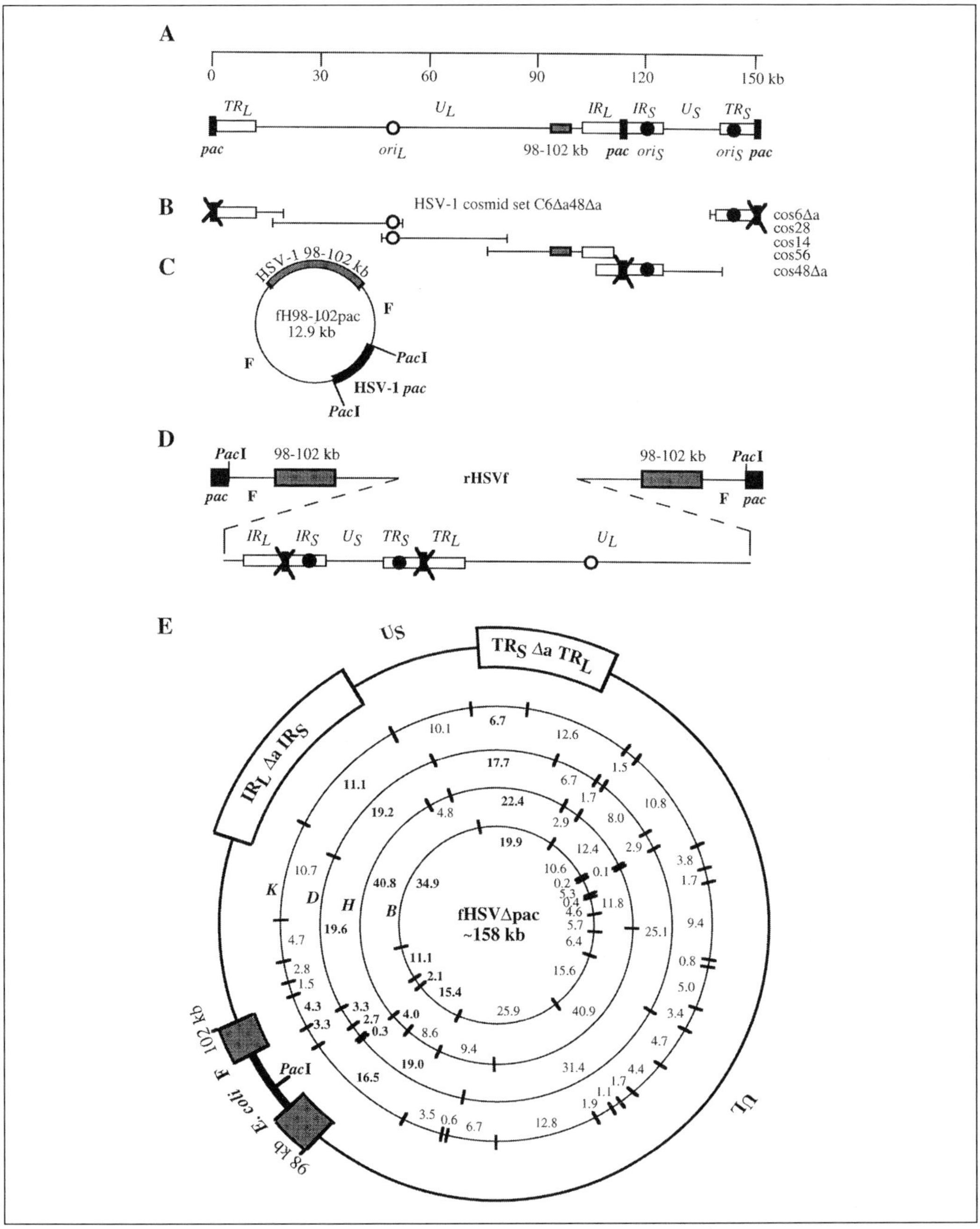

Figure 3. Cloning of the HSV-1 genome as a BAC in *E. coli*. Schematic diagrams of (A) the HSV-1 genome and (B) cosmid set C6Δa48Δa. (C) Transfer plasmid fH98-102pac, which is based on the single-copy F-plasmid from *E. coli*, contains the sequences between nucleotides 98 968 and 102 732 of the HSV-1 genome to facilitate homologous recombination with HSV-1 DNA, and the HSV-1 *pac* signal flanked by recognition sites for *Pac*I, a restriction endonuclease that does not cut HSV-1 DNA. (D) Hypothetical structure of the replication- and packaging-competent virus genome (rHSVf) generated via homologous recombination in Vero 2-2 cells co-transfected with transfer plasmid fH98-102pac and the five clones of cosmid set C6Δa48Δa. DNA isolated from rHSVf was cleaved with *Pac*I; a fragment of approximately 160 kb was isolated, self-ligated, and then electroporated into *E. coli* DH10B. (E) Clones that contained the HSV-1 genome deleted for the *pac* signals (fHSVΔpac) were identified by restriction endonuclease analysis. The sites for *Kpn*I *(K)*, *Dra*I *(D)*, *Hin*dIII *(H)*, and *Bgl*II *(B)*, and the resulting fragment sizes are indicated. Reprinted with permission from Hum. Gene Ther. (70).

Vector stocks prepared with fHSVΔpac were expected to be contaminated with large numbers of replication-competent helper viruses since these can easily be generated via a single recombination event between the *ori*$_S$ sequences present on both fHSVΔpac and amplicon. However, contamination with such viruses was surprisingly low, less than 100 plaque-forming units (pfu) per mL, perhaps due to a limited packaging capacity of the HSV-1 capsid because even smaller numbers of replication-competent helper viruses (<10 pfu/mL) were obtained with larger amplicons (approximately 15 kb; Y. Saeki et al., unpublished data). Vector stocks produced by using fHSVΔpac helper DNA efficiently transduced neural cells in culture and in the rat brain, as well as hepatocytes in the rat liver, without causing apparent cytopathic effects (Figures 4–6) (70). Nevertheless, the possibility of vector stocks being contaminated with replication-competent helper virus must be eliminated entirely (e.g., by deleting one or several essential HSV-1 genes from fHSVΔpac). Such targeted modifications of fHSVΔpac can be achieved efficiently through *recA*-mediated homologous recombination in *E. coli* (Y. Saeki et al., unpublished data) (57,61,95).

HSV-1 AMPLICON AND HYBRID AMPLICON VECTORS

Components of the Standard HSV-1 Amplicon

Standard HSV-1 amplicon vectors contain three types of genetic elements: *(i)* a prokaryotic plasmid backbone with an *E. coli* origin of DNA replication and an antibiotic resistance gene; *(ii)* sequences

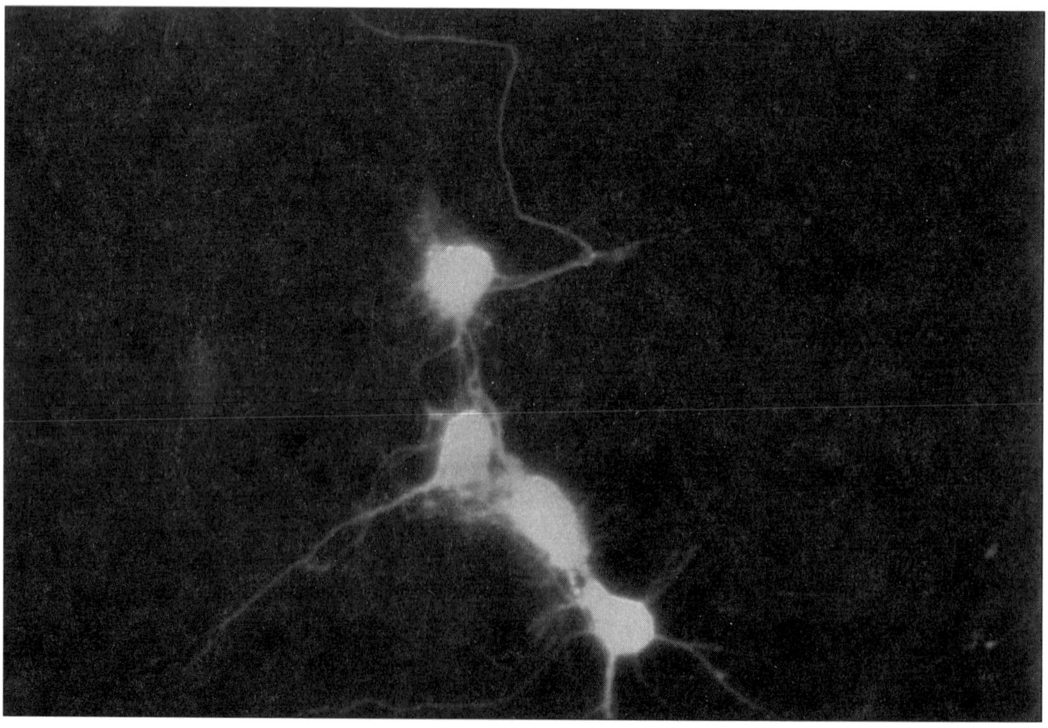

Figure 4. Photomicrograph showing GFP cells. Cultured rat cortical neurons were infected with stocks of HSV-1 amplicon vector pHSVGFP that have been prepared by using fHSVΔpac as the replication-competent, packaging-defective HSV-1 helper genome. pHSVGFP expresses the gene for GFP from the HSV-1 *IE4/5* promoter. The cultures were examined 24 hours after infection using a fluorescence microscope. Original magnification is 20×. (**See color plate A1.**)

from HSV-1, including *ori* and *pac*, to support amplicon replication and packaging in the presence of helper functions; and *(iii)* a transcription unit with one or more genes of interest (Figure 7A) (20,32,42).

In general, amplicons use pBR322-based or pUC-based backbones, which support medium to high copy number plasmid propagation and are therefore not suitable to explore the full transgene capacity of HSV-1. The use of low- or single-copy plasmids as backbones for amplicons may increase genetic stability and allow the packaging of very large transgenes (C. Fraefel et al., unpublished data).

The HSV-1 genome contains two different origins of DNA replication, ori_S and ori_L (see Figure 1A), both of which have been found in naturally occurring defective genomes and were used in amplicon

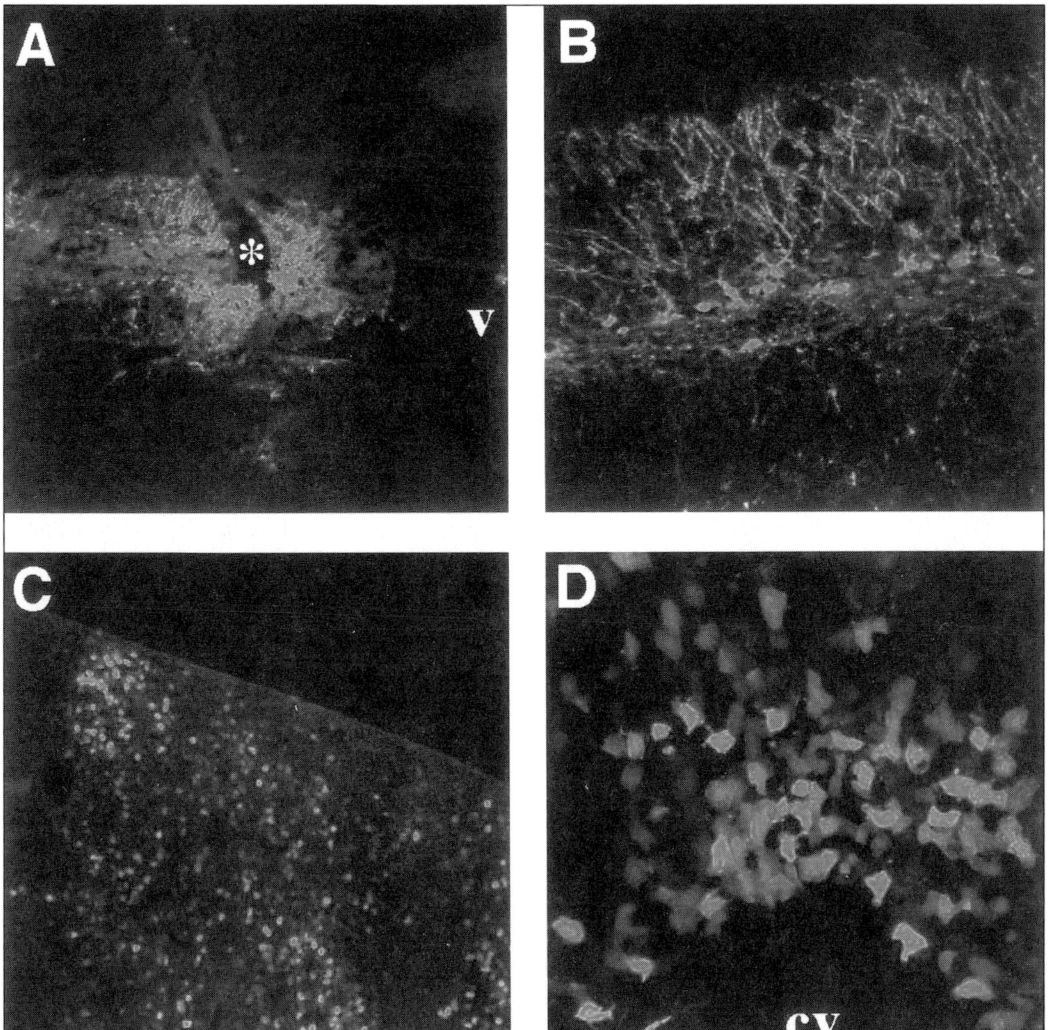

Figure 5. Photomicrographs showing GFP cells on day 4 following injection of pHSVGFP amplicon vector into rat hippocampus (7.5 × 10^5 tu in 10 µL; A and B) or rat liver (7.5 × 10^6 tu in 100 µL; C and D). Needle tract (*), third ventricle (v), and central vein (cv) are indicated. Original magnifications are 5× (A and C) and 20× (B and D). Reprinted with permission from Hum. Gene Ther. (70). **(See color plate A2.)**

vectors (24,48,77). However, the ori_L sequence contains an AT rich, 144-bp perfect palindrome and is unstable in bacteria (65,91); by contrast, the smaller, 90-bp ori_S sequence, which contains an almost perfect 45-bp palindrome, has proven more stable in bacteria and is typically used in amplicons (83). In the HSV-1 genome, ori_S is flanked by two promoters that control, in opposite directions, the expression of the *IE3* (ICP4) and *IE4/5* (ICP22/47) genes. Both promoters contain Sp1 binding sites and TAATGARAT consensus sequences that respond to the virion tegument protein VP16, while also enhancing virus DNA replication (79,93). A major concern for the inclusion of the *IE3* and/or *IE4/5* promoters as part of the amplicon ori_S fragment is, especially when cell type-specific or inducible transgene expression is desired, that the viral promoters can interfere with other promoters located on the same construct. The design of minimal or artificial origins of DNA replication that support efficient vector amplification would reduce or eliminate both interference between promoters, as well as recombinations between ori_S sequences present in amplicons and the packaging-defective HSV-1 helper genomes.

The *pac* signals are located within the *a* repeats at the junctions between the long and the short segment and at both termini of the HSV-1 genome (see Figure 1A). They consist of alternating repeats and unique sequences of 250–500 bp, depending on the virus strain, and have the following configuration: direct repeat *(DR) 1* (20 bp) - unique sequence *(U) b* (65 bp) - *DR2* (12 bp × 19–23) - *DR4* (37 bp × 2–3) - *Uc* (59 bp) - *DR1* (13,14). The repeat nature of these sequences and additional recombinational hot spots may contribute to their instability in *E. coli* and in mammalian cells (87). Of note, the promoter of the *γ34.5* gene is located within the *a* repeat and thus could also be used to express a transgene from the amplicon.

Although the absolute transgene capacity has not yet been determined, amplicon plasmids could express several transgenes simultaneously. Each transgene cassette consists of a viral or cellular promoter, the cDNA encoding a gene of interest, and a polyadenylation signal. Because amplicon DNA is packaged as a concatamer of approximately 152 kb, depending on the size of the seed amplicon, each vector particle can deliver several copies of the transgene(s) into a target cell. Upon infection of

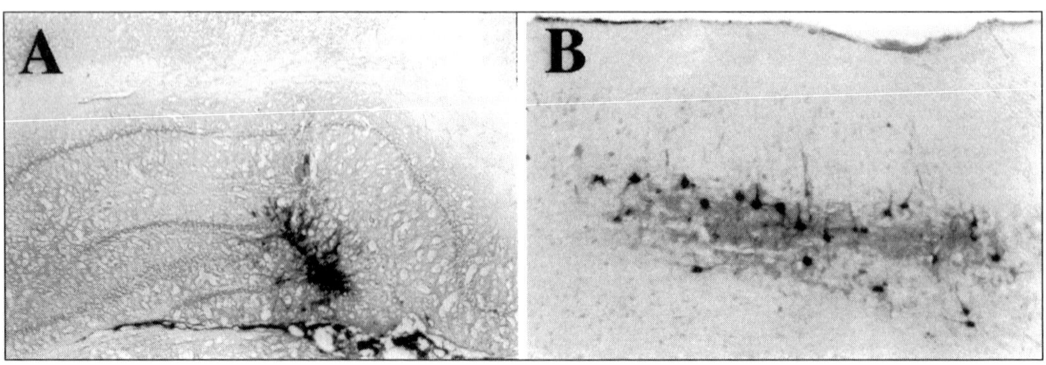

Figure 6. Photomicrographs showing β-galactosidase-positive cells in the rat brain. HSV/EBV hybrid amplicon pREHcagLacZ, which expresses the *E. coli lacZ* gene from the chicken β-actin promoter, was packaged into HSV-1 particles by using fHSVΔpac helper DNA, and then injected into the hippocampus (1×10^6 tu in 10 μL). At day 3 after injection, the rat was killed and brain sections were incubated with X-gal solution to stain β-galactosidase-positive cells. (A) Intense blue staining was observed in the hippocampal region of the injected side. (B) Some of the cortical neurons adjacent to the needle track were also β-galactosidase-positive. Magnification is 4× (A) and 10× (B). (**See color plate A3.**)

a cell, vector capsids are transported to the nuclear pores, and the concatameric amplicon genomes are released into the nucleus, circularize, and express the transgene(s). However, the circularized episomal amplicon DNA can be lost easily during cell division, and moreover, transgene expression mediated by standard amplicon vectors is usually transient in dividing and nondividing cells, particularly when viral promoters are used. Therefore, several attempts have been made both to stabilize the amplicon vector genome and to increase the efficiency of long-term gene expression.

Hybrid Amplicon Vectors

All currently available vector systems, both viral and nonviral, have limitations (see Table 1). HSV-1-based vectors exploit most of the advantageous features of their parent virus, including: *(i)* the ability to transduce most cell types, *(ii)* the large transgene capacity, *(iii)* the high stability of the virion, and *(iv)* the relatively high titers (6,20,30). However, they have also retained major drawbacks. Recombinant HSV-1 vectors, although replication defective in noncomplementing cells, express many viral genes that induce cytopathic effects and immune responses. Moreover, transgene expression from most viral or cell type-specific promoters is transient. HSV-1 amplicons, on the other hand, do not express any viral genes, but their replication and packaging into virions depend on helper functions (77). If these helper functions are provided by replication-defective mutants of HSV-1, the resulting vector stocks are contaminated with at least as much helper virus as packaged amplicon vector, and consequently, also induce cytopathic effects and immune responses (26,33,35,36,94). While these problems are largely reduced if the helper functions are provided by a packaging-defective HSV-1 genome, which results in vector stocks that are essentially free of contami-

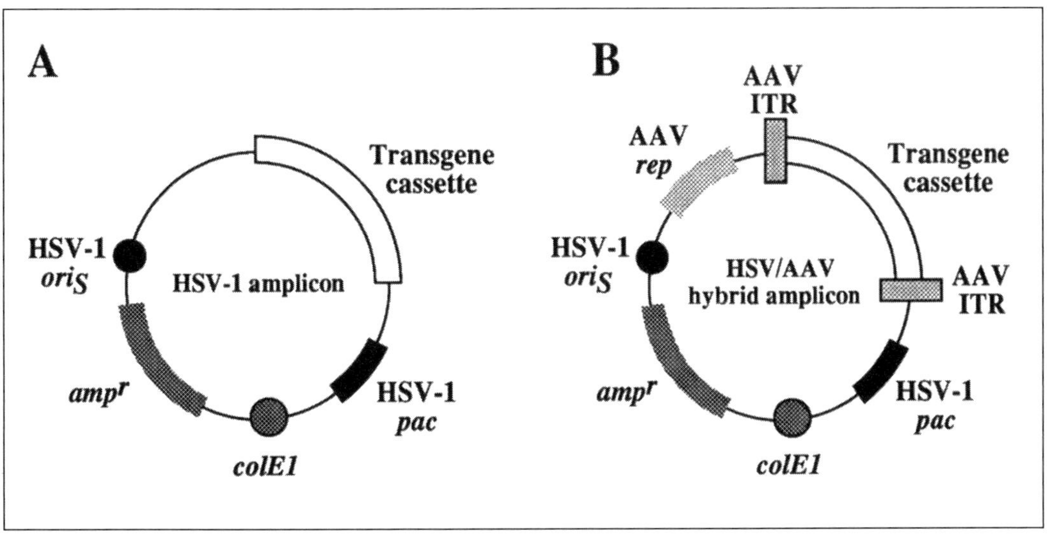

Figure 7. Amplicon structures. (A) The standard HSV-1 amplicon is composed of three types of genetic elements: *(i)* sequences from bacteria, including an origin of DNA replication *(colE1)* and an antibiotic resistance gene *(amp^r)*, which allow propagation of plasmid DNA in *E. coli*; *(ii)* sequences from HSV-1, in particular an origin of DNA replication *(ori$_S$)* and a DNA cleavage/packaging signal *(pac)*, which support replication of amplicon DNA and subsequent packaging into HSV-1 particles in mammalian cells in the presence of helper functions; and *(iii)* a transgene cassette with one or more genes of interest. (B) In addition to the standard amplicon elements, HSV/AAV hybrid amplicons contain the AAV *rep* genes and a transgene cassette that is flanked by AAV ITRs.

Table 1. Properties of Selected Vector Systems

Vector System		Advantages	Disadvantages
HSV-1	recombinant	broad host range, dividing and nondividing cells, large transgene capacity (~30 kb)	immunogenic, toxic, instability of transgene expression
	amplicon helper virus-dependent	as above (transgene capacity theoretically up to 150 kb)	as above
	helper virus-free	as above, low toxicity and immunogenicity	vector production depends on transfection of cells, instability of transgene expression
Retrovirus	MoMLV-based	integration into host genome	transduction limited to dividing cells, random integration, limited transgene capacity (~8 kb)
	HIV-based	integration into genome of dividing and nondividing cells	HIV-1-based, random integration, limited transgene capacity (~8 kb)
Adenovirus	E1, E2, E4 deletions	broad host range, dividing and nondividing cells	immunogenic, toxic, instability of transgene expression, limited transgene capacity (~8 kb)
	elimination of all virus genes (gutless)	as above, large transgene capacity (~30 kb)	vector production depends on helper virus
AAV		integration into genome of dividing and nondividing cells nonimmunogenic, nontoxic	limited transgene capacity (~4.5 kb)
Nonviral	liposomes, particle bombardment, etc.	safety, low toxicity	limited transduction efficiency, instability of transgene expression

nating helper virus, the second major problem of HSV-1–based vectors, the inefficiency of long-term transgene expression, remains (2,20,22,70).

Adeno-associated virus (AAV)-based vectors are created by substituting a transgene cassette for the viral *rep* and *cap* genes, which are flanked by the AAV inverted terminal repeats (ITRs). AAV vectors are distinguished by their capability to either persist in an episomal form (19,69) or integrate into the host cell genome, preferentially into the AAVS1 site on human chromosome 19q13.3-qter (3,39,46,85). Furthermore, AAV vectors have a wide host range that includes dividing and nondividing cells. However, the transgene capacity of only approximately 4.5 kb dramatically limits the usefulness of AAV vectors.

To overcome some of the deficits of various virus- and nonvirus-based vector systems, hybrid vectors have been designed as the new generation of gene transfer vehicles (see Table 2). Optimally, hybrid vectors combine the advantageous components of their parent vector systems while eliminating the disadvantages. The HSV/AAV hybrid vector represents one example of a successful combination of genetic elements from two different viruses (37). In addition to the standard HSV-1 amplicon elements, HSV/AAV hybrid amplicons incorporate sequences from AAV that allow for stable maintenance of the transgene, in particular the ITR and the *rep* genes (Figure 7B). Because HSV/AAV hybrid vectors can be packaged into HSV-1 virions, they conserve the numerous advantages of HSV-1 as a vector system, including the high efficiency of gene transfer, the large transgene capacity, and the availability of a helper virus-free packaging system. Once the hybrid vector genome reaches the host cell nucleus, it has the potential to act like an AAV vector with *rep*-mediated amplification, and eventually, genomic integration of the ITR-flanked transgene cassette specifically at the AAVS1 site on human chromosome 19 (Figure 8). The functionality of the HSV-1 and AAV elements in the context of the hybrid vector has been demonstrated. The vector can efficiently be packaged into HSV-1 virions by using helper virus-dependent or helper virus-free packaging systems, and the resulting vector stocks can transduce many different cell types, support amplification of the ITR-flanked transgene cassette, and allow for long-term gene expression in dividing and nondividing cells (21,37). Although the specificity of integration at AAVS1 remains to be demonstrated conclusively by fluorescence in situ hybridization, polymerase chain reaction (PCR), and genomic Southern analyses, the efficiency of integration of the transgene cassette into the genome of human cells is surprisingly high. Depending on the cell line, 20%–30% of the cells initially transduced with a hybrid vector that contained an ITR-flanked transgene cassette consisting of the genes for green fluorescent protein (GFP) and neomycin resistance formed stable, neomycin-resistant and green fluorescent colonies. In comparison, only approximately 4% of the cells transduced with a standard amplicon vector formed colonies under these conditions (Y. Saeki and C. Fraefel, unpublished data). With the goal to increase the stability of the transgene cassette, AAV ITR and *rep* genes have been incorporated also into adenovirus (Ad) vectors and baculovirus vectors (18,63).

A hybrid vector of similar rationale has been developed by inserting components from EBV into the HSV-1 amplicon. EBV-based vectors, so-called replicons, are bacterial plasmids containing sequences from EBV, in particular the *EBNA-1* gene and the latent origin of DNA replication, *oriP*, and up to 180 kb of foreign DNA. *EBNA-1* and *oriP* are sufficient to maintain the replicon as a circular episome that replicates and segregates to daughter cells in phase with the cellular genome (25,58, 84,92,96). Because replicons are packaged

Table 2. Properties of Selected Hybrid Vector Systems

Hybrid Vector	Packaging Functions	Heterologous Elements	Intent of Design	Reference(s)
HSV/AAV	HSV-1 amplicon	AAV ITRs and *rep*	Long-term gene expression, transgene amplification, site-specific integration in dividing and nondividing cells	37
HSV-EBV	HSV-1 amplicon	EBV *oriP* and *EBNA-1*	Long-term gene expression, episomal retention and replication	90
HSV/MoMLV	HSV-1 amplicon	MoMLV *gag-pol-env*	Platform for production of retrovirus vectors	72
Lentivirus/VSV-G	HIV vector	VSV G protein	Broad host range, integration in dividing and nondividing cells, high titers	60
MoMLV/VSV-G	MoMLV vector	VSV G protein	Broad host range, high titers	7
Ad/AAV	Ad vector	AAV ITRs and *rep*	Long-term gene expression, transgene amplification, site-specific integration in dividing and nondividing cells	18
Ad/MoMLV	Ad vector	MoMLV *gag-pol-env* MoMLV LTRs	Platform for production of retrovirus vectors	17
Baculovirus/VSV-G	Baculovirus	VSV G protein	Broader host range	4
Baculorivus/AAV	Baculovirus	AAV ITRs and *rep*	Long-term gene expression, transgene amplification, site-specific integration in dividing and nondividing cells	63
Nonviral/Viral:				
Liposome/VSV-G	Liposomes	VSV G protein	Increased transduction efficiency	1
Liposome/HVJ	Liposomes	HVJ fusion protein	Increased transduction efficiency	38,71

into EBV virions, their host range is limited to human epithelial, muscle and some hematopoietic cells, as well as B and T cells. By inserting the EBV *EBNA-1* and *oriP* elements into an HSV-1 amplicon, a hybrid vector has been developed that replicates episomally and supports long-term transgene expression in a wide range of host cells (89,90).

Several vectors, including those based on human immunodeficiency virus type 1 (HIV-1), Moloney murine leukemia virus (MoMLV), and baculovirus have been pseudotyped with the vesicular stomatitis virus (VSV-G) glycoprotein to increase both the transduction efficiency and the host range (4,7,60). Combinations of elements from nonviral and viral delivery systems also exist. To improve the efficiency of liposome-mediated gene transfer, liposomes have been complexed with the VSV-G glycoprotein (1), or alternatively, with the fusion proteins of hemagglutinating virus of Japan (HVJ; Sendai virus) (38,71). Some hybrid vectors have been developed to serve as platforms for the production of other vectors, including HSV-1 amplicons for the production of AAV vectors (11) and both HSV-1 amplicons and Ad vectors for the production of retrovirus vectors (17,72).

In summary, hybrid vectors hold the promise to combine efficient, nontoxic gene

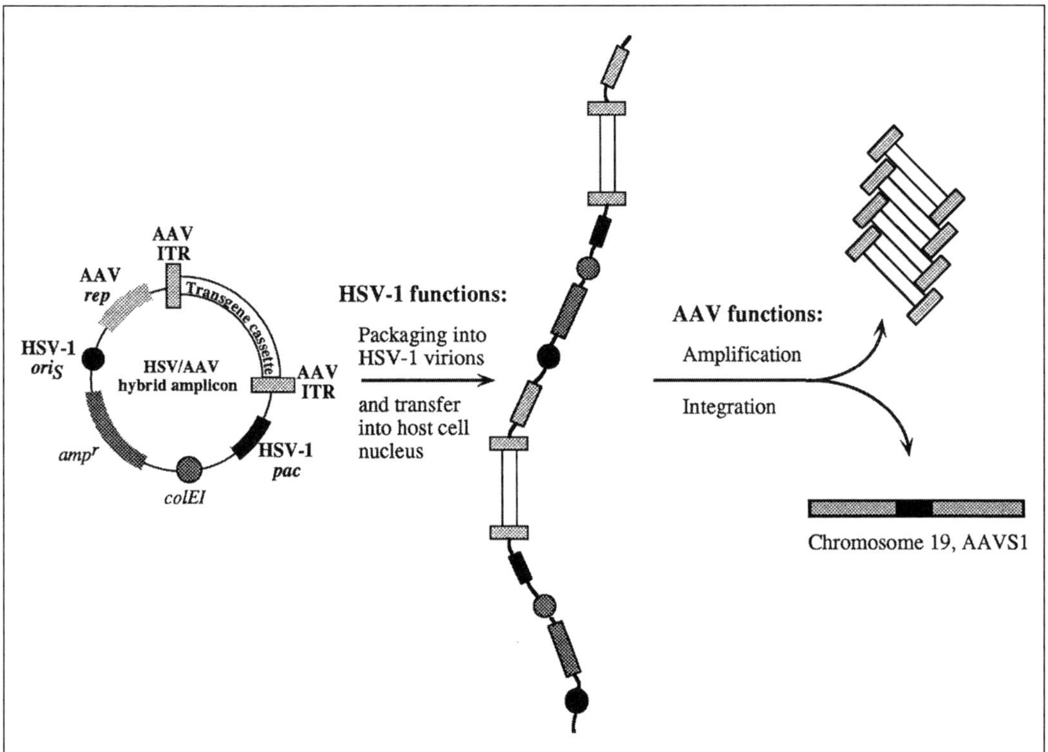

Figure 8. Schematic diagram of proposed HSV/AAV hybrid amplicon vector functions. HSV/AAV hybrid amplicon DNA is replicated and packaged as a concatamer of approximately 152 kb into HSV-1 virions by using helper virus-dependent or helper virus-free packaging systems. After delivery into the target cell nucleus, the vector-encoded Rep proteins can mediate amplification and genomic integration of the ITR-flanked transgene cassette, potentially into the AAVS1 site on human chromosome 19. The *rep* genes are placed outside of the ITR-flanked transgene cassette to avoid their amplification and integration into the host genome. Because the *rep* genes are expressed from their own promoters, feedback inhibition reduces the levels of the Rep proteins and consequently reduces both Rep-associated toxicity and the chance for rescuing integrated transgene cassettes. Moreover, the *rep* genes should eventually be degraded, together with the amplicon backbone.

delivery and long-term transgene expression, a prerequisite for successful gene therapy.

PERSPECTIVES

In spite of new packaging strategies and amplicon designs, some aspects of the amplicon system still need to be further addressed, such as vector titers, large-scale production of vector stocks, toxicity, and safety issues. Because the current helper virus-free packaging methods depend on transient co-transfection of amplicon DNA and replication-competent, packaging-defective HSV-1 helper genomes, the transfection efficiencies directly influence the titers of the resulting amplicon vector stocks. Therefore, isolation of highly transfectable cell lines and the development of more efficient transfection protocols, which use viral fusion proteins (1,38,71) and/or DNA-binding nuclear proteins (38), will have a direct effect on amplicon titers. However, as an ultimate goal, a helper virus-free packaging cell line will eliminate many difficulties with large-scale production of vector stocks. The development of such cell lines depends on efficient means to control viral gene expression and replication. BACs containing the HSV-1 genome, such as fHSVΔpac, will be invaluable tools in achieving these goals.

Toxic effects caused by amplicon vector-mediated gene transfer have been largely reduced by introduction of the helper virus-free packaging system. However, certain virion tegument proteins, such as VHS and VP16, are co-introduced into a target cell with the vector particle and may cause some cytopathic effects, at least at high moi. VHS induces nonspecific degradation of mRNA in infected cells (41,62). VP16 is a transcriptional transactivator of viral *IE* promoters, but it can also alter cellular gene expression patterns as well as vector transgene expression. While both proteins contain domains that are important for the virion structure, the host shut-off domain of VHS and the transactivating domain of VP16 are both nonessential for virus replication and can therefore be modified or deleted.

The development of strategies that eliminate all possibilities for the generation of replication-competent helper virus is imperative. One approach is to reduce the possibility of recombination between amplicon DNA and replication-competent, packaging-defective helper HSV-1 genomes by eliminating sequence homologies. Another approach would be to prevent the helper HSV-1 DNA from being packaged even after the acquisition of a functional *pac* signal through recombination with the amplicon. This may be achieved by increasing the size of fHSVΔpac (Y. Saeki et al., unpublished data) (70). Alternatively, the use of packaging-defective helper HSV-1 genomes that can replicate exclusively in complementing cells may also increase the safety of the vector system (Y. Saeki et al., unpublished data).

One of the most unique and attractive features of HSV-1 amplicon vectors is their large transgene capacity of theoretically up to 150 kb, although inserts of more than 20 kb have not been reported to date. The recent development of an HSV-1 amplicon vector based on a single-copy, F-plasmid backbone (f-amplicon) will allow us to assess both the transgene capacity of the amplicon and the genetic stability of large inserts during replication and packaging (C. Fraefel et al., unpublished data). For some applications in gene therapy, it would be desirable to regulate the transgene expression in a cell type-specific manner. In such situations, the amplicon could accommodate large transcription units with native promoters and adequate flanking sequences to allow correct regulation of transgene expression. The large capacity of the amplicon also supports the incorporation of genetic elements from other virus vectors, such as the AAV ITRs and *rep* genes or EBV *oriP* and *EBNA-1*, to specifically alter the properties of the vector and to promote genetic stability and long-term gene expression.

In summary, the large capacity and flexibility of the HSV-1 amplicon will facilitate the development of new, customized, synthetic, hybrid, and artificial vectors for specific applications in basic research and gene therapy alike.

PROTOCOLS

Packaging of HSV-1 Amplicon Vectors by Using Replication-Competent, Packaging-Defective HSV-1 Genomes

The helper virus-free packaging of amplicons into virions is based on co-transfection of cells with amplicon DNA and replication-competent, packaging-defective HSV-1 helper DNA (see Figure 2, B and C). There are two types of replication-competent, packaging-defective HSV-1 genomes available: *(i)* cosmid sets, such as C6Δa48Δa (Figure 1C); and (ii) BAC, such as fHSVΔpac (Figure 3). The following protocol describes *(i)* the preparation of cosmid DNA and BAC DNA, *(ii)* the transfection of cells using LIPOFECTAMINE™ (Life Technologies, Rockville, MD, USA), and the *(iii)* harvest and *(iv)* titration of vector stocks.

Preparation of DNA from Cosmid Set C6Δa48Δa and BAC fHSVΔpac

1a. Prepare cosmid DNA from 300-mL cultures (for each of the five clones from cosmid set C6Δa48Δa) grown overnight at 37°C in SOB medium (SOB per liter: 20 g Bacto-tryptone, 5 g Bacto-yeast extract, 0.5 g NaCl, 10 mL of 250 mM KCl, 5 mL of 2 M $MgCl_2$ containing 50 μg/mL ampicillin. Extract and purify DNA by alkaline lysis and QIAGEN TIP-500 columns according to the instructions of the manufacturer (Qiagen, Valencia, CA, USA).

1b. Prepare fHSVΔpac DNA from 1 L cultures grown overnight at 37°C in LB medium containing 12.5 μg/mL chloramphenicol. Extract DNA by alkaline lysis and purify over QIAGEN TIP-500 columns and CsCl equilibrium centrifugation.

2. Determine the absorbance of the DNA preparations at 260 nm (A_{260}) and 280 nm (A_{280}) using a UV spectrophotometer and calculate the DNA concentration in μg/μL. A value of 1.0 for A_{260} is equivalent to 50 μg/mL of double-stranded DNA; the ratio between A_{260} and A_{280}, which is an indicator for the purity of the DNA preparation, should be larger than 1.8.

3. To excise the HSV-1 inserts from the cosmid backbones (this step is not necessary for fHSVΔpac), pool 10 μg of each of the five clones from step 1a in a microcentrifuge tube and digest with 50 U of *Pac*I restriction endonuclease (New England Biolabs, Beverly, MA, USA) in a volume of 100 μL for at least 3 hours at 37°C. Confirm the completion of the digest by electrophoresis of an aliquot (1–2 μL) of the reaction mixture on a 0.4% agarose gel. Extract the DNA in the reaction mixture with phenol/chloroform and chloroform, and precipitate with ethanol. Resuspend the DNA in 100 μL distilled water and measure the DNA concentration.

Co-transfection of Amplicon DNA and Packaging-Defective HSV-1 Helper DNA

Any cell line that is well transfectable

and that supports HSV-1 replication can be used for packaging of amplicon vectors. The following procedure, however, has been optimized for Vero 2-2 cells (76).

4. Vero 2-2 cells are maintained in DMEM supplemented with 10% fetal bovine serum (FBS) and 500 μg/mL G418. On the day before transfection, trypsinize, count, and plate the cells at 1.2×10^6 cells/60-mm tissue culture dish in 3 mL DMEM/10% FBS.
5. The following day, prepare two 15-mL conical tubes each containing 100 μL optiMEM I™ (Life Technologies). To one tube, add 0.6 μg amplicon DNA and either 2 μg of the *Pac*I-digested cosmid DNA from set C6Δa48Δa or 1.5 μg of fHSVΔpac DNA. Add 12 μL LIPOFECTAMINE to the other tube. Combine the contents of the two tubes, mix, and incubate for 45 minutes at room temperature.
6. Wash the cultures from step 4 once with optiMEM I. Add 1.1 mL optiMEM I to the tube containing the DNA-LIPOFECTAMINE transfection mixture. Aspirate the medium from the culture, add the transfection mixture, and incubate for 5.5 hours at 37°C.
7. Wash the culture three times with optiMEM I, add 3.5 mL DMEM/6% FBS and incubate 2.5 days in a humidified, 37°C CO_2 incubator. As a control, use an amplicon that expresses the gene for GFP to monitor transfection efficiency.

Harvest Vector Stocks

8. To harvest vector particles, scrape the cells from step 7 into the medium, transfer the suspension to a 15-mL conical tube, and freeze-thaw three times using a dry ice/ethanol bath and a 37°C water bath. Sonicate the suspension for 20 seconds with 20% output energy, remove cell debris by centrifugation at 1400× *g* and 4°C, and filter the supernatant through a 0.45-μm syringe-tip filter. After removing a sample for titration, divide the vector stock into aliquots, freeze in a dry ice/ethanol bath, and store at -80°C. Alternatively, purify and concentrate the vector stock before storage.
9. To purify and concentrate the vector stock from step 8, prepare a sucrose gradient (with 7 mL 60% sucrose, 7 mL 30% sucrose, and 3 mL 10% sucrose) in a Beckman Ultra-Clear™ 25 × 89-mm centrifuge tube (Beckman-Coulter, Fullerton, CA, USA). Add the vector stock from step 8 (up to 20 mL) carefully on top of the gradient and centrifuge for 2 hours at 100 000× *g* and 4°C using a Beckman SW40 rotor.
10. Collect the virus band at the interphase between the 60% and 30% sucrose layers after aspirating the 10% and 30% layers. Transfer to a Beckman Ultra-Clear 14 × 95-mm centrifuge tube, add 15 mL phosphate-buffered saline (PBS), and pellet vector particles for 1 hour at 100 000× *g* and 4°C using a Beckman SW40 rotor.
11. Resuspend the pellet in, for example, 200 μL PBS containing 10% sucrose. After removing a sample for titration, divide the purified and concentrated vector stock into aliquots, freeze in a dry ice/ethanol bath, and store at -80°C.

Titration of Vector Stocks

12. Plate cells (e.g., Vero and BHK) in a 24-well tissue culture plate at 1 ×

10^5 cells in 0.5 mL DMEM/2% FBS/well.

13. The following day (when the cells are confluent), remove the medium and add 0.1, 1, or 5 µL of the samples collected at steps 8 and 11 in 250 µL DMEM/2% FBS. Incubate in a humidified, 37°C CO_2 incubator for 2 hours. Remove the inoculum, wash the cells twice with PBS, add 250 µL DMEM/2% FBS, and continue incubation for ≥20 hours.
14. Fix the cells for 20 minutes at room temperature with 250 µL of 4% paraformaldehyde (fixation is not necessary for detection of GFP). Wash the fixed cells three times with PBS and proceed, depending on the transgene, with a detection protocol such as green fluorescence, X-Gal staining, or immunocytochemical staining using an antibody.

15a. For detection of GFP, examine the culture from step 14 (before or after fixation) using an inverted fluorescence microscope. Count the green fluorescent cells and determine the vector titer in tu/mL.

15b. For detection of β-galactosidase activity, add 250 µL of X-Gal staining solution (20 mmol/L $K_3Fe(CN)_6$, 20 mmol/L $K_4Fe(CN)_6$, 2 mmol/L $MgCl_2$, in PBS; just before use, add 20 µL/mL of a 50 mg/mL 5-bromo-4-chloro-3-indolyl-β-D-galactopyranoside solution) to the cells from step 14 and incubate at 37°C for 4 to 12 hours in the dark. Wash the cells three times with PBS to stop the staining reaction, count the blue cells under an inverted light microscope, and determine the vector titer in tu/mL.

15c. For immunocytochemical detection of the transgene product, add 250 µL of GST solution (2% goat serum and 0.2% Triton® X-100 in PBS) to the cells from step 14 and incubate for 30 minutes at room temperature to block nonspecific binding and to permeabilize the cell membrane. Replace the blocking solution with the primary antibody (diluted in 250 µL GST solution) and incubate overnight at 4°C. Wash the cells three times with PBS, add the conjugated secondary antibody (diluted in GST solution), and incubate for 4 hours at room temperature. Wash the cells three times with PBS and develop, depending on the antibody conjugation, according to an appropriate visualization protocol. Count the positive cells under an inverted light microscope and determine the vector titer in tu/mL.

ACKNOWLEDGMENTS

The authors thank Tomotsugu Ichikawa for collaborative insights, and Martin Schwyzer and Phung Lang for critical readings of the manuscript. Supported by NIH NCI P01CA69246 (Y.S.), and the Swiss National Science Foundation (C.F.).

REFERENCES

1. **Abe, A., A. Miyanohara, and T. Friedman.** 1998. Enhanced gene transfer with fusogenic liposomes containing vesicular stomatitis virus G glycoprotein. J. Virol. *72*:6159-6163.
2. **Aboody-Guterman, K.S., P.A. Pechan, N.G. Rainov, M. Sena-Esteves, A. Jacobs, E.Y. Snyder, P. Wild, E. Schraner, K. Tobler, X.O. Breakefield, and C. Fraefel.** 1997. Green fluorescent protein as a reporter for retrovirus and helper virus-free HSV-1 amplicon vector-mediated gene transfer into neural cells in culture and *in vivo*. NeuroReport *8*:3801-3808.
3. **Balague, C., M. Kalla, and W.W. Zhang.** 1997. Adeno-associated virus Rep78 protein and terminal repeats enhance integration of DNA sequences into the cellular genome. J. Virol. *71*:3299-3306.
4. **Barsoum, J., R. Brown, M. McKee, and F.M. Boyce.** 1997. Efficient transduction of mammalian cells by a

recombinant baculovirus having the vesicular stomatitis virus G glycoprotein. Hum. Gene Ther. *8*:2011-2018.

5. **Bataille, D. and A.L. Epstein.** 1997. Equimolar generation of the four possible arrangements of adjacent L components in herpes simplex virus type 1 replicative intermediates. J. Virol. *71*:7736-7743.

6. **Breakefield, X.O., C.M. Kramm, E.A. Chiocca, and P.A. Pechan.** 1995. Herpes simplex virus vectors for tumor therapy, p. 41-56. *In* R.E. Sobol and K.J. Scanlon (Eds.), The Internet Book of Gene Therapy: Cancer Gene Therapeutics. Appleton and Lange, Stamford, CT.

7. **Burns, J.C., T. Friedmann, W. Driever, M. Burrascano, and J.K. Yee.** 1993. Vesicular stomatitis virus G glycoprotein pseudotyped retroviral vectors: concentration to very high titer and efficient gene transfer into mammalian and non-mammalian cells. Proc. Natl. Acad. Sci. USA *90*:8033-8037.

8. **Cai, W., B. Gu, and S. Person.** 1988. Role of glycoprotein B of herpes simplex virus type 1 in viral entry and cell fusion. J. Virol. *62*:2596-2604.

9. **Campadelli-Fiume, G., D. Stirpe, A. Boscaro, E. Avitabile, L. Foa-Tomasi, D. Barker, and B. Roizman.** 1990. Glycoprotein C dependent attachment of herpes simplex virus to susceptible cells leading to productive infection. Virology *178*:213-222.

10. **Chou, J., E.R. Kern, R.J. Whitley, and B. Roizman.** 1990. Mapping of herpes simplex virus 1 neurovirulence to $\gamma_1$34.5, a gene nonessential for growth in cell culture. Science *252*:1262-1266.

11. **Conway, J.E., S. Zolothukin, N. Muzycka, G.S. Hayward, and B.J. Byrne.** 1997. Recombinant adeno-associated virus type 2 replication and packaging is entirely supported by a herpes simplex virus type 1 amplicon expressing Rep and Cap. J. Virol. *71*:8780-8789.

12. **Cunningham, C. and A.J. Davison.** 1993. A cosmid-based system for constructing mutants of herpes simplex virus type 1. Virology *197*:116-124.

13. **Davidson, A.J. and N.M. Wilkie.** 1981. Nucleotide sequence of the joint between the L and S segments of herpes simplex virus type 1 and 2. J. Gen. Virol. *55*:315-331.

14. **Deiss, L.P., J. Chou, and N. Frenkel.** 1986. Functional domain within the a sequence involved in the cleavage-packaging of herpes simplex virus DNA. J. Virol. *59*:605-618.

15. **Delecluse, H.-J., T. Hilsendegen, D. Pich, R. Zeidler, and W. Hammerschmidt.** 1998. Propagation and recovery of intact, infectious Epstein-Barr virus from prokaryotic to human cells. Proc. Natl. Acad. Sci. USA *95*:8245-8250.

16. **Desai, P.G., P.A. Schaffer, and A.C. Minson.** 1988. Excretion of non-infectious virus particles lacking glycoprotein H by a temperature-sensitive mutant of herpes simplex virus type 1: evidence that gH is essential for virion infectivity. J. Gen. Virol. *69*:1147-1156.

17. **Feng, M., W.H. Jackson, Jr., C.K. Goldman, C. Rancourt, M. Wang, S.K. Dusing, G. Siegal, and D.T. Curiel.** 1997. Stable in vivo gene transduction via a novel adenoviral/retroviral chimeric vector. Nat. Biotechnol. *15*:866-870.

18. **Fisher, K.J., W.M. Kelley, J.F. Burda, and J.M. Wilson.** 1996. A novel adenovirus-adeno-associated virus hybrid vector that displays efficient rescue and delivery of the AAV genome. Hum. Gene Ther. *7*:2079-2987.

19. **Flotte, T.R., S.A. Afione, and P.L. Zeitlin.** 1994. Adeno-associated virus vector gene expression occurs in nondividing cells in the absence of vector DNA integration. Am. J. Respir. Cell Mol. Biol. *11*:517-521.

20. **Fraefel, C., X.O. Breakefield, and D.R. Jacoby.** 1998. HSV-1 amplicon, p. 63-82. *In* E.A. Chiocca and X.O. Breakefield (Eds.), Gene Therapy for neurological disorders and brain tumors. Humana Press, Totowa, NJ.

21. **Fraefel, C., D.R. Jacoby, C. Lage, H. Hilderbrand, J.Y. Chou, F.W. Alt, X.O. Breakefield, and J.A. Majzoub.** 1997. Efficient gene transfer into hepatocytes mediated by helper virus-free HSV/AAV hybrid vectors. Mol. Med. *3*:813-825.

22. **Fraefel, C., S. Song, F. Lim, P. Lang, L. Yu, Y. Wang, P. Wild, and A.I. Geller.** 1996. Helper virus-free transfer of herpes simplex virus type 1 plasmid vectors into neural cells. J. Virol. *70*:7190-7197.

23. **Frenkel, N.** 1981. Defective interfering herpesviruses, p. 99-120. *In* A.J. Nahmias, W.R. Dowdle and R.S. Schinazy (Eds.), The Human Herpesviruses—An Interdisciplinary Prospective. Elsevier/North-Holland, New York, NY.

24. **Frenkel, N., H. Locker, B. Batterson, G.S. Hayward, and B. Roizman.** 1976. Anatomy of herpes simplex virus. VI. Defective DNA originates from the S component. J. Virol. *20*:527-531.

25. **Gahn, T.A. and C.L. Schildkraut.** 1989. The Epstein-Barr virus origin of plasmid replication, oriP, contains both the initiation and termination sites of DNA replication. Cell *58*:527-535.

26. **Geller, A.I. and X.O. Breakefield.** 1988. A defective HSV-1 vector expresses *Escherichia coli* beta-galactosidase in cultured peripheral neurons. Science *241*:1667-1669.

27. **Geller, A.I., K. Keyomarski, J. Bryan, and A.B. Pardee.** 1990. An efficient deletion mutant packaging system for defective HSV-1 vectors; potential applications to neuronal physiology and human gene therapy. Proc. Natl. Acad. Sci. USA *87*:8950-8954.

28. **Geraghty, R.J., C. Krummenacher, G.H. Cohen, R.J. Eisenberg, and P.G. Spear.** 1998. Entry of alphaherpesviruses mediated by poliovirus receptor-related protein 1 and poliovirus receptor. Science *280*:1618-1620.

29. **Gilbert, W. and D. Dressler.** 1968 DNA replication: the rolling-circle model. Cold Spring Harbor Symp. Quant. Biol. *33*:473-484.

30. **Glorioso, J.C., N.A. DeLuca, and D.J. Fink.** 1995. Development and application of herpes simplex virus vectors for human gene therapy. Annu. Rev. Microbiol. *49*:675-710.

31. **Herold, B.C., D. WuDunn, N. Soltys, and P.G. Spear.** 1991. Glycoprotein C of herpes simplex virus type 1 plays a principal role in the adsorption of virus to cells and in infectivity. J. Virol. *65*:1090-1098.

32. **Ho, D.Y.** 1994. Amplicon-based herpes simplex virus vectors. Meth. Cell. Biol. *43*:191-210.

33. **Ho, D.Y., S.L. Fink, M.S. Lawrence, T.J. Meier, T.C. Saydam, R. Dash, and R.M. Sapolski.** 1995. Herpes simplex virus vector system: analysis of its in vivo and in vitro cytopathic effects. J. Neurosci. Methods *57*:205-215.
34. **Jacob, R.J. and B. Roizman.** 1977. Anatomy of herpes simplex virus DNA. VIII. Properties of the replicating DNA. J. Virol. *23*:394-411.
35. **Johnson, P.A., A. Miyanohara, F. Levine, T. Cahill, and T. Friedman.** 1992. Cytotoxicity of a replication-defective mutant of herpes simplex virus type 1. J. Virol. *66*:2952-2965.
36. **Johnson, P.A., K. Yoshida, F.H. Gage, and T. Friedman.** 1992. Effects of gene transfer into cultured CNS neurons with a replication-defective herpes simplex virus type 1 vector. Mol. Brain. Res. *12*:95-102.
37. **Johnston, K.M., D. Jacoby, P. Pechan, C. Fraefel, P. Borghesani, D. Schuback, R.J. Dunn, F.I. Smith, and X.O. Breakefield.** 1997. HSV/AAV hybrid amplicon vectors extend transgene expression in human glioma cells. Hum. Gene Ther. *8*:359-370.
38. **Kaneda, Y., K. Iwai, and T. Uchida.** 1989. Increased expression of DNA cointroduced with nuclear protein in adult rat liver. Science *243*:375-378.
39. **Kotin, R.M., R.M. Linden, and K.I. Berns.** 1992. Characterization of a preferred site on human chromosome 19q for integration of adeno-associated virus DNA by non-homologous recombination. EMBO J. *11*:5071-5078.
40. **Kwong, A.D. and N. Frenkel.** 1985. The herpes simplex virus amplicon IV. Efficient expression of a chimeric chicken ovalbumin gene amplified within defective virus genome. Virology *142*:421-425.
41. **Kwong, A.D. and N. Frenkel.** 1989. The herpes simplex virus virion host shutoff function. J. Virol. *63*:4834-4839.
42. **Kwong, A.D. and N. Frenkel.** 1995. Biology of herpes simplex virus (HSV) defective viruses and development of the amplicon system. p. 25-42. *In* M.G. Kaplitt and A.D. Loewy (Eds.), Viral vectors: Gene therapy and neuroscience applications. Academic Press, New York.
43. **Laquerre, S., R. Argnani, D.B. Anderson, S. Zucchini, R. Manservigi, and J.C. Glorioso.** 1998. Heparan sulfate proteoglycan binding by herpes simplex virus type 1 glycoproteins B and C, which differ in their contributions to virus attachment, penetration, and cell-to-cell spread. J. Virol. *72*:6119-6130.
44. **Ligas, M.W. and D.C. Johnson.** 1988. A herpes simplex virus mutant in which glycoprotein D sequences are replaced by β-galactosidase sequences binds to but is unable to penetrate into cells. J. Virol. *62*:1486-1494.
45. **Lim, F., D. Hartley, P. Starr, P. Long, S. Song, L. Yu, Y. Wang, and A.I. Geller.** 1996. Generation of high-titer defective HSV-1 vectors using an IE2 deletion mutant and quantitative study of expression in cultured cortical cells. BioTechniques *20*:458-469.
46. **Linden, R.M., P. Ward, C. Giraud, E. Wincour, and K.I. Berns.** 1996. Site-specific integration by adeno-associated virus. Proc. Natl. Acad. Sci. USA *93*:11288-11294.
47. **Locker, H. and N. Frenkel.** 1979. Structure and origin of defective genomes contained in serially passaged herpes simplex virus type 1 (Justin). J. Virol. *29*:1065-1077.
48. **Locker, H., N. Frenkel, and I. Halliburton.** 1982. Structure and expression of class II defective herpes simplex virus genomes encoding infected cell polypeptide number 8. J. Virol. *43*:574-593.
49. **Lockshon, D. and D.A. Galloway.** 1986. Cloning and characterization of OriL2, a large palindromic replication origin of herpes simplex virus type 2. J. Virol. *58*:513-521.
50. **Longnecker, R., B. Roizman, and B. Meignier.** 1988. Herpes simplex viruses as vectors: properties of a prototype vaccine strain suitable for use as a vector, p.68. *In* Y. Gluzman, S.H. Hughes (Eds.), Viral Vectors. CSH Laboratory Press, Cold Spring Harbor, NY.
51. **Luckow, V.A., S.C. Lee, G.F. Barry, and P.O. Olins.** 1993. Efficient generation of infectious recombinat baculoviruses by site-specific transposone-mediated insertion of foreign genes into a baculovirus genome propagated in *Escherichia coli*. J. Virol. *67*:4566-4579.
52. **Lukonis, C.J., J. Burkham, and S.K. Weller.** 1997. Herpes simplex virus type 1 prereplicative sites are a heterologous population: only a subset are likely to be precursors to replication compartments. J. Virol. *71*:4771-4781.
53. **Lycke, E., M. Johansson, B. Svennerholm, and U. Lindahl.** 1991. Binding of herpes simplex virus to cellular heparan sulfate, an initial step in the adsorption process. J. Gen. Virol. *72*:1131-1137.
54. **Martin, D.W. and P.C. Weber.** 1996. The *a* sequence is dispensable for isomerization of the herpes simplex virus type 1 genome. J. Virol. *70*:8801-8812.
55. **McCarthy, A.M., L. McMahan, and P.A. Schaffer.** 1989. Herpes simplex virus type 1 ICP27 deletion mutants exhibit altered patterns of transcription and are DNA deficient. J. Virol. *63*:18-27.
56. **McGeoch, D.J., M.A. Dalrymple, A.J. Davidson, A. Dolan, M.C. Frame, D. McNab, L.J. Perry, J.E. Scott, and P. Taylor.** 1988. The complete DNA sequence of the long unique region in the genome of herpes simplex virus type 1. J. Gen Virol. *69*:1531-1574.
57. **Messerle, M., I. Crnkovic, W. Hammerschmidt, H. Ziegler, and U.H. Koszinowski.** 1997. Cloning and mutagenesis of a herpesvirus genome as an infectious bacterial artificial chromosome. Proc. Natl. Acad. Sci. USA *94*:14759-14763.
58. **Middleton, T. and B. Sugden.** 1994. Retention of plasmid DNA in mammalian cells is enhanced by binding of the Epstein-Barr virus replication protein EBNA1. J. Virol. *68*:4067-4071.
59. **Montgomery, R.I., M.S. Warner, B.J. Lum, and P.G. Spear.** 1996. Herpes simplex virus-1 entry into cells mediated by a novel member of the TNF/NGF receptor family. Cell *87*:427-436.
60. **Naldini, L., U. Blomer, P. Gallay, D. Ory, R. Mulligan, F.H. Gage, I.M. Verma, and D. Trono.** 1996. In vivo delivery and stable transduction of nondividing cells by a lentiviral vector. Science *272*:263-267.
61. **O'Connor, M., M. Peifer, and W. Bender.** 1989. Construction of large DNA segments in *Escherichia coli*. Science *244*:1307-1312.

62.**Oroskar, A. and G.S. Read.** 1989. Control of mRNA stability by the virion host shutoff function of herpes simplex virus. J. Virol. *63*:1897-1906.

63.**Palombo, F., A. Monciotti, A. Recchia, R. Cortese, G. Ciliberto, and N. LaMonica.** 1998. Site-specific integration in mammalian cells mediated by a new hybrid baculovirus-adeno-associated virus vector. J. Virol. *72*:5025-5034.

64.**Perry, L.J. and D.J. McGeoch.** 1988. The DNA sequences of the long repeat region and adjoining parts of the long unique region in the genome of herpes simplex virus type 1. J. Gen. Virol. *69*:2831-2864.

65.**Polvino-Bodnar, M., P.K. Orberg, and P.A. Schaffer.** 1987. Herpes simplex virus type 1 ori_L is not required for virus replication or for the establishment and reactivation of latent infection in mice. J. Virol. *61*:3528-3535.

66.**Preston, C.** 1979. Control of herpes simplex virus type 1 mRNA synthesis in cells infected with wild type or the temperature sensitive mutant tsK. J. Virol. *29*:257-284.

67.**Roizman, B.** 1996. The function of herpes simplex virus genes: a primer for genetic engineering of novel vectors. Proc. Natl. Acad. Sci. USA *93*:11307-11312.

68.**Roizman, B. and A.E. Sears.** 1996. Herpes simplex viruses and their replication, p. 2231-2295. *In* B.N. Fields, D.M. Knipe, and P.M. Howley (Eds.), Fields Virology, Vol. 2 (3rd ed). Lippincott-Raven, Philadelphia.

69.**Russel, D.W., A.D. Miller, and I.E. Alexander.** 1994. Adeno-associated virus vectors preferentially transduce cells in S phase. Proc. Natl. Acad. Sci. USA *91*:8915-8919.

70.**Saeki, Y., T. Ichikawa, A. Saeki, E.A. Chiocca, K. Tobler, M. Ackermann, X.O. Breakefield, and C. Fraefel.** 1998. Herpes simplex virus type 1 DNA amplified as bacterial artificial chromosome in *Escherichia coli*: rescue of replication-competent virus progeny and packaging of amplicon vectors. Hum. Gene Ther. *9*:2787-2794.

71.**Saeki, Y., N. Matsumoto, Y. Nakano, M. Mori, K. Awai, and Y. Kaneda.** 1997. Development and characterization of cationic liposomes conjugated with HVJ (Sendai virus): reciprocal effect of cationic lipid for in vitro and in vivo gene transfer. Hum. Gene Ther. *8*:2133-2141.

72.**Savard, N., F.-L. Cosset, and A.L. Epstein.** 1997. Defective herpes simplex virus type 1 vectors harboring gag, pol, and env genes can be used to rescue defective retrovirus vectors. J. Virol. *71*:4111-4117.

73.**Severini, A., D.G. Scraba, and D.L. Tyrrell.** 1996. Branched structures in the intracellular DNA of herpes simplex virus type 1. J. Virol. *70*:3169-3175.

74.**Shieh, M.-T., D. WuDunn, R.I. Montgomery, J.D. Esko, and P.G. Spear.** 1992. Cell surface receptors for herpes simplex virus are heparan sulfate proteoglycans. J. Cell Biol. *116*:1273-1281.

75.**Shizuya, H., B. Birren, U.-J. Kim, V. Mancino, T. Slepak, Y. Tachiiri, and M. Simon.** 1992. Cloning and stable maintenance of 300-kilobase-pair fragments of human DNA in *Escherichia coli* using an F-fator-based vector. Proc. Natl. Acad. Sci. USA *89*:8794-8797.

76.**Smith, I.L., M.A. Hardwicke, and R.M. Sandri-Goldin.** 1992. Evidence that the herpes simplex virus immediate early protein ICP27 acts post-transcriptionally during infection to regulate gene expression. Virology *186*:74-86.

77.**Spaete, R.R. and N. Frenkel.** 1982. The herpes simplex virus amplicon: a new eucaryotic defective-virus cloning-amplifying vector. Cell *30*:295-304.

78.**Stavropoulos, T.A. and C.A. Strathdee.** 1998. An enhanced packaging system for helper-dependent herpes simplex virus vectors. J. Virol. *72*:7137-7143.

79.**Stern, S., M. Tanaka, and W. Herr.** 1989. The oct-1 homeo domain directs formation of multiprotein-DNA complex with the HSV transactivator VP16. Nature *341*:624-630.

80.**Steven, A.C. and P.G. Spear.** 1997. Herpesvirus capsid assembly and envelopment, p. 312-351. *In* W. Chiu, R. Burnett, and R. Garcea (Eds.), Structural Biology of Viruses. Oxford University Press, New York.

81.**Stevens, J.G.** 1975. Latent herpes simplex virus and the nervous system. Curr. Top. Microbiol. Immunol. *70*:31-50.

82.**Stow, N.D.** 1982. Localization of an origin of DNA replication in the TRs/IRs repeated region of the herpes simplex virus type 1 genome. EMBO J. *1*:863-867.

83.**Stow, N. and E.C. McMonagle.** 1983. Characterization of TRs/IRs origin of DNA replication of herpes simplex virus type 1. Virology *130*:427-438.

84.**Sun, T.-Q., E. Livanos, and J.-M. Vos.** 1996. Engineering a mini-herpesvirus as a general strategy to transduce up to 180 kb of functional self-replicating human mini-chromosomes. Gene Ther. *3*:1081-1088.

85.**Surosky, R.T., M. Urabe, S.G. Godwin, S.A. McQuiston, G.J. Kurtzman, K. Ozawa, and G. Natsoulis.** 1997. Adeno-associated virus rep proteins target DNA sequences to a unique locus in the human genome. J. Virol. *71*:7951-7957.

86.**Terry-Allison, T., R.I. Montgomery, C. Whitbeck, R. Xu, G.H. Cohen, R.J. Eisenberg, and P.G. Spear.** 1998. HveA (herpesvirus entry mediator A), a coreceptor for herpes simplex virus entry, also participates in virus-induced cell fusion. J. Virol. *72*:5802-5810.

87.**Umene, K.** 1993. Herpes simplex virus type 1 variant *a* sequence generated by recombination and breakage of the *a* sequence in defined regions, including the one involved in recombination. J. Virol. *67*:5685-5691.

88.**Vlazny, D.A., A. Kwong, and N. Frenkel.** 1982. Site specific cleavage-packaging of herpes simplex virus DNA and the selective maturation of nucleocapsids containing full-length viral DNA. Proc. Natl. Acad. Sci. USA *79*:1423-1427.

89.**Wang, S., S. Di, W.-B. Young, C. Jacobson, and C. Link, Jr.** 1997. A novel herpesvirus amplicon system for in vivo gene delivery. Gene Ther. *4*:1132-1141.

90.**Wang, S. and J.-M. Vos.** 1996. A hybrid herpes virus infectious vector based on Epstein-Barr virus and herpes simplex virus type 1 for gene transfer into human cells in vitro and in vivo. J. Virol. *70*:8422-8430.

91.**Weller, S.K., A. Spadaro, J.E. Schaffer, A.W. Murray, A.M. Maxam, and P.A. Schaffer.** 1985. Cloning, sequencing, and functional analysis of oriL, a herpes simplex virus type 1 origin of DNA synthesis. Mol. Cell. Biol. *5*:930-942.

92. **Wohlgemuth, J.G., S.H. Kang, G.H. Bulboaca, K.A. Nawotka, and M.P. Calos.** 1996. Long-term gene expression from autonomously replicating vectors in mammalian cells. Gene Ther. *3*:503-512.

93. **Wong, S.W. and P.A. Schaffer.** 1991. Elements in the transcriptional regulatory region flanking herpes simplex virus type 1 ori_S stimulate origin function. J. Virol. *65*:2601-2611.

94. **Wood, M.J.A., A.P. Byrnes, D.W. Pfaff, S.D. Rabkin, and H.M. Charlton.** 1994. Inflammatory effects of gene transfer into the CNS with defective HSV-1 vectors. Gene Ther. *1*:283-291.

95. **Yang, X.W., P. Model, and N. Heintz.** 1997. Homologous recombination based modification in *Escherichia coli* and germline transmisssion in transgenic mice of a bacterial artificial chromosome. Nat. Biotechnol. *15*:859-865.

96. **Yates, J., N. Warren, D. Reisman, and B. Sugden.** 1984. A cis-acting element from the Epstein-Barr viral genome that permits stable replication of recombinant plasmids in latently infected cells. Proc. Natl. Acad. Sci. USA *81*:3806-3810.

97. **Zhang, X., S. Efstathiou, and A. Simmons.** 1994. Identification of novel herpes simplex virus replicative intermediates by field inversion gel electrophoresis: implications for viral DNA amplification strategies. Virology *202*:530-539.

14 Stable Gene Transfer Using Hybrid Herpes Simplex-Epstein-Barr Virus Amplicon Vectors

Tom A. Stavropoulos and Craig A. Strathdee
Gene Therapy and Molecular Virology Group, The John P. Robarts Research Institute, and Department of Microbiology and Immunology, University of Western Ontario, London, ON, Canada

INTRODUCTION

Herpesviruses have many favorable properties that allow them to be exploited as gene delivery and expression platforms for human gene therapy. For example, they are able to establish persistent and in many cases lifelong infection of their hosts even in the face of a vigorous and prolonged antiviral immune response. The viral genome is maintained as an episome in infected cells, such that gene expression is independent of position effects, and there is no risk of insertional mutagenesis of the host cell genome. Finally, although the viral genome itself is very complex and difficult to manipulate, the replication strategy utilized by herpesviruses is fully compatible with the use of conventional plasmid vectors. Our laboratory is investigating the utility of two herpesviruses to serve as gene delivery and expression platforms, the herpes simplex virus (HSV) and the Epstein-Barr virus (EBV).

HELPER-DEPENDENT HSV VECTORS

HSV is a neurotropic herpesvirus that can establish lifelong infections in its human host through latent maintenance in the ganglia of sensory neurons. The virus is relatively well characterized, with a genome of approximately 155 kb that is maintained as a concatamerized circular or linear episome in infected cells (37,38). Because of their wide host range, efficient infection, long-term persistence, capacity to accommodate large amounts of foreign DNA, and ability to deliver genes to postmitotic cell types, considerable effort has been expended to develop gene transfer vectors that can exploit the natural biology of HSV (4,13,25). Two different types of HSV-based vectors have been developed, which can be defined as helper-independent and helper-dependent (HD) according to their packaging requirements. Inasmuch as both types are designed to take advantage of

Viral Vectors: Basic Science and Gene Therapy
Edited by A. Cid-Arregui and A. García-Carrancá

HSV's ability to enter into a latent life cycle in neuronal cell types, an additional advantage is that HSV is very efficient at infecting most other mammalian cell types regardless of their mitotic state or degree of differentiation. Although the later stages of viral entry into the cell are only beginning to be defined (24), the initial stages involve a fairly nonspecific binding to various cell surface glycosaminoglycans such as heparan sulfate (1,6,51). There is no mammalian cell line that HSV cannot bind to (54)—a fact that reflects the ubiquitous nature of this receptor and that holds great promise for the future application of HSV vectors in diverse gene delivery and expression strategies.

Helper-independent HSV vectors contain the complete viral genome but have deletions in one or more essential viral genes and can be used to express three or more cDNAs by replacing other nonessential viral genes (5,23). Helper-independent HSV vectors thus have a replication-defective viral phenotype and can establish a productive infection only in complementing cell lines that express the corresponding deleted viral gene (8). The major problem associated with this class of vectors in general is that there is leaky or inappropriate expression of some of the immediate early and early HSV genes in noncomplementing cell types; this phenomenon ultimately results in the death of the infected cell even though no viral replication has occurred (18–20,40,41,46). Since the focus of this chapter is on HD HSV vectors, helper-independent vectors will not be discussed further.

HD HSV vectors (commonly known as HSV amplicons) contain only the *cis* elements required for HSV replication and packaging (11,43). Because the vector backbone is only a small fraction of the size of the HSV genome, usually less than 10%, HSV amplicons have the potential to express a large number of genes or cDNAs. Moreover, since they contain no viral-encoded genes, these vectors have the potential to provide long-term gene expression in transduced cells by obviating the latency-associated shutdown of HSV gene expression. Therefore, they are well positioned to take full advantage of the ability of HSV to infect virtually every type of cell (5,11,15,43).

The prototypical HSV amplicon vector is pHSVlac (11) (Figure 1). There are two functional components with respect to the ability of this vector to be packaged into infectious viral particles: *(i)* the *a* sequences from the HSV genome, which are repetitive sequences that contain the conserved *pac1* and *pac2* packaging elements; and *(ii)* an origin of replication, ori_S, derived from the U_S region of the HSV genome. Efficient recognition of the ori_S element by the HSV replication machinery is the critical factor that limits the ability of the amplicon vector to compete with the HSV genome during growth of the virus (57). The ori_S is colinear with the *IE4* promoter element in the viral genome, and this configuration has been preserved in the pHSVlac vector to maximize its packaging efficiency. The *IE4* promoter is normally transactivated by VP16 during growth of the virus, and because VP16 is a component of the virion that is introduced into infected cells along with the amplicon vector, this promoter can provide relatively high levels of cDNA expression (40). VP16 activity is lost after a few days, however, and in its absence, the *IE4* promoter is relatively weak. The *IE4* promoter is thus not well suited for providing consistently high levels of cDNA expression. A number of HSV amplicon vectors have been developed to facilitate the use of alternative promoter elements to control cDNA expression. One such vector developed by our laboratory, designated pHEX, is designed such that it conforms to a fully modular format (Figure 1). Specifically, the pHEX vector has a configuration similar to that of pHSVlac but is designed

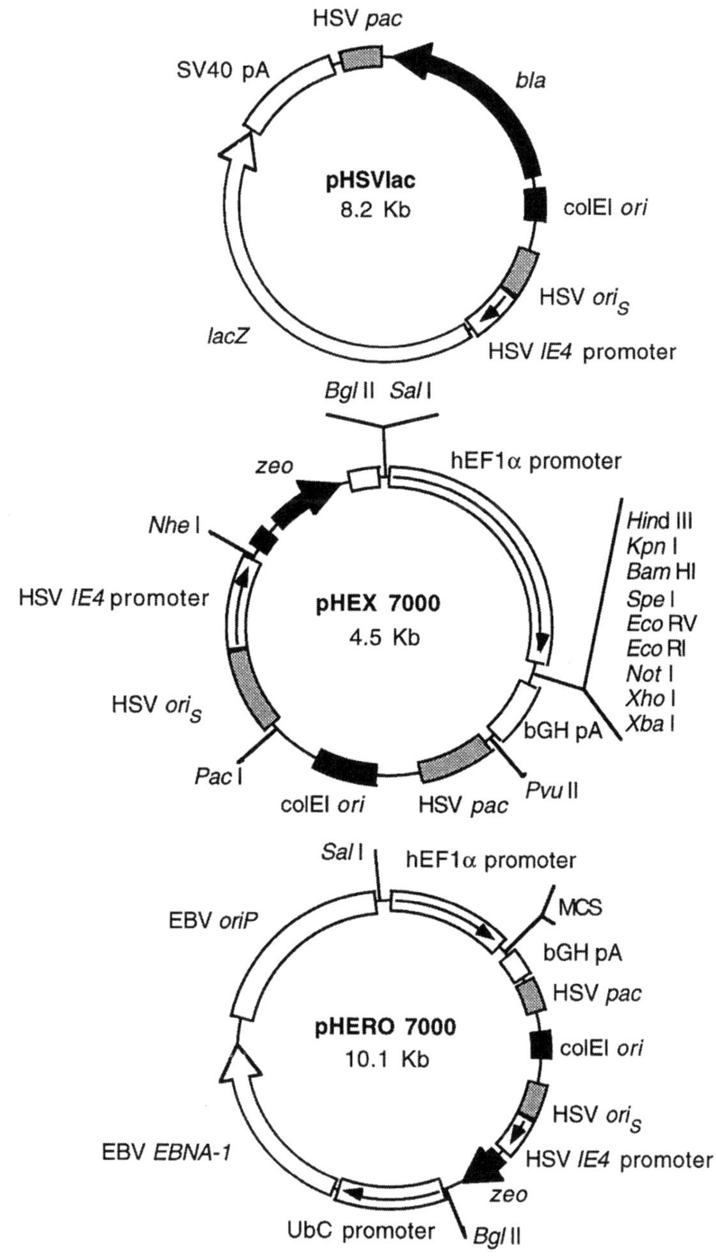

Figure 1. Configuration of the pHSVlac, pHEX, and pHERO vectors. The vectors contain identical *pac*, *ori*$_S$, and *IE4* promoter elements derived from the HSV-1 genome. Note that the latter two are colinear to maximize the replicative efficiency of the vectors during packaging. The *IE4* promoter drives expression of the bacterial β-galactosidase reporter gene in pHSVlac and the *zeo* selectable marker in pHEX and pHERO. The pHEX and pHERO vectors utilize the human elongation factor 1α promoter and bovine growth hormone polyadenylation signal to control cDNA expression. The pHERO vector contains the EBV episomal replicon, consisting of the *oriP* element and *EBNA-1* gene driven by the human ubiquitin C promoter, but is otherwise identical to the pHEX vector.

to facilitate easy subcloning of cDNAs to be expressed as well as to incorporate alternative promoter elements.

TRANSFECTION-BASED PACKAGING SYSTEMS FOR HSV AMPLICONS

The first practical system for packaging HSV amplicons into infectious viral particles utilized an *ICP4*-deleted, replication-defective HSV as a helper virus to provide the necessary viral replication and packaging functions (12). In this system, the amplicon vector is transfected into an *ICP4*-complementing cell line, which is then superinfected with the helper virus and the resulting infectious particles harvested following growth of the virus. Although this approach can yield titers of recombinant virus approaching 1×10^9 plaque-forming units (pfu)/mL, unless there is a mechanism in place to provide for the selective replication and packaging of the amplicon vector over that of the helper virus, the transducing lysates produced are heavily contaminated with helper virus (12,28). Even when selective replication and packaging of the amplicon vector is provided, the helper virus represents 10% to 25% of the total yield of virus produced (32,62). Similar to the situation with helper-independent vectors, this helper virus contamination leads to significant delivery-associated cytotoxicity in transduced cells. Although this problem can potentially be eliminated through the use of less cytotoxic forms of HSV to package amplicon vectors, such mutants typically grow much more slowly, and the yield of packaged amplicon is correspondingly lower.

A variety of second-generation packaging systems have been developed for HD HSV vectors that are able to yield transducing lysates free of helper virus contamination, thus making these the packaging systems of choice for HSV amplicons. The first successful system was centered around a set of five overlapping HSV-1 cosmid clones that together encode the complete wild-type (*wt*) viral genome (10). By specifically deleting the *pac* element in the repetitive *a* sequences of two of the cosmids, the cosmid set is rendered packaging defective, and when co-transfected into mammalian cells along with an HSV amplicon vector, only the latter is packaged into mature viral particles.

Enhanced versions of the cosmid-based packaging system have now been developed for use with HD HSV vectors. Although the specific details differ for each system, they are all based on the use of a single bacmid clone (bacterial artificial chromosome [BAC]-based plasmid) that contains the entire HSV-1 genome from which the *pac* elements are specifically deleted. The infectious bacmid that was generated by our laboratory, designated pBAC-V2, has an insertional mutation in the *vhs* gene (44). Similar bacmid constructs have been described that have insertional mutations in the *tk* and U_L46/U_L47 genes, respectively (16,39). Since these genes are all nonessential for replication in cultured cells, the respective bacmid clones are thus replication competent, but packaging defective, and can be used to provide all of the helper functions required to package HSV amplicon vectors (Figure 2). Although there is a modest increase in packaged amplicon compared with that generated using the cosmid-based system (16,39,44), the major advantage gained through the bacmid-based system is its inherent simplicity. Specifically, the bacmid system requires the propagation of a single bacterial clone, and following DNA purification, this can be transfected as supercoiled DNA. The cosmid system requires the propagation of five bacterial clones, some of which are unstable, and each must be linearized by restriction enzyme digestion and then repurified before use.

In either the cosmid- or bacmid-based system, the yield of infectious particles produced remains at least two orders of magnitude lower than that obtained using intact HSV particles. The inability to grow to similar titers likely stems from limitations inherent to transfection and not the sequence of the viral genome. Specifically, HSV normally establishes three distinct consecutive waves of gene expression during lytic replication, the first of which is initiated by accessory tegument proteins associated with the HSV particle (37,38). Since the transfection-based system uses only naked DNA and thus lacks these proteins, it is innately inefficient at initiating the first wave of gene expression in the viral lytic cycle. Additionally, transfection-based methods can never ensure every cell receives at least one helper virus genome, which limits the total number of cells

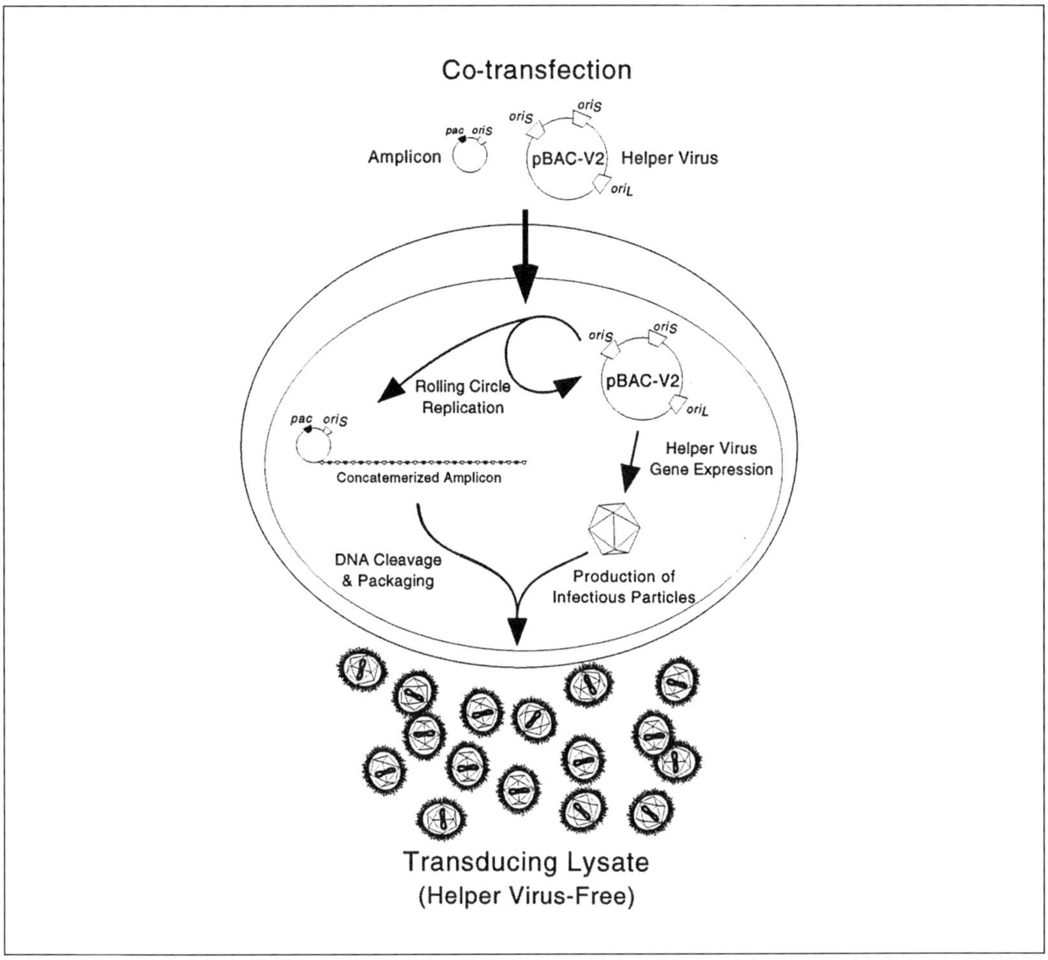

Figure 2. Transfection-based packaging of amplicon vectors using an HSV bacmid. The packaging strategy is based on the ability of an amplicon vector to be incorporated into infectious viral particles using a replication-competent but packaging-deficient HSV helper virus. The cloned HSV helper virus genome in the pBAC-V2 bacmid contains one ori_L and two ori_S elements, but no *pac* elements, whereas the amplicon vector contains a single ori_S and *pac* element. Following co-transfection into HEK293T cells, the helper virus genome initiates its program of gene expression to produce infectious viral particles. The helper virus genome and the amplicon vector are both replicated in this process. Only the latter, however, is packaged into maturing virus particles. Barring any recombination events between the pBAC-V2 bacmid and the amplicon vector, the resulting transducing lysates are free of contaminating helper virus.

undergoing lytic infection and reduces the corresponding titer of packaged amplicon vector. Our laboratory is currently investigating a variety of methods for increasing the efficiency of the transfection-based system, by using, for example, improved transfection methodologies that are compatible with large bacmid constructs, and by including additional plasmid constructs in the transfection that express some of the necessary HSV tegument proteins.

BACMID DNA PURIFICATION

The bacmid-based transfection system relies on the ability to generate high-quality DNA preparations to maximize transfection efficiencies. Because the construct involved is so large, special precautions must be observed to limit contamination with bacterial debris, especially lipopolysaccharide. It is recommended that a frozen glycerol stock of bacteria transformed with the pBAC-V2 bacmid construct be maintained and used for streaking out fresh plates for large-scale DNA preparations. Even though the cloned HSV genome is stable in the bacmid vector, this approach will minimize the outgrowth of rearranged clones. Because the pBAC-V2 construct is approximately 160 kb in size, to avoid shearing, the purified DNA must be pipetted with a wide-bore pipet tip and never mixed vigorously since we find that linear DNA is more than 100-fold less efficient at packaging than supercoiled DNA.

Since the virus genome is propagated as a bacmid at a copy number of 1 to 2 in bacteria, it is difficult to obtain large amounts of high-quality DNA for transfection. It is recommended that the DNA be streaked out on a fresh plate of pBAC-V2 from a frozen glycerol stock and grown overnight at 37°C on an LB plate supplemented with 12.5 µg/mL chloramphenicol plate overnight at 37°C. From this freshly streaked plate, 2.5 L of LB medium supplemented with 12.5 µg/mL chloramphenicol should be inoculated with a single colony of bacteria and grown overnight at 37°C with shaking. Our preferred method to isolate pBAC-V2 bacmid DNA is through the commercially available EndoFree™ Plasmid Mega Kit (Qiagen, Valencia, CA, USA). Although similar purification methods are marketed by other manufacturers, we are able to obtain consistent yields of high-quality DNA using this kit. Following isolation, the DNA pellet was resuspended in 1.5 mL of 10 mmol/L Tris-HCl (pH 8.5), quantitated using $OD_{260/280}$ and stored at -20°C at a concentration of 0.2 µg/µL in small aliquots until use to minimize degradation caused by successive freeze-thaw cycles. Although some user optimization is required in each laboratory (e.g., we have found that using elution buffer prewarmed to 65°C greatly increases the yield of DNA without any loss in quality), it is relatively simple to obtain yields of 300 to 450 µg from a 2.5-L culture.

HSV AMPLICON PACKAGING PROTOCOLS

It is recommended that individual laboratories develop their own optimized transfection protocols for use with the HSV bacmids, and, in fact, we evaluate new transfection methods on a continuing basis. The current method of choice in our laboratory utilizes the GenePORTER lipid (Gene Therapy Systems, San Diego, CA, USA). A protocol for one small-scale transfection/packaging reaction using this lipid is presented in Table 1. To upscale, either the number of transfections should be increased (multiple plates) and/or a larger culture dish used. Transfection conditions will have to reoptimized for the latter. Although packaging reactions can be car-

Table 1. Protocol for Packaging HSV Amplicon Vectors

1. Plate 293T cells in a P60 culture dish (60 mm) the day before so that the cells are 90% to 95% confluent at the time of transfection. Note that 293T cells require 400 μg/mL Geneticin® (Life Technologies, Rockville, MD, USA).
2. Pipet 1 mL of serum-free media into each of two 5-mL tubes.
3. With a wide-bore pipet tip, add 5 μg pBAC-V2 and 2 μg amplicon vector to one of the tubes. Avoid vigorous mixing or pipeting.
4. Add 35 μL GenePORTER™ lipid (Gene Therapy Systems, San Diego, CA, USA) to the second tube containing 1 mL serum-free media and mix.
5. Combine the two solutions by slowly adding the DNA to the GenePORTER lipid tube using a wide-bore pipet tip and while moving the tip around to avoid precipitating the DNA. Mix and let stand for 20 min at room temperature to allow for complex formation. If a precipitate forms, discard the tube and start over.
6. Remove media from cells and wash once with prewarmed serum-free media. Using a wide-bore tip, add the DNA/GenePORTER complexes to the cells evenly and swirl to distribute. Return to incubator for 3 to 5 h.
7. Add 2 mL of Dulbecco's modified Eagle medium (DMEM) containing 20% fetal bovine serum (FBS) to the culture to stop the transfection reaction, and then return to the incubator overnight. We also add N,N′-hexamethylene-bis-acetamide (HMBA) to a final concentration of 2 mmol/L to stimulate immediate early gene expression from pBAC-V2 at this point.
8. The next day, replace media with 5 mL DMEM plus 10% FBS and 2 mmol/L HMBA, then return it to incubator for a further 2 to 3 days (i.e., until cells show significant cytopathic effect).
9. Harvest packaged vector by scraping the cells into a 15-mL tube. Wash any residual cells from the plate with 5 mL cold phosphate-buffered saline (PBS). The lysate is kept on ice from this point forward.
10. Centrifuge at 1000× *g* at 4°C for 10 minutes, and aspirate all but approximately 500 μL of media.
11. Mix vigorously to resuspend cells, then lyse by sonication to release the infectious particles from the cellular membranes. Sonication should always be done on ice and for 4 cycles of 30 s on followed by 30 s off.
12. Centrifuge at 1000× *g* at 4°C for 10 minutes to pellet debris, and then collect supernatant. Repeat. Store supernatant at -70°C as a transducing lysate in suitable aliquots until use because repeated freeze-thaw cycles lower the titer of packaged amplicon. Depending on the vector being packaged, our typical yield for a single 60-mm culture dish ranges from 1×10^4 to 1×10^6 transducing units (tu).

ried out in any cell line permissive to HSV-1, it is important to choose one that can be highly transfected. It is possible to obtain approximately 90% transfection efficiencies using 293T cells, the packaging cell line of choice in our laboratory.

The results of a typical packaging reaction comparing the relative packaging efficiency of the pHEX and pHSVlac vectors are presented in Table 2. Despite the fact that the two vectors are the same relative size and contain identical ori_S and *pac* elements derived from the HSV genome, pHSVlac is packaged more efficiently than pHEX. A similar relationship has been observed between pHSVlac and other HSV

Table 2. Packaging Efficiency of HSV Amplicon Vectors

Vector Construct[a]	β-Gal+ Titer[b]	HSV Titer[c]
pHSVlac	2×10^5	<10
pHEX 7000β	8×10^4	<10
pHERO 7000β	2×10^4	<10

[a]All vectors express the bacterial β-galactosidase reporter gene.

[b]The β-gal+ titer was obtained by X-Gal histochemical staining two days after infecting Vero cells and represents the number of β-galactosidase tu/mL of lysate.

[c]The HSV titer was obtained by screening for plaque formation on Vero cells and represents the number of replication-competent infectious viral units/mL of lysate. This is the background reversion rate of the packaging-defective HSV helper virus caused by recombination with the amplicon vector.

amplicon vectors, and although the underlying basis for this is not fully understood, it is likely that the efficient function of the respective HSV elements is somewhat context specific. The reduction in packaged amplicon yield using pHEX is small, however, and this slight disadvantage is offset by the increased flexibility of the vector.

AMPLICON PREREPLICATION INCREASES PACKAGING EFFICIENCY

Various modifications to the amplicon vector can have a dramatic effect on packaging efficiency. In addition to the activity of the ori_S, a critical factor is the ability of the concatamerized vector to be incorporated into the HSV capsid, which normally accommodates approximately 150 kb of DNA between the *pac* elements. Variations of this length can lead to inefficient packaging because of the failure of the virus particle to mature. The size of the vector should therefore be as close to an even multiple of 150 kb as possible to optimally position the *pac* elements during packaging and maximize virus production. Furthermore, because the HSV bacmid packaging system does not select for replication of the amplicon vector, the yield of packaged amplicon vector is somewhat limited by ineffective competition with the helper virus for virus-specific replication factors. We have seen this with a number of our vectors and can compensate for this inefficiency by increasing the copy number of the amplicon vector relative to that of the helper virus, thereby ensuring that subsequent replication is skewed to favor the vector rather than the helper virus. The approach is to introduce the simian virus 40 (SV40) *ori* as an HSV-independent replicon in T-antigen-positive cells to afford such a competitive advantage to the amplicon and thus increase the overall yield of packaged vector particles (53).

The ability of prereplication to increase the yield of packaged vector is somewhat context specific and is dependent on sequences in the amplicon and/or a specific HSV amplicon. For example, the prereplication by the SV40 *ori* element may only be effective under those conditions in which replication of the HSV helper virus (and thus the HSV amplicon) are suboptimal, as is the case when a replication-defective or packaging-deficient HSV helper virus is used (58). In support of this view, we have incorporated the minimal SV40 *ori* element into other HSV amplicon vectors that our laboratory has engineered and have found that it is more effective in those vectors that are packaged relatively inefficiently in comparison to pHSVlac and can increase the yields of packaged vector by as much as tenfold in such cases (T.A. Stavropoulos and C.A. Strathdee, unpublished data).

EBV VECTORS

EBV is a lymphotropic human herpes virus that causes lifelong latent infections in humans and is endemic to the global population with over 95% of adults being seropositive (21,34). The virus has a very high tissue tropism and can only infect and replicate in two cell types. A productive lytic infection occurs in epithelial cells, whereas a latent infection occurs in B lymphocytes, which are transformed into dividing blast cells as a result (33). Lymphoblasts can proliferate indefinitely in culture and carry multiple copies of the approximately 170 kb EBV genome, which replicates episomally at low copy number and only once per cell cycle. The viral genes that control latent replication are clearly distinct from those contributing to cellular transformation, productive viral replication, and cell lysis. Specifically, the virus expresses 11 genes during latency, encompassing six nuclear antigens (*EBNA-1, -2, -3A, -3B, -3C,* and *-LP*), three membrane proteins (*LMP-1, -2A,* and *-2B*), and two small, nonpolyadenylated RNAs (*EBER-1* and *-2*) (29,42). Of these, only *EBNA-1* is required for episomal replication and maintenance of the viral genome during latency.

The *EBNA-1* gene encodes a multifunctional nuclear phosphoprotein that binds to *cis* sequences in the viral genome located in the *oriP* element (48,50). The EBNA-1/*oriP* system functions as the episomal origin of replication as well as a nuclear matrix attachment region for the EBV genome. It also acts as an enhancer element that facilitates *trans*-activation of other EBV latency-associated genes (3,7). Bacterial plasmids that contain *EBNA-1* and *oriP* are not limited to replication in lymphoblast or epithelial cells (49,59,61), and this observation has been exploited in the development of a variety of EBV-based vectors for cDNA and genomic DNA expression (14,52,56). For the most part, these vectors share common characteristics and contain all of the elements required for efficient selection and maintenance in vitro: *(i)* their replicative fidelity is quite high; *(ii)* they can support stable gene expression in both mitotic and postmitotic human cell types for extended periods; and *(iii)* they are only rarely associated with genome-altering integration events (56,60). Because the virus uses the EBNA-1/*oriP* system to persist indefinitely in vivo (30), this implies that EBV-based vectors will share similar characteristics, and their future clinical application to provide for long-term expression in gene-based therapies is very promising.

Unlike HSV, which does not express any viral genes during latency, EBV has developed a strategy of immunoevasion during latent infection that is well suited to future clinical application. Latently infected and growth-transformed B cells stimulate a cytotoxic T cell response towards eight of the nine EBV proteins, but not to EBNA-1 (35,36,45). The virus is able to restrict expression to *EBNA-1* in some latently infected cells, and this pattern of expression is not involved in growth transformation. Therefore, only EBV-transformed B cells that express immunogenic proteins are recognized and eliminated by the immune system (22). This strong selection for *EBNA-1*-restricted expression, coupled with the observation that a subpopulation of B cells expressing only *EBNA-1* can be detected in seropositive individuals, suggests that this is the cellular compartment in which EBV persists during latent infection without detection by the immune system. The mechanism underlying the poor antigenicity of EBNA-1 is now partially understood. The protein contains a unique glycine-alanine copolymer domain that composes approximately one-third of its length, which acts in *cis* to prevent processing of EBNA-1 by the proteosome and thus prevents presentation of any EBNA-1-derived peptides through the major histocompatibility complex (MHC) class I (2,26,27). As a result of this, infected cells that express the bulk of the viral genes enter either the transforming or productive viral replication cycles and are rapidly eliminated from the host through an MHC class I-restricted cytotoxic T cell response. However, cells that are latently infected and express only the viral *EBNA-1* are not altered in any detectable way, maintain the EBV genome for an indefinite period, and are completely tolerated by the host (31).

HYBRID EBV-HSV VECTORS FACILITATE HIGH-EFFICIENCY GENE DELIVERY AND STABLE GENE EXPRESSION

The principal disadvantage associated with the use of EBV-based vectors is that they rely on physical delivery systems to facilitate entry into target cells. Although such methods are acceptable for use with cultured cells and perhaps even amenable for use with ex vivo approaches, they are in general too inefficient for use in vivo. In theory, it should be possible to adapt the HSV packaging system for use with EBV-based vectors because the viruses involved have compatible genomic organization and utilize a similar DNA replication and packaging process during productive infection. A number of EBV-HSV hybrid vectors have been developed based on this rationale (47,55). The hybrid vector developed in our laboratory is designated pHERO and contains the *EBNA-1* and *oriP* sequences of EBV as well as the ori_S and *pac* elements from HSV (Figure 1). Note that the HSV ori_S is colinear with the HSV *IE4* promoter element and that the latter is used to drive expression of the *zeo* selectable marker. This configuration is identical to that of the pHEX vector and closely resembles that of pHSVlac, which is packaged very efficiently relative to other HSV amplicon vectors. A second important feature of the vector is that the bacterial *zeo* gene that encodes resistance to the commercially available antibiotic Zeocin™ (Invitrogen, Carlsbad, CA, USA) was used as a selectable marker. The *zeo* gene is expressed from a hybrid promoter that is active in bacteria as well as in mammalian cells, and no additional markers are required for selection.

EBV-HSV hybrid vectors can be packaged into infectious particles using the same protocols as a conventional HSV amplicon vector. It is possible to achieve transducing lysates of greater than 1×10^5 infectious particles/mL using the pHERO vector in conjunction with the pBAC-V2 bacmid-based packaging system (Table 2). The titers obtained with pHERO are only 10% to 20% that of titers obtained with pHSVlac and pHEX, however, despite the fact that pHERO contains identical HSV packaging elements. The higher efficiency of pHSVlac over pHERO is likely related to the differences in size between the two vectors. Specifically, the packaging efficiency of each vector in this system is a function of its ability to compete with the packaging-defective HSV genome for the viral polymerase during rolling-circle DNA replication and also the ease with which the subsequent concatamerized vector can be cleaved at the *pac* elements for packaging into the viral capsid.

The ability of the hybrid pHERO vector to provide for stable gene expression has been established in two complemen-

Table 3. Stable Transduction of Reporter Gene Expression in Mitotically Active Cells

Vector Construct[a]	β-Gal+ Titer[b]	ZeoR Titer[c]	HSV Titer[d]
pHERO400β	1.4×10^5	8.0×10^4	<10
pHSVlac	4.5×10^6	Not done	< 10

[a]All vectors express the bacterial β-galactosidase reporter gene.

[b]The β-gal+ titer was obtained by X-Gal histochemical staining of HEK293 cells two days after infection with serial tenfold dilutions of the vector stock and counting the positive cells. It represents the number of β-galactosidase tu/mL of lysate. This is the transient transducing titer.

[c]The ZeoR titer was obtained by replating the transduced HEK293 cells one day after infection into a P100 culture dish, selecting cultures for 14 days with 250 μg/mL Zeocin and counting growing colonies. It represents the number of *zeo* tu/mL of lysate. All ZeoR colonies were positive for β-galactosidase activity by X-Gal histochemical staining. This is the stable transducing titer.

[d]The HSV titer was obtained by screening for plaque formation on Vero cells and represents the number of replication-competent infectious viral units/mL of lysate. This is the background reversion rate of the packaging-defective HSV helper virus caused by recombination with the amplicon vector.

tary assay systems. The first system involved analysis of replicational stability following transduction into mitotically active cells. Briefly, the pHSVlac and pHERO vectors were packaged into infectious particles, and the transducing titers were determined using HEK293 cells (human kidney) (Table 3). The transient transducing titer was determined by staining for reporter activity after two days postinfection and provides an index of the total amount of vector particles in the volume of lysate used for transduction. The stable transducing titer was determined by replating the transduced cells and selecting in medium with Zeocin for the outgrowth of colonies (stable clones). This provides an index of the ability of the vector to replicate in the transduced cells. The transient and stable transducing titers are similar for the pHERO vector, which indicates that the EBV components are fully functional in terms of replicating the concatamerized vector and maintaining it in the nucleus. Moreover, because stable clones can be selected out, the pHERO vector confers long-term gene expression of the *zeo* selectable marker. Similarly, a nonselected β-galactosidase reporter gene is also stably expressed, suggesting that the EBV components can maintain the transduced pHERO vector in transcriptionally active regions of the nucleus such that extended expression is possible using any promoter.

As demonstrated using postmitotic cells, the ability of EBV-HSV hybrid vectors to facilitate long-term transduced gene expression is independent of the replicative functions of the EBV components of the vector. Although neurons represent a natural cell type to target for such experiments, a strength of the HSV gene delivery system is that it can be used to target gene expression to a wide variety of cell types. For example, cultured human primary myoblasts can be stimulated to differentiate and fuse into myofibers through serum withdrawal and are amenable to transduction by HSV amplicons at all points in this process. The pHERO vector can facilitate transduced reporter expression for extended periods in differentiated myoblasts, as demonstrated using the jellyfish green fluorescent protein (GFP) reporter gene

(Figure 3). GFP expression can be detected in live cells; it is possible to follow the course of transduced reporter gene expression in individually transduced myofibers. Although expression levels decreased somewhat after the first week in culture, GFP activity was clearly detectable throughout the course of the experiment, indicating that the EBV components of the pHERO vector function independently of mitotic status in stabilizing gene expression from an episomal platform. The ability of EBV-HSV hybrid vectors to direct long-term stable gene expression in vivo is currently under investigation.

CONCLUSION

There is no one particular viral vector that possesses all of the features required to serve as an ideal platform for providing efficient gene delivery and long-term gene expression. With respect to the use of HSV as a vector, for example, the virus is able to establish lifelong persistent infections of its host although the mechanism used is likely not suitable for exploitation in the development of vectors for therapeutic application. Specifically, HSV can remain undetected by the immune system only in its latent state, and although the underlying basis of this is not fully understood, it is clear that the viral genome supports very little gene expression during latency. Thus, inasmuch as HD HSV vectors are well positioned to serve as generalized vectors for gene delivery by exploiting the high infectivity of the virus, they are not well suited to serve as platforms for long-term gene expression.

An effective solution to this problem is to engineer a hybrid viral vector that contains suitable components of selected viruses such that the weaknesses of one component are specifically addressed as strengths of the other (17). Because of their inherent flexibility, HD HSV vectors are at the forefront of the development of hybrid viral vectors. Two different configurations have been described to date, incorporating components of the adeno-associated virus (AAV) (9) or EBV (47,55), respectively, and it is likely that other configurations are also in development.

The full power of HD HSV vectors will

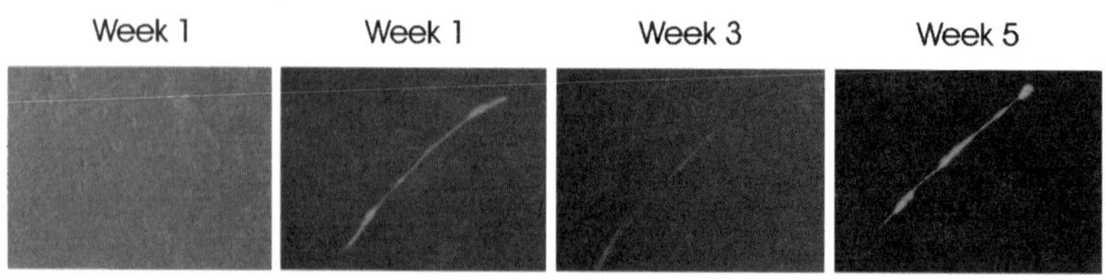

Figure 3. Stable transduction of reporter gene expression in postmitotic cells. Differentiated human myoblasts were transduced with infectious particles containing a pHERO vector expressing the jellyfish GFP at an moi of 0.1, and reporter gene expression was monitored in live cells. GFP activity could be detected throughout the full course of the experiment, as indicated by the micrographs of a representative myofiber assayed at weeks 1, 3, and 5. The first panel shows a bright field view of the cultured cells at week 1, and the remaining three panels show the same field viewed under fluorescent microscopy. After five weeks, the culture was no longer viable, and the experiment was terminated. (**See color plate A3.**)

be realized only if a packaging system can be developed that yields high titers of packaged vector material that is free of contaminating helper virus. The transfection-based packaging systems described in this chapter and elsewhere in this volume are limited in terms of the quality and amount of material that can be produced. They represent only the first step in the development of such a system. Future developments must therefore move beyond such simple transfection-based systems if HSV amplicons are to be applied in the clinic.

REFERENCES

1. **Banfield, B.W., Y. Leduc, L. Esford, K. Schubert, and F. Tufaro.** 1995. Sequential isolation of proteoglycan synthesis mutants by using herpes simplex virus as a selective agent: evidence for a proteoglycan-independent virus entry pathway. J. Virol. *69*:3290-3298.
2. **Blake, N., S. Lee, I. Redchenko, W. Thomas, N. Steven, A. Leese, P. Steigerwald-Mullen, M.G. Kurilla, L. Frappier, and A. Rickinson.** 1997. Human CD8+ T cell responses to EBV EBNA1: HLA class I presentation of the (Gly-Ala)-containing protein requires exogenous processing. Immunity *7*:791-802.
3. **Bochkarev, A., J.A. Barwell, R.A. Pfuetzner, W. Furey, Jr., A.M. Edwards, and L. Frappier.** 1995. Crystal structure of the DNA-binding domain of the Epstein-Barr virus origin-binding protein EBNA 1. Cell *83*:39-46.
4. **Breakefield, X.O. and N.A. DeLuca.** 1991. Herpes simplex virus for gene delivery to neurons. New Biol. *3*:203-218.
5. **Chiocca, E.A., B.B. Choi, W.Z. Cai, N.A. DeLuca, P.A. Schaffer, M. DiFiglia, X.O. Breakefield, and R.L. Martuza.** 1990. Transfer and expression of the *lacZ* gene in rat brain neurons mediated by herpes simplex virus mutants. New Biol. *2*:739-746.
6. **Dyer, A.P., B.W. Banfield, D. Martindale, D.M. Spannier, and F. Tufaro.** 1997. Dextran sulfate can act as an artificial receptor to mediate a type-specific herpes simplex virus infection via glycoprotein B. J. Virol. *71*:191-198.
7. **Edwards, A.M., A. Bochkarev, and L. Frappier.** 1998. Origin DNA-binding proteins. Curr. Opin. Struct. Biol. *8*:49-53.
8. **Fink, D.J., N.A. DeLuca, W.F. Goins, and J.C. Glorioso.** 1996. Gene transfer to neurons using herpes simplex virus-based vectors. Annu. Rev. Neurosci. *19*:265-287.
9. **Fraefel, C., D.R. Jacoby, C. Lage, H. Hilderbrand, J.Y. Chou, F.W. Alt, X.O. Breakefield, and J.A. Majzoub.** 1997. Gene transfer into hepatocytes mediated by helper virus-free HSV/AAV hybrid vectors. Mol. Med. *3*:813-825.
10. **Fraefel, C., S. Song, F. Lim, P. Lang, L. Yu, Y. Wang, P. Wild, and A.I. Geller.** 1996. Helper virus-free transfer of herpes simplex virus type 1 plasmid vectors into neural cells. J. Virol. *70*:7190-7197.
11. **Geller, A.I. and X.O. Breakefield.** 1988. A defective HSV-1 vector expresses *Escherichia coli* β-galactosidase in cultured peripheral neurons. Science *241*:1667-1669.
12. **Geller, A.I., K. Keyomarsi, J. Bryan, and A.B. Pardee.** 1990. An efficient deletion mutant packaging system for defective herpes simplex virus vectors: potential applications to human gene therapy and neuronal physiology. Proc. Natl. Acad. Sci. USA *87*:8950-8954.
13. **Glorioso, J.C., W.F. Goins, D.J. Fink, and N.A. DeLuca.** 1994. Herpes simplex virus vectors and gene transfer to brain. Dev. Biol. Stand. *82*:79-87.
14. **Groger, R.K., D.M. Morrow, and M.L. Tykocinski.** 1989. Directional antisense and sense cDNA cloning using Epstein-Barr virus episomal expression vectors. Gene *81*:285-294.
15. **Ho, D.** 1994. Amplicon-based herpes simplex virus vectors. Meth. Cell Biol. *43*:191-210.
16. **Horsburgh, B.C., M.M. Hubinette, D. Qiang, M.L.E. MacDonald, and F. Tufaro.** 1999. Allele replacement: an application that permits rapid manipulation of herpes simplex virus type 1 genomes. Gene Ther. *6*:922-930.
17. **Jacoby, D.R., C. Fraefel, and X.O. Breakefield.** 1997. Hybrid vectors: a new generation of virus-based vectors designed to control the cellular fate of delivered genes. Gene Ther. *4*:1281-1283.
18. **Johnson, P.A., A. Miyanohara, F. Levine, T. Cahill, and T. Friedmann.** 1992. Cytotoxicity of a replication-defective mutant of herpes simplex virus type 1. J. Virol. *66*:2952-2965.
19. **Johnson, P.A., M.J. Wang, and T. Friedmann.** 1994. Improved cell survival by the reduction of immediate-early gene expression in replication-defective mutants of herpes simplex virus type 1 but not by mutation of the virion host shutoff function. J. Virol. *68*:6347-6362.
20. **Johnson, P.A., K. Yoshida, F.H. Gage, and T. Friedmann.** 1992. Effects of gene transfer into cultured CNS neurons with a replication- defective herpes simplex virus type 1 vector. Brain Res. Mol. Brain Res. *12*:95-102.
21. **Kieff, E.** 1996. Epstein-Barr virus and its replication, p. 2343-2396. *In* B.N. Fields, D.M. Knipe, and P.M. Howley (Eds.), Fields Virology, 3rd ed. Lippincott-Raven, New York.
22. **Klein, G.** 1994. Epstein-Barr virus strategy in normal and neoplastic B cells. Cell *77*:791-793.
23. **Krisky, D.M., P.C. Marconi, T. Oligino, R.J. Rouse, D.J. Fink, and J.C. Glorioso.** 1997. Rapid method for construction of recombinant HSV gene transfer vectors. Gene Ther. *4*:1120-1125.
24. **Krummenacher, C., A.V. Nicola, J.C. Whitbeck, H. Lou, W. Hou, J.D. Lambris, R.J. Geraghty, P.G. Spear, G.H. Cohen, and R.J. Eisenberg.** 1998. Herpes simplex virus glycoprotein D can bind to poliovirus receptor-related protein 1 or herpesvirus entry mediator, two structurally unrelated mediators

of virus entry. J. Virol. *72*:7064-7074.

25. **Leib, D.A. and P.D. Olivo.** 1993. Gene delivery to neurons: is herpes simplex virus the right tool for the job? Bioessays *15*:547-554.

26. **Levitskaya, J., M. Coram, V. Levitsky, S. Imreh, P.M. Steigerwald-Mullen, G. Klein, M.G. Kurilla, and M.G. Masucci.** 1995. Inhibition of antigen processing by the internal repeat region of the Epstein-Barr virus nuclear antigen-1. Nature *375*:685-688.

27. **Levitskaya, J., A. Sharipo, A. Leonchiks, A. Ciechanover, and M.G. Masucci.** 1997. Inhibition of ubiquitin/proteasome-dependent protein degradation by the Gly-Ala repeat domain of the Epstein-Barr virus nuclear antigen 1. Proc. Natl. Acad. Sci. USA *94*:12616-12621.

28. **Lim, F., D. Hartley, P. Starr, P. Lang, S. Song, L. Yu, Y. Wang, and A.I. Geller.** 1996. Generation of high-titer defective HSV-1 vectors using an *IE2* deletion mutant and quantitative study of expression in cultured cortical cells. BioTechniques *20*:460-469.

29. **Middleton, T., T.A. Gahn, J.M. Martin, and B. Sugden.** 1991. Immortalizing genes of Epstein-Barr virus. Adv. Virus Res. *40*:19-55.

30. **Middleton, T. and B. Sugden.** 1994. Retention of plasmid DNA in mammalian cells is enhanced by binding of the Epstein-Barr virus replication protein EBNA1. J. Virol. *68*:4067-4071.

31. **Mukherjee, S., P. Trivedi, D.M. Dorfman, G. Klein, and A. Townsend.** 1998. Murine cytotoxic T lymphocytes recognize an epitope in an EBNA-1 fragment, but fail to lyse EBNA-1-expressing mouse cells. J. Exp Med. *187*:445-450.

32. **Pechan, P.A., M. Fotaki, R.L. Thompson, R. Dunn, M. Chase, E.A. Chiocca, and X.O. Breakefield.** 1996. A novel 'piggyback' packaging system for herpes simplex virus amplicon vectors. Hum. Gene Ther. *7*:2003-2013.

33. **Rickinson, A.B.** 1998. Epstein-Barr virus in action *in vivo*. N. Engl. J. Med. *338*:1461-1463.

34. **Rickinson, A.B. and E. Kieff.** 1996. Epstein-Barr virus, p. 2397-2446. *In* B.N. Fields, D.M. Knipe, and P.M. Howley (Eds.), Fields Virology, 3rd ed. Lippincott-Raven, New York.

35. **Rickinson, A.B., S.P. Lee, and N.M. Steven.** 1996. Cytotoxic T lymphocyte responses to Epstein-Barr virus. Curr. Opin. Immunol. *8*:492-497.

36. **Rickinson, A.B. and D.J. Moss.** 1997. Human cytotoxic T lymphocyte responses to Epstein-Barr virus infection. Annu. Rev. Immunol. *15*:405-431.

37. **Roizman, B.** 1996. The function of herpes simplex virus genes: a primer for genetic engineering of novel vectors. Proc. Natl. Acad. Sci. USA *93*:11307-11312.

38. **Roizman, B. and A.B. Sears.** 1996. Herpes simplex viruses and their replication, p. 2231-2295. *In* B.N. Fields, D.M. Knipe, and P.M. Howley (Eds.), Fields Virology, 3rd ed. Lippincott-Raven, New York.

39. **Saeki, Y., T. Ichikawa, A. Saeki, E.A. Chiocca, K. Tobler, M. Ackermann, X.O. Breakefield, and C. Fraefel.** 1998. Herpes simplex virus type 1 DNA amplified as bacterial artificial chromosome in *Escherichia coli*: rescue of replication-competent virus progeny and packaging of amplicon vectors. Hum. Gene Ther. *9*:2787-2794.

40. **Samaniego, L.A., L. Neiderhiser, and N.A. DeLuca.** 1998. Persistence and expression of the herpes simplex virus genome in the absence of immediate-early proteins. J. Virol. *72*:3307-3320.

41. **Samaniego, L.A., N. Wu, and N.A. DeLuca.** 1997. The herpes simplex virus immediate-early protein ICP0 affects transcription from the viral genome and infected-cell survival in the absence of ICP4 and ICP27. J. Virol. *71*:4614-4625.

42. **Sample, C. and E. Kieff.** 1991. Molecular basis for Epstein-Barr virus induced pathogenesis and disease. Springer Semin. Immunopathol. *13*:133-146.

43. **Spaete, R.R. and N. Frenkel.** 1985. The herpes simplex virus amplicon: analyses of *cis*-acting replication functions. Proc. Natl. Acad. Sci. USA *82*:694-698.

44. **Stavropoulos, T.A. and C.A. Strathdee.** 1998. An enhanced packaging system for helper-dependent herpes simplex virus vectors. J. Virol. *72*:7137-7143.

45. **Steven, N.M., N.E. Annels, A. Kumar, A.M. Leese, M.G. Kurilla, and A.B. Rickinson.** 1997. Immediate early and early lytic cycle proteins are frequent targets of the Epstein-Barr virus-induced cytotoxic T cell response. J. Exp. Med. *185*:1605-1617.

46. **Stevens, J.G.** 1989. Human herpesviruses: a consideration of the latent state. Microbiol. Rev. *53*:318-332.

47. **Strathdee, C.A.** 1996. Development of a hybrid herpesvirus-based vector for expression in mammalian tissues [abstr W31-8]. *In* 15th Annual Meeting of the American Society for Virology, London, Ontario, Canada.

48. **Su, W., T. Middleton, B. Sugden, and H. Echols.** 1991. DNA looping between the origin of replication of Epstein-Barr virus and its enhancer site: stabilization of an origin complex with Epstein-Barr nuclear antigen 1. Proc. Natl. Acad. Sci. USA *88*:10870-10874.

49. **Sugden, B., K. Marsh, and J. Yates.** 1985. A vector that replicates as a plasmid and can be efficiently selected in B-lymphoblasts transformed by Epstein-Barr virus. Mol. Cell. Biol. *5*:410-413.

50. **Summers, H., J.A. Barwell, R.A. Pfuetzner, A.M. Edwards, and L. Frappier.** 1996. Cooperative assembly of EBNA1 on the Epstein-Barr virus latent origin of replication. J. Virol. *70*:1228-1231.

51. **Tal-Singer, R., C. Peng, M. Ponce De Leon, W.R. Abrams, B.W. Banfield, F. Tufaro, G.H. Cohen, and R.J. Eisenberg.** 1995. Interaction of herpes simplex virus glycoprotein gC with mammalian cell surface molecules. J. Virol. *69*:4471-4483.

52. **Tomiyasu, K., E. Satoh, Y. Oda, K. Nishizaki, M. Kondo, J. Imanishi, and O. Mazda.** 1998. Gene transfer *in vitro* and *in vivo* with Epstein-Barr virus-based episomal vector results in markedly high transient expression in rodent cells. Biochem. Biophys. Res. Commun. *253*:733-738.

53. **Tsui, L.C., M.L. Breitman, L. Siminovitch, and M. Buchwald.** 1982. Persistence of freely replicating SV40 recombinant molecules carrying a selectable marker in permissive simian cells. Cell *30*:499-508.

54. **Tufaro, F.** 1997. Virus entry: two receptors are better than one. Trends Microbiol. *5*:257-258.

55. **Wang, S. and J.M. Vos.** 1996. A hybrid herpesvirus infectious vector based on Epstein-Barr virus and her-

pes simplex virus type 1 for gene transfer into human cells *in vitro* and *in vivo*. J. Virol. *70*:8422-8430.

56. **Wendelburg, B.J. and J.M. Vos.** 1998. An enhanced EBNA1 variant with reduced IR3 domain for long-term episomal maintenance and transgene expression of *oriP*-based plasmids in human cells. Gene Ther. *5*:1389-1399.

57. **Wong, S.W. and P.A. Schaffer.** 1991. Elements in the transcriptional regulatory region flanking herpes simplex virus type 1 ori_S stimulate origin function. J. Virol. *65*:2601-2611.

58. **Wu, X., Y. Leduc, M. Cynader, and F. Tufaro.** 1995. Examination of conditions affecting the efficiency of HVS-1 amplicon packaging. J. Virol. Methods *52*:219-229.

59. **Yates, J., N. Warren, D. Reisman, and B. Sugden.** 1984. A *cis*-acting element from the Epstein-Barr viral genome that permits stable replication of recombinant plasmids in latently infected cells. Proc. Natl. Acad. Sci. USA *81*:3806-3810.

60. **Yates, J.L. and N. Guan.** 1991. Epstein-Barr virus-derived plasmids replicate only once per cell cycle and are not amplified after entry into cells. J. Virol. *65*:483-488.

61. **Yates, J.L., N. Warren, and B. Sugden.** 1985. Stable replication of plasmids derived from Epstein-Barr virus in various mammalian cells. Nature *313*:812-815.

62. **Zhang, X., H. O'Shea, C. Entwisle, M. Boursnell, S. Efstathiou, and S. Inglis.** 1998. An efficient selection system for packaging herpes simplex virus amplicons. J. Gen. Virol. *79*:125-131.

15 Herpesvirus/Retrovirus Chimeras for Gene Delivery

Cécile Zaupa and Alberto L. Epstein
Center for Molecular and Cellular Genetics, UMR 5534–CNRS, University Claude Bernard Lyon I, Villeurbanne, France

INTRODUCTION

Gene delivery systems for gene therapy must fit two major requirements: *(i)* they must be absolutely safe for the receiving organism, and *(ii)* they must be efficient enough to induce a significant therapeutic modification of the disease. These two requirements are often contradictory since in many cases improving the safety of the vector results in a reduction of its therapeutic usefulness. Efficient therapy of many pathological conditions requires a genetic modification of such a high number of cells that it could hardly be achieved without using replication-competent vectors. These are able to amplify themselves within the target tissue and to deliver the therapeutic gene into cells that have escaped from input vector infection. But replicative vectors, even when replication-compromised or attenuated, may be dangerous, not only because most of them will kill the infected cells but also because they can spread out from the target tissue and endanger the survival of the whole organism.

On the other hand, replication-incompetent (defective) vectors, which can only replicate in specially constructed *trans*-complementing cell lines, will not spread out from the target tissue. Provided that most toxic functions have been deleted from these vectors, cells will survive the infection and durably express the therapeutic transgene. The main inconvenience of these vectors is, however, that the number of cells they can infect is limited. Cells that have not been reached by the input vector will not experience the therapeutic action of the transgene, except in a few cases, as illustrated by the bystander effect of some antitumor strategies, or when the transgenic protein is secreted from the genetically manipulated cell.

To explore whether such a contradictory situation could be resolved, we have started to develop a new strategy, based on the use of defective chimerical vectors. The main concept is to use the genome of a large defective vector (herpes simplex virus type 1 [HSV-1]) to carry the genes required for the de novo synthesis of a small defective vector (derived from Moloney murine leukemia retrovirus [MoMLV]). According

Viral Vectors: Basic Science and Gene Therapy
Edited by A. Cid-Arregui and A. García-Carrancá

to this strategy, cells infected with the defective HSV-1 vector will express the MoMLV vector genes and will thus induce the cells to produce in situ and release defective MoMLV vector particles during several days. In other words, each cell infected with the HSV-1 vector would become a relatively stable factory for production of retrovirus vectors. MoMLV vectors would in this way infect many more cells in the target tissue, where they will express the transgenic function. Since both vectors are defective, however, the system will not spread further. The strategy requires the therapeutic transgene to be carried and expressed by the retrovirus vector genome, but depending on the promoter sequences employed, it may be also expressed by the HSV-1 vector.

Two other potential uses of such a hybrid vector system can be envisioned. The first would be to overcome some of the major drawbacks of standard systems utilized to produce retrovirus vectors, such as low-vector titer production, generation of recombinant vectors that may be replication competent and the fact that those systems are mainly suitable for use in cell cultures only. This hybrid vector system might also be used for vaccines. Several vectors, especially those derived from poxviruses (17,19), have already been used as vaccines, for the expression of antigens belonging to different classes of viruses. The possibility of inducing in situ synthesis not just of a single antigenic protein, but of a whole set of proteins that make a complete virus structure, which would be antigenically indistinguishable from the target virus structure, could represent a significant improvement in the design of vector vaccines.

Several other chimerical virus or virus-derived systems, in which a large genome vector provides the genes for the in situ synthesis of a smaller genome vector, have recently been described. For example, Semliki Forest virus-derived replication-competent RNA was introduced into host cells and used to produce recombinant retroviral particles, including particles carrying intron-containing genes (12,13,28). Furthermore, an adenovirus (AD)/retrovirus chimerical system consisting of two different elements was also described. The first one encodes retroviral-packaging functions, while the second encodes a defective retroviral genome. If both types of AD vectors coinfect a target cell, the system allows in vivo gene transduction (6).

A different AD/retrovirus chimerical system, consisting of three different elements (by dissociation of *gag-pol* and *env* transcription units) has also been reported (5). More recently, a poxviral/retroviral chimera that allows cytoplasmic production of transducing defective retroviral particles was described (10). Finally, an HSV-1/adeno-associated vector (AAV) chimera has been used to mediate amplification and genomic integration of the AAV element, which resulted in the improvement of gene transfer and expression into target cells (8,11).

THE HSV-1/MoMLV CHIMERAS

The final goal of our program is to construct a single HSV-1-defective vector harboring the complete set of genes required for the de novo synthesis of a defective retrovirus vector. We expect these vectors to induce the in situ production of infectious, though defective, retrovirus vectors upon infection of cell types containing no retrovirus endogenous or exogenous genetic elements, both in vivo and in vitro.

However, we have first characterized this hybrid vector system in two preliminary and complementary steps. One of them involved the construction of an HSV-1 vector carrying MoMLV *gag*, *pol*, and *env* genes (the human cytomegalovirus (HCMV)-GPE vector). The other one involved the construction of an HSV-1 vector carrying a reporter gene (*lacZ*) flanked by the complete set of *cis*-acting sequences required for the fulfillment of a MoMLV retroviral cycle (i.e., long terminal repeats [LTRs], the primer binding sites, [PBSs], the polypurine tracts [PPTs], and the packaging signal E) (the LTR-*lacZ* vector). As we describe below, both the HCMV-GPE vector (24) and the LTR-*lacZ* vector (unpublished results) induced the production of defective MoMLV vectors upon infection of cells harboring the complementary set of retrovirus sequences, relative to those carried by the herpes vectors.

HSV-1 Amplicon Plasmids and Vectors

The HSV-1 vectors employed in these studies were amplicon vectors (25). These vectors have two major advantages compared with HSV-1 recombinant vectors (22). The first and most important is that amplicon particles are nontoxic for the infected cells since they carry no HSV-1 viral genes. Second, the repetitive character of the genome carried by a single amplicon particle vector ensures the introduction of multiple copies of the transgenic transcription unit per infected cell, likely resulting in strong expression.

Amplicon vectors derive from a plasmid containing one origin of viral DNA replication (ori_S) and one packaging sequences (*pac*) from HSV-1, in addition to the transgenic transcription unit. Because the amplicon does not encode any further viral sequences, it can only be grown in the presence of *trans*-acting functions provided by a helper virus. In the presence of a helper virus, amplicon DNA is amplified into a head-to-tail concatamer that is then packaged into defective HSV-1 particles, up to the genome size (152 kb).

The two transgenic transcription units carried by the amplicon plasmids employed in these studies are shown in Figure 1. Amplicon plasmid pA-HCMV-GPE harbors the ecotropic MoMLV *gag-pol-env* transcription unit, under the control of the HCMV major immediate early enhancer/promoter and the HSV-1 thymidine kinase polyadenylation sequence (polyA_nTK). The modifications introduced in the transgenic sequence, which derives from plasmid pCrip (3), should ensure that the retrovirus transcription unit is expressed as an immediate early gene during HSV-1 infection. Compared to the MoMLV genome, the HCMV-GPE transcription unit lacks PBSs and PPTs, both LTRs and the packaging sequence E. Only the donor and acceptor splicing sites were conserved to allow *env* gene expression. The design of this defective retrovirus genome was adopted to abolish or minimize its packaging, reverse transcription, and integration.

Amplicon plasmid pA-LTR-LacZ (Figure 1) derives from pIRV-LacZ (16) and bears a transgenic defective MoMLV retrovirus vector (i.e., a packageable *lacZ* gene, surrounded by two MoM-LV LTRs). The *lacZ* open reading frame (ORF) is placed under the control of the actin promoter. In addition, this plasmid contains the Neo^R gene, expressed under the control of the 5′ LTR enhancer/promoter, conferring resistance to the drug G418.

Using the aforementioned amplicon plasmids, we have prepared stocks of both HCMV-GPE and LTR-LacZ amplicon vectors. In the first series of studies, both vectors were generated using replication-defective HSV-1 mutants as helpers (9). The viral mutant employed as helper, HSV-1 D30EBA (21), carries deletions in both copies of a gene essential for virus replication (*IE3* gene, encoding ICP4 protein). This virus can support amplicon amplification and packaging in cells that complement the missing function, such as M64A cells, which are BHK-21-derived cells expressing ICP4 (4). To prepare our amplicon stocks, 5 mg of either HCMV-GPE or LTR-LacZ amplicon plasmids were

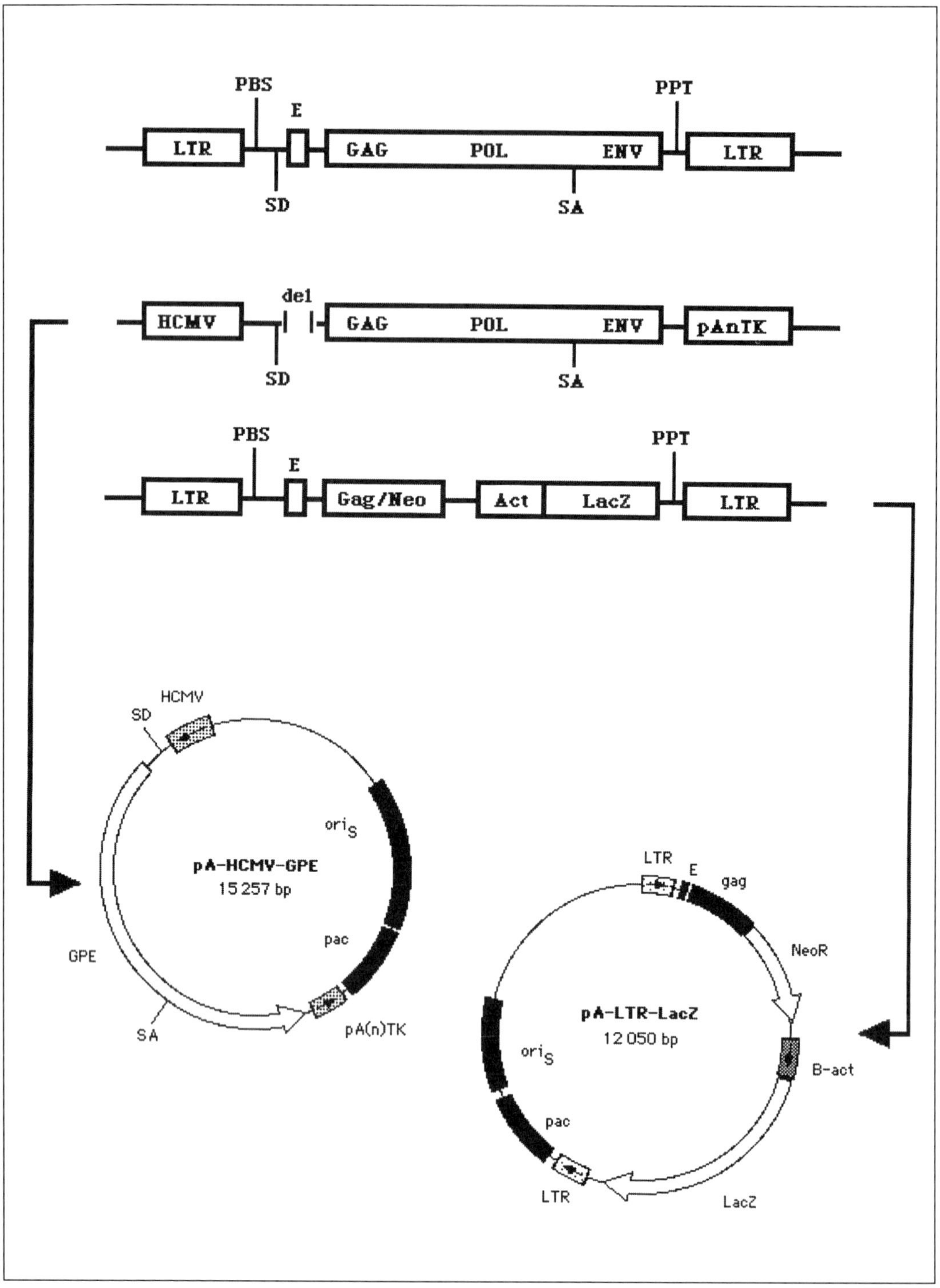

Figure 1. Structure of the transcription units carried by the amplicon vectors. The upper line shows the structure of the MoMLV genome, from which derive the transcription units carried by the chimerical vectors. The middle line shows the transcription unit carried out by pA-HCMV-GPE amplicon plasmid, whereas the bottom line shows the transcription unit carried out by pA-LTR-LacZ amplicon plasmid.

transfected into M64A cells, and the transfected cells were superinfected with D30EBA helper virus at 1 plaque-forming unit (pfu)/cell the following day (Figure 2). When total cytopathic effect was observed, cells were frozen and thawed three times to release the infectious virus. The vector populations were then serially passaged onto fresh M64A cells by using a 1:2 dilution to increase the vector-to-helper ratio (25). Titers of the contaminating helper virus in the viral stocks were determined by plaque assays performed on M64A cells. To estimate the titer of the amplicon vector LTR-LacZ, which expresses β-galactosidase (β-gal) activity, nonpermissive Vero cells were infected with serial dilutions of the vector stocks. The following day, the infected cell cultures were fixed and β-gal-positive cells were scored using X-Gal. To estimate the amount of the HCMV-GPE amplicons, which express no reporter genes, the relative amounts of amplicon vector DNA were determined by blot hybridization, in comparison with helper virus DNA, by using a ^{32}P-labeled probe corresponding to cloned ori_S sequences. The ^{32}P-labeled probe recognizes both helper and amplicon vector DNAs. The titers of the HCMV-GPE vectors present in the viral stocks were then estimated by the helper virus titers as well as the ratios of the amplicon DNA to the helper virus DNA (amplicon-to-helper ratios). As a control, the same procedure was performed with the amplicon vector pA-SF1 (15), which carries the *lacZ* gene and can be titrated independently by staining β-gal-positive cells with X-Gal. Following these procedures, we observed that the titers of both types of amplicon vectors varied according to the serial passages (both ranging from 1×10^5 to 1×10^7 transducing units (tu)/mL), while the titer of the helper virus remained relatively constant during the passages (ranging from 1×10^7 to 5×10^7 pfu/mL in both cases). In the assays described below, we used vector stocks that had an estimated vector-to-helper ratio of about 1:5 (24).

Deletion mutant packaging systems produce relatively high amplicon titers, but the vector preparations are heavily contaminated with defective helper particles, which are predominant in the vector stocks. On the other hand, we had already shown that the contaminating helper particles, even if defective, can negatively affect transgenic expression from the amplicon genome in cells that were co-infected with both types of particles (15). To avoid the possible cytotoxic and adverse effects induced by the contaminating helper particles, we have recently generated amplicon stocks using a helper virus-free packaging system (8). This system utilizes transient co-transfection of amplicon DNA with a set of five cosmids that overlap and represent the entire HSV-1 genome (2), but were mutated to remove the packaging signals. After co-transfection, the virus genomes that are potentially reconstituted, via homologous recombination, are not packageable, but can still provide all the helper functions required for replication and packaging of co-transfected amplicon plasmids. The resulting amplicon stocks are virtually free from detectable helper virus, but have considerably lower titers than amplicon stocks prepared using replication-defective helper HSV-1. In our hands, the titers of the pure amplicon stocks ranged from 5×10^3 to 8×10^4 pfu/cell.

Rescue of Retroviral Vectors using HCMV-GPE Amplicons

Experimental protocol. The experimental protocol employed for testing HCMV-GPE vectors, which is shown in Figure 2, is based on the infection of cells providing the complementary retrovirus element. To study whether HCMV-GPE amplicons (either helper-free or helper-contaminated) could rescue LacZ retroviral vectors, approximately 1×10^6 TE-Lac2 cells (27), a

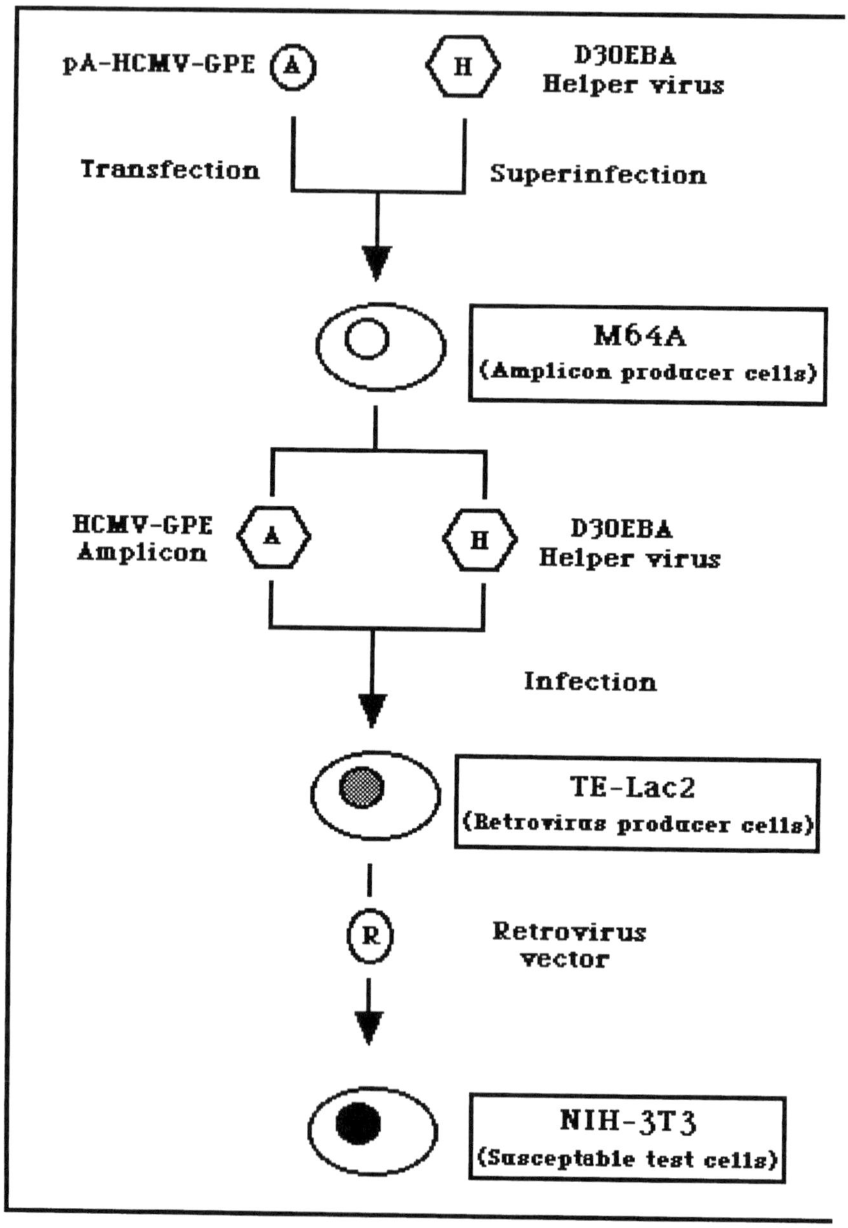

Figure 2. Production of helper-contaminated HCMV-GPE amplicons and rescue of retrovirus vectors using these amplicons. M64 cells, a BHK-21-derived cell line expressing the *IE3* gene of HSV-1, were transfected with 5 μg of pA-HCMV-GPE amplicon plasmid. The following day, the transfected cells were superinfected at 1 pfu/cell, with helper HSV-1. The strain of HSV-1 used as helper is D30EBA, a defective strain lacking the essential *IE3* gene. The absolute amount of input vector particles thus corresponded to 2×10^4. At 48 hours postinfection, the supernatant of infected TE-Lac2 cells, containing the putative retrovirus vector particles, were collected and filtered. Aliquots of this supernatant were then used to infect susceptible NIH-3T3 cells. Infected NIH-3T3 cells were fixed and stained with X-Gal 48 hours later, and β-gal-positive foci were scored.

human tumor cell line carrying the proviral form of a defective retroviral vector expressing *lacZ*, were infected with HCMV-GPE amplicons at a multiplicity of infection (moi) of 0.02 amplicon particles/cell. This corresponded to an moi of 0.1 for the helper particles, in the case of helper-contaminated stocks. Cell culture supernatants were collected at different times after infection, filtered with a 0.2-μm (pore size) filter, and centrifuged at low speed to eliminate cell debris. To check whether these supernatants contained retrovirus vectors that were able to transduce β-gal activity to susceptible cells, and also to estimate the amount of infectious retrovirus vectors present in the supernatants, aliquots of filtered supernatants were overlaid on either murine NIH-3T3 or human TE-671 target cells in the presence of polybrene. β-gal expression was checked 48 hours later. As shown in Figure 3, NIH-3T3 cells (Figure 3A), but not TE-671 cells (Figure 3B), displayed *lacZ* colonies. Following a similar experimental protocol, we later demonstrated that infection of MΦ9 cells, a murine monocyte-macrophage cell line, with HCMV-GPE amplicons, also resulted in the rescue of the defective retroviral LacZ vector carried by these cells (20). These tests thus demonstrated that HCMV-GPE amplicons could rescue the defective retrovirus vectors expressed from the genomes of two different cell types.

Controls

Several supplementary control experiments were carried out to confirm the specificity of our results (24). First, production of defective retroviral particles was observed neither in noninfected TE-Lac2 cells (Figure 3C) nor in the same cells infected with HSV-1 helper virus alone (Figure 3D).

Second, two neutralization experiments were performed to further explore this novel vector-generating system. In the first one, designed to assess whether the rescue of LacZ retrovirus particles from TE-Lac2 cells was exclusively induced by the HSV-1 amplicon vector (and not by naked amplicon DNA or by retroviral transfer of GPE function), HCMV-GPE amplicons were treated with antibodies directed against either HSV-1 or MoMLV envelope glycoproteins before infecting TE-Lac2 cells. As expected, anti-HSV-1 neutralizing antibodies could block the generation of LacZ retrovirus particles if added before the infection of TE-Lac2 cells, while nonimmune or neutralizing antibodies specific for MoMLV or RD114 gp70-SU protein did not block the generation of retrovirus particles at this step (24).

The second neutralization assay was designed to confirm whether the LacZ activity transduced to NIH-3T3 cells by the supernatants of HCMV-GPE amplicon-infected TE-Lac2 cells resulted from ecotropic retroviral vectors (as expected from the fact that the *env* gene carried by the HCMV-GPE amplicons is itself ecotropic). The supernatants were treated either with antibodies against the ecotropic envelope glycoprotein or with antibodies against RD114 retrovirus, which belong to a distinct receptor group. Only the antibodies specific to the ecotropic glycoprotein showed neutralization activity when added before infection of NIH-3T3 cells. Furthermore, as described above, the retrovirus vector particles produced by TE-Lac2 cells could transduce β-gal activity to murine NIH-3T3 cells (Figure 3A), but not to human TE-671 cells (Figure 3B), the parental cell line from which both TE-Lac2 and TE-GPE cells were generated.

Third, we confirmed that the retroviral RNA molecules packaged by the retrovirus particles synthesized in HCMV-GPE-infected TE-Lac2 cells were identical to the mRNA expressed in these cells before infection. For this, TE-Lac2 cell culture supernatants, collected 48 hours after

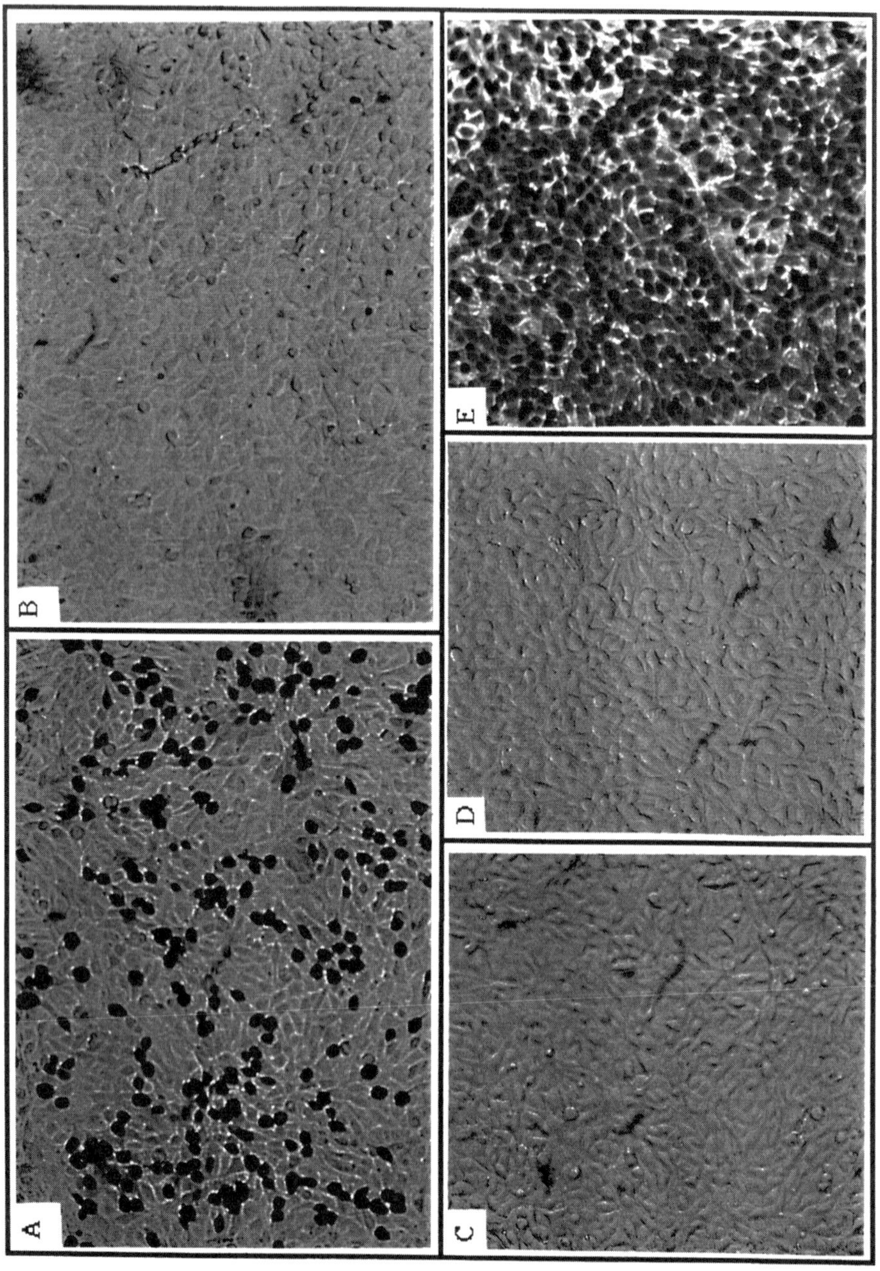

Figure 3. Specificity of the retrovirus vector rescue by HCMV-GPE infection of TE-Lac2 cells. Murine NIH-3T3 cells (A) or human TE-671 cells (B) were infected with supernatants taken from HCMV-GPE-infected TE-Lac2 cells. Only the murine cells (A) show foci of β-gal-expressing cells, indicating that, as expected, the rescued retrovirus was ecotropic. As controls, NIH-3T3 cells infected with supernatants taken from noninfected TE-Lac2 cells (C), or with supernatants taken from TE-Lac2 cells that were infected with helper virus alone (D), showed no β-gal activity. (E) NIH3T3 cells were infected with supernatants taken from a clonal retrovirus vector producer cell line. (**See color plate A4.**)

infection with the amplicon vector, were filtered as described and concentrated by ultracentrifugation, at 35 000 rpm (Sorvall RC 5C PLUS™; Sorvall, Newton, CT, USA). Putative retroviral pellets were analyzed by reverse transcription polymerase chain reaction (RT-PCR) using oligonucleotides specific for the LacZ retroviral vector. The expected fragment was detected by RT-PCR, but not by PCR. This fragment was found to be indistinguishable from both the RT-PCR band generated by total RNA extracted from noninfected TE-Lac2 cells, and the PCR band generated by a plasmid harboring the LacZ retrovirus vector. No DNA fragments were detected after RT-PCR with supernatants taken from noninfected TE-Lac2 cells or from TE-Lac2 cells infected with helper virus alone. Thus, the LacZ-defective retroviral genome expressed by TE-Lac2 cells could be packaged only after infection of these cells with the HCMV-GPE amplicons.

Finally, to study whether the supernatants of the HCMV-GPE amplicon-infected TE-LacZ were devoid of replication-competent retroviral particles that might have arisen from recombination events, aliquots of these supernatants were overlaid on 3T3-LacZ cells (27), which express a defective LacZ retroviral vector. The supernatants of infected 3T3-LacZ cells were tested after three days for the presence of retroviral vectors able to transduce β-gal activity to NIH-3T3 cells. No rescue of LacZ vectors could be observed, thereby suggesting that the retrovirus vectors produced by the HCMV-GPE amplicon infection were devoid of replication-competent retroviruses.

Rescue of Retroviral Vectors Using LTR-LacZ Amplicons

More recently, we have started to explore the complementary vector system, using an experimental protocol similar to that described for the HCMV-GPE vectors. The only difference is that, in this case, TE-GPE cells (1), a packaging cell line expressing MoMLV *gag*, *pol*, and *env* proteins, were infected with LTR-LacZ amplicon stocks (either helper-free or helper-contaminated). As before, cells were infected at an moi of 0.02 amplicon particles. We observed (Figure 4B) that infection of these packaging cells with LTR-LacZ amplicons resulted in the rescue of the defective genomic retrovirus RNA expressed from the amplicon genome, with the concomitant synthesis and release of defective retrovirus vectors. These vectors were able to transduce β-gal activity to susceptible NIH-3T3 cells (unpublished results).

Quantitative Considerations: Kinetics of Retroviral Vector Production

Using either HCMV-GPE vectors or LTR-LacZ vectors, we observed that the amount of retroviral vectors produced by the infected cells increased with the dose of the input amplicons. On the other hand, at any amplicon input dose, the ratio of total retroviral vectors produced by the cell culture, per amplicon-infected cell (i.e., the absolute amount of input amplicon particles) remained close to 1, suggesting that each infected cell was able to produce a vector. It is interesting to note that the amount of retrovirus vectors produced per infected cell in our system is similar to those obtained using either a poxvirus/retrovirus (10) or an AD/retrovirus (5,6) chimera. Furthermore, the most popular clonal cell lines also produce no more than 1 retroviral tu/cell (some 1×10^6 tu/mL are released by 1×10^6 clonal cells), which is similar to the values obtained with the chimerical systems. It will be interesting to know whether the ratio of retrovirus vector produced per infected cell could be increased by infecting cells at higher moi.

The presence of defective helper parti-

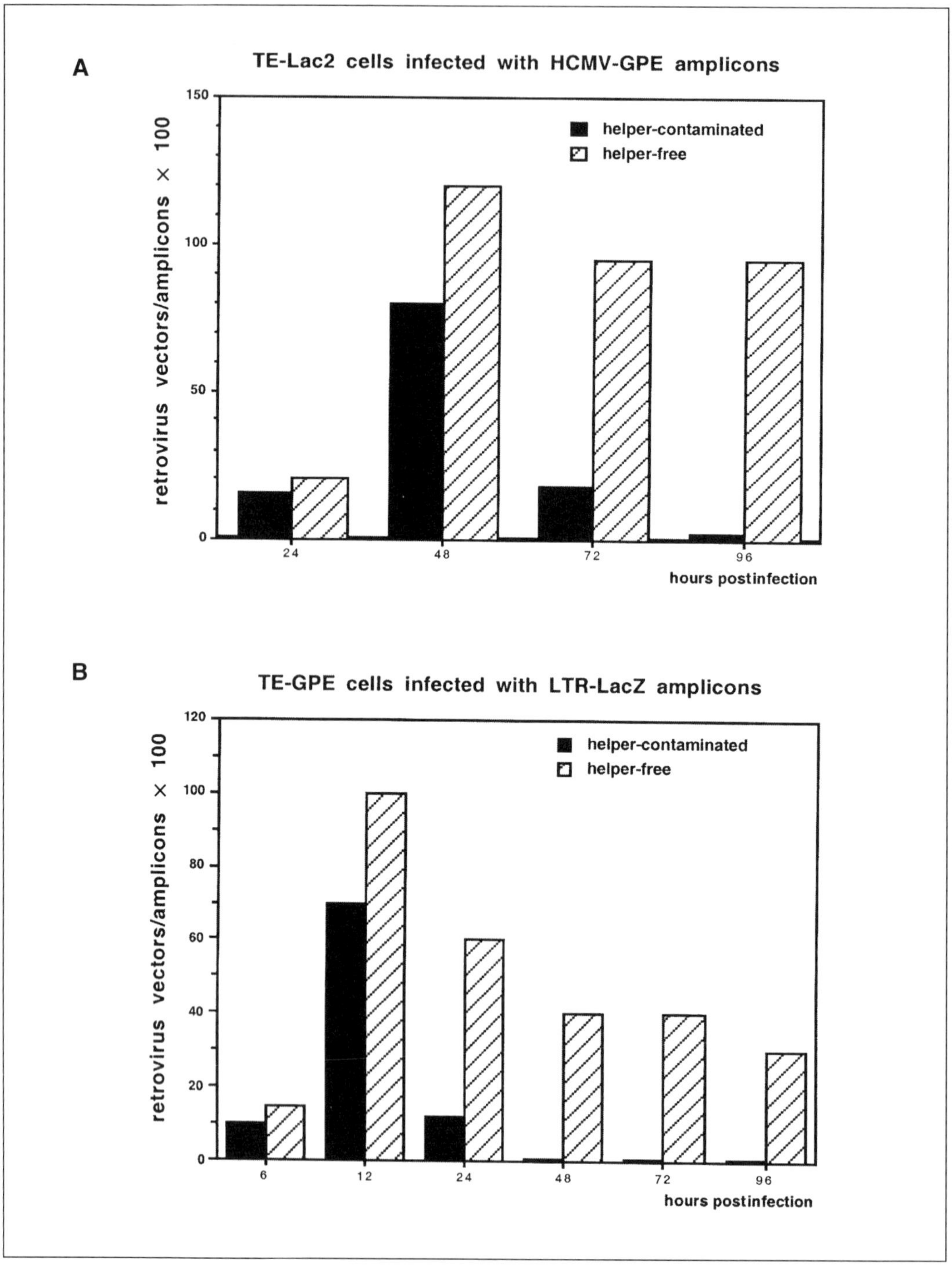

Figure 4. Production kinetics of retrovirus vectors through infection of cells with the chimerical vectors. TE-Lac2 cells were infected with HCMV-GPR amplicons (A), whereas TE-GPE cells were infected with LTR-LacZ amplicons (B). In these experiments, some 1×10^4 amplicon particles (with or without contaminating helper particles) were introduced into each cell culture. The moi was, in all cases, 0.01 vector particles/cell. At different times postinfection, cell culture media were harvested, filtered, and assayed for the presence of ecotropic retrovirus particles, which are able to transduce β-gal activity to NIH-3T3 cells. Cell foci that stained blue after X-Gal staining were counted. Results were plotted as the amount of retrovirus vectors produced per amplicon-infected cell.

cles in the amplicon stocks strongly interfered with the rescue of retrovirus vectors (24). More precisely, while increasing the input dose of a particular amplicon stock, we noticed that as the moi of the helper component of the stocks became close to 1 pfu/cell, the amount of retroviral vector released by the infected cells started to decrease. Moreover, adding extra helper particles to a vector stock, thus increasing the helper-to-vector ratio of that particular virus population, resulted in inhibition of retrovirus production. This is the reason why cells should be infected at low moi when helper-contaminated stocks of amplicons are used.

Once the cosmid system for producing helper-free amplicons became available (7), we decided to use this system to prepare new amplicon stocks and to compare their abilities to rescue retrovirus vectors, relative to those of helper-contaminated stocks. As shown in Figure 4, helper-free stocks of both HCMV-GPE (A) and LTR-LacZ vectors (B) induced higher titers of retrovirus vectors. In addition, the release of MoMLV vectors remained stable for at least five days, whereas with helper-contaminated stocks, MoMLV vectors were difficult to detect at three days postinfection. The main limit of the helper-free system is that, although pure, the titers of the amplicon vectors obtained are considerably lower than those present in helper-contaminated stocks (not exceeding 1×10^5 amplicon particles/mL). This fact prevented the cells from being infected at high moi.

Interestingly, the kinetics of production of retrovirus particles varied according to the type of amplicon vector employed, but not according to the presence or absence of helper particles. In TE-Lac2 cells infected with HCMV-GPE vectors, retrovirus production reached a maximum around 48 hours postinfection. In the case of TE-GPE cells infected with LTR-LacZ vectors, there was a peak of retrovirus production at about 12 hours postinfection, to decrease thereafter. The most likely reason for this difference is that TE-GPE cells are already assembling empty retrovirus capsids and that the only element provided by the LTR-LacZ amplicons are transcripts of packageable MoMLV genomic RNA. However, in the case of amplicon-infected TE-Lac2 cells, all steps required to give rise to retrovirus production (i.e., transcription, splicing, translation, posttranslational modifications, and assembly) must occur to incorporate the genomic RNA that is being transcribed from the proviral vector integrated in the cellular chromosomes.

PERSPECTIVES

The next step in the strategy described here is to induce de novo retrovirus vector synthesis and release in cells that do not contain retrovirus elements. This could be achieved either by co-infecting naive cells with both HCMV-GPE and LTR-LacZ amplicons or by constructing a new amplicon vector that expresses both retroviral elements simultaneously. We are currently exploring both avenues.

Preliminary results indicate that co-infection of cells containing no retroviral elements with a helper-contaminated HCMV-GPE and a helper-free LTR-LacZ vector result in de novo synthesis and release of retrovirus vectors, which are able to transduce β-gal activity to NIH-3T3 cells. However, co-infection of cells with helper-free stocks of both HCMV-GPE and LTR-LacZ vectors have failed to induce production of significant amounts of retrovirus vectors. This failure may be related to the low titers of our pure (helper-free) amplicon stocks, thereby resulting in a low probability for a single cell to become simultaneously infected by the two complementary amplicons. It is thus crucial to develop methods to improve the titers of the vector stocks. Very recently, a new helper system, based on a bacterial artificial chromosome carrying the entire HSV-1 genome, except for the packaging signals, has been described (23,26) (see also Chapters 13 and 14). We are currently developing an alternative system based on the intracellular deletion of HSV-1 packaging signals flanked by two parallel loxP sites, in cells expressing the P1 phage-*Cre* recombinase (14). We are confident that one or the other system will bring a quantitatively significant improvement to the current methods for amplicon production.

At the same time, we are constructing new amplicon vectors derived from plasmoviruses (18). A plasmovirus is a plasmid that contains, in a disorganized manner, a whole set of retroviral genes and *cis*-acting sequences required for generating defective retrovirus vectors upon transfection of naive cells. The cloning of an amplicon module to one such plasmovirus plasmid is already finished, and we will soon be able to test whether infection of naive cells with such a chimeric amplicon will result in the de novo synthesis of retrovirus vectors.

A second challenge will be to put the system described here, as well as the other chimerical systems, to work efficiently in vivo by infected animal models. At least two different strategies can be conceived using chimerical systems, which could result in an improvement of transducing efficiencies. As described above, a way to pursue this goal is the simultaneous introduction into naive cells of all genetic elements required for the synthesis of a complete retrovirus vector, either using two complementary vectors or a single vector carrying a plasmovirus-like genome. Another way would be to first infect tumor cells with a single defective retrovirus vector that expresses, for example, the herpetic thymidine kinase gene, as it is currently being done by several teams, and later on superinfect the tumor cells carrying integrated the retrovirus vector genome with HCMV-GPE amplicons. Infection of such tumor cells with HCMV-GPE amplicon vectors, which could be done either before or after a first wave of ganciclovir treatment, should result in rescue of retrovirus vectors and subsequent infection with retrovirus vectors of tumor cells that escaped the former infection.

ACKNOWLEDGMENTS

We wish to thank the Association pour la Recherche sur le Cancer and the Ligue National Contre le Cancer for funding most of the research described in this review. C. Zaupa has received a fellowship from the Ministère de l'Education Supérieure et la Recherche Scientifique, France. We thank Emiliano Rial Verde for critical reading of this manuscript.

REFERENCES

1. **Cosset, F.L., Y. Takeuchi, J.L. Battini, R.A. Weiss, and M.K.L. Collins.** 1995. High titer packaging cells producing recombinant retroviruses resistant to human serum. J. Virol. *69*:7430-7436.
2. **Cunningham, C. and A.J. Davison.** 1993. A cosmid-based system for constructing mutants of herpes simplex virus type 1. Virology *197*:116-124.
3. **Danos, O. and R.C. Mulligan.** 1988. Safe and efficient generation of recombinant retroviruses with amphotropic and ecotropic host ranges. Proc. Natl. Acad. Sci. USA *85*:6460-6464.
4. **Davidson, I. and N.D. Stow.** 1985. Expression of an immediate early polypeptide and activation of a viral origin of DNA replication in cells containing a fragment of herpes simplex virus DNA. Virology *141*:77-88.
5. **Duisit, G., A. Salvetti, P. Moullier, and F.L. Cosset.** 1999. Functional characterization of adenoviral/retroviral chimeric vectors and their use for efficient screening of retroviral producer cell lines. Hum. Gene Ther. *10*:189-200.
6. **Feng, M., W.H. Jackson, C.K. Goldman, C. Rancourt, M. Wang, S.K. Dusing, G. Siegal, and D.T. Curiel.** 1997. Stable in vivo gene transduction via a novel adenoviral/retroviral chimeric vector. Nat. Biotechnol. *15*:866-870.
7. **Fraefel, C., D.R. Jacoby, C. Lage, H. Hilderbrand, J.Y. Chou, F.W. Alt, X.O. Breakefield, and J.A. Majzoub.** 1997. Gene transfer into hepatocytes mediated

by helper virus-free HSV/AAV hybrid vectors. Mol. Med. *3*:813-825.

8. **Fraefel, C., S. Song, F. Lim, P. Lang, L. Yu, Y. Wang, P. Wild, and A.I. Geller.** 1996. Helper virus-free transfer of herpes simplex virus type 1 plasmid vectors into neural cells. J. Virol. *70*:7190-7197.

9. **Geller, A.I., K. Keyomarsi, J. Bryan, and A.B. Pardee.** 1990. An efficient deletion mutant packaging system for defective herpes simplex virus vectors: potential applications to human gene therapy and neuronal physiology. Proc. Natl. Acad. Sci. USA *87*:8950-8954.

10. **Holzer, G.W., J.A. Mayhofer, W. Gritschenberger, F. Dorner, and F.G. Falkner.** 1999. Poxviral/retroviral chimeric vectors allow cytoplasmic production of transducing defective retroviral proteins. Virology *253*:107-114.

11. **Johnston, K.M., D. Jacoby, P. Pechan, C. Fraefel, P. Borghesani, D. Schuback, R.J. Dunn, F.I. Smith, and X.O. Breakefield.** 1997. HSV/AAV hybrid amplicon vectors extend transgene expression in human glioma cells. Hum. Gene Ther. *8*:359-370.

12. **Li, K.J. and H. Garoff.** 1996. Production of infectious recombinant Moloney murine leukemia virus particles in BHK cells using Semliki forest virus-derived RNA expression vectors. Proc. Natl. Acad. Sci. USA *93*:11658-11663

13. **Li, K.J. and H. Garoff.** 1998. Packaging of intron-containing genes into retrovirus vectors by alphavirus vectors. Proc. Natl. Acad. Sci. USA *95*:3650-3654.

14. **Logvinoff, C. and A.L. Epstein.** 2000. Genetic engineering of herpes simplex virus and vectors carrying loxP sites in cells expressing Cre recombinase. Virology *267*:102-110.

15. **Lowenstein, P.R., S. Fournel, D. Bain, P. Tomasec, P. Clissold, M.G. Castro, and A.L. Epstein.** 1994. Herpes simplex virus 1 (HSV-1) helper co-infection affects the distribution of an amplicon encoded protein in glia. NeuroReport *5*:1625-1630.

16. **Morgenstern, J.P. and H. Land.** 1990. Advanced mammalian gene transfer: high titer retroviral vectors with multiple drug selection markers and a complementary helper-free packaging cell line. Nucleic Acid Res. *18*:3587-3596.

17. **Moss, B.** 1996. Genetically engineered poxviruses for recombinant gene expression, vaccination and safety. Proc. Natl. Acad. Sci. USA *93*:11341-11348.

18. **Noguiez-Hellin, P., M. Robert-Le Meur, J. Salzmann, and D. Klatzmann.** 1996. Plasmoviruses: non-viral/viral vectors for gene therapy. Proc. Natl. Acad. Sci. USA *93*:4175-4180.

19. **Paoletti, E.** 1996. Application of pox virus vectors to vaccination: an update. Proc. Natl. Acad. Sci. USA *93*:11349-11353.

20. **Parrish, E., E. Peltékian, G. Dickson, A.L. Epstein, and L. Garcia.** 1999. Cell engineering for muscle gene therapy: extemporaneous production of retroviral vector packaging macrophages using defective herpes simplex virus type 1 vectors harboring gag, pol, env genes. Cytotechnology *30*:173-180.

21. **Paterson, T. and R.D. Everett.** 1990. A prominent serine-rich region in Vmw175, the major transcriptional regulator protein of herpes simplex virus type 1, is not essential for virus growth in tissue culture. J. Gen. Virol. *71*:1775-1783.

22. **Post, L. and B. Roizman.** 1981. A generalized technique for deletion of specific genes in large genomes: alpha 22 gene of herpes simplex virus 1 is not essential for growth. Cell *25*:227-232.

23. **Saeki, Y., T. Ichikawa, A. Saeki, E.A. Chiocca, K. Tobler, M. Ackerman, X.O. Breakefield, and C. Fraefel.** 1998. Herpes simplex virus type 1 DNA amplified as bacterial artificial chromosome in *Escherichia coli*: rescue of replicative-competent virus progeny and packaging of amplicon vectors. Hum. Gene Ther. *9*:2787-2794.

24. **Savard, N., F.L. Cosset, and A.L. Epstein.** 1997. Defective herpes simplex virus type 1 vectors harboring *gag*, *pol*, and *env* genes can be used to rescue defective retrovirus vectors. J. Virol. *71*:4111-4117.

25. **Spaete, R.R. and N. Frenkel.** 1982. The herpes simplex virus amplicon: a new eucaryotic defective-virus cloning-amplifying vector. Cell *30*:295-304.

26. **Stavropoulos, T.A. and C.A. Strathdee.** 1998. An enhanced packaging system for helper-dependent herpes simplex virus vectors. J. Virol. *72*:7137-7143.

27. **Takeuchi, Y., F.L. Cosset, P.J. Lachmann, H. Okada, R.A. Weiss, and M.K.L. Collins.** 1994. Type C retrovirus inactivation by human complement is determined by both the viral genome and producer cell. J. Virol. *68*:8001-8007.

28. **Wahlfors, J.J., K.G. Xanthopoulos, and R.A. Morgan.** 1997. Semliki forest virus-mediated production of retroviral vector RNA in retroviral packaging cells. Hum. Gene Ther. *8*:2031-2041.

16 Gene Transfer into Brain Slices from Mouse Embryos Using Herpes Simplex Amplicons

Angel Cid-Arregui[1,2] and Wulf Haubensak[2]
[1]Institute of Applied Tumor Virology, German Cancer Research Center, Heidelberg; [2]Max Planck Institute of Molecular Cell Biology and Genetics, Dresden, Germany

INTRODUCTION

As new generations of viral vectors are being developed and improved to eliminate cytotoxicity and other side effects of the parental viruses, they become more applicable to fine studies on fundamental aspects of developmental neurobiology. This chapter describes the application of herpes simplex virus type 1 (HSV-1) amplicons to target gene expression to the mouse neuroepithelium. The chapter includes a brief review of the basic aspects of neurogenesis, methods for culturing and infecting embryonic brain slices, and some illustrative results.

THE NEUROEPITHELIAL CELLS

The mammalian central and peripheral nervous systems derive from a specialized region along the dorsal midline of the embryo known as the neurectoderm (12,16,17). As it thickens, the neurectoderm forms the neural plate, whose edges elevate progressively to form a neural groove, bounded laterally by walls of raised neural folds whose edges form the neural crest. Starting from the cervical region, the neural folds fuse together progressively in the rostral and caudal directions, thereby transforming the neural plate into the neural tube. Further expansion of the rostral region gives rise to the forebrain, midbrain, and hindbrain. The caudal, unexpanded part of the neural tube will form the spinal cord.

Histologically, during the initial stages, the neural plate and the neural tube consist of a simple columnar epithelium (called neuroepithelium or ventricular zone) that, as cells proliferate, becomes a pseudostratified epithelium containing several layers of nuclei. Neuroepithelial (NE) cells have an elongated shape, perpendicular to the lumen of the neural tube, with a cytoplasmic process in contact with the basal lamina, and another process extending toward the lumen, where it projects microvilli and joins laterally to neighboring luminal processes by junctional complexes (Figure 1).

Analysis of cell proliferation in the neuroepithelium using [^{3}H]-thymidine has shown that the nuclei in the S phase lie in

Viral Vectors: Basic Science and Gene Therapy
Edited by A. Cid-Arregui and A. García-Carrancá

the outer half of the layer (Figure 1A) (26). As the cells enter the G_2 phase, their nuclei migrate toward the lumen, a moment when the cells lose contact with the basal lamina and round off, leading to the M phase. After mitosis (G_1 phase), the daughter cells, while remaining anchored to the lumen, elongate and expand to again reach the basal lamina, and their nuclei migrate in the same direction to repeat the cycle. This movement of the nuclei during the cell cycle is known as interkinetic migration (27). The total duration of the mitotic cycle varies along the neural tube and with the age of the embryo from 8 to 24 hours and longer (16), and it lengthens linearly during development (18).

Proliferation and Differentiation of NE Cells

From early stages in the development of the nervous system, some NE cells do not reenter the cell cycle. Instead, they lose contact with the ventricle, start to differentiate into neurons or glial cells, and migrate to the margins of the ventricular zone to form a new layer (i.e., the mantle zone) (Figure 1, B and C) (8,15). Proliferation of NE cells and generation of neurons and glia cells persist throughout development. In regions where more nerve cells should form, like the cerebral hemispheres, a second proliferative layer appears next to the NE (i.e., the subventricular zone) in which mitotic activity persists even after cessation of proliferation in the NE layer (2,25). In the mature nervous system, the neuroepithelium remains as a thin layer of ependymal cells lining the ventricular system of the brain and the spinal cord.

From a cell biology point of view, the dynamic polarization of nerve cells (26) is reflected in their NE progenitors, whose radial disposition implies three distinct plasma membrane domains (Figure 1): two small patches (one facing the ventricle and another in contact with the basal lamina) and a large columnar region immersed in the interstitium. Moreover, the interkinetic movements of nuclei and the changes in shape of the NE cells during the cell cycle require polarization of certain cytoplasmic components, as has been shown for microtubules and microfilaments (21). Finally, the postulated asymmetric mode of division of NE cells (Figure 1B) is likely to require a refined polarization of cytoplasmic and plasma membrane components of the progenitor cell during mitosis.

Data on proliferation and differentiation in the vertebrate nervous system have accumulated during the last century, mainly from histologic, genetic, autoradiographic, and cell lineage studies. From this data, some generalizations can be made: *(i)* All NE cells are capable of mitosis, since they incorporate [^{3}H]-thymidine into their DNA (6). *(ii)* In most regions, the first neurons and the first glial cells arise at the same time (Reference 16, and references therein). *(iii)* All nerve cells have to migrate from the NE region where they are generated to their final destination. In most regions they follow an inside-out pattern by which neurons originating later migrate past those formed earlier (1).

Fundamental aspects of NE cell proliferation remain unknown, in particular: *(i)* the mechanisms that maintain NE cell division and control the duration of the cell cycle in the neuroepithelium; *(ii)* the factors that induce cessation of DNA synthesis in certain NE cells and bring them to a permanent postmitotic state; *(iii)* whether the two daughter cells from an NE cell division may undergo different fates, and *(iv)* if so, the mechanisms underlying such asymmetric divisions. Progress in unraveling the molecular mechanisms underlying the polarization and asymmetric division of NE cells has been burdened by the complexity of these phenomena, which are difficult to approach experimentally in vivo. However, progressive refinement of strategies and techniques is facilitating studies that address the above questions. One such strategy is the use of viral vectors. Thus, for instance, retroviral vectors have been exploited to target expression of marker genes to the neuroepithelium for cell lineage analyses (11,23). Here

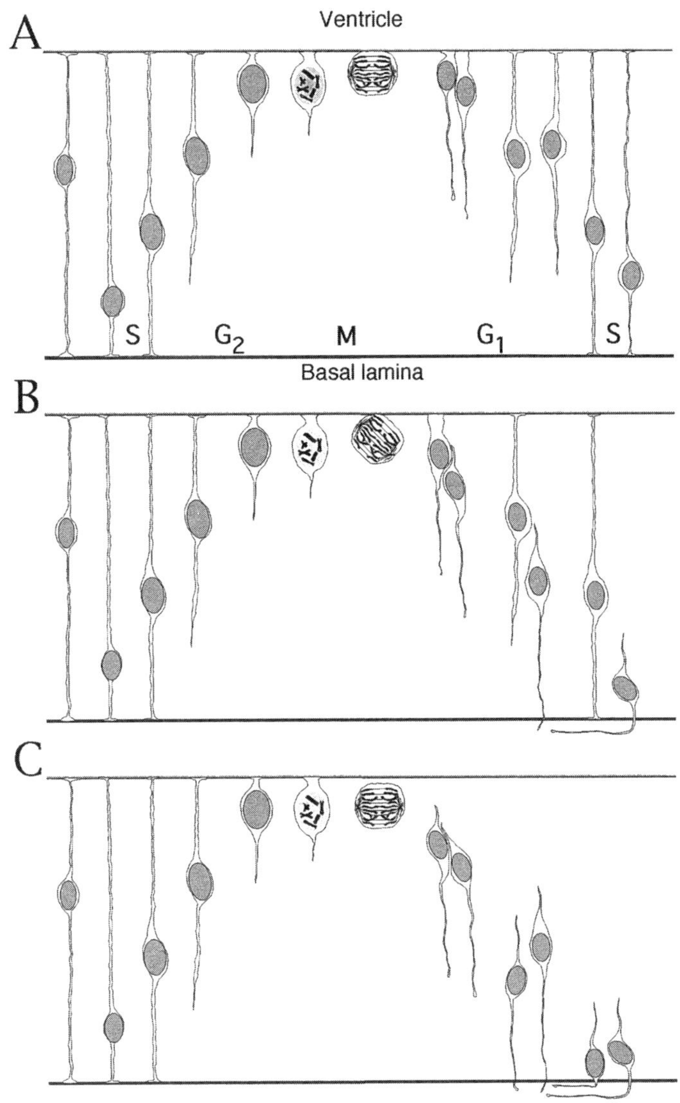

Figure 1. Schematic representation of the interkinetic migration of nuclei and the different modes of division of the NE germinal cells in the vertebrate neural tube. (A) Movement of the nucleus of a single NE cell during the mitotic cycle: DNA replication is initiated in the way of the nucleus toward the basal lamina and continues while it moves back to the lumen (ventricle); migration is completed during the G_2 phase, and mitosis occurs at the luminal side; in G1, the two daughter cells expand and their nuclei start to migrate outwards. During the early stages of the neural tube, the daughter cells reenter the cell cycle so that the number of germinal cells increases logarithmically (symmetrical mitosis, proliferative mode of division). (B) During neurogenesis, part of the germinal cells undergo mitosis that generates one postmitotic daughter cell and one germinal cell (asymmetrical mitosis, stem cell mode). (C) Later in development, an increasing number of germinal cells enter their terminal cycles to produce two postmitotic cells (symmetrical mitosis, terminal mode). Adapted from J. Comp. Neurol. (25).

we describe the use of HSV-1 amplicons to target expression of recombinant proteins to NE cells in vitro, whose fate can be followed in short-term slices and primary explants from brains of mouse embryos.

GENE TARGETING TO NE CELLS IN VITRO

The limited knowledge of the molecular mechanisms that control NE cell polarity and proliferation may be explained by the lack of cell type-specific markers that could help study the progenitor NE cells and their progeny. Attempts to overcome this problem have been made, in vitro and in vivo, by labeling NE cells in at least three different ways: *(i)* using lipophilic dyes, like DiI (6), *(ii)* neural cell lineage in transgenic mice expressing *lacZ* (31), and *(iii)* genetic marking of NE cells with retroviral vectors that express reporter genes. The latter has proven useful for tracing cell lineages in the nervous system (11,14,20,24,32). Of these, the use of viral vectors seems most convenient for studies that require gene targeting of single NE cells rather than overall expression of a transgene. However, retroviral vectors have some limitations, such as the difficulty in purifying adequately high titers of vector and their inability to infect nondividing cells. These limitations make it necessary to explore other vector systems. As shown below, the HSV-1 amplicon system is proving useful for targeting high levels of ectopic gene expression to NE cells in short-term studies in vitro.

Gene Transfer into Acute Brain Slices

Slice sections (50–500 μm) of living nervous tissue can be maintained in vitro under conditions that preserve cell viability and tissue architecture in a way which closely resembles that observed in situ (9,10,29). Short-term (acute) slice cultures of embryonic brains have proven to be a valuable system for studying a variety of molecular and cellular aspects of NE cells, such as asymmetric cell division (4), migration (5), and signaling through intracellular calcium fluctuations (22).

Detailed protocols for organotypic slice cultures of the embryonic neocortex have been described elsewhere (10,13,29).

In an attempt to study polarity and proliferation aspects of the NE cells in vitro in the context of the NE tissue, we have chosen HSV-1 vectors to infect slices of mouse telencephalon at embryonic day 15 (E15), when intense cortical neurogenesis is taking place (1). HSV-1 amplicons had been applied previously to transduce hippocampal slice cultures of postnatal rats (3), although their use in slices of developing brains had not been documented. Therefore, as a first step, we utilized HSV-1 amplicons (helper-containing stocks; see below) expressing either the bacterial β-galactosidase or the jellyfish green fluorescent protein (GFP) to determine the susceptibility of NE cells to infection and the effects of these vectors on cell and tissue viability.

Brain slices from E15 embryos were infected and cultured for 5 and 20 hours. Under the conditions described below, infection seemed not to alter the morphology of the tissue (Figure 2, A-C), which was comparable to that of noninfected slices (data not shown). Thus, for instance, the expression pattern of class III β-tubulin, a neuron-specific isotype whose expression is initiated immediately after the NE cells complete their final mitotic division (19), resembles that seen in situ (i.e., it is not expressed in the NE cell layer but in the neuronal layers above it) (Figure 2, A–C). Fluorescence microscopy of NE cells expressing GFP showed that they had their characteristic bipolar shape (Figure 2, B and D) (16,26,27). In addition, their nuclei incorporated bromodeoxyuridine (BrdU) (Figure 2, F–G), suggesting that

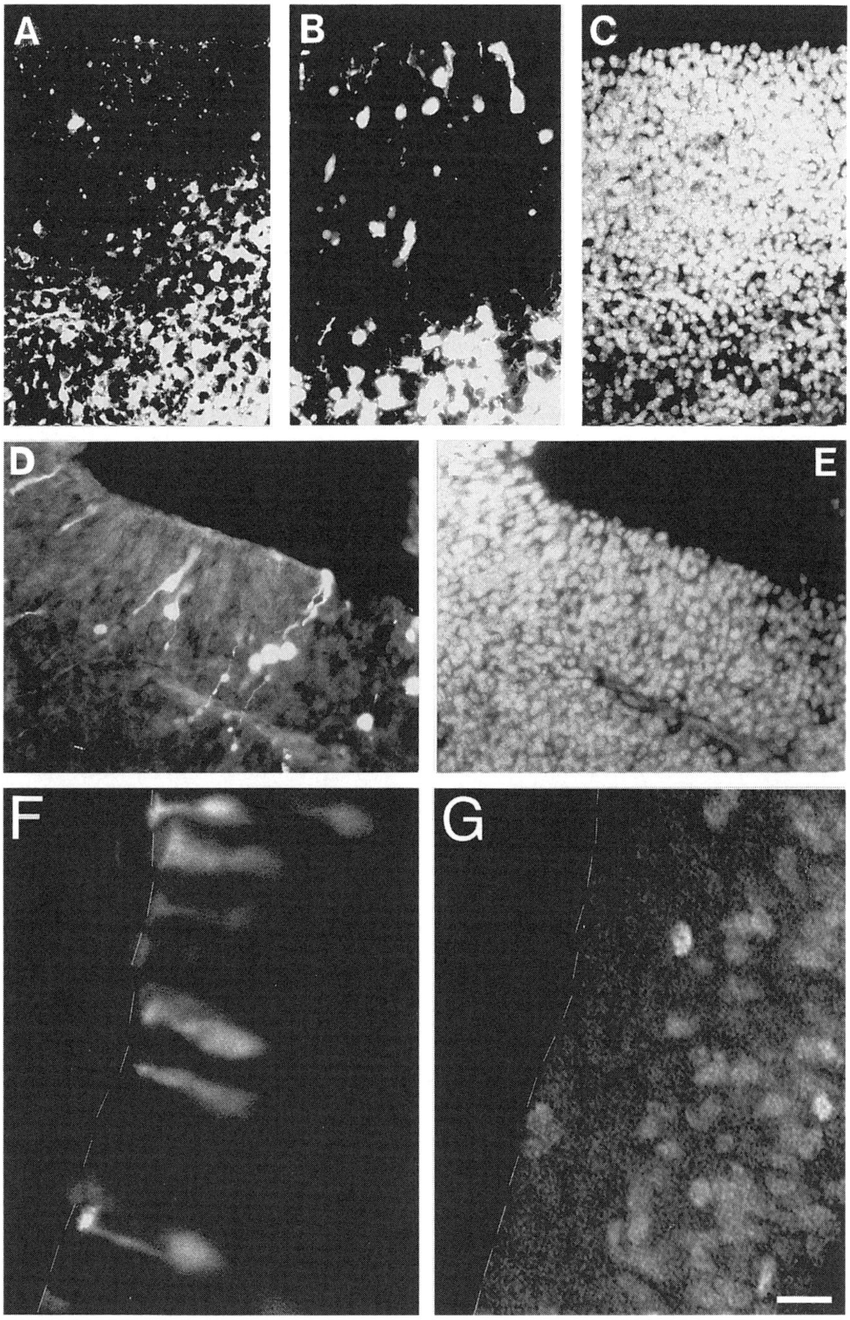

Figure 2. Representative slice sections from the brains of E15 mouse embryos infected with HSVEGFP amplicon. (A–C) Low-magnification view of a slice that was infected with amplicon for 1 hour, transferred to fresh medium without virus and incubated for 20 hours. The tissue was then fixed, embedded, and snap-frozen. Sections (10 μm) were stained with anticlass III β-tubulin (A) and anti-GFP (B) antibodies. The nuclei were counterstained with DAPI (C). Note that in the NE layer in panel A there is no expression of the β-tubulin neuronal marker, and that GFP-expressing cells are scattered throughout the tissue. (D–E) Higher magnification of the NE layer of an infected slide showing the bipolar shape of a GFP expressing cell. Some processes are labeled whose cell bodies are not visible, and vice versa. (F–G) Confocal image of a sectioned slice showing expression of GFP in NE cells (F) and nuclei in S phase in the NE cell layer labeled with BrdU (G). BrdU was added to the medium for the last 30 min of culture. Panels F and G correspond to two consecutive sections. Bar, 80 μm (A–C), 40 μm (D–E), and 20 μm (F–G).

the slices maintained the pattern of cellular proliferation seen in situ. Strong reporter expression was seen 5 hours after infection. After 20 hours, infected cells expressing GFP were seen not only in the NE layer and the margins of the slices that were exposed to the virus, but also deep into the neuronal layers expressing class III β-tubulin (Figure 2B). These cells, far from the surface of the slices and hence not accessible to the virus, could have been derived from cells primarily infected in the NE layer, which migrated later to such upper layers. Migration of neurons from the NE layer upwards has been estimated to occur in vivo at an average speed of 4 µm/h (8,25), but is much faster in vitro (10-100 µm/h) (6,16). Alternatively, some viral particles could have been transported intracellularly by the NE cells and released at their basal pole, although this pathway would imply that the virus could pass through the basal lamina.

An alternative strategy to the infection of slices is the infection of the NE layer of the forebrain by injection of virus suspension into the telencephalic vesicle of living embryos followed by incubation, in vitro or in uterus, prior to sectioning the brains into slices. An advantage of this procedure is that the neuroepithelium is the only tissue exposed to the virus. A drawback may be the difficulty in monitoring the NE cells at early postinfection stages.

CONCLUSION

In summary, we have found that: *(i)* HSV-1 amplicons are able to infect NE cells efficiently in vitro; *(ii)* the HSV-1 immediate early 4/5 promoter, used by these vectors to express the foreign gene, drives high levels of reporter gene expression; and *(iii)* NE cells proliferate and/or differentiate normally after infection. Therefore, HSV-1 amplicons are a valid system for targeting expression to NE cells, which can be applied to monitor division, differentiation, and migration of these cells in short-term slices and explants from the developing forebrain.

PROTOCOLS

Materials

1. Culture dishes: normal petri dishes (3- or 6-cm ø), dishes (3-cm ø) with glass bottom surface for microscopy.
2. Dissection material: Fine forceps (DuMont™ #5 or #55; Fine Science Tools, Foster City, CA, USA), blunt forceps (for holding uteri), and small scissors (for dissecting uteri).
3. Microscopes: dissection stereomicroscope (Olympus SZX12 [Melville, NY, USA] or equivalent) and epifluorescence microscope (Carl Zeiss Axiophot™ [Thornwood, NY, USA] or equivalent).
4. Glass (autoclaved) and plastic (UV-irradiated) Pasteur pipets.
5. Plastic tubes (e.g., 15-mL Falcon polypropylene tubes or equivalent).
6. Vibrating blade microtome: Leica VT1000S (Deerfield, IL, USA) with stainless steel razor blades.

Reagents and Media

1. Antibodies: anti-BrdU (monoclonal; Roche Molecular Biochemicals, Indianapolis, IN, USA), anti-β-III tubulin (monoclonal; Sigma, St. Louis, MO, USA), anti-GFP (polyclonal; Clontech, Palo Alto, CA, USA), anti-HSV-1 (Dako, Carpinteria, CA, USA), monoclonal and polyclonal antimouse and antirabbit secondary antibodies conjugated with fluorescein isothyocyanate and rhodamine (Molecular Probes, Eugene, OR, USA).

2. Blocking solution: 2% fetal calf serum (FCS)/2% bovine serum albumin (BSA)/0.2% gelatin in phosphate-buffered saline (PBS).
3. Bromodeoxyuridine (BrdU): stock 5 mg/mL in PBS (16 mmol/L).
4. Cyanoacrylate-based glue (e.g., Krazy Glue®, Roticoll®, or other similar glue).
5. DAPI (Molecular Probes): stock 1 mg/mL in water, stored at -20°C. Working dilution should be 1 µg/mL.
6. L-glutamine, 200 mmol/L, 100×, (Life Technologies, Rockville, MD, USA).
7. Laminin, 1 mg/mL (Sigma).
8. Mowiol (Calbiochem-Novabiochem, La Jolla, CA, USA); Prepare a stock as follows: add 10 g to 40 mL PBS and stir for 2 hours. Then, add 20 mL glycerin and stir for 10 minutes. Centrifuge at 3000 rpm (800× *g*) for 15 minutes, make aliquots of 1 mL, and store at -20°C.
9. N-2 Supplement, 100× (Life Technologies).
10. Neurobasal (NB) medium (Life Technologies).
11. Paraformaldehyde (PFA): Prepare a stock of 4% PFA and 4% sucrose as follows: weigh 8 g PFA and 8 g sucrose in a beaker, fill up to 180 mL with water and heat to 56° to 60°C. Add a small amount of phenol red and a few drops of 1 mol/L NaOH (one per minute) until the solution clears. Let cool to room temperature. Add 20 mL 10× PBS to a final volume of 200 mL. Adjust pH to 7.0 to 7.5, if necessary. Aliquot and store at -20°C.
12. PB1 medium (28): 136.9 mmol/L NaCl, 2.68 mmol/L KCl, 0.49 mmol/L $MgCl_2 \times 6H_2O$, 8.04 mmol/L $Na_2HPO_4 \times 12\ H_2O$, 1.47 mmol/L KH_2PO_4, 0.9 mmol/L $CaCl_2 \times 2H_2O$, 5.56 mmol/L glucose, 0.33 mmol/L sodium pyruvate, 4.35 mmol/L sodium lactate. Adjust pH to 7.3 to 7.4 and sterilize by filtering through 0.45-µm membrane filter.
13. PBS: 137 mmol/L NaCl, 2.7 mmol/L KCl, 8 mmol/L $Na_2HPO_4 \times 2\ H_2O$, 1.5 mmol/L KH_2PO_4, 1 mmol/L $MgCl_2$, and 2 mmol/L $CaCl_2$.
14. Poly-L-lysine: stock 100 µg/mL in water (MW 70 000–150 000; Sigma)
15. TESPA (3-aminopropyl triethoxysilane; Sigma); Working solution: 2% (vol/vol) TESPA in 100% acetone. Coat slides by incubating 10 seconds into each: *(i)* 2% TESPA solution; *(ii)* twice in 100% acetone; *(iii)* deionized water. Dry at 37°C for 4 hours or at room temperature overnight. Store at room temperature protected from the dust.
16. Tissue freezing medium.

Preparation of HSV-1 Amplicons

The HSV-1 amplicons used here to infect explants of mouse embryo brains (HSV-EGFP and HSV-LacZ), and the protocols for purification and titration of vector stocks are described in Chapter 12. Typically, the titer of the amplicon stocks (containing helper virus) was in the range of 1×10^7 infectious units (iu), and the amplicon-to-helper ratio was approximately 1:1. Infections of living brain sections were carried out by incubating 5 to 10 slices (200 µm thick) or the explanted NE tissue (see below) in 1 mL of conditioned medium containing about 1 to 5×10^5 iu.

Preparation, Infection, and Culture of Embryonic Forebrain Slices

The experiments described here were performed with 200-µm slices. However, it should be noted that for studies that require direct observation of infected cells, for example, to follow their migration by time-lapse video fluorescence microscopy, the thickness of the slices should be less than 100 μm to minimize light scatter and thus enhance cell visualization. The ability of current microscopy methods to resolve detailed cellular features is limited to 40 to 50 μm from the slice surface.

For better preservation of tissue, dissection of embryos and sectioning of brains should be carried out as quickly as possible and at low temperature, using an ice-cold platform under the stereomicroscope and cold PB1 medium. Routinely, three to five brains are sectioned at a time.

1. Kill a pregnant mouse of the desired strain (e.g., CD1) by cervical dislocation at embryonic day 15. (Mice are 0.5 days pregnant on the day vaginal plug is detected.)
2. Open the abdominal walls, dissect the uterus and transfer it into a 10-cm ø dish containing 10 mL of cold PB1.
3. Using fine scissors, cut the uterine wall longitudinally by its inner side (where the placentas are implanted), taking care not to damage the embryos. Cut three embryos with the scissors transversally through the neck, and transfer the heads into a new dish filled with cold PB1 medium. Carefully, take out the brains by cutting the skull along the midline, without damaging the cerebral hemispheres, and separating the edges with fine forceps until the brain can be released by pushing gently from the frontal side with a fine spatula or using blunt forceps.
4. Transfer the three brains onto a specimen plate (Figure 3) with a perforated spoon so that they lie on their basal sides. Align them close to the center with the olfactory bulb pointing to the margin of the plate and the hindbrain to the center. Using a razor blade, section the hindbrains to leave a flat surface. Wipe excess PB1 medium on the plate. Put three drops of cyanoacrylate-based glue in a line parallel to the three brains, but separated 1 to 2 mm from them. With the help of a spatula, lift the brains, one by one, from below the olfactory bulbs until they lie on the hindbrain side on top of the glue (sections will be frontal). The brains can be better handled and glued if the specimen plate is at room temperature because the glue starts to polymerize in contact with water condensed on the plate when it is cold. This step should be performed very quickly to prevent the tissue from drying.
5. Transfer the specimen plate with the brains immediately to the microtome buffer tray (filled with cold PB1) and fix it so that the line formed by the brains is parallel to the edge of the razor blade (Figure 3). Cut slice sections 100 to 200 μm thick. The microtome can be operated at different adjustable settings according to the type of tissue. For E15 brains we used the following settings: knife angle, 5 to 7 degrees; sectioning speed, 0.2 mm/s; oscillation frequency, 80 mHz; oscillation amplitude, 0.6 mm.
6. Transfer slices to a 6-cm ø dish filled with cold PB1 and select those of intact appearance under the dissection microscope. Remove fragments of meninges that remain attached to the slices.
7. Transfer slices carefully into an appropriate culture 3-cm ø dish (normal, or with glass bottom surface dishes, according to the nature of the experiment) coated with PLL and laminin (first PLL 100 μg/mL in PBS, and then laminin 10 μg/mL in PBS; coat dishes by pipetting 1 mL of the respective solution and incubating

the dish for 30 to 60 minutes at room temperature, then wash twice with PBS).

8. Cover the slices with 0.5 mL of culture medium (NB + N2 supplement/10% rat or mouse heat-inactivated serum) and transfer to the incubator (36°C/5% CO_2).

Infection of Slices

Infection is more efficient if the tissue is incubated with virus diluted in serum-free medium.

1. Immediately after sectioning or at different times in culture, the slices can be infected by removing the medium carefully with a Pasteur pipet from the dish and replacing it with PB1 medium containing the appropriate amount of virus. Return the dish to the incubator for 1 hour.
2. Remove infecting medium and wash twice with PB1. Then, add 0.5 mL of fresh medium and incubate for a convenient time (several hours to 24 hours).

BrdU Labeling of S-Phase Cells

BrdU is a thymidine analog that is incorporated specifically into DNA during DNA synthesis. Apparently, it has little or no effect on cell viability or growth, and it seems not to prevent or delay nerve cell differentiation in vivo (14,31). BrdU has a short metabolic half-life and therefore can be used to label only those cells that are synthesizing DNA at the time of its addition to the medium. Small amounts of BrdU incorporated into cellular DNA can be detected with monoclonal antibodies.

S-phase NE cells of slices maintained in vitro can be labeled at any time by adding 10 to 15 µmol/L BrdU to the culture medium and incubating for 15 to 30 minutes to allow for BrdU uptake and incorporation into DNA. The tissue can be stained immediately or, if convenient, the BrdU pulse can be followed by a chase. BrdU-labeled cells are immunostained in sections of fixed tissue as follows:

1. Fix slices in 4% PFA/4% sucrose/PBS at 4°C for 1 hour on a rocking platform.

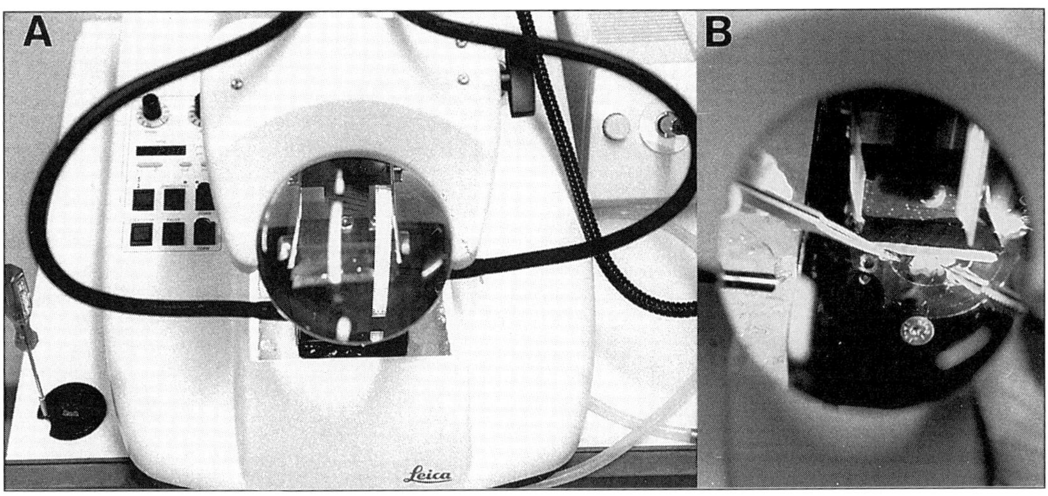

Figure 3. Vibrating blade microtome (Leica VT100S) used to produce slice sections from living mouse embryo brains. (A) Microtome overview. (B) Detail of the buffer tray viewed through the lens during sectioning.

2. Wash slides with PBS and saturate in 30% sucrose/PBS at 4°C for 30 minutes (or until they sink). Embed in tissue freezing medium and snap-freeze in liquid nitrogen.
3. Cut 10- to 20-μm sections and place them on TESPA-coated slides.
4. Permeabilize sections with 0.3% Triton X-100™ in PBS for 5 minutes.
5. Reduce background fluorescence by quenching free aldehydes in 50 mmol/L NH_4Cl/PBS for 10 minutes.
6. Denature DNA in 2 *N* HCl for 30 minutes at 37°C.
7. Neutralize in 0.5 mol/L Tris HCl, pH 8.5, for 10 minutes.
8. Wash three times with PBS for 2 minutes each time.
9. Incubate sections in blocking solution for 30 minutes at room temperature.
10. Replace blocking solution with blocking solution/PBS (1:10) containing 5 μg/mL of anti-BrdU antibodies, and incubate for 1 hour at room temperature.
11. Wash three times with PBS, 2 min each.
12. Incubate in blocking solution containing the appropriate dilution of secondary antibody. Incubate for 1 hour at room temperature.
13. Wash three times with PBS for 2 minutes each time. Mount with a drop of Mowiol.

Notes:

1. If double immunodetection of BrdU and a cellular antigen is required, reduce the time of incubation in HCl to 10 to 15 minutes. Blocking and incubation with primary and secondary antibodies are performed the same way but must include the appropriate antibodies.
2. Triton X-100 can interfere with the detection of some antigens (in particular, integral membrane proteins). In such cases, permeabilization with this detergent should be shorter (1 minute) or performed by passing the tissue through an ethanol series (1 minute into each solution 25%, 50%, 70%, 90%, 100%, 90%, 70%, 50%, and 25% ethanol; finally, rehydrate 1 minute in PBS and proceed with immunofluorescence).
3. Infected cells expressing GFP can be followed in vivo by time-lapse video microscopy; the reader interested in this technique should consult specialized literature [e.g., Reference 4].
4. Alternatively, or complementarily to the culture, the slices can be fixed, snap-frozen and sectioned with a cryostat, and expression of the transgene analyzed by immunostaining or by in situ hybridization.

ACKNOWLEDGEMENTS

We thank J. Koch and B. Oback for helpful comments on the manuscript, and W.B. Huttner for stimulating discussions and support. A.C.-A. gratefully acknowledges H. zur Hausen for continuous support and interest in this work.

REFERENCES

1. **Angevine, J.B. and R.L. Sidman.** 1961. Autoradiographic study of the cell migration during histogenesis of cerebral cortex in the mouse. Nature *192*:766-768.
2. **Bayer, S.A. and J. Altman.** 1991. Neocortical Development. Raven Press, New York.
3. **Bergold, P., P. Casaccia-Bonnefil, Z. Xiu-Liu, and H. Federoff.** 1993. Transsynaptic neuronal loss induced in hippocampal slice cultures by a herpes simplex virus vector expressing the GluR6 subunit of the kainate receptor. Proc. Natl. Acad. Sci. USA *90*:6165-6169.
4. **Bradke, F. and G.C. Dotti.** 1998. Videomicroscopy of living microinjected hippocampal neurons in culture, p. 81-94. *In* A. Cid-Arregui and A. Garcia-Arranca (Eds.), Microinjection and Transgenesis: Strategies and Protocols. Springer-Verlag, Berlin.

5.**Chenn, A. and S.K. McConnell.** 1995. Cleavage orientation and the asymmetric inheritance of notch1 immunoreactivity in mammalian neurogenesis. Cell *82*:631-641.

6.**Fishell, G., C.A. Mason, and M.E. Hatten.** 1993. Dispersion of neural progenitors within the germinal zones of the forebrain. Nature *362*:636-638.

7.**Fujita, S.** 1964. Analysis of neuron differentiation in the central nervous system by tritiated thymidine autoradiography. J. Comp. Neurol. *122*:311-328.

8.**Fujita, S.** 1966. Application of light and electron microscopy to the study of the cytogenesis in the forebrain, p. 180-196. *In* R. Hassler and H. Stephen (Eds.), Evolution of the Forebrain. Plenum Press, New York.

9.**Gahwiler, B.H.** 1981. Organotypic monolayer cultures of nervous tissue. J. Neurosci. Meth. 4:329-342.

10.**Gähwiler, B.H., M. Capogna, D. Debanne, R.A. McKinney, and S.M. Thompson.** 1997. Organotypic slice cultures: a technique has come of age. Trends Neurosci. *20*:471-477.

11.**Galileo, D.S., G.E. Gray, G.C. Owens, J. Majors, and J.R. Sanes.** 1990. Neurons and glia arise from a common progenitor in chicken optic tectum: demonstration with two retroviruses and cell type-specific antibodies. Proc. Natl. Acad. Sci. USA *87*:458-462.

12.**Gilbert, S.F.** 1994. Developmental Biology, 4th ed. Sinauer Associates, Sunderland.

13.**Haydar, T.F., L.L. Bambrick, B.K. Krueger, and P. Rakic.** 1999. Organotypic slice cultures for analysis of proliferation, cell death, and migration in the embryonic neocortex. Brain Res. Protocols *4*:425-437.

14.**Huszar, D., A. Sharpe, S. Hashmi, B. Bouchard, A. Houghton, and R. Jaenisch.** 1991. Generation of pigmented stripes in albino mice by retroviral marking of neural crest melanoblasts. Development *113*:653-660.

15.**Jacobson, M.** 1968. Cessation of DNA synthesis in retinal ganglion cells correlated with the time of specification of their central connections. Dev. Biol. *17*:219-232.

16.**Jacobson, M.** 1991. Developmental Neurobiology, 3rd ed. Plenum Press, New York.

17.**Jones, E.G. and W.M. Cowan.** 1983. The nervous tissue. *In* L. Weiss (Ed.), Cell and Tissue Biology, 5th ed. Elsevier Science Publisher, New York.

18.**Kauffman, S.L.** 1968. Lengthening of the generation cycle during embryonic differentiation of the mouse neural tube. Exp. Cell Res. *49*:420-424.

19.**Lee, M.K., J.B. Tuttle, L.I. Rebhuhn, D.W. Cleveland, and A. Frankfurter.** 1990. The expression and posttranslational modification a neuron-specific β-tubulin isotype during chick embryongenesis. Cell Motil. Cytoskeleton *17*:118-132.

20.**Luskin, M.B., A.L. Pearlman, and J.R. Sanes.** 1988. Cell lineage in the cerebral cortex of the mouse studied in vivo and in vitro with a recombinant retrovirus. Neuron *1*:635-647.

21.**Messier, P.E. and C. Auclair.** 1974. Effect of cytochalasin B on interkinetic nuclear migration in the chick embryo. Dev. Biol. *36*:218-223.

22.**Owens, D.F. and A.R. Kriegstein.** 1998. Patterns of intracellular calcium fluctuation in precursor cells of the neocortical ventricular zone. J. Neurosci. *18*:5374-5388.

23.**Price, J. and L. Thurlow.** 1989. Cell lineage in the rat cerebral cortex: a study using retroviral-mediated gene transfer. Development *104*:473-482.

24.**Price, J., D. Turner, and C. Cepko.** 1987. Lineage analysis in the vertebrate nervous system by retrovirus-mediated gene transfer. Proc. Natl. Acad. Sci. USA *84*:156-160.

25.**Rakic, P.** 1974. Neurons in Rhesus monkey visual cortex: systematic relation between time of origin and eventual disposition. Science *183*:425-427.

26.**Ramón y Cajal, S.** 1995. Histology of the Nervous System of Man and Vertebrates. Oxford University Press, New York.

27.**Sauer, F.C.** 1935. Mitosis in the neural tube. J. Comp. Neurol. *62*:377-405.

28.**Sidman, R.L., I.L. Miale, and N. Feder.** 1959. Cell proliferation and migration in the primitive ependymal zone: an autoradiographic study of histogenesis in the nervous system. Exp. Neurol. *1*:322-333.

29.**Stoppini, L., P.A. Buchs, and D. Muller.** 1991. A simple method for organotypic cultures of nervous tissue. J. Neurosci. Methods *37*:173-182.

30.**Sturm, K. and P.P.L. Tam.** 1993. Isolation and culture of whole postimplantation embryos and germ layer derivatives, p. 164-90. *In* P.M. Wassarman and M.L. DePanphilis (Ed.), Methods in Enzymology: Guide to Techniques in Mouse Development, Vol. 225. Academic Press, New York.

31.**Tan, S.-S. and S. Breen.** 1993. Radial mosaicism and tangential cell dispersion both contribute to mouse neocortical development. Nature *362*:638-640.

32.**Walsh, C. and C. L. Cepko.** 1992. Widespread dispersion of neuronal clones across functional regions of the cerebral cortex. Science *255*:434-440.

33.**Wilt, F.H. and M. Anderson.** 1972. The action of 5-bromodeoxyuridine on differentiation. Dev. Biol. *28*:443-447.

17 Gene Transfer into Cultured Postimplantation Mouse Embryos Using Herpes Simplex Amplicons

Björn Oback[1,2], Angel Cid-Arregui[1,3], and Wieland B. Huttner[1]
[1]Max Planck Institute of Molecular Cell Biology and Genetics, Dresden, and Department of Neurobiology, University of Heidelberg, Heidelberg, Germany; [2]Reproductive Technologies, AgResearch, Ruakura Research Centre, Hamilton, New Zealand; [3]Institute of Applied Tumor Virology, German Cancer Research Center, Heidelberg, Germany

INTRODUCTION

For many years, simple and reproducible techniques for the culture of postimplantation rodent embryos have been available (9,24). These methods support growth and development that approximate that of embryos in utero. The possibility of maintaining and manipulating embryos in vitro is particularly attractive to developmental biologists who are interested in early mammalian morphogenesis and organogenesis. In this chapter, we focus on the application of whole-embryo culture to study early development of the central nervous system in the mouse. We have combined whole-embryo culture with herpes simplex virus (HSV)-based viral gene transfer to functionally characterize proteins involved in early vertebrate neurogenesis.

Using whole-embryo culture for the expression of the gene of interest has several advantages over other biological systems, such as cell culture approaches or transgenic animals. It allows the study of early embryonic cells and tissues in their natural three-dimensional microenvironment, thereby preserving cell-to-cell contacts, cell polarity, contact to basal lamina, and other physiological parameters. Therefore, whole-embryo culture provides data that are more readily transferable to the in vivo situation; it is thus superior to cell lines, primary cultures of dissociated cells, or organotypic cultures. Expression of the gene of interest can be accomplished at a defined time and location, which reduces the risk of compensatory, pleiotropic, or embryonic lethal effects often observed with transgenic animals. Once established, the method is relatively easy to handle and fast, requiring only 2 to 3 days between start of the experiment and evaluation of the resulting phenotype. At the time of

Viral Vectors: Basic Science and Gene Therapy
Edited by A. Cid-Arregui and A. García-Carrancá

microinoculation of the HSV amplicon vector (between embryonic day [E] 7 and E9), most organ primordia originate (27) and are therefore amenable to virus infection. Some tissues (e.g., neural plate, neural crest, and heart) are directly facing the amniotic cavity and can be targeted simply by injecting virus into the amniotic fluid.

HSV-1 Amplicons

Amplicons are plasmids that contain an HSV origin of replication and an HSV packaging signal in addition to the gene of interest and the sequences required for replication/selection in bacteria (12,26) (see Chapter 12, Construction of Defective Herpes Simplex Viral Vectors and Helper-Dependent Amplicons). In the presence of viral protein provided by a mutated helper virus, the amplicon will package into infectious viral particles that transmit the gene of interest to its target cell. We have focused on using recombinant amplicon vectors to mediate gene transfer into neuroepithelial precursor cells of the neural plate and early neural tube stage.

When performing infections for the first time, it is advisable to use a viral vector carrying an easily detectable reporter gene. We have used a *lacZ*-coding HSV (13) to establish and optimize amplicon-based gene transfer into cultured mouse embryos. E7.5–E8.5 embryos were infected with HSV*lacZ* by injection into the amniotic cavity, cultured for 3 to 48 hours, sectioned, and analyzed for expression of the transgene, either by staining with X-Gal (5-bromo-4-chloro-3-indolyl-β-D-galactoside) or immunocytochemistry for β-galactosidase (β-gal) (see below).

After the initial characterization with HSV*lacZ*, we used the amplicon system for functional assays of proteins expressed during early mouse neurogenesis, namely the antiproliferative protein TIS21 (17). Different versions of TIS21 were subcloned into the pHSVpUC plasmid (13), packaged into infectious virus particles, and ectopically expressed in the neuroepithelium.

Procedures used in our laboratory for packaging of HSV amplicons, virus purification, determination of amplicon, and helper virus titer are described in detail in Chapter 12.

WHOLE-EMBRYO CULTURE

Before starting any microinjection experiments, it is useful to first optimize the conditions for culturing whole embryos. In principle, mouse embryos between early gastrulation (E6.5) and 40-somite, hind limb-bud stage (E10.5), can be kept in culture under the conditions described below. However, embryos explanted from the late primitive streak (E7.5) to early somite stage (E8.5) consistently give the best results and develop outside the uterus at the same rate as in vivo for about two days.

Principal features of whole embryo culture conditions are: *(i)* rat or mouse serum supplemented with chemically defined media, *(ii)* continuous rotation at physiological pH (7.3–7.4) and temperature (37°C), and *(iii)* gas conditions that are varied according to the developmental stage of the embryos.

Materials

- Blunt forceps (for holding uterus)
- Dissection stereomicroscope (Olympus SZX12 (Olympus, Melville, NY, USA) or equivalent)
- Fine forceps (DuMont™ #5 or #55; Fine Science Tools, Foster City, CA, USA) for removing decidua and Reichert's membrane)
- Gas bottle with appropriate mixture (see below)
- Plastic tubes (e.g., 50-mL FALCON®

polypropylene tubes (Becton-Dickinson, Franklin Lakes, NJ, USA) or equivalent)
- Plastic Petri dishes (ø 60 or 100 mm)
- Plastic transfer pipets (3 mL)
- Roller apparatus
- Small scissors (for opening abdomen and removing uterus)

Reagents, Solutions, and Media

- Dulbecco's modified Eagle's medium (DMEM) (containing 24 mmol/L $NaHCO_3$ and 1 g/L glucose)
- Immediately centrifuged (IC) serum (rat and/or mouse)
- L-glutamine 200 mmol/L (100×)
- Phosphate-buffered dissection medium [PB1 (32)] dissection medium consisting of 150 mmol/L NaCl, 5.4 mmol/L KCl, 2 mmol/L $CaCl_2$, 1 mmol/L $MgCl_2$, 8 mmol/L Na_2HPO_4, 1.5 mmol/L KH_2PO_4, 0.3 mmol/L sodium pyruvate, 1 g/L glucose, 5 mmol/L HEPES, and 10% fetal calf serum (FCS) (v/v). Make up 1 L solution in water, add 100 U/mL penicillin/streptomycin, filter through 0.22-μm Millipore filter (Millipore, Bedford, MA, USA), aliquot, and freeze at -20°C.
- Penicillin/streptomycin consisting of 100× stock solution (10 000 U penicillin/10 000 μg streptomycin/mL)

Dissection of Embryos for Microinjection and Culture

A major difficulty in working with mammalian embryos is that they implant and develop inside the uterus and cannot simply be flushed out for experiments. Detailed protocols for explanting early somite-stage embryos have been previously described (9,16). All explanting instruments must be clean, and media and dishes must be sterile. Embryos are dissected in 60-mm petri dishes containing freshly thawed PB1 medium (32) at room temperature. Briefly, the uterus, decidua, and Reichert's membrane are removed, while the ectoplacental cone and all extraembryonic membranes are left intact and cultured with the embryo.

1. Kill a pregnant female by cervical dislocation. Soak the abdominal skin with 70% ethanol and make a large V-shaped incision (Figure 1A). Grasp one end of the uterus near the ovary with fine forceps and lift up. Trim away the fat and mesentery with fine scissors until the cervix is reached. Cut through it without separating the two uterine horns and trim the other horn until the whole uterus can be removed.
2. Rinse the uterus in PB1 and remove residual fat and mesentery. Transfer the uterus to a new dish with fresh PB1 and conduct all subsequent steps under the dissection microscope. Open the uterus with two pairs of fine forceps by tearing it along the antimesometrial side (Figure 1B). Keep the tips of the two forceps close together to prevent any excessive strain on the embryos, which could damage them.
3. Gently isolate the decidual swellings by sliding with closed forceps along the inner uterine wall (Figure 1C). Transfer them to a new dish containing PB1.
4. Open the decidua capsularis by grasping it with one pair of forceps and piercing it with the other one as shown in Figure 1D. Slowly open the pair of forceps, thereby separating the upper region of the decidua in halves. The reddish ectoplacental cone at the upper end of the embryo should now become visible.
5. Hold each half of the decidua carefully with a pair of forceps and tear it completely apart (Figure 1E). The embryo will most often remain in one half.
6. Grasp the embryo at the tip of the ectoplacental cone and remove it from the decidua mass. If the conceptus is very sticky, it may be necessary to sever the con-

nections between embryonic trophoblast and maternal decidua with the second pair of forceps (Figure 1F). Transfer all embryos to a new dish with a plastic transfer pipet.

7. The last step is to remove the thin, acellular Reichert's membrane. Reichert's membrane does not grow in culture and would otherwise block in vitro development. Moreover, it is hard to penetrate and would interfere with subsequent microinjection. Reichert's membrane is clearly visible at the embryonic pole of the conceptus and can be torn open with two pairs of fine watchmaker's forceps (Figure 1G).

Care must be taken during the whole procedure not to compress the fluid-filled cavities or damage the surrounding yolk sac because this will cause leakage of fluid and eventual collapse of the embryo. Any damaged, retarded, squashed, or otherwise abnormal embryos are discarded. Intact embryos are microinjected and cultured for the desired time.

Culture Protocols

Culture Media

Embryos are cultured in DMEM supplemented with either 50% rat or mouse serum (DR50 or DM50) or a combination of 50% rat and 25% mouse serum (DRM75). Commercially available sera lack some yet unidentified embryotrophic activities and are not recommended. Serum must be prepared by immediate centrifugation (i.e., formation of the fibrin clot occurs during centrifugation, and the clot is squeezed afterwards), often referred to as IC serum (28), followed by heat inac-

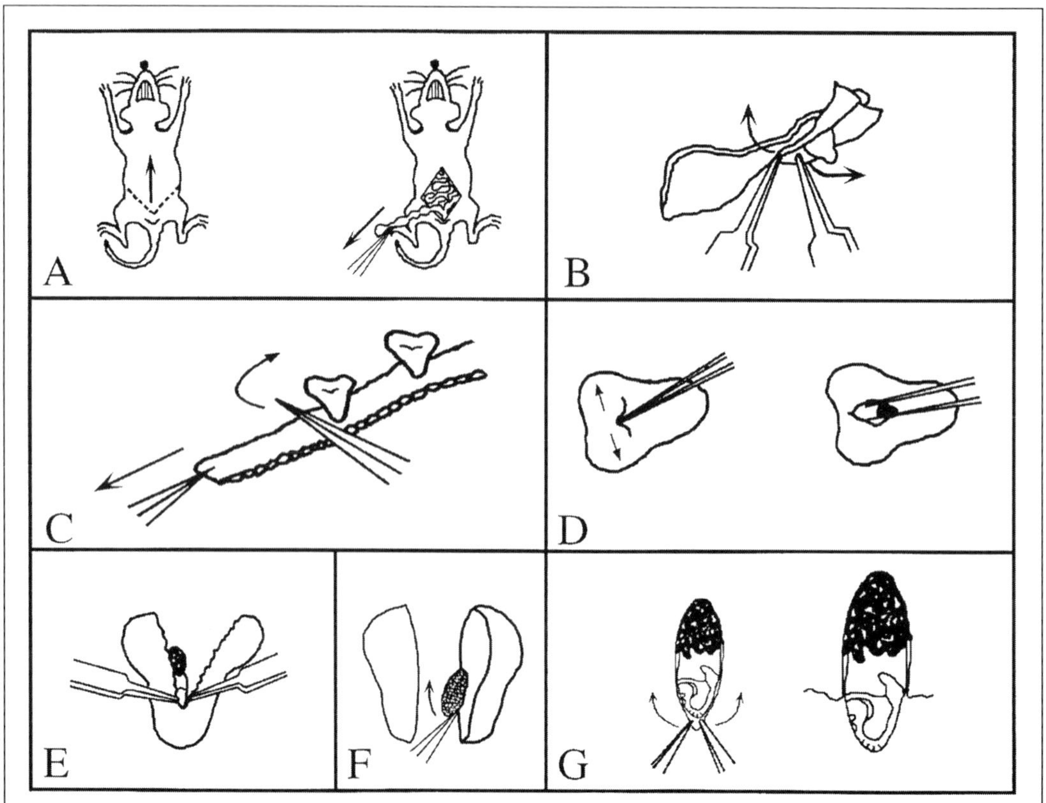

Figure 1. Dissection of embryos for microinjection and culture. Steps A–G are described in more detail in the text.

tivation at 56°C for 30 minutes (with the cap of the tube partly loosened to allow anesthetic ether to evaporate). Serum can be stored at -20°C (for up to six months) or at -70°C (for up to one year). For more comprehensive descriptions of how to prepare IC serum, see references 9 and 16.

To prepare the culture medium, thaw an aliquot of serum and mix 1:1 (DR50) or 3:1 (DRM75) with DMEM containing 24 mmol/L $NaHCO_3$, 4 mmol/L L-glutamine, 1 g/L glucose, and 100 U/mL penicillin/streptomycin. Four milliliters of medium is sufficient to culture eight E7.5 or four E8.0–E8.5 embryos for 36 hours.

Culture Equipment

Embryos are cultured in rotating tubes. Any sterile tube that can be made gas-tight and is free of detergents or any other toxic substances will suffice. Disposable plastic tubes (e.g., 50-mL polypropylene Falcon tubes) are preferable over reusable glass bottles since they do not require any thorough cleaning steps. The roller apparatus consists of multiple parallel horizontal rollers rotating at around 30 rpm (Figure 2A). It may be constructed in a workshop or bought (from neoLab Migge, Heidelberg, Germany). To ensure a sufficient depth of medium, the whole roller apparatus should be tilted at a 10°–20° angle.

Gas Conditions

Different gas mixtures are needed for different embyonic stages, but at all stages, the mixture must contain 5% CO_2 to maintain physiological pH (7.3–7.4). Older embryos require higher O_2 levels, increasing from 5% at early somite stage (0–10 somites) to 20% at 10 to 20 somites, to 40% at 20 to 35 somites, and to 95% O_2 for embryos with more than 35 somites. These mixtures contain N_2 to make up the final volume and are commercially available. The appropriate gas mixture is supplied directly into the culture tube for about 1 minute either via a 1-mL syringe (without piston) fitted to a 0.22-µm filter unit or simply via a cotton-plugged Pasteur pipet. After gassing, the tubes are tightly sealed with Parafilm and continously rotated on the roller apparatus within a 37°C incubator.

Cultures of early somite stage embryos start in 5% O_2. After 18 to 22 hours, the concentration should be raised to 20%. For cultures rotating in a 5% CO_2 incubator, regassing can be done by simply opening the tube for some minutes.

Culture Period

Embryos develop normally for two days in roller bottle cultures with one additional gassing (as above) when explanted from early somite stage. After about 48 hours, however, they begin to display obvious developmental retardation and morphological anomalies. Using more sophisticated culture techniques (e.g., constant gassing or renewal of culture medium), the culture period can be extended for up to 72 hours (29), but we have not analyzed HSV infections after more than 48 hours.

MICROINJECTION OF MOUSE EMBRYOS

Materials

- Base plate (Leitz [Bensheim, Germany] or equivalent), where the microinjection setup is mounted
- Depression slide injection chamber
- Glass capillaries with filament, (Fa Hilgenberg GmbH, Malsfeld, Germany)
- Glass cutter
- Inverted light microscope (Leica DM [Bensheim, Germany] IRB/E or equivalent)

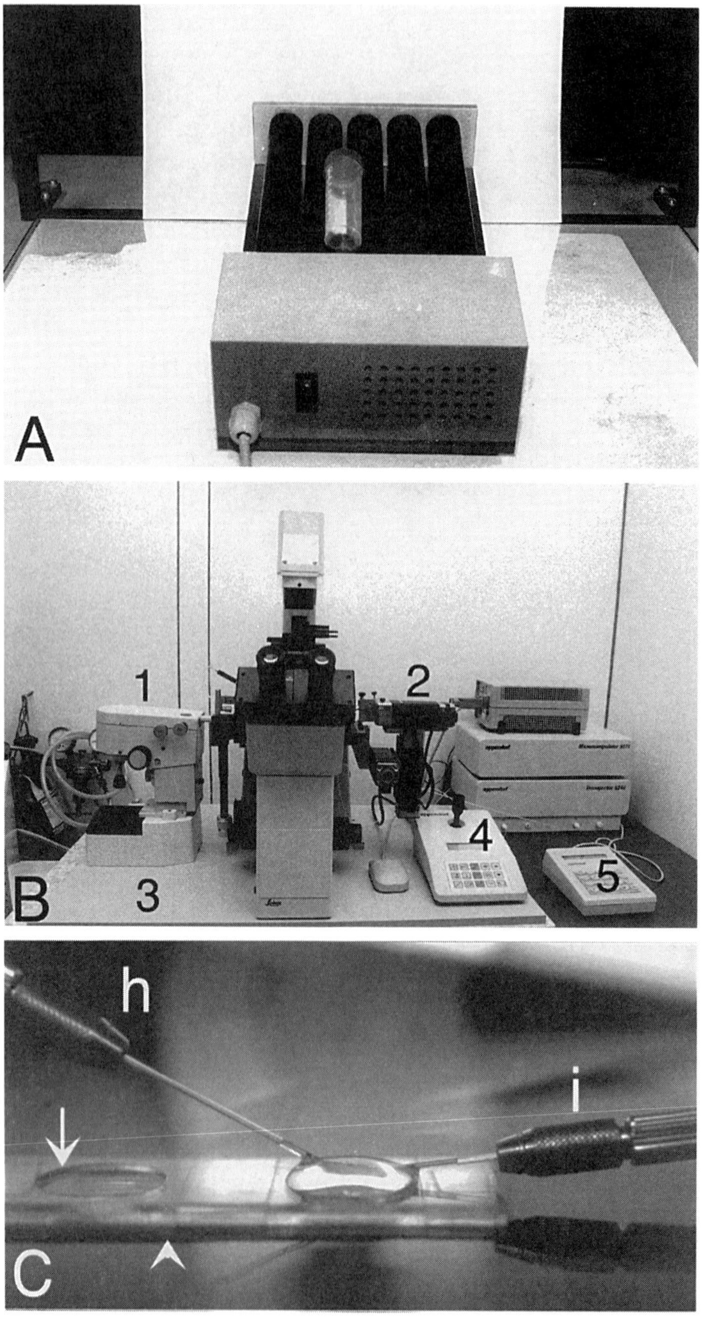

Figure 2. Components of whole embryo culture and microinjection setup. (A) Roller apparatus for rotating tube cultures. It consists of multiple parallel horizontal rollers rotating at around 30 rpm. To ensure a sufficient depth of medium, the whole apparatus should be tilted at an 10° to 20° angle. (B) 1: Manual micromanipulator (Leitz) for the holding capillary, fixed to a custom-made metal base plate. 2: Motorized micromanipulator (Eppendorf 5171) for the injection capillary, fixed to the microscope stage. 3. Vibration-free stand. 4: Joystick control board for motorized micromanipulator. 5: Control keypad for pneumatic transjector (Eppendorf 5246). It controls the output parameters for both microinjection and holding capillaries. (C) Injection chamber. A plexiglass form (arrow) is glued to a depression glass slide (arrowhead) with silicon grease and tape to form two injection chambers. H, holding capillary; I, injection capillary.

- Manual micromanipulator (Leitz)
- Motorized micromanipulator (Eppendorf, model 5171 [Hamburg, Germany])
- Pipet puller (e.g., P-87, Sutter Instrument, Novato, CA, USA)
- Spirit burner
- Transjector (Eppendorf, model 5246 plus)
- Vibration-free stand

Preparation of Capillaries

Two types of capillaries are used: holding and injection capillaries (8). They are made of borosilicate glass and have an inner filament (Hilgenberg or Malsfeld). The tubing for injection capillaries should be 100 mm long with a 1.0-mm outer diameter and an 0.8-mm inner diameter. The tubing for holding capillaries should be 150 mm long with 1.0-mm outer diameter and an 0.8-mm inner diameter. Most horizontal micropipet pullers do not allow one to pull capillaries of the relatively large tip diameter needed for embryo injection. Therefore, capillaries are first prepulled. The following instructions are given for usage of a Flaming/Brown micropipet puller (Model P-87; Sutter Instrument Co.).

1. Set the starting heat value to the ramp test value of the capillary tubing.

2. Set the time and pull parameters to 0 and the vel parameter to a low value (between 10 and 30). These settings will produce a long and gradually tapering tip. A gradual taper will provide a relatively long section of the desired diameter so that a single pipet can be broken repeatedly until the correct shape is attained.

3. After a number of capillaries have been pulled, they are broken manually with a razor blade or scalpel on a solid surface (e.g., metal plate) to give an even tip with an inner diameter of 10 to 30 μm. Usually the break is perpendicular, and there is no need to further bevel or flame-polish these needles.

Holding pipets require no pulling step. Instead the tubing is cut with a glass cutter in halves and the ends polished for a few seconds in the outer flame cone of a spirit lamp or a self-made microburner (i.e., a Bunsen burner where the outlet has been replaced by silicone tubing and a hypodermic needle to create a microflame). They should give a smooth and even tip of 100- to 300-μm inner diameter. Injection capillaries are straight, but for holding capillaries it is necessary to introduce a slight bend (about 5 mm above the tip) so that the mouth of the pipet reaches the embryo almost horizontally. To achieve this, the tip is held in the blue inner flame cone, and the capillary subsequently bends where it contacts the hot outer flame. The final angle should be 5° to 10°.

Check the tips of both injection and holding capillaries under a microscope (50–100× magnification) for correct size and cracks or edges. Discard all needles with jagged edges. Capillaries should be handled with care and stored under dry and dust-free conditions. UV-sterilize for 10 minutes before use.

Microinjection of Virus

Injections require a micromanipulation setup similar to the one shown in Figure 2B. The essential components are a pneumatic transjector, two micromanipulators for holding and injecting embryos, respectively, that are connected to a fixed-stage inverted microscope with long working distance objectives, and a condensor.

Replication-deficient HSV-1 amplicons can be transmitted in liquids. Virus-loaded injection capillaries can suddenly become loose at increasing injection pressure and can be ejected from the holder because of the blow-pipe effect. Always check that the capillary fits in the grip head and is tightly fastened. Never direct the capillary at people. Any spill of virus must be immediately treated with ethanol or detergent. See Chapter 37 for safety considerations.

Before starting the dissection of embryos, all tools for injection should be installed properly. Choose the working magnification (usually 2.5–5× objective) and set up brightfield illumination according to Köhler. Insert holding and injection needle in their capillary holder, align them in parallel just above the injection chamber, and adjust the right focal plane. The whole procedure is carried out at room temperature under normal atmospheric conditions.

1. Embryos are transferred with a clean plastic pipet in a drop of PB1 medium in a depression slide injection chamber (Figure 2C). The chamber consists of a depression slide with a custom-made plexiglass form (1 mm thick) glued on top of it. The plexiglass form has the same dimensions as the glass slide and a hole of the same diameter (15 mm) where the depression is. It is fixed on the glass slide with highly viscous silicon grease to prevent leakage from the chamber. Four to eight embryos per chamber can be injected in about 1 mL medium.

2. Prepare the injection sample. Virus suitable for microinjection should be purified according to the procedures described in Chapter 16, Gene Transfer into Brain Slices from Mouse Embryos Using Herpes Simplex Amplicons, including centrifugation through a sucrose step gradient. Virus aliquots (5–10 µL in Hank's balanced salt solution [10% sucrose/1% BSA]) are thawed and kept on ice until use. Injection capillaries are filled immediately before injection to minimize nonspecific adherence of virus particles to the charged glass capillary. They are backfilled through the rear open end down near the tip using a sterile microloader (Eppendorf) and avoiding air bubbles. The thin inner filament supports transport of the sample solution to the micropipet tip by capillary forces. Typically loaded volumes are 2 to 5 µL.

3. The embryo is gently fixed with the holding capillary using low negative pressure (P_{-} = -0.3 to -3.0 hPa) and lifted up in the same focal plane as the injection needle. The embryo is best fixed at the ectoplacental cone since this creates the least damage.

4. Carefully penetrate yolk sac and amnion with the injection needle without touching the embryo (Figure 3). The tip of the needle should always be visible inside the amniotic cavity. Perform injection using maximal pressure ("clean" button = 7000 hPa for the Eppendorf system) for 2 to 4 seconds. *You should see the amnion expanding during injection. The injection needle can be moved inside the amniotic cavity to a certain extent, but must not touch the embryo proper. Do not overlook that there may be factors inside tissue fluids (e.g., amnion fluid or ventricular fluid) that inhibit or destroy viral infectivity. Therefore, it is important to deliver the virus directly to the area containing the highest percentage of target cells.* The injection capillary is retracted carefully, the embryo is released from the holding capillary (P_{+} = 5.0 hPa) and transferred to a dish with fresh PB1. When all embryos are injected, they are transferred to the tube with culture medium, gassed appropriately, sealed, and placed on a roller culture apparatus.

The injection setup described here is suitable for all embryonic stages up to E8.5. Older embryos are best manipulated using a standard dissection microscope and less precise manual micromanipulators (e.g., Singer Mk1 [Singer Instruments, Somerset, UK]).

Quantification of Injected Sample Volume

The main factors that influence the injected volume are: *(i)* injection pressure, *(ii)* injection time, *(iii)* inner diameter of the capillary, *(iv)* viscosity of the injection sample, and *(v)* inner resistence of the injected embryos. Injection pressure and injection time can be reproduced with high reliability by the transjector. The inner diameter can be reproduced by the puller. Injection sample viscosity and inner resistence of the injected embryos can vary. Therefore, it is advisable for a given capillary, solution, and embryonic stage to calibrate the system. This can be most precisely done by co-injection with a radioactive tracer.

1. Dilute ^{35}S-methionine (10 µCi/µL) 1 × 10^3 to 1 × 10^4 in the sample. Inject as usual.

2. After injection, embryos are washed, homogenized with a yellow tip in 500 µL phosphate-buffered saline (PBS), boiled for 5 minutes, and counted in a liquid scintillation counter. Using a standard calibration curve, the injected volume can be calculated. Typical volumes are 50 to 200 nL for 0 to 5 somite stage embryos. This corresponds to about 10% of the amniotic cavity volume. Higher volumes have been found to interfere with normal development.

EVALUATION OF INJECTED EMBRYOS

The protocols given below include modifications of standard histological procedures that were found to improve the morphology of early embryonic tissue.

Additional Materials and Reagents

- Cryostat (Leica Frigocut 2800 or equivalent)
- Embedding molds
- Isopentane
- Microscope slides
- Moviol (Calbiochem-Novabiochem, San Diego, CA, USA)
- PAP-Pen (Zymed Laboratories Inc.)
- Paraformaldehyde (PFA) 4%/sucrose 4%. To make, weigh out 8 g PFA and 8 g sucrose, fill up to 180 mL with water and heat to 56°–60°C. Add a small amount of phenol red and a few drops of 1 mol/L NaOH and wait until solution clears. Cool to room temperature. Add 20 mL 10× PBS to final volume of 200 mL. Final pH

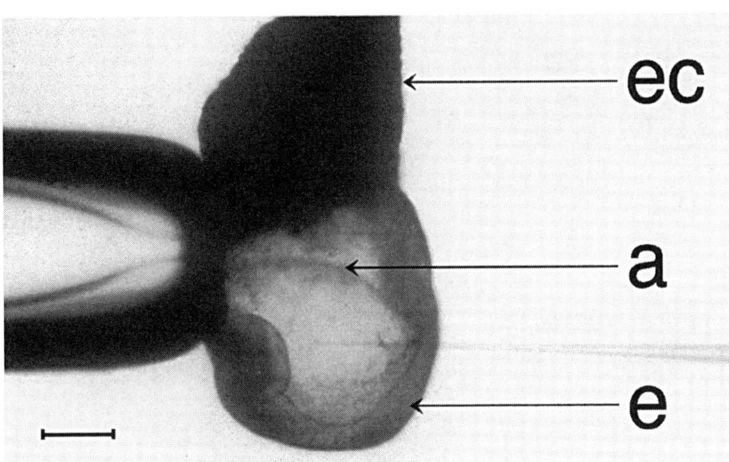

Figure 3. Microinjection of an E8.0 embryo. The embryo is gently held near the ectoplacental cone (ec). The injection needle penetrates yolk sac and amnion (a) without touching the embryo proper (e). Bar = 150 µm.

should be 7.0–7.5. Aliquot and store at -20°C. Including sucrose in the fixing solution improves morphological preservation of the tissue.

- To make PBS, combine 137 mmol/L NaCl, 2.7 mmol/L KCl, 8 mmol/L Na_2HPO_4, and 1.5 mM KH_2PO_4
- To make staining solution, combine 20 mmol/L $K_3Fe(CN_6)$, 20 mmol/L $K_4Fe(CN_6)$, 2 mmol/L $MgSO_4$, 100 mmol/L Na_2HPO_4, 0.02% Nonidet P-40, and 0.01% sodium deoxycholate. Adjust pH to 7.3–7.4 with HCl. Because solution is light-sensitive, keep in a dark bottle at 4°C for at least one year.
- Sucrose 30% (w/v) in PBS
- TESPA (3-aminopropyltriethoxy-silane; Sigma Chemical, St. Louis, MO, USA)
- Tissue-Tek™ (OCT compound; Sakura Finetek USA, Torrance, CA, USA)
- X-Gal stock solution (40 mg/mL in dimethylformamide; keep away from light at -20°C)

Assessment of In Vitro Development

Morphological development of embryos is studied by comparing their anatomy with that of embryos that have developed for an equivalent period of time in utero. The gross morphological assessment is made from fresh samples to detect all abnormalities and retardations resulting from culture (Figure 4). The results are summarized using a semiquantitative scoring scheme, such as the Brown-Fabro system (6,7). Various scoring schemes have been devised, but they are all similiar. They list many different embryonic features, such as development of various organ primordia, degree of embryonic flexion, function of the circulatory system, number of somites, and give a score for each stage. From the total score for all features, the overall stage of development is estimated.

Routine histology is carried out for a more detailed examination (i.e., scoring for cell density, mitotic index, and dead cell index) using a nuclear stain (e.g., DAPI or

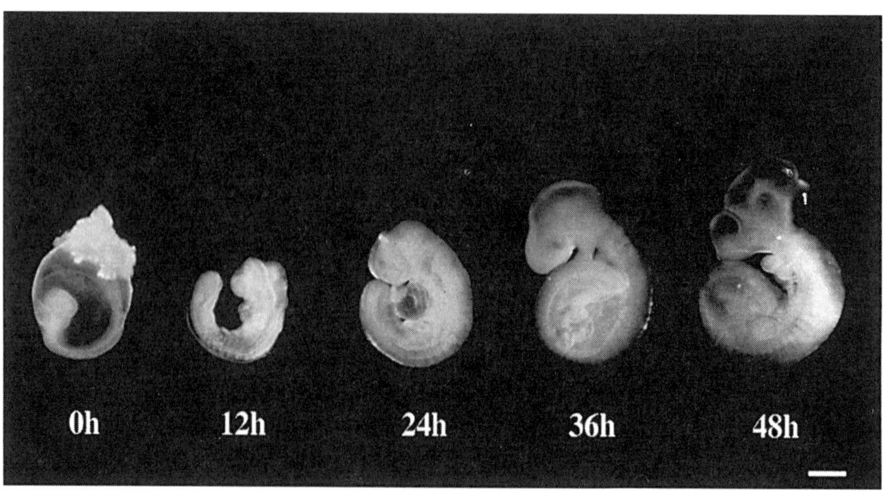

Figure 4. Development of mouse embryos in vitro. E8.0 embryos were cultivated in rotating tubes in a 1:2:1 mixture of DMEM, rat serum, and mouse serum (DRM75). After 12, 24, 36, and 48 hours, embryos were fixed and photographed. After 20 hours in culture, the oxygen concentration in the tubes was raised to 20%. Bar = 300 µm.

SYTOX®; Molecular Probes, Leiden, The Netherlands). Gross morphology and histology are not informative enough to ensure that the developmental process under investigation is also faithfully reproduced at the molecular level. Before starting gene transfer studies in vitro, it should therefore always be clarified first (e.g., by in situ hybridizations or immunocytochemistry) whether all appropriate candidate genes for a particular experiment are still expressed at the correct time and location under culture conditions.

X-Gal Staining Assay on Whole-Mount Embryos

Detection of β-gal activity in infected cells in situ can be done directly on frozen sections using the histochemical X-Gal stain. It is easier, however, to first perform a whole-mount, X-Gal stain and section the embryo afterwards. The substrate X-Gal is hydrolyzed by β-galactosidase to generate galactose and soluble indoxyl molecules, which in turn are oxidized to insoluble indigo.

1. Fix embryos in 4% PFA/4% sucrose for 30 minutes to 2 hours at 4°C.
2. Wash 3× for 10 minutes in PBS containing 2 mmol/L $MgCl_2$ and 5 mmol/L ethyleneglycotetraacetic acid (EGTA), pH 7.4. (For X-Gal assays, always include 2 mmol/L $MgCl_2$ and 5 mmol/L EGTA in fixing and washing solutions to enhance staining intensity. Magnesium is a cofactor for the enzyme and EGTA and specifically chelates Ca^{2+} ions, which would compete with Mg^{2+} for the active center of the enzyme.)
3. Prewarm staining solution at 37°C. Immediately before the staining reaction dilute 40× X-Gal stock in staining solution to 1 mg/mL. Stain for 6 to 24 hours at 37°C. If background problems occur, stain at 30°C. If solutions are not warm, precipitate forms. (Potassium ferrocyanide and potassium ferricyanide act as a redox system to increase the rate of conversion of the soluble, and therefore diffusible, indoxyl molecules to the insoluble indigo form, thereby enhancing localization of the blue precipitate.)

Cryostat Sectioning

1. Fix embryos in 4% PFA (w/v)/4% (w/v) sucrose for 30 minutes for 2 hours at 4°C. If embryos were already stained as whole mounts with X-Gal, fixation time can be reduced to 15 minutes.
2. Wash 3× for 10 min in PBS.
3. Equilibrate embryos in 30% (w/v) sucrose solution (as a cryoprotector) for about 1–2 hours (until embryos sink down).
4. Put embryo into embedding mold filled with Tissue-Tek and orient in the desired position under a dissection microscope.
5. Place a beaker with isopentane on dry ice for 10 minutes. Freeze embryo by slowly immersing in cold isopentane.
6. Cut 5- to 20-μm sections on TESPA-coated glass slides. For TESPA-coating, clean slides with Kimwipes®, immerse for 2 minutes in acetone, immerse for 5 minutes in 2% (v/v) TESPA solution in acetone, briefly rinse in water, and air dry at room temperature.

5\. Air dry sections for at least 30 minutes and proceed with immunocytochemistry.

Immunocytochemistry on Frozen Sections

1. Encircle sections with a PAP-Pen. Postfix for 10 minutes in 4% PFA/4% sucrose.
2. Wash 5 minutes in PBS.
3. Permeabilize in 0.1% Triton® X-100 in PBS for 10 minutes.
4. Quench free aldehydes with 10 mmol/L NH_4Cl in PBS for 10 minutes.

5. Wash 1× in PBS.

6. Block in 5% bovine serum albumin, 5% fetal calf serum in PBS for at least 20 minutes. Prepare the first antibody dilutions in this blocking solution.

7. Place 5–10 µL of first antibody on each section. Make a wet chamber by lining a 24 × 24 cm-square dish (or equivalent) with two sheets of filter paper and add sufficient water to moisten the filter paper. Leave overnight at 4°C (or for 30 minutes at 37°C).

8. Wash 5 minutes in PBS.

9. Add secondary antibody for 30 minutes at 37°C.

10. Wash 2× for 5 minutes in PBS.

11. Put briefly in water. Place coverslips on slides in minimal amount of Moviol and secure the coverslips with nail polish. Keep slides in the dark at 4°C for conservation. Specimens should not be allowed to dry out at any stage in the procedure.

GENE TRANSFER INTO EMBRYOS USING HSV AMPLICONS

Amplicon Titer

The titer, expressed as infectious units/mL (iu/mL), varies and depends on a combination of factors. The vector, the insert(s), the packaging line, the methods of transfection, and harvesting all contribute to the final titer. Since the volume that can be safely delivered into the amniotic cavity is quite small (50–200 nL), it is important to generate high-titer stocks (1 × 10^6 to 1 × 10^7 iu/mL). Using 5*dl*1.2 helper virus together with the permissive packaging cell line (Vero 2-2) and the procedures described in Chapter 12, such titers should be readily achievable.

Target Cell Specificity

Earlier studies showed that HSV is capable of infecting many differentiated cell types in adult mice and is not restricted to skin epithelial or neural host cells as in naturally occuring infections (26). We infected early embryonic stages (E7.5–E8.5) with HSV*lacZ* amplicons for 12 hours and prepared serial sections to determine the susceptibility of cells from different regions and stages for HSV amplicons. Sections were stained with antibodies against HSV-1 coat proteins (DAKO, Carpinteria, CA, USA), and β-galactosidase (monoclonal; clone no. GAL-13; Sigma). For comparison, sections were also stained with X-Gal. Cells from various stages and regions along the anterior-posterior axis were highly susceptible to viral infection and strongly expressed the reporter gene *lacZ*. HSV showed no target cell specificity and infected all directly accessible cell types (e.g., neuroepithelial, primitive streak, heart, surface ectoderm, amnion, allantois, and yolk sac cells). Cell types that were not directly accessible to the virus in intact embryos could be successfully infected in primary embryonic cultures (e.g., mesodermal cells from the perineural mesenchyme; note absence of staining in Figure 6, B, D, and F).

The histochemical detection of β-gal with X-Gal is the most widely used method for rapid and easy visualization of whole-mount embryos and sections. Therefore, it may be of methodical interest that on cryosections of formaldehyde-fixed mouse embryos the detection of β-gal expression by anti-β-gal antibodies was consistently more sensitive than the X-Gal staining procedure (Figure 5). Only 30% to 50% of immunoreactive neuroepithelial cells were detectable by X-Gal staining, a discrepancy not observed with cell culture cells subjected to the same staining protocols (unpublished data). Some cellular structures (e.g., the thin processes of neuroepithelial cells) often stained particularly poorly with X-Gal.

Efficiency of Infection

Infection efficiency, defined as the percentage of β-gal-positive cells per total cell number, depends on the target tissue, the titer of amplicons, the amount of injected virus, and the method used to detect β-gal (see above). It can be close to 100% for small embryonic cell populations, such as amnion or heart cells. For neuroepithelial cells, it was estimated to be in the range of 30% to 50% (using β-gal immunocytochemistry; see Figure 6). Injection of diluted virus stocks offers the possibility of manipulating gene expression in only a small subset of cells within a given cell population. This is not possible with transgenic animals even when tissue-specific promoters are being exploited.

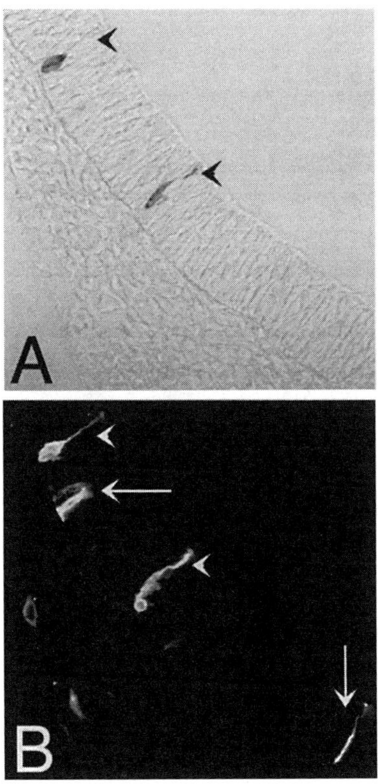

Figure 5. Comparison of β-gal activity versus β-gal immunoreactivity. E8.0 embryos were inoculated with HSV*lacZ*, cultured for 24 hours, fixed and stained with X-Gal (A). Positive embryos were cryosectioned and stained with an anti-β-gal antibody (B). Arrowheads mark neuroepithelial cells, which are both X-Gal stained and immunoreactive. Arrows in panel B mark cells only detectable by immunocytochemistry for β-gal.

Time Course of Infection

During infection, the HSV amplicon enters the host cell nucleus, where it stays episomal and neither replicates nor stably integrates into the host cell chromosome. Thus, in contrast to retroviral vectors, HSV-introduced gene(s) will be transmitted to only one daughter cell each time the cell divides (provided that there are no multiple infections per cell). We injected E8.0 embryos with HSV*lacZ* amplicons and fixed them after 3, 12, 24, and 48 hours. The presence of the β-gal marker gene was already clearly detectable after 3 hours (Figure 6B). While the percentage of β-gal-expressing cells per section declined with time (Figure 6B, D, and F), most of the β-gal-positive cells detected after 24 hours (Figure 6F) and 48 hours (not shown) showed a level of β-gal immunoreactivity (or staining) that was similar to that observed at earlier points in time. These observations can be explained by the ongoing proliferation of neuroepithelial cells during the culture period. Since HSV amplicons do not integrate into the host cell genome, two populations of β-gal-positive cells can be predicted. First, neuroepithelial cells that still contain the amplicon, continue to express β-gal and decrease in percentage because upon cell division the HSV amplicon is inherited by only one of the daughter cells. The brightly β-gal-positive cells observed after 24 and 48 hours presumably corresponded to this cell population. Second, neuroepithelial cells that do not contain the amplicon, but upon cell division inherit a certain, variable amount of β-gal. If the mother cell still contains an amplicon, the amplicon-lacking daughter cell should inherit approximately half the amount of β-gal. If the mother cells already lacks an amplicon, the amount of β-gal in the daughter cells will be less because of its progressive dilution. Given the relatively short cell cycle length of neuroepithelial cells at this stage of development (6–8 hours), the amount of β-gal in cells generated by division of amplicon-lacking cells will be reduced to about 6% to 12% of the initial value after 24 hours and to about 0.4% to 1.5% after 48 hours. The more weakly β-gal-positive cells observed after 24 and 48 hours and the cells in which β-gal levels were below the limit of detection

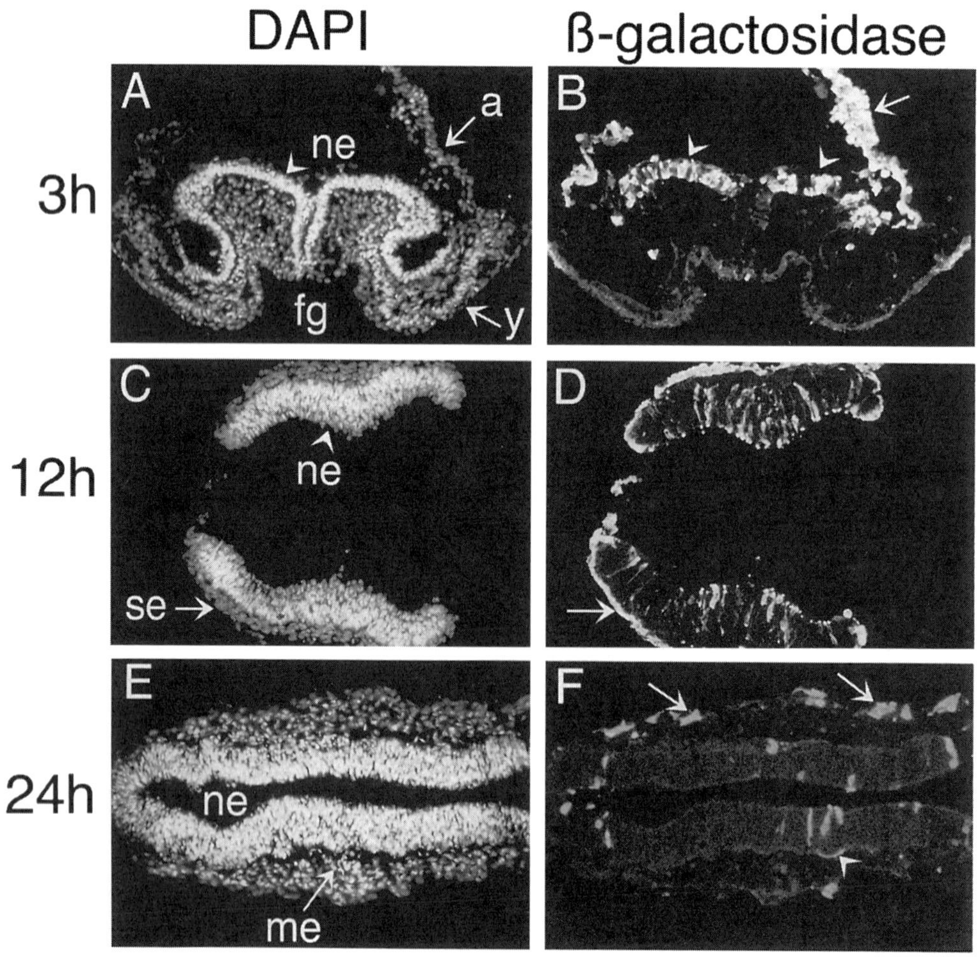

Figure 6. Time course of β-gal expression in HSV*lacZ*-infected embryos. E8.0 embryos were injected with HSV*lacZ*, cultured, fixed, horizontally sectioned after 3, 12, and 24 hours, and stained with an antibody against β-gal. (A and B) Three hours after injection, expression of β-gal was detectable in the neuroepithelium (arrowheads) and amnion (arrows). Weak staining of yolk sac (y) and foregut (fg) in panel B is due to cross-reactivity of the secondary antibody. (C and D) After 12 hours, strong staining is observed in the midbrain neuroepithelium (arrowheads) and surface ectoderm (arrows). (E and F) After 24 hours, the number of positive cells in the midbrain neuroepithelium and surface ectoderm (arrows) has decreased. The arrowhead in (F) marks a putative neuron. ne, neuroepithelium; se, surface ectoderm; y, yolk sac; a, amnion; fg, foregut; me, mesenchyme. (A, C, and E) Counterstain of nuclei with DAPI. In panels A and B, anterior is top; in panels C–F, anterior is to the left.

presumably reflect this progressive dilution of the marker protein.

Side Effects of HSV Infection on Development

Replication-defective HSV vectors have been used for a wide range of in vitro and in vivo gene transfer studies (10,11,13,15). These experiments have also shown some limitations of the current packaging systems, namely cytopathic side effects associated with the helper virus. Cytopathic effects may result both from proteins present in the helper HSV particles and from expression of remaining HSV immediate early genes. In addition, there is a risk of recombination between amplicon, helper virus genome, and complementing sequences in the packaging cell, thereby resulting in the potential generation of replication-competent virus. Frequency of recombination after repeated passages has been estimated as 1×10^{-6} (22); therefore, for normal virus titers (1×10^{5} to 1×10^{7} colony-forming units [cfu]/mL) only 0.1 to 10 replication-competent infectious virus particles/mL can be expected. Because the volume of injected virus is very small (<1 μL), the risk is low that these HSV revertants contribute to the cytotoxicity.

We first examined whether herpesvirus infection affects normal embryonic development on a gross morphological and histological level and found that this was not the case. Infected embryos developed comparably to noninfected embryos. To clarify whether infection affects the normal time course of gene expression during neurogenesis, cryosections along the anterior-posterior axis were prepared. Sections were stained for the candidate gene β-III-tubulin, an early marker for neurons (21), and TIS21, a marker for neuron-generating neuroepithelial cells (17). In HSV*lacZ*-infected neuroepithelial cells, as determined by anti-β-gal staining, there was no premature expression of β-III-tubulin or TIS21 after 12 or 24 hours of infection. Therefore, infection with HSV per se did not influence the timing of neuronal differentiation. Furthermore, 24 hours after injection, the normal onset of endogenous β-III-tubulin and TIS21 expression in neighboring uninfected cells was not affected in HSV*lacZ*-infected embryos.

After longer incubation times (36–48 hours), however, viral infection led to increasing cytotoxicity and cell death (as shown by increasing numbers of DAPI-stained fragmented nuclei as compared to uninfected control embryos).

Experiments using other HSV amplicons (e.g., HSV-TIS21) gave similar results as HSV*lacZ* with respect to viral titers, cell specificity, infection efficiency, and time course of viral expression (Figure 7).

Understanding the developmental effect(s) of a gene of interest depends very much on the quality of the assay. For morphological assays (e.g., up- or downregulation of downstream acting candidate genes in the target cells), embryo culture is a suitable system. For biochemical assays, it is less appropriate because the number of transgenic cells will often not suffice for standard biochemical procedures.

Other Applications of Microinjection in Cultured Embryos

The combination of microinjection technique and whole-embryo culture can be exploited in different ways to manipulate early postimplantation mouse embryos:

- microsurgical perturbation using transplantation of single cells or tissues with microcapillaries (3,5,18)
- lineage tracing using horseradish peroxidase (HRP), wheat germ agglutinin, or fluorescent vital dyes, such as DiI or DiO (19,20,25,30)
- injection of enzymes that degrade extracellular matrix molecules (23,31)

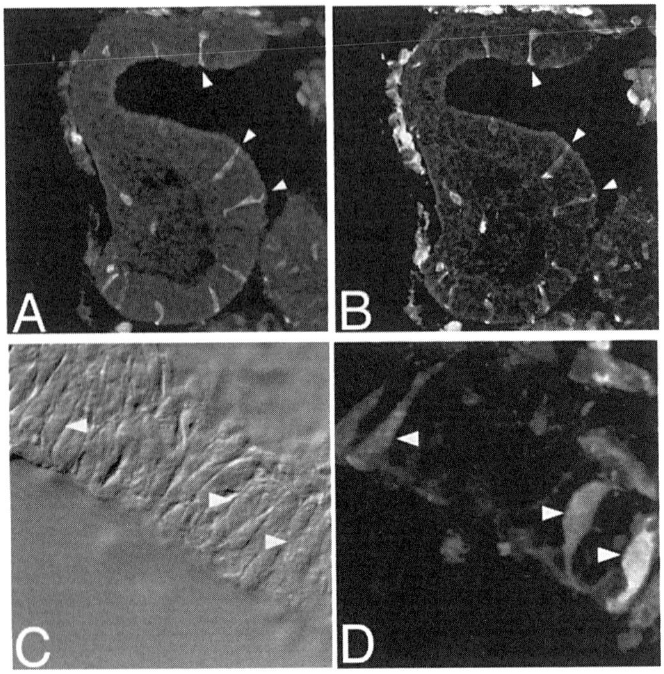

Figure 7. Overexpression of TIS21 in the neuroepithelium. E8.0 embryos were injected with HSV amplicons carrying the myc-tagged TIS21-cDNA and sectioned after 12 hours (A and B) and 24 hours (C and D). Arrowheads mark infected neuropithelial cells. (A and B) Horizontal section through the forebrain/hindbrain region, stained with antibodies against TIS21 (A) and the myc-epitope (B). (C and D) Higher magnification of a sagittal section through the midbrain, stained with anti-TIS21 antibody (D) and control Nomarski-image (C).

- HRP injections to demonstrate the loss of functional tight junctions in the neuroepithelium preceding neurogenesis (1)
- injection of vesicular stomatitis virus and fowl plague virus to investigate changes in apico-basal polarity of neuroepithelial cells (2)
- injection of antisense oligonucleotides (4)

These examples show how useful the whole-embryo culture system can be for studying various questions in developmental cell biology on a single-cell level.

ACKNOWLEDGMENTS

Björn Oback was supported by a fellowship of the Studienstiftung des Deutschen Volkes and Wieland B. Huttner was supported by a grant from the Deutsche Forschungsgemeinschaft (SFB 317).

REFERENCES

1. **Aaku-Saraste, E., A. Hellwig, and W.B. Huttner.** 1996. Loss of occludin and functional tight junctions, but not ZO-1, during neural tube closure-remodeling of the neuroepithelium prior to neurogenesis. Dev. Biol. *180*:664-679.
2. **Aaku-Saraste, E., B. Oback, A. Hellwig, and W.B. Huttner.** 1997. Neuroepithelial cells downregulate their plasma membrane polarity prior to neural tube closure and neurogenesis. Mech. Dev. *69*:71-81.
3. **Arkell, R. and R.S. Beddington.** 1997. BMP-7 influences pattern and growth of the developing hindbrain of mouse embryos. Development *124*:1-12.
4. **Augustine, K., E.T. Liu, and T.W. Sadler.** 1993. Antisense attenuation of *Wnt-1* and *Wnt-3a* expression in whole-embryo culture reveals roles for these genes in craniofacial, spinal cord and cardiac morphogenesis.

Dev. Genet. *14*:500-520.

5. **Beddington, R.S.** 1994. Induction of a second neural axis by the mouse node. Development *120*:613-620.

6. **Brown, N.A.** 1990. Assessment of development, p. 93-108. *In* A.J. Copp and D.L. Cockroft (Eds.), Postimplantation Mammalian Embryos, IRL-Press, Oxford.

7. **Brown, N.A. and S. Fabro.** 1981. Quantitation of rat embryonic development *in-vitro*: a morphological scoring system. Teratology *24*:65-78.

8. **Cid-Arregui, A., C. Santana, M. Guido, N. Morales-Peza, M.V. Juarez, R. Wepf, and A. Garcia-Carranca.** 1998. Protocols for the manual and automatic microinjection of somatic cells in culture and for the analysis of microinjected cells, p. 3-30. *In* A. Cid-Arregui and A. Garcia-Carranca (Eds.), Microinjection and Transgenesis. Strategies and Protocols. Springer-Verlag, Berlin.

9. **Cockroft, D.L.** 1990. Dissection and culture of postimplantation embryos, p. 15-40. *In* A.J. Copp and D.L. Cockroft (Eds.), Postimplantation Mammalian Embryos. IRL-Press, Oxford.

10. **During, M.J., J.R. Naegele, K.L. O'Malley, and A.I. Geller.** 1994. Long-term behavioral recovery in parkinsonian rats by an HSV vector expressing tyrosine hydroxylase. Science *266*:1399-1403.

11. **Federoff, H.J., M.D. Geschwind, A.I. Geller, and J.A. Kessler.** 1992. Expression of nerve growth factor in vivo from a defective herpes simplex virus 1 vector prevents effects of axotomy on sympathetic ganglia. Proc. Natl. Acad. Sci. USA *89*:1636-1640.

12. **Fink, D.J., N.A. DeLuca, W.F. Goins, and J.C. Glorioso.** 1996. Gene transfer to neurons using herpes-simplex virus-based vectors. Annu. Rev. Neurosci. *19*:265-287.

13. **Geller, A.I. and X.O. Breakefield.** 1988. A defective HSV-1 vector expresses Escherichia coli beta-galactosidase in cultured peripheral neurons. Science *241*:1667-1669.

14. **Geller, A.I., M.J. During, J.W. Haycock, A. Freese, and R. Neve.** 1993. Long-term increases in neurotransmitter release from neuronal cells expressing a constitutively active adenylate cyclase from a herpes simplex virus type 1 vector. Proc. Natl. Acad. Sci. USA *90*:7603-7607.

15. **Geller, A.I., M.J. During, Y.J. Oh, A. Freese, and K. O'Malley.** 1995. An HSV-1 vector expressing tyrosine hydroxylase causes production and release of L-dopa from cultured rat striatal cells. J. Neurochem. *64*: 487-496.

16. **Hogan, B., R.S. Beddington, F. Costantini, and E. Lacy.** 1994. Manipulating the Mouse Embryo: A Laboratory Manual. CSH Laboratory Press, Cold Spring Harbor, NY.

17. **Iacopetti, P., M. Michelini, I. Stuckmann, B. Oback, E. Aaku-Saraste, and W.B.H. Huttner.** 1999. Expression of the antiproliferative gene TIS21 at the onset of neurogenesis identifies single neuroepithelial cells that switch from proliferative to neuron-generating division. Proc. Natl. Acad. Sci. USA *96*:4639-4644.

18. **Jaenisch, R.** 1985. Mammalian neural crest cells participate in normal embryonic development on microinjection into post-implantation mouse embryos. Nature *318*:181-183.

19. **Lawson, K.A., J.J. Meneses, and R.A. Pedersen.** 1986. Cell fate and cell lineage in the endoderm of the presomite mouse embryo, studied with an intracellular tracer. Dev. Biol. *115*:325-339.

20. **Lawson, K.A. and R.A. Pedersen.** 1987. Cell fate, morphogenetic movement and population kinetics of embryonic endoderm at the time of germ layer formation in the mouse. Development *101*:627-652.

21. **Lee, M.K., J.B. Tuttle, L.I. Rebhuhn, D.W. Cleveland, and A. Frankfurter.** 1990. The expression and posttranslational modification of a neuron-specific β-tubulin isotype during chick embryongenesis. Cell. Motil. Cytoskeleton *17*:118-132.

22. **Lim, F., D. Hartley, P. Starr, P. Lang, S. Song, L. Yu, Y. Wang, and A.I. Geller.** 1996. Generation of high-titer defective HSV-1 vectors using an IE 2 deletion mutant and quantitative study of expression in cultured cortical cells. BioTechniques *20*:460-469.

23. **Morriss-Kay, G. and F. Tuckett.** 1989. Immunohistochemical localisation of chondroitin sulphate proteoglycans and the effects of chondroitinase ABC in 9- to 11-day rat embryos. Development *106*:787-798.

24. **New, D.A.T.** 1978. Whole-embryo culture and the study of mammalian embryos during embryogenesis. Biol. Rev. *53*:81-122.

25. **Quinlan, G.A., E.A. Williams, S.-S. Tan, and P.L. Tam.** 1995. Neuroectodermal fate of epiblast cells in the distal region of the mouse egg cylinder: implication for body plan organization during early embryongenesis. Development *121*:87-98.

26. **Roizman, B. and A.E. Sears.** 1996. Herpes simplex viruses and their replication, p. 2231-2295. *In* B.N. Fields, D.M. Knipe, P. Howley, R.M. Chanock, M.S. Hirsch, J.L. Melnick, T.P. Monath, and B. Roizman (Eds.), Fields Virology. Lippincott-Raven, Philadelphia.

27. **Rugh, R.** 1968. The Mouse. Its Reproduction and Development. Burgess Publishing, Minneapolis.

28. **Steele, C.E.** 1972. Improved development of "rat egg-cylinders" *in vitro* as a result of fusion of the heart primordia. Nat. New Biol. *237*:150-151.

29. **Sturm, K. and P.P.L. Tam.** 1993. Isolation and culture of whole postimplantation embryos and germ layer derivatives, p. 164-190. *In* P.M. Wassarman and M.L. DePamphilis (Eds.), Guide to Techniques in Mouse Development, vol. 225. Academic Press, San Diego.

30. **Tan, S.S. and G.M. Morriss-Kay.** 1986. Analysis of cranial neural crest cell migration and early fates in postimplantation rat embryos. J. Embryol. Exp. *98*:21-58.

31. **Tuckett, F. and G.M. Morriss-Kay.** 1989. Heparitinase treatment of rat embryos during cranial neurulation. Anat. Embryol. *180*:393-400.

32. **Whittingham, D.G. and R.G. Wales.** 1969. Storage of two-cell mouse embryos *in vitro*. Aust. J. Biol. Sci. *22*:1065-1068.

18 The Use of Herpes Simplex Vectors to Study Alzheimer's Disease Neurodegeneration

Rachael L. Neve[1], Donna L. McPhie[1], and Sherry Bursztajn[2]
[1]Department of Psychiatry and Program in Neuroscience, Harvard Medical School, McLean Hospital, Belmont, MA; [2]Biomedical Research Institute, Louisiana State University Medical Center, Shreveport, LA, USA

INTRODUCTION

All individuals with Alzheimer's disease (AD) experience a progressive loss of cognitive function that results from a neurodegenerative process characterized classically by the deposition of β-amyloid (Aβ) in plaques and in the cerebrovasculature, and the formation of neurofibrillary tangles in neurons. Additional pathological hallmarks of AD include degeneration of synapses and decreases in cell density in distinct regions of the brain. AD does not have a simple etiology. It can occur as a sporadic event; it can result from the possession of an extra copy of chromosome 21 (Down's syndrome); or it can be caused by mutations in the amyloid precursor protein (APP) gene on chromosomes 21 or by mutations in the presenilin genes on chromsomes 1 and 14. Such familial AD (FAD) mutations cause less than 10% of all cases of AD; but because they are the only known causes of AD, researchers have exploited the existence of these mutants by expressing them in cells to obtain data on the mechanisms by which they may kill neurons in AD.

At least two hypotheses concerning the etiology of AD have arisen as a result of these FAD mutant studies. The amyloid hypothesis proposes that plaque-like amyloid depositions or partially aggregated soluble Aβ trigger a neurotoxic cascade, thereby causing neurodegeneration and AD (10,16). This theory is based on in vitro studies that suggest that Aβ is toxic to neurons (20) and on the measurements of increased release of Aβ by cells that express FAD mutant genes (10,16,21). While this amyloid hypothesis is attractive, molecular mechanisms other than those mediated by extracellular Aβ could also lead to neuronal death and impaired cognition in AD. We and others have proposed a modified version of the amyloid hypothesis that postulates that the primary contributor to the etiology of AD is the C-terminal 100-amino acid fragment of the amyloid precursor protein (APP-C100), which includes the 42-amino acid Aβ peptide and 58 adjacent amino acids in the carboxyl-terminus of APP. The genetic and biochemical data that support this hypothesis (14) are that *(i)* APP-C100 is both neuro-

Viral Vectors: Basic Science and Gene Therapy
Edited by A. Cid-Arregui and A. García-Carrancá

toxic and amyloidogenic, and *(ii)* expression of APP-C100 in vivo causes neuropathology, including neurodegeneration and cognitive dysfunction, that is similar in many ways to that in AD. One prediction of this hypothesis is that levels of APP-C100 would be elevated in neurons that express FAD mutants of APP. Such an increase had not been detected in transfected cell lines overexpressing the London mutants of APP. However, the metabolism of APP in neurons, rather than in cell lines, is of particular relevance to AD. To identify metabolic fragments of APP that are generated by neurons in FAD, we used a herpes simplex virus (HSV) vector to express five different FAD mutants of APP in primary neurons in culture. As we describe below and in McPhie et al. (13), all of the mutants caused abnormal intracellular increases in APP-C100, as well as increased secretion of Aβ.

A second hypothesis guiding research on FAD is that regulation of apoptosis by the presenilins PS-1 and PS-2 may be disrupted in AD. It has been reported that overexpression of these genes in transfected cell lines can cause apoptosis (11) or result in an increased susceptibility to apoptosis (3,7–9,19), with the effects of FAD mutants of these genes being even more prominent. However, it is not known whether the overexpression of wild-type (wt) or FAD mutant PS-1 in postmitotic cortical neurons results in DNA damage or apoptosis. To gain a better understanding of the role of these genes in neuronal apoptosis, we used HSV vectors (1) to overexpress wt and mutant PS-1 cDNAs in primary cortical neurons. We discovered that apoptosis is not enhanced in neuronal cells infected with these vectors and that, on the contrary, expression of wt or FAD mutant PS-1 in neurons protects against apoptosis. These data indicate that the previously reported enhancement of apoptosis by presenilins may be cell type dependent.

NEURONAL EXPRESSION OF FAD MUTANTS OF APP CAUSES INTRACELLULAR ACCUMULATION OF APP-C100

We prepared (13) replication-defective HSV vectors that express human APP-695 and APP-751, the five known FAD mutants of each (the Swedish mutation, $APP_{K595N,M596L}$ [numbering based on the APP-695 spliced form] at the N-terminal end of Aβ, the three London mutations $APP_{V642F,G\ or\ I}$ carboxyl-terminal to Aβ, and the mutation APP_{A617G}, which can cause AD or Dutch hereditary cerebral hemorrhage with amyloidosis [HCHWA-D]), and APP-C100 in the amplicon expression vector pHSVPrpUC (12). The resultant recombinant plasmids were packaged into virus particles in the packaging line Vero 2-2, using the IE2 deletion mutant 5dl1.2 as described (12). The titer of the helper virus component of each stock was 1 to 1.2×10^6 plaque forming units (pfu)/mL on Vero 2-2 cells. The titer of the recombinant virus component of each stock, as assayed by expression of the exogenous gene in PC12 cells, was consistently 3×10^7 infectious units (iu)/mL; the HSVlac negative control was 9×10^7 iu/mL (as assayed by immunocytochemistry with an antibody to *E. coli* β-galactosidase). A second lot of each virus was prepared, in which the titers of the viruses were uniformly 2×10^7 iu/mL with the exception of HSVlac, which was 8×10^7/mL. The titer of the helper virus component of each was 1 to 1.2×10^6 pfu/mL.

Twenty microliters (multiplicity of infection [moi] of about 0.1) of each virus was used to infect six-day-old primary cortical cultures dissociated from E21 rat embryos, plated at a density of 5×10^6 viable cells per 10-cm poly-D-lysine-coated dish in neurobasal medium supplemented with B27 (both from Life Technologies, Rockville, MD, USA), 10% fetal bovine serum, and 5% horse serum. Sixteen hours later (prior to the appearance of C100-induced neurotoxicity), the infected cells were harvested and their proteins analyzed by immunoblot analysis. The results of a typical experiment (Figure 1) reveal that a 12-kDa fragment that co-migrates with a fragment representing the carboxyl-terminal 100 amino acids of APP (APP-C100, expressed from a HSV/C100 vector) is detected at low levels in cell lysates from neurons infected with APP-695 or APP-751. Below it (p10) is the fragment presumably generated by the constitutive α secretase cleavage of APP between lys16 and leu17 in the Aβ sequence. In Figure 1, the amount of the fragment co-migrating with APP-C100 appears to be increased several fold in cell lysates from neurons infected with HSV/APP FAD mutant recombinants relative to those infected with wt recombinants.

This result was verified by quantification of the data (13). The Swedish mutation, $APP_{K595N,M595L}$, caused a sevenfold increase in accumulation of the C100-sized fragment over wt in APP-695 and an eightfold increase in APP-751. Each of the other mutations except for the V642I mutation in APP-695 caused increased levels of the C100-sized fragment in neurons, with the increases achieving significance for V642I in APP-751, V642F in APP-695, V642G in APP-695, and A617G (HCHWA) in APP-751, and near significance for the remaining mutations.

The primary cultures used for these experiments contained glial as well as neuronal cells. To test whether the abnormal processing of APP observed in primary neuronal cultures infected with HSV/FAD-APP vectors occurs in neurons or the

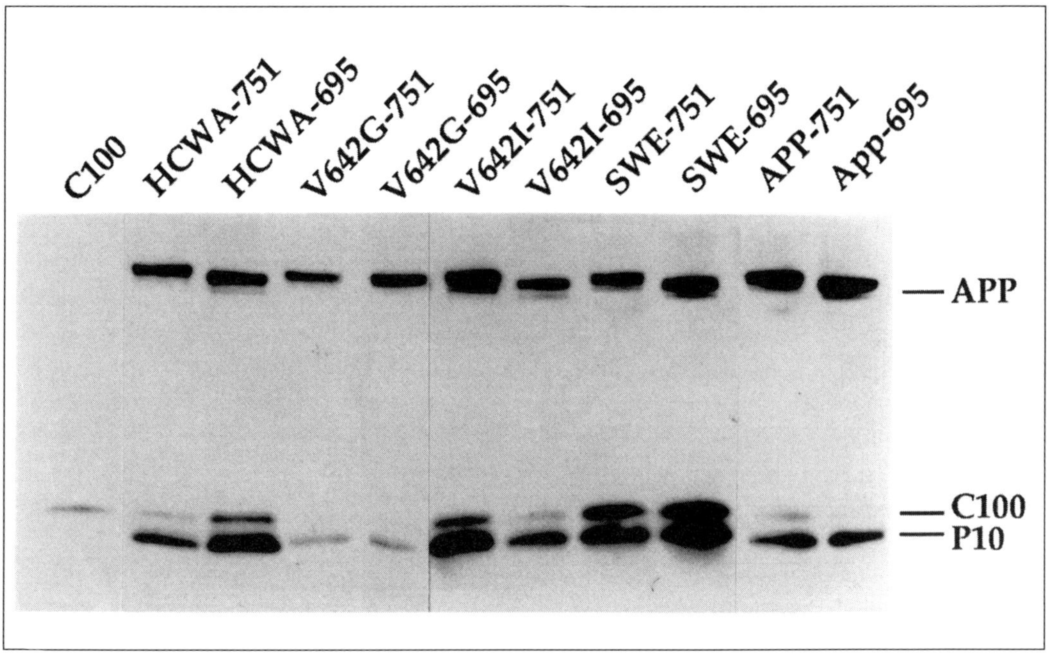

Figure 1. Immunoblot analysis of C-terminal fragments of APP present in neurons infected with HSV vectors expressing wt and FAD mutants of APP. The blot was probed with C8, a polyclonal antibody directed to the C-terminal 20 amino acids of APP. A fragment that co-migrates with C100 (as expressed in the rat neurons infected with HSV/C100; see lane labeled C100) is detected at low levels in cell lysates from neurons infected with APP-695 or APP-751. Below it (P10) is the fragment that is presumably generated by the cleavage of APP between lys16 and leu17 in the Aβ sequence. It is clear that the amount of the fragment co-migrating with C100 is increased several fold in cell lysates from neurons infected with HSV vectors expressing FAD mutants of APP.

glia, we generated primary astrocyte cultures, which were infected with the HSV vectors after seven days in vitro. Although infection of these cells by the HSV vectors was efficient, and resulted in the production of APP and P10 (Figure 2), C100-sized fragments were not detectable in astrocyte cultures infected with the HSV/FAD-APP vectors. These data suggest that the abnormal accumulation of APP-C100 fragments in neuronal cultures infected with HSV/FAD-APP vectors is due to abnormal processing that occurs in neurons rather than in astrocytes.

It had been reported that cell cultures transfected with vectors expressing FAD mutants of APP display increased secretion of $A\beta_{(1\text{-}42)}$ (10,16,21). To ascertain whether the same holds true for neurons infected with FAD APP vectors, levels of $A\beta_{(1\text{-}40)}$ and $A\beta_{(1\text{-}42)}$ were quantified (13) in conditioned media from cells infected with HSV/FAD-APP, HSV/C100, and control HSV vectors. Comparison of cultures infected with HSV vectors expressing wt APP, with those infected with HSV vectors expressing FAD APP mutations revealed that the Swedish mutation in both APP-695 and APP-751 caused an approximately 20-fold increase in the secretion of Aβ relative to that caused by vectors expressing wt APP, while the V642I mutation in APP-751 increased secretion of Aβ over 30-fold. Interestingly, and in contrast to results obtained with transfected cell lines, the ratio of secreted $A\beta_{(1\text{-}42)}$ to $A\beta_{(1\text{-}40)}$ was not altered significantly in neurons infected with HSV/FAD-APP

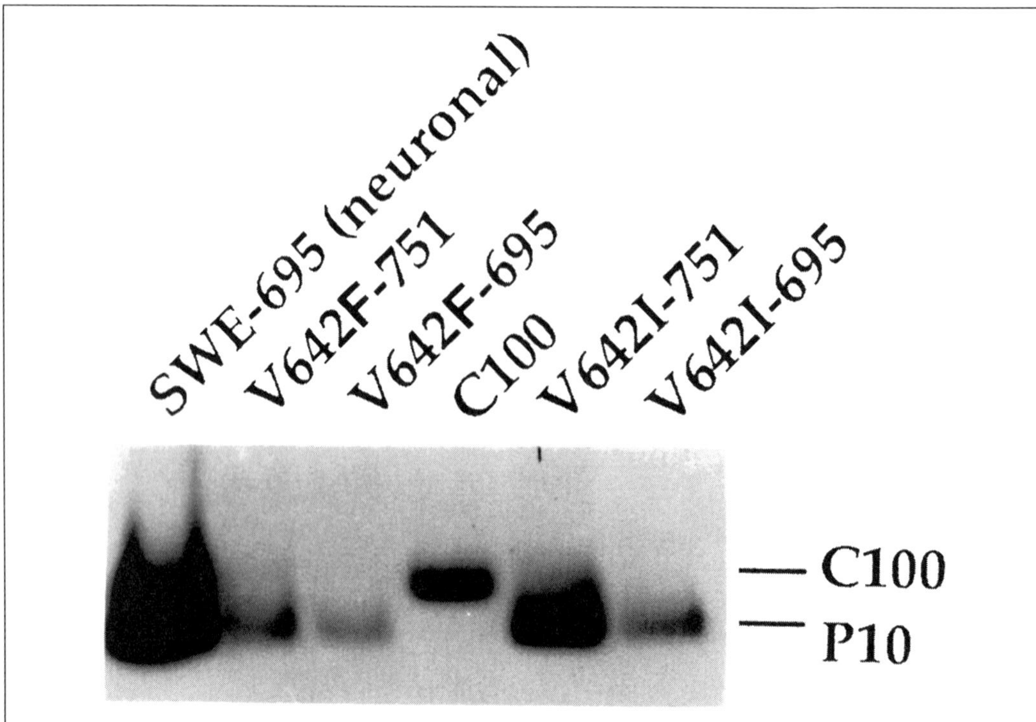

Figure 2. Immunoblot analysis of lysates from astrocytes infected with representative HSV vectors expressing wt and FAD mutants of APP. The blot was probed with an antibody to the C terminus of APP. Note that P10 but not C100-sized fragments accumulate in astrocytes infected with FAD mutants of APP, whereas infection of mixed neuronal-glial cortical cultures with the Swedish mutant of APP (SWE-695) and with the same 642 mutants of APP (Figure 1) resulted in the accumulation of large amounts of both P10 and C100-sized fragments.

vectors, although it was significantly increased in HSV/C100-infected cultures compared with HSV/lac-infected cultures.

OVEREXPRESSION IN NEURONS OF PS-1 OR A FAD MUTANT OF PS-1 DOES NOT ENHANCE APOPTOSIS

We prepared (1) replication-defective HSV vectors expressing the wt (HSV/PS-1) and mutant (HSV/A246E-PS-1) human PS-1 cDNAs and *E. coli* β-galactosidase (HSV/lac; negative control) in the expression vector pHSVPrpUC as described (12). Some experiments utilized an HSV vector that expresses N-terminal myc epitope (EQKLISEEDL)-tagged PS-1 cDNA. HSV/APP-695 was prepared as described (13). The titer of the helper virus component of each stock was 1 to 1.2 × 10^6 pfu/mL on Vero 2-2 cells. The titer of the recombinant virus component of each stock, as assayed by expression of the exogenous gene in PC12 cells, was consistently 3 × 10^7 iu/mL.

Cortices from embryonic day 18 (E18) mice (Charles River, Wilmington, MA, USA) were dissociated mechanically in serum-free neurobasal medium supplemented with B27. Cells were plated at a density of 3 × 10^6 cells/mL in poly-L-ornithine-coated 35-mm culture dishes and were maintained in neurobasal medium supplemented with B27 and 1% each fetal bovine and horse serum. Cells were infected with HSV recombinants five days after plating.

HSV/lac, expressing the *E. coli* β-galactosidase gene, was used to determine the optimal conditions for gene transfer into the mouse cortical neurons. E17-18 neurons were infected with HSV/lac at five days in vitro using a range of moi (0.1–5.0) as described (13). Twelve hours after infection, X-Gal histochemistry was performed to visualize cells expressing the β-galactosidase transgene. Infection with HSV/lac at an moi of 2 yielded an efficiency of expression that was very near that of cultures infected with HSV/lac at an moi of 5. Quantitation of β-galactosidase-positive cells showed that at an moi of 0.1 almost no cells that express LacZ were detected, whereas at moi of 1, 2, and 5, 85%, 90%, and 95%, respectively, of cells were LacZ positive. A similar efficiency of expression was seen when the cultures were infected with HSV/PS-1 or HSV/A246E-PS-1.

To assess potential cytotoxicity in response to infection of neurons with HSV recombinants that express β-galactosidase, PS-1 or A246-PS-1, cortical neurons infected at a range of moi were assayed for trypan blue exclusion and mitochondrial function. Cortical neurons infected with HSV/PS-1 or HSV/A246E-PS-1 and incubated with 0.2% trypan blue 24 hours after infection showed no staining with the dye, which is taken up only by cells with compromised membranes. Trypan blue exclusion does not rule out apoptosis since loss of membrane integrity occurs very late during apoptosis, but it does indicate that the cells are viable.

Further confirmation of the health of the cells was seen in microscopic images (1) that showed phase bright cells with long neuronal processes. In an independent assay for the viability of the HSV-infected cells, we quantified cellular survival following HSV/PS-1 infection, using the 3-[4,5-dimethylthiozol-2-yl]-2,5-diphenyl tetrazolium bromide (MTT) metabolic assay. Neurons were infected in parallel with vehicle (mock) or with HSV vectors expressing wt or A246E mutated PS-1, or LacZ, at an moi of 0.1 to 5. No significant differences between groups in cellular survival was observed up to 72 hours. It can be concluded, therefore, that overexpression of wt or mutated PS-1 via HSV vectors up to an moi of 5 is not sufficient to induce cell death in cortical neurons.

We carried out four different quantitative assays (1) to determine whether PS-1 or FAD mutant PS-1 overexpression, 24 and 48 hours following infection at an moi of 2, induces single- or double-stranded DNA breaks as seen in apoptosis. These assays included *(i)* determination of the amount of cytoplasmic histone-associated DNA fragments by a photometric enzyme immunoassay, *(ii)* detection of DNA single-strand breaks using a Klenow fragment end-labeling technique, *(iii)* detection of DNA double-strand breaks using the TUNEL (terminal deoxynucleotidyl transferase [TdT]-mediated dUTP nick end-label) in situ apoptosis assay, and *(iv)* determination of frequency of apoptotic nuclear morphology using bisbenzimide. All of these assays yielded the same result, which is that apoptosis was not induced in mouse neurons infected with HSV vectors that overexpress wt or mutant PS-1. Interestingly, apoptosis was induced in mouse neurons infected with HSV vectors that overexpress wt APP-695.

Enhancement of apoptosis by wt and mutant presenilins has been reported by a number of laboratories. To determine whether presenilins predispose to apoptosis in neurons, we tested the effect of overexpression of PS-1 or A246E-mutated PS-1 on neurons treated with the apoptosis-inducing agents etoposide and staurosporine. Both etoposide and staurosporine induced a significant increase in DNA fragmentation in cortical cells that were mock-infected or infected with HSV/lac. This increase in DNA fragmentation was prevented if the cells were infected with HSV/PS-1 or with HSV/A246E-PS-1 before being exposed to the apoptosis-inducing agents. Thus, in contrast to the data obtained with transfected cell lines, PS-1 and a FAD mutant of PS-1 actually protect neurons from experimentally induced apoptosis.

UTILITY OF HSV VECTORS FOR STUDYING MECHANISMS OF AD NEURODEGENERATION

We have described experiments using HSV vectors that have revealed that the behavior of FAD mutant proteins in neurons differs from their behavior in cell lines. First, we showed important differences in the processing of FAD mutants of APP in primary neurons relative to cell lines (13). Expression of V642 mutations of APP in non-neuronal mammalian cells had not been observed to cause changes in APP-C100 levels (2,5); this observation suggests that the presumedly increased β-secretase processing of these mutants observed in our study occurs primarily or exclusively in neurons. In contrast to data from an earlier study showing that neuroblastoma cells that express the V642I mutation secrete an altered ratio of $A\beta_{(1-42)}$ to $A\beta_{(1-40)}$, but show no change in the absolute level of Aβ (18), our data suggest that the same mutation expressed in neurons causes an overall increase in Aβ secretion, but no change in the ratio of $A\beta_{(1-42)}$ to $A\beta_{(1-40)}$. The results most likely reflect a difference in APP processing in primary neurons relative to cell lines.

Multiple laboratories have reported evidence for apoptosis, or programmed cell death, in pathological tissues obtained from the brains of AD patients (7,17). Therefore, it is interesting that overexpression of PS-1 and PS-2 in transfected cell lines has been reported to result in an increased susceptibility to apoptosis (3,8,19), possibly by perturbing cellular calcium regulation and promoting oxidative stress (7,9). The question of whether these genes induce apoptosis in primary neurons, however, had remained unanswered. It was an important question because PS-1 function may differ in primary neurons from cell lines. PS-1 is reported to be expressed primarily in central nervous system neurons in the brain, suggesting that this protein may perform a neuron-specific function (4). In fact, in AD, neurons that express PS-1 antigen are less vulnerable to the disease than are neurons that do not express it (6), and inhibition of PS-1 expression results in apoptosis (15), which suggests a protective role for this protein.

We therefore sought (1) to determine whether overexpression of wt or A246E mutated PS-1 induces DNA damage or apoptosis in primary mouse neurons in culture. We used HSV vectors expressing wt or A246E mutant PS-1, as well as wt APP-695 and appropriate controls, to determine whether overexpression of these proteins leads to enhanced apoptosis in primary cortical neurons. We showed that cortical neurons infected with HSV vectors are viable and that when we infect at an moi of 2, most cortical neurons express the appropriate HSV-encoded protein. We also demonstrated that the overexpression of wt or A246E mutant PS-1 does not result in enhanced apoptosis. Finally, we showed that overexpression of PS-1 or PS-1-A246E in cortical neurons suppresses apoptosis induced by etoposide or staurosporine. Thus, in contrast to results that have implicated presenilins in enhancing apoptosis or in rendering cell lines more susceptible to apoptosis, our results using HSV infection of cortical neurons show that overexpression of wt or FAD-mutated PS-1 is not sufficient by itself to induce apoptosis. In fact, it protects cortical neurons from experimentally induced apoptosis. The ability of PS-1 to induce apoptosis appears to be cell type-specific.

In aggregate, our data demonstrate a difference in APP processing in primary neurons compared to cell lines and suggest that the role of PS-1 in apoptosis is very different in neurons relative to dividing cells. These data may have important implications for the pathogenesis in AD, in which neurons are differentially affected. The power of using HSV vectors to deliver genes into neurons for the study of mechanisms of neurodegeneration is highlighted by these experiments.

REFERENCES

1. **Bursztajn, S., R. DeSouza, D.L. McPhie, S.A. Berman, J. Shioi, N.K. Robakis, and R.L. Neve.** 1998. Overexpression in neurons of human presenilin-1 or a presenilin-1 familial Alzheimer disease mutant does not enhance apoptosis. J. Neurosci. *18*:9790-9799.
2. **Cai, X.-D., T.E. Golde, and S.G. Younkin.** 1993. Release of excess amyloid β protein from a mutant amyloid β protein precursor. Science *259*:514-516.
3. **Deng, G., C.J. Pike, and C.W. Cotman.** 1996. Alzheimer-associated presenilin-2 confers increased sensitivity to apoptosis in PC12 cells. FEBS Lett. *397*:50-54.
4. **Elder, G.A., N. Tezapsidis, J. Carter, J. Shioi, C. Bouras, H.C. Li, J.M. Johnson, S. Efthimiopoulos, V.L. Friedrich, Jr., and N.K. Robakis.** 1996. Identification and neuron specific expression of the PS-1/presenilin I protein in human and rodent brains. J. Neurosci. Res. *45*:308-320.
5. **Felsenstein, K.M. and L. Lewis-Higgins.** 1993. Processing of the β-amyloid precursor protein carrying the familial, Dutch-type, and a novel recombinant C-terminal mutation. Neurosci. Lett. *152*:185-189.
6. **Giannakopoulos, P., C. Bouras, E. Kovari, J. Shioi, N. Tezapsidis, P.R. Hof, and N.K. Robakis.** 1997. Presenilin-1-immunoreactive neurons are preserved in late-onset Alzheimer's disease. Am. J. Pathol. *150*:429-436.
7. **Guo, Q., W. Fu, J. Xie, H. Luo, S.F. Sells, J.W. Geddes, V. Bondada, V.M. Rangnekar, and M.P. Mattson.** 1998. Par-4 is a mediator of neuronal degeneration associated with the pathogenesis of Alzheimer disease. Nat. Med. *4*:957-963.
8. **Guo, Q., K. Furukawa, B.L. Sopher, D.G. Pham, N. Robinson, G.M. Martin, and M.P. Mattson.** 1996. Alzheimer's PS-1 mutation perturbs calcium homeostasis and sensitizes PC12 cells to death induced by amyloid β-peptide. Neuroreport *8*:379-383.
9. **Guo, Q., B.L. Sopher, K. Furukawa, D.G. Pham, N. Robinson, G.M. Martin, and M.P. Mattson.** 1997. Alzheimer's presenilin mutation sensitizes neural cells to apoptosis induced by trophic factor withdrawal and amyloid beta peptide: involvement of calcium and oxyradicals. J. Neurosci. *17*:4212-4222.
10. **Hardy, J.** 1997. The Alzheimer family of diseases: many etiologies, one pathogenesis? Proc. Natl. Acad. Sci. USA *94*:2095-2097.
11. **Janicki, S. and M.J. Monteiro.** 1997. Increased apoptosis arising from increased expression of Alzheimer's disease associated presenilin-2 mutation (N141I). J. Cell Biol. *139*:485-495.
12. **Lim, F. and R.L. Neve.** 1999. Generation of high-titer defective HSV-1 vectors, p. 4.13.1–4.13.17. *In* Cur-

rent Protocols in Neuroscience. Greene Publishing Associates and Wiley-Interscience, New York.

13.**McPhie, D.L., R.K.K. Lee, E.B. Eckman, D.H. Olstein, S.P. Durham, D. Yager, S.G. Younkin, R.J. Wurtman, and R.L. Neve.** 1997. Neuronal expression of β-amyloid precursor protein Alzheimer mutations causes intracellular accumulation of a C-terminal fragment containing both the amyloid β and cytoplasmic domains. J. Biol. Chem. *272*:24743-24746.

14.**Neve, R.L.** 1997. Carboxyl-terminal fragments of the Alzheimer amyloid precursor protein: a new target for drug development. Alzheimer Dis. Invest. Drugs Res. Alert *2*:553-556.

15.**Roperch, J.P., V. Alvaro, S. Peieur, M. Tuynder, M. Nemani, F. Lethrosne, L. Piouffre, M.C. Gendron, D. Israeli, J. Dausset, et al.** 1998. Inhibition of presenilin 1 expression is promoted by p53 and p21WAF-1 and results in apoptosis and tumor suppression. Nat. Med. *4*:835-838.

16.**Selkoe, D.J.** 1996. Amyloid β-protein and the genetics of Alzheimer's disease. J. Biol. Chem. *271*:18295-18298.

17.**Su, J.H., A.J. Anderson, B.J. Cummings, and C.W. Cotman.** 1994. Immunohistochemical evidence for apoptosis in Alzheimer's disease. Neuroreport *20*:2529-2533.

18.**Suzuki, N., T.T. Cheung, X.-D. Cai, A. Odaka, L. Otvos, Jr., C. Eckman, T.E. Golde, and S.G. Younkin.** 1994. An increased percentage of long amyloid beta protein secreted by familial amyloid β protein precursor (βAPP717) mutants. Science *264*:1336-1340.

19.**Wolozin, B., K. Iwasaki, P. Vito, J.K. Ganjei, E. Lacana, T. Sunderland, B. Zhao, J.W. Kusiak, W. Wasco, and L. D'Adamio.** 1996. Participation of presenilin 2 in apoptosis enhanced basal activity conferred by an Alzheimer mutation. Science *274*:1710-1713.

20.**Yankner, B.A., L.K. Duffy, and D.A. Kirschner.** 1990. Neurotrophic and neurotoxic effects of amyloid β-protein: reversal by tachykinin neuropeptides. Science *250*:279-282.

21.**Younkin, S.G.** 1995. Evidence that A42 is the real culprit in Alzheimer's disease. Ann. Neurol. *37*:287-288.

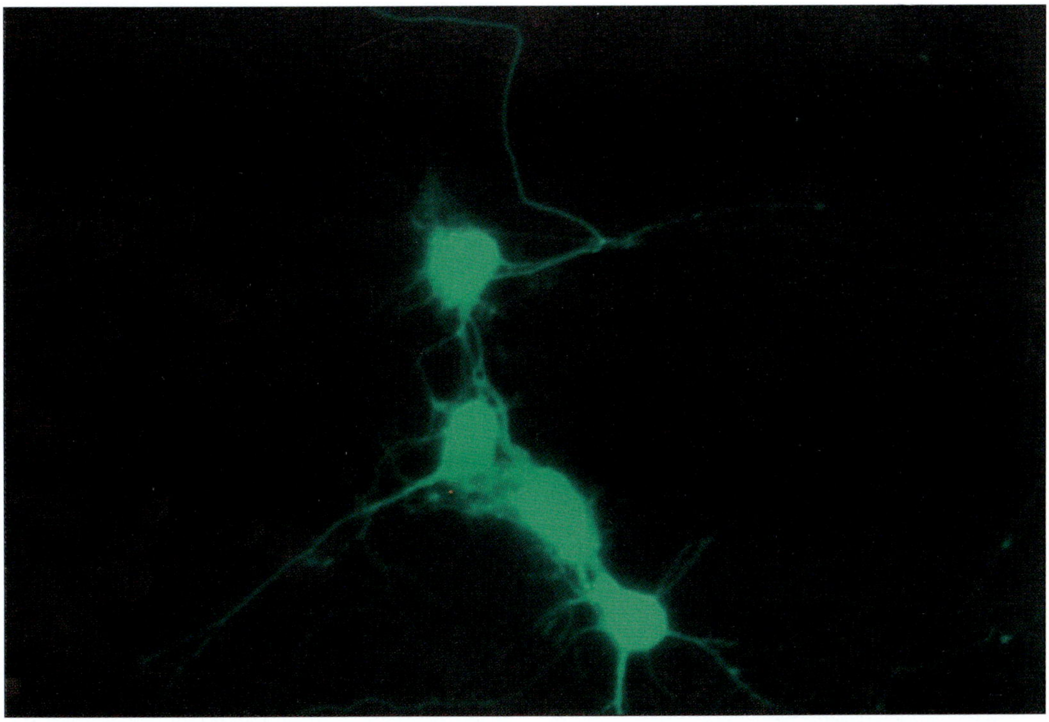

Chapter 13, Figure 4. Photomicrograph showing GFP cells. Cultured rat cortical neurons were infected with stocks of HSV-1 amplicon vector pHSVGFP that have been prepared by using fHSVΔpac as the replication-competent, packaging-defective HSV-1 helper genome. pHSVGFP expresses the gene for GFP from the HSV-1 *IE4/5* promoter. The cultures were examined 24 hours after infection using a fluorescence microscope. Original magnification is 20×.

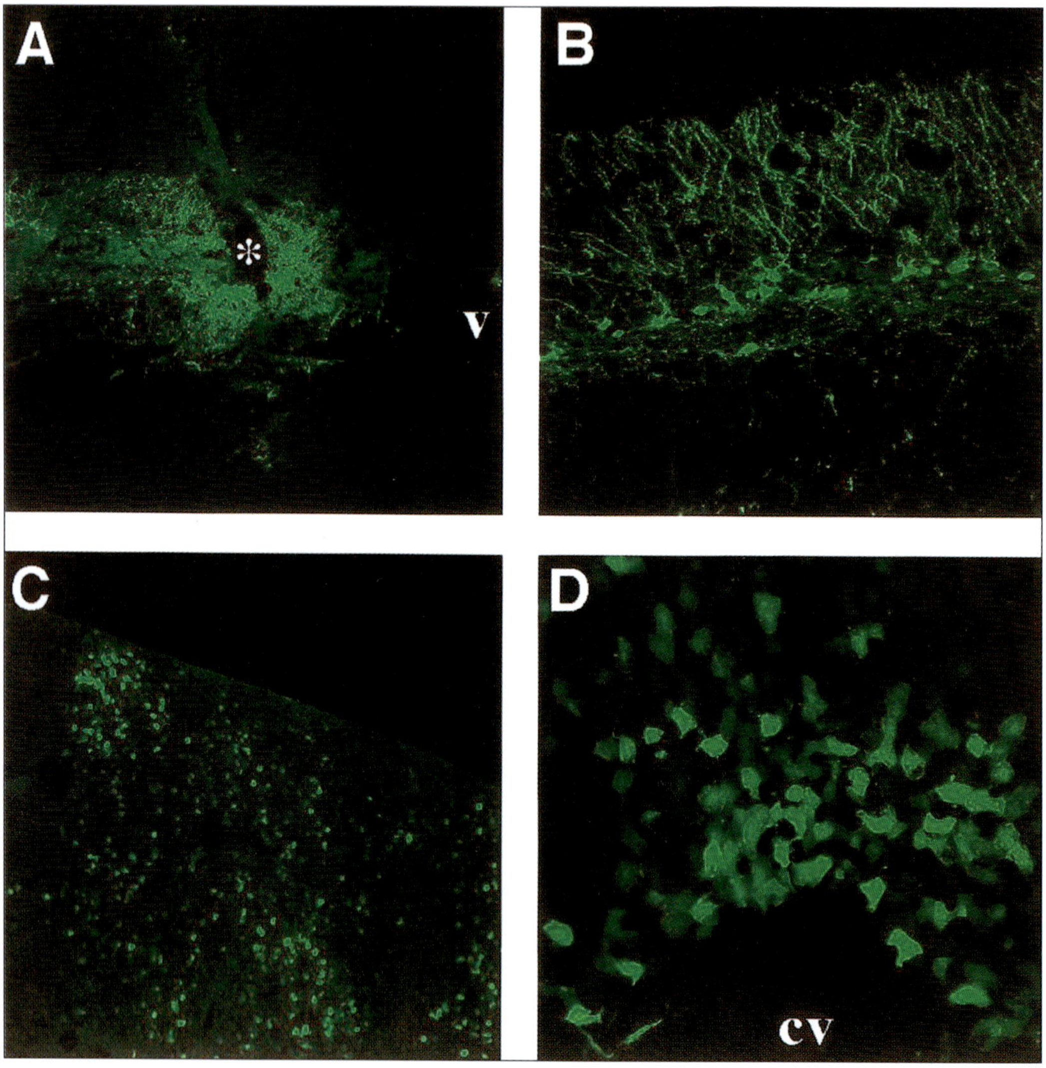

Chapter 13, Figure 5. Photomicrographs showing GFP cells on day 4 following injection of pHSVGFP amplicon vector into rat hippocampus (7.5×10^5 **tu in 10 μL;** A **and** B) **or rat liver** (7.5×10^6 **tu in 100 μL; C and D**). Needle tract (*), third ventricle (v), and central vein (cv) are indicated. Original magnifications are 5× (A and C) and 20× (B and D). Reprinted with permission from Hum. Gene Ther. (70).

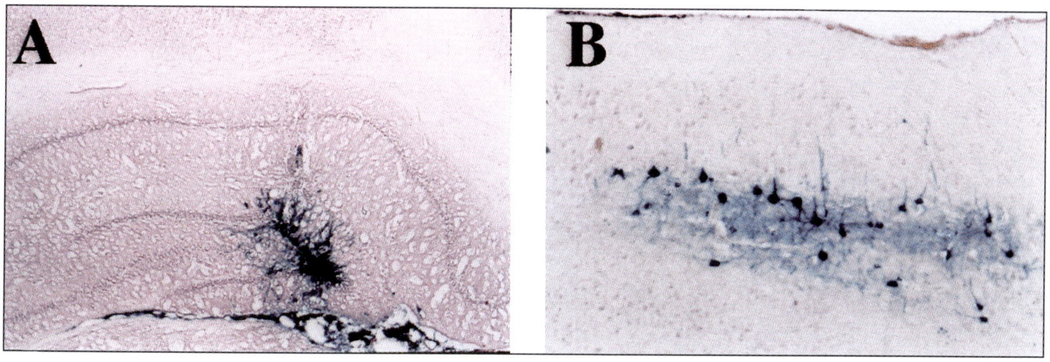

Chapter 13, Figure 6. Photomicrographs showing β-galactosidase-positive cells in the rat brain. HSV/EBV hybrid amplicon pREHcagLacZ, which expresses the *E. coli lacZ* gene from the chicken β-actin promoter, was packaged into HSV-1 particles by using fHSVΔpac helper DNA, and then injected into the hippocampus (1×10^6 tu in 10 μL). At day 3 after injection, the rat was killed and brain sections were incubated with X-gal solution to stain β-galactosidase-positive cells. (A) Intense blue staining was observed in the hippocampal region of the injected side. (B) Some of the cortical neurons adjacent to the needle track were also β-galactosidase-positive. Magnification is 4× (A) and 10× (B).

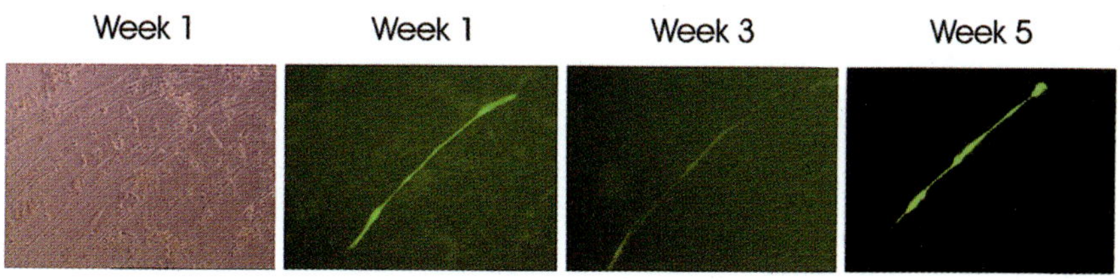

Chapter 14, Figure 3. Stable transduction of reporter gene expression in postmitotic cells. Differentiated human myoblasts were transduced with infectious particles containing a pHERO vector expressing the jellyfish GFP at an moi of 0.1, and reporter gene expression was monitored in live cells. GFP activity could be detected throughout the full course of the experiment, as indicated by the micrographs of a representative myofiber assayed at weeks 1, 3, and 5. The first panel shows a bright field view of the cultured cells at week 1, and the remaining three panels show the same field viewed under fluorescent microscopy. After five weeks, the culture was no longer viable, and the experiment was terminated.

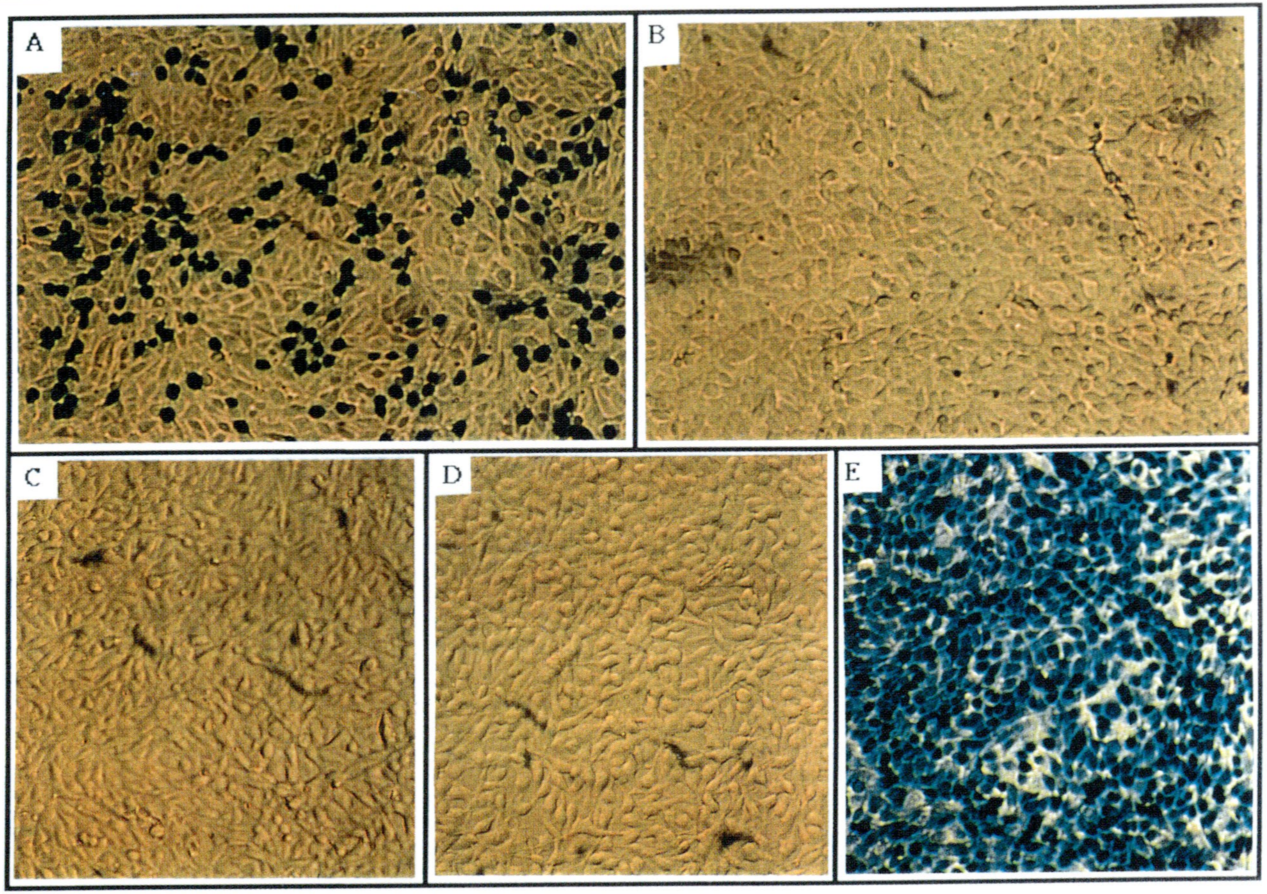

Chapter 15, Figure 3. Specificity of the retrovirus vector rescue by HCMV-GPE infection of TE-Lac2 cells. Murine NIH-3T3 cells (A) or human TE-671 cells (B) were infected with supernatants taken from HCMV-GPE-infected TE-Lac2 cells. Only the murine cells (A) show foci of β-gal-expressing cells, indicating that, as expected, the rescued retrovirus was ecotropic. As controls, NIH-3T3 cells infected with supernatants taken from noninfected TE-Lac2 cells (C), or with supernatants taken from TE-Lac2 cells that were infected with helper virus alone (D), showed no β-gal activity. (E) NIH3T3 cells were infected with supernatants taken from a clonal retrovirus vector producer cell line.

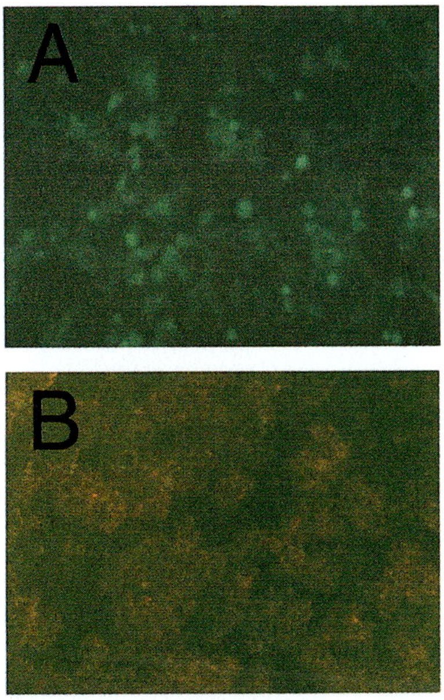

Chapter 21, Figure 6. Transformed 293 cells derived from infection with recombinant AmEPV-TKUF5. (A) GFP-positive, G418-resistant cells. (B) Nonfluorescent, nontransformed control 293 cells.

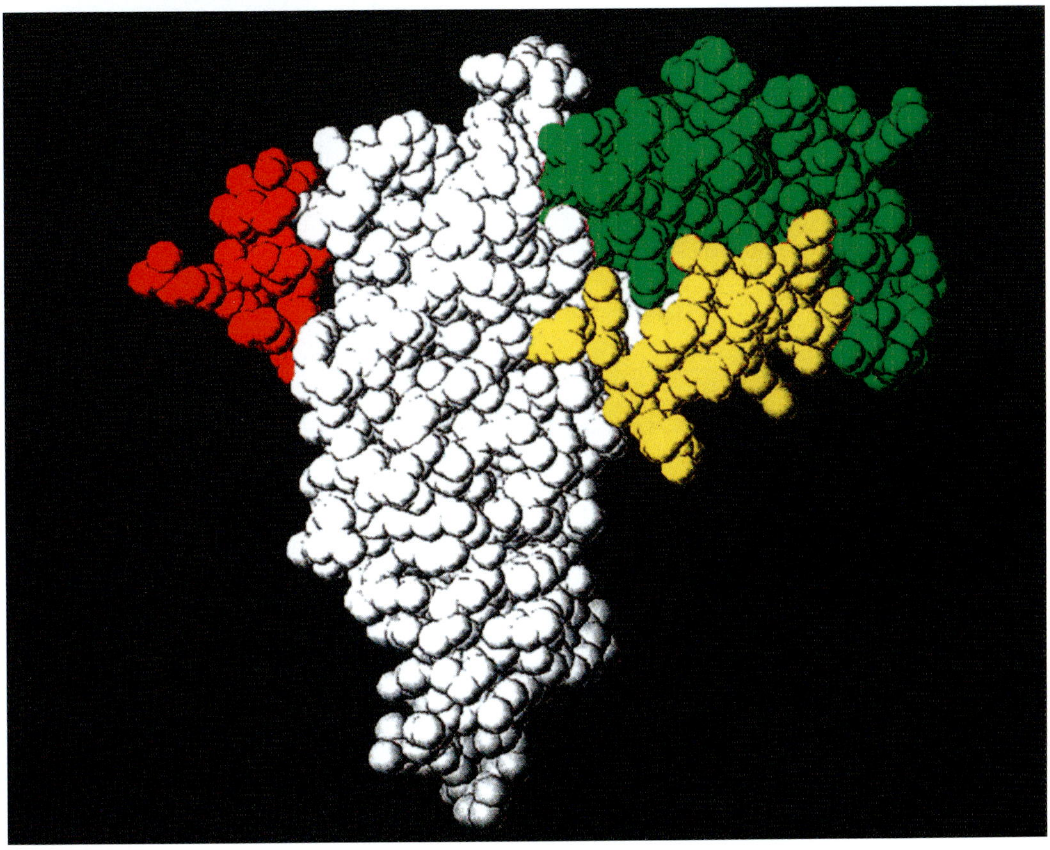

Chapter 23, Figure 2. The structure of the receptor-binding domain of FrMLV Env (27). VRA is in green, VRB in yellow, and VRC in red.

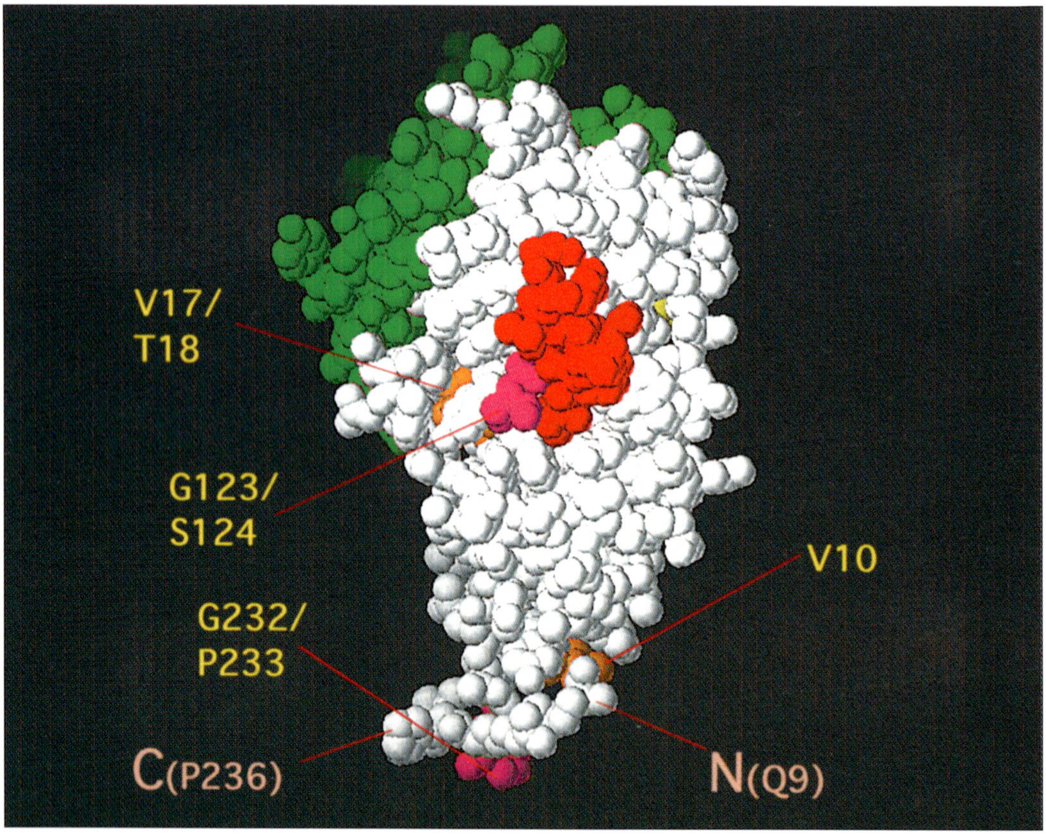

Chapter 23, Figure 3. Locations of insertions of some novel cell-targeting ligands (orange) and linker insertions (magenta) within the ecotropic Env receptor-binding domain. VRC is facing out from the page. VRA, B, and C are in green, yellow, and red, respectively. VRB is barely visible from this angle. The locations of the amino-terminal Q9 (N) and carboxy-terminal P236 (C) residues of the resolved crystal structure are indicated. FrMLV amino acid numbers are indicated for the homologous residues of MoMLV Env that are discussed in the text.

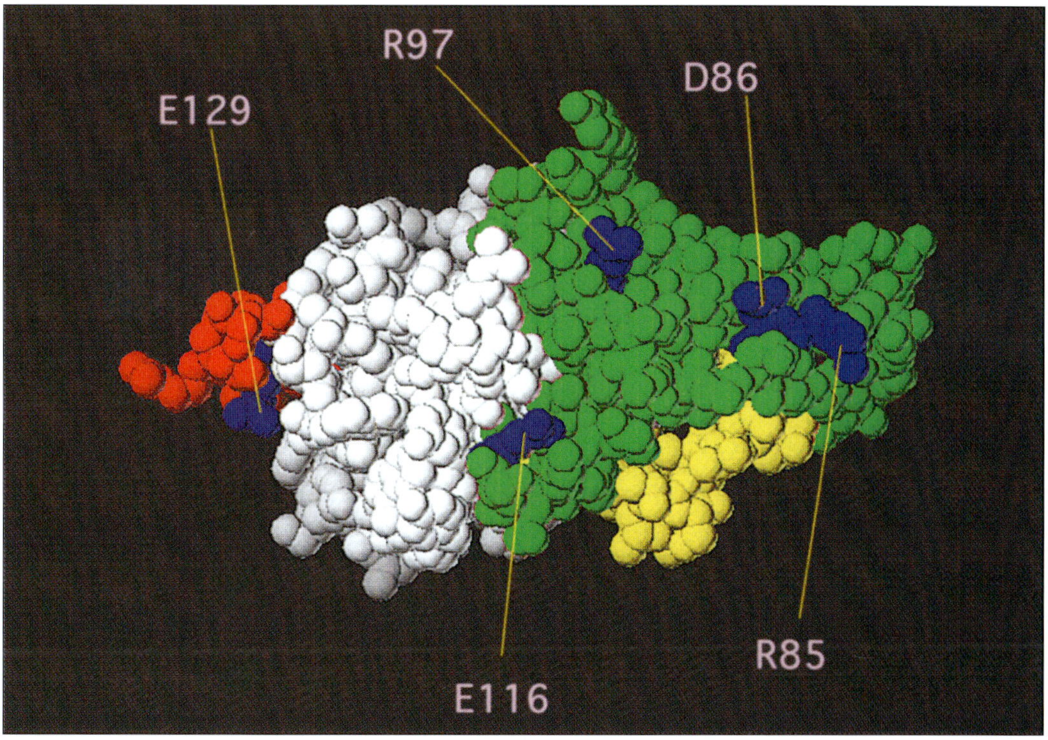

Chapter 23, **Figure 4. Amino acid residues shown to be important for Env-receptor interactions.** View is from the top of the structure shown in Figure 2. VRA, B, and C are in green, yellow, and red, respectively. Critical residues discussed in the text are indicated in blue. FrMLV amino acid numbers are indicated for the homologous residues of MoMLV residues that are discussed in the text.

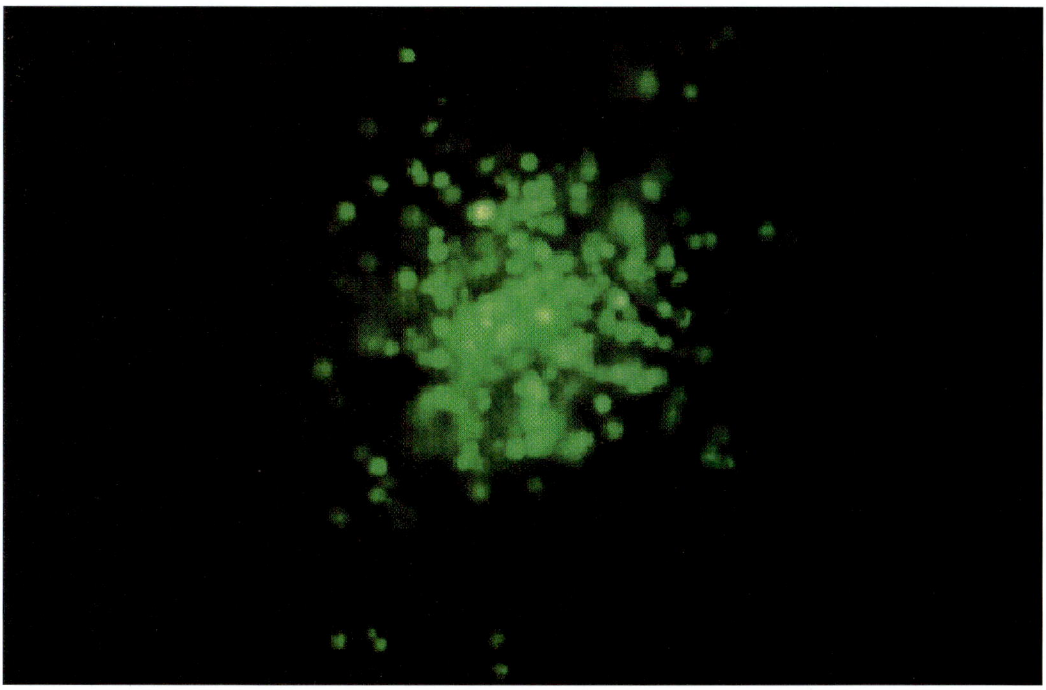

Chapter 27, **Figure 3. Cord blood-derived CFU-GM expressing EGFP (400× optical magnification).**

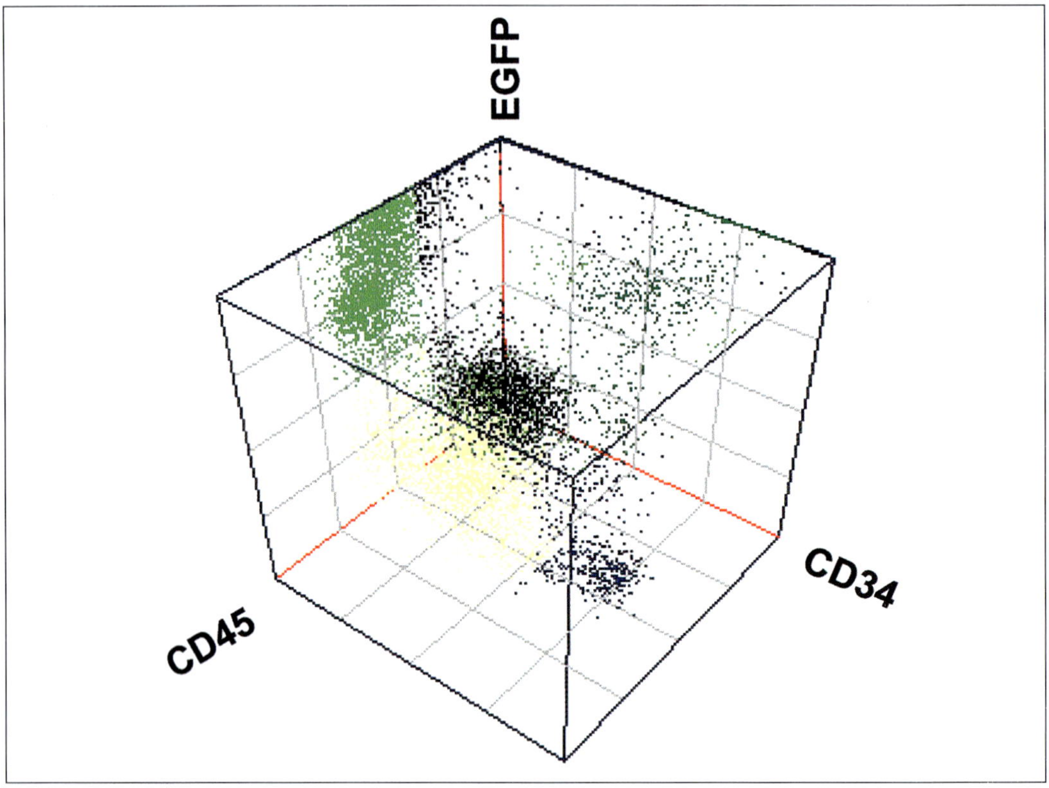

Chapter 27, Figure 5. EGFP expression in human CD45⁺ and CD45⁺CD34⁺ cells obtained from a BM biopsy of NOD-SCID mice at 30 days after transplantation. BM cells were labeled with specific antobodies for the CD45 and CD34 human antigens and analyzed by flow cytometry. Yellow: $CD45^+/CD34^-$ cells negative to EGFP; blue: $CD45^+/CD34^+$ cells negative to EGFP; light green: $CD45^+/CD34^-$ cells positive to EGFP; dark green: human $CD34^+$ cells expressing the EGFP protein. See text for information on percentage values.

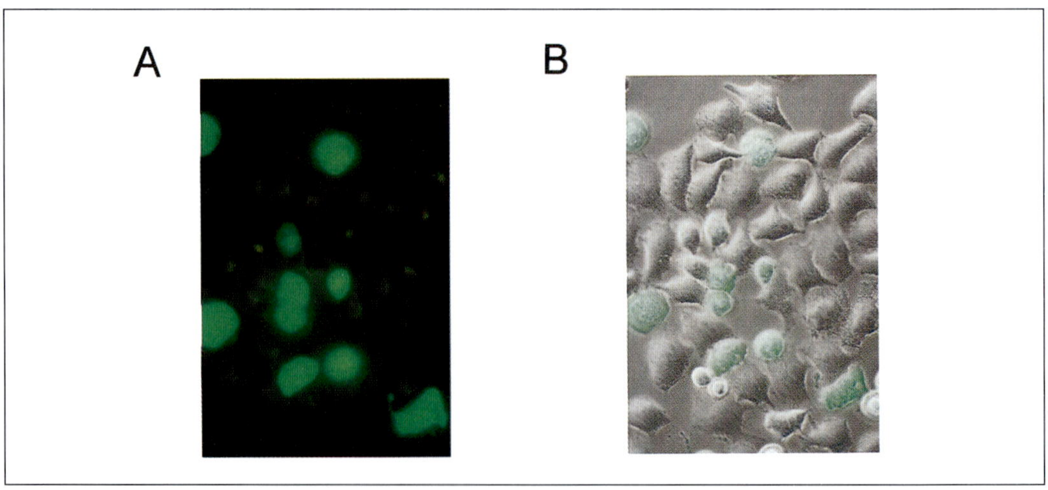

Chapter 33, Figure 3. HeLa cells infected with poliovirus polyprotein fusion vector expressing GFP (85). (A) Fluorescence microscopy. Cells were seeded on coverslips, infected, and mounted after 7 hours in PBS (unfixed) and photographed immediately. (B) Double exposure photography. Fluorescence microscopy was performed and the first exposure was made. Then, without rewinding the film and moving the object, a second exposure of the brightfield image was taken. This method allows fluorescing and non-fluorescing cells to be visualized on the same photograph. *Note:* The image as it appears is not actually observed under the microscope.

FAMILY: Herpesviridae
SUBFAMILY: Gammaherpesvirinae
GENUS: Rhadinovirus

19

Gamma-2-Herpesvirus Vectors for Investigation of Viral Pathogenesis and Experimental Gene Transfer

S. Monroe Duboise[1] and Jae U. Jung[2]

[1]Department of Applied Medical Sciences, University of Southern Maine, Portland, ME; [2]Department of Microbiology and Molecular Genetics, New England Regional Primate Research Center, Harvard Medical School, Southborough, MA, USA

INTRODUCTION

Gamma-2-herpesviruses or rhadinoviruses have coevolved in intimate association with multicellular hosts and establish lifelong persistent infection in balance with host innate and immune defenses. Rhadinovirus genome sequences reveal extensive evolutionary molecular piracy and subversion of cellular mechanisms of immune recognition and response, cell cycle control, complement-mediated killing, and apoptotic response. Understanding of gammaherpesvirus persistence and modulation of host responses may contribute to gene transfer vector designs that prolong persistence and transgene expression while minimizing undesired inflammatory responses. The abilities of rhadinoviruses to accommodate substantial foreign genetic material and to persist in episomal form during latent infection are useful vector attributes. Viral modulators of host responses, notably cellular gene homologs within rhadinovirus genomes, are genetic resources potentially adaptable to therapeutic applications in diverse vector systems. Understanding rhadinovirus-associated hyperplastic or neoplastic disease in immunocompromised or genetically susceptible hosts is a challenge that accompanies development of rhadinovirus-based gene transfer vectors and is a necessary prerequisite for their clinical or veterinary application. This report presents illustrative manipulations of a rhadinovirus genome that have been used to delete and replace viral genes during investigations of the biology and pathogenesis in nonhuman primates of a prototypic rhadinovirus, herpesvirus saimiri (HVS). Limited gene transfer-based study of the Kaposi's sarcoma-associated herpesvirus (KSHV), the only recognized human rhadinovirus, is also discussed. Recombinant rhadinoviruses constructed to elucidate specific gene functions are contributing fundamental understanding of viral pathogenesis and are establishing the foundation for first-generation rhadinovirus gene transfer vectors.

Viral Vectors: Basic Science and Gene Therapy
Edited by A. Cid-Arregui and A. García-Carrancá

The family Herpesviridae is large, diverse, and includes viruses with distinct tissue tropisms (100). Herpesviruses characteristically establish persistent or latent infections in which viral gene expression is severely limited and viral clearance by host immune defenses is efficiently evaded (92). Viruses of the subfamily Gammaherpesvirinae infect and establish latency in lymphoblastoid cells and typically, during acute infection or following reactivation from latency, also produce lytic infection in other cell types that can include epithelial, fibroblastic, endothelial, and monocytic cells (10,12,38,91,114,115). Latent or persistent infection of nonlymphoid cells has also been reported (113,126). Comparisons of nucleotide sequence homology and genomic organization of gammaherpesviruses have been used to distinguish the two genera *Lymphocryptovirus* (gamma-1-herpesviruses) and *Rhadinovirus* (gamma-2-herpesviruses) (76,77). Epstein-Barr virus (EBV) or human herpesvirus 4 (HHV-4) is the etiologic agent of infectious mononucleosis and has also been linked to Burkitt's lymphoma, nasopharyngeal carcinoma, and possibly to Hodgkin's disease (72,96). The relevance of gamma-2-herpesviruses to human health was unconfirmed prior to the discovery of KSHV, which is formally designated human herpesvirus 8 (HHV-8) (18,19). This human rhadinovirus has been implicated in the etiology of Kaposi's sarcoma, body cavity-based lymphomas, and in multicentric Castleman's disease (108).

While there has been active development of EBV-based gene transfer vectors (41,61,62,98) and extensive development and utilization of herpes simplex virus (HSV) vectors (22,45,46,47,51,64,74), little attention has been given to development of rhadinovirus vectors. Development and use of gamma-2-herpesvirus vectors have been limited to investigations of rhadinovirus pathogenesis using HVS, which are reviewed in this report. HVS is a rhadinovirus that is apparently nonpathogenic in its natural host, squirrel monkeys (Saimiri sciureus), but is highly oncogenic, rapidly producing fatal T-cell lymphomas, in other New World primates (3,24,26,37) and is capable of transforming human T lymphocytes in culture (7,9,13). Other rhadinoviruses of nonhuman primates (55), including those recently isolated from rhesus macaques (27,102), and several other viruses that can infect domesticated animals (15,36,116) also provide useful models. Another rhadinovirus, first isolated in 1980 (11), that is increasingly being investigated is murine gammaherpesvirus 68 (MHV-68), which establishes latent infection in B lymphocytes (34,110,115, 120) and also in some nonlymphoid cells, notably lung epithelial cells (113,126). Interest in MHV-68 has been stimulated by its genomic similarity to KSHV, HVS, and to a lesser extent to EBV (3,34, 104,122), its association with lymphoproliferative disease resembling infectious mononucleosis (29,119), and by the opportunities it provides to study rhadinovirus biology and pathogenesis in genetically defined mouse models (17,35,110, 122). MHV-68 infection of mice with specific gene knockouts are providing insight into the immune effectors required for control of lytic infection and cellular requirements for establishment of viral latency (110,112,126,127). MHV-68 infection of mice lacking either interferon gamma or receptors for this cytokine fail to control acute infection of arterial smooth muscle and thus provide an interesting model of herpesvirus-induced vascular disease (127). With the MHV-68 genome fully sequenced (122), this virus, which is capable of highly productive lytic growth in various continuous cell culture lines, is readily manipulated genetically using the same methods that are discussed below for producing HVS recombinant vectors.

GENE TRANSFER POTENTIAL OF RHADINOVIRUSES

Gene transfer methods are being designed for a variety of therapeutic purposes including vaccination, immunomodulation, replacement of defective genes in somatic cells, and delivery of lethal genes to neoplastic tissues (97,101,111,121). The extent to which viral methods for gene delivery have dominated the gene transfer literature reflects the viruses being elegant mechanisms for gene transfer resulting from innumerable experiments in the natural laboratory of coevolution with multicellular hosts. The challenges of viral evolution have parallels in the design requirements for targeted gene delivery systems. Appropriate targeting and regulation of gene expression, and some degree of control of interference from host innate responses and immune defenses, are common requirements of viral evolution and successful gene transfer vector design. Herpesviruses have evolved in particularly close and prolonged association with their hosts (77). The complex DNA genomes of herpesviruses exhibit highly regulated gene expression and contain numerous accessory genes unnecessary for viral replication, but relevant to successful virus-host interactions. Extensive studies of HSV, for example, have shown that 45 of the 83 open reading frames (ORFs) that encode viral genes are unnecessary for viral replication in at least some cells in culture (99). While herpesvirus genetic complexity is a daunting characteristic for vector design, the genetic endowment of herpesviruses does allow many choices of whether to remove or exploit accessory genes. Consequently, thorough investigation and understanding of the molecular cell biology and immunology of virus interactions with hosts are expected to be of fundamental importance for the design of sophisticated gene transfer methods. Vector design options within the context of gamma-2-herpesvirus biology and general methods for manipulating rhadinovirus genomes are discussed in this report.

Development of gamma-2-herpesvirus gene transfer vectors will potentially provide new therapeutic tools that may target either cell types in which latent infection is established such as lymphoid cells or the various cell types in which acute or reactivated lytic infection can occur. Initially, however, the most appropriate use of rhadinovirus recombinant vectors is in the investigation of the complex biology and pathogenesis of these viruses. Studies reviewed in this report will illustrate that rhadinovirus gene transfer methods can serve as powerful experimental tools. While both HVS and MHV-68 are capable of infecting human cells, KSHV-based gene transfer vectors may be more likely to be developed for human applications. Development of KSHV vectors will be greatly facilitated if difficulties in cultivating the virus in vitro at high titers (94,95) can be circumvented. Hyperplastic or neoplastic disease associated with KSHV is highly correlated with human immunodeficiency virus type 1 infection (18,43,80). Whether KSHV is truly an oncogenic virus is controversial (42). KSHV, unlike EBV, is not ubiquitous in human populations (80, 108). KSHV-based vectors could, therefore, be applied with some advantages to an immunologically naive population. Replication-defective vector development for immunoprophylaxis similar to designs that have been explored for HSV (22,82,87) may also be considered. Rhadinoviruses of veterinary importance such as the alcephaline herpesvirus, equine herpesvirus 2, and bovine herpesvirus 4 could be manipulated to produce multiantigen veterinary vaccine vectors.

General methods for producing rhadinovirus-based gene transfer vectors have been developed and will be discussed.

Beyond the basic methods outlined in this report, many of the lessons learned and vector designs that have been incorporated into the more extensively studied EBV and HSV vectors are equally applicable to use in rhadinovirus vectors. Using rhadinoviruses, it will be possible to develop Replication-defective vectors similar to those studied extensively using HSV (46,74,82,87) or amplicons containing *cis*-regulated elements and limited viral sequences needed for episomal maintenance that previously have been derived from EBV, HSV, or even both in combination (45,62,78,98,123,125,129). Attenuated replication-competent herpesvirus vectors have also been considered useful for certain applications (133). The HVS prototypic vectors that have been described (30–32) and will be discussed in this report are replication competent, but contain deletions or mutations in genes critical to pathogenesis. Replication-defective rhadinovirus vectors can be produced by similar methods that rely additionally on viral cultivation in complementing cell lines that express essential viral gene products.

GENE TRANSFER APPLICATIONS OF RHADINOVIRUS MODULATORS OF HOST RESPONSES

Efficient delivery and appropriate expression of a therapeutic transgene are typical goals of gene transfer vector design. Difficulties involving targeting, levels of expression, sustainability of expression, regulation of expression, and undesired triggering of host responses have plagued gene transfer efforts (14,121). The bidirectional interaction between viral vectors and their hosts requires careful attention when prolonged gene transfer is the goal. Design of effective vectors may require inclusion of mechanisms for evasion of host responses either by stealth or by regulation of innate and immune responses. These critical requirements for effective gene transfer are also requirements for the lifelong infections that herpesviruses routinely and, in some cases, ubiquitously accomplish. Herpesvirus latency is unparalleled as an example of stealth. Coevolution with host species and possession of large DNA genomes capable of accommodating many accessory genes have also provided herpesviruses with a formidable array of genes with products that directly target host innate cellular and immune responses or target signal transduction pathways relevant to cellular proliferation (1,2,4,8,20, 21,52,56,58,59,67,68,86,88,90,117,130, 131). Complete genomic sequences reveal that this is particularly true of gammaherpesviruses, which infect and persist in lymphoid cells (3,36,104,108,116,122).

The extensive rhadinovirus repertoire of genes that target host signal transduction, cell cycle control, apoptosis, the complement cascade, and immune responses are summarized in Table 1. While some of this impressive genetic endowment contributes directly to the persistence of viral infection, it is also clear that lymphoproliferative disorders and neoplastic or hyperplastic pathology are potential outcomes of extensive viral-induced dysregulation. Nevertheless, as rhadinovirus-induced pathogenesis is more thoroughly understood and replication-defective vectors are produced, it may be possible to adapt portions of the rhadinovirus genetic armamentarium for judicious use in gene transfer applications either within rhadinovirus-based vectors or in other vector systems. For example, viral genes limiting apoptosis or complement-mediated killing could contribute to prolonging transgene expression or to producing therapeutic effects as has been observed in experiments in which Bcl-2 expression was elevated (71,73). If viral genes that alter host signaling and innate and

Table 1. Representative Genes of HVS, KSHV, and MHV-68 that are Homologous to Cellular Genes or that are Known to Modulate Host Responses or Cellular Signaling

	Virus	References
Bcl-2 homolog	K,M,S	20,86,122
CD59 homolog	S	3,4,103
Complement control protein homolog	K,M,S	2,39,122
Cyclin D homolog	K,M,S	59,70,106,122
Dihydrofolate reductase (DHFR)	K,S	3,104
FLICE Inhibitory Protein (vFLIP)	K,S	117
Interferon regulatory factor	K	44,69,134
Interleukin-6 homolog	K	80,104
Interleukin-8 receptor homolog	K,M,S	1,5,6,93,122
Interleukin-17 homolog	S	130
K1	K	68,81,108
Macrophage inhibitory protein-1 alpha homolog	K	88,104
Macrophage inhibitory protein-1 beta homolog	K	88,104
Mouse mammary tumor virus superantigen homolog	S	31,131
Ox2 homolog	K	88,104
Saimiri transformation-associated protein (STP)	S	52,53,54,60,63,84
Serpin homolog	M	109,110,122
Thymidylate synthetase (TS)	K,S	50,66,104
Tyrosine kinase interacting protein (Tip)	S	8,30,33,48,57,58

K, KSHV; M, MHV-68; S, HVS

immune responses are regarded as part of a pharmacopeia for gene transfer technology, then rhadinovirus genomes may prove to be significant resources. Access to these resources, however, depends on intensive study of rhadinovirus biology and pathogenesis. The studies of HVS that are discussed below illustrate that the rhadinovirus gene transfer vector development and studies of viral pathogenesis can be parallel and complementary processes.

GENE TRANSFER-BASED STUDIES OF HVS PATHOGENESIS

HVS can be propagated lytically at high titers in owl monkey kidney cells. This characteristic along with its dramatic pathogenesis, a fulminant T-cell lymphoma in New World primates such as common marmosets, provides an excellent system for studying rhadinovirus biology both in vivo and in vitro. The ease of propagation, which is also characteristic of some of the other rhadinoviruses such as MHV-68, makes this virus particularly amenable to genetic manipulation. Nevertheless, isolation of viral mutants or recombinants without the use of genetic markers for selection or identification of recombinants is a highly labor-intensive effort to detect rare recombination events (25,28,85). Consequently, the introduction of a method that facilitates identification of viral recombinants by incorporation of a transgene encoding a reporter (e.g., the secreted engineered alkaline phosphatase [SEAP] or green fluorescent protein [GFP]) provided a significant methodological advance (32). Further advantage was provided by flanking control elements of the reporter expression cassette and the reporter gene itself with restriction enzyme recognition sites that were not found elsewhere in the viral genome. These provided sites for direct cloning into the recombinant virion DNA, thereby allowing convenient removal or replacement of either the reporter gene or the entire expression cassette. The same goals were also attainable by use of standard recombination methods followed by selection and isolation of recombinant virus by assaying for loss of reporter gene expression. Verification of the desired recombination finally depends on polymerase chain reaction (PCR) analysis, hybridization studies, and nucleotide sequence analysis. This method has been used to introduce deletions (30), point mutations (33), and gene replacements (49,68) into the bicistronic transcriptional unit located at the left end of the unique coding sequence in the genome of highly oncogenic HVS-C488. Further refinement of these methods to permit precise gene replacement beginning with the initiation codon of a viral ORF has been demonstrated in a study of HVS orf14, which encodes a viral superantigen related to mouse mammary tumor virus superantigens (31). The precision of this replacement simplifies construction of recombinant virus in which transgene expression can be driven by endogenous viral control elements rather than by a foreign promoter.

These studies have provided significant insight into the role of selected viral genes in the pathogenesis and persistence of HVS. The contributions of the saimiri transformation-associated protein (STP-C488) (30), the tyrosine kinase interacting protein (Tip) (30,33), and the orf14 superantigen (31) to oncogenesis and viral persistence were assessed through in vitro and in vivo studies of infection with HVS recombinants with specific gene deletions. Each of these genes proved to be essential to oncogenesis. Cocultivation of owl monkey kidney cell monolayers with serially diluted peripheral blood mononuclear cells (PBMC) collected periodically from marmosets infected with each recombinant virus was used to quantitate the persistence in vivo of each recombinant virus. HVS recombinants lacking either STP-C488 or Tip could be detected in PBMCs of infected marmosets for at least five months postinfection with the higher viral loads being observed for the virus with Tip deleted. HVS with deletion of the orf14 superantigen could not be detected in infected marmosets beyond four to eight weeks postinfection. Sera from marmosets infected with recombinant HVS expressing a SEAP reporter gene were also tested for SEAP production. Serum of those animals infected with HVS lacking Tip contained detectable levels of SEAP at one month postinfection (30).

Beyond these specific gene deletion studies, HVS recombinants with specific genes replaced by reporter expression cassettes have also provided the starting material for additional gene replacements expected to be informative with respect to specific gene functions and the pathogenesis of the virus. The same methods are applicable to the introduction of any transgene of choice. The STP-C488 oncogene product has been shown to associate with cellular ras (56). Virion DNA from the HVS recombinant with the STP-C488 open reading frame replaced by a SEAP expression cassette was used for isolation of a HVS recombinant that expresses ras rather than STP-C488 (49). This resulted in restoration of the oncogenicity of the virus. A similar replacement of STP-C488 was also done using the K1 ORF of KSHV, which is the left-most ORF in the unique coding sequence of the KSHV genome, a position analogous to that of STP in HVS (68). Restoration of oncogenicity in the K1-expressing HVS recombinant provided the first in vivo demonstration of the transforming potential of the KSHV K1 gene product. Similarly, the HVS orf14 superantigen has been replaced by a mouse mammary tumor virus superantigen that restored transforming capacity to the virus with orf14 deleted (unpublished data). This method was also applied to further examination of the function of Tip, which is known to associate with the nonreceptor tyrosine kinase Lck, a member of the Src kinase family that is prominently expressed in T-lymphocytes (33). A SH3 binding domain of Tip necessary for the association with Lck had been previously identified (57). Using site-directed mutagenesis, a Tip gene with mutated SH3 binding domain was isolated to produce a Tip product incapable of binding Lck. When recombinant HVS expressing this mSH3B Tip instead of wild-type Tip was isolated and used for in vitro and in vivo studies, the association of Tip with Lck was shown to be unnecessary for oncogenesis (33). In fact, the mSH3B Tip virus produced enhanced lymphoid cell activation, an observation consistent with other mutational studies, suggesting that the association of Tip and Lck serves a signal-blocking function (48,58).

GENERAL PROTOCOL FOR DELETION AND REPLACEMENT OF RHADINOVIRUS GENES

The general protocol below is applicable to any rhadinovirus (and indeed, to any herpesvirus) that can be efficiently cultivated in a cultured cell line. The methods are highly dependent on knowledge of viral nucleotide sequences and are, therefore, most readily applied to rhadinovirus genomes that have been fully sequenced. Construction of recombinant viruses with specific gene deletions, point mutations, or transgene insertions as described previously (25,30–33,48,68) depends on low frequency in vitro homologous recombination events between co-transfected virion DNA and linearized plasmid constructs containing sequences to be inserted (such as altered viral sequences and/or reporter or other transgene expression cassettes) flanked on each side by hundreds of base pairs of unaltered viral sequence. Unless an inserted transgene can be conveniently assayed during isolation of recombinant virus, the process is facilitated by using a reporter gene cassette that expresses a readily assayed marker such as SEAP or GFP. Use of a reporter system may add recombination steps in some cases, but will greatly facilitate the detection and isolation of desired recombinants. The essential process of producing a specific rhadinovirus recombinant in which a SEAP reporter cassette has been inserted in a precise genomic location is diagrammed in Figure 1 using the ORF 14 superantigen gene deletion (31) as an example. The unique restriction enzyme recognition sites shown flanking the SEAP expression cassette provide convenient cloning sites for replacement of the reporter cassette with any other transgene of choice and can be used for direct cloning into virion DNA as previously demonstrated (32). For studies of specific deletion mutations, these restriction sites permit convenient removal of the foreign reporter cassette by restriction enzyme digestion of the recombinant virion DNA followed by ligation and transfection of permissive cultured cells. SEAP expression can either be driven by an exogenous promoter provided in the plasmid construct or by an endogenous viral promoter if the precise gene replacement method described for study of HVS orf14 is used (31). Whether endogenous or exogenous control elements are employed, the genomic locus chosen for transgene expression is expected to be critical for level and persistence of expression and can only be ascertained empirically and through understanding of

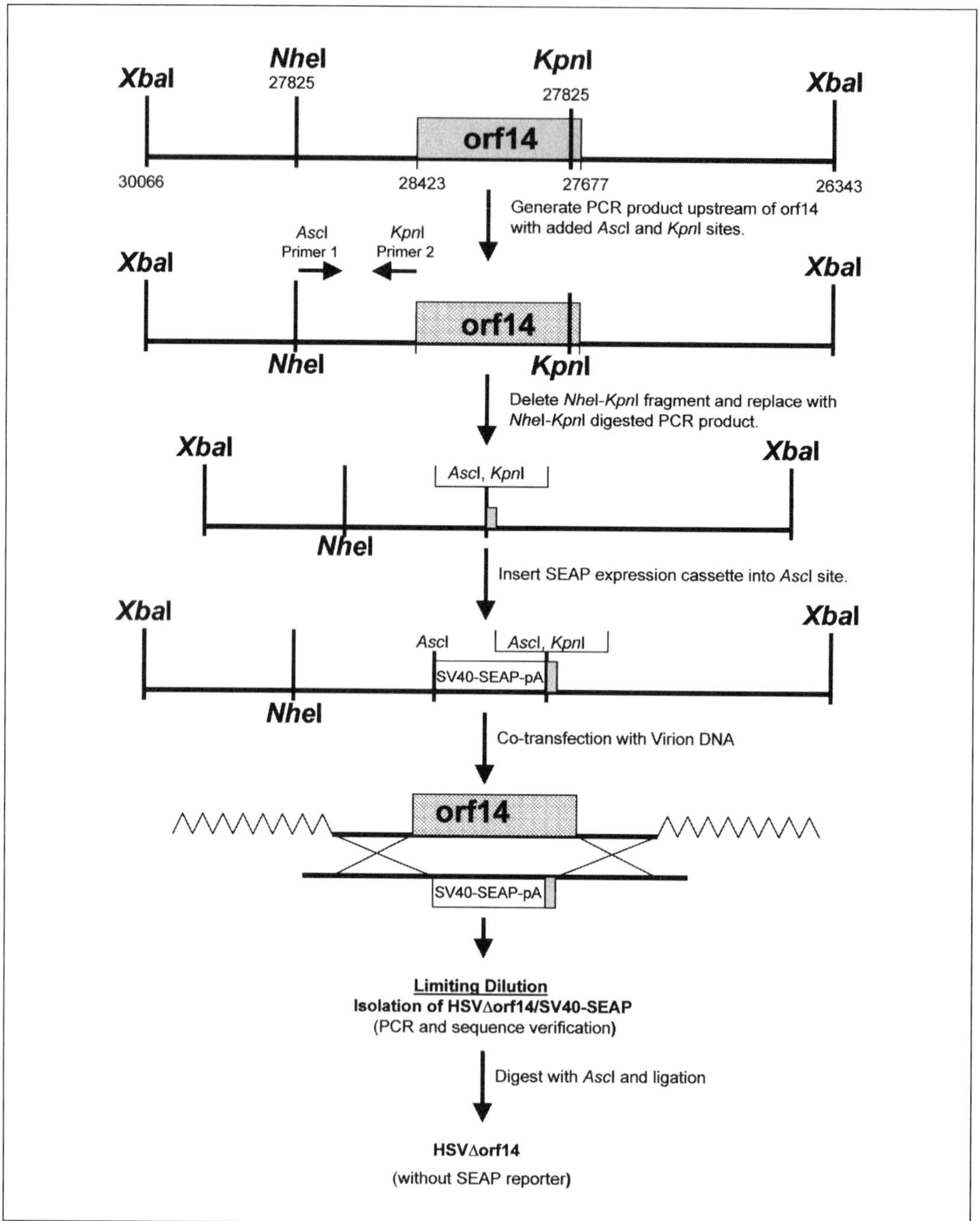

Figure 1. Construction of HVS recombinants. The method used for construction of a HVS recombinant with orf14, the viral superantigen gene, deleted is shown (31). PCR amplification of the sequence upstream of orf14 using primers adding *Asc*I and *Kpn*I restriction sites allows removal of a *Nhe*I-*Kpn*I fragment and precise replacement of only the sequence upstream of and orf14 deletion. The *Asc*I site allows insertion of the simian virus 40 (SV40) early promoter driven SEAP expression cassette in place of orf14. The plasmid construct is then co-transfected along with HVS virion DNA into owl monkey kidney cells to allow homologous recombination and isolation of SEAP expressing recombinant virus. After repeated limiting dilution purification of the virus with orf14 deleted, virion DNA can be isolated, digested with *Asc*I to remove the reporter expression cassette and then be ligated and transfected into owl monkey kidney cells to isolate virus with orf14 deleted and with the reporter cassette also removed. Since *Asc*I sites are not found elsewhere in the viral genome, the site can be used for direct cloning of any gene of interest into the viral genome (32).

the biology of the virus. Once optimal expression characteristics are determined using recombinants expressing SEAP or other reporters, virion DNA from these recombinants can be used for recombinations that will result in a rhadinovirus vector containing any transgene of choice as shown in Figure 2, again using the HVS orf14 deletion virus as an example. In this recombination process, loss of reporter gene expression serves as a marker for detection and isolation of desired recombinants. Major steps of a generalized protocol for producing a rhadinovirus-based gene transfer vector are described below.

Virion DNA Isolation

Purified or highly enriched preparations of intact virion DNA are required for use in transfections and as a source of viral DNA fragments for plasmid clones used in production of transgene expression constructs suitable for homologous recombination with virion DNA. The following is a convenient method of virion DNA isolation that has been adequate for providing large quantities of intact HVS DNA suitable for transfection and for cloning. The procedure should work well for any rhadinovirus that is released at high titers during lytic infection of cultured cells. Infected cultures with nearly confluent lysis are cleared of intact cells and cell debris by low-speed centrifugation. Then, intact virus is pelleted from the culture supernatant fraction by centrifugation at 40 000 × *g* for 2 hours. To disrupt the pelleted virus and digest proteinaceous components, the viral pellet is resuspended in lysis buffer containing 10 mmol/L Tris (pH 8.5), 1 mmol/L EDTA, 1% sarkosyl, and 0.1 mg/mL proteinase K, and then incubated for 2 hours at 60°C. Virion DNA is then extracted in the aqueous phase first with an equal volume of phenol and then twice with chloroform. All manipulation of the DNA is done using pipet tips that are cut to reduce the shearing forces to which the large viral DNA molecules are subjected. Virion DNA prepared in this fashion is typically sufficiently pure and intact for all necessary cloning and transfection procedures.

Reporter Gene and Transgene Plasmid Construction

For production of HVS recombinants, reporter plasmids containing either SEAP or GFP expression cassettes flanked by various restriction enzyme sites (useful for insertion into viral sequences) were generated using commercially available plasmids combined with PCR-amplified reporter genes plus the desired restriction sites (32). While these reporter plasmids were designed for flexibility of use and may be applicable to making other rhadinovirus recombinants, the design considerations were determined significantly by specific HVS sequences. Thus, the need for modifications for optimal use in other rhadinovirus systems is probable. An important design feature was the addition of unique restriction sites (not present in virion DNA) flanking the entire expression cassette along with other unique sites flanking only the reporter gene itself. This design allows maximum flexibility either for use in direct cloning into recombinant rhadinoviruses (32) or for use in exchanging reporter genes for other transgenes within the plasmids used for viral recombination.

Plasmids for recombination following co-transfection with virion DNA need to contain not only the reporter or transgene but also must contain sufficient unaltered viral DNA sequence to allow the double-crossover homologous recombination that results in the desired insertion into the recombinant viral genome. Extensive homologous sequences of a kilobase or more and flanking the foreign insert are desirable to increase the frequency of the desired recombination event. Thus, the plasmids used for recombination generally contain several kilobases of viral DNA into which the reporter gene or other transgene can be cloned. Removal of viral DNA fragments followed by replacement with PCR fragments containing needed restriction sites is sometimes a useful strategy for increasing precision of placement of a transgene construct (31).

Recombination of Transgene Plasmids with Virion DNA

Mixed transfection of cells permissive for viral infection with infectious virion DNA and linearized plasmid that contains specific mutations and/or insertions is a commonly used method for generation of herpesvirus recombinants and has been described previously using HVS (25,32).

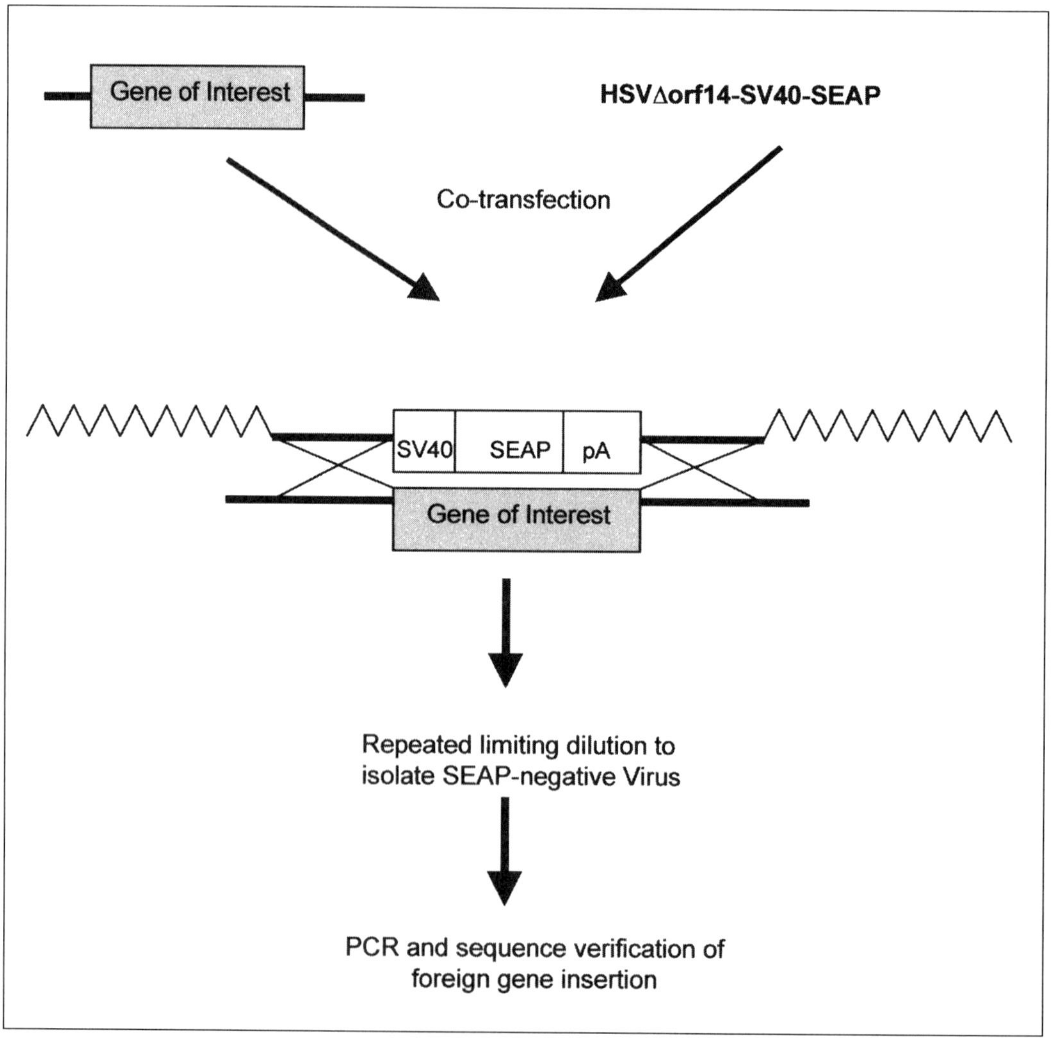

Figure 2. Construction of recombinant HVS with orf14 replaced with any gene of interest. A gene of interest, flanked by sequences homologous to those surrounding the SEAP expression cassette in the virion DNA of the HVS orf14 deletion recombinant, can be inserted by co-transfection of plasmid and virion DNA allowing homologous recombination to replace the reporter cassette. The recombinant can be isolated by limiting dilution purification of SEAP-negative virus.

Various transfection protocols may provide satisfactory results, but we have typically used Ca^{2+} coprecipitation of virion and linearized plasmid DNA at a molar ratio of 1:200 with absolute quantities being determined empirically for any given virion DNA preparation. Transfected cells are then incubated appropriately and monitored for appearance of cytopathic effect. When a reporter construct is being inserted into the virus, a crude assessment of whether successful recombination has occurred can be obtained by assaying for reporter gene expression at a point in time beyond optimal transient expression (i.e., at four or more days after transfection).

The HVS recombinants previously produced were replication competent. Some recombinants designed for safe transgene delivery, however, would be expected to contain specific defects that render the virus incapable of replication in the absence of a complementing gene product. To supply this needed complementation, stable permissive cell lines expressing the needed gene product must be produced by transfection of expression plasmid. Complementing viral genes that may yield products which either adversely affect growth or are toxic to the cells may be provided in an inducible context such as in the pIND vector that requires the presence of an ecdysone analog to induce gene expression (89). Such a cell line could be maintained without the inducer until the complementing gene product is needed during production of viral recombinants.

Isolation of Viral Recombinants

Virus resulting from co-transfection of virion DNA and linearized plasmid constructs is expected to include recombinants at low frequency. Thus, isolation, purification, and verification of the desired recombinant virus are needed. Repeated plaque purification is an option with some rhadinoviruses, while others may require repeated limiting dilution purification. Use of a reporter gene cassette greatly facilitates this isolation process. If the desired recombinant is expected to produce a reporter gene product (as shown in Figure 1) or, conversely, if the desired recombination results in replacement of a reporter cassette (as shown in Figure 2), then acquisition or loss of reporter expression provides the necessary selection factor for the desired recombinant rhadinovirus. After repeated purification of the recombinant virus, the purity of the virus and correctness of the recombination can be verified by appropriate analyses that may include sequence analysis, PCR, hybridization, and Western blot analysis. When this verification is obtained, in vitro and in vivo studies of the recombinant virus and of the level and duration of transgene expression can commence.

FUTURE DIRECTIONS FOR STUDIES USING RHADINOVIRUS-BASED GENE TRANSFER

Gene transfer methods using herpesvirus vectors may contribute to effective treatment of inherited and acquired disease. Stringent requirements and challenges must be met, however, for the potential therapeutic value of gene transfer technology to be realized (14,46,97,101,111,121). Physiologically appropriate transgene expression requires that delivery be targeted, gene expression be sustained and regulated, and unwanted host responses be avoided. Progress in meeting these formidable challenges is likely to be achieved by a careful and incremental process of investigation using appropriate model systems. In the case of gammaherpesvirus-based vectors, there is the additional challenge of understanding the complex biology and pathogenic potential of the viruses. As with other gene transfer vectors, ensuring safety of vectors through attenuation or replication defects is of great importance in the design of rhadinovirus-based vectors. Studies of the molecular basis of HVS pathogenesis have contributed to the development of HVS gene transfer methods and demonstrate the synergy between these pursuits.

HVS provides opportunities to study the basis of dramatic rhadinovirus-induced pathology as well as gene transfer in nonhuman primate models. The recently described rhadinoviruses infecting rhesus macaques (27,102) and MHV-68 (122) provide B-lymphotropic rhadinoviruses with significant genetic and biological similarity to KSHV. MHV-68 and genetically defined mouse models provide particularly powerful experimental systems for exploration of rhadinovirus biology and for testing of rhadinovirus-based gene transfer methods. Murine models for testing gene delivery and therapeutic efficacy clearly offer substantial advantages over other animal models for defined experimental testing. Numerous inbred strains and genetically unique animals such as knockout mice are particularly valuable assets. Murine models are clearly the most advanced animal models available with respect to knowledge of genetics and immunology. Reagents for genetic, biochemical, and immunological investigations are generally more available for mouse models than for any other system. The favorable economics of using a small animal model for testing gene transfer methods is also an obvious advantage. The advantages of mouse models for testing of gene transfer methods also apply equally to performing the needed analysis of the pathogenic mechanisms of the viruses that may be adapted for gene transfer applications.

Gene transfer technology may provide methods of delivery of immunoprotective antigens for vaccination against a variety of infectious agents. Live attenuated viral vaccines, which have a history dating back to Jenner's vaccination against smallpox, are generally without equal for stimulation of cell-mediated immunity and long-lasting immunoprotection. Use of vaccinia virus recombinants for vaccination has shown considerable promise and potentially can simultaneously deliver multiple antigens (83). Herpesviruses including rhadinoviruses have similar potential. The potential for addressing some of the most challenging immunoprotection problems (e.g., vaccination against agents of malaria) using viral vectors expressing multiple antigens has been discussed and demonstrated (65,118). Engineered vaccines expressed in viral vectors may be particularly useful in a veterinary context for immunizing animals that are economically or socially important. Viral delivery systems for veterinary vaccines may also have significant impact on human health when directed toward breaking the chain of transmission of zoonotic agents by vaccination of animals that are potential reservoir hosts.

Designing appropriate quantitative and temporal regulation of transgene expression and ensuring safety are fundamental problems that must be addressed in construction of virus-based vaccines and gene transfer vectors. While attenuating mutations that do not preclude viral replication have been the basis for successful vaccines, higher levels of safety may be achieved by using replication-defective viruses that can infect target cells and produce immunogenic or therapeutic molecules but are unable to produce infective virus particles except in cultured cell lines that have been engineered for stable expression of missing viral components (22,82,87). Incorporation into gene transfer vectors of a variety of pharmacologically regulatable gene expression systems such as those controlled by tetracycline (75), rapamycin (132), or steroid hormone analogs (23,124,128) may provide opportunities to produce viral vectors that not only have regulatable transgene expression but also have conditional replication competence. Since some replication-defective herpesviruses are nevertheless capable of persistence (46,74, 105), it may be possible to produce conditionally replication-defective herpesvirus vectors that combine the advantages of replication-competent vectors with the safety advantages of replication-defective vectors. The extent to which viral genes are expressed by replication-defective herpesvirus vectors is highly dependent on which essential gene is deleted or mutated. Deletion of one of the immediate early genes that activate expression of other viral genes, for example, will greatly limit the expression of most viral genes (105). Substitution of regulatable control elements for the promoter of an essential rhadinovirus gene such as an immediate early transactivator could result in a virus that is replication dependent on the presence of an inducer. Replication-defective HSV vectors have been extensively characterized (22,46,74,82,87,105). Some essential genes that have been deleted from HVS vectors, such as ICP8 and ICP27, have homologs in rhadinoviruses that might be targeted for either deletion or regulation in the design of safe rhadinovirus vectors.

Development of rhadinovirus-based gene transfer vectors has proceeded through only the most preliminary stages at present. The tools described in this report have only been applied in the context of exploring viral pathogenesis. While highly modified episomal amplicon vectors have been produced using EBV and HSV (16,61,62,98,107,129) as well as replication-defective viruses, no such vectors have been reported in a rhadinovirus system. Nevertheless, the potential of rhadinoviruses is certainly no less than any of the other herpesvirus-based gene transfer vector systems. As suggested above, these viruses possess a rich endowment of genes with products that modulate cellular signaling and host innate and immune responses that may have potential to contribute not only to rhadinovirus-based gene transfer but also to other therapeutic gene delivery systems. These genetic resources as well as concepts such as creation of chimeric vectors (40,97) await development in synergy with fundamental investigation of rhadinovirus biology.

ACKNOWLEDGMENT

We thank Paul E. Richardson for assistance with manuscript preparation.

REFERENCES

1. **Ahuja, S.K. and P.M. Murphy.** 1996. Molecular piracy of mammalian interleukin-8 receptor type B by herpesvirus saimiri. J. Biol. Chem. *268*:20691-20694.
2. **Albrecht, J.-C., B. Fleckenstein.** 1992. New member of the multigene family of complement control proteins in herpesvirus saimiri. J. Virol. *66*:3937-3940.
3. **Albrecht, J.-C., J. Nicholas, D. Biller, K.R. Cameron, B. Biesinger, C. Newman, S. Wittmann, M.A. Craxton, H. Coleman, B. Fleckenstein, and R.W. Honess.** 1992. Primary structure of the herpesvirus saimiri genome. J. Virol. *66*:5047-5058.
4. **Albrecht, J.-C., J. Nicholas, K.R. Cameron, C. Newman, B. Fleckenstein, and R.W. Honess.** 1992. Herpesvirus saimiri has a gene specifying a homologue of the cellular membrane glycoprotein CD59. Virology *190*:527-530.
5. **Arvanitakis, L., E. Geras-Raaka, A. Varma, M.C. Gershengorn, and E. Cesarman.** 1997. Human herpesvirus KSHV encodes a constitutively active G-protein-coupled receptor linked to cell proliferation. Nature *385*:347-350.
6. **Bais, C., B. Santomasso, O. Coso, L. Arvanitakis, E.G. Raaka, J. Silvio Gutkind, A.S. Asch, E. Cesarman, M.C. Gerhengorn, and E.A. Mesri.** 1998. G-protein-coupled receptor of Kaposi's sarcoma-associated herpesvirus is a viral oncogene and angiogenesis activator. Nature *391*:86-89.
7. **Biesinger, B., I. Muller-Fleckenstein, B. Simmer, G. Lang, S. Wittman, E. Platzer, R.C. Desrosiers, and B. Fleckenstein.** 1992. Stable growth transformation of human T lymphocytes by herpesvirus saimiri. Proc. Natl. Acad. Sci. USA *89*:3116-3119.
8. **Biesinger, B., A.Y. Tsygankov, H. Fickenscher, F. Emmrich, B. Fleckenstein, J.B. Bolen, and B.M. Broker.** 1995. The product of the herpesvirus saimiri ORF1 (Tip) interacts with T cell specific kinase p56lck in transformed cells. J. Biol. Chem. *270*:4729-4734.

9. **Berend, K.R., J.U. Jung, T.J. Boyle, J.M. DiMaio, S.A. Mungal, R.C. Desrosiers, and H.K. Lyerly.** 1993. Phenotypic and functional consequences of herpesvirus saimiri infection of human CD8+ cytotoxic T lymphocytes. J.Virol. *67*:6317-6321.

10. **Blasig, C., C. Zeitz, B. Haar, F. Neipel, S. Esser, N.H. Brockmeyer, E. Tschachler, S. Colombini, B. Ensoli, and M. Sturzl.** 1997. Monocytes in Kaposi's sarcoma lesions are productively infected by human herpesvirus 8. J. Virol. *71*:7963-7968.

11. **Blaskovic, D., M. Stancekova, J. Svobodova, and J. Mistrikova.** 1980. Isolation of 5 strains of herpesviruses from two species of free living small rodents. Acta Virol. *24*:468.

12. **Boshoff, C., T.F. Schulz, M.M. Kennedy, A.K. Graham, C. Fisher, A. Thomas, J. O'D. McGee, R.A. Weiss, and J.J. O'Leary.** 1995. Kaposi's sarcoma-associated herpesvirus infects endothelial and spindle cells. Nat. Med. *1*:1274-1278.

13. **Broker, B.M., A.Y. Tsygankov, I. Muller-Fleckenstein, A.H. Guse, N.A. Chitaev, B. Biesinger, B. Fleckenstein, and F. Emmrich.** 1993. Immortalization of human T cell clones by herpesvirus saimiri. J. Immunol. *151*:1184-1192.

14. **Bromberg, J.S., L.A. Debruyne, and L. Qin.** 1998. Interactions between the immune system and gene therapy vectors: Bidirectional regulation of response and expression. Adv. Immunol. *69*:353-409.

15. **Bublot, M., P. Lomonte, A.S. Lequarre, J.C. Albrecht, J. Nicholas, B. Fleckenstein, P.P. Pastoret, and E. Thiry.** 1992. Genetic relationships between bovine herpesvirus 4 and the gammaherpesviruses Epstein-Barr virus and herpesvirus saimiri. Virology *190*:654-665.

16. **Calos, M.P.** 1996. The potential of extrachromosomal replicating vectors for gene therapy. Trends Genet. *12*:463-466.

17. **Cardin, R.D., J.W. Brooks, S.R. Sarawar, and P.C. Doherty.** 1996. Progressive loss of CD8+ T cell-mediated control of a gammaherpesvirus in the absence of CD4+ T cells. J. Exp. Med. *184*:863-871.

18. **Cesarman, E., Y. Chang, P.S. Moore, J.W. Said, and D.M. Knowles.** 1996. Kaposi's sarcoma-associated herpesvirus-like DNA sequences in AIDS-related body-cavity-based lymphomas. N. Engl. J. Med. *332*: 1186-1191.

19. **Chang, Y., E. Cesarman, M.S. Pessin, F. Lee, J. Culpepper, D.M. Knowles, and P.S. Moore.** 1994. Identification of herpesvirus-like DNA sequences in AIDS-associated Kaposi's sarcoma. Science *266*:1865-1869.

20. **Cheng, E.H.-Y., J. Nicholas, D.S. Bellows, G.S. Hayward, H.-G. Guo, M. Reitz, and J.M. Hardwick.** 1997. A Bcl-2 homolog encoded by Kaposi sarcoma-associated virus, human herpesvirus 8, inhibits apoptosis but does not heterodimerize with Bax or Bak. Proc. Natl. Acad. Sci. USA *94*:690-694.

21. **Collette, Y. and D. Olive.** 1997. Non-receptor protein tyrosine kinases as immune targets of viruses. Immunol. Today *18*:393-400.

22. **Da Costa, X.J., N. Bourne, L.R. Stanberry, and D.M. Knipe.** 1997. Construction and characterization of a replication-defective herpes simplex virus 2 ICP8 mutant strain and its use in immunization studies in a guinea pig model of genital disease. Virology *232*:1-12.

23. **Delort, J.P. and M.R. Capecchi.** 1996. TAXI/UAS: a molecular switch to control expression of genes in vivo. Hum. Gene Ther. *7*:809-820.

24. **Desrosiers, R.C., A. Bakker, J. Kamine, L.A. Falk, R.D. Hunt, and N.W. King.** 1985. A region of the herpesvirus saimiri genome required for oncogenicity. Science *228*:184-187.

25. **Desrosiers, R.C., R.L. Burghoff, A. Bakker, and J. Kamine.** 1984. Construction of replication-competent herpesvirus saimiri deletion mutants. J. Virol. *49*:343-348.

26. **Desrosiers, R.C. and L.A. Falk.** 1982. Herpesvirus saimiri strain variability. J. Virol. *43*:352-356.

27. **Desrosiers, R.C., V.G. Sasseville, S.C. Czajak, X. Zhang, K.G. Mansfield, A. Kaur, R.P. Johnson, A.A. Lackner, and J.U. Jung.** 1997. A herpesvirus of rhesus monkeys related to the human Kaposi's sarcoma-associated herpesvirus. J. Virol. *71*:9764-9769.

28. **Desrosiers, R.C., D.P. Silva, L.M. Waldron, and N.L. Letvin.** 1986. Nononcogenic deletion mutants of herpesvirus saimiri are defective for in vitro immortalization. J. Virol. *57*: 701-705.

29. **Doherty, P.C., R.A. Tripp, A.M. Hamilton-Easton, R.D. Cardin, D.L. Woodland, and M.A. Blackman.** 1997. Tuning into immunological dissonance: an experimental model for infectious mononucleosis. Curr. Opin. Immunol. *9*:477-483.

30. **Duboise, S.M., J. Guo, S. Czajak, R.C. Desrosiers, and J.U. Jung.** 1998. STP and tip are essential for herpesvirus saimiri oncogenicity. J. Virol. *72*:1308-1313.

31. **Duboise, S.M., J. Guo, S. Czajak, H. Lee, R. Veazey, R.C. Desrosiers, and J.U. Jung.** 1998. A role for herpesvirus saimiri orf14 in transformation and persistent infection. J. Virol. *72*:6770-6776.

32. **Duboise, S.M., J. Guo, R.C. Desrosiers, and J.U. Jung.** 1996. Use of virion DNA as a cloning vector for the construction of mutant and recombinant herpesviruses. Proc. Natl. Acad. Sci. USA *93*:11389-11394.

33. **Duboise, S.M., H. Lee, J. Guo, J.-K. Choi, S. Czajak, M. Simon, R.C. Desrosiers, and J.U. Jung.** 1997. Mutation of the lck-binding motif of tip enhances lymphoid cell activation by herpesvirus saimiri. J. Virol. *72*:2607-2614.

34. **Efstathiou, S., Y.M. Ho, S. Hall, C.J. Styles, S.D. Scott, and U.A. Gompels.** 1990. Murine herpesvirus 68 is genetically related to the gammaherpesviruses Epstein-Barr virus and herpesvirus saimiri. J. Gen. Virol. *71*:1365-1372.

35. **Ehtisham, E., N.P. Sunil-Chandra, and A.A. Nash.** 1993. Pathogenesis of murine gammaherpesvirus infection in mice deficient in CD4 and CD8 T cells. J. Virol. *67*:5247-5252.

36. **Ensser, A., R. Pflanz, and B. Fleckenstein.** 1997. Primary structure of the alcelaphine herpesvirus 1 genome. J. Virol. *71*:6517-6525.

37. **Falk, L.A., L.G. Wolfe, and F. Deinhardt.** 1972. Isolation of herpesvirus saimiri from blood of squirrel monkeys (Saimiri sciureus). J. Natl. Cancer Inst. *48*:

1499-1505.

38.**Flore, O., S. Rafii, S. Ely, J.J. O'Leary, E.M. Hyjek, and E. Cesarman.** 1998. Transformation of primary human endothelial cells by Kaposi's sarcoma-associated herpesvirus. Nature *394*:588-592.

39.**Fodor, W.L., S.A. Rollins, S. Bianco-Caron, R.P. Rother, E.R. Guilmette, W.V. Burton, J.-C. Albrecht, B. Fleckenstein, and S.P. Squinto.** 1995. The complement control protein homolog of herpesvirus saimiri regulates serum complement by inhibiting C3 convertase activity. J. Virol. *69*:3889-3892.

40. **Fraefel, C., D.R. Jacoby, C. Lage, H. Hilderbrand, J. Y. Chou, F.W. Alt, X.O. Breakefield, and J.A. Majzoub.** 1997. Gene transfer into hepatocytes mediated by helper virus-free HSV/AAV hybrid vectors. Mol. Med. *3*:813-825.

41.**Franken, M., A. Estabrooks, L. Cavacini, B. Sherburne, F. Wang, and D.T. Scadden.** 1996. Epstein-Barr virus-driven gene therapy for EBV-related lymphomas. Nat. Med. *2*:1379-1382.

42.**Gallo, R.C.** 1998. The enigmas of Kaposi's sarcoma. Science *282*:1837-1839.

43.**Ganem, D.** 1996. Human herpesvirus 8 and the biology of Kaposi's sarcoma. Semin. Virol. *7*:325-332.

44.**Gao, S.J., C. Boshoff, S. Jayachandra, R.A. Weiss, Y. Chang, and P.S. Moore.** 1997. KSHV ORF K9 (vIRF) is an oncogene which inhibits the interferon signaling pathway. Oncogene *15*:1979-1985.

45. **Geller, A.I.** 1997. Herpes simplex virus-1 plasmid vectors for gene transfer into neurons. Adv. Neurol. *72*:143-148.

46.**Glorioso, J.C., W.F. Goins, M.C. Schmidt, T. Oligino, D.M. Krisky, P.C. Marconi, J.D. Cavalcoli, R. Ramakrishnan, P.L. Poliani, and D.J. Fink.** 1997. Engineering herpes simplex virus vectors for human gene therapy. Adv. Pharmacol. *40*:103-136

47.**Goins, W.F., K.A. Lee, J.D. Cavalcoli, M.E. O'Malley, S.T. DeKosky, D.J. Fink, and J.C. Glorioso.** 1999. Herpes simplex virus type 1 vector-mediated expression of nerve growth factor protects dorsal root ganglion neurons from peroxide toxicity. J. Virol. *73*:519-532.

48.**Guo, J., M. Duboise, H. Lee, M. Li, J.-K. Choi, M. Rosenzweig, and J.U. Jung.** 1997. Enhanced down-regulation of lck-mediated signal transduction by a Y114 mutation of herpesvirus saimiri tip. J. Virol. *71*:7092-7096.

49.**Guo, J., K. Williams, S.M. Duboise, L. Alexander, R. Veazey, and J.U. Jung.** 1998. Substitution of ras for the herpesvirus saimiri STP oncogene in lymphocyte transformation. J. Virol. *72*:3698-3704.

50.**Honess, R.W., W. Bodemer, K.R. Cameron, H.H. Niller, B. Fleckenstein, and R.E. Randall.** 1986. The A+T-rich genome of herpesvirus saimiri contains a highly conserved gene for thymidylate synthase. Proc. Natl. Acad. Sci. USA *11*:3604-3608.

51.**Huard, J., D. Krisky, T. Oligino, P. Marconi, C.S. Day, S.C. Watkins, and J.C. Glorioso.** 1997. Gene transfer to muscle using herpes simplex virus-based vectors. Neuromusc. Disord. *7*:299-313.

52.**Jung, J.U. and R.C. Desrosiers.** 1991. Identification and characterization of the herpesvirus saimiri oncoprotein, STP-C488. J. Virol. *65*:6953-6960.

53.**Jung, J.U. and R.C. Desrosiers.** 1992. Herpesvirus saimiri oncogene STP-C488 encodes a phosphoprotein. J. Virol. *66*:1777-1780.

54.**Jung, J.U. and R.C. Desrosiers.** 1994. Distinct functional domains of STP-C488 of herpesvirus saimiri. Virology *204*:751-758.

55.**Jung, J.U. and R.C. Desrosiers.** 1994. Herpesvirus saimiri and ateles, p. 614-622. *In* R. Webster and A. Granoff (Eds.), Encyclopedia of Virology. Saunders Scientific Publications, Philadelphia.

56.**Jung, J.U. and R.C. Desrosiers.** 1995. Association of the viral oncoprotein STP-C488 with cellular ras. Mol. Cell. Biol. *15*:6506-6512.

57.**Jung, J.U., S.M. Lang, U. Friedrich, T. Jun, T.M. Roberts, R.C. Desrosiers, and B. Biesinger.** 1995. Identification of lck binding elements in Tip of herpesvirus saimiri. J. Biol. Chem. *270*:20660-20667.

58.**Jung, J.U., S.M. Lang, T. Jun, T.M. Roberts, A. Veillette, and R.C. Desrosiers.** 1995. Downregulation of Lck-mediated signal transduction by tip of herpesvirus saimiri. J. Virol. *69*:7814-7822.

59.**Jung, J.U., M. Stäger, and R.C. Desrosiers.** 1994. Virus encoded cyclin. Mol. Cell. Biol. *14*:7235-7244.

60.**Jung, J.U., J.J. Trimble, N.W. King, B. Biesinger, B.W. Fleckenstein, and R.C. Desrosiers.** 1991. Identification of transforming genes of subgroup A and C strains of herpesvirus saimiri. Proc. Natl. Acad. Sci. USA *88*:7051-7055.

61.**Kenney, S., J.Q. Ge, E.M. Westphal, and J. Olsen.** 1998. Gene therapy strategies for treating Epstein-Barr virus-associated lymphomas: comparison of two different Epstein-Barr virus-based vectors. Hum. Gen. Ther. *9*:1131-1141.

62.**Kirchmaier, A.L. and B. Sugden.** 1995. Plasmid maintenance of derivatives of oriP of Epstein-Barr Virus. J. Virol. *69*:1280-1283.

63.**Kretschmer, C., C. Murphy, B. Biesinger, J. Beckers, H. Fickenscher, T. Kirchner, B. Fleckenstein, and U. Ruther.** 1996. A herpes saimiri oncogene causing peripheral T-cell lymphoma in transgenic mice. Oncogene *12*:1609-1616.

64.**Krisky, D.M., P.C. Marconi, T. Oligino, R.J.D. Rouse, D.J. Fink, and J.C. Glorioso.** 1997. Rapid method for construction of recombinant HSV gene transfer vectors. Gene Ther. *4*:1120-1125.

65.**Lanar, D.E., J.A. Tine, C. Detaisne, M.C. Seguin, W.I. Cox, J.P. Winslow, L.A. Ware, E.B. Kauffman, D. Gordon, W.R. Ballou, et al.** 1996. Attenuated vaccinia virus-circumsporozoite protein recombinants confer protection against rodent malaria. Infect. Immun. *64*:1666-1671.

66.**Lang, G. and B. Fleckenstein.** 1990. Trans-activation of the thymidylate synthase promoter of herpesvirus saimiri. J. Virol. *64*:5333-5341.

67.**Lee, H., J.J. Trimble, D.-W. Yoon, D. Regier, R.C. Desrosiers, and J.U. Jung.** 1997. Genetic variation of herpesvirus saimiri subgroup A transforming protein and it association with cellular src. J. Virol. *71*:3817-3825.

68.**Lee, H., R. Veazey, K. Williams, M. Li, J. Guo, F. Neipel, B. Fleckenstein, A. Lackner, R.C. Desrosiers, and J.U. Jung.** 1998. Deregulation of cell growth by the K1 gene of Kaposi's sarcoma-associated herpesvirus. Nat. Med. *4*:435-440.

69.**Li, M., H. Lee, J. Guo, F. Neipel, B. Fleckenstein, K. Ozato, and J.U. Jung.** 1998. Kaposi's sarcoma-associated herpesvirus viral interferon regulatory factor. J. Virol. *72*:5433-5440.

70.**Li, M., H. Lee, D.W. Yoon, J.C. Albrecht, B. Fleckenstein, F. Neipel, and J.U. Jung.** 1997. Kaposi's sarcoma-associated herpesvirus encodes a functional cyclin. J. Virol. *71*:1984-1991.

71.**Lieber, A., C.-Y. He, L. Meuse, C. Himeda, C. Wilson, and M.A. Kay.** 1998. Inhibition of NF-kappaB activation in combination with Bcl-2 expression allows for persistence of first-generation adenovirus vectors in the mouse liver. J. Virol. *72*:9267-9277.

72.**Liebowitz, D.** 1998. Pathogenesis of Epstein-Barr virus, p. 175-199. *In* D.J. McCance (Ed.), Human Tumor Viruses. ASM Press, Washington, DC.

73.**Linnik, M.D., P. Zahos, M.D. Geschwind, and H.J. Federoff.** 1995. Expression of bcl-2 from a defective herpes simplex virus-1 vector limits neuronal death in focal cerebral ischemia. Stroke *26*:1670-1674.

74.**Marconi, P., D. Krisky, T. Oligino, P.L. Poliani, R. Ramakrisnan, W.F. Goins, D.J. Fink, and J.C. Glorioso.** 1996. Replication-defective herpes simplex virus vectors for gene transfer in vivo. Proc. Natl. Acad. Sci. USA *93*:11319-11320.

75.**Massie, B., F. Couture, L. Lamoureux, D.D. Mosser, C. Guilbault, P. Jolicoeur, F. Belanger, and Y. Langelier.** 1998. Inducible overexpression of a toxic protein by and adenovirus vector with a tetracycline-regulatable expression cassette. J. Virol. *72*:2289-2296.

76.**Mayo, M.A. and C.R. Pringle.** 1998. Virus taxonomy-1997. J. Gen. Virol. *79*:649-657.

77.**McGeoch, D.J., S. Cook, A. Dolan, F.E. Jamieson, and E.A. Telford.** 1995. Molecular phylogeny and evolutionary timescale for the family of mammalian herpesviruses. J. Mol. Biol. *247*:443-458.

78.**Mesri, E.A., H.J. Federoff, and M. Brownlee.** 1995. Expression of vascular endothelial growth factor from a defective herpes simplex virus type 1 amplicon vector induces angiogenesis in mice. Circ. Res. *76*:161-167.

79.**Moore, P.S., C. Boshoff, R.A. Weiss, and Y. Chang.** 1996. Molecular mimicry of human cytokine and cytokine response pathway genes by KSHV. Science *274*:1739-1744.

80.**Moore, P.S. and Y. Chang.** 1998. Kaposi's sarcoma (KS), KS-associated herpesvirus, and the criteria for causality in the age of molecular biology. Am. J. Epidemiol. *147*:217-221.

81.**Moore, P.S. and Y. Chang.** 1998. Kaposi's sarcoma-associated herpesvirus-encoded oncogenes and oncogenesis. J. Natl. Cancer Inst. Monogr. *23*:65-71.

82.**Morrison, L.A. and D.M. Knipe.** 1994. Immunization with replication-defective mutants of herpes simplex virus type 1: sites of immune intervention in pathogenesis of challenge virus infection. J. Virol. *68*:689-696.

83.**Moss, B.** 1996. Genetically engineered pox viruses for recombinant gene expression, vaccination, and safety. Proc. Natl. Acad. Sci. USA *93*:11341-11348.

84.**Murphy, C., C. Kretschmer, B. Biesinger, J. Beckers, J. Jung, R.C. Desrosiers, H.K. Muller-Hermelink, B.W. Fleckenstein, and U. Ruther.** 1994. Epithelial tumors induced by a herpesvirus oncogene in transgenic mice. Oncogene *9*:221-226.

85.**Murty, S.C.S., J.J. Trimble, and R.C. Desrosiers.** 1989. Deletion mutants of herpesvirus saimiri define an open reading frame necessary for transformation. J. Virol. *63*:3307-3314.

86.**Nava, V.E., E.H.-Y. Cheng, M. Veliuona, S. Zou, R.J. Clem, M.L. Mayer, and J.M. Hardwick.** 1997. Herpesvirus saimiri encodes a functional homolog of the humanbcl-2 oncogene. J. Virol. *71*:4118-4122.

87.**Nguyen, L.H., D.M. Knipe, and R.W. Finberg.** 1992. Replication-defective mutants of herpes simplex virus (HSV) induce cellular immunity and protect against lethal HSV infection. J. Virol. *66*:7067-7072.

88.**Nicholas, J., V. Ruvolo, J. Zong, D. Ciufo, H.G. Guo, M.S. Reitz, and G.S. Hayward.** 1997. A single 13 kilobase divergent locus in the Kaposi sarcoma-associated herpesvirus (human herpesvirus 8) genome contains nine open reading frames that are homologous to or related to cellular proteins. J. Virol. *71*:1963-1974.

89.**No, D., T.-P. Yao, and R.M. Evans.** 1996. Ecdysone-inducible gene expression in mammalian cells and transgenic mice. Proc. Natl. Acad. Sci. USA *93*:3346-3351.

90.**O'Brien, V.** 1998. Viruses and apoptosis. J. Gen. Virol. *79*:1833-1845.

91.**Panyutich, E.A., J.W. Said, and S.A. Miles.** 1998. Infection of primary dermal microvascular endothelial cells by Kaposi's sarcoma-associated herpesvirus. AIDS *26*:467-472.

92.**Ploegh, H.L.** 1998. Viral strategies of immune evasion. Science *280*:248-253.

93.**Raaka, E.G., L. Arvanitakis, C. Bais, E. Cesarman, E.A. Mesri, and M.C. Gershengorn.** 1998. Inhibition of constitutive signaling of Kaposi's sarcoma-associated herpesvirus G-protein coupled receptor by protein kinases in mammalian cells in culture. J. Exp. Med. *187*:801-806.

94.**Renne, R., D. Blackbourn, D. Whitby, J. Levy, and D. Ganem.** 1998. Limited transmission of Kaposi's sarcoma-associated herpevirus in cultured cells. J. Virol. *72*:5182-5188.

95.**Renne, R., W. Zhong, B. Herndier, M. McGrath, N. Abbey, D. Kedes, and D. Ganem.** 1996. Lytic growth of Kaposi's sarcoma-associated herpesvirus (human herpesvirus 8) in culture. Nat. Med. *2*:342-346.

96.**Rickinson, A.B. and E. Kieff.** 1996. Epstein-Barr virus, p. 2397-2446. *In* B.N. Fields, D.M. Knipe, and P.M. Howley (Eds.), Fields Virology. Lippencott-Raven, Philadelphia.

97.**Robbins, P.D., H. Tahara, and S.C. Ghivizzani.** 1998. Viral vectors for gene therapy. Trends Biotechnol. *16*:35-40.

98.**Robertson, E.S., T. Ooka, and E.D. Kieff.** 1996. Epstein-Barr virus vectors for gene delivery to B lymphocytes. Proc. Natl. Acad. Sci. USA *93*:11334-11340.

99.**Roizman, B.** 1996. The function of herpes simplex virus genes: a primer for genetic engineering of novel vectors. Proc. Natl. Acad. Sci. USA *93*:11307-11312.

100.**Roizmann, B., R.C. Desrosiers, B. Fleckenstein, C.**

Lopez, A.C. Minson, and M.J. Studdert. 1992. The family *Herpesviridae*: an update. Arch. Virol. *123*: 425-449.

101.**Rolph, M.S. and I.A. Ramshaw.** 1997. Recombinant viruses as vaccines and immunological tools. Curr. Opin. Immunol. *9*:517-524.

102.**Rose, T.M., K.B. Strand, E.R. Schultz, G. Schaefer, G.W. Rankin, Jr., M.E. Thouless, C.-C. Tsai, and M.L. Bosch.** 1997. Identification of two homologs of the Kaposi's sarcoma-associated herpesvirus (human herpesvirus 8) in retroperitoneal fibromatosis of different macaque species. J. Virol. *71*:4138-4144.

103.**Rother, R.P., S.A. Rollins, W.L. Fodor, J.-C. Albrecht, E. Setter, B. Fleckenstein, and S.P. Squinto.** 1994. Inhibition of complement-mediated cytolysis by the terminal complement inhibitor of herpesvirus saimiri. J. Virol. *68*:730-737.

104.**Russo, J.J., R.A. Bohenzky, M.-C. Chien, J. Chen, M. Yan, D. Maddalena, J.P. Parry, D. Peruzzi, I.S. Edelman, Y. Chang, and P.S. Moore.** 1996. Nucleotide sequence of the Kaposi sarcoma-associated herpesvirus (HHV8). Proc. Natl. Acad. Sci. USA *93*:14862-14867.

105.**Samanieg, L.A., L. Neiderhiser, and N.A. DeLuca.** 1998. Persistence and expression of the herpes simplex virus genome in the absence of immediate-early proteins. J. Virol. *72*:3307-3320.

106.**Sarid, R., J.S. Wiezorek, P.S. Moore, and Y. Chang.** 1999. Characterization and cell cycle regulation of the major Kaposi's sarcoma-associated herpesvirus (human herpesvirus 8) latent genes and their promoter. J. Virol. *73*:1438-1446.

107.**Satoh, E., M. Osawa, K. Tomiyasu, H. Hirai, C. Shimazaki, Y. Oda, M. Nakagawa, M. Kondo, S. Kinoshita, O. Mazda, and J. Imanishi.** 1997. Efficient gene transduction by Epstein-Barr-virus-based vectors coupled with cationic liposome and HVJ-liposome. Biochem. Biophys. Res. Commun. *238*:795-799.

108.**Schultz, T.F.** 1998. Kaposi's sarcoma-associated herpesvirus (human herpesvirus-8). J. Gen. Virol. *79*: 1573-1591.

109.**Simas, J.P., R.J. Bowden, V. Paige, and S. Efstathiou.** 1998. Four tRNA-like sequences and a serpin homologue encoded by murine gammaherpesvirus 68 are dispensable for lytic replication *in vitro* and latency *in vivo*. J. Gen. Virol. *79*:149-153.

110.**Simas, J.P. and S. Efstathiou.** 1998. Murine gammaherpesvirus 68: a model for the study of gammaherpesvirus pathogenesis. Trends Microbiol. *6*:276-282.

111.**Smith, A.E.** 1995. Viral Vectors in Gene Therapy. Ann. Rev. Microbiol. *49*:807-838.

112.**Stevenson, P.G., G.T. Belz, J.D. Altman, and P.C. Doherty.** 1998. Virus specific CD8+ T cell numbers are maintained during gammaherpesvirus reactivation in CD4- deficient mice. Proc. Natl. Acad. Sci. USA *95*:15565-15570.

113.**Stewart, J.P., E.J. Usherwood, A. Ross, H. Dyson, and T. Nash.** 1998. Lung epithelial cells are a major site of murine gammaherpesvirus persistence. J. Exp. Med. *187*:1941-1951.

114.**Sunil-Chandra, N.P., S. Efstathiou, J. Arno, and A.A. Nash.** 1992. Virological and pathological features of mice infected with murine gammaherpesvirus 68. J. Gen. Virol. *73*:2347-2356.

115.**Sunil-Chandra, N.P., E. Efstathiou, and A.A. Nash.** 1992. Murine gammaherpesvirus 68 establishes a latent infection in mouse B lymphocytes in vivo. J. Gen. Virol. *73*:3275-3279.

116.**Telford, E.A., M.S. Watson, H.C. Aird, J. Perry, and A.J. Davidson.** 1995. The DNA sequence of equine herpesvirus 2. J. Mol. Biol. *249*:520-528.

117.**Thome, M., P. Schneider, K. Hofmann, H. Fickenscher, E. Meinl, F. Neipel, C. Mattmann, K. Burns, J.L. Bodmer, M. Schroter, et al.** 1997. Viral FLICE-inhibitory proteins (FLIPs) prevent apoptosis induced by death receptors. Nature *386*:517-521.

118.**Tine, J.A., D.E. Lanar, D.M. Smith, B.T. Wellde, P. Schultheiss, L.A. Ware, E.B. Kaufmann, R.A. Wirtz, C. Detaisne, G. Hui, et al.** 1996. NYVAC-Pf7—a poxvirus-vectored, multiantigen, multistage vaccine candidate for *Plasmodium falciparum* malaria. Infect. Immun. *64*:3833-3844.

119.**Tripp, R.A., A.M. Hamilton-Easton, R.D. Cardin, P. Nguyen, F.G. Behm, D.L. Woodland, P.C. Doherty, and M.A. Blackman.** 1997. Pathogenesis of an infectious mononucleosis-like disease induced by a murine gammaherpesvirus: role for a viral superantigen? J. Exp. Med. *185*:1641-1650.

120.**Usherwood, E.J., J.P. Stewart, and A.A. Nash.** 1996. Characterization of tumor cell lines derived from Murine gammaherpesvirus-68-infected mice. J. Virol. *70*:6516-6518.

121.**Verma, I.M. and N. Somia.** 1997. Gene therapy—promises, problems, and prospects. Nature *389*:239-242.

122.**Virgin, H.W., P. Latreille, P. Wamsley, K. Hallsworth, K.E. Weck, A.J. Dal Canto, and S.H. Speck.** 1997. Complete sequence and genomic analysis of murine gammaherpesvirus 68. J. Virol. *71*:5894-5902.

123.**Wang, S., S. Di, W.B. Young, C. Jacobson, and C.J. Link, Jr.** 1997. A novel herpesvirus amplicon system for in vivo gene delivery. Gene Ther. *4*:1132-1141.

124.**Wang, Y., B.W. O'Malley, Jr., S.Y. Tsai, and B.W. O'Malley.** 1994. A regulatory system for use in gene transfer. Proc. Natl. Acad. Sci. USA *91*:8180-8184.

125.**Wang, S. and J.M. Vos.** 1996. A hybrid herpesvirus infectious vector based on Epstein-Barr virus and herpes simplex virus type I for gene transfer into human cells in vitro and in vivo. J. Virol. *70*:8422-8430.

126.**Weck, K.E., M.L. Barkson, L.I. Yoo, S.H. Speck, and H.W. Virgin.** 1996. Mature B cells are required for acute splenic infection but not for establishment of latency, by murine gammaherpesvirus 68. J. Virol. *70*:6775-6780.

127.**Weck, K.E., A.J. Dal Canto, J.D. Gould, A.K. O'Guin, K.A. Roth, J.E. Saffitz, S.H. Speck, and H.W. Virgin.** 1997. Murine gammaherpesvirus 68 causes severe large-vessel arteritis in mice lacking interferon gamma responsiveness: a new model for virus-induced vascular disease. Nat. Med. *3*:1346-1353.

128.**White, J.H. 1997.** Modified steroid receptors and steroid-inducible promoters as genetic switches for gene therapy. Adv. Pharmacol. *40*:339-367.

129.**Wohlgemuth, J.G., S.H. Kang, G.H. Bulboaca, K.A. Nawotka, and M.P. Calos.** 1996. Long-term gene expression from autonomously replicating vectors in mammalian cells. Gene Ther. *3*:503-512.

130.**Yao, Z., W.C. Fanslow, M.F. Seldin, A.M. Rousseau, S.L. Painter, M.R. Comeau, J.I. Cohen, and M.K. Spriggs.** 1995. Herpesvirus saimiri encodes a new cytokine IL-17, which binds to a novel cytokine receptor. Immunity *3*:811-821.

131.**Yao, Z., E. Maraskovsky, M.K. Spriggs, J.I. Cohen, R.J. Armitage, and M.R. Alderson.** 1996. Herpesvirus saimiri open reading frame 14, a protein encoded by a T lymphotropic herpesvirus, binds to MHC Class II molecules and stimulates T cell proliferation. J. Immunol. *156*:3260-3266.

132.**Ye, X., V.M. Rivera, P. Zoltick, F. Cerasoli, Jr., M.A. Schnell, G. Gao, J.V. Hughes, M. Gilman, and J.M. Wilson.** 1999. Regulated delivery of therapeutic proteins after in vivo somatic cell gene transfer. Science *283*:88-91.

133.**Yoon, S.S., N.M. Carroll, E.A. Chiocca, and K.K. Tanabe.** 1998. Cancer gene therapy using a replication-competent herpes simplex type 1 vector. Ann. Surg. *228*:366-374.

134.**Zimring, J.C., S. Goodbourn, and M.K. Offermann.** 1998. Human herpesvirus 8 encodes an interferon regulatory factor (IRF) homolog that represses IRF-1-mediated transcription. J. Virol. *72*:701-707.

FAMILY: Herpesviridae
SUBFAMILY: Gammaherpesvirinae
GENUS: Epstein-Barr Virus

20 Application of Epstein-Barr Virus and its Genetic Elements to Gene Therapy

Osam Mazda
Department of Microbiology, Kyoto Prefectural University of Medicine, Kamikyo, Kyoto, Japan

INTRODUCTION

The Epstein-Barr virus (EBV) is a B-lymphocyte-trophic human herpesvirus with a genome that is a double-stranded circular DNA of approximately 172 kb in length (36). EBV is carried in a latent, nonsymptomatic state by more than 90% of the adult population worldwide. The primary silent infection is generally established early in childhood and persists throughout life without causing any serious disease. On the other hand, EBV is also the causative agent of infectious mononucleosis as well as various malignancies in humans. Therefore, EBV may require certain risk factor(s), such as those of immunologic, genetic, and environmental origin, for its oncogenesis.

Infection of human resting B cells with EBV in vitro results in the generation of immortalized lymphoblastoid cell lines, in which multiple copies of the latent EBV genome are maintained as episomes (latent infection). Taking advantage of its efficient infectivity and latency mechanisms, EBV and its genetic elements could be applicable to various gene therapy strategies. These can be classified into several categories, such as utilization of plasmid vectors carrying certain genetic elements of EBV and the generation of infectious recombinant EBV.

THE EBNA1/*ori*P SYSTEM

The Epstein-Barr nuclear antigen 1 (EBNA1)/*ori*P system plays a crucial role in the replication and maintenance of the EBV genome in latent EBV-infected cells. EBNA1 is a nuclear phosphoprotein with DNA-binding properties. The consensus binding sequence has been identified (3,42,43). The protein binds to DNA as a homodimer, and both DNA-binding and dimerization domains map to the C terminus of the protein (2,5). Binding to *ori*P in a sequence-specific manner, EBNA1 mediates replication of the episome in synchrony with chromosomal DNA during early S phase (1,17,63). EBNA1 can also bind to RNA and could control gene expression at the posttranscriptional level (50).

Viral Vectors: Basic Science and Gene Therapy
Edited by A. Cid-Arregui and A. García-Carrancá
©2000 Eaton Publishing, Natick, MA

The *ori*P consists of two elements, which are both essential and sufficient for replication. One element, the dyad symmetry, spans approximately 120 bp, containing four EBNA1-binding sites. The dyad symmetry coincides with the origin of bidirectional replication (14). The other element is referred to as family of repeats and is composed of a family of approximately 20 members of 30-bp tandem repeats, each containing an EBNA1 binding site. The 30-bp repeats act in a highly cooperative manner to activate replication (44,62), their function being independent of orientation and distance from the dyad symmetry. The family of repeats also acts as an EBNA1-dependent transcriptional enhancer (43,51) and performs a role in the episomal maintenance of viral DNA (6,29,44).

It is well known that plasmids carrying *ori*P exclusively replicate in primate cells and that rodent cells are nonpermissive in terms of DNA replication, even in the presence of EBNA1 (28,64). Therefore, some primate-specific factor(s) may cooperate with EBNA1 to mediate plasmid replication. Because EBNA1 lacks DNA helicase activity (13,37), it seems reasonable to assume that some human endogenous enzyme(s) with DNA helicase activity may play a crucial role as cofactor(s).

On the other hand, the EBNA1 has some additional functions. Through binding to *ori*P, EBNA1 facilitates nuclear localization of the plasmid (2,11), anchorage of plasmid to nuclear matrix (22), and transcriptional upregulation of viral genes (15,40,43,51,62). Due to these multiple functions of EBNA1, one can obtain both high transient expression and long-term maintenance of the plasmid by transfecting human cells with a plasmid vector carrying EBNA1 and *ori*P (Figure 1).

We have recently found that high-level transient expression can be observed not only in human cells but also in rodent cells (54). Several rodent cell lines were transfected with a plasmid carrying *ori*P, EBNA1 and *Escherichia coli* β-galactosidase (β-gal) genes (pSES.β), or another plasmid carrying the β-gal gene alone (pS.β). Three days later, β-gal activity was measured to assess the transient expression. The pSES.β-transfected cells exhibited more than tenfold higher β-gal activity compared with those transfected with pS.β. Co-transfection experiments with a plasmid carrying β-gal gene and an EBNA1-bearing plasmid indicated that higher transient expression was obtained when both *ori*P and EBNA1 were provided in *cis* and *trans*, respectively (Figure 2). The results

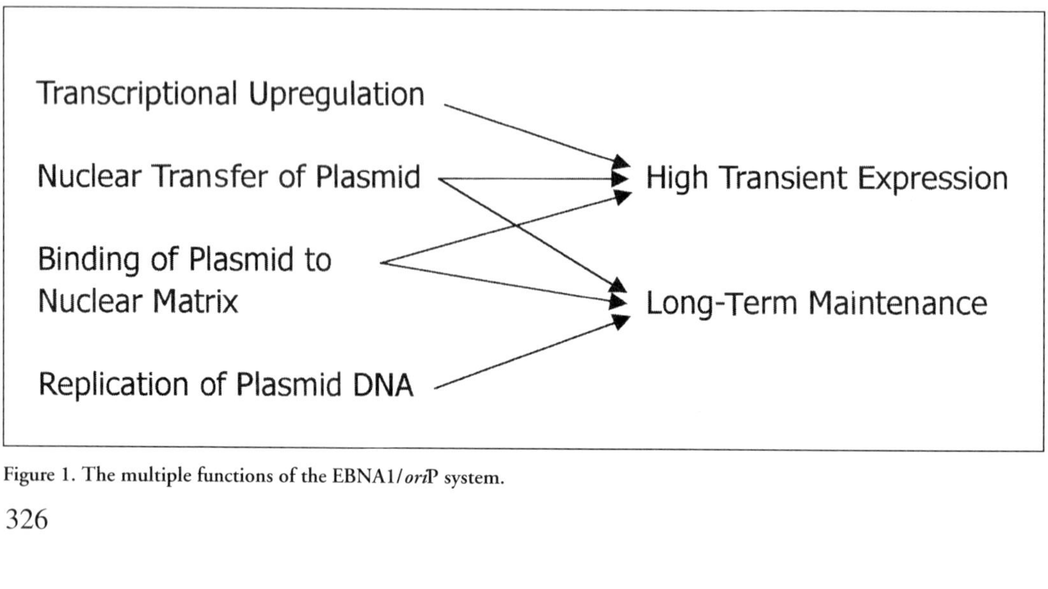

Figure 1. The multiple functions of the EBNA1/*ori*P system.

strongly suggest that cellular factor(s) required for transient expression and long-term maintenance are different; the former is available in both rodent and primate cells, while the latter is exclusively present in primate cells. The results also suggest that rodent models can be utilized to estimate the effectiveness and toxicity of the EBV-based plasmid vector systems in vivo.

THE EBV-BASED PLASMID VECTORS

Yates et al. were the first to describe plasmid vectors carrying the EBNA1 gene and *ori*P (EBV-based episomal vector, or EBV-based plasmid vector) (64). Modified in various ways, these vectors have been utilized to transfer genes into various cell lines in vitro. More recently, they have been applied to experimental gene therapy strategies.

We have estimated the transfection/expression efficiencies of EBV-plasmid vectors (34). EBNA1-positive or EBNA1-negative cell lines were transfected by electroporation with a plasmid carrying the EBNA1 gene and *ori*P, plus the neomycin resistance (*neo*R) and the murine CD8α genes as markers (Table 1). As a control, a plasmid carrying the *ori*P, *neo*R and CD8α genes, but not the EBNA1 gene, was transfected in parallel. Three days later, the cells expressing murine CD8α were analyzed by flow cytometry to estimate the transient expression efficiency. By a limiting dilution

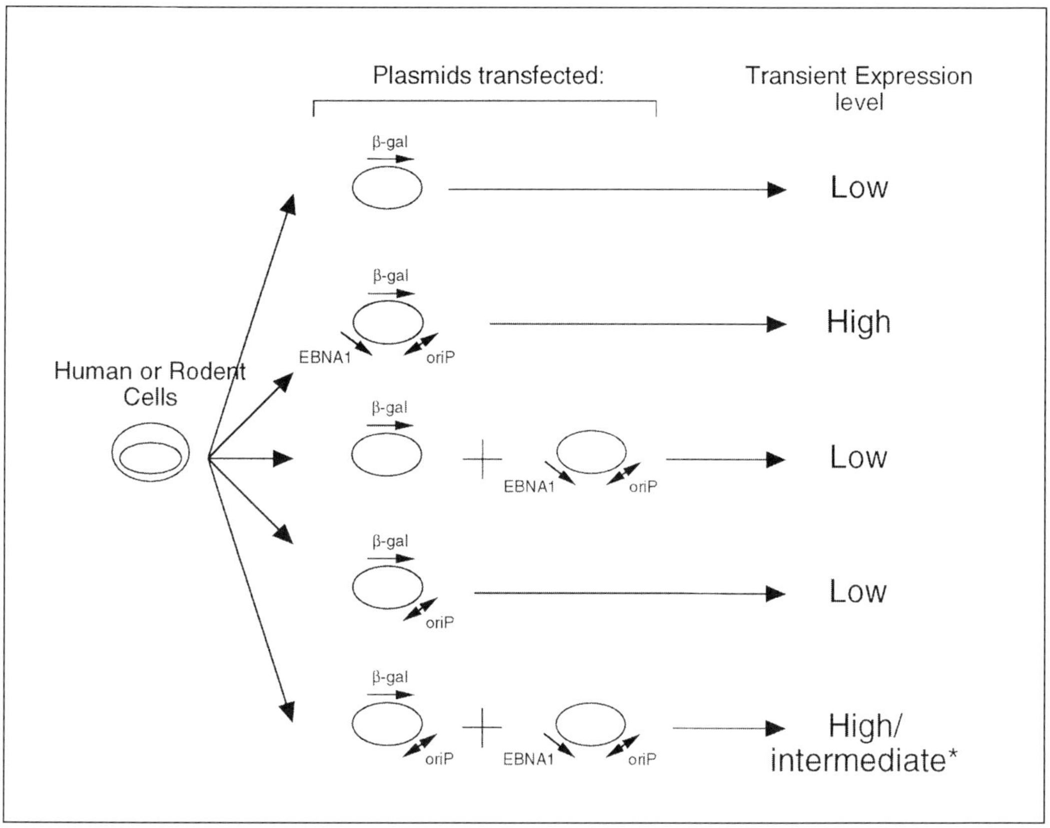

Figure 2. Schematic representation of co-transfection experiments (54). HeLa, a human cervical carcinoma cell line, or H9C2, a rat fibroblastoid cell line, was transfected with indicated plasmids by electroporation, and three days later, expression level of β-gal was assayed. *Expression level correlated with the amount of the plasmid carrying EBNA1 gene and *ori*P.

Table 1. Transfection/Expression Efficiencies with EBV-Based Plasmid Vectors (33)

Cells	Plasmid	% CD8α-positive cells on day 3[a]	% G418-resistant cells generated[b]
K562	-	<1	<0.002
	pNS.CD8α	<1	0.05
	pNSES.CD8α	40	29
KE	-	<1	<0.002
	pNSES.CD8α	100	100

K562 or KE cells were transfected with 20 μg of the indicated plasmids by electroporation. KE is a K562-derived cell line constitutively expressing EBNA1. pNS.CD8α is a plasmid carrying *neo*^{R} and murine CD8α genes as markers. Another plasmid, pNSES.CD8α, carries *neo*^{R}, murine CD8α and EBNA1 genes.

[a]Percent CD8α-positive cells on day 3 after the transfection. Cells were stained by phycoerythrin-conjugated anti-CD8α antibody and analyzed by flow cytometry.

[b]Percent of cells that gave rise to G418-resistant transfectants. Cells were harvested on day 3 after transfection, seeded into 96-well plates at limited cell numbers, and cultured with 1 mg/mL of G418. Fourteen days later, number of wells in which the cells had been dead was counted, and the precursor cell frequency was calculated based on the Poisson probability distribution.

analysis, we also assessed the percentages of transfected cells that could give rise to neomycin-resistant stable transformants. Our results showed that the transfection/expression efficiencies with an EBV-based plasmid vector are much higher than those obtained with a conventional plasmid, both in transient and long-term expression tests.

The EBV-plasmid vectors must be combined with some gene delivery vehicles (nonviral vectors), so that the plasmids can be delivered into target cells of interest. Experimental gene therapy strategies in which the EBV-plasmids are engaged can be classified into different categories on the basis of the delivery system used.

Transfection of the EBV-Based Plasmid Vectors In Vitro by Electroporation and Calcium Phosphate Precipitation Methods

Electroporation and calcium phosphate precipitation are classical and popular methods to transfect plasmid DNA into cultured cells. These procedures are simple and easy, giving highly reproducible results. Lei et al. transferred the cystic fibrosis transmembrane conductance regulator (*CFTR*) gene into an airway epithelial cell line by the calcium phosphate method (30). They observed functional expression of the transgene, although the intensity and duration of expression from their EBV-based plasmid were not compared with those of a non-EBV-based one.

Muecke et al. transfected a Burkitt's lymphoma cell line with an EBV vector by electroporation to obtain stable transformants that express cytokine genes, (i.e., granulocyte-macrophage colony-stimulating factor [*GM-CSF*], interleukin [*IL*]-6, and tumor necrosis factor alpha [*TNFα*]) (39). Finke et al. transfected melanoma cells by electroporation with an EBV-plasmid vector carrying the *IL-7* gene. The resultant transfectants were effectively killed by autologous or allogenic cytotoxic T cells in vitro (10).

The disadvantage of these delivery methods is that they cannot be applied to direct in vivo gene transfer. In case of ex vivo gene transfer, however, electroporation may be a feasible method—a typical example being gene transfer into hematopoietic progenitor cells.

We have reported successful transfer of an adenosine deaminase (*ADA*) gene into human immature hematopoietic progenitor cells (47). Immobilized peripheral blood CD34+ cells were obtained from cancer patients who had received high-dose chemotherapy and administration with recombinant granulocyte-colony-stimulating factor (G-CSF). The peripheral blood CD34+ cells were transfected with EBV-based plasmid vectors carrying human *ADA* gene expression unit and thereafter allowed to produce colony-forming units in culture (CFU-c). Reverse transcription-polymerase chain reaction (RT-PCR) analysis of individual colonies revealed that 37% to 100% of CFU-c were actually transduced by the system. The ADA activity elevated 1.5- to 2-fold in cultured CD34+ cells as well as pooled CFU-c colonies. It is not yet clear how long the transgene expression could persist in vivo, in cases when the transfected cells were transplanted into an animal.

One problem with electroporation is cellular toxicity. In our hands, cell viability decreased to 50% to 70% after the pulse. This, however, may not cause any practical problem because the number of peripheral blood progenitor cells harvested from a donor is usually much higher than that transplanted to a recipient. Therefore, electroporation may be a useful method for ex vivo gene transfer into hematopoietic progenitor cells, provided that transfer efficiencies can be improved by using EBV-plasmid vectors instead of conventional plasmid vectors.

Gene Transfer In Vitro Using EBV-Based Plasmid Vectors Supported by Synthetic Nonviral Vectors

Cationic liposomes, cationic polymers, and other synthetic vectors are means to transfer plasmid DNA into cells both in vitro and in vivo. They could make EBV-plasmid vectors available to a broader application range.

We transferred marker genes into primary cells of human origin (e.g., skin fibroblasts and bone marrow cells) by combining the EBV-plasmid vectors with cationic liposomes (EBV/lipoplex) or hemagglutinating virus of Japan (HVJ)-liposomes (48). In all cases, transfection with EBV-plasmid vectors resulted in significant expression of marker genes, while expression obtained with conventional plasmid vectors used in parallel was very faint, if observed at all. When primary fibroblasts were transfected, cationic liposomes gave stronger expression of the marker gene than did HVJ-liposomes. In contrast, bone marrow cells were more effectively transfected by HVJ-liposomes than by cationic liposomes. Therefore, distinct types of cells may require different delivery systems to be transduced optimally.

We also utilized EBV-plasmid vectors for suicide gene transfer into tumor cells in vitro (19). Cationic liposomes and polyamidoamine (PAMAM) dendrimer were employed as delivery vehicles. The PAMAM dendrimer is a branched spherical polymer carrying positive charge on its surface and has been reported to be a safe and efficient vehicle for plasmid DNA transfer (41,53). Three human hepatocellular carcinoma (HCC) cell lines were transfected with EBV-plasmid vectors or conventional plasmid vectors carrying β-gal genes. In our hands, the PAMAM dendrimer gave higher transfection efficiencies than cationic liposomes, in terms of β-gal expression in transient experiments. When the herpes simplex virus type I thymidine kinase (HSV1-Tk) gene contained in an EBV-based plasmid (pSES.Tk) was transferred by means of the PAMAM dendrimer, transfected HCC cell lines showed several hundredfold higher susceptibility to ganciclovir (GCV) than mock transfected cells. In contrast, transfection with a conventional plasmid vector (pS.Tk) did not significantly affect, or elevated only slightly, the GCV susceptibility of the cells. More importantly, the pSES.Tk-transfected cells were effectively killed by day nine with a clinically feasible concentration of GCV (25 μmol/L), whereas pS.Tk-transfected cells partially survived under the same culture conditions. These results strongly suggest the effectiveness of the EBV-plasmid vector coupled with PAMAM dendrimer (EBV/polyplex) in suicide gene therapy of cancer. More recently, we have succeeded in in vivo suicide gene therapy by using this system.

Gene Transfer In Vivo Using EBV-Based Plasmid Vectors

The EBV-plasmid vectors are also useful for in vivo gene transduction into various organs. Wang and Becker constructed EBV-based plasmid vectors carrying a fibroblast growth factor (*FGF*) receptor gene or a basic *FGF* cDNA in antisense orientation (55). These plasmids were coupled with cationic liposomes and injected into tumors that had been established in nude mice by implanting cells of a human melanoma cell line. They found drastic regression of tumors caused by suppression of intratumoral angiogenesis (55).

Saeki et al. employed HVJ-liposomes to deliver EBV-based plasmid vectors into the rat liver (46). They found that infusion of the EBV-vector/HVJ-liposomes via the portal vein led to significant expression of marker genes (β-gal and firefly luciferase genes) in hepatocytes. In addition, longer reporter expression was obtained by infusion of EBV-vector/HVJ-liposomes than by infusion of a conventional plasmid vector/HVJ-liposomes.

In our experiments, we focused on the heart as the target organ. Cardiac muscle cells have been reported to take up naked plasmid DNA and express transgenes (32). An EBV-based plasmid vector or a conventional plasmid vector carrying a β-gal gene as marker was injected into rat hearts under the guidance of echocardiography (54). On day eight after injection, rats were sacrificed and their hearts were examined after X-*gal* staining. Stronger expression of the marker gene was observed in cardiomyocytes injected with the EBV-plasmid vector than those given injection of a conventional plasmid vector. Recently, we transferred the β2-adrenergic receptor gene into myocardium of cardiomyopathic hamsters by injecting naked EBV-plasmid DNA. We observed significant elevation of the catecholamine-induced increase of cardiac output and fractional shortening, as revealed by echocardiographic examination of the failing heart (K. Tomiyasu et al., unpublished data).

Further, we have utilized a gene gun to directly transfer the EBV plasmid vector into rat heart (Nishizaki et al., manuscript submitted). Gold particles of 1.0 or 1.5 μm in diameter were coated with either an EBV-based plasmid vector or a conventional plasmid vector carrying β-gal gene. A gene gun (model TF-1) designed by Nihon Medical & Chemical Instruments, Co. Ltd. (Osaka, Japan) was used. Being accelerated by helium gas pressure, the particles were bombarded into the left ventricular wall exposed through an anterothoracotomy. The particle-mediated transfer of the EBV-plasmid resulted in more sustained expression of the marker gene in cardiomyocytes, compared with that of the conventional plasmid vectors. The advantage of the gene gun is that gene expression is topologically restricted to cells in the bombarded area.

More recently, we engaged the EBV/polyplex system (see above) to in vivo suicide gene therapy of cancer. An EBV-plasmid vector or a conventional plasmid vector carrying *HSV1-Tk* gene was combined with PAMAM dendrimer, and the resultant complex was injected into tumors that had been established in severe combined immunodeficiency (SCID) mice. Systemic administration of GCV resulted in drastic suppression of tumor growth and prolonged survival of mice. In contrast, injection of the conventional plasmid vector/dendrimer only partially suppressed the tumor growth and did not influence survival of mice. The results demonstrate the potential application of the EBV/polyplex but not the conventional plasmid vector/dendrimer to suicide gene therapy of cancer (33).

HUMAN ARTIFICIAL CHROMOSOMES BASED ON THE LATENT INFECTION MACHINERY OF EBV

An artificial chromosome is a large-sized DNA that replicates autonomously in a cell cycle-coordinated manner and segregates appropriately into the daughter cells. It may be an ideal tool for gene therapy, offering a potential solution to eliminate positional effects and alleviate size constraints. Since the EBNA1/*ori*P system enables a plasmid to be retained in human cells for the long-term, the EBNA1 and *ori*P are being utilized to engineer human artificial chromosomes.

Calos' group has shown autonomous replication of plasmids carrying EBNA1, family of repeats, and a human DNA, but not dyad symmetry (20,28,60). They demonstrated that any human DNA fragment larger than approximately 10 kb could support efficient autonomous replication in human cells. The capability of a fragment to be an origin appeared to be solely dependent on the length rather than sequence of the fragment. Probably, the signals that direct initiation of DNA replication have a low degree of sequence specificity. Of particular interest is that the chimeric constructs are capable of replicating not only in human cells but also in rodent cells in which a plasmid carrying EBNA1 and whole *ori*P (dyad symmetry and family of repeats) cannot replicate (27).

Vos and collaborators have constructed a human artificial episomal chromosome (HAEC) library of 1500 independent clones carrying random human genomic DNA with average sizes of 60 to 330 kb (52). The constructs were introduced into human cells that express EBNA1. The HAEC library propagates stably as circular minichromosomes in the cells. The entire human β-globin gene locus including its locus control region was also inserted (58). The HAEC, 200 kb in size, persisted as a structurally intact episome in EBNA1-positive cells for up to one year. Significant expression of the human β-globin gene was detectable in this long-term culture.

RECOMBINANT EBV VECTORS FOR GENE THERAPY

Some reports have shown the successful gene delivery into lymphoid cell lines by means of infectious recombinant EBV. $CD21^+$ cells should be used as target cells because this cell surface molecule is the receptor for the EBV.

Banerjee et al. devised a system to generate recombinant EBV and transduced $CD21^+$ B lymphoblastoid cells (4). Their constructs included the *ori*P, the origin of lytic replication (*ori*Lyt), and the terminal repeat sequence. These plasmids were transfected into a helper cell line that cleaves the vector DNA and packages it into viral particles. The β-gal and Fanconi anemia group C (*FA-C*) genes were successfully transferred in vitro by infecting $CD21^+$ B lymphoblastoid cell lines with the recombinant virus.

Shimizu et al. made Akata cells generate recombinant EBV (49). The Akata⁻ is an EBV-nonproducing clone derived from an EBV highly producing cell line (Akata). The *neo*^{R} gene was first introduced into Akata cells to generate recombinant EBV in addition to a wild-type virus. Akata⁻ cells were infected with the mixture of viruses, and G418 selection was performed. High titers of recombinant virus were obtained from the resultant G418-resistant clones, while the wild-type virus was undetectable. Interestingly, treatment of these cells with anti-immunoglobulin antibody induced production of a higher titer of recombinant EBV.

Robertson et al. used Raji, a Burkitt's lymphoma cell line, as the producer cells (45). The EBV genome in Raji cells lacks most of the EBNA3 open reading frame (orf). Therefore, Raji cells are not able to produce endogenous virus that could propagate in newly infected cells.

Kenney et al. aimed at targeting EBV-associated lymphoma cells (24). Their construct contained the *ori*Lyt, terminal repeat, family of repeats, and the *Bam*HI Z fragment leftward orf-1 (*BZLF1*) gene. The *BZLF1* gene product induces expression of EBV proteins that are required for replication of *ori*Lyt. Being transfected into an EBV-positive cell line by electroporation, the plasmid induced lytic EBV infection and was packaged into a virus particle. The generated recombinant virus could subsequently be transmitted to other $CD21^+$ EBV-positive cells, in which another round of replication occurred. The cytosine deaminase (*CD*) gene was transferred using this system, resulting in specific killing of EBV-positive cell lines in vitro.

Recombinant EBV systems could provide the tools for useful strategies for gene transfer into $CD21^+$ B lineage cells. The principal limitations of these systems are their restricted host range, the possible generation of replication-competent recombinants, and the potent oncogenicity of the virus.

TARGETING EBV-ASSOCIATED TUMORS BY USING THE GENETIC ELEMENTS FROM EBV

EBV is the etiologic agent of various malignancies in humans. These include Burkitt's lymphoma, nasopharyngeal carcinoma, Hodgkin's disease, gastric carcinoma, lymphoepithelial-like carcinomas, and a variety of tumors occurring in the immunocompromised host (30).

In EBV-associated tumor cells, EBNA1 is constitutively expressed, so that the EBV episome is stably maintained. In contrast, most normal cells in human do not express EBNA1, except for a very small percentage of B cells (approximately one out of 1×10^6 cells) and pharyngeal epithelial cells.

We and others have reported the targeted killing of EBV-associated tumor cells based on the particular features of the tumors (9,21,23,24). As mentioned above, a plasmid vector carrying *ori*P but not EBNA1, permits strong and persistant expression of a transgene in EBNA1-positive cells, while in EBNA1-negative cells, the same plasmid vector gives only marginal expression (34,35). When EBNA1-positive and EBNA1-negative cells were transfected with plasmid vectors carrying a suicide gene (*HSV1-Tk* or *CD* gene) and *ori*P but not EBNA1, only EBNA1-positive cells were rendered highly susceptible to the corresponding prodrug (GCV or 5-fluorocytosine) in vitro. (Representative data of ours are shown in Table 2.) In these studies, the plasmids were transfected by electroporation, except that we used not only electroporation but also the HVJ-liposome, which is capable of transferring genes in vivo (8).

Franken et al. presented a different strategy (12). EBNA2 is a sequence-specific DNA binding protein and plays essential roles in transformation of EBV-infected cells. In their construct, the hygromycin-thymidine kinase fusion gene was driven by EBNA2-dependent EBV BamC promoter. An

Table 2. Targeted Killing of EBNA1-Positive Cells In Vitro (21)

Cells	ID$_{50}$ to GCV (μmol/L)			
	Mock	pNS.CD8α	pO.TkA	pHSV-106
Raji	500	500	3	20
K562	100	100	100	100
KE	100	100	5	100

Raji (an EBNA1-positive Burkitt's lymphoma cell line), K562 (an EBNA1-negative leukemic cell line), or KE (an EBNA1-positive transfectant derived from K562) was transfected by electroporation with pNS.CD8α (a plasmid vector that does not carry HSV1-*Tk* gene), pO.TkA (a plasmid vector that carries *ori*P and HSV1-*Tk* gene) or pHSV-106 (a plasmid vector that carries HSV1-*Tk* gene but not *ori*P). Three days later, cells were cultured in the presence of various concentrations of GCV, and cell viability was assessed by an Alamar Blue assay. The concentration of GCV at which 50% of the cells were killed is indicated as ID$_{50}$ to GCV.

EBNA2-positive cell line was transfected with this construct by electroporation, followed by selection with hygromycin B. The resultant stable transfectants were implanted into SCID mice, and their susceptibility to GCV was demonstrated in vivo.

The latent membrane protein 1 (LMP1) is another EBV protein responsible for B cell transformation. LMP1 binds to NF-kappa B binding sites within the HIV1 long terminal repeat. Wu et al. constructed a plasmid vector in which HIV1 long terminal repeat drives the *HSV1-Tk* gene (61). Stable transformants established from a nasopharyngeal carcinoma cell line exhibited high susceptibility to GCV when implanted into SCID mice.

Kenney et al. devised a recombinant EBV vector to target EBV-associated tumors (24) (see previous section).

USE OF EBNA1/*ori*P TO OBTAIN HIGH-TITER RETROVIRUS AND HSV VECTORS

Although retroviral vectors are the most widely used vehicles for gene therapy, their use is hampered by several technical limitations. One of the major limiting aspects is the difficulty of producing high titers of retrovirus. This problem could be overcome by virtue of the EBNA1/*ori*P system. Kinsella and Nolan (25), and Grignani et al. added EBNA1 and *ori*P to retroviral constructs and transfected these into retrovirus packaging cell lines. The EBNA1/*ori*P system permitted a rapid and long-lasting production of high titers (as high as 1×10^7 cfu/mL) of recombinant retrovirus.

The EBNA1/*ori*P system is also being combined with another viral expression system: herpesvirus vectors. Wang and Vos constructed miniviral vectors, which carry HSV-1 lytic replication origin (*ori*S) and HSV packaging sequence as well as EBNA1 and *ori*P (57). The *cis*-acting elements of HSV-1 allowed vector replication and packaging in an IE3 gene-expressing cell line, infected with a helper virus carrying a deleted IE3 gene. The EBNA1/*ori*P enabled episomal maintenance of the construct, thereby allowing its efficient replication and packaging into HSV-1 capsids. By using the resultant high-titer recombinant HSV, they succeeded in transducing β-gal gene into cultured human cell lines in vitro as well as human tumor cells established in nude mice in vivo. Wang et al. next employed a modified EBNA1 to improve the plasmid stability in helper cells. They could transduce marker genes into neurons, by stereotactically injecting the recombinant HSV into rat brains (56).

PERSPECTIVES

As demonstrated in different experimental systems, the EBV-based plasmid vectors may provide powerful tools for gene therapy of various diseases. To make the EBV-plasmid vectors available for practical use, some problems must be solved. Among them, the safety issue is particularly important.

EBV is an oncogenic virus, and hence potential oncogenicity by EBNA1 cannot be ruled out. This issue represents a strong impediment for its use in gene therapy. In vitro transformation experiments have demonstrated that EBNA2, LMP1, and other latent gene products are a prerequisite for transformation of B cells (7,18). EBNA1 has not been reported to possess transforming activity by itself. Recently, Wilson et al. established transgenic mouse lines carrying the EBNA1 gene driven by an immunoglobulin heavy chain gene enhancer and found that these mice develop B cell lymphomas (59). However, epidemiological studies suggest that EBNA1 expression per se does not lead to tumor development. Therefore, more careful analyses should be performed to assess whether human somatic cells could become cancerous in vivo just by transferring to them the EBNA1 gene.

The EBNA1 is expressed in latently EBV-infected B lymphocytes that persist for life in healthy virus carriers. Cells expressing EBNA1 cannot be recognized by major histocompatibility complex (MHC) class I-restricted cytotoxic T lymphocytes. This is because the EBNA1 interferes with the antigen processing machinery in antigen-presenting cells (26, 38). Roughly one-third of the EBNA1 is composed of a polymer of glycine and alanine. These glycine-alanine repeats may provide a *cis*-acting inhibitory signal that interferes with antigen processing and MHC class I-restricted presentation (31). Therefore, the EBV-based plasmid vectors may not induce any harmful immune response or inflammation in humans.

Concerning recombinant EBV vectors, the safety issue is more complicated because viral genes other than EBNA1 are also present in these systems. Potential oncogenicity and immunogenicity of each gene product must be assessed. Moreover, to avoid any possible generation of replication-competent recombinants, the constructs and producer cells need to be carefully designed.

Among all the trials mentioned in this review, the most feasible strategy is probably the targeted killing of EBV-associated tumors by means of an *ori*P-bearing plasmid vector (9,21,23,24). In those constructs, *ori*P is the only genetic element derived from EBV. Since *ori*P does not encode any functional protein, no viral proteins will be produced.

The artificial chromosome can be an ultimate tool for gene therapy. It is an ideal carrier of exogenous genes of interest, allowing long-term expression without distortion of structure and function of host cell chromosomal DNA. Because human autonomous replicating sequences are not available yet, utilization of genetic elements from EBV may greatly help engineer artificial chromosomes.

Utilization of an appropriate delivery system is quite important for both EBV-based plasmid vectors and human artificial chromosomes. The most suitable vehicle should be chosen, conditioned, and optimized for each application, taking into account the target cell type, route, and method of administration, DNA dose to be transferred, and intensity and duration of the therapeutic gene required. Of course, possible toxicity of the delivery vehicle itself is also a crucial factor. We are currently testing various synthetic macromolecules and combining them with EBV-based plasmid vectors in gene transfer experiments. Further improvement of the delivery systems, especially those applicable to gene transfer in vivo, could make the EBV-based plasmid vectors really attractive for clinical use.

ACKNOWLEDGMENTS

Our studies were supported by a grant-in-aid for scientific research from the Ministry of Education, Science, Sports and Culture, a Health Sciences Research Grant (Research on Eye and Ear Sciences, Immunology, Allergy and Organ Transplantation) and grants from the Japan Heart Foundation, IBM Japan Research and the Japan Foundation for Cardiovascular Research.

REFERENCES

1. **Adams, A.** 1987. Replication of latent Epstein-Barr virus genomes in Raji cells. J. Virol. *61*:1743-1746.
2. **Ambinder, R.F., M. Mullen, Y. Chang, G.S. Hayward, and S.D. Hayward.** 1991. Functional domains of Epstein-Barr virus nuclear antigen EBNA-1. J. Virol. *65*:1466-1478.
3. **Ambinder, R.F., W.A. Shah, D.R. Rawlins, G.S. Hayward, and S.D. Hayward.** 1990. Definition of the sequence requirements for binding of the EBNA-1 protein to its palindromic target sites in Epstein-Barr virus DNA. J. Virol. *64*:2369-2379.
4. **Banerjee, S., E. Livanos, and J.-M.H. Vos.** 1995. Therapeutic gene delivery in human B-lymphoblastoid cells by engineered non-transforming infectious Epstein-Barr virus. Nat. Med. *1*:1303-1308.
5. **Chen, M.R., J. Zong, and S.D. Hayward.** 1994. Delineation of a 16 amino acid sequence that forms a core DNA recognition motif in the Epstein-Barr virus EBNA-1 protein. Virology *205*:486-495.
6. **Chittenden, T., S. Lupton, and A.J. Levine.** 1989. Functional limits of *oriP*, the Epstein-barr virus plasmid origin of replication. J. Virol. *63*:3016-3025.
7. **Cohen, J.I., F. Wang, J. Mannick, and E. Kieff.** 1989. Epstein-Barr virus nuclear protein 2 is a key deterinant of lymphocyte transformation. Proc. Natl. Acad. Sci. USA *86*:9558-9562.
8. **Dzau, V.J., M.J. Mann, R. Morishita, and Y. Kaneda.** 1996. Fusigenic viral liposome for gene therapy in cardiovascular diseases. Proc. Natl. Acad. Sci. USA *93*:11421-11425.
9. **Evans, T.J., L. Brooks, and P.J. Farrell.** 1997. A strategy for specific targeting of therapeutic agents to tumour cells of virus-associated cancers. Gene Ther. *4*:264-267.
10. **Finke, S., B. Trojaneck, P. Moeller, D. Schadendorf, A. Neubauer, D. Huhn, and I.G.H. Schmidt-Wolf.** 1997. Increase of cytotoxic sensitivity of primary human melanoma cells transfected with the interleukin-7 gene to autologous and allogeneic immunologic effector cells. Cancer Gene Ther. *4*:260-268.
11. **Fischer, N., E. Kremmer, G. Lautscham, N. Mueller-Lantzsch, and F.A. Graesser.** 1997. Epstein-Barr virus nuclear antigen 1 forms a complex with the nuclear transporter Karyopherin a2. J. Biol. Chem. *272*:3999-4005.
12. **Franken, M., A. Estabrooks, L. Canacini, B. Sherburne, F. Wang, and D.T. Scadden.** 1996. Epstein-Barr virus-driven gene therapy for EBV-related lymphomas. Nat. Med. *2*:1379-1382.
13. **Frappier, L. and M. O'Donnel.** 1991. Overproduction, purification, and characterization of EBNA-1, the origin binding protein of Epstein-Barr virus. J. Biol. Chem. *266*:7819-7826.
14. **Gahn, T.A. and C.L. Schildkraut.** 1989. The Epstein-Barr virus origin of plasmid replication, *oriP*, contains both the initiation and termination sites of DNA replication. Cell *58*:527-535.
15. **Gahn, T.A. and B. Sugden.** 1995. An EBNA-1-dependent enhancer acts from a distance of 10 kilobase pairs to increase expression of the Epstein-Barr virus LMP gene. J. Virol. *69*:2633-2636.
16. **Grignani, F., T. Kinsella, A. Mencarelli, M. Valtieri, D. Riganelli, F. Grignani, L. Lanfrancone, C. Peschle, G.P. Nolan, and P.G. Pelicci.** 1998. High-efficiency gene transfer and selection of human hematopoietic progenitor cells with a hybrid EBV-retroviral vector expressing the green fluorescence protein. Cancer Res. *58*:14-19.
17. **Haase, S.B. and M.P. Calos.** 1991. Replication control of autonomously replicating human sequences. Nucleic Acids Res. *19*:5053-5058.
18. **Hammerschmidt, W. and B. Sugden.** 1989. Genetic analysis of immortalizing functions of Epstein-Barr virus in human B lymphocytes. Nat. *340*:393-397.
19. **Harada, Y., M. Iwai, S. Tanaka, T. Okanoue, K. Kashima, H. Maruyama-Tabata, H. Hirai, E. Satoh, J. Imanishi, and O. Mazda.** 2000. Highly efficient suicide gene expression in hepatocellular carcinoma cells by Epstein-Barr virus-based plasmid vectors combined with polyamidoamine dendrimer. Cancer Gene Ther. *7*:27-36.
20. **Heinzel, S., P.J. Krysan, C.T. Tran, and M.P. Calos.** 1991. Autonomous DNA replication in human cells is affected by the size and the source of the DNA. Mol. Cell. Biol. *11*:2263-2272.
21. **Hirai, H., E. Satoh, M. Osawa, T. Inaba, C. Shi-**

mazaki, S. Kinoshita, M. Nakagawa, O. Mazda, and J. Imanishi. 1997. Use of Epstein-Barr virus-based vector/hemagglutinating virus of Japan-liposome complex vector for targeted gene therapy of Epstein-Barr virus-associated neoplasms. Biophys. Biochem. Res. Commun. *241*:112-118.

22.**Jankelevich, S., J.L. Kolman, J.W. Bodnar, and G. Miller.** 1992. A nuclear matrix attachment region organized the Epstein-Barr viral plasmid in Raji cells into a single DNA domain. EMBO J. *11*:1165-1176.

23.**Judde, J.-G., G. Spangler, I. Magrath, and K. Bhatia.** 1996. Use of Epstein-Barr virus nuclear antigen-1 in targeted therapy of EBV-associated neoplasia. Hum. Gene Ther. *7*:647-653.

24.**Kenney, S., J.-Q. Ge, E.M. Westphal, and J. Olsen.** 1998. Gene therapy strategies for treating Epstein-Barr virus-associated lymphomas: comparison of two different Epstein-Barr virus-based vectors. Hum. Gene Ther. *9*:1131-1141.

25.**Kinsella, T.M. and G.P. Nolan.** 1996. Episomal vectors rapidly and stably produce high-titer recombinant retrovirus. Hum. Gene Ther. *7*:1405-1413.

26.**Klein, G.** 1994. Epstein-Barr virus strategy in normal and neoplastic B cells. Cell *77*:791-793.

27.**Krysan, P.J. and M.P. Calos.** 1993. Epstein-Barr virus-based vectors that replicate in rodent cells. Gene *136*:137-143.

28.**Krysan, P.J., S.B. Haase, and M.P. Calos.** 1989. Isolation of human sequences that replicate autonomously in human cells. Mol. Cell. Biol. *9*:1026-1033.

29.**Lei, D.C., K. Kunzelmann, T. Koslowsky, M.J. Yezzi, L.C. Escobar, Z. Xu, A.R. Ellison, J.M. Rommens, L.-C. Tsui, M. Tykocinski, and D.C. Gruenert.** 1996. Episomal expression of wild-type CFTR corrects cAMP-dependent chloride transport in respiratory epithelial cells. Gene Ther. *3*:427-436.

30.**Levin, L.I. and P.H. Levine.** 1998. The epidemiology of Epstein-Barr virus-associated human cancers, p. 51-74. *In* T. Osata, K. Takada, and M. Tokunaga (Eds.), Epstein-Barr Virus and Human Cancer. Japan Scientific Societies Press, Tokyo.

31.**Levitskaya, J., M. Coram, V. Levitsky, S. Imreh, P.M. Steigerwald-Mullen, G. Klein, M.G. Kurilla, and M.G. Masucci.** 1995. Inhibition of antigen processing by the internal repeat region of the Epstein-Barr virus nuclear antigen-1. Nature *375*:685-688.

32.**Lin, H., M.S. Parmacek, G. Morle, S. Bolling, and J.M. Leiden.** 1990. Expression of recombinant genes in myocardium *in vivo* after direct injection of DNA. Circulation *82*:2217-2221.

33.**Maruyama-Tabata, H., Y. Harada, T. Matsumura, E. Satoh, F. Cui, M. Iwai, M. Kita, S. Hibi, J. Imanishi, T. Sawada, and O. Mazda.** 2000. Effective suicide gene therapy in vivo by EBV-based plasmid vector coupled with polyamidoamine dendimer. Gene Ther. *7*:53-60.

34.**Mazda, O., E. Satoh, K. Yasutomi, and J. Imanishi.** 1997. Extremely efficient gene transfection by Epstein-Barr virus vectors into lympho-hematopoietic cell lines. J. Immunol. Methods *204*:143-152.

35.**Mazda, O., K. Teshigawara, S. Fujimoto, N. Hattori, Y.-M. Dou, and Y. Katsura.** 1994. A reporter system using a flow cytometer to detect promoter/enhancer activity in lymphoid cell lines. J. Immunol. Methods *169*:53-61.

36.**Middleton, T., T.A. Gahn, J.M. Martin, and B. Sugden.** 1991. Immortalizing genes of Epstein-Barr virus. Adv. Virus Res. *40*:19-55.

37.**Middleton, T. and B. Sugden.** 1992. EBNA1 can link the enhancer element to the initiator element of the Epstein-Barr virus plasmid origin of DNA replication. J. Virol. *66*:489-495.

38.**Moss, D.J., S.R. Burrows, A. Suhrbier, and R. Khanna.** 1994. Potential antigenic targets on Epstein-Barr virus-associated tumours and the host response. Ciba Found. Symp. *187*:4-13.

39.**Muecke, S., A. Polack, M. Pawlita, D. Zehnpfenning, N. Massoudi, H. Massoudi, H. Bohlen, W. Doerfler, G. Bornkamm, V. Diehl, and J. Wolf.** 1997. Suitability of Epstein-Barr virus-based episomal vectors for expression of cytokine genes in human lymphoma cells. Gene Ther. *4*:82-92.

40.**Puglielli, M.T., M. Woisetschlaeger, and S.H. Speck.** 1996. *ori*P is essential for EBNA gene promoter activity in Epstein-Barr virus-immortalized lymphoblastoid cell lines. J. Virol. 70:5758-5768.

41.**Qin, L., D.R. Pahud, Y. Ding, A.U. Bielinska, J.F. Kukowska-Latallo, J.R. Baker, and J.S. Bromberg.** 1998. Efficient transfer of genes into murine cardiac grafts by Starburst polyamidoamine dendrimers. Gene Ther. *9*:553-560.

42.**Rawlins, D.R., G. Milman, S.D. Hayward, and G.S. Hayward.** 1985. Sequence-specific DNA binding of the Epstein-Barr virus nuclear antigen (EBNA-1) to clustered sites in the plasmid maintenance region. Cell *42*:859-868.

43.**Reisman, D. and B. Sugden.** 1986. *trans* activation of an Epstein-Barr viral transcriptional enhancer by the Epstein-Barr viral nuclear antigen 1. Mol. Cell. Biol. *6*:3838-3846.

44.**Reisman, D., J. Yates, and B. Sugden.** 1985. A putative origin of replication of plasmids derived from Epstein-Barr virus is composed of two cis-acting components. Mol. Cell. Biol. *5*:1822-1832.

45.**Robertson, E.S., T. Ooka, and E.D. Kieff.** 1996. Epstein-Barr virus vectors for gene delivery to B lymphocytes. Proc. Natl. Acad. Sci. USA *93*:11334-11340.

46.**Saeki, Y., M. Wataya-Kaneda, K. Tanaka, and Y. Kaneda.** 1998. Sustained transgene expression in vitro and in vivo using an Epstein-Barr virus replicon vector system combined with HVJ-liposome. Gene Ther. *5*:1031-1037.

47.**Satoh, E., H. Hirai, T. Inaba, C. Shimazaki, M. Nakagawa, J. Imanishi, and O. Mazda.** 1998. Successful transfer of adenosine deaminase (ADA) gene in vitro into human peripheral blood $CD34^+$ cells by transfecting Epstein-Barr virus (EBV)-based episomal vectors. FEBS Lett. *441*:39-42.

48.**Satoh, E., M. Osawa, K. Tomiyasu, H. Hirai, C. Shimazaki, Y. Oda, M. Nakagawa, M. Kondo, S. Kinoshita, O. Mazda, and J. Imanishi.** 1997. Efficient gene transduction by Epstein-Barr virus (EBV)-based vectors coupled with cationic liposome and HVJ-liposome. Biochem. Biophys. Res. Commun. *238*:795-799.

49. **Shimizu, N., H. Yoshiyama, and K. Takada.** 1996. Clonal propagation of Epstein-Barr virus (EBV) recombinants in EBV-negative Akata cells. J. Virol. *70*:7260-7263.

50. **Snudden, D.K., J. Hearing, P.R. Smith, F.A. Graesser, and B.E. Griffin.** 1994. EBNA-1, the major nuclear antigen of Epstein-Barr virus, resembles 'RGG' RNA binding proteins. EMBO J. *13*:4840-4847.

51. **Sugden, B. and N. Warren.** 1989. A promoter of Epstein-Barr virus that can function during latent infection can be transactivated by EBNA1, a viral protein required for viral DNA replication during latent infection. J. Virol. *63*:2644-2649.

52. **Sun, T.-Q., D.A. Fenstermacher, and J.-M.H. Vos.** 1994. Human artificial episomal chromosomes for cloning large DNA fragments in human cells. Nat. Genet. *8*:33-41.

53. **Tang, M.X., C.T. Redemann, and F.C. Szoka.** 1996. In vitro gene delivery by degraded polyamidoamine dendrimers. Bioconjug. Chem. *7*:703-714.

54. **Tomiyasu, K.-I., E. Satoh, Y. Oda, K. Nishizaki, M. Kondo, J. Imanishi, and O. Mazda.** 1998. Gene transfer in vitro and in vivo with Epstein-Barr virus-based episomal vector results in markedly high transient expression in rodent cells. Biochem. Biophys. Res. Commun. *253*:733-738.

55. **Wang, Y. and D. Becker.** 1997. Antisense targeting of basic fibroblast growth factor and fibroblast growth factor receptor-1 in human melanomas blocks intratumoral angiogenisis and tumor growth. Nat. Med. *3*:887-893.

56. **Wang, S., S. Di, W.-B. Young, C. Jacobson, and C.J.L. Link, Jr.** 1997. A novel herpesvirus amplicon system for in vivo gene delivery. Gene Ther. *4*:1132-1141.

57. **Wang, S. and J.M. Vos.** 1996. A hybrid herpesvirus infectious vector based on Epstein-Barr virus and herpes simplex virus type 1 for gene transfer into human cells in vitro and in vivo. J. Virol. *70*:8422-8430.

58. **Westphal, E.M., H. Sierakowska, E. Livanos, R. Kole, and J.-M. Vos.** 1998. A system for shuttling 200-kb BAC-PC clones into human cells: stable extrachromosomal persistence and long-term ectopic gene activation. Hum. Gene Ther. *9*:1863-1873.

59. **Wilson, J.B., J.L. Bell, and A.J. Levine.** 1996. Expression of Epstein-Barr virus nuclear antigen-1 induces B cell neoplasia in transgenic mice. EMBO J. *15*:3117-3126.

60. **Wohlgemuth, J.G., S.H. Kang, G.H. Bulboaca, K.A. Nawotka, and M.P. Calos.** 1996. Long-term gene expression from autonomously replicating vectors in mammalian cells. Gene Ther. *3*:503-512.

61. **Wu, C.J., C.Y. Leu, S.T. Liu, K.P. Chow, C.L. Meng, and Y.S. Chang.** 1998. Transcriptional activation of NF-kappa B activity by Epstein-Barr virus (EBV) LMP1 as a selective therapeutic strategy for EBV-associated diseases. Gene Ther. *5*:905-912.

62. **Wysokenski, D.A. and J. Yates.** 1989. Multiple EBNA-1 binding sites are required to form an EBNA-1-dependent enhancer and to activate a minimal replicative origin within *ori*P of Epstein-Barr virus. J. Virol. *63*:2657-2666.

63. **Yates, J.L. and N. Guan.** 1991. Epstein-Barr virus-derived plasmids replicate only once per cell cycle and are not amplified after entry into cells. J. Virol. *65*:483-488.

64. **Yates, J.L., N. Warren, and B. Sugden.** 1985. Stable replication of plasmids derived from Epstein-Barr virus in various mammalian cells. Nature *313*:812-815.

FAMILY: Poxviridae
SUBFAMILY: Entomopoxvirinae
GENUS: Entomopoxvirus B

21 Entomopoxvirus Vectors

Alison L. Bawden, Yi Li, Kathryn J. Glassberg, and Richard W. Moyer
Department of Molecular Genetics and Microbiology, University of Florida College of Medicine, Gainesville, FL, USA

INTRODUCTION

Until recently, complex insect viruses (entomoviruses) had not been considered for use as possible viral gene therapy vectors. In the past, studies of entomoviruses have mainly concentrated on their use as biopesticides, expression systems, or taxonomic novelties to compare with their mammalian virus counterparts. We will show in this chapter that newly discovered properties of an insect poxvirus, *Amsacta moorei* entomopoxvirus (AmEPV), allow it to be considered as a legitimate and feasible candidate for development as a gene therapy vector. Two key findings that support this thesis are: *(i)* mammalian cells survive AmEPV infection following expression of early viral genes, and *(ii)* vertebrate cells can be stably transformed by AmEPV, a property that directly stems from the cells' surviving and continuing to grow despite AmEPV infection. These properties of AmEPV will be discussed in further detail in subsequent sections of this chapter.

The family *Poxviridae* comprises two subfamilies, the *Chordopoxvirinae* (vertebrate) and the *Entomopoxvirinae* (insect) viruses (EPVs). EPVs were first discovered in the early 1960s and have subsequently been shown to have a worldwide distribution. The subfamily contains three genera: A, B, and C, which infect beetles (Coleoptera), butterflies, moths, or grasshoppers, and midge flies, respectively (27). It should be recognized that classification within the three EPV genera is based solely on morphological and host range criteria and not molecular properties. Indeed, it is now clear that the group B viruses of butterflies and moths (Lepidoptera) and grasshoppers (Orthoptera) are quite distinct from one another (1). AmEPV was originally isolated in India from the red hairy caterpillar, and it is the prototype virus of this group. This is primarily because of its ability to be easily grown in cultured insect cells, although certain *Choristoneura* and *Heliothis* EPVs have also been shown to replicate in cell cultures at low levels (8,24).

EPVs are the most distant relatives of mammalian poxviruses and exhibit both similarities to and differences from the more commonly studied chordopoxviruses, such as vaccinia virus (VV). Similarities include morphology, a large linear, double-stranded genome (225 kb for AmEPV and 190 kb for VV), common transcriptional regulation sequence motifs, nonspliced transcripts, and a cytoplasmic site of replication. Differences include the GC content of the viral DNA (a low 18% for AmEPV

Viral Vectors: Basic Science and Gene Therapy
Edited by A. Cid-Arregui and A. García-Carrancá

and 37% for VV), optimal growth temperatures (28°C for AmEPV and 37°C for VV), and host range. AmEPV does not replicate in vertebrate cells, and VV does not replicate in insect cells although both viruses enter their respective nonpermissive cells and initiate a replicative cycle (20).

Generally, growth of AmEPV in insect cell cultures is similar to that of vertebrate poxviruses in mammalian cells. Receptors mediating poxvirus attachment and entry appear to be widespread and common since EPVs infect vertebrate cells and VV infects insect cells (21,23). It is assumed by analogy with the vertebrate poxviruses that AmEPV gene expression patterns can be classified as early, intermediate, and late, but experimental data is minimal (39). EPVs have been shown to contain vertebrate poxvirus promoter elements and early transcription termination motifs (1,14). The most unique feature of poxvirus replication is the development mostly, if not exclusively, within the cytoplasm. As a consequence of cytoplasmic development, EPV promoters (like those of vertebrate poxviruses) are recognized only by the virally encoded transcription system. The general availability of poxvirus-specific promoters, coupled with exclusion of the nuclear transcription apparatus, are major advantages for engineering and control of foreign gene expression related to gene therapy applications.

EPVs, like VV, contain a number of genes that are nonessential for growth in cell culture. Two examples are the thymidine kinase (*TK*) and spheroidin genes. The spheroidin gene can be viewed as a counterpart to the polyhedrin and A-type (*ATI*) occlusion genes of baculoviruses and cowpox viruses, respectively. VV also contains an *ATI* gene, but it is defective. Spheroidin is the most abundantly expressed AmEPV gene and serves to occlude infectious virions within an environmentally resistant occlusion body. Both the AmEPV *TK* and spheroidin gene can readily serve as sites for insertion and expression of foreign genes by utilizing standard plasmid-mediated recombination.

As stated earlier, there are, however, two unique features of AmEPV *not* shared by VV that argue favorably for the use of AmEPV as a novel vector for delivery of foreign genes to mammalian cells and tissues. The first is that entomopoxvirus infection of vertebrate cells is abortive and cells survive. AmEPV enters vertebrate cells, and early genes are expressed before the infection aborts. Neither viral DNA synthesis nor late gene expression is observed. Hence, AmEPV can be used for nonlethal, transient expression of foreign genes in mammalian systems. Second, although it is not necessarily intuitive for a cytoplasmic virus, we have shown that AmEPV can be used to stably transform cells because AmEPV infection of vertebrate cells is nonlethal. Hence, AmEPV also has potential use for the stable introduction and long-term expression of genes in mammalian cells. Since there is little technical information published on the handling and manipulation of AmEPV, in this chapter, we first summarize the basic techniques and strategies utilized by our laboratory for manipulation of this virus. We then discuss experiments that directly illustrate the potential of AmEPV as a gene therapy vector.

BASIC METHODOLOGY FOR THE GROWTH AND TITERING OF AmEPV

Growth and Amplification of AmEPV

AmEPV productively infects *Lymantria dispar*-derived IPLB-LD-652 (LD) cells. LD cells were maintained at 28°C in a 1:1 ratio of TC-100 medium (Life Technologies, Rockville, MD, USA) to EX-CELL™ 401 medium (JRH Biosciences, Lenexa,

KN, USA) supplemented with 10% fetal bovine serum (FBS), 50 U/mL penicillin, and 50 µg/mL streptomycin (1:1 Tris EDTA [TE]). To amplify virus, cells grown in 150-mm dishes are inoculated at a multiplicity of infection (moi) of 0.01 in sufficient medium (5 mL) to cover the surface of the tissue culture vessel. The cells should be no more than 70% confluent. After adsorption at 28°C for 2 hours, 25 mL medium is added, and the infections are incubated for four to six days at 28°C. The infection is considered complete when most cells become occlusion body-positive as seen by light microscopy (i.e., when refractile occlusion bodies can be seen). For recombinant viruses in which *lacZ* has been inserted into either the *TK* or the spheroidin locus, infection is monitored by in situ staining of infected cells with 1 mg/mL 5-bromo-4-chloro-3-indolyl-β-D-galactopyranoside (X-Gal), 4 mmol/L potassium ferricyanide, 4 mmol/L potassium ferrocyanide, and 2 mmol/L $MgCl_2$ in phosphate-buffered saline (PBS) (140 mmol/L NaCl, 2.7 mmol/L KCl, 10 mmol/L Na_2PO_4, 1.8 mmol/L KH_2PO_4, pH 7.4). This stain allows β-galactosidase (β-gal)-producing infected cells to be visualized by the appearance of blue color. Though numbers can vary, an AmEPV infection at an moi of 0.01 generates a net yield of approximately 1.5 plaque-forming units (pfu) based on the total number of cells and results in a 100-fold increase in pfu. Titers are routinely between 1×10^5 and 1×10^6 pfu/mL. Cells infected at a higher moi can generate higher yields/cell (approximately 10), but the net increase in virus over input is lower.

Partial Purification and Concentration of Virus

Some experiments necessitate higher concentrations of virus. The procedure that follows typically renders semipure virus stocks at titers greater than 1×10^8 pfu/mL. AmEPV-infected cells are harvested by scraping and centrifuged at 700× *g* for 15 minutes to remove cells. The supernatant is then subjected to ultracentrifugation at 70 000× *g* for 2 hours to pellet the extracellular virus. The virus is resuspended in an appropriate amount of PBS (typically 100 µL/40 mL supernatant), and the titer is determined by plaque assay. Total yield is typically 50% of input virus.

Plaque Assay

Virus to be titered is subjected to tenfold serial dilutions in 1:1 TE medium. LD-652 cells are plated at 70% confluency in six-well dishes, each having a 34.6-mm diameter (roughly 1.4×10^6 cells/well). Once the cells have adhered, the medium is removed, and 0.5 mL diluted virus is added to the wells. After adsorption at 28°C for 2 hours, the inoculum is removed, and 2.5 mL of overlay is added to each well. The overlay is a 2:1 ratio of 1.33× TC-100 medium (containing 14% fetal calf serum) and 4% sterile low-melting point agarose, equilibrated to 42°C and mixed just before addition to the monolayer.

For a spheroidin-positive virus, visible plaques appear and are counted one week after infection (Figure 1). Spheroidin-negative virus plaques are much more difficult to visualize; hence, most spheroidin-negative viruses have been engineered to contain *lacZ*. Such viruses can be readily visualized by staining with X-Gal as follows: a liquid overlay of 400 µg X-Gal in 50 µL total solution is spread over the agarose in each well of the plaque assay three to four days after infection. Plaques appear as blue patches of infected cells and are counted one week after infection.

Comet-like plaques of wild-type AmEPV, in the absence of an agarose overlay can be more rapidly visualized by immunostaining (39). Plaque assays are prepared as above, except that a liquid overlay of medium replaces the agarose overlay. Infected cell monolayers are air-dried three days after infection and fixed in acid alcohol (95% ethanol [EtOH], 5% glacial acetic acid) for 30 minutes. After equilibration with Tris-buffered saline (TBS; 0.02 mol/L Tris (pH 7.4) and 0.15 mol/L NaCl), the cells are incubated in TBS-Block (0.5% w/v blocking reagent in TBS) (Roche Molecular Biochemicals, Mannheim Germany) for 1 hour at room temperature to prevent nonspecific antibody binding. The primary antibody (rabbit anti-AmEPV occlusion body antiserum) (15) or secondary antibody (goat anti-rabbit conjugated to alkaline phosphatase) (Fisher Scientific, Pittsburgh, PA, USA) are both diluted in TBS-Block. Antibody reactions and color development are performed as previously described (16).

Isolation of AmEPV Genomic DNA

LD-652 cells are infected with AmEPV at an moi of 0.01. The infections (cells and medium) are harvested by scraping and centrifuged at 700× *g* for 15 minutes to remove cells. The supernatant is centrifuged at 39 000× *g* for 30 minutes to pellet extracellular virus. The viral pellet is resuspended in deionized water (100 μL for 40 mL supernatant). DNase-free RNase is added to the resuspended viral pellet at a final concentration of 50 μg/mL and incubated at 37°C for 30 minutes. The virus sample is then heated to 50°C, and an equal volume of lysis buffer (100 mmol/L Tris [pH 8.0], 10 mmol/L EDTA, 54% sucrose, 2% SDS and 10 mmol/L β-mercaptoethanol) is added to the sample. Proteinase K is then added to a final concentration of 0.6 mg/mL, and the viral lysate is incubated overnight at 50°C. The lysate is extracted three times with 50:49:1 phenol:chloroform:isoamyl alcohol and once with chloroform, and the DNA is precipitated in 0.4 mol/L LiCl and 95% ethanol. This procedure typically yields 2 μg of genomic AmEPV DNA/1 × 10^7 infected cells.

THE PREPARATION OF RECOMBINANT AmEPVs

The strategy for generation of recombinant viruses is identical to that used for VV and takes advantage of the high levels of recombination with transfected plasmids mediated by these viruses. The basic procedure utilizes transfection of AmEPV-infected cells with an appropriately designed shuttle vector. Insertion of foreign genes occurs within a nonessential gene (spheroidin or *TK*). Because of the cytoplasmic nature of AmEPV, it is necessary to place all foreign genes under the control of an AmEPV (early or late) poxvirus promoter. Recombinants are selected and subjected to three rounds of plaque purification before use.

Shuttle Vector Plasmid Construction

Following the sequence determination of two nonessential genes, *TK* and spher-

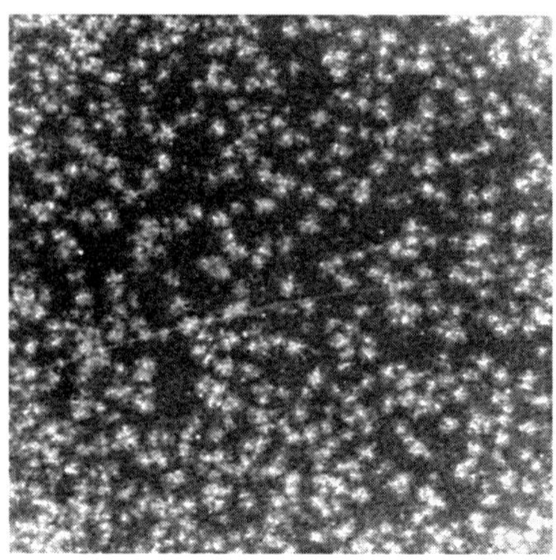

Figure 1. A monolayer of infected LD-652 cells, seven days after infection showing wild-type AmEPV plaques. Plaques appear as small, pale foci on the cell background.

oidin (12,14), we were able to create shuttle vector plasmids for the generation of recombinant AmEPV viruses. The shuttle vectors are described below.

The TK Insertion Site Shuttle Vector

Oligonucleotide primers were used to polymerase chain reaction (PCR)-amplify a 748-bp fragment of downstream *TK* flanking sequence from plasmid pMEGTK-1 (12). Another set of oligonucleotide primers was used to PCR-amplify a 663-bp fragment of *TK* upstream flanking sequence from pMEGTK-1. These two fragments were separately inserted into pBluescript® I SK(+) (Stratagene, La Jolla, CA, USA) to produce pDUTK (21). Foreign genes were then cloned within the *TK* flanks to generate shuttle vectors for the generation of recombinants (21).

The Spheroidin Insertion Site Shuttle Vector

Oligonucleotides were used to PCR-amplify a 1046-bp fragment of upstream spheroidin flanking sequence from plasmid pRH512 (14). In addition to 1023 bp of sequence upstream of the spheroidin gene, this fragment contained the starting ATG of the spheroidin coding sequence and 20 bp following the ATG. Another set of oligonucleotide primers was used to amplify a 998-bp fragment of downstream spheroidin flanking sequence from pRH512. These two fragments were separately inserted into pBluescript I SK(+) to produce pDU20 (13). Subsequent constructs were cloned within the spheroidin flanks to generate various shuttle vectors for the generation of recombinants (13).

AmEPV Early-Promoter Constructs

Promoters for early poxvirus genes are active prior to viral DNA replication. We have utilized two early EPV promoters in our constructs. The first, an AmEPV early strong promoter (*esp*), was derived from a strongly expressed 42-kDa early protein (21). The second promoter was derived from the early expressed fusolin (*fus*) gene as described and provided by Dr. M. Bergoin (9). These promoters have been used to regulate reporter genes (e.g., *lacZ* and *gfp*). The appropriately regulated genes are then placed within shuttle vectors and transfected into infected cells to produce recombinant viruses. The shuttle vector pTK-fus*lacZ* was constructed by PCR amplification of the MmEPV fusolin early promoter from pHF51 (provided by Dr. M. Bergoin) and insertion into pDUTK; *lacZ* was subcloned from pMC1871 (Amersham Pharmacia Biotech, Piscataway, NJ, USA) as described (21). The shuttle vector pTK-esp*lacZ* (pTK-42k*lacZ*) was constructed by cloning the PCR-amplified *esp* promoter into pTK-fus*lacZ* after excision of the fusolin promoter (21). To construct pTK-esp*gfp*, a green fluorescent protein gene (*gfp*) was PCR-amplified from the pTR-UF5 plasmid (Vector Core, University of Florida, Gainesville, FL, USA) (18) and cloned into pTK-esp*lacZ* replacing the *esp*lacZ cassette as described (21).

AmEPV Late-Promoter Constructs

We have used the spheroidin (*sph*) promoter as an example of an AmEPV strong late promoter. This promoter has two rather unexpected properties: *(i)* the *sph* promoter appears to be insect cell specific and functions very poorly in vertebrate cells (13), and *(ii)* unlike previously described late poxvirus promoters, we have found in insect cells that expression is further enhanced by including the 20 bp downstream of the TAAATG promoter in the reporter gene constructs (22). pDU20-*lacZ* was created by insertion of *lacZ* into the *Bam*HI site of pDU20. The final reporter contains 1046 bp of potential *sph* promoter sequence plus 20 bp additional downstream spheroidin coding sequence following the TAAATG sequence before fusion to lacZ (13). pDU2*lacZ* was constructed using the same strategy as that for pDU20*lacZ*, except that only 2 bp spheroidin coding sequence follows the translation-starting TAAATG before fusion to *lacZ* (13,22). We have also constructed and used the cowpox virus late *ATI* gene promoter to drive *lacZ*, which functions well in both insect and vertebrate cells (22).

Construct Driven by Pol II-Specific Promoters

An AmEPV construct containing reporter genes driven by *Pol*II rather than poxvirus promoters has also been prepared based on the plasmid pTR-UF5 (Vector Core) (18), which is a plasmid containing two mammalian reporter genes: *(i)* a gene coding for neomycin resistance (*neo*^{R}) driven by the herpes simplex virus (HSV) *TK* promoter and *(ii) gfp* driven by the immediate early promoter/enhancer from cytomegalovirus (CMV). The dual gene cassette is flanked by the adeno-associated virus (AAV) inverted terminal repeats (ITRs). pTR-UF5 was digested with *Sal*I to remove two *Pst*I sites then religated to form pTRUF5*Sal*I. This construct was then digested with *Pst*I, and the fragment containing the two reporter genes was inserted into the *Pst*I site of pDUTK to produce pTKUF5 *Sal*I. This construct was then digested with *Sal*I, and the previously removed *Sal*I fragment was reinserted into the construct to produce pTKUF5 (Figure 2).

Generation and Selection of Recombinant AmEPV

Neither the spheroidin gene nor the *TK* gene is required for propagation of AmEPV in cell culture (29) (R.W. Moyer, Y. Li, and A. Bawden, unpublished data), providing sites for insertion of foreign genes by homologous recombination. Following transfection of AmEPV-infected LD-652 cells, all foreign genes are inserted into either the *TK* or the spheroidin locus (13).

Transfection of Infected Cells

LD-652 cells (1.4 × 10^6 cells, 70% confluent in a 34.6-mm dish) are infected with AmEPV at an moi of 5 pfu/cell in a volume of 1 mL. Two hours postinfection, the

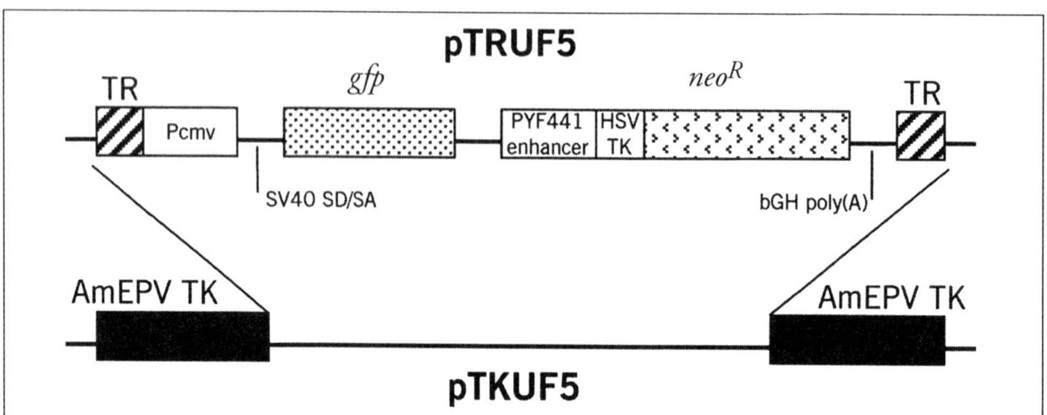

Figure 2. Diagram of plasmid pTR-UF5, supplied by the University of Florida Vector Core, and used to construct recombinant AmEPV for mammalian cell transformation experiments. Pertinent elements include: TRs (ITRs of AAV); *neo*^{R} expressed from a CMV immediate early promoter/enhancer (Pcmv); and *gfp* expressed from the HSV-TK promoter.

inoculum is aspirated, and 1 mL of transfection mixture and DNA is added.

Transfection mixture and DNA is prepared by separately combining 20 µL Lipofectin® (Life Technologies) and 80 µL 1:1 TE media without FBS, and 5 µg shuttle vector plasmid DNA in a volume of 100 µL of the same media. The mixture is incubated at room temperature for 15 minutes. The concentrated transfection mixture is then diluted by addition of 800 µL of 1:1 TE without FBS and added to cells. After 6 hours, the transfection mixture is removed and replaced with 2 mL 1:1 TE with 10% FBS. Three days postinfection, the supernatant is harvested by centrifugation at 200× *g*.

Selection of AmEPV Recombinants

For selection of recombinants inserted into the *TK* gene, a TK(-) cell line, C11.3, was derived by serial passage of LD-652 cells in increasing concentrations of 5-bromo-2′-deoxyuridine (BudR) (10 µg/mL increasing increments of BudR at intervals of five weeks over one year). C11.3 cells are maintained in 1:1 TE medium supplemented with 100 µg/mL BudR (21). Supernatants harvested from the infection/transfection described above are plaqued on C11.3 cells in the presence of 100 µg/mL BudR. Only virus lacking a functional *TK* gene (i.e., recombinant virus) will grow in the presence of BudR. For selection of recombinants where insertion is within the spheroidin gene (where *lacZ* has been incorporated into the shuttle plasmid), recombinant virus plaques are identified following staining with X-Gal as described above or by the selection of nonrefractile plaques.

Purification of Recombinants

Dilutions at 1×10^{-2} to 1×10^{-4} of the infected cell supernatant are assayed by repeated plaquing (described above). Individual plaques are isolated and diluted in 1 mL 1:1 TE medium plus 10% FBS. Isolates are replaqued at dilutions of 1×10^{-1} to 1×10^{-3}. The plaque purification is repeated three to four times before plaque expansion for the propagation of larger stocks. These viruses can be stored at 4°C for two to three months or at -80°C for long-term storage.

THE USE OF AmEPV AS AN EXPRESSION VECTOR

There are basically three applications of EPVs: *(i)* as a system for the expression of high levels of foreign proteins, *(ii)* for the transient expression of foreign genes in mammalian cells, and *(iii)* for the stable transformation of vertebrate cells for the long-term expression of foreign proteins.

Foreign Gene Expression in Permissive Insect Cells

We find that the level of foreign gene expression within infected insect cells when driven by either of two late promoters (the cowpox virus *ATI* or spheroidin +20 promoters) equals or exceeds that of VV or baculoviruses on a per cell basis (22). The technology needs no elaboration, because once suitable recombinant viruses are constructed, only appropriate infection of insect cells is necessary.

Transient Expression in Vertebrate Cells

The normal host range of AmEPV is limited to Lepidoptera, and early experiments attempting to infect vertebrate cells with AmEPV indicated no obvious deleterious effects on the cells (11). Given the general promiscuity of poxviruses in the binding and entering of cells and the similarity of the AmEPV life cycle to that of vaccinia, we had reason to believe AmEPV would infect and enter vertebrate as well as insect cells. AmEPV recombinants were constructed carrying the *lacZ* reporter gene regulated by either of two early AmEPV promoters, the late spheroidin promoter or the *ATI* promoter from cowpox (TK-fus-*lacZ*, TK-esp*lacZ*, SPH(20)*lacZ* and TK-ATI*lacZ*, described above). When mammalian CV-1 cells were infected with the recombinant viruses at an moi of 10 pfu/cell, those cells infected with AmEPV recombinants in which *lacZ* was regulated by either the MmEPV fusolin or the AmEPV 42kD protein early promoters, expressed β-gal (Figure 3, D and E) (21). No expression was observed for the recombinants where *lacZ* was under control of either of the two late promoters (Figure 3, B and C). Thus, AmEPV enters vertebrate cells and undergoes at least a partial uncoating, which allows early, prereplicative viral genes to be expressed. The infection is blocked thereafter, and neither viral DNA replication nor late protein synthesis is observed (21). Early expression results from the fact that AmEPV, like other poxviruses, packages the enzymes necessary for early gene transcription within the virion. However, if vertebrate cells are coinfected with both VV and AmEPV, late promoters within AmEPV are rescued and activated (21) suggesting that vaccinia can provide factors *in trans* that are needed for the infection to progress and activate the late promoters.

While determining the basis of host range restriction is difficult, the cytoplasmic nature of AmEPV coupled with a

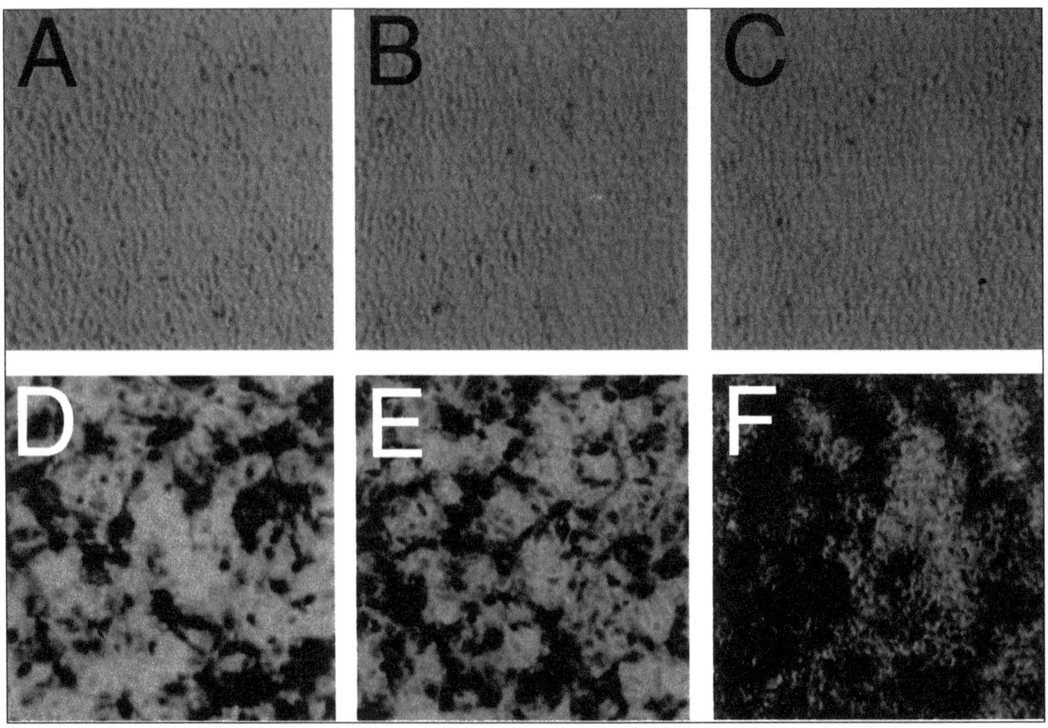

Figure 3. Expression of *lacZ* in recombinant AmEPV-infected mammalian cells. CV-1 cells were mock infected (A) or infected with various AmEPV *lacZ* recombinants in which *lacZ* was under the control of the cowpox virus late *ATI* gene promoter (B), the late AmEPV spheroidin promoter (C), the *Melolontha melonontha* early *fus* promoter (D) or the AmEPV early *esp* promoter (E). Infection of human Huh-7 liver cells with the AmEPV TK*esp*-lacZ recombinant is also shown as an additional control (F). The infected cell monolayers were stained with X-gal 24 hours after infection.

virus-encoded transcription and replication machinery offers major advantages for vector design. We can be fairly certain that late genes are not transcribed because of the lack of β-gal expression from late promoters in vertebrate cells and because DNA synthesis, a requirement for late mRNA synthesis, does not occur. It is quite possible that incomplete uncoating of the virus leads to the block in gene expression.

Normally, uncoating of poxviruses occurs in two discrete steps. Upon entry into cells, virions are sufficiently permeabilized to allow early gene transcription from the viral core. Early proteins allow the complete uncoating of the core to allow transcription of the later classes of genes following interaction of newly synthesized DNA with intermediate and late transcription factors. The uncoating of vertebrate poxviruses has been thoroughly studied (17, 34,40), and uncoating intermediates have been identified through differential centrifugation of cellular extracts infected with labeled virus. A viral activity specifically required for the second stage of uncoating has been identified (31). By analogy with VV, AmEPV might be expected to encode a similar uncoating factor. If so, then one would not necessarily expect a cell line-specific block in uncoating unless this uncoating protein acts in conjunction with cellular components. Should we find that the particle *is* uncoated, then AmEPV fails to express other genes in vertebrate cells required for the infection to continue. Host range restriction of another insect virus family, the *Baculoviridae*, has received considerable attention. For baculoviruses, blockage may occur at many stages during the activation of late or very late genes after viral DNA enters the nucleus (4,5,25,26). The questions related to host range specificity are outside the scope of this review, but the limit of expression to those genes under the control of early promoters following infection of vertebrate cells is a key property of the virus that ultimately makes it a potential candidate as a vector for transient expression in vertebrate cells.

The most novel feature of this system is the survival and continued growth of the infected vertebrate cells. We immediately noticed that there were no observable cytopathic effects in the infected CV-1 cells. Survival and growth of the infected cells was shown following infection with recombinant AmEPV TK-esp*gfp*. Initially, individual fluorescent cells resulted, which over a period of two to three days, divided to form fluorescent microclusters of cells (Figure 4). There is no other known precedent of cells surviving a poxvirus infection. The fact that mammalian cells survive suggests that AmEPV offers the potential for a highly efficient, nontoxic method of foreign gene delivery into vertebrate cells for transient expression of foreign genes, after which the cells continue to grow unabated. While initial observations were made using CV-1 cells, these results have been extended to many other cell lines. In general, lymphocytic cells are more resistant to infection (21).

AmEPV-Mediated Gene Expression in the Mouse

To examine the feasibility of AmEPV to deliver and express foreign genes in vivo, we examined the effects of injection of TK-esp*lacZ* and SPH(20)*lacZ* into mouse muscle. Approximately 2×10^6 pfu (100 μL) of each virus was injected into the muscle of the hind leg of two separate mice; as an additional control, we also injected 100 μL PBS into a third mouse. Two days postinfection, the mice were sacrificed, and the muscle was excised into small pieces, fixed in a 3% formaldehyde solution for 30 minutes, and stained with X-Gal. Extensive β-gal expression occurred in the muscle from the TK-esp*lacZ*-infected mouse (Figure 5). No expression was seen in either the SPH(20)*lacZ* or the PBS control (data not shown). Thus, consistent with our observations of infected mammalian cells in culture, AmEPV can also enter cells in vivo and allow early, but not late, expression of a reporter gene.

The Control of AmEPV-Induced Inflammation

One concern with complex viral vectors is the potential for the unintended induction of inflammatory and immunological responses following administration. In studies with adenovirus (Ad), inflammation and immunogenicity to the virus and to virus-infected cells has limited transgene expression and the utility of this approach to treat chronic illnesses. Inflammation is initially characterized by perivascular and peribronchiolar inflammatory cell infiltration. Neutrophils and later macrophages and lymphocytes frequent the site of the infected area. Specific cytokines can also be measured as an index of the inflammatory response (10,28). The early response to Ad infection consists of diffuse cellular infiltration of peribronchiolar and alveolar regions associated with the appearance of several classes of proinflammatory cytokines (10, 28). These include TNF-α, IL-1, IL-6, and IL-8 (KC/GRO in the mouse). There is considerable experimental evidence from rodents that demonstrates that these classes of cytokines, and in particular TNF-α and IL-8 (or KC/GRO), play central roles in the recruitment and activation of inflammatory cell populations in the lung.

It is likely with a virus as complex as AmEPV that unintended inflammation will result when the virus is introduced in vivo. However, vertebrate poxviruses may serve as a source of genes to provide a solution to this problem. There have been a variety of vertebrate poxvirus-encoded

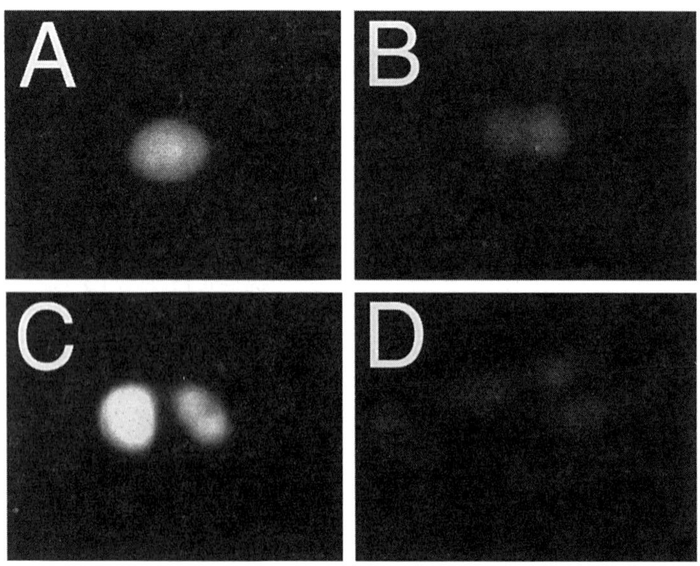

Figure 4. The survival of mammalian cells following infection by recombinant AmEPV TK*esp-gfp*. Subconfluent CV-1 cells were infected with AmEPV TK*esp-gfp* at an moi of 1 pfu/cell. The individual fluorescent cells were located and followed over a period of two to three days and periodically photographed with a fluorescent microscope. One fluorescent cell, identified 18 hours after infection (A), had divided into two cells by 24 (B) to 26 (C) hours after infection and by 50 hours had become a small cluster of dividing cells (D).

secreted virokines and viroceptors described including IFN-α/β, IFN-γ, TNF, and IL-1, as well as chemokine receptors (3, 37,38). Should inflammation be observed, cloning any or all of these genes into AmEPV to control any AmEPV-induced inflammation is quite plausible.

THE USE OF AmEPV TO STABLY TRANSFORM MAMMALIAN CELLS

Gene therapy is an attractive concept that offers the possibility of applications designed to treat a number of human diseases. Recently, a number of virus-based vectors have been investigated as a means of delivery of genes of therapeutic value. Each of the intensively studied major candidate virus vectors has been shown to possess distinct advantages and disadvantages. For example, retroviruses are able to mediate stable transduction of a gene of interest into cells, but generally have a limited host range, require actively dividing cells, and result in random integration (36). Ad vectors infect cells efficiently and express large amounts of gene products, but transfer genes in an unstable manner and, as discussed previously, elicit an unintended immune response (7). AAV is not pathogenic and preferentially integrates its DNA into a defined region of the human genome (19,33). However, AAV vectors suffer from a limited packaging capacity of approximately 4.7 kb for foreign insert DNA. Herpesvirus vectors, which have a large capacity for foreign DNA, and lentivirus vectors, which infect nondividing cells, are also under development.

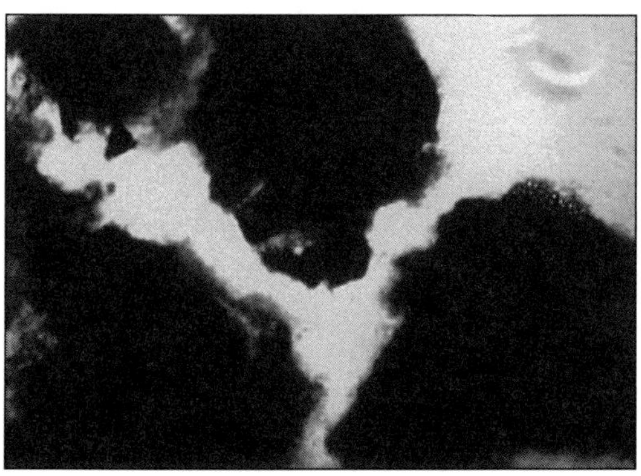

Figure 5. AmEPV-mediated β-gal expression in the muscle of mouse. 2×10^6 pfu (100 μL) of recombinant AmEPV-*esplacZ* were injected into the muscle of the hind leg of a mouse. As a control, mice were injected with the same amount of recombinant AmEPV-*SPHlacZ* or 100 μL PBS. Two days later, the mice were sacrificed, the muscle was excised from the injected area, cut into small pieces, fixed with 3% formaldehyde for 30 minutes, and then stained with X-gal. The muscle injected with recombinant AmEPV pTK-*esplacZ* showed β-gal expression. No β-gal expression was observed in control mice (data not shown).

EPVs have virtually no size constraints for foreign gene insertion(s). We show here the feasibility of using AmEPV to permanently transform cells leading to long-term expression of foreign genes. The fact that vertebrate cells could survive AmEPV infection implied that a recombinant AmEPV could be used to deliver foreign genes of therapeutic value to vertebrate cells, provided the viral DNA from the abortive infection could enter the nucleus and recombine with chromosomal DNA. It has been shown earlier that vaccinia DNA can transform cells and be stably incorporated into the nucleus of the cell (32).

Considerations in the Design of AmEPV Vectors for Stable Transformation

Poxviruses are cytoplasmic, and therefore poxvirus promoters are recognized only by the poxvirus-encoded RNA polymerase and not by RNA polymerase II of the host cell. Since we wished to demonstrate the ability of AmEPV to mediate stable transformation of mammalian cells, we constructed a recombinant AmEPV containing a cassette in which *gfp* and neo^R were cloned under the control of promoters recognized by the *cellular* (not poxvirus) RNA polymerase (Figure 6). The promoters chosen were the CMV immediate early and herpesvirus *TK* gene promoters, respectively, and the cassette was flanked by the ITR sequences of AAV.

Although the exact mechanism of AAV site-specific integration is unknown, two required viral components have been identified (35). These are the ITR sequences of AAV DNA and the Rep 78/68 proteins. The ITRs comprise two 145-nucleotide elements located at either end of the AAV genome. ITR sequences enclosing marker genes have been shown to allow a lower level of random genome integration when compared with the levels of specific integration observed when genes encoding the Rep 78/68 proteins are also included in constructs (2).

Selection of Stable Recombinants

When vertebrate cells are transfected with plasmid pTK-UF5, and selected with G418 (250 μg/mL) 24 hours later, colonies of resistant, fluorescent vertebrate cells are observed after about 10 days of selection. In contrast, colonies from cells infected with AmEPVpTK-UF5 at an moi of 5 were G418-resistant, but not initially GFP-positive. Approximately 20 G418-resistant clones can be isolated following infection of 1×10^6 cells (one 35-mm dish). The efficiency of transformation is only five- to tenfold less than that for optimized plasmid-mediated transfections. These AmEPV-derived colonies were observed to initially grow more slowly in the presence of G-418 than those produced from a plasmid transfection. Most likely, the explanation for these results is that the *gfp* and *neo*^{R} gene copy number in AmEPV-derived colonies is less than that for clones derived from cells transfected with the pTKUF5 plasmid. This explanation is further supported by the observation that the AmEPV-derived colonies gradually became more resistant to G418 and that GFP expression, while delayed, is ultimately observed. Fluorescent cells became both more numerous and brighter, consistent with a gradual increase in GFP concentration with time. Ultimately, all cells in each clonal isolate were GFP-positive, as shown in Figure 6.

We are currently characterizing these cell lines, which have been grown reliably for multiple generations. Our unpublished data suggests random integration of the marker cassette into the cellular genome. Inclusion of the AAV *rep* gene in future constructs is anticipated to provide directed insertion into chromosome 19 (33).

While similar observations have been noted for baculoviruses, it must be remembered that these viruses replicate in the nucleus of the host cell. Recently, baculoviruses containing AAV DNA elements (the ITR regions) (6) and expressing the AAV *rep* gene have been shown to yield site-specific integration and transformation (30). Despite the rather divergent developmental

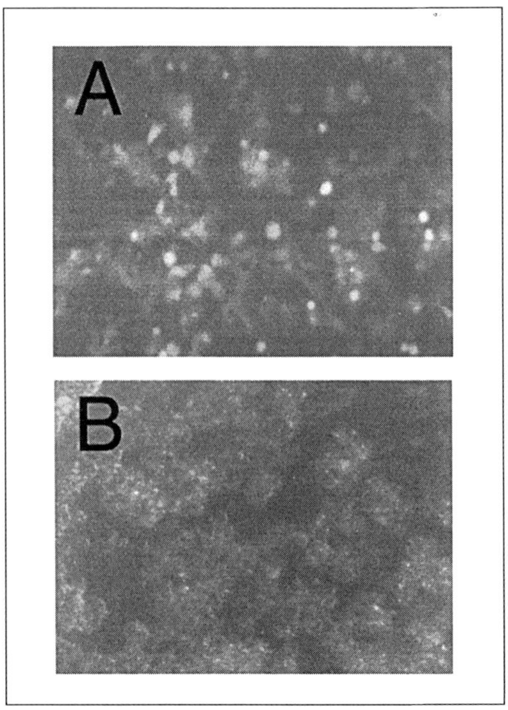

Figure 6. Transformed 293 cells derived from infection with recombinant AmEPV-TKUF5. (A) GFP-positive, G418-resistant cells. (B) Nonfluorescent, nontransformed control 293 cells. (**See color plate A5.**)

strategies for the two viruses, the number of transformants that arise through AmEPV or baculovirus infections are similar.

FUTURE POTENTIAL USES OF AmEPV VECTORS

There are several rather obvious potential applications of AmEPV vectors as well as accompanying difficulties to consider. The first application takes advantage of the ability of the virus to rapidly synthesize proteins regulated by early promoters for a short period of time within vertebrate cells, after which the infection aborts and cell growth/function continues as before. As one specific example, it is possible to imagine administration of AmEPV recombinant viruses via aerosols to relieve acute, episodic pulmonary inflammation when topical, temporary, rapid synthesis of anti-inflammatory proteins is required. However, the use of AmEPV for these purposes will require a thorough evaluation of the inflammation induced by the vector itself and testing of various recombinant proteins that might be incorporated into the vector to negate the unintended inflammation.

Another use of AmEPV vectors would be as a means to generate stable transformants, or to serve as a launchpad for other vectors (e.g., AAV or retroviruses). First, using AAV as an example, as mentioned above there is the potential advantage of using protein and nucleic acid elements from AAV flanked by their own ITRs to target integration of a given gene specifically into chromosome 19. An extension is a launchpad vector, in which AmEPV would be used to generate and/or deliver a second viral vector. Another example would be AAV, which depends on in vitro packaging systems to produce a suitable vector. While excellent progress has been made with such packaging systems, the procedure is not straightforward. It should be possible to clone portions, or an entire AAV recombinant, into AmEPV under the control of early AAV promoters together with a gene of interest flanked by AAV ITRs. While the gene or cassette of interest should be flanked by the AAV ITRs, the AAV-encoded genes (such as *rep*), under control of early AmEPV promoters, need not be contiguously cloned. This would allow for maximum flexibility in genetic manipulations. Infection with AmEPV would then generate all the elements *in trans* needed for either AAV formation or AAV-mediated transduction without the need for in vitro packaging.

From the data presented in this chapter, we have shown AmEPV to be a legitimate and feasible gene therapy agent. The two key findings that allow this are *(i)* mammalian cells survive AmEPV infection following expression of early genes, and *(ii)* vertebrate cells can be transformed in a stable manner by AmEPV, a property that directly stems from the cells remaining alive despite AmEPV infection.

ACKNOWLEDGMENTS

Work carried out in this chapter was supported by USDA grant no. 97-35302-4431 and NSF grant IBN 9406471.

REFERENCES

1. **Afonso, C.L., E.R. Tulman, Z. Lu, E. Oma, G.F. Kutish, and D.L. Rock.** 1999. The genome of *Melanoplus sanguinipes* Entomopoxvirus. J. Virol. *73*:533-552.
2. **Balague, C., M. Kalla, and W.-W. Zhang.** 1997. Adeno-associated virus Rep78 protein and terminal repeats enhance integration of DNA sequences into the cellular genome. J. Virol. *71*:3299-3306.
3. **Barry, M. and G. McFadden.** 1997. Virus encoded cytokines and cytokine receptors. Parasitology *115*:S89-100.
4. **Carbonell, L.F., M.J. Klowden, and L.K. Miller.** 1985. Baculovirus-mediated expression of bacterial genes in dipteran and mammalian cells. J. Virol. *56*:153-160.

5. **Carbonell, L.F. and L.K. Miller.** 1987. Baculovirus interaction with nontarget organisms: a virus-borne reporter gene is not expressed in two mammalian cell lines. Appl. Environ. Microbiol. *53*:1412-1417.

6. **Condreay, J.P., S.M. Witherspoon, W.C. Clay, and T.A. Kost.** 1999. Transient and stable gene expression in mammalian cells transduced with a recombinant baculovirus vector. Proc. Natl. Acad. Sci. USA *96*:127-132.

7. **Engelhardt, J.F., X. Ye, B. Doranz, and J.M. Wilson.** 1994. Ablation of E2A in recombinant adenoviruses improves transgene persistence and decreases inflammatory response in mouse liver. Proc. Natl. Acad. Sci. USA *91*:6196-6200.

8. **Fernon, C.A., A.P. Vera, R. Crnov, J. Lai-Fook, R.J. Osborne, and D.J. Dall.** 1995. Replication of *Heliothis armigera* entomopoxvirus *in vitro*. J. Invert. Pathol. *66*:216-223.

9. **Gauthier, L., F. Cousserans, J.C. Veyrunes, and M. Bergoin.** 1995. The *Melolontha melolontha* entomopoxvirus (MmEPV) fusolin is related to the fusolins of lepidopteran EPVs and to the 37K baculovirus glycoprotein. Virology *208*:427-436.

10. **Ginsberg, H.S., L.L. Moldawer, P.B. Sehgal, M. Redington, P.L. Kilian, R.M. Chanock, and G.A. Prince.** 1991. A mouse model for investigating the molecular pathogenesis of adenovirus pneumonia. Proc. Natl. Acad. Sci. USA *88*:1651-1655.

11. **Granados, R.R.** 1981. Entomopoxvirus infection in insects, p. 101-126. *In* E.W. Davidson (Ed.), Pathogenesis of Invertebrate Microbial Diseases. Allanheld, Osmun & Co., Totowa, NJ.

12. **Gruidl, M.E., R.L. Hall, and R.W. Moyer.** 1992. Mapping and molecular characterization of a functional thymidine kinase from *Amsacta moorei* entomopoxvirus. Virology *186*:507-516.

13. **Hall, R.L., Y. Li, J.A. Feller, and R.W. Moyer.** 1996. The *Amsacta moorei* entomopoxvirus spheroidin gene is improperly transcribed in vertebrate poxviruses. Virology *224*:427-436.

14. **Hall, R.L. and R.W. Moyer.** 1991. Identification, cloning, and sequencing of a fragment of *Amsacta moorei* entomopoxvirus DNA containing the spheroidin gene and three vaccinia virus-related open reading frames. J. Virol. *65*:6516-6527.

15. **Hall, R.L. and R.W. Moyer.** 1993. Identification of an *Amsacta* spheroidin-like protein within the occlusion bodies of *Choristoneura* entomopoxviruses. Virology *192*:179-187.

16. **Harlow, E. and D. Lane.** 1998. Laboratory Antibodies—A Laboratory Manual, p. 396-400. CSH Press, Cold Spring Harbor, NY.

17. **Holowczak, J.A.** 1972. Uncoating of poxviruses. I. Detection and characterization of subviral particles in the uncoating process. Virology *50*:216-232.

18. **Klein, R.L., E.M. Meyer, A.L. Peel, S. Zolotukhin, C. Meyers, N. Muzyczka, and M.A. King.** 1998. Neuron-specific transduction in the rat septohippocampal or nigrostriatal pathway by recombinant adeno-associated virus vectors. Exp. Neurol. *150*:183-194.

19. **Kotin, R.M., J.C. Menninger, D.C. Ward, and K.I. Berns.** 1991. Mapping and direct visualization of a region-specific viral DNA integration site on chromosome 19q13-qter. Genomics *10*:831-834.

20. **Langridge, W.H.** 1983. Detection of *Amsacta moorei* entomopoxvirus and vaccinia virus proteins in cell cultures restrictive for poxvirus multiplication. J. Invert. Pathol. *42*:77-82.

21. **Li, Y., R.L. Hall, and R.W. Moyer.** 1997. Transient, nonlethal expression of genes in vertebrate cells by recombinant entomopoxviruses. J. Virol. *71*:9557-9562.

22. **Li, Y., R.L. Hall, S. Yuan, and R.W. Moyer.** 1998. High-level expression of *Amsacta moorei* entomopoxvirus *spheroidin* depends on sequences within the gene. J. Gen. Virol. *79*:613-622.

23. **Li, Y., S. Yuan, and R.W. Moyer.** 1998. The non-permissive infection of insect (gypsy moth) LD-652 cells by vaccinia virus. Virology *248*:74-82.

24. **Lytvyn, V., Y. Fortin, M. Banville, B. Arif, and C. Richardson.** 1992. Comparison of the thymidine kinase genes from three entomopoxviruses. J. Gen. Virol. *73*:3235-3240.

25. **Morris, T.D. and L.K. Miller.** 1992. Promoter influence on baculovirus-mediated gene expression in permissive and nonpermissive insect cell lines. J. Virol. *66*:7397-7405.

26. **Morris, T.D. and L.K. Miller.** 1993. Characterization of productive and non-productive AcMNPV infection in selected insect cell lines. Virology *197*:339-348.

27. **Moyer, R.W.** 1994. Entomopoxviruses, p. 392-397. *In* R.G. Webster and A. Granoff (Eds.), Encyclopedia of Virology. Academic Press Ltd., London.

28. **Noah, T.L., I.A. Wortman, P.C. Hu, M.W. Leigh, and R.C. Boucher.** 1996. Cytokine production by cultured human bronchial epithelial cells infected with a replication-deficient adenoviral gene transfer vector or wild-type adenovirus type 5. Am. J. Respir. Cell Mol. Biol. *14*:417-424.

29. **Palmer, C.P., D.P. Miller, S.A. Marlow, L.E. Wilson, A.M. Lawrie, and L.A. King.** 1995. Genetic modification of an entomopoxvirus: deletion of the spheroidin gene does not affect virus replication *in vitro*. J. Gen. Virol. *76*:15-23.

30. **Palombo, F., A. Monciotti, A. Recchia, R. Cortese, G. Ciliberto, and N. La Monica.** 1998. Site-specific integration in mammalian cells mediated by a new hybrid baculovirus adeno-associated virus vector. J. Virol. *72*:5025-5034.

31. **Pedley, C.B. and R.J. Cooper.** 1987. The assay, purification and properties of vaccinia virus-induced uncoating protein. J. Gen. Virol. *68*:1021-1028.

32. **Pellicer, A. and M. Esteban.** 1982. Gene transfer, stability, and biochemical properties of animal cells transformed with vaccinia DNA. Virology *122*:363-380.

33. **Samulski, R.J., X. Zhu, X. Xiao, J.D. Brook, D.E. Housman, N. Epstein, and L.A. Hunter.** 1991. Targeted integration of adeno-associated virus (AAV) into human chromosome 19. EMBO J. *10*:3941-3950.

34. **Sarov, I. and W.K. Joklik.** 1972. Characterization of intermediates in the uncoating of vaccinia virus DNA. Virology *50*:593-602.

35. **Shelling, A.N. and M. Smith.** 1994. Targeted inte-

gration of transfected and infected adeno-associated virus vectors containing the neomycin resistance gene. Gene Ther. *1*:165-169.
36. **Smith, A.E.** 1995. Viral vectors in gene therapy. Annu. Rev. Microbiol. *49*:807-838.
37. **Smith, G.L., J.A. Symons, A. Khanna, A. Vanderplasschen, and A. Alcami.** 1997. Vaccinia virus immune evasion. Immunol. Rev. *159*:137-154.
38. **Turner, P.C. and R.W. Moyer.** 1998. Control of apoptosis by poxviruses. Semin. Virol. *8*:453-469.
39. **Winter, J., R.L. Hall, and R.W. Moyer.** 1995. The effect of inhibitors on the growth of the entomopoxvirus from *Amsacta moorei* in *Lymantria dispar* (gypsy moth) cells. Virology *211*:462-473.
40. **Zaslavsky, V.** 1985. Uncoating of vaccinia virus. J. Virol. *55*:352-356.

FAMILY: Baculoviridae
GENUS: Nucleopolyhedrovirus (Baculovirus)

22

High-Efficiency Transduction of Mammalian Cells Using Baculovirus: Practical Aspects

Frederick M. Boyce and Elizabeth A. Franco
Massachusetts General Hospital East, Charlestown, MA, USA

INTRODUCTION

Baculoviral vectors are commonly used for expression of proteins in insect cells (8,10). However, it has recently been shown that baculoviruses can also be used to deliver genes to animal cells (1–4,7,9). Because baculoviruses are relatively easy to grow and manipulate, are able to accommodate large inserts (>20 kb), are able to grow to high titers (>1 × 10^8 plaque-forming units [pfu]/mL), and can be concentrated to even higher titers (>1 × 10^{10} pfu/mL), they are an attractive vector for gene delivery in vitro. Furthermore, because there are no vertebrate homologs of baculoviruses, there is little chance of recombination with endogenous viruses after transduction of mammalian cells. Since the majority of baculoviral genes do not contain TATA boxes in their promoters, they are not transcribed by mammalian RNA polymerase II after transduction of mammalian cells. Thus, although baculoviral vectors are able to transduce mammalian cells, they are not able to transcribe their genomes efficiently or to replicate in mammalian cells. Although not all mammalian cells are susceptible to infection with baculoviruses, many commonly used cell lines, such as 293 cells, and even primary cultures of some cell types may be transduced efficiently. Taken together, these properties of baculoviruses suggest that they may be useful tools for genetic manipulation of mammalian cells for research purposes, and for efficient expression of large inserts or of toxic genes that cannot be easily propagated in conventional animal viral vectors. One growing use of baculoviruses is as a method of producing animal viral vectors for gene therapy experiments. Thus, the methods described in this chapter may be useful for readers working with the other vector systems described in this book.

To allow expression of an exogenous gene in mammalian cells after baculoviral transduction, a promoter active in mammalian cells must be properly positioned upstream of the gene (2,4). In addition, mammalian splice and polyadenylation sequences are usually added downstream to allow efficient mRNA formation. Conventional baculoviral vectors used for production in insect cells usually contain the viral

Viral Vectors: Basic Science and Gene Therapy
Edited by A. Cid-Arregui and A. García-Carrancá

polyhedrin promoter, which is not active in mammalian cells (2). Thus, the vectors commonly used for insect cell expression cannot be used for expression in mammalian cells without modification. However, there is at least one company (Novagen, Madison, WI, USA) that now sells baculoviral vectors designed for expression in mammalian cells.

The most commonly used method of creating recombinant baculoviruses involves cloning the gene to be expressed into a transfer vector that contains the proper transcriptional elements flanked by sequences derived from the baculoviral genome. Usually, these sequences derive from the polyhedrin region of the genome and allow homologous recombination into the polyhedrin gene after co-transfection with viral DNA that has been linearized at the polyhedrin locus. This procedure normally results in greater than 95% recombinant viruses in the transfection supernatant. These viruses can be purified by sequential plaque isolations or can be used as an impure preparation. An alternative method of recombination into the baculoviral genome during propagation as a bacterial artificial chromosome-based plasmid (bacmid) has been described that may obviate the need for extensive plaque purification (5).

In this chapter, we describe the methods used in our laboratory to create recombinant baculoviruses and their subsequent use for transduction of mammalian cells. These methods are modifications of standard baculoviral techniques, and we are grateful for the efforts of the baculovirologists who have developed them. This chapter is written as a starting point for nonbaculovirologists who may wish to use baculoviral vectors to transduce mammalian cells. We have attempted to emphasize modifications to standard procedures that we have found to be useful or that are not well described in other publications. We suggest that the reader also consult more comprehensive baculoviral methods books that may provide much more detail. One book that we have found particularly useful also has an excellent review of baculoviral biology (6).

PROTOCOLS

Media and Cells Used for Propagation of Baculoviruses

The insect cell lines used for propagation of the baculovirus, AcNPV, are derived from *Spodoptera frugiperda*, the fall armyworm. We have found that the highest titers are obtained by growth of the virus in Sf9 cells, while Sf21 cells yield better results when used for plaque assays. Both cell lines are grown at 27°C in low-temperature incubators. Because the media used to grow the cells are buffered by phosphate rather than CO_2, a CO_2 incubator is not necessary. The cells can be grown in regular plastic tissue culture flasks, but for convenience and scale-up are often propagated in spinner flasks.

A large variety of specialized media have been developed for the propagation of insect cells and for baculovirus expression. The most commonly used medium, which we refer to as "supplemented Grace's," consists of Grace's medium supplemented with lactalbumin hydrosylate (3300 mg/L) and yeastolate (3300 mg/L). We routinely propagate our cells in this medium in the presence of 10% fetal bovine serum (FBS) that has been heat-inactivated by treatment at 55°C for 30 minutes. Heat inactivation is used to destroy serum complement, which severely inhibits baculovirus growth. Additional components added to the complete medium include a surfactant, Pluronic F-68 (final concentration of 0.1%), which is necessary for cells grown in spinner culture, and antibiotics. For antibiotics, penicillin/streptomycin is added at half the normal dose used for mammalian cells (final concentration of 50 U/mL penicillin and 50 μg/mL streptomycin). For diluting virus or cells, Grace's medium lacking supplements or serum can be used for economy. The virus can also be grown in cells adapted to serum-free medium, such as Sf-900-II (Life Technologies, Rockville, MD, USA), although we do not routinely do so.

Generation of Recombinant Baculovirus by Transfection of Insect Cells

Materials

- Sf9 cells
- Recombinant baculoviral transfer vector containing gene of interest
- Linear baculoviral DNA
- Lipofectin® (Life Technologies)
- Sterile water
- Polystyrene tubes (Falcon #2054)
- 25-cm^2 tissue culture flasks
- Grace's medium, unsupplemented and serum-free Grace's medium, supplemented and containing 10% fetal calf serum (heat-inactivated)
- Tabletop centrifuge
- Low-temperature incubator (27°C)

Procedure

1. Seed one 25-cm^2 flask with 3×10^6 Sf9 cells for each transfection. This is done by counting the cells with a hemacytometer and adding the appropriate amount of cells to a flask containing 5 mL serum-free medium. Allow to attach 1 hour on a level surface. We use a plexiglas leveling table on the benchtop for this purpose.

2. Meanwhile, prepare DNA/Lipofectin® complexes. First, a solution containing the transfer vector and linear viral DNA is prepared. Dilute 2 μg transfer vector DNA and 0.5 μg linear baculoviral DNA into a total volume of 40 μL in sterile water in a sterile 5-mL polystyrene tube. Pipet the linear viral DNA slowly to avoid shearing. Next, prepare the diluted Lipofectin in a separate tube. Mix 15 μL Lipofectin with 30 μL water in a sterile 5-mL polystyrene tube. If multiple transfections are to be performed, the amounts should be scaled up accordingly. Finally, add 40 μL diluted Lipofectin to each DNA mixture. Mix gently and let stand at room temperature for 15 minutes.

3. Gently rinse the flask of cells twice, 5 mL each, with serum-free Grace's medium. Add 2 mL serum-free medium to the flask.

4. Add DNA/Lipofectin mixture (80 μL) to the flask of cells. Mix gently and return to incubator for 4 to 8 hours.

5. Remove DNA/Lipofectin medium from the cells by gentle aspiration and refeed cells with 5 mL complete medium; return to incubator.

6. Harvest virus 48 hours after transfection by gently removing the medium containing the virus and placing it into a 15-mL polystyrene centrifuge tube. Spin out any contaminating cells by centrifugation at 2000 rpm for 5 minutes in a tabletop centrifuge and remove the viral supernatant to a fresh tube. Store this 48-hour supernatant at 4°C until it is used for plaque purification. Refeed the flask with 5 mL complete medium and let infection continue for an additional five days. Harvest this seven-day supernatant similarly and save for a backup in case sufficient viral titer is not found in the 48-hour supernatant.

Notes

Although some manufacturers recommend performing controls by transfection with control transfer vectors and/or linear viral DNA alone, we have not found it necessary to do so routinely. In addition, while some manufacturers recommend using greater amounts of the somewhat expensive linear viral DNA, we have not found any disadvantage to using lesser amounts. We use Lipofectin because it is more economical than some other cationic lipids, but other cationic lipid transfection reagents, such as those packaged in some linear viral DNA kits, may be substituted.

Plaque Purification

Materials

- Water bath equilibrated to 42°C
- Sf21 cells
- 10-cm, tissue culture-grade petri dishes
- 5-mL polystyrene tubes (Falcon #2054)
- 50-mL polypropylene tubes
- 3% (w/v) SeaPlaque™ agarose (FMC Bioproducts, Rockland, ME, USA) in water, made sterile by autoclaving
- Grace's medium (2×), supplemented and containing 20% FBS and antibiotics
- Grace's medium, unsupplemented
- Plastic box for humidity

Procedure

1. For each virus to be plaque-purified, prepare four petri dishes so that a range of viral dilutions may be plated. Each petri dish is first filled with 10 mL serum-free medium so that the dish is wetted and cells will attach evenly. Next, determine the density of the cell culture by counting a portion on a hemocytometer. If there are many clumps of cells, allow these to settle out by gravity in a 50-mL tube for 5 minutes, remove the upper portion containing single cells, and recount. Seed exactly 4×10^6 Sf21 cells per 10-cm petri dish by adding the appropriate volume of cells to each dish. Getting the right density of cells is critical to forming a good monolayer with visible plaques. Too many cells will give plaques that are very small, while too few cells will allow gaps in the monolayer, which may be confused with plaques. Rock the dishes to distribute the cells (do not swirl). Allow to attach at room temperature on a level surface for 1 hour. We use a plexiglas leveling table on the benchtop for this purpose.

2. Perform serial tenfold dilutions of virus in Grace's unsupplemented medium in polystyrene tubes. Use 100 µL 48-hour supernatant virus in the first tube and dilute so that subsequent tubes contain 10, 1, and 0.1 µL of virus, with a volume of 3 mL for each tube. (During the secondary or tertiary rounds of purification, use 100, 10, 1, and 0.1 µL of plaque eluate in a total of 3 mL Grace's unsupplemented medium.)

3. Gently remove medium from petri dishes. Add diluted virus to dish (3 mL/dish) and allow infection for 1 hour at room temperature on a level surface with gentle rocking every 15 minutes.

4. Meanwhile, melt the 3% agarose in a microwave. Since it is important that the agarose is thoroughly dissolved, use ample swirling and reheating. Pipet 25 mL agarose solution into a 50-mL polypropylene tube. Place the tube in a water bath equilibrated to 42°C. Similarly, pipet 25 mL complete 2× medium into a 50-mL tube and equilibrate at 42°C.

5. After 1 hour of infection, carefully aspirate the medium from each petri dish. Because it is important to remove all the liquid, tilt each dish and allow to drain briefly before aspiration. Working quickly, remove agarose and 2× medium from water bath, mix using a 25-mL pipet, and overlay each dish with 12.5 mL agarose/medium mixture. Set mixture aside in the tissue culture hood with the lid cracked open for a few minutes, then cover, and leave undisturbed until agarose solidifies (about 20 minutes). Do not let the plates dry too much in the hood, or a proper monolayer will not form. Wrap plates tightly with Parafilm and invert so that the agarose layer is on top. Put plates in a humidified box in the incubator. The humidified box can be prepared by lining a plastic shoebox with blotting paper

dampened with water and covering the floor of the box with plastic wrap or Parafilm. We add a small amount of water bath antifungal solution to inhibit growth of mold in the box. The box is placed in the 27°C incubator and left undisturbed for four to five days.

6. Plaques should become visible after four to five days, but may take up to seven days. Plaques are visualized as holes in the monolayer and are most easily seen by looking up through the dish toward a fluorescent ceiling light. Under microscopic observation, plaques will have a central area of cell death with surrounding areas of swollen, well-delineated cells. Mark each plaque by drawing a circle around it on the dish. Pick plaques from the plates containing the highest dilutions (lowest numbers of plaques) so that contamination from other plaques is minimized. Plaques are picked by aspiration of the agarose overlaying the marked plaque using a sterile Pasteur pipet. The agarose plug is then deposited into 1 mL sterile serum-free medium in a cryovial. Use a separate pipet for each plaque. Allow the virus to diffuse from the agarose for 1 hour or longer before proceeding to the next round of purification or amplification. We usually pick ten plaques during the first round of plaque purification and after screening proceed with purification of one or two positive viral clones.

Notes

Most protocols employ smaller (35-mm or 60-mm) tissue culture dishes. However, we have found that there is an edge effect that occurs around the periphery of the dish that may result in monolayer loss or hard-to-visualize plaques. The use of larger dishes minimizes this effect. To obtain an even monolayer of cells, it is critical to have sufficient volume in the dishes during plating. Hence, do not use less than 10 mL medium during plating. If the agarose layer is allowed to dry too much, the cells will take on a pancake-like appearance and will not form a proper monolayer. To be certain that the final viral stock is pure, we usually perform three rounds of plaque purification for each virus before amplification.

Variation: Titering Virus by Plaque Assay

The viral concentration, or titer, can be determined by quantitative plaque assay of a known volume of virus stock. This is done similarly to that described above for purification, except that it is usually necessary to dilute the virus more extensively. For titering purposes, the plaques may be more easily visualized by staining with MTT (Thiazolyl blue; Sigma, St. Louis, MO, USA) after they have formed five days after infection. This is done by overlaying a second agarose layer composed of 4 mL 3% agarose, 4 mL 2× medium (serum-free, unsupplemented is fine), and 0.4 mL MTT (10 mg/mL in water) per plate. The MTT overlay is prepared by mixing solutions prewarmed to 42°C as done for the original overlay. After the MTT overlay has solidified, the plates are returned to the incubator briefly. After 1 hour, plaques are visible as clear holes in the cellular monolayer stained red by the MTT.

Screening of Recombinant Plaques by Polymerase Chain Reaction (PCR)

Materials

- 96-well tissue culture plate (Falcon # 3072)
- Sf9 cells
- Grace's medium with supplements and serum
- 0.5 M NaOH
- 5 mol/L ammonium acetate
- 0.1× Tris EDTA (TE) (1 mmol/L Tris [pH 7.5] and 0.1 mmol/L EDTA)
- PCR reagents and thermocycler

Procedure

1. Allow virus to diffuse from the agarose plug for at least 1 hour.

2. Seed a 96-well dish with 1×10^4 Sf9 cells/well. This is most easily done by diluting cells to 1×10^5 cells/mL and then seeding 100 µL/well using the eight-channel pipettor. Seed enough wells for all of the plaques to be screened, at least one positive and negative control well, and an uninfected well. Allow cells to attach for 30 to 60 minutes.

3. Carefully aspirate the medium from each well using a multichannel pipettor. Add 100 µL each primary plaque eluate to each well. Include an eluate from an unrelated virus and uninfected as well as negative controls. If possible, add a positive control virus. (The 48-hour supernatant can be used if no other positive control is available.) Allow the virus to adsorb for 1 to 2 hours in the 27°C incubator.

4. Remove the virus from the wells using a multichannel pipettor, being careful that there is no contamination between wells. Refeed the wells with 100 µL complete medium using the multichannel pipettor. Carefully wrap the plate with Parafilm and place in a humidified box in the incubator. (Be very careful not to move the plate suddenly to avoid cross-contamination.) Allow infection to proceed for 48 to 72 hours, but harvest before widespread cell death occurs.

5. When cells are all infected (usually about 72 hours after infection), aspirate the medium using a multichannel pipetter. Sterile technique does not need to be used from this point on, but be very careful to avoid cross-contamination. Lyse the cells by addition of 200 µL 0.5 mol/L NaOH. Allow lysis to occur for 1 minute.

6. Using the multichannel pipettor, neutralize the lysis by addition of 40 µL 5 mol/L ammonium acetate. Mix the lysate gently using the tips of the pipettor. (The lysate may still be somewhat basic at this point.) Store lysate at -20°C if not used immediately for PCR.

7. Dilute 5 µL lysate in 95 µL 0.1× TE in 96-well plates.

8. Set up PCR in a 96-well PCR plate using 5 µL diluted lysate in a final reaction volume of 15 µL. Performing PCR for 30 cycles will usually give strong signals in positives and weak or no background signals in negative controls.

Notes

Ideally, primary screening should be done by transducing mammalian cells with an amplified (P1) stock of virus and assaying for function (e.g., Western blot analysis for expression of the desired protein). However, we include this protocol here because it can be used for any virus constructed and can be used as a preliminary screening procedure. In our experience, PCR screening gives a significant number (perhaps 30%) of false negatives. Usually, there are sufficient numbers of positive viral clones so that this is not a problem. If the number of cycles is increased beyond what is recommended, it is possible to obtain false positives in the screen. Therefore, it is important to include both negative and positive controls in each set of reactions.

Amplification of Baculovirus Stocks

Once the virus stock is plaque-purified, several rounds of amplifications are performed to give large volumes of high-titer viral stock. These successive amplifications are done using a low (<1.0) multiplicity of infection (moi) so that defective viruses will not be propagated.

P1 Amplification

Seed a 25-cm^2 tissue culture flask with 2.5×10^6 Sf9 cells. Remove medium and refeed with 5 mL complete medium containing serum. Add 100 μL viral eluate. The P1 virus can be harvested when widespread cell death occurs, usually about seven days later. The viral supernatant is separated from cellular debris by low-speed centrifugation (e.g., 2000 rpm for 5 minutes in a tabletop centrifuge) and stored at 4°C. We always store 1 mL P1 amp of each virus at -80°C as backup protection against loss of the stock in the future.

P2 Amplification

Grow a 100-mL spinner flask containing 50 mL Sf9 cells to a density of 1×10^6 cells/mL. Add 0.25 mL P1 virus stock. Allow infection to proceed for seven days, harvest virus supernatant by centrifugation, and store in a sterile 50-mL polypropylene tube at 4°C.

Concentration and Purification of Baculovirus

Materials

- Sf9 Cells and medium
- Spinner flask (500 mL or larger)
- 25 × 89-mm SW28 Ultra-Clear™ tubes (Beckman Coulter, Fullerton, CA, USA)
- SW28 rotor and ultracentrifuge
- 2× phosphate-buffered saline (PBS)
- 27% (w/v) sucrose in PBS
- 20% (w/v) sucrose in 2× PBS
- 60% (w/v) sucrose in 2× PBS

Procedure

1. Infect 200 mL Sf9 cells at a density of 1 to 1.5×10^6 cells/mL with virus at an moi of 1. Typically, this is approximately 2 mL virus stock at a titer of 1×10^8 pfu/mL for 200 mL cells at a density of 1×10^6 cells/mL). Allow infection to progress in suspension culture for five to seven days until most cells appear very sick or dead.

2. Harvest virus supernatant by centrifugation at 2000 rpm for 5 minutes to pellet cellular debris. The virus may be stored at 4°C in the dark until pelleting.

3. Load SW28 tubes with 33 mL cleared viral supernatant per tube with six tubes total. Underlay virus with 3 mL 27% (w/v) sucrose in PBS. Make sure tubes are balanced ±10 mg by redistributing virus or adding PBS.

4. Pellet virus at 24 000 rpm for 75 minutes. Aspirate off the medium and sucrose supernatant. The pelleted virus is visualized as a dime-sized, thin, and white-to-brown pellet.

5. Add 0.3 mL/tube of 2× PBS/tube. Cover with Parafilm and allow to resuspend overnight at 4°C. Pool virus into a 15-mL polystyrene (clear) tube.

6. Prepare two sucrose gradients in SW28 Ultra-Clear tubes: First, fill each tube with precisely 18 mL 60% (w/v) sucrose in 2× PBS. Make sure sucrose solutions are at room temperature before using. Next, slowly overlay with 18 mL 20% sucrose in 2× PBS. The two solutions should be separate layers and should not intermix when they are layered. Tightly wrap the top of each tube with Parafilm. Place the tubes in a secure, well-fitting rack, and gently lower the tubes so that they are horizontal to allow diffusion of the sucrose. Let diffuse at room temperature for 2.5 hours, and then carefully raise them to the vertical position.

7. Centrifuge resuspended virus at 2000 rpm for 5 minutes to pellet any clumps. Carefully overlay each gradient tube with 0.9 to 1.5 mL resuspended virus. Weigh each tube to be certain it is balanced ±10 mg. Spin in SW28 rotor at 27 000 rpm for 2.5 hours at 4°C.

8. The virus forms a sharp opalescent band about 60% of the way down the tube. Using a pipet inserted from the top collect approximately 5 mL of the viral band.

9. Concentrate the purified, banded

virus now by repelleting. Place up to 10 mL sucrose-banded virus into a SW28 tube and fill the remainder of the tube to within 5 mm of the top using 2× PBS. Cover the tube tightly with Parafilm and mix well by inversion. Make a balance tube using the same amount of virus and PBS so that the tubes are balanced ±10 mg. Pellet virus at 24 000 rpm and aspirate supernatant. Resuspend the pellet in 0.2 mL 2× PBS/ tube overnight at 4°C. If desired, the virus may be sterilized by passage through a 0.45-μm filter although some loss of titer (up to 50%) may occur.

Notes

Concentrated baculovirus tends to clump. We have found that clumping is minimized by resuspension of the virus in 2× PBS rather than isotonic saline. The virus can also be resuspended by sonication, although this may cause some loss of titer. To do this, we use a Model 550 sonic dismembranator (Fisher Scientific, Pittsburgh, PA, USA) equipped with a cup horn. The virus (0.5–2.5 mL) is placed in a polystyrene tube and sonicated at setting 2, 50% pulse, for 2 minutes. Gentle mixing of the viral clumps should be visible during sonication. Since overly aggressive sonication may lead to loss in titer, reduce the power level if vigorous bubbling occurs. For virus that is to be stored for long periods, we incorporate antibiotics in the resuspension buffer and/or filter-sterilize the final preparation. Freezing of the virus leads to clumping.

Transduction of Cultured Mammalian Cells Using Baculovirus

Procedure

1. Twenty-four hours before transduction, trypsinize recipient cells (e.g., Huh7 or 293 cells) and seed 5×10^5 cells/60-mm petri dish. For HepG2 cells, use double this amount because they tend to spread less. Allow cells to attach and grow overnight.

2. Remove medium and refeed with 1 mL serum-free medium (e.g., Dulbecco's modified Eagle's medium [DMEM]). Add virus to the cells and rock gently to mix. For maximal transduction, an moi of 100 or more is used. For example, 10 μL of 1×10^{10} pfu/mL viral stock would be used to give an moi of 100. (This assumes the number of cells doubles overnight). Because small volumes of virus in 2× PBS are typically used, it is not necessary to dilute the virus stock to 1× PBS before use. Return the dish to the CO_2 incubator and rock gently every 15 minutes.

3. After one hour, add 4 mL complete medium containing serum and return the dish to the incubator. The virus solution does not need to be removed.

4. Harvest the cells 24 to 48 hours after transduction.

Notes

The optimal medium and conditions for transduction may vary among cell lines. In general, we have found that it is not necessary to remove the virus from the cells after infection because longer exposure to the virus may increase the transduction efficiency by up to twofold. It is also possible to transduce cells using unpurified viral supernatants. However, the moi may be limited by the volume of medium that can be added. We have found that most mammalian cell lines can be exposed to solutions containing up to 50% Grace's medium (and 50% normal mammalian medium, such as DMEM) during the hour of infection, although we usually keep the volume of virus added to 10% of the total volume, if possible.

REFERENCES

1. **Barsoum, J., R. Brown, M. McKee, and F.M. Boyce.** 1997. Efficient transduction of mammalian cells by a

recombinant baculovirus having the vesicular stomatitis virus G glycoprotein. Hum. Gene Ther. *8*:2011-2018.

2. **Boyce, F.M. and N.L.R. Bucher.** 1996. Baculovirus-mediated gene transfer into mammalian cells. Proc. Natl. Acad. Sci. USA *93*:2348-2352.
3. **Condreay, J.P., S.M. Witherspoon, W.C. Clay, and T.A. Kost.** 1999. Transient and stable gene expression in mammalian cells transduced with a recombinant baculovirus vector. Proc. Natl. Acad. Sci. USA *96*:127-132.
4. **Hofmann, C., V. Sandig, G. Jennings, M. Rudolph, P. Schlag, and M. Strauss.** 1995. Efficient gene transfer into human hepatocytes by baculovirus vectors. Proc. Natl. Acad. Sci. USA *92*:10099-10103.
5. **Luckow, V.A., S.C. Lee, G.F. Barry, and P.O. Olins.** 1993. Efficient generation of infectious recombinant baculoviruses by site-specific transposon-mediated insertion of foreign genes into a baculovirus genome propagated in *Escherichia coli.* J. Virol. *67*:4566-4579.
6. **O'Reilly, D.R., L.K. Miller, and V.A. Luckow.** 1992. Baculovirus Expression Vectors: A Laboratory Manual. W.H. Freeman and Company, New York.
7. **Palombo, F., A. Monciotti, A. Recchia, R. Cortese, G. Ciliberto, and N. La Monica.** 1998. Site-specific integration in mammalian cells mediated by a new hybrid baculovirus-adeno-associated virus vector. J. Virol. *72*:5025-5034.
8. **Pennock, G.D., C. Shoemaker, and L.K. Miller.** 1984. Strong and regulated expression of *Escherichia coli* β-galactosidase in insect cells with a baculovirus vector. Mol. Cell. Biol. *4*:399-406.
9. **Shoji, I., H. Aizaki, H. Tani, K. Ishii, T. Chiba, I. Saito, T. Miyama, and Y. Matsuura.** 1997. Efficient gene transfer into various mammalian cells, including non-hepatic cells, by baculovirus vectors. J. Gen. Virol. *78*:2657-2664.
10. **Smith, G.E., M.J. Fraser, and M.D. Summers.** 1983. Production of human β-interferon in insect cells infected with a baculovirus expression vector. Mol. Cell. Biol. *3*:2156-2165.

Part II: RNA Virus Vectors

Introduction to Part II: RNA Virus Vectors

Angel Cid-Arregui

This section (Chapters 23–35) describes transduction vectors and modified viruses generated from retroviruses (Chapters 23–31), influenza virus (Chapter 32), poliovirus (Chapter 33) and alphaviruses (Chapters 34 and 35). This introduction contains a taxonomic description of the parental RNA viruses (see also virus images on the web sites recommended at the end of this section).

Family: Retroviridae
Genus: Oncovirinae (MLV)
Genus: Lentivirinae (Human Immunodeficiency Virus [HIV])
Genus: Spumavirinae (Human and Simian Foamy Viruses [HFV, SFV])

Virion: The virion is spherical, enveloped, and 80–130 nm in diameter. It comprises a ribonucleoprotein core (nucleoid) enclosed in an outer envelope made of a lipid bilayer, with spikes (8 nm in diameter) of variable lengths projecting from the outer surface. According to their morphology, retroviruses are classified into four categories: (A) A-type particles occur as intracellular forms and are not infective; they are 60–90 nm in diameter and have and electron-lucent core surrounded by a double shell. They can be intracytoplasmic (precursor forms to B-type of mouse mammary tumor virus [MMTV]) or intracisternal (likely, prototypes to C-type or D-type particles). (B) B-type budding particles have a toroidal (doughnut-shaped) core 75 nm in diameter, and long spikes. Extracellular B-type particles (mature forms) are 120–130 nm in diameter, and have an electron-dense nucleoid located eccentrically within the enveloped particle. (C) The majority of retroviruses form particles of C-type. The main feature of this class is the absence of recognizable intracytoplasmic precursor forms. No viral structures are observed in the cytoplasm until budding commences at the plasma membrane in the form of an electron-dense structure that becomes the core of the particle. As virion formation proceeds, the core forms a sphere with an electron-lucent center while the plasma membrane surrounding it begins to pinch off. Extracellular virions (80–110 nm in diameter) contain centrally located electron-lucent or electron-dense cores (immature and mature particles, respectively). Spikes are less well defined than those of B-type. (D) D-type particles have two forms: intracellular (immature) particles are ring-shaped, 60–90 nm in diameter, most abundant near the plasma membrane; and extracellular (mature) particles, 100–120 nm in

Viral Vectors: Basic Science and Gene Therapy
Edited by A. Cid-Arregui and A. García-Carrancá

diameter, containing a central electron-dense core and shorter spikes than B-type.

Virion proteins include nucleoprotein bound to the RNA genome, reverse transcriptase, protease and integrase enzymes, capsid and matrix proteins, and two envelope proteins. All retroviruses have similar overall chemical composition (by weight): 60% protein; 35% lipids, derived from the host-cell plasma membrane; and 4% carbohydrates, associated with Env proteins. Virion Mr: 150–180 × 10^6. Buoyant density in CsCl: 1.16–1.21 g/mL; in sucrose: 1.16–1.18 g/mL.

Physicochemical properties: Virions are relatively resistant to UV- and γ-irradiation, but sensitive to heat (56°C, 30 min), detergents, lipid solvents, and formaldehyde.

Genome: The genome is a homodimer of a positive-sense, linear, single-stranded RNA of 7000–11 000 nucleotides (60S-70S). Both monomers contain a 5′-cap structure and a 3′-poly(A) tail and are bound noncovalently by hydrogen bonds. A molecule of tRNA is associated to a complementary region near the 5′-end of each monomer. This tRNA is the primer on which the viral reverse transcriptase initiates DNA synthesis. The retroviral genome is composed of noncoding sequences located in terminal regions, which serve as recognition signals for DNA and RNA synthesis, and of coding sequences lying in the internal region. Noncoding sequences include (from 5′ to 3′):

1. A terminally redundant sequence of variable length (12–235 bases) located after the capping structure at the 5′-end.
2. A unique sequence near the 5′-end (U_5), which is the first part of the genome being copied into DNA during reverse transcription and becomes the 3′ end of the long terminal repeat (LTR). The U_5 region is involved in reverse transcription and integration into the host genome.
3. The primer binding site (PB, 18 nucleotides) is a sequence complementary to the 3′ terminus of the tRNA.
4. The untranslated sequence between PB and the first open reading frame (*gag*). It contains the packaging signal (Ψ or E). In addition, it contains the donor site for the generation of spliced subgenomic mRNAs (in general only for messages encoding the Env protein).
5. The 3′ untranslated region, involved in transport of unspliced mRNA.
6. The polypurine tract (nine or more A and G bases), containing the initiation site for synthesis of the plus strand DNA.
7. The U_3 region, which forms the 5′ sequence of the LTR, is involved in virus replication, integration and regulation of transcription.
8. Preceding the poly(A) tract at the 3′-end, there is another copy of the R sequence.

Proteins: Most retroviruses have four genes named, from 5′ to 3′, *gag, pro, pol,* and *env.* An mRNA comprising the whole viral genome is used for translation of the *gag, pro, and pol* products. The *gag* gene is translated into a polyprotein precursor, which is cleaved to yield three proteins: the matrix (MA), capsid (CA), and nucleic acid-binding (NC) proteins. The *pro* gene encodes the protease that cleaves the *gag* and *pol* polyproteins. The *pol* gene encodes the reverse transcriptase and the integrase proteins. The *env* gene encodes a precursor form of the two envelope glycoproteins: the larger is a surface protein (SU) responsible for interaction with cell-surface receptors; the smaller is a transmembrane protein that is required for fusion of virion and cell plasma membranes.

Some virus genera encode additional proteins: *(i)* Lentiviruses, human immunodeficiency virus (HIV) and its relatives, encode at least seven additional products: Tat (stimulates transcription from the HIV LTR), Rev (regulator of viral gene expression), Nef (localizes to the inner side of the plasma membrane where it causes removal of CD4, the HIV receptor; it may also activate T cells), Vif (necessary for production of infective virus), Vpu (involved in budding and release of virions), and Vpr and Vpx (unknown function). *(ii)* Spumaviruses encode several proteins in the region between Env and U_3: Bel1 (transcriptional transactivator), Bel2 (unknown function), Bel3 (unknown function), and Bet (fusion protein).

Replication: After penetration of the virion core into the cell, replication of viral RNA is initiated within the core structure by reverse transcription of RNA into cDNA. As cDNA synthesis proceeds, the template RNA is being degraded by the RNase activity of the reverse transcriptase. The RNA fragments generated during degradation are used to prime synthesis of positive-sense cDNA on the negative-sense cDNA transcripts. The DNA, still complexed with virion proteins, is transported to the nucleus. The circular, double-stranded DNA generated from the viral genome, containing the long terminal repeats, is then integrated into the host genome. Cellular RNA polymerase II produces viral RNA from the integrated provirus. These transcripts are processed to new viral genomes and to viral mRNAs which are translated into virion proteins. Nucleocapsids are assembled by interaction of Gag precursor molecules with a unit genome (through NC), with neighboring Gag precursors (through CA) and with the cell membrane (through MA). Interactions with Pro and Pol bring these proteins into the virion. Late proteolytic processing of the polyproteins by the PR protein leaves the virion prepared for its functions once it is in the proper environment. Finally, virions pinch off from the plasma membrane.

Family: Orhomyxoviridae
Genus: Influenza Virus (types A, B, and C)

Virion: The virion comprises enveloped particles, 80–120 nm in diameter, with a helical nucleocapsid 9 nm in diameter with transcriptase activity and a lipid bilayer with approximately 500 spikes (10–14 nm in length, 4–6 nm in diameter) protruding from its surface. There are two types of spikes: rod-shaped and mushroom-shaped, made of trimeric hemagglutinin (HA) and of tetrameric neuraminidase (NA), respectively. The influenza A virion contains the following proteins: *(i)* three integral membrane proteins: HA, NA, and M2; *(ii)* underlying the lipid membrane is the viral matrix protein (M1), the most abundant virion protein; *(iii)* inside the virion are the eight ribonucleoprotein complexes (RNPs), each made of an RNA molecule and four protein species: NP, the predominant subunit of the nucleocapsid, and the replicase/transcriptase complex (PB1, PB2, and PA); and *(iv)* the NS2 protein. Virion composition: 70% protein, 20% lipid derived form the host cell plasma membrane, and 5%–8% carbohydrate in the form of *N*-glycosidic side chains of glycoproteins, glycolipids, and mucopolysaccharides. Virion Mr: 250×10^6. Buoyant density in sucrose: 1.19 g/mL.

Physicochemical properties: Virions are sensitive to heat, UV- and γ-irradiation, lipid solvents, nonionic detergents, oxidizing agents, and formaldehyde.

Genome: The viral genome consists of negative-sense, linear, single-stranded RNA (10 000-13 600 nucleotides; influenza A: 13588 nucleotides) divided into six (Thogoto virus), seven (influenza virus C, Dhori virus), or eight (influenza virus A and B) segments of 900–2350 nucleotides in length. The first 13 nucleotides at the 5′-end and the last 12 at the 3′-end are conserved in all RNA segments and may function as promoters during viral transcription.

Proteins: The coding assignment for influenza A is as follows: segment 1 codes for PB2 (recognizes and binds to capped RNA fragments derived from host-cell

RNA transcripts; endonuclease activity); 2 for PB1 (catalyzes nucleotide polymerization and chain elongation); 3 for PA (unknown function); 4 for HA (major surface glycoprotein and antigenic determinant; binds sialic acid; fusion activity at acidic pH); 5 for NP (binds to viral RNA to form coiled RNP; involved in switch from mRNA to virion RNA synthesis); 6 for NA (surface glycoprotein, antigenic determinant receptor destroying activity); 7 for M1, (unknown function, interacts with other virion proteins) and M2 (integral membrane protein, ion channel activity to proton and Na^+ ions); and 8 for NS1 (predominantly nuclear; inhibits pre-mRNA splicing and nuclear export of poly(A)$^+$-mRNA) and NS2 (unknown function; cytoplasmic and nuclear location). Upon synthesis in the cytoplasm, PB1, PB2, and PA migrate to the nucleus where they form a transcriptase and replicase complex. Viral mRNA synthesis in the nucleus requires initiation by capped RNA fragments derived from host-RNA polymerase II transcripts.

Replication: Replication occurs in the nucleus (RNA synthesis and accumulation of N, M, and NS1 proteins) and the cytoplasm. Replication of virion RNA requires de novo synthesis of viral RNA and comprises three sequential steps: *(i)* switch from mRNA to template RNA synthesis; *(ii)* synthesis of template RNA (full-length copies of the virion RNA segments); and *(iii)* synthesis of multiple copies of virion RNAs. Closely related viruses, such as influenza A virus infecting humans and birds, can interchange genes during co-infection, thus generating new virus strains, a mechanism that is believed to account for the periodic pandemic waves of influenza. Virion assembly takes place at discrete patches of the plasma membrane containing viral envelope proteins. It is unclear whether the eight RNA segments of the genome are assembled together by virtue of a selective mechanism or if segmented genomes are packaged randomly.

Family: Picornaviridae
Genus: Enterovirus (Poliovirus)

Virion: The virion is a nonenveloped particle of icosahedral symmetry (T = 1), 24–30 nm in diameter, made of 60 subunits (protomers), each composed of a copy of the four capsid proteins (VP1, VP2, VP3, and VP4). In addition, virions contain a single copy of a small protein (VPg) covalently linked to the 5′-end of the RNA genome. Virion Mr: 8–9 × 10^6. Buoyant density in CsCl: 1.33–1.45 g/mL.

Physicochemical properties: The virus is sensitive to heat-oxidizing agents and formaldehyde. Virions are insensitive to nonionic detergents and organic solvents, and are stable at pH 3–9.

Genome: The genome is a single molecule of positive-sense, linear, single-stranded RNA of 7200–8450 nucleotides (Mahoney strain of type 1 poliovirus, 7433 nucleotides). The RNA carries a small protein (VPg) bound to its 5′ terminus that contains a conserved nontranslated region (600–1200 bases), and contains a nontranslated 3′-end (47–126 bases) followed by a poly(A) tract of variable length (35–100 bases, poliovirus: 62 bases).

Proteins: The genome contains a single open reading frame encoding a long polypeptide chain, the polyprotein, which is cleaved during translation by three virus-coded proteinases to generate 11 proteins (from the N to the C terminus of the polyprotein): VP4, VP2, VP3, VP1, 2A, 2B, 2C, 3A, 3B, 3C, and 3D. The first fragment released from the nascent chain is P1, the capsid precursor protein containing VP0 (precursor form to VP4 and VP2), VP3, and VP1. The next fragment is P2 (precursor to 2ABC), and the last P3 (3ABCD). The main function of protein 2A is to cleave P1 away from the nascent chain; in addition, 2A is involved in shut-off of host-cell protein synthesis. 2B and 2C are involved in RNA synthesis. 3B (VPg) is derived from a precursor 3AB, which is thought to be involved in initiating RNA synthesis. 3D catalyzes elongation of nascent RNA chains from viral RNA templates. 3C, or its precursor 3CD, catalyzes most cleavages in the polyprotein.

Replication: Replication and virion assembly occur entirely in the cytoplasm. The viral genome is used to form complementary minus-strand RNA from which new plus-strands are synthesized on smooth endoplasmic reticulum. Some of the newly synthesized plus-strand RNA form additional replication centers, while others are packaged into virions. Generation of infective particles depends on a late cleavage of VP0 to form mature protomers. Mature virus particles form crystals in the cytoplasm before being released by cell lysis caused by infection.

Family: Togaviridae
Genus: Alphaviruses (Semliki Forest Virus, Sindbis Virus)

Virion: The virion is a spherical enveloped particle, 70 nm in diameter. The nucleocapsid consists of an RNA molecule packed into an icosahedral shell (T = 4) made of 240 copies of a single species of capsid protein. Surrounding this complex there is a lipid bilayer derived from the host-cell plasma membrane, which contains embedded 80 trimers of glycoprotein spikes. Each spike is a heterotrimer consisting of two integral membrane proteins (E1 and E2) and a surface protein (E3), arranged in an icosahedral symmetry (T = 4). Spikes are held together by E1–E1 interactions in the core of the trimer. Virion Mr: approximately 52×10^6. Buoyant density in sucrose: 1.20 g/mL.

Physicochemical properties: The viruses are sensitive to heat, detergents, and organic solvents. Virions are very stable at pH 7–8 but inactivate rapidly at acidic pH.

Genome: The viral genome consists of a nonsegmented single molecule of linear, positive-sense, single-stranded RNA (42S) of about 12 000 nucleotides. The RNA is capped at the 5′-end and contains a poly(A) tract at the 3′-end. During replication, a subgenomic (26S) RNA is synthesized, which is also capped and polyadenylated and is the mRNA for the viral structural proteins. Four regions of the viral genome are highly conserved among alphaviruses: *(i)* the 5′ terminus; *(ii)* 51 nucleotides near the 5′-end; *(iii)* the junction region (21 nucleotides) the nonstructural and structural coding parts of the genome, including the start of the 26S RNA; and *(iv)* 19 nucleotides at the 3′ terminus. These four regions seem to be replicase binding sites and to function as promoter elements (e.g., the junction region contains the minimal promoter for transcription of the subgenomic RNA). A region within the NSP1 coding sequence binds the capsid protein and is required for packaging of newly synthesized viral genomes into virions. Transcription and replication of alphavirus RNA occurs on cellular membranes.

Proteins: Translation of the virion RNA is initiated at a single AUG near the 5′-end and proceeds through the first two thirds of the genome until a stop codon located at the beginning of the subgenomic 26S RNA sequence. The resulting polyprotein is cleaved to yield the four nonstructural proteins (NSP1–4). NSP1 is a methyltransferase with capping activity, which is also involved in initiation of minus-strand synthesis. NSP2 is the protease that cleaves the polyprotein. NSP3 is a phosphoprotein of unknown function. NSP4 is the viral RNA polymerase. The 26S RNA is translated into a polyprotein, which is co- and post-translationally cleaved into the structural proteins in the order C, p62 (precursor to E3 and E2), 6K, and E1. C is the capsid protein. E3, E2, and E1 form the viral spikes. 6K seems to be involved in envelopment and budding of the viral particle.

Replication: Replication and virion assembly take place in the cytoplasm. As production of the NSP proteins proceeds, the viral genome begins to be used as template for the synthesis of full-length antisense strands, which in turn serve as templates for replication of the genomic RNA. In addition, antisense strands are used to produce the subgenomic 26S RNA. Unidentified host-cell proteins are involved in the synthesis of viral RNA. Assembly of nucleocapsids involves a positive selection of the genomic RNA, since subgenomic RNA is not encapsidated although it is present in molar excess in the infected cell. At the plasma membrane, preformed nucleocapsids interact with the viral transmembrane glycoproteins, causing the membrane to bend around them until it fuses to release viral particles. The E2 carboxy-terminal domain, which binds the capsid protein, is essential for budding. In insect cells, budding occurs intracellularly and virions need to be transported in vesicles to the cell surface where they are released.

Alphaviruses have a wide host range. They replicate in many cell types of different species ranging from insects to humans. The infection of vertebrate cells by alphaviruses causes shut-down of host-cell macromolecular synthesis resulting in cell lysis. In mosquito cells, these viruses establish a nonlytic infection. They are arboviruses ("arthropod-born viruses"). In nature, these viruses are transmitted between susceptible vertebrates by hematophagous arthropods. Both Sindbis and SFV undergo an enzootic cycle involving wild rodents, births and subhuman primates, with mosquitoes as vectors. SFV undergo first isolated from *Aedimorphus abnormalis* in Uganda in 1942. SFV strains have been found in Africa, India, and Southeast Asia.

RECOMMENDED READING

Coffin, J.M., S.H. Hughes, and H.E. Varmus (Eds.). 1997. Retroviruses. CSH Laboratory Press, Cold Spring Harbor, NY.

Fields, B.N., D.M. Knipe, P.M. Howley, R.M. Chanock, J.L. Melnick, T.P. Monath, B. Roizman, and S.E. Straus (Eds.). 1996. Fundamental Virology, 3rd. ed. Lippincott-Raven Publishers, New York.

Webster, R.G. and A. Granoff (Eds.). 1994. Encyclopedia of Virology. Academic Press, San Diego.

Weiss, R., N. Teich, H. Varmus, and J. Coffin (Eds.). 1984. RNA Tumor Viruses. Molecular Biology of Tumor Viruses. CSH Laboratory Press, Cold Spring Harbor, New York.

RECOMMENDED URLs

www.tulane.edu/~dmsander/garryfavweb.html (Provides abundant information on virology. It contains listings of virology-related web sites of interest).

www.virology.net/Big_Virology/BVHomePage.html (The big picture book of viruses. It also serves as a resource for virus taxonomy).

FAMILY: Retroviridae
GENUS: Oncovirinae (MLV)

23 Strategies and Mechanisms for Retrovirus Retargeting

Keith Bupp and Monica Roth
Department of Biochemistry, Robert Wood Johnson Medical School, University of Medicine and Dentistry of New Jersey, Piscataway, NJ, USA

INTRODUCTION

Targeting a retroviral gene delivery vehicle specifically to the tissue where expression of a transgene is required is a critical aspect of gene therapy. In order to retarget a retrovirus to a tissue of interest, three basic strategies have been widely used to alter retroviral interactions with cellular receptors: *(i)* complete substitution of the envelope glycoprotein (Env) by another protein; *(ii)* incorporation of an antibody fragment into Env; and *(iii)* incorporation of a cell-binding ligand into Env.

In this chapter, we summarize the accumulated data for results employing these strategies. The findings indicate that fusing a cell-binding ligand or single-chain antibody fragment to Env only rarely leads to reasonably efficient retargeted gene transfer. The lack of success using these approaches may be due to structural perturbations of the Env molecule as a result of fusion to the novel targeting domain. An alternative approach to avoid this problem is the use of random amino acid substitutions within the receptor-binding site of an Env molecule followed by screening for mutants with the desired targeting properties. This approach is conceptually analogous to phage display technology and should be useful in retargeting retroviruses. Improvements in the entry properties of such selected mutants should also be improved by passage of a virus containing the novel binding site in culture.

VIRUS ENTRY

Retroviral entry involves the interaction between the viral envelope proteins and the host cell receptors. For the murine leukemia viruses (MLVs), five different classes have been identified based on superinfection interference assays. They are ecotropic, amphotropic, polytropic, xenotropic, and 10A1 (11,74,75). Except for the xenotropic class, all of the MLVs infect mouse cells such as NIH 3T3 cells (11,74,75). Ecotropic retroviruses such as Friend MLV (FrMLV) and Moloney MLV (MoMLV) only infect mouse and rat cells, whereas the other classes have broader host ranges (89).

Viral Vectors: Basic Science and Gene Therapy
Edited by A. Cid-Arregui and A. García-Carrancá

The receptors for three of these classes have been identified and are all multiple membrane-passing proteins (3,59,64,93). While the ecotropic receptor is a cationic amino acid transporter (47,95), the amphotropic receptor is a phosphate symporter termed Pit2 (45,93,96). 10A1 virus binds to Pit2 as well as another phosphate symporter, Pit1 (60). Receptors for the polytropic and xenotropic classes have not yet been identified. Since most retroviral retargeting experiments have employed the ecotropic and amphotropic *env* genes, these two are discussed in detail in this chapter.

The viral Env consists of two proteins, the surface (SU) and transmembrane (TM) proteins, which are cleaved from a common precursor (Figure 1). Extensive analyses have localized the receptor-binding domain to the N-terminal half of the SU protein (5–8,21,35,38,68,71). This is followed by a proline-rich region, which is proposed to behave either as a hinge or trigger to communicate receptor binding to the fusion machinery (30). Ecotropic Env fusion is triggered by low pH in most cell types, and amphotropic Env fusion is pH independent (58). The C-terminus of the SU is highly conserved and interacts with the TM protein (65,72). TM contains the fusion peptide (F) and a coiled-coil region, which is analogous to that found in influenza hemagglutinin (28), followed by the transmembrane and cytoplasmic domains. Amino acid sequence comparisons of MLVs have indicated that the regions of highest variability are located within the receptor-binding domain (7).

The crystal structure of the FrMLV SU receptor-binding domain has been solved (27) and sheds light on the variable regions. Throughout this chapter, we refer to this fragment and analogous fragments as the receptor-binding domain, although the region of the molecule actually involved in receptor contacts only makes up a small portion of this fragment. We term this smaller region the receptor-binding site. Three variable regions within the receptor-binding domain have been identified as named variable regions (VR) A, B, and C. VRC is spatially distinct from VRA and VRB (Figure 2) and has been proposed to be involved in the trimerization interface of SU molecules (26). As discussed below, VRA and VRB contribute in varying degrees to the formation of the receptor-binding site. Although separated in the primary sequence, their interrelation within the three-dimensional structure is convincing (Figure 2).

As shown in Table 1, the sizes of the VRAs and VRBs from different retroviral isolates show great variation (7,27,49). The ecotropic MLV envelopes, such as MoMLV and FrMLV, contain the largest and most complex VRA and a small VRB. In contrast, the amphotropic MLV Env contains a smaller and less complex VRA and a larger VRB. Phylogenetically related retroviral isolates that show strong conservation to

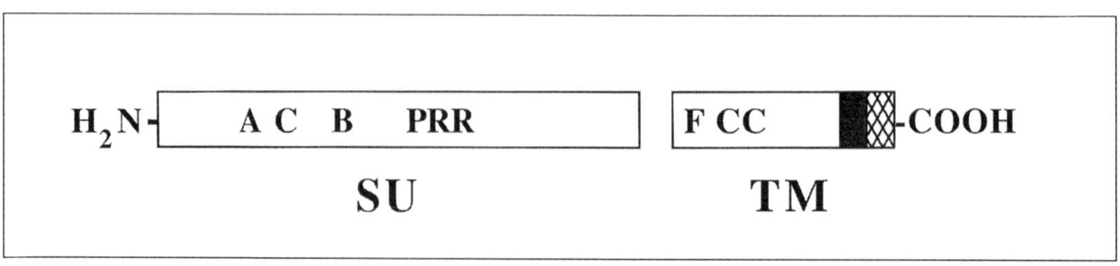

Figure 1. Linear structure of the MLV Env proteins. The SU and TM proteins are indicated. Variable regions A, B and C are labeled as well as the proline-rich region (PRR) within SU. The putative fusion peptide (F) and the coiled-coil (CC) region are indicated within TM. The membrane spanning portion of TM is shaded. The cytoplasmic domain is cross-hatched. For FrMLV, SU is 445 amino acids long, and TM is 200 amino acids long.

MLVs in the C-terminus of SU and TM represent a third pattern. Feline leukemia virus-A (FeLV-A), FeLV-C, and rat leukemia virus (RaLV) Env proteins have minimal VRB sequences (23,49,77). Amino acids that would be expected to be important for interactions with the receptors for these viruses should therefore lie exclusively within the VRA domain. We discuss below the possible advantages of using an FeLV Env molecule for retargeting purposes.

For MuLV SU, only the structure of the N-terminal half of FrMLV has been determined (27). The interactions of this domain with the rest of the SU protein and the TM protein have not been characterized, nor are the requirements for entry of the MLV envelope proteins fully understood. One approach our laboratory has taken toward this understanding is to examine the mutations generated within chimeric amphotropic/ecotropic Env proteins of replication-competent viruses (71). The chimeras that result in infectious virus all contain amphotropic/ecotropic junctions located at or downstream from the proline-rich region. The high degree of misincorporation by reverse transcriptase generates a pool of mutations from which the replicating virus can select to improve the chimeric Env proteins generated in vitro. Interestingly, analyses of chimeras that contained junctions within the C terminus of SU revealed compensatory mutations located along a single face of the N-terminal receptor-binding domain (66). These results imply that the N and C termini of the SU protein interact.

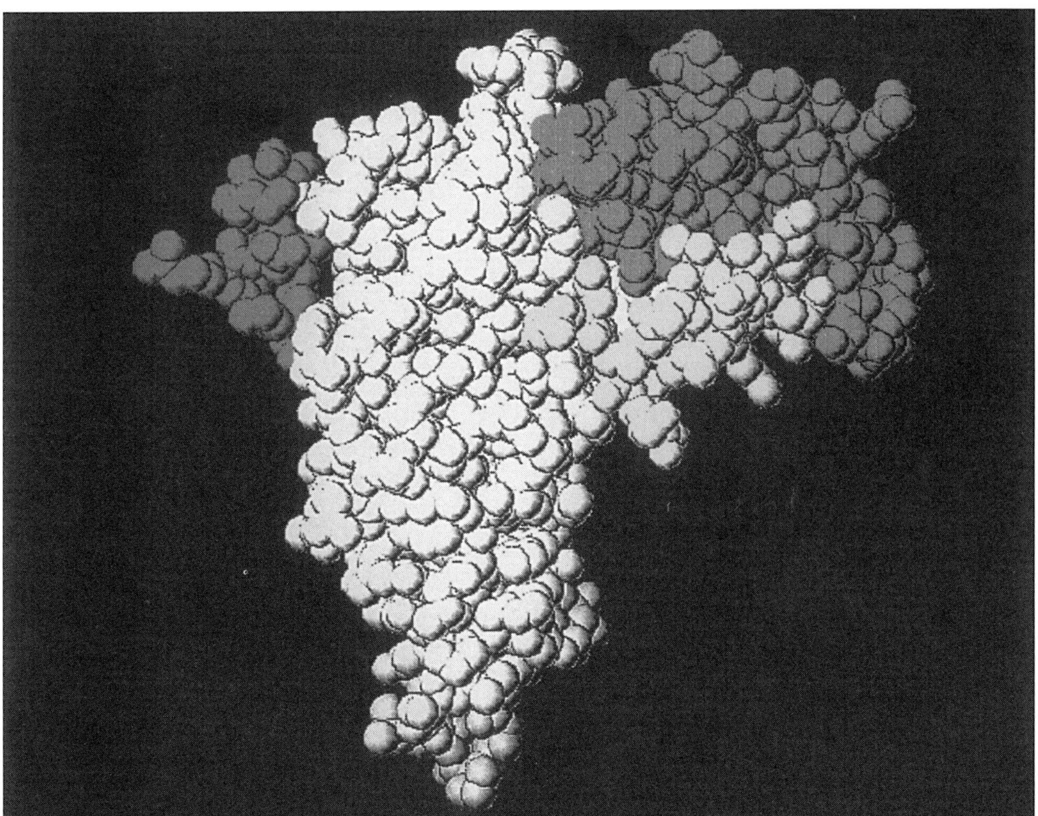

Figure 2. The structure of the receptor-binding domain of FrMLV Env (27). VRA is in green, VRB in yellow, and VRC in red. (See color plate A5.)

Table 1. Comparison of Env Variable Regions

Virus	# of Amino Acids in Variable Regions		
	VRA	VRB	VRC
MoMLV	65	11	9
4070A amphotropic	29	30	6
FeLV-A	39	6	11
RaLV	21	3	6

Boundaries of the variable regions are as defined in (7) except that the carboxy terminus of VRA has been redefined as VRC (27). Sequences are those described in (7) with the addition of RaLV (49).

For productive retargeting of an envelope protein to an alternative receptor, the basic structure and functions of the envelope proteins should be maintained. By analogy to influenza hemagglutinin, multiple conformations of Env have been proposed to be required during the process of virus-cell fusion (28). The Env protein on the virus must present the receptor-binding site while maintaining the fusion domain in a buried state. The fusion-competent state is only revealed upon receptor binding and is triggered through complex conformational changes in both the SU and TM proteins. Retargeted envelopes must be able to change the receptor-binding amino acids and yet be able to effect the communication required for triggering and activation of membrane fusion upon binding.

Several different systems have been used to test the abilities of various surface proteins to mediate retroviral entry (19,20,52, 85). Most of them measure the ability of these proteins to mediate a single round of infection as measured by the transfer (transduction) of a marker gene from a donor to an acceptor cell. Generally, an expression plasmid encoding the protein to be tested is transfected into a cell line stably expressing viral core particles that are the products encoded by the retroviral *gag* and *pol* genes. Expressing the protein on the surface of the virion without incorporating its gene into the particle in this manner is termed pseudotyping. The retroviral core-producing cell line also frequently carries a drug resistance and/or other marker enzyme gene that can be packaged into a retrovirus and transduced into an acceptor cell (19). In some cases, a wild-type *env* gene may also be stably expressed by the core-producing cell line (20,52). A number of variations of this general strategy exist. For example, the *gag* and *pol* genes, the transferable marker gene, and recombinant surface protein may all be transfected simultaneously (85). Approximately one to two days after transfection of the SU protein gene into the core-producing cell line, the supernatant that contains viral particles is harvested, filtered, and used to inoculate acceptor cells. The efficiency of transfer can then be quantitated by counting drug-resistant colonies or the number of cells staining positive for a marker enzyme such as β-galactosidase or alkaline phosphatase.

ENV REPLACEMENT

One relatively straightforward technique for targeting a retrovirus to a specific cell type is to completely replace the Env protein with another protein known to promote fusion. A number of surface proteins from other viruses have been used to pseudotype retroviruses. For example, the vesicular stomatitis virus (VSV) G protein mediates cellular entry via a phospholipid receptor rather than a specific protein and can target retroviral vectors to a wide variety of cell types (2,14,24,66; see also Chapter 24). Since the G protein is a single TM protein, it is more stable than Env and permits the concentration of pseudotyped retroviruses (14). Other viral SU proteins that may be useful in retroviral targeting caused by the ubiquitous presence of their receptors are influenza and fowl plague virus hemagglutinins (22,37).

Two strategies have been used to target retroviral vectors for the development of anti-HIV (human immunodeficiency virus) therapies. One approach is to target a retrovirus to HIV-infected cells by pseudotyping the retrovirus with CD4 and an HIV-1 coreceptor (25). The retrovirus was able to enter cells via the HIV-1 Env expressed on the cell surface. A complementary strategy is to target CD4+ cells by pseudotyping a retrovirus with HIV-1 Env (82). This entails the deletion of most of the cytoplasmic domain of HIV-1 TM (82). Similar minor modifications might also be necessary with other foreign glycoproteins that are unable to pseudotype a retrovirus in their native forms.

Two other viral glycoproteins with interesting cell type-specific targeting potentials have recently been used to pseudotype retroviruses. The Ebola virus glycoprotein may be useful for targeting because it is relatively specific for endothelial cells (98,101). The Sendai virus fusion (SV-F) protein may be useful for targeting to the liver since it binds to the asialoglycoprotein receptor, although titers of retroviral particles pseudotyped with SV-F are reportedly low (86).

ATTACHMENT OF ANTIBODIES AND OTHER CELL-BINDING LIGANDS TO ENV

A number of different antibodies or cell-binding ligands have been attached to Env in an effort to promote retroviral entry via the receptors targeted by the Env-attached molecule. These approaches have met with only limited success and, in most cases, the additional presence of a wild-type Env was required to observe any level of marker gene transduction. The wild-type Env presumably forms mixed oligomers with the hybrid protein (76,102) and could compensate for defects in the hybrid such as cell surface transport (42,43) or cellular entry (54).

Single-chain antigen-binding fragments of antibodies (single-chain Abs) that have been tested include anticarcinoembryonic antigen (CEA) (48), anti-CD34 (10), anti-ErbB2 (81) anti-high-molecular-weight melanoma-associated antigen (54), anti-low-density lipoprotein (LDL) receptor (84), and anti-major histocompatibility complex class I (53), as well as others (15,16,41,79). Other ligands tested have included the receptor-binding domain of amphotropic Env (17,91,92), a collagen-binding domain from von Willebrand clotting factor (32), epidermal growth factor (EGF) (17,29,63, 70), erythropoietin (EPO) (44), hepatocyte growth factor (62), heregulin (36), stem cell factor (SCF) (29,100), and an RGD-containing peptide (90).

Some of these hybrid Env proteins failed to be transported to the cell surface and/or failed to be incorporated into virions (10, 81,100). Other hybrid proteins that failed to mediate transduction were incorporated into virions that were demonstrated to bind the cells to which they were targeted (1,17,92). The block to efficient transduction of marker genes in those cases was apparently due to failure at a subsequent stage of fusion and entry.

Knowledge of the structure of the receptor-binding domain of the FrMLV Env protein allows some analysis of the initial constructs used to present alternative binding domains. One of the key questions is where to insert the alternative binding domain into the SU/TM complex. Several positions have already been tested. One site is at the beginning of the proline-rich region, replacing the viral receptor-binding domain with an alternative domain. For all MLVs, there is a strong conservation of an ordered proline sequence GPR(L/V/I)PIGPNPV, marking the N-terminus of the proline-rich region (30). Inserting erythropoietin (EPO) or heregulin eight amino acids upstream of this sequence in place of the normal receptor-binding domain led to transduction in the presence of wild-type MoMLV Env (36,44). However, a number of single-chain Ab-Env fusions located at various positions between VRB and the proline-rich region led to the production of proteins that were not transported to the cell surface and therefore not incorporated into virions (10).

Several targeting domains have been attached at position 1 or 6 of MoMLV SU (1,10,17,29,48,53,54,63,70,81,91,92). Positions 1 and 6 are not within the crystallized structure of the FrMLV receptor-binding domain. However, they would be located near the base of the structure, which begins with Q9 (Figure 3). Mutational analyses have indicated that there is some degree of tolerance for changes in this region. A linker insertion at amino acid positions 3/4 had little effect on viral viability (30). However, deletion of residues 2 through 7 decreased transduction efficiency by two orders of magnitude, and the additional deletion of residue 8 reduced transduction by a further four orders of magnitude (5). Substitution of another amino acid for the highly conserved residue H8 also abolished infectivity (5). None of the changes in this amino-terminal region affected protein incorporation into virions or cell binding (5).

Insertion of targeting ligands at position 1 or 6 might be expected to be suboptimal for their presentation because of steric hindrance from the Env protein. However, in some cases, linkers added to create spacing between the ligand and Env did improve transduction efficiencies to nearly 1×10^5/mL. The degree of success depended on the specific hybrid (91,92), and the presence of spacers did not always lead to successful transduction (54,81). The ligand can also interfere with the normal functioning of the Env protein to which it is fused (1). Without such spacers, transduction efficiencies of chimeric env genes with ligands attached at positions 1 or 6 led to significantly lower titers than with spacers (10,48,53,91). This supports the notion that ligand insertion at the base of the receptor-binding domain results in adverse steric effects.

Ligands and antibody fragments have also been inserted at amino acids V10 and V17/T18 of MoMLV Env (Figure 3). The level of successful transduction depended on the specific construct, and there was no obvious systematic influence from the position of insertion. Residue V10 is located at the base of the receptor-binding domain. An insertion of the 161-amino-acid-long murine stem cell factor at this position enabled a low level of transduction (approximately 1×10^2/mL) via its receptor c-Kit only in the presence of chloroquine to prevent lysosomal degradation. However, the insertion of granulocyte-macrophage colony-stimulating factor at this position prevented the hybrid protein from being transported to the surface of the cell (100).

Residues V17 and T18 are buried within the 1β sheet, and an insertion at this location would appear to be disruptive to the overall structure of the receptor-binding domain. A hybrid protein containing an insertion of an anti-CD34 single-chain Ab at this position was not, in fact, incorporated into virions even in the presence of wild-type Env (10). However, the insertion of an anti-LDL receptor single-chain Ab (83) or a small peptide containing a collagen-binding domain (32) at the same position did lead to successful transduction of a marker gene. In the case of the anti-LDL receptor-Env hybrid, the titer on HeLa cells, which express LDL receptors and are normally uninfectable by MoMLV, was 1×10^4/mL only when the wild-type MoMLV Env was co-expressed. The function of the hybrid Env protein disrupted by the single-chain Ab and compensated by the presence of the wild-type Env was not determined. In the other construct, the influence of the collagen-binding domain on transduction was indirect. The virus was first adsorbed to a collagen-coated well, MoMLV-susceptible NIH 3T3 cells were added, and infection then proceeded via the ecotropic receptor. Virus that expressed wild-type Env was washed off the collagen-coated wells, and, therefore, no transduction was observed with this control. Successful transduction mediated by the collagen-binding domain-containing Env did not require the presence of a wild-type Env. The absence of such a requirement may be due to the small size of the collagen-binding peptide, which did not disrupt the Env structure as the bulky anti-LDL receptor insertion did. This conclusion was borne out by the nearly wild-type titers obtained on NIH 3T3 cells when infected via the ecotropic receptor by the collagen-binding, domain-containing virus (32).

A number of other nondisruptive insertion sites have been identified through mutagenesis studies. Most tetrapeptide insertions within the receptor-binding domain of MoMLV Env made by Gray and Roth (30) had deleterious effects on viral

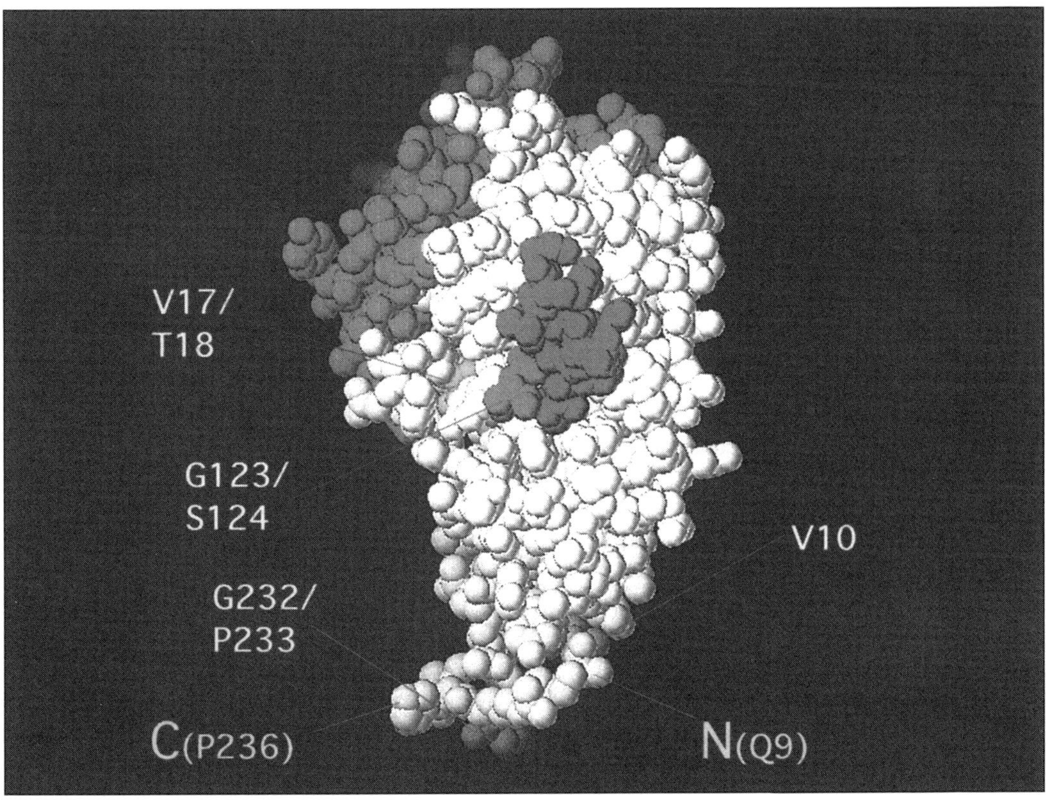

Figure 3. Locations of insertions of some novel cell-targeting ligands (orange) and linker insertions (magenta) within the ecotropic Env receptor-binding domain. VRC is facing out from the page. VRA, B, and C are in green, yellow, and red, respectively. VRB is barely visible from this angle. The locations of the amino-terminal Q9 (N) and carboxy-terminal P236 (C) residues of the resolved crystal structure are indicated. FrMLV amino acid numbers are indicated for the homologous residues of MoMLV Env that are discussed in the text. (**See color plate A6.**)

viability. The exceptions (Figure 3) were at positions 3/4 of the N terminus of the mature protein, 121/122 at the N-terminal edge of VRC (corresponding to positions 123/124 of FrMLV Env), and 230/231 (corresponding to positions 232/233 of FrMLV Env) at the N-terminal edge of the proline-rich region. Whereas the insertions at 3/4 and 121/122 exhibited nearly wild-type levels of virus production, the 230/231 insertion was temperature sensitive. This was due to increased shedding of the SU portion of Env. A number of different insertions within the proline-rich region of FrMLV Env (46) or an ecotropic/amphotropic hybrid Env (99) did not significantly interfere with transduction via the retroviral receptor (99) or viral viability (46) even when the insertions were up to 252 amino acids long (46). These results suggest regions where cell-targeting ligands might be inserted, although none of them is spatially proximal to the receptor-binding residues identified in other experiments (Figures 3 and 4).

In some cases, successful entry of a retargeted retrovirus could be attributed to the presence of wild-type Env-binding function, expressed either as a separate Env protein within a mixed oligomer or as a part of the hybrid protein itself. Binding of these Env molecules to the endogenous retroviral receptor sometimes occurred after initial binding via the targeting ligand (32,54,91). Thus, the targeting ligand served to concentrate the virus on the surface of the cell; this concentration facilitated the subsequent binding and entry via the normal route. Hall et al. used this strategy to target the collagen-binding domain MoMLV Env chimera discussed above to regions of wound repair in injured rat carotid artery (32). Transduction of the endothelial and smooth muscle cells in these regions was presumably mediated by the ecotropic receptor.

In a variation of this mechanism, retroviral particles displaying EGF-amphotropic Env chimeras did bind to EGF receptors. However, this led to inactivation of the virus within lysosomes (17). Protease-sensitive linkers were therefore engineered between the EGF and amphotropic Env domains of such a hybrid (63,70). Cleavage of this linker either by exogenously added or cellular proteases after virus binding to cells via the EGF receptor allowed transduction via the amphotropic receptor (63,70). This strategy has been discussed in greater detail elsewhere (18) and has been extended to stem cell factor-Env constructs (29).

In other cases, fusion and entry have been observed even in the absence of the ecotropic Env receptor on human cells (17,48,53,84,100). In these cases, internalization of the virus may lead to transport to an intracellular compartment where a fusion event occurs. Since ecotropic Env-mediated entry is pH dependent (58), it might be expected that internalization of a retrovirus that expresses ecotropic Env or an ecotropic Env derivative could lead to fusion once inside a low-pH compartment. However, in two studies (17,100), the addition of chloroquine to raise the pH within intracellular compartments actually resulted in increased transduction titers. Thus, abrogation of lysosomal degradation seemed to be more important in those cases than a requirement for low endosomal pH.

Single-chain Abs have also been inserted at several different positions within the SU protein of the avian spleen necrosis virus (SNV). Initial experiments in this system gave only low levels of transduction (15). In those cases, there was little difference in transduction efficiencies between the different sites used for insertion. More recently, direct fusions to the N-terminus of the TM protein have been principally employed. Successful transduction in this system almost always depends on the presence of a functional wild-type Env and titers of 1×10^5/mL or more have been attained (16,41). Successful transduction was also observed when avian leukosis virus (ALV) was retargeted to EGF receptor-expressing cells using a soluble receptor-ligand fusion protein (83). The fusion protein contained the cellular receptor for ALV attached to EGF. Preincubation of EGF-receptor-expressing cells with the soluble receptor-ligand, followed by incubation with virus led to EGF-receptor-dependent transduction of a drug resistance marker.

ALTERATION OF RECEPTOR BINDING SITES IN ENV

So far, retargeting a retrovirus using nonviral proteins as targeting ligands has been relatively unsuccessful. Part of the reason for the lack of success may be due to the rather significant changes made in Env that involve cumbersome additions or substitutions of novel binding domains. Whereas Env proteins have evolved specifically for binding, fusion, and entry into infectable cells, antibodies and other cell-binding ligands evolved to serve quite different functions; thus the failure of these proteins to mediate efficient fusion and entry of virions is not too surprising.

In an effort to make less radical changes in Env structure, preserving its functionality while altering its receptor-binding capabilities, it may be useful to take into account data that delineate the residues responsible for receptor targeting. By knowing precisely which residues are involved in receptor targeting, one could theoretically change them in order to retarget the retrovirus to a novel receptor. An analogy could be made to antibodies in which the antigen recognition site is basically defined by a small number of hypervariable amino acid residues (4,94).

Recently, researchers have been using phage display technology to select antibodies in vitro that express the highest affinity for an antigen of interest. Phage display entails the creation of libraries of mutated antibody genes that have incorporated engineered random amino acid sequences into their variable domains. Selective amplification after affinity purification of phage-expressing antibodies that contain new variable domains results in the isolation of antibodies with high affinities for a specific ligand. Since the locations of variable regions of the antibody are already known and can consequently be substituted by novel sequences, such a strategy is feasible and has been widely used (31,39). Data concerning the receptor-binding sites of retroviruses have been accumulating for several years. The availability of the crystal structure of the receptor-binding domain of FrMLV allows a better understanding of the results obtained from a number of experiments.

In this section, we analyze and compare data from a number of experiments to narrow down the regions responsible for interactions between Env and its receptor. Since creating a library of *env* genes with randomized sequences requires that a short linear section of the protein be identified for randomization, the ability to limit the binding site to only a few residues is important. Variable regions of ecotropic, amphotropic, and FeLV are discussed below with the object of identifying such short linear regions. By making substitutions within a region already known to be important for interactions of Env with its receptor, it is hoped that only the binding function of the protein will be altered, thereby leaving fusion functions intact. Replacing the wild-type binding site with a novel binding site should also place the new binding site in an optimal location for its presentation to the cellular receptor.

Ecotropic Retroviruses

Receptor-Binding Domain

FrMLV and MoMLV are members of the ecotropic class of murine retroviruses.

They are capable of infecting mouse and rat cells via the ecotropic receptor (3), which is a cationic amino acid transporter (47,95). The three-dimensional structure of a receptor-binding domain of FrMLV Env has been determined at 2.0 Å resolution (27). This fragment encompasses the amino-terminal 236 amino acid residues of the mature Env glycoprotein upstream of the proline-rich region. A number of studies have shown that the amino termini of ecotropic Env proteins encompassing this region are responsible for receptor binding (5,6,21,38,71). The receptor-binding domain contains VRA, B, and C (Figure 2), which are obvious potential determinants of receptor choice. Experimental data discussed below support this idea.

The smallest ecotropic-determining region delineated was shown by Morgan et al. (61) to be located within the N-terminal 88 amino acids of MoMLV Env including approximately half of VRA. In that study, the interference sensitivities of virus particles pseudotyped with MoMLV/4070A (amphotropic) Env hybrid constructs were tested. Although the construct containing only 88 amino acids of ecotropic sequence retained ecotropic specificity, the titers of particles pseudotyped with it were nearly two orders of magnitude lower than titers obtained using wild-type pseudotyped particles. This suggests that more carboxy terminal residues including VRB might also be involved in binding and entry or that the conformation of the hybrid protein was somehow altered.

Amino Acid Substitutions

Within the receptor-binding domain, changes in amino acids E116 and E129 have been shown to play a role in altering cell tropism (55,57,88) (Figure 4). This was determined by studying the properties of chimeric recombinants constructed between the neuropathogenic Friend variant PVC-211 and FrMLV clone 57. The double change E116 to G and E129 to K allowed increased infectability of capillary endothelial (CE) cells from rat brain, kidney, and liver when compared to FrMLV clone 57 (55,56). These changes also allowed infection of hamster cells (57), which are not normally infectable by FrMLV. Whether the newly acquired tropism is due to altered binding or a postbinding event has not yet been determined. Interference studies demonstrated that PVC-211 used the ecotropic receptor on mouse 3T3 and Rat-1 cells. The receptor utilized on hamster cells or CE cells has not been well characterized (60,97). It should be noted that a different control strain of FrMLV (FB29), which retains E at both 116 and 129, was also capable of infecting rat brain CE cells in vitro (55) and mouse brain CE cells in vivo (69). The determinants of tropism are therefore not limited to these two residues.

E116 and E129 are found within the hypervariable regions of VRA and VRC, respectively. Examination of the crystal structure (27) reveals that E116 is located on a face of Env, which includes the residues R85, D86, and R97, which have been shown to be important for Env-receptor interactions (Figure 4). It has been proposed that the VRC domain is involved in Env trimerization (26). The contribution of E129 to receptor choice is, therefore, presumably not due to an alteration in the receptor-binding site, but rather an indirect mechanism, such as the regulation of subunit interactions. It is possible that the altered binding properties of the E116G mutation alone may not be optimal for entry into CE cells and must be compensated by the E129K mutation to be effective. This interpretation suggests that *(i)* alterations in the vicinity of E116 should alter receptor specificity and *(ii)* allowing an engineered retroviral Env to evolve into an improved structure during infection can be advantageous. The second concept is consistent with results obtained in our laboratory concerning acquired Env mutations that improve viral viability (66).

Bae et al. (5) examined the effects of substitutions and deletions of the R residues at positions 83 and 95 within VRA of MoMLV Env (R85 and R97, respectively, of FrMLV Env) (Figure 4). Whereas substitutions at 83 had no effect on transduction, most Env mutants containing substitutions at 95 failed to be incorporated into virions. The R83 deletion and the R95D change each led to decreases in transduction efficiency of an order of magnitude. Combining the two changes completely eliminated transduction because the virions no longer bound to cells, although the protein was incorporated into virions.

MacKrell et al. (51) generated a large number of point mutations within the hypervariable regions of VRA and VRB of MoMLV Env. The most interesting effect was a D to K change at position 84 (D86 of FrMLV Env) (Figure 4) within VRA, which eliminated binding and transduction although Env incorporation into virions was normal. Although most other mutations had no effect on transduction efficiency, some in both VRA and VRB had small negative effects, suggesting direct or indirect contributions to receptor binding.

Figure 4 indicates on the three-dimensional structure that the residues R85, D86, and R97 form a small domain on one face of Env. One of the mutated residues of PVC-211 shown to be important in CE cell-tropism (E116) is also located on this face. Additional residues in this region that may be important for Env-receptor interactions remain to be identified. For the ecotropic Env, the residues

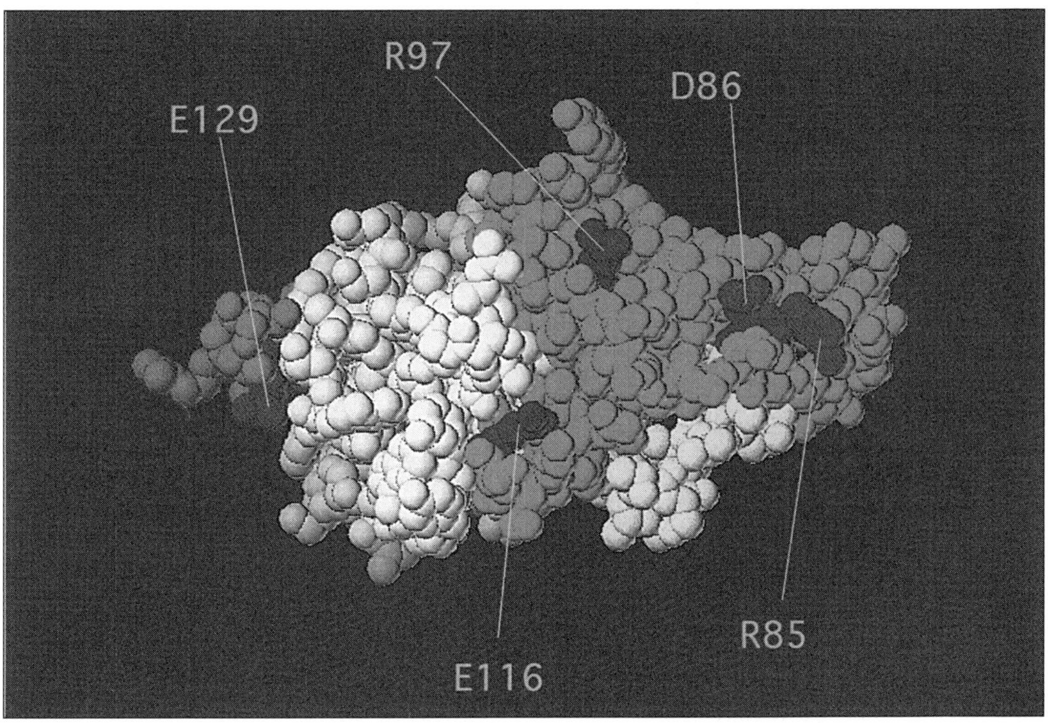

Figure 4. Amino acid residues shown to be important for Env-receptor interactions. View is from the top of the structure shown in Figure 2. VRA, B, and C are in green, yellow, and red, respectively. Critical residues discussed in the text are indicated in blue. FrMLV amino acid numbers are indicated for the homologous residues of MoMLV residues that are discussed in the text. (**See color plate A7.**)

that constitute this face of the molecule present reasonable targets for substitution by random amino acids. From a display library of *env* genes containing random amino acids within this presumed receptor-binding site, one should be able to select genes that encode Env proteins with novel binding specificities. Since the substitutions are located at the normal receptor-binding site, there should be minimal deleterious effect on other functions of the molecule such as trimerization and the mechanics of fusion.

Amphotropic MLV and 10A1

Receptor-Binding Domain

The amphotropic Env protein is highly divergent from the ecotropic Env protein within VRA, B, and C. While a high-resolution crystal structure is not yet available for amphotropic Env, such a structure might be expected to be similar to FrMLV Env over conserved regions of the molecule, but quite different over VRA, B, and C. Although the amphotropic and ecotropic Env proteins recognize different receptors (45,47,93,95, 96), evidence indicates that VRA and VRB make direct or indirect contributions to receptor binding and specificity for both.

As with FrMLV Env, an N-terminal fragment of the amphotropic 4070A Env upstream of the proline-rich region contains the receptor-binding domain. The 208 N-terminal amino acids of amphotropic Env were sufficient to block entry when a soluble preparation was added to cells before challenge by amphotropic 4070A pseudotyped particles (8). (Several different numbering systems have been used to indicate amino acid positions in the studies cited below. For ease of discussion, all amino acid numbers have been assigned here with number 1 as the N-terminal amino acid of the predicted mature protein in Reference 67.)

Analyses of hybrid recombinants between amphotropic 4070A, 10A1, polytropic, and xenotropic Env segments indicated that sequences upstream of amino acid 167 (amphotropic) were sufficient to determine infection via the amphotropic receptor (7,35,68). Amino acid 167 is located just downstream of VRB. The amphotropic VRB could be replaced by VRB from either polytropic or xenotropic Env, but the hybrid still mediated transduction via the amphotropic receptor on NIH 3T3 cells as shown by interference assays (7). This suggested that VRA sequences alone might be responsible for amphotropic receptor choice. However, lower transduction efficiencies on NIH 3T3 cells and amphotropic-susceptible D17 canine cells by the same VRB replacement chimeras (7) indicated that VRB sequences also make a direct or indirect contribution to amphotropic receptor binding.

Tests of viable viruses containing hybrid amphotropic/ecotropic Env constructs led to the conclusion that the 209 N-terminal amino acids were sufficient to confer amphotropism (71). Sequencing of viral DNA from D17 canine cell cultures that had been infected for several weeks with the hybrid construct containing the 209 N-terminal amino acids of mature amphotropic Env revealed the acquisition of two amino acid changes within VRA (66). One was E50K, and the other was G70R. Further analyses of other hybrid constructs in that study led to the determination that the G70R change could lead to improved viability of a hybrid virus. Since this adaptive mutation arose in several different hybrids passaged in D17 cells, it is possible that it may improve interaction with the receptor present on those cells. Hybrids with other fusion junctions were also able to improve their viability in this and/or other cell types by the acquisition of amino acid changes in other regions of the molecule. For retroviral retargeting applications, these data indicate that passage of an engineered virus in culture can improve its infection efficiency of specific cell types.

Localized Mutagenesis

Several mutations in VRA and VRB of amphotropic 4070A Env have been generated to further define a receptor-binding site. With one exception, none of the point mutations constructed by Han et al. between positions E50 and F152 throughout VRA, B, and C had a very significant effect on transduction efficiency (34). The sole exception led to a defect in SU/TM cleavage. The absence of effect may indicate that disruption of the interaction of a single amino acid of amphotropic Env with its receptor is insufficient to eliminate receptor binding.

A number of constructs containing polytropic substitutions over a similar region of amphotropic 4070A Env either retained amphotropism or completely eliminated transduction of NIH 3T3 cells, which are infectable by both amphotropic and polytropic retroviruses (34). All of these Env constructs were incorporated into viral particles. The ability of a fusion-defective, ecotropic Env mutant to functionally complement the noninfective amphotropic mutant containing a polytropic substitution at positions 48 to 74 (Figure 5) indicated that the fusion function of this mutant was not affected. Presumably, the complementing ecotropic Env provided the binding function, while the fusion function was provided by the amphotropic mutant within a mixed oligomer. Therefore, this mutant may have disrupted interactions with the amphotropic receptor Pit2. The deletion of residues 54 to 61 from the wild-type sequence affected transduction efficiency by an order of magnitude. Substitution of residues 63 to 74 with polytropic sequences also led to a 95% decrease in transduction efficiency of NIH 3T3 cells. These results suggest that a combination of residues from the region 54 to 74 play a role in amphotropic receptor binding (Figure 5).

Battini et al. (9) made substitutions of short sequences within VRA and VRB of amphotropic 4070A Env with an 11 amino acid tag from VSV to analyze functional domains of the Env protein. One substitution within VRA and the two substitutions within VRB had only small effects on the transduction efficiency of NIH 3T3 cells, although the substitution of the amino terminal residues of VRB (129–141) resulted in a protein that could not confer interference to infection by amphotropic virus. Another substitution within VRA spanning residues 66 to 78 inhibited SU/TM cleavage. The third substitution within VRA (50–63) eliminated infection, binding of virus to cells, and interference capability without affecting incorporation into virions, although processing was temperature sensitive. These results suggested that residues within the regions 50 to 63 of VRA and 129 to 141 of VRB (underlined in Figure 5) play a role in amphotropic Env interactions with its receptor, Pit2.

Tailor and Kabat (87) constructed several different hybrid Env proteins between amphotropic 4070A and the closely related FeLV-B (see under "Feline Leukemia Virus"). Whereas FeLV-B can use human Pit1 as a receptor, the amphotropic virus uses Pit2. They showed that both VRA and VRB of amphotropic Env make contributions to receptor choice. They also demonstrated that Y60 and V61 (Figure 5) within the amphotropic VRA play important roles in mediating transduction when tested on a chimeric Pit1/Pit2 receptor.

The *env* genes of amphotropic 4070A and 10A1 retroviruses are closely related (66). The receptors for both are phosphate symporters (45,93,96). But while 10A1 can use both the Pit1 and Pit2 symporters, 4070A can only use Pit2 (60,97). Han et al. (33) were able to switch the cellular specificity of amphotropic 4070A Env to that of the closely related 10A1 by making the triple change A69G/Q72K/V137M (Figure 5). Whereas the wild-type 4070A Env cannot mediate transduction of 10A1-infectable CHO-K1 cells, the triple mutant is able to mediate transduction of this cell type. Interference experiments indicate that the triple mutant uses Pit1 on NIH 3T3 cells (33,97). Amino acids A69 and Q72 are located within VRA; V137 is in VRB. The Q72K/V137M mutant Env-mediated transduction less efficiently. No single change alone altered the tropism toward CHO-K1 cells. Interestingly, sequences beyond VRB downstream of amino acid 158 also seem to play a role in determining receptor specificity since a hybrid amphotropic 4070A/10A1 Env containing 4070A sequences from the N terminus through the VRB domain can infect CHO-K1 cells at a very low level (33). This could indicate that sequences within the proline-rich region or further downstream play a role in cell type specificity.

Feline Leukemia Virus

FeLV isolates are divided into the three subgroups (A, B, and C) based on receptor usage. Whereas subgroup A preferentially infects for feline cells, subgroups B and C can infect cells from other species (40,78,80). By using chimeric Env molecules between A and C viruses, it was shown that receptor specificity segregated with the N-terminal 107 amino acids of FeLV Env (13) and more specifically the Vr1 region (73,78), which corresponds to the N-terminal portion of VRA of MLV. Replacement of the FeLV-A sequence shown in Figure 6 by the corresponding FeLV-C sequence changed the receptor tropism of the virus from A to C. Attempts to more narrowly define the subgroup-determining domain led to inactivation of the virus. Within the C subgroup, a significant amount of variation exists over this region, indicating that more than one sequence there permits C receptor interaction (78). Interestingly, FeLV Env molecules from subgroups A and C contain very few amino acid residues over a region that would correspond to VRB of MLV (Table 1) (23,77).

The subgroup B FeLV Env is closely related to amphotropic 4070A Env. It has shorter VRA and VRC regions and a longer VRB than members of subgroups A and C, which makes it more like amphotropic Env than Env proteins from the other FeLV subgroups. Using chimeras between either FeLV-B and amphotropic Env (87) or FeLV-A and FeLV-B (12), it

```
                46      50          60              70    74
                        .           .               .
VRA   4070A     D L V G E E W D P S D Q E P Y V G Y G C K Y P A G R Q R T
      Mo-MCF        I   D D     E T - - - - - -   L     R T   G     K   A

                129                140               150             157
                  .                  .                 .
VRB   4070A     P W D T G C S K V A C G P C Y D L S K V S N S F Q G A T R
```

Figure 5. Determinants of receptor specificity for amphotropic 4070A virus. The amino acid sequences of amphotropic 4070A Env VRA from residue 46 to 74 and VRB from 129 to 157 are shown. Differences between amphotropic Env and MoMCF polytropic Env within VRA are shown below the amphotropic sequence. Hyphens denote the absence of residues. Substitution of amphotropic sequences by polytropic sequences within this region affects interaction with the amphotropic receptor (34). Regions affecting receptor binding when substituted by a VSV tag are underlined (9). Y60 and V61 (bold) have been shown to be important for infection mediated by amphotropic/FeLV-B Env hybrids (86). Changes A69G, Q72K, and V137M (bold) alter the host range from amphotropic to 10A1 (33). See text for further details.

was shown that both VRA and VRB make contributions to receptor recognition by FeLV-B. F60 and P61 of FeLV-B VRA were demonstrated to be important determinants of receptor choice (87).

CONCLUSION

The alteration of receptor specificity by the attachment of an antibody fragment or cell-binding ligand to Env has met with limited success. Further work needs to be done to characterize the interactions between binding domains and fusion domains present on separate monomers within mixed oligomers (76,102). Increased understanding from such studies should allow improvements in the efficiency of fusion after cell binding.

Pseudotyping a retrovirus with a surface protein from another virus or a receptor for another virus by using one of the transduction systems described at the beginning of this chapter provides a simple way of redirecting a retroviral gene therapy vector to a different cell type. It takes advantage of the fact that viral surface glycoproteins already have predetermined cellular specificities. Since the pseudotyping protein is already known to promote binding and entry of another virus, problems concerning the manufacture of an artificial, fusion-competent protein with a novel binding specificity can be circumvented by employing such proteins.

The recent elucidation of the three-dimensional structure of the receptor-binding domain of FrMLV Env aids in the understanding of results from a number of experimental studies that have characterized Env-receptor interactions. The contributions of both VRA and VRB to receptor specificity are readily explainable by the close proximity of these two domains in the molecule. The experimental data discussed above indicate that multiple amino acid residues are important for binding to the receptor. Some of these residues are indicated in Figure 4. A region on the surface of the Env molecule delineated by these residues is presumably responsible for interactions with the receptor. Residues implicated in receptor binding by amphotropic 4070A and FeLV Env proteins are indicated in Figures 5 and 6, respectively.

Given the difficulty of constructing highly functional hybrid Env proteins that contain bulky retargeting domains, alternative approaches need to be attempted. These could involve making small localized changes in the receptor-binding site of Env. The binding site information discussed above provides a useful starting point for making such changes.

One strategy for remodeling the receptor-binding site for retargeting involves the creation of a library of *env* genes that have incorporated random amino acid residues at positions determined to be important for receptor specificity. This would then be

FeLV A	^{48} V G D T W E P I V L N P T N V K H G A R Y S S S K Y G C K^{76}
FeLV C	^{48} V G D T W E P M A P D P - - - R S W A R Y S S S T H G C K^{73}
4070A	^{48} V G E E W D P S D Q E P - - - - - Y V G Y - - - - - G C K^{66}

Figure 6. Determinants of receptor specificity for a C-subgroup FeLV. Shown are sequences of FeLV-A strain A/Glasgow (top) and FeLV-C strain C/FZ215 (middle) within the N-terminal portions of Vr1. Hyphens denote the absence of residues. Replacing the A/Glasgow sequence in this region by the C/FZ215 sequence switches the tropism of the virus from A to C (77). A portion of the VRA region of amphotropic 4070A Env is shown for comparison (bottom).

followed by screening for mutant Env proteins that target desired cell types. A basic outline for such a strategy is depicted in Figure 7. In this approach, the *env* genes containing random nucleotide sequences in the receptor targeting site (*env**) are carried on a packageable retroviral cassette along with a drug resistance marker (Figure 7A). First, plasmids containing this cassette are transfected into cells expressing retroviral core particles (Figure 7B). Virions produced from these cells will express mutant

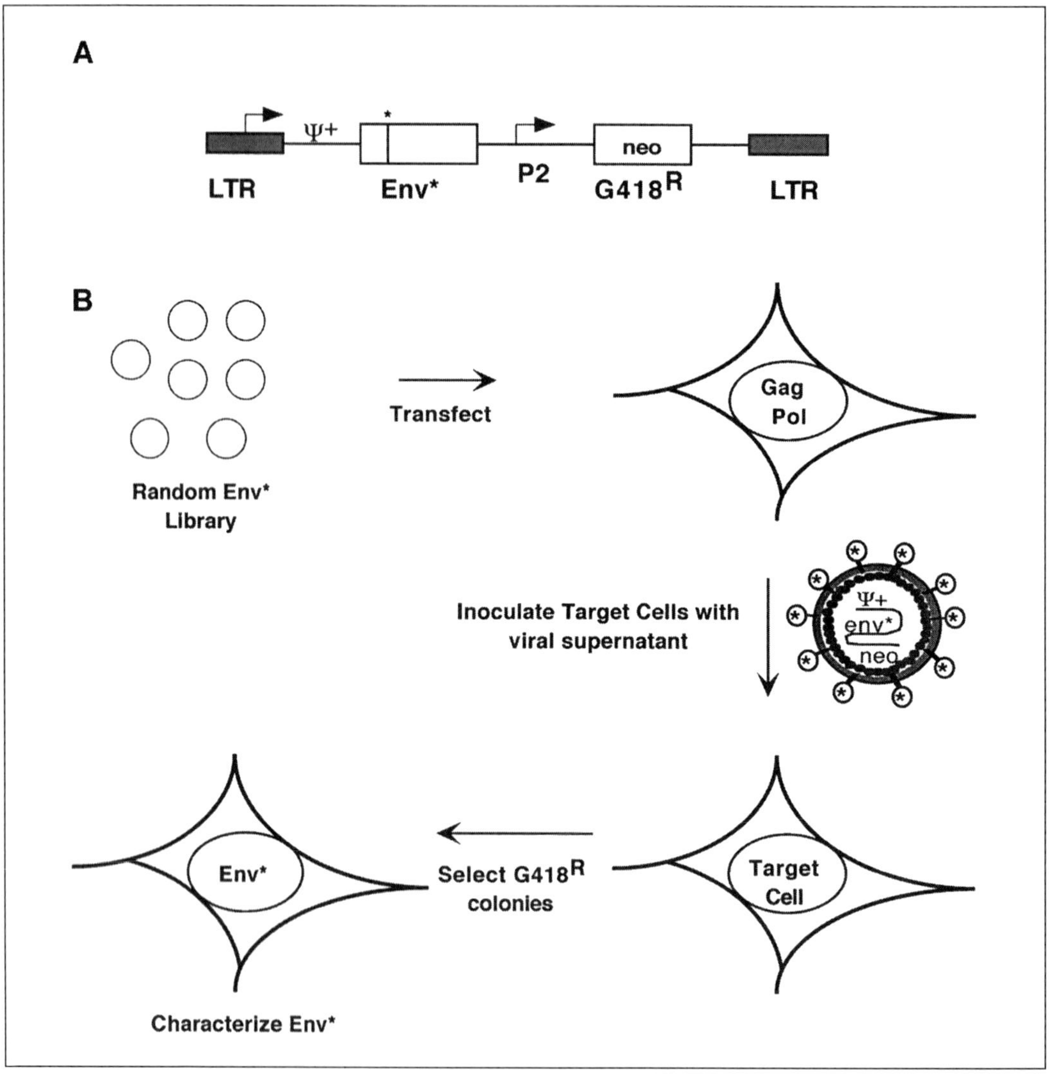

Figure 7. Use of a random Env library to select Env proteins with novel targeting properties. (A) Schematic diagram of a retroviral vector expressing Env proteins containing random amino acid sequences within the receptor targeting site (Env*). The position of the random sequence is denoted by an asterisk. The Env* proteins are expressed from the retroviral long terminal repeat (LTR) promoter. The G418 resistance gene is expressed from a second promoter (P2). The presence of the Ψ sequence allows the transcript to be packaged into a retrovirus. (B) To select Env* proteins capable of mediating binding and entry of cells to be targeted, a library of plasmids containing the construct shown in A are transfected into cells that express the retroviral core proteins Gag and Pol. Viral supernatants from these transfected cells are harvested and used to inoculate the cells to be targeted. If an Env* protein contains an amino acid sequence within its receptor targeting site that allows it to mediate viral entry, then a retroviral transcript containing that Env* variant gene will be transferred to the targeted cells and give rise to $G418^R$ colonies. The novel Env* gene can then be further characterized.

Env* proteins on their surface and will carry the cassettes. Env* proteins that contain appropriate novel amino acid sequences will mediate binding and entry into the desired cell type. Incorporation of the cassette into the genome of the infected cell will lead to the acquisition of drug resistance. The Env* protein present within such a drug-resistant clone can then be further characterized. This strategy thus allows the direct selection of fully functional novel Env proteins. Variations of this strategy are also possible.

One of the advantages of this approach over using a preselected ligand with known cell-binding properties is that it does not require a predetermined conformationally active orientation to fit onto the Env protein. It also does not necessarily entail foreknowledge of a cell type-specific receptor. One technical limitation of the strategy is that only a single linear region of the Env sequence can be randomized to create a library. Since VRA and VRB interact directly or indirectly with the receptor in both amphotropic and ecotropic Env, randomization of sequences would be more effective in an Env molecule that contains minimal VRB sequences such as FeLV-A. The goal of this strategy differs from that of using phage display to isolate novel antibodies with high affinity for specific ligands in that the Env protein does not simply supply a binding function. However, the successful use of a random library to alter the physical properties of an enzyme (50) suggests that it should succeed.

Note added in proof: cDNA clones encoding receptors for xenotropic and polytropic retroviruses have now been isolated and characterized in the following publications: Battini, J.-L., J.E.L. Rasko, and A.D. Miller. 1999. A human cell surface receptor for xenotropic and polytropic murine leukemia viruses: possible role in G protein-coupled signal transduction. Proc. Natl. Acad. Sci. USA *96*:1385-1390. C.S. Tailor, A. Nouri, C.G. Lee, C. Kozak, and D. Kabat. 1999. Cloning and characterization of a cell surface receptor for xenotropic and polytropic murine leukemia viruses. Proc. Natl. Acad. Sci. USA *96*:927-932.

ACKNOWLEDGMENTS

We would like to thank Victor Stollar and members of our laboratory for useful comments on the manuscript. We would also like to thank Paula Cannon for providing us with results prior to publication. This research is supported by Grant No. RO1CA49932 from the NIH.

REFERENCES

1. **Ager, S., B.H.K. Nilson, F.J. Morling, K.W. Peng, F.-L. Cosset, and S.J. Russell.** 1996. Retroviral display of antibody fragments; interdomain spacing strongly influences vector infectivity. Hum. Gene Ther. *7*:2157-2164.
2. **Akkina, R.K., R.M. Walton, M.L. Chen, Q.-X. Li, V. Planelles, and I.S.Y. Chen.** 1996. High-efficiency gene transfer into CD34+ cells with a human immunodeficiency virus type 1-based retroviral vector pseudotyped with vesicular stomatitis virus envelope glycoprotein G. J. Virol. *70*:2581-2585.
3. **Albritton, L.M., L. Tseng, D. Scadden, and J.M. Cunningham.** 1989. A putative murine ecotropic retrovirus receptor gene encodes a multiple membrane-spanning protein and confers susceptibility to virus infection. Cell *57*:659-666.
4. **Alzari, P.M., M.-B. Lascombe, and R.J. Poljak.** 1988. Three-dimensional structure of antibodies. Ann. Rev. Immunol. *6*:555-580.
5. **Bae, Y., S.M. Kingsman, and A.J. Kingsman.** 1997. Functional dissection of the Moloney murine leukemia virus envelope protein gp70. J. Virol. *71*:2092-2099.
6. **Battini, J.-L., O. Danos, and JM. Heard.** 1995. Receptor-binding domain of murine leukemia virus envelope glycoproteins. J. Virol. 69:713-719.
7. **Battini, J.-L., J.M. Heard, and O. Danos.** 1992. Receptor choice determinants in the envelope glycoproteins of amphotropic, xenotropic, and polytropic murine leukemia viruses. J. Virol. *66*:1468-1474.
8. **Battini, J.-L., P. Rodrigues, R. Muller, O. Danos, and J.M. Heard.** 1996. Receptor-binding properties of a purified fragment of the 4070A amphotropic murine leukemia virus envelope glycoprotein. J. Virol. *70*: 4387-4393.
9. **Battini, J.-L., O. Danos, and J.M. Heard.** 1998. Def-

inition of a 14-amino-acid peptide essential for the interaction between the murine leukemia virus amphotropic envelope glycoprotein and its receptor. J. Virol. *72*:428-435.

10. **Benedict, C.A., R.Y.M. Tun, D.B. Rubinstein, T. Guillaume, P.M. Cannon, and W.F. Anderson.** 1999. Targeting retroviral vectors to CD34-expressing cells: binding to CD34 does not catalyze virus-cell fusion. Hum. Gene Ther. *10*:545-557.
11. **Besmer, P. and D. Baltimore.** 1977. Mechanism of restriction of ecotropic and xenotropic murine leukemia viruses and formation of pseudotypes between the two viruses. J. Virol. *21*:965-973.
12. **Boomer, S., M. Eiden, C.C. Burns, and J. Overbaugh.** 1997. Three distinct envelope domains, variably present in subgroup B feline leukemia virus recombinants, mediate Pit1 and Pit2 receptor recognition. J. Virol. *71*:8116-8123.
13. **Brojatsch, J., B.S. Kristal, G.A. Viglianti, R. Khiroya, E.A. Hoover, and J.I. Mullins.** 1992. Feline leukemia virus subgroup C phenotype evolves through distinct alterations near the N terminus of the envelope surface glycoprotein. Proc. Natl. Acad. Sci. USA *89*:8457-8461.
14. **Burns, J.C., T. Friedman, W. Driever, M. Burrascano, and J.-K. Yee.** 1993. Vesicular stomatitis virus G glycoprotein pseudotyped retroviral vectors: concentration to very high titer and efficient gene transfer into mammalian and nonmammalian cells. Proc. Natl. Acad. Sci. USA *90*:8033-8037.
15. **Chu, T.-H.T. and R. Dornburg.** 1995. Retroviral vector particles displaying the antigen-binding site of an antibody enable cell-type-specific gene transfer. J. Virol. *69*:2659-2663.
16. **Chu, T.-H.T. and R. Dornburg.** 1997. Toward highly efficient cell-type-specific gene transfer with retroviral vectors displaying single-chain antibodies. J. Virol. *71*:720-725.
17. **Cosset, F.-L., F.J. Morling, Y. Takeuchi, R.A. Weiss, M.K.L. Collins, and S.J. Russell.** 1995. Retroviral retargeting by envelopes expressing an N-terminal binding domain. J. Virol. *69*:6314-6322.
18. **Cosset, F.-L. and S.J. Russell.** 1996. Targeting retrovirus entry. Gene Ther. *3*:946-956.
19. **Cosset, F.-L., Y. Takeuchi, J.-L. Battini, R.A. Weiss, and M.K.L. Collins.** 1995. High-titer packaging cells producing recombinant retroviruses resistant to human serum. J. Virol. *69*:7430-7436.
20. **Danos, O. and R.C. Mulligan.** 1988. Safe and efficient generation of recombinant retroviruses with amphotropic and ecotropic host ranges. Proc. Natl. Acad. Sci. USA *85*:6460-6464.
21. **Davey, R.A., C.A. Hamson, J.J. Healy, and J.M. Cunningham.** 1997. *In vitro* binding of purified murine ecotropic retrovirus envelope surface protein to its receptor, MCAT-1. J. Virol. *71*:8096-8102.
22. **Dong, J., M.G. Roth, and E. Hunter.** 1992. A chimeric avian retrovirus containing the influenza virus hemagglutinin gene has an expanded host range. J. Virol. *66*:7374-7382.
23. **Donohue, P.R., E.A. Hoover, G.A. Beltz, N. Riedel, V.M. Hirsch, J. Overbaugh, and J.I. Mullins.** 1988. Strong sequence conservation among horizontally transmissible, minimally pathogenic feline leukemia viruses. J. Virol. *62*:722-731.
24. **Emi, N., T. Friedmann, and J.-K. Yee.** 1991. Pseudotype formation of murine leukemia virus with the G protein of vesicular stomatitis virus. J. Virol. *65*:1202-1207.
25. **Endres, M.J., S. Jaffer, B. Haggarty, J.D. Turner, B.J. Doranz, P.J. O' Brien, D.L. Kolson, and J.A. Hoxie.** 1997. Targeting of HIV- and SIV-infected cells by CD4-chemokine receptor pseudotypes. Science *278*:1462-1464.
26. **Fass, D.** 1997. The protein structures underlying receptor binding and membane fusion of ecotropic murine leukemia viruses [PhD thesis]. Boston, MA: Massachusetts Institute of Technology.
27. **Fass, D., R.A. Davey, C.A. Hamson, P.S. Kim, J.M. Cunningham, and J.M. Berger.** 1997. Structure of a murine leukemia virus receptor-binding glycoprotein at 2.0 Angstrom resolution. Science *277*:1662-1666.
28. **Fass, D., S.C. Harrison, and P.S. Kim.** 1996. Retrovirus envelope domain at 1.7 Å resolution. Nat. Struct. Biol. *3*:465-469.
29. **Fielding, A.K., M. Maurice, F.J. Morling, F.-L. Cosset, and S.J. Russell.** 1998. Inverse targeting of retroviral vectors: selective gene transfer in a mixed population of hematopoietic and nonhematopoietic cells. Blood *91*:1802-1809.
30. **Gray, K.D. and M.J. Roth.** 1993. Mutational analysis of the envelope gene of Moloney murine leukemia virus. J. Virol. *67*:3489-3496.
31. **Griffiths, A.D. and A.R. Duncan.** 1998. Strategies for selection of antibodies by phage display. Curr. Opin. Biotechnol. *9*:102-108.
32. **Hall, F.L., E.M. Gordon, L. Wu, N.L. Zhu, M.J. Skotzko, V.A. Starnes, and W.F. Anderson.** 1997. Targeting retroviral vectors to vascular lesions by genetic engineering of the Mo-MuLV gp70 envelope protein. Hum. Gene Ther. *8*:2183-2192.
33. **Han, J.-Y., P.M. Cannon, K.-M. Lai, Y. Zhao, M.V. Eiden, and W.F. Anderson.** 1997. Identification of envelope protein residues required for the expanded host range of 10A1 murine leukemia virus. J. Virol. *71*:8103-8108.
34. **Han, J.-Y., Y. Zhao, W.F. Anderson, and P.M. Cannon.** 1998. Role of variable regions A and B in receptor binding domain of amphotropic murine leukemia virus envelope protein. J. Virol. *72*:9101-9108.
35. **Han, L., T. Hofmann, Y. Chiang, and W.F. Anderson.** 1995. Chimeric envelope glycoproteins constructed between amphotropic and xenotropic murine leukemia retroviruses. Somatic Cell Mol. Genet. *21*:205-214.
36. **Han, X., N. Kasahara, and Y.W. Kan.** 1995. Ligand-directed retroviral targeting of human breast cancer cells. Proc. Natl. Acad. Sci. USA *92*:9747-9751.
37. **Hatziioannou, T., S. Valsesia-Wittmann, S.J. Russell, and F.-L. Cosset.** 1998. Incorporation of fowl plague virus hemagglutinin into murine leukemia virus particles and analysis of the infectivity of the pseudotyped retroviruses. J. Virol. *72*:5313-5317.
38. **Heard, J.M. and O. Danos.** 1991. An amino-terminal fragment of the Friend murine leukemia virus envelope glycoprotein binds the ecotropic receptor. J.

Virol. *65*:4026-4032.
39.**Hoogenboom, H.R.** 1997. Designing and optimizing library selection strategies for generating high-affinity antibodies. Trends Biotechnol. *15*:62-70.
40.**Jarrett, O., H.M. Laird, and D. Hay.** 1973. Determinants of the host range of feline leukaemia viruses. J. Gen. Virol. *20*:169-175.
41.**Jiang, A., T.-H.T. Chu, F. Nocken, K. Cichutek, and R. Dornburg.** 1998. Cell-type-specific gene transfer into human cells with retroviral vectors that display single-chain antibodies. J. Virol. *72*:10148-10156.
42.**Kasahara, N.** 1994. Cell-specific targeting of retroviral vectors via ligand-receptor interaction [PhD thesis]. San Francisco, CA: University of California.
43.**Kasahara, N., A. Dozy, and Y.W. Kan.** 1995. Targeting retroviral vectors to specific cells. Science *269*:417.
44.**Kasahara, N., A.M. Dozy, and Y.W. Kan.** 1994. Tissue-specific targeting of retroviral vectors through ligand-receptor interactions. Science *266*:1373-1375.
45.**Kavanaugh, M.P., D.G. Miller, W. Zhang, W. Law, S.L. Kozak, D. Kabat, and A.D. Miller.** 1994. Cell-surface receptors for gibbon ape leukemia virus and amphotropic murine retrovirus are inducible sodium-dependent phosphate symporters. Proc. Natl. Acad. Sci. USA *91*:7071-7075.
46.**Kayman, S.C., H. Park, M. Saxon, and A. Pinter.** 1999. The hypervariable domain of the murine leukemia virus surface protein tolerates large insertions and deletions, enabling development of a retroviral particle display system. J. Virol. *73*:1802-1808.
47.**Kim, J.W., E.I. Closs, L.M. Albritton, and J.M. Cunningham.** 1991. Transport of cationic amino acids by the mouse ecotropic retrovirus receptor. Nature *352*:725-728.
48.**Konishi, H., T. Ochiya, K.A. Chester, R.H.J. Begent, T. Muto, T. Sugimura, and M. Terada.** 1998. Targeting strategy for gene delivery to carcinoembryonic antigen-producing cancer cells by retrovirus displaying a single-chain variable fragment antibody. Hum. Gene Ther. *9*:235-248.
49.**Lee, S.-Y., T.M. Howard, and S. Rasheed.** 1998. Genetic analysis of the rat leukemia virus: influence of viral sequences in transduction of the c-*ras* proto-oncogene and expression of its transforming activity. J. Virol. *72*:9906-9917.
50.**MacBeath, G., P. Kast, and D. Hilvert.** 1998. Redesigning enzyme topology by directed evolution. Science *279*:1958-1961.
51.**MacKrell, A.J., N.W. Soong, C.M. Curtis, and W.F. Anderson.** 1996. Identification of a subdomain in the Moloney murine leukemia virus envelope protein involved in receptor binding. J. Virol. *70*:1768-1774.
52.**Mann, R., R.C. Mulligan, and D. Baltimore.** 1983. Construction of a retrovirus packaging mutant and its use to produce helper-free defective retrovirus. Cell *33*:153-159.
53.**Marin, M., D. Noel, S. Valsesia-Wittman, F. Brockly, M. Etienne-Julan, S. Russell, F.-L. Cosset, and M. Piechaczyk.** 1996. Targeted infection of human cells via major histocompatibility complex class I molecules by Moloney murine leukemia virus-derived viruses displaying single-chain antibody fragment-envelope fusion proteins. J. Virol. *70*:2957-2962.
54.**Martin, F., J. Kupsch, Y. Takeuchi, S. Russell, F.-L. Cosset, and M. Collins.** 1998. Retroviral vector targeting to melanoma cells by single-chain antibody incorporation in envelope. Hum. Gene Ther. *9*:737-746.
55.**Masuda, M., C.A. Hanson, W.G. Alvord, P.M. Hoffman, S.K. Ruscetti, and M. Masuda.** 1996. Effects of subtle changes in the SU protein of ecotropic murine leukemia virus on its brain capillary endothelial cell tropism and interference properties. Virology *215*:142-151.
56.**Masuda, M., C.A. Hanson, N.V. Dugger, D.S. Robbins, S.G. Wilt, S.K. Ruscetti, and P.M. Hoffman.** 1997. Capillary endothelial cell tropism of PVC-211 murine leukemia virus and its application for gene transduction. J. Virol. *71*:6168-6173.
57.**Masuda, M., M. Masuda, C.A. Hanson, P.M. Hoffman, and S.K. Ruscetti.** 1996. Analysis of the unique hamster cell tropism of ecotropic murine leukemia virus PVC-211. J. Virol. *70*:8534-8539.
58.**McClure, M.O., M.A. Sommerfelt, M. Marsh, and R.A. Weiss.** 1990. The pH independence of mammalian retrovirus infection. J. Gen. Virol. *71*:767-773.
59.**Miller, D.G., R.H. Edwards, and A.D. Miller.** 1994. Cloning of the cellular receptor for amphotropic murine retroviruses reveals homology to that for gibbon ape leukemia virus. Proc. Natl. Acad. Sci. USA *91*:78-82.
60.**Miller, D.G. and A.D. Miller.** 1994. A family of retroviruses that utilize related phosphate transporters for cell entry. J. Virol. *68*:8270-8276.
61.**Morgan, R.A., O. Nussbaum, D.D. Muenchau, L. Shu, L. Couture, and W.F. Anderson.** 1993. Analysis of the functional and host range-determining regions of the murine ecotropic and amphotropic retrovirus envelope proteins. J. Virol. *67*:4712-4721.
62.**Nguyen, T.H., J.-C. Pages, D. Farge, P. Briand, and A. Weber.** 1998. Amphotropic retroviral vectors displaying hepatocyte growth factor-envelope fusion proteins improve transduction efficiency of primary hepatocytes. Hum. Gene Ther. *9*:2469-2479.
63.**Nilson, B.H.K., F.J. Morling, F.-L. Cosset, and S.J. Russell.** 1996. Targeting of retroviral vectors through protease-substrate interactions. Gene Ther. *3*:280-286.
64.**O'Hara, B., S.V. Johann, H.P. Klinger, D.G. Blair, H. Rubinson, K.J. Dunn, P. Sass, S.M. Vitek, and T. Robins.** 1990. Characterization of a human gene conferring sensitivity to infection by gibbon ape leukemia virus. Cell Growth Differ. *1*:119-127.
65.**Opstelten, D.-J.E., M. Wallin, and H. Garoff.** 1998. Moloney murine leukemia virus envelope protein subunits, gp70 and Pr15E, form a stable disulfide-linked complex. J. Virol. *72*:6537-6545.
66.**O'Reilly, L. and M.J. Roth.** 2000. Second-site changes affect viability of amphotropic/ecotropic chimeric enveloped murine leukemia viruses. J. Virol. *74*:899-913.
67.**Ott, D., R. Friedrich, and A. Rein.** 1990. Sequence analysis of amphotropic and 10A1 murine leukemia viruses: close relationship to mink cell focus-inducing viruses. J. Virol. *64*:757-766.

68. **Ott, D. and A. Rein.** 1992. Basis for receptor specificity of nonecotropic murine leukemia virus surface glycoprotein gp70 SU. J. Virol. *66*:4632-4638.

69. **Park, B.H., B. Matuschke, E. Lavi, and G.N. Gaulton.** 1994. A point mutation in the *env* gene of a murine leukemia virus induces syncytium formation and neurologic disease. J. Virol. *68*:7516-7524.

70. **Peng, K.-W., F.J. Morling, F.-L. Cosset, G. Murphy, and S.J. Russell.** 1997. A gene delivery system activatable by disease-associated matrix metalloproteinases. Hum. Gene Ther. *8*:729-738.

71. **Peredo, C., L. O'Reilly, K. Gray, and M.J. Roth.** 1996. Characterization of chimeras between the ecotropic Moloney murine leukemia virus and the amphotropic 4070A envelope proteins. J. Virol. *70*:3142-3152.

72. **Pinter, A., R. Kopelman, Z. Li, S.C. Kayman, and D.A. Sanders.** 1997. Localization of the labile disulfide bond between SU and TM of the murine leukemia virus envelope protein complex to a highly conserved CWLC motif in SU that resembles the active-site sequence of thiol-disulfide exchange enzymes. J. Virol. *71*:8073-8077.

73. **Ramsey, I.K., N. Spibey, and O. Jarrett.** 1998. The receptor binding site of feline leukemia virus surface glycoprotein is distinct from the site involved in virus neutralization. J. Virol. *72*:3268-3277.

74. **Rein, A.** 1982. Interference grouping of murine leukemia viruses: a distinct receptor for the MCF-recombinant viruses in mouse cells. Virology *120*:251-257.

75. **Rein, A. and A. Schultz.** 1984. Different recombinant murine leukemia viruses use different cell surface receptors. Virology *136*:144-152.

76. **Rein, A., C. Yang, J.A. Haynes, J. Mirro, and R.W. Compans.** 1998. Evidence for cooperation between murine leukemia virus env molecules in mixed oligomers. J. Virol. *72*:3432-3435.

77. **Riedel, N., E.A. Hoover, R.E. Dornsife, and J.I. Mullins.** 1988. Pathogenic and host range determinants of the feline aplastic anemia retrovirus. Proc. Natl. Acad. Sci. USA *85*:2758-2762.

78. **Rigby, M.A., J.L. Rojko, M.A. Stewart, G.J. Kociba, C.M. Cheney, L.J. Rezanka, L.E. Mathes, J.R. Hartke, O. Jarrett, and J.C. Neil.** 1992. Partial dissociation of subgroup C phenotype and *in vivo* behaviour in feline leukaemia viruses with chimeric envelope genes. J. Gen. Virol. *73*:2839-2847.

79. **Russell, S.J., R.E. Hawkins, and G. Winter.** 1993. Retroviral vectors displaying functional antibody fragments. Nucleic Acids Res. *21*:1081-1085.

80. **Sarma, P.S., T. Log, D. Jain, P.R. Hill, and R.J. Huebner.** 1975. Differential host range of viruses of feline leukemia-sarcoma complex. Virology *64*:438-446.

81. **Schnierle, B.S., D. Moritz, M. Jeschke, and B. Groner.** 1996. Expression of chimeric envelope proteins in helper cell lines and integration into Moloney murine leukemia virus particles. Gene Ther. *3*:334-342.

82. **Schnierle, B.S., J. Stitz, V. Bosch, F. Nocken, H. Merget-Millitzer, M. Engelstadter, R. Kurth, B. Groner, and K. Cichutek.** 1997. Pseudotyping of murine leukemia virus with the envelope glycoproteins of HIV generates a retroviral vector with specificity of infection for CD4-expressing cells. Proc. Natl. Acad. Sci. USA *94*:8640-8645.

83. **Snitkovsky, S. and J.A.T. Young.** 1998. Cell-specific viral targeting mediated by a soluble retroviral receptor-ligand fusion protein. Proc. Natl. Acad. Sci. USA *95*:7063-7068.

84. **Somia, N.V., M. Zoppe, and I.M. Verma.** 1995. Generation of targeted retroviral vectors by using single-chain variable fragment: an approach to in vivo gene delivery. Proc. Natl. Acad. Sci. USA *92*:7570-7574.

85. **Soneoka, Y., P.M. Cannon, E.E. Ramsdale, J.C. Griffiths, G. Romano, S.M. Kingsman, and A.J. Kingsman.** 1995. A transient three-plasmid expression system for the production of high titer retroviral vectors. Nucleic Acids Res. *23*:628-633.

86. **Spiegel, M., M. Bitzer, A. Schenk, H. Rossman, W.J. Neubert, U. Seidler, M. Gregor, and U. Lauer.** 1998. Pseudotype formation of Moloney murine leukemia virus with Sendai virus glycoprotein F. J. Virol. *72*:5296-5302.

87. **Tailor, C.S. and D. Kabat.** 1997. Variable regions A and B in the envelope glycoproteins of feline leukemia virus subgroup B and amphotropic murine leukemia virus interact with discrete receptor domains. J. Virol. *71*:9383-9391.

88. **Tanaka, A., K. Oka, K. Tanaka, A. Jinno, S.K. Ruscetti, and K. Kai.** 1998. The entire nucleotide sequence of Friend-related and paralysis-inducing PVC-441 murine leukemia virus (MuLV) and its comparison with those of PVC-211 MuLV and Friend MuLV. J. Virol. *72*:3423-3426.

89. **Teich, N.** 1984. Taxonomy of retroviruses, p. 25-207. *In* R. Weiss, N. Teich, H. Varmus, and J. Coffin (Eds.), RNA Tumor Viruses, 2nd ed., vol. 1. CSH Laboratory Press, Cold Spring Harbor, NY.

90. **Valsesia-Wittman, S., A. Drynda, G. Deleage, M. Aumailley, J.-M. Heard, O. Danos, G. Verdier, and F.-L. Cosset.** 1994. Modifications in the binding domain of avian retrovirus envelope protein to redirect the host range of retroviral vectors. J. Virol. *68*:4609-4619.

91. **Valsesia-Wittman, S., F.J. Morling, T. Hatziioannou, S.J. Russell, and F.-L. Cosset.** 1997. Receptor co-operation in retrovirus entry: recruitment of an auxiliary entry mechanism after retargeted binding. EMBO J. *16*:1214-1223.

92. **Valsesia-Wittman, S., F.J. Morling, B.H.K. Nilson, Y. Takeuchi, S.J. Russell, and F.-L. Cosset.** 1996. Improvement of retroviral retargeting by using amino acid spacers between an additional binding domain and the N terminus of Moloney murine leukemia virus SU. J. Virol. *70*:2059-2064.

93. **van Zeijl, M., S.V. Johann, E. Closs, J. Cunningham, R. Eddy, T.B. Shows, and B. O'Hara.** 1994. A human amphotropic retrovirus receptor is a second member of the gibbon ape leukemia virus receptor family. Proc. Natl. Acad. Sci. USA *91*:1168-1172.

94. **Wagner, S.D. and M.S. Neuberger.** 1996. Somatic hypermutation of immunoglobulin genes. Ann. Rev. Immunol. *14*:441-457.

95. **Wang, H., M.P. Kavanaugh, R.A. North, and D.**

Kabat. 1991. Transport of cationic amino acids by the mouse ecotropic retrovirus receptor. Nature *352*:729-731.

96. **Wilson, C.A., M.V. Eiden, W.B. Anderson, C. Lehel, and Z. Olah.** 1995. The dual-function hamster receptor for amphotropic murine leukemia virus (MuLV), 10A1 MuLV and gibbon ape leukemia virus is a phosphate symporter. J. Virol. *69*:534-537.

97. **Wilson, C.A., K.B. Farrell, and M.V. Eiden.** 1994. Properies of a unique form of the murine amphotropic leukemia virus receptor expressed on hamster cells. J. Virol. *68*:8270-8276.

98. **Wool-Lewis, R.J. and P. Bates.** 1998. Characterization of Ebola virus entry by using pseudotyped viruses: identification of receptor-deficient cell lines. J. Virol. *72*:3155-3160.

99. **Wu, B.W., P.M. Cannon, E.M. Gordon, F.L. Hall, and W.F. Anderson.** 1998. Characterization of the proline-rich region of murine leukemia virus envelope protein. J. Virol. *72*:5383-5391.

100. **Yajima, T., T. Kanda, K. Yoshiike, and Y. Kitamura.** 1998. Retroviral vector targeting human cells via c-Kit-stem cell factor interaction. Hum. Gene Ther. *9*:779-787.

101. **Yang, Z.-y., R. Delgado, L. Wu, R.F. Todd, E.G. Nabel, A. Sanchez, and G.J. Nabel.** 1998. Distinct cellular interactions of secreted and transmembrane Ebola virus glycoproteins. Science *279*:1034-1037.

102. **Zhao, Y., S. Lee, and W.F. Anderson.** 1997. Functional interactions between monomers of the retroviral envelope protein complex. J. Virol. *71*:6967-6972.

24 Efficient Production of VSV-G-Pseudotyped Retrovirus Vectors Using Stable Packaging Cell Lines

Hideo Iba[1], Tohru Arai[1,2] and Motoyasu Ui[1]
[1]*Department of Gene Regulation, Institute of Medical Science, University of Tokyo, Tokyo;* [2]*Tsukuba Research Laboratories, Eisai Co., Ltd, Tsukuba, Japan*

INTRODUCTION

Development of New Retrovirus Vectors

Vectors based on murine leukemia virus (MLV) expressing amphotropic envelope glycoprotein (Env) have been used as powerful tools for gene transfer in human gene therapy and in basic research (18). Several laboratories, however, have shown that amphotropic vectors are not able to introduce transgenes in vitro into entire populations of fibroblasts and several other cell lines even when transduction is performed at high titers of infectious vector particles (7,17). The reasons for this phenomenon are not yet fully understood, although possible explanations include the presence of inhibitory factors in amphotropic virus stocks or heterogeneity of the cell cycle phase that the host cells are experiencing at the time of infection (2,3,7,17). This feature of amphotropic vectors makes them disadvantageous in practice, especially if gene transfer into entire populations of human cells in culture is required, either for basic research purposes or for applications to gene therapy of human cancers.

Advantages of VSV-G-Pseudotyped Vectors

Previous studies have shown that when MLV-based retrovirus vectors are pseudotyped with the G protein of vesicular stomatitis virus (VSV-G), a single-stranded RNA virus belonging to the rhabdoviridae family (27), they exhibit a much broader host range than vectors using the conventional amphotropic Env (4,8,15,22,29). This probably results because the receptors for VSV-G include ubiquitous anionic phospholipids such as phosphatidylserine (15,21). Furthermore, unlike amphotropic vectors, the stocks of VSV-G-pseudotyped virus can be concentrated about 1000 times by ultracentrifugation (4,29) because of the physical stability of the pseudotyped virus particles.

In our studies, the transduction efficiency of a VSV-G-pseudotyped vector encoding a β-galactosidase that contained a

Viral Vectors: Basic Science and Gene Therapy
Edited by A. Cid-Arregui and A. García-Carrancá

Table 1. Transduction into the Entire Cellular Population by VSV-G-Pseudotyped Retrovirus

Human Solid Tumor Cell Lines	
MIA PaCa-II, PSN1	pancreatic cancer
A549, H1299, NCI-H23, NCI-H460	lung cancer
MDA-MB	breast cancer
HCT-116, Colo320DM, HT29, SW480, SW620	colon cancer
U373MG	glioblastoma
HepG2	hepatoma
U2-OS, SAOS-2	osteogenic sarcinoma
HT1080, PtG-S2	fibrosarcoma
Cells from Other Species	
NIH3T3	mouse fibroblasts
ATDC5	mouse chondrogenic EC cells
3Y1	rat fibroblasts
MDCK	dog kidney cells
rat embryonic fibroblasts	
rabbit primary chondrocytes	
chicken primary fibroblasts	

nuclear localization signal was higher in human tumor cell lines, murine fibroblasts, and several other mammalian and avian cell lines, as compared with that of an amphotropic vector carrying the same marker. The ratio of cells transduced with the VSV-G-pseudotyped vector corresponded closely to $1\text{-}e^{-moi}$, as predicted from a Poisson distribution of transduction to the entire cellular population (2); this was not the case for the amphotropic vector. Here, multiplicity of infection (moi) is defined as the ratio of input infectious units (iu) (titrated on the corresponding cell line) to the number of cells used for the transduction. At high moi (more than 3), the VSV-G-pseudotyped vector induced lacZ expression in about 95% of the culture population in all cell lines examined (Table 1). In a single transduction event, multiple copies of viral DNA could be easily transferred as provirus into the host genome. Further, cells such as NIH 3T3 and 3Y1 could be grown for more than two weeks without any loss of lacZ-expressing cells, thereby confirming that the observed gene transfer was stable and was not pseudotransduction as reported previously in other studies (9,14). The transduction efficiency of amphotropic vectors, however, has been reported not to be dose dependent and to reach a plateau or even decrease, especially at high moi (2,17). The tight dose-dependent transduction by VSV-G-pseudotyped vectors should be highly advantageous for gene transfer into entire populations of human tumor cell lines at an appropriate dose (2).

Strategies for Production of VSV-G-Pseudotyped Vectors

The cytotoxicity of VSV-G protein expression for most mammalian cells imposes a significant growth disadvantage to the transduced cells. Thus, while titers of pseudotyped vector stocks of 1×10^5 to 1×10^6 iu/mL have been obtained after transient expression of the VSV-G gene by transfection into cell lines that constitutively express the *gag* and *pol* genes, these systems would not be suitable for reproducible, large-scale preparations of certified vectors (i.e., those of tested quality that show no contamination by replication-competent retrovirus, for example). Because of the cytotoxicity of VSV-G, however, the generation of stable packaging cell lines for production of such pseudotyped vectors has been difficult in spite of their potential advantages for gene transfer.

We have recently developed a new system suitable for large-scale production of VSV-G-pseudotypes (1,2,26). To this end, we first constructed prepackaging cell lines that express the *gag* and *pol* genes and harbor a completely silent *VSV-G* gene. The prepackaging cell lines were designed to be converted into packaging cell lines that produce high levels of VSV-G. Such conversion was carried out via *lox*P-specific recombination, for which the Cre recombinase was expressed in the prepackaging cells using a replication-defective adenovirus (Ad) vector. High-titer virus stocks of VSV-G-pseudotyped retrovirus (1×10^6 iu/mL) were reproducibly produced after three days following Ad infection. Neither RCR nor Ad vector was detected in purified virus stocks of 1×10^7 iu of VSV-G-pseudotyped vector prepared by this method. In this chapter, we will describe the Cre-*lox*P mediated system for VSV-G-pseudotyped vector production.

THE PtG-S2 PREPACKAGING CELL LINE

FLY cells, a derivative of the HT1080 (human fibrosarcoma) cell line, express the *gag* and *pol* genes of Moloney murine leukemia retrovirus (MoMLV) at high levels and have been used as efficient packaging cells after introduction of the *env* gene of either amphotropic virus or RD114 (a feline endogenous retrovirus) (6). We made use of the FLY cell line for the generation of a new prepackaging cell line, PtG-S2 (Figure 1), which harbors an inducible transcriptional unit for the *VSV-G* gene (pCALNdLG). In the pCALNdLG construct, the *VSV-G* gene is preceded by a neo^R gene flanked by a tandem pair of *lox*P sequences. The transcription of neo^R is designed to be driven from a CAG (10,19) promoter present upstream of the 5′-*lox*P sequence and to be terminated by a polyadenylation signal located upstream of the 3′-*lox*P sequence. Therefore, the *VSV-G* gene is not transcribed from the CAG promoter in the prepackaging form of PtG-S2. After site-specific recombination between the pair of *lox*P sequences by Cre recombinase, the neo^R gene and the poly A signal are removed from the chromosomal DNA as a closed circular molecule, and the CAG promoter becomes available for active transcription of the *VSV-G* gene (Figure 2) to produce VSV-G pseudotypes.

For the introduction of a Cre recombinase gene into the prepackaging cells to convert them into packaging cells, we have chosen a replication-defective Ad vector, AxCANCre, which encodes Cre recombinase containing a nuclear localization signal (10,12). This vector has been shown to efficiently infect human cells, to express very high levels of the exogenous gene within a short period after infection, and to cause efficient excision of DNA fragments contained between a pair of *lox*P sequences.

Introduction of Vector DNA into PtG-S2 Cells

In the first step for retrovirus production, the MLV-based vector DNA should be introduced into PtG-S2 cells. If MLV-based vectors carrying amphotropic Env are available, these could be easily transduced into PtG-S2 cells for their conversion into VSV-G-pseudotyped vectors. The PtG-S2 cell line is highly susceptible to amphotropic vectors. Alternatively, proviral vector DNA could be introduced into PtG-S2 by transfection of its DNA. In this case, cloning of transfectants that express vector RNA at high levels would be helpful to prepare high-titer virus stocks after transduction with the vector that expresses Cre recombinase.

Conditions for Optimal Production of Pseudotyped Virus by PtG-S2 Cells

PtG-S2 cells were transduced with a MLV-based vector, MFGnlslacZ, with amphotropic Env at an moi of 3. Mixed populations of transduced cells were used for screening of optimal conditions for virus production. Our results can be summarized as follows (1):

1. Before introduction of Cre recombinase, the culture did not produce any transmissible viral particles, even though the cells were kept growing for more than nine months. Such a culture, however, was shown to retain full vector-producing activity after transduction with the Cre recombinase-expressing vector.

2. From the third day after infection with Ad vector carrying the Cre recombinase gene, production of high levels of VSV-G-pseudotyped vector (approximate-

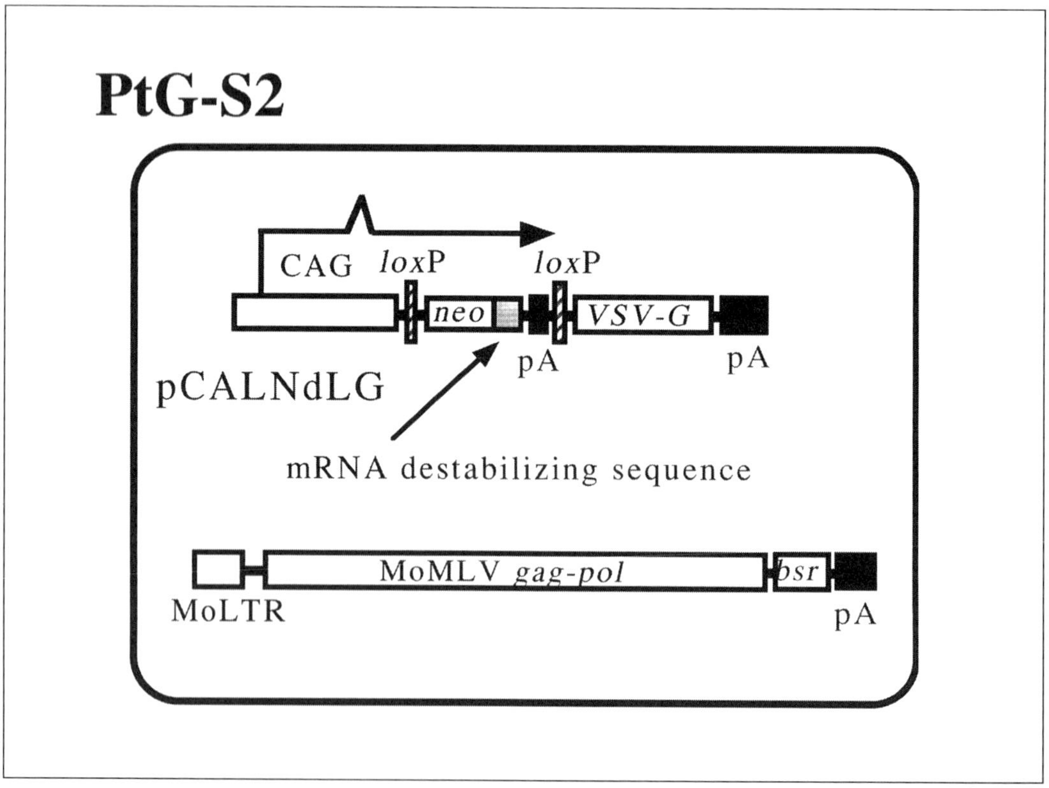

Figure 1. The prepackaging cell line PtG-S2 (1). PtG-S2 was generated by transfecting the plasmid pCALNdLG into FLY cells (a derivative of the human fibrosarcoma cell line HT1080) expressing high levels of the *gag* and *pol* genes from MoMLV. In the prepackaging cell line, the neo^R gene is transcribed from a CAG promoter, while the *VSV-G* gene is completely silent because the RNA transcript terminates just before the beginning of its coding sequence. Arrows indicate the predicted transcripts. The 3′-untranslated region of the neo^R transcripts from pCALNdLG includes an mRNA destabilizing sequence originated from the 3′-noncoding sequence of the chicken c-*fos* gene to reduce the relative amount of neo^R transcripts. When FLY is exposed to G418 after transfection with pCALNdLG, it is possible to efficiently select cell clones in which the synthesis of transcripts from the CAG promoter is sufficiently active so as to keep the unstable neo^R transcript at the high levels required for survival. The PtG-S2 clone was selected among 250 clones analyzed because of its higher capacity for virus production.

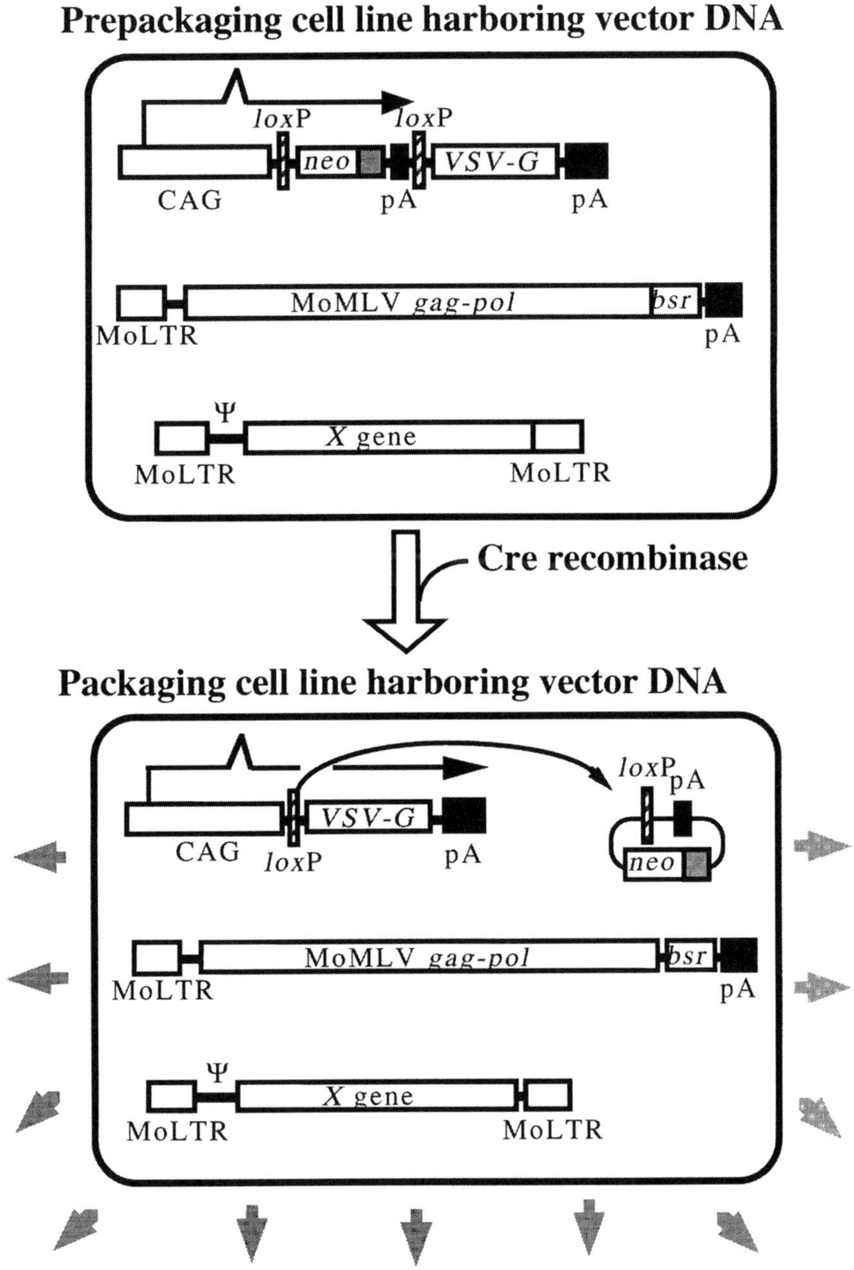

Figure 2. Schematic representation of the conversion from a prepackaging to a packaging cell line (modified from Reference 1). A PtG-S2 cell line harboring proviral DNA was generated by either transduction of amphotropic vector or DNA transfection. The Cre recombinase excises the *neo*^{R} gene, the mRNA destabilizing signal, and the polyA signal by site-specific recombination between the two *lox*P sequences and thus converts the prepackaging cell line into a packaging cell line within four days after infection with adenovirus carrying the Cre recombinase gene. In the packaging cell line, the *VSV-G* gene is transcribed from the same promoter used previously for the *neo*^{R} expression. MoLTR, MoMLV long terminal repeat; MoMLV *gag-pol*, MoMLV *gag* and *pol* genes; *bsr*, blasticidin resistance gene; pA, polyadenylation signal; Ψ, packaging signal of retrovirus vector.

ly 1 × 10^6 iu/mL) was observed, which continued for the next three to four days.

3. The structure of the pCALNdLG DNA construct in the host genome was fully processed as expected within four days after Cre recombinase introduction. Excision of the DNA fragment flanked by the *lox*P sequences was accompanied by the loss of *neo*^{R} expression and also by the induction of high levels of expression of VSV-G.

4. The moi of AxCANCre virus required to obtain high levels of production of pseudotyped retrovirus was about 3 to 15. Higher moi could be cytotoxic for PtG-S2. The kinetics and yield of virus was not significantly different if the cells were maintained at 37°C or at 32°C. However, a growth temperature of 32°C seemed preferable for better reproducibility.

PROTOCOLS

Transduction of Amphotropic Vectors into PtG-S2 Cells

Reagents

- Culture medium for PtG-S2 and its derivatives (for routine cell growth): Dulbecco's modified Eagle's medium (DMEM) (high-glucose) (Life Technologies, Rockville, MD, USA), 10% fetal calf serum (FCS) (Life Technologies), 4 µg/mL blasicidin S (Funakoshi, Tokyo, Japan; KK-400), 1 mg/mL G418 (Life Technologies 11811-031), penicillin, and streptomycin
- Polybrene: Stock solution (10 mg/mL in phosphate-buffered saline [PBS])

Procedure

1. Seed 7.5 × 10^5 PtG-S2 cells on a 90-mm plate.

2. The next day, add polybrene to a final concentration of 8 µg/mL and infect with amphotropic vectors using an moi of 3. Susceptibility of PtG-S2 to amphotropic vector is exceptionally high (2), and more than 70% of the population can be transduced without selection.

Production of Virus: Small-Scale/Large-Scale Purification

Reagents

- Culture medium for virus production: DMEM (high-glucose), 10% FCS, penicillin, and streptomycin
- AxCANCre (10,12) (replication-defective Ad vector) carrying the Cre recombinase gene (1 × 10^5 to 4 × 10^8 plaque-forming units [pfu]/mL)

Procedures

Small-scale preparation of VSV-G-pseudotyped vectors. This protocol is useful for checking the efficiency of virus production from an isolated (or cloned) prepackaging cell line harboring a vector DNA encoding gene X (PtG-S2-X).

Day 1

1. Seed PtG-S2-X at 4 × 10^4 cells/24-well plate (500 µL/well) in medium for virus production and incubate at 37°C/5% CO_2. This medium is used throughout the procedures described in this section. (PtG-S2-X loses neomycin resistance after Cre recombinase expression.)

Day 2

1. Remove medium.

2. Introduce the Cre recombinase gene into PtG-S2-X cells by infecting them with AxCANCre at an moi of 3 to 10 (1.2 to 4 × 10^5 pfu). The total volume of culture in each well of a 24-well plate should be adjusted to 150 µL with medium.

3. Rock 24-well plates gently every 10 minutes so that the medium is distributed evenly and cells do not dry up.

4. After 1 hour, add fresh medium (350 µL/well) and continue incubation at 37°C.

Day 3

1. Wash the culture three times with medium to minimize the amount of remaining Ad vector.

2. Add fresh medium (500 µL/well) and incubate further at 37°C.

Day 4

1. Wash the cells three times with medium. Add fresh medium (500 µL/well).

2. Shift incubation temperature to 32°C.

Day 5 to Day 8 (or Day n)

1. Collect medium containing pseudotyped virus into Eppendorf® tubes with screw caps harboring O-rings (Sarstedt, Nümbrecht, Germany).

2. Add fresh medium (500 µL/well) to the culture.

3. Centrifuge Eppendorf tubes at 5000 rpm (2000× *g*) for 30 minutes. Transfer supernatants to serum tubes and store them at -80°C.

4. Continue recovering pseudotyped virus vectors once a day.

Day n

1. Determine virus titers using 3Y1 (rat fibroblast) cells as indicator. (NIH 3T3 cells can also be used for titration and should yield similar titer values.)

Large-Scale Preparation of VSV-G-Pseudotyped Vectors

Day 1

1. Seed PtG-S2 cells at a density of 1.0×10^6 cells/90-mm dish in medium for virus production (10 mL medium/dish) and incubate at 37°C/5% CO_2. (Using 80 plates, a stock of about 1×10^9 iu of concentrated virus stock can be obtained.)

Day 2

1. Replace medium with 2 mL fresh medium.

2. Add AxCANCre at an moi of 3 to 10 and incubate at 37°C/5% CO_2.

3. For efficient infection, rock gently every 10 minutes for a total of 1 hour.

4. Add fresh medium (8 mL/plate) and continue incubation at 37°C.

Day 3

1. Wash cells three times with medium to remove remaining Ad vector as much as possible.

2. Add fresh medium (10 mL/plate) and incubate further at 37°C.

Day 4

1. Wash cells and add fresh medium as on day 3.

2. Shift incubation temperature to 32°C.

Day 5 to Day 8 (or Day n)

1. Collect medium every 24 hours, filtrate immediately through 0.45-µm filters (0.45-nm HV-filters; Millipore, Bedford, MA, USA) to remove cell debris. Cell debris could also be removed by low-speed centrifugation (1500× *g* for 30 min).

Day n

1. Titrate virus stocks using 3Y1 cells as indicator. Pool virus stocks showing higher titers and concentrate virus as described in the next section.

Concentration of Pseudotyped Vectors Ultracentrifugation/Low-Speed Centrifugation

Before the concentration of virus by centrifugation, it is essential to remove the remaining cell debris either by low-speed centrifugation or by filtration as described above. Either ultracentrifugation or low-speed centrifugation can be used for virus purification with a recovery of virus titer ranging from 30% to 80% in a single step. Low-speed centrifugation (6000× *g* for 16 hours) can also be used for concentration and may be more convenient as a first step in large-scale preparations. Virus stocks (approximately 1×10^6 iu/mL) of VSV-G-pseudotyped vectors prepared as described under "Production of Virus: Small-Scale/Large-Scale Purification" can be concentrated up to 1000-fold by two successive ultracentrifugation steps (30–200-fold concentration in one cycle of ultracentrifugation).

Ultracentrifugation

1. Pool virus stocks prepared as described under "Production of Virus: Small-Scale/Large-Scale Purification" (approximately 37 mL) into 40-mL ultraclean tubes (Beckman Coulter, Fullerton, CA, USA) and centrifuge in a SW28 rotor at 19500 rpm (70000× *g*) for 100 minutes at 4°C.
2. Remove supernatants by decantation.
3. Resuspend pellet in 0.35 mL ice-cold DMEM (high-glucose, without FCS).
4. Transfer virus suspension to a serum tube and keep frozen at -80°C.

Low-Speed Centrifugation

1. Pool virus stocks (up to 200 mL) into 290-mL bottles (Nalge Nunc International, Rochester, NY, USA) and centrifuge at 6000× *g* for 16 hours at 4°C using a TA-9 rotor in a SRX-200 (Tomy, Tokyo, Japan) or similar centrifuge. (A centrifuge with swing bucket rotor may give higher yields of infectious particles.)
2. Remove supernatants by decantation.
3. Add 3 mL of ice-cold DMEM (high-glucose, without FCS) to the pellet of virus particles and rock for 1 to 2 days at 4°C.
4. Filtrate through a 0.2-µm filter (EB-DISK 25; Kanto Chemical, Tokyo, Japan).
5. Dispense into serum tubes and keep at -80°C.

Large-Scale Preparation of Ad Vector Carrying the *Cre* Recombinase Gene (11,12)

Amplification of Ad Vector

1. Seed 293 cells (5 × 10^6 cells in 15 mL DMEM containing 10% FCS) into collagen-coated T75 flask (75-mL; Corning, Corning, NY, USA).
2. On the next day, remove the medium and add 1.9 mL fresh medium plus 100 µL Ad stock (first seed). (It is recommended that the volume of seed added is not higher than 1/20 of the infection volume.)
3. Rock the flask gently every 10 minutes to prevent cells from drying.
4. After 1 hour, add 14 mL fresh medium and culture for three days at 37°C.
5. When most 293 cells have lysed as a result of amplification of the Ad, collect both medium and cell debris (second seed). Cells remaining on the flask can be collected by gentle pipetting. If cells did not lyse, the seed of Ad vector should be replaced by a new one.
6. Sonicate the collected medium containing dead cells for efficient recovery of Ad from cells. Sonicator must be sterile and designed not to disperse Ad (closed-box type). Never use tip-type sonicators. Ad is infectious through aerosol contamination! If such a sonicator is not available, break down cells by repeated freeze-and-thaw cycles six times using a 37°C water bath and a dry ice-ethanol bath instead.
7. Centrifuge at 3000 rpm for 10 min (1500× *g*) at 4°C to remove cell debris. Recover supernatants, freeze using a dry ice-acetone bath and keep at -80°C.
8. Determine the titers of Ad stocks as described elsewhere (11) and check their quality by restriction enzyme digestion.

DISCUSSION AND PERSPECTIVES

In this chapter, we have described a system for producing VSV-G-pseudotyped vectors using stable helper cell lines. Considering the cytotoxicity of the VSV-G protein for the packaging cells, its expression was designed to be regulated in an all-or-none manner by making use of the Cre-*lox*P system. Some other stable packaging cell lines (5,20) or virus-producing cell lines (28) for VSV-G pseudotypes have been established using tetracycline-modulated promoters to drive VSV-G expression, which have been reported to produce high titers of pseudotyped vector (1×10^6 to 1×10^7 iu/mL). However, those stable cell lines produced low titers of pseudotypes (1×10^1 to 1×10^2 iu/mL [5], and 1×10^3 to 1×10^4 iu/mL [28]) even in the presence of transcriptional repressors such as tetracycline.

In the case of a natural retrovirus, cells producing an Env acquire resistance to the retrovirus with the same Env (a phenomenon known as interference). We have found, however, that VSV-G-producing cells are fully susceptible to VSV-G-pseudotyped retroviruses (1). This might be partly explained because the VSV-G receptor contains ubiquitous anionic phospholipids (15,21). This means that a pseudotyped particle can infect the cell that produced it, and further that the packaging cells can accumulate virus vector during culture for virus production by a self-Ping-Pong mechanism. Therefore, some packaging cell lines inducible for VSV-G expression would have the disadvantage of possible long-lasting, leaky production of pseudotyped vectors, which might lead to the production of replication-competent virus before the induction of virus propagation.

The vector-producing system described here would be also applicable to other types of vector preparation. For example, virus vectors pseudotyped with proteins other than VSV-G could be produced using transcriptional units similar to pCALNdLG. In addition, a producing system for VSV-G-pseudotyped lentivirus vector could be generated using pCALNdLG, which may be useful for gene transfer into nondividing cells.

Retrovirus produced from human-derived cell lines have been reported to be much more resistant to the neutralizing effect of human serum than those produced by conventional murine-derived packaging cell lines, which is probably due to the humanized glycosylation pattern of the virus envelope protein (23–25). PtG-S2 cells originate from a human cell line, and VSV-G-pseudotypes produced with this system were in fact quite stable in human fresh blood (1). This property sould be also advantageous when considering the in vivo application of the vector to human gene therapy.

The development of retrovirus vectors that express tumor suppressor genes with cytotoxic or cytostatic potential would be essential for gene therapy of human cancers. However, it is quite difficult to establish stable packaging cell lines producing such retrovirus vectors because those genes also affect the growth of the packaging cell lines. To overcome this problem, we have developed new MLV-based vectors, in which expression of an exogenous gene with cytotoxic activity is completely silent, but can be induced upon Cre recombinase introduction (26). We have generated a PtG-S2 cell line harboring one such new vector carrying a *p53* cDNA as an exogenous gene. Recombination worked as expected, and we have obtained high titers of *p53*-expressing virus stocks, which can induce G1 arrest in 100% of a population of transduced human cells (26).

ACKNOWLEDGMENTS

We are grateful to Dr. I. Saito who kindly introduced us to the protocols for Ad vector production. We also thank M. Takada for her excellent technical support.

REFERENCES

1. **Arai, T., K. Matsumoto, K. Saitoh, M. Ui, T. Ito, M. Murakami, Y. Kanegae, I. Saito, F.L. Cosset, Y. Takeuchi, and H. Iba.** 1998. A new system for stringent, high-titer vesicular stomatitis virus G protein-pseudotyped retrovirus vector induction by introduction of Cre recombinase into stable prepackaging cell lines. J. Virol. *72*:1115-1121.

2. **Arai, T., M. Takada, M. Ui, and H. Iba.** 1999. Dose-dependent transduction of vesicular stomatitis virus G protein-pseudotyped retrovirus vector into human solid tumor cell lines and murine fibroblasts. Virology *260*:109-115.

3. **Batra, R.K., J.C. Olsen, D.K. Hoganson, B. Caterson, and R.C. Boucher.** 1997. Retroviral gene transfer is inhibited by chondroitin sulfate proteoglycans/glycosaminoglycans in malignant pleural effusions. J. Biol. Chem. *272*:11736-11743.

4. **Burns, J.C., T. Friedmann, W. Driever, M. Burrascano, and J.K. Yee.** 1993. Vesicular stomatitis virus G glycoprotein pseudotyped retroviral vectors: concentration to very high titer and efficient gene transfer into mammalian and nonmammalian cells. Proc. Natl. Acad. Sci. USA *90*:8033-8037.

5. **Chen, S.T., A. Iida, L. Guo, T. Friedmann, and J.K. Yee.** 1996. Generation of packaging cell lines for pseudotyped retroviral vectors of the G protein of vesicular stomatitis virus by using a modified tetracycline inducible system. Proc. Natl. Acad. Sci. USA *93*:10057-10062.

6. **Cosset, F.L., Y. Takeuchi, J.L. Battini, R.A. Weiss, and M.K.L. Collins.** 1995. High-titer packaging cells producing recombinant retroviruses resistant to human serum. J. Virol. *69*:7430-7436.

7. **Doux, J.M.L., J.R. Morgan, R.G. Snow, and M.L. Yarmush.** 1996. Proteoglycans secreted by packaging cell lines inhibit retrovirus infection. J. Virol. *70*:6468-6473.

8. **Emi, N., T. Friedmann, and J.K. Yee.** 1991. Pseudotype formation of murine leukemia virus with the G protein of vesicular stomatitis virus. J. Virol. *65*:1202-1207.

9. **Gallardo, H.F., C. Tan, D. Ory, and M. Sadelain.** 1997. Recombinant retroviruses pseudotyped with the vesicular stomatitis virus G glycoprotein mediate both stable gene transfer and pseudotransduction in human peripheral blood lymphocytes. Blood *90*:952-957.

10. Kanegae, Y., G. Lee, Y. Sato, M. Tanaka, M. Nakai, T. Sakaki, S. Sugano, and I. Saito. 1995. Efficient gene activation in mammalian cells by using recombinant adenovirus expressing site-specific Cre recombinase. Nucleic Acids Res. *23*:3816-3821.

11. **Kanegae, Y., M. Makimura, and I. Saito.** 1994. A simple and efficient method for purification of infectious recombinant adenovirus. Jpn. J. Med. Sci. Biol. *47*:157-166.

12. **Kanegae, Y., K. Takamori, Y. Sato, G. Lee, M. Nakai, and I. Saito.** 1996. Efficient gene activation system on mammalian cell chromosomes using recombinant adenovirus producing Cre recombinase. Gene *181*:207-212.

13. **Kavanaugh, M.P., D.G. Miller, W. Zhang, W. Law, S.L. Kozak, D. Kabat, and A.D. Miller.** 1994. Cell-surface receptors for gibbon ape leukemia virus and amphotropic murine retrovirus are inducible sodium-dependent phosphate symporters. Proc. Natl. Acad. Sci. USA *91*:7071-7075.

14. **Liu, M.L., B.L. Winther, and M.A. Kay.** 1996. Pseudotransduction of hepatocytes by using concentrated pseudotyped vesicular stomatitis virus G glycoprotein (VSV-G)-Moloney murine leukemia virus-derived retrovirus vectors: comparison of VSV-G and amphotropic vectors for hepatic gene transfer. J. Virol. *70*:2497-2502.

15. **Mastromarino, P., C. Conti, P. Goldoni, B. Hauttecoeur, and N. Orsi.** 1987. Characterization of membrane components of the erythrocyte involved in vesicular stomatitis virus attachment and fusion at acidic pH. J. Gen. Virol. *68*:2359-2369.

16. **Matsubara, T., R.W. Beeman, H. Shike, N.J. Besansky, O. Mukabayire, S. Higgs, A.A. James, and J.C. Burns.** 1996. Pantropic retroviral vectors integrate and express in cells of the malaria mosquito, *Anopheles gambiae*. Proc. Natl. Acad. Sci. USA *93*:6181-6185.

17. **Morgan, J.R., J.M. LeDoux, R.G. Snow, R.G. Tompkins, and M.L. Yarmush.** 1995. Retrovirus infection: effect of time and target cell number. J. Virol. *69*:6994-7000.

18. **Naviaux, R.K., E. Costanzi, M. Haas, and I.M. Verma.** 1996. The pCL vector system: rapid production of helper-free, high-titer, recombinant retroviruses. J. Virol. *70*:5701-5705.

19. **Niwa, H., K. Yamamura, and J. Miyazaki.** 1991. Efficient selection for high-expression transfectants with a novel eukaryotic vector. Gene *108*:193-199.

20. **Ory, D.S., B.A. Neugeboren, and R.C. Mulligan.** 1996. A stable human-derived packaging cell line for production of high titer retrovirus/vesicular stomatitis virus G pseudotypes. Proc. Natl. Acad. Sci. USA *93*:11400-11406.

21. **Schlegel, R., T.S. Tralka, M.C. Willingham, and I. Pastan.** 1983. Inhibition of VSV binding and infectivity by phosphatidylserine: is phosphatidylserine a VSV-binding site? Cell *32*:639-646.

22. **Sharma, S., M. Cantwell, T.J. Kipps, and T. Friedmann.** 1996. Efficient infection of a human T-cell line and of human primary peripheral blood leukocytes with a pseudotyped retrovirus vector. Proc. Natl. Acad. Sci. USA *93*:11842-11847.

23. **Takeuchi, Y., F.L. Cosset, P.J. Lachmann, H. Okada, R.A. Weiss, and M.K.L. Collins.** 1994. Type C retrovirus inactivation by human complement is determined by both the viral genome and producer cell. J. Virol. *68*:8001-8007.

24. **Takeuchi, Y., S.-H. Liong, P.D. Bieniasz, U. Jäger, C.D. Porter, T. Friedmann, M.O. McClure, and R.A. Weiss.** 1997. Sensitization of rhabdo-, lenti-, and spumaviruses to human serum by galactosyl (1-3) galactosylation. J. Virol. *71*:6174-6178.

25. **Takeuchi, Y., C.D. Porter, K.M. Strahan, A.F. Preece, K. Gustafsson, F.L. Cosset, R.A. Weiss, and M.K. Collins.** 1996. Sensitization of cells and retroviruses to human serum by (1-3) galactosyltransferase. Nature *379*:85-88.

26. **Ui, M., M. Takada, T. Arai, K. Matsumoto, K. Yamada, T. Nakahata, T. Nishiwaki, Y. Furukawa, T. Tokino, Y. Nakamura, and H. Iba.** 1999. Retrovirus vectors designed for efficient transduction of cytotoxic or cytostatic genes. Gene Ther. *6*:1670-1678.

27. **Weiss, R.A., D. Boettiger, and H.M. Murphy.** 1977. Pseudotypes of avian sarcoma viruses with the envelope properties of vesicular stomatitis virus. Virology *76*:808-825.

28. **Yang, Y., E.F. Vanin, M.A. Whitt, M. Fornerod, R. Zwart, R.D. Schneiderman, G. Grosveld, and A.W. Nienhuis.** 1995. Inducible, high-level production of infectious murine leukemia retroviral vector particles pseudotyped with vesicular stomatitis virus G envelope protein. Hum. Gene Ther. *6*:1203-1213.
29. **Yee, J.K., A. Miyanohara, P. LaPorte, K. Bouic, J.C. Burns, and T. Friedmann.** 1994. A general method for the generation of high-titer, pantropic retroviral vectors: highly efficient infection of primary hepatocytes. Proc. Natl. Acad. Sci. USA *91*:9564-9568.

25 Utility of Scaffold Attachment Regions for Enhanced Retroviral Vector Expression in Human Hematopoietic Cells

Ivan Plavec and Manju Agarwal
SyStemix, Inc., Palo Alto, CA, USA

INTRODUCTION

For hematopoietic cells, and in particular hematopoietic stem/progenitor cells (HSPC), retroviral vectors are the gene transfer modality of choice because integration of retrovirally transduced genes has the potential to promote long-term persistence of a transgene in vivo (2,49). However, sustained and high-level transgene expression remains a major technical challenge for the development of effective gene therapies. In our laboratory we have discovered that scaffold attachment region (SAR) elements can enhance vector expression in hematopoietic cells. The effect of an SAR element is on the level as well as on the persistence of transgene expression and does not depend on the vector backbone. We believe that SARs can be broadly utilized to improve performance of the existing retroviral vectors. This article summarizes our current data on expression of SAR-containing vectors in human hematopoietic cells.

AN OVERVIEW OF CURRENT APPROACHES FOR IMPROVING RETROVIRAL VECTOR EXPRESSION

Moloney murine leukemia virus (MoMLV)-based vectors are commonly used in preclinical experiments and have also been utilized in phase I clinical studies (13,49). In general, one can readily detect vector expression in HSPCs a few days after transduction, but the expression in differentiated cells is less efficient. For example, in T cells, vector expression depends on the activation status; it is high in activated cells but low in resting cells (14,57). Expression is also down-regulated when HSPCs differentiate along myeloid pathways (44). De novo methylation of the MoMLV long terminal repeats (LTR) (16,62), negative regulatory transcription factors that bind to the LTR and the primer-binding site (23,56), and position effects contribute to poor expression in vivo.

Viral Vectors: Basic Science and Gene Therapy
Edited by A. Cid-Arregui and A. García-Carrancá

Several different types of retroviral vectors were developed with the aim of improving expression (Figure 1). In one type of vector (type 1 in Figure 1), the gene of interest is expressed from the viral LTR. These vectors differ from the MoMLV vectors in the LTR promoter/en hancer region and the primer-binding site. They are derived from the myeloproliferative sarcoma virus (MPSV), murine stem cell virus (MSCV), spleen focus-forming virus (SFFV), and dl587rev virus (5,17,32,54,61,62). The main advantage of these vectors is that they express better than the MoMLV vectors in mouse bone marrow transplantation models. All of the type 1 vectors have the packaging sequence derived from the MoMLV-based LxSN vector (50). The protein expression depends on the splicing that occurs between the genuine MoMLV splice donor and the cryptic splice acceptor located at the very end of the packaging sequence. Although most of the genes will be expressed from these vectors, the efficiency of splicing is unpredictable and is influenced by the sequences cloned into the vector (I. Plavec and M. Agarwal, unpublished data). The splicing problem has been addressed in the type 2 vectors (Figure 1). The example of this type is the MFG vector (21). The MFG vector contains the genuine MoMLV splice donor and the splice acceptor, which the wild-type virus uses to express the *env* gene. Furthermore, the cDNA is inserted in frame with the *env* ATG to maximize mRNA translation (Figure 1).

In the type 3 group of vectors, the gene of interest is expressed from an internal promoter (Figure 1). The promoters that were successfully used to express genes in hematopoietic cells include the phosphoglycerate kinase gene promoter (43) and

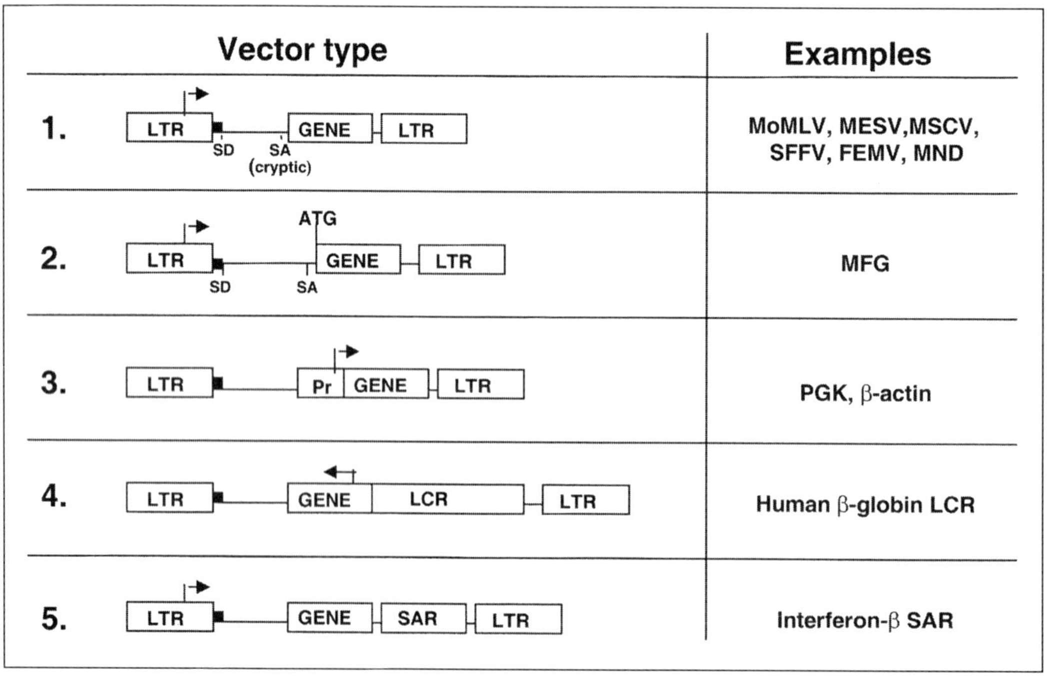

Figure 1. An overview of current retroviral vector designs (not drawn to scale). LTR, long terminal repeat; SD, splice donor; SA, splice acceptor; MoMLV, Moloney murine leukemia virus; MESV, murine embryonic stem cell virus; MSCV, murine stem cell virus; SFFV, spleen focus-forming virus; Pr, promoter; LCR, locus control region; SAR, scaffold attachment region; PGK, phosphoglycerate kinase promoter. Arrows indicate orientation of transcription. Small black boxes adjacent to the 5′ LTRs indicate primer binding site sequence.

the β-actin gene promoter (71). However, the activity of many promoters is suppressed when cloned into a retroviral vector, and the final choice of the promoter has to be empirically defined. The examples of promoters that did not perform well are the simian virus 40 (SV40) immediate early (70), the immunoglobulin μ or κ-chain (30); the methallothionein (47); and the CD11a, CD11b, and CD18 gene promoters (4).

All of the above vectors have been demonstrated to express in hematopoietic cells. However, they are subject to position-dependent variation in expression. An approach to solve this problem is to use a locus control region (LCR) or SAR elements (type 4 and type 5 vectors in Figure 1). These elements have the potential to confer high-level, position-independent expression to a transgene (9,40). Several LCR elements have been described to date: the β-globin LCR (26), the α-globin LCR (33), the CD2 LCR (25), the T cell receptor α/δ LCR (19), the chicken lysozyme LCR (10), the apolipoprotein E/C-I LCR (18). Unfortunately, the available LCR elements have not been used much, primarily for technical reasons. First, the LCR elements are often too large to be packaged into retroviral vectors, and reducing the size can result in a loss of activity (53,58,63). Second, at least with the human β-globin LCR, it has proven difficult to generate stable and high-titer retroviral vectors (53,58). Once these technical problems are solved, the LCR elements will certainly find broader application.

SARs

In contrast to the LCRs, SAR sequences are relatively small (0.5–2 kb) and can easily be integrated into a retroviral vector. SARs are several hundred base pairs long and AT rich (approximately 70%). The sequences are defined by high-affinity binding to an isolated nuclear scaffold or nuclear matrix in vitro (9). Different SAR elements share some common structural features, but to date no consensus sequence has been identified (12). SAR elements can enhance expression of heterologous genes in transfection experiments in vitro and in transgenic mice (39,41,48,59,68). In some instances, it has been reported that SAR elements can also confer position-independent expression to a linked transgene (39,48). How the SAR elements enhance transgene expression is not well understood. The current hypothesis is that SARs define boundaries of independent chromatin domains and allow access of transcription factors to the enhancer/promoter sequences within the domain (8,35).

SAR-CONTAINING RETROVIRAL VECTORS

The Human Interferon-β SAR Improves Retroviral Vector Expression in Human Primary T Cells

We have used primary T cell culture to assay for the effect of SAR on retroviral vector expression. Primary T cells were isolated from peripheral blood of healthy donors and activated with allogeneic feeder cells, phytohemagglutinin (PHA) and interleukin-2 (IL-2). Upon activation, cells start to proliferate and express specific antigens (e.g., CD25). The cells are then maintained in the medium with low concentration IL-2 (20–40 U/mL). Under such conditions, at three to four days after activation cells are actively dividing, and CD25 expression is at a maximum with greater than 95% $CD25^+$ cells (Figures 2 and 3). By days 10 to 14, cells cease to proliferate, and the $CD25^+$ expression is down-regulated (<50% $CD25^+$ cells), reflecting a mitotically resting state of the cells (Figures 2 and 3). Kinetic analysis of CD25 expression in the

T-cell culture is shown in Figure 2. These cells can be restimulated for another activation cycle.

The retroviral vectors that we used were derived from the MoMLV-based LMiLy vector (57). The LMiLy vector (Figure 4) encodes two genes from one bicistronic mRNA transcript: the *RevM10* gene (45) and the *Lyt-2* surface marker (mouse CD8 α′ chain) (67). Translation of the Lyt-2 protein is mediated by the internal ribosomal entry site (IRES) of the human encephalomyocarditis virus (EMCV) (34) and therefore is linked to RevM10 protein expression (11,66). Standard retroviral vectors such as LMiLy, in which the gene of interest is expressed from the viral LTR, are transcribed at high levels in activated T cells, but the expression is significantly reduced in resting cells (Figures 2 and 3). This is true for the MoMLV, MPSV, and murine embryonic stem cell virus (MESV)-based vectors (Figure 2), indicating that simply changing the LTR and the primer binding site may not be sufficient to rescue expression.

In sharp contrast, insertion of an SAR sequence into a vector significantly improves vector expression (Figure 3). Initial SAR-containing vectors were generated by inserting the human interferon-β SAR (IFN-SAR) sequences into the LTR of the LMiLy vector. The example of such double-copy type vectors (29) is the LMiLy2S vector (Figure 4). Following transduction with the LMiLy2S, the 3′ LTR SAR sequence is duplicated in the 5′ LTR, generating an integrated provirus that is bordered by two SAR sequences. In the experiment shown in Figure 3, there are 2.6-fold more transgene-positive resting T cells with the LMiLy2S vector compared with the control LMiLy vector. Overall, in multiple experiments, we observed 2.4 ± 0.9-fold

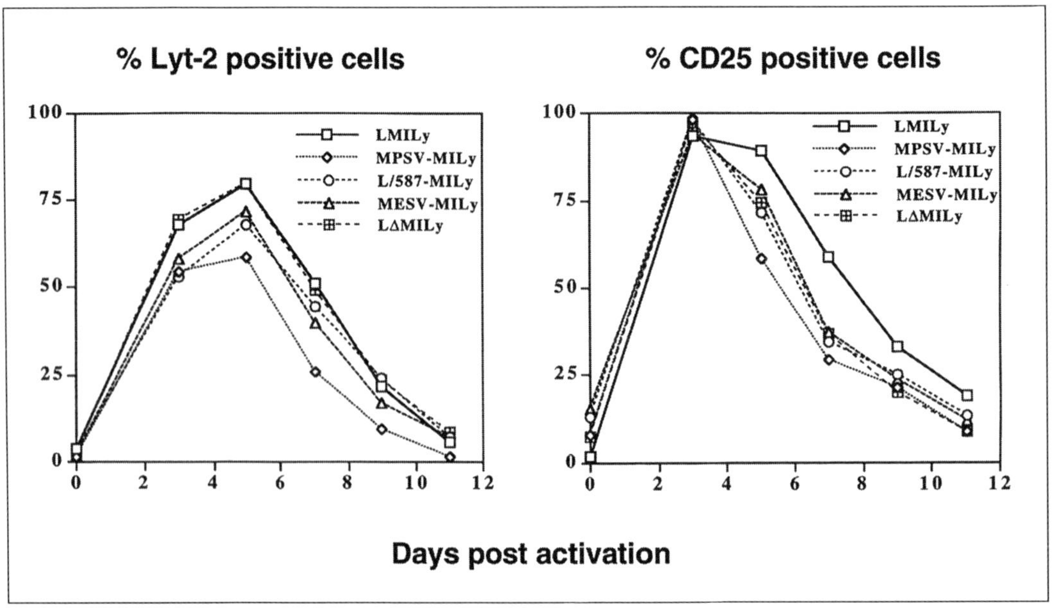

Figure 2. Standard retroviral vectors are poorly expressed in resting T cells. All vectors used in this experiment were derived from the LMiLy vector (Figure 4). MPSV-MiLy has the myeloproliferative sarcoma virus LTR (7), L/587-MiLy has the primer binding site from the dl587 virus (17), and the MESV-MiLy has both the MPSV LTR and the dl587 PBS (30). PBL-derived CD4+ T cells were transduced with the various vectors (indicated in legend) and enriched for Lyt-2 surface protein as described in the text. Fully resting cells (as determined by staining for the CD25 marker) were stimulated with PHA and feeder cells, and Lyt-2 and CD25 expression were measured every two to three days over a period of 11 days. A representative experiment is shown. Results were reproduced using four separate tissues.

more expressing cells with the LMiLy2S than with the LMiLy vector (Table 1). The effect of SAR on vector expression was most pronounced in the resting CD25$^-$ T cells in which the LMiLy2S vector showed 5.7 ± 3.4-fold more expressing cells than the vector without the SAR (Table 2). We were concerned that the observed improved expression with the LMiLy2S vector may be a result of disruption of the negative control region (NCR) (23), which is located at the 5′ end of the LTR. To

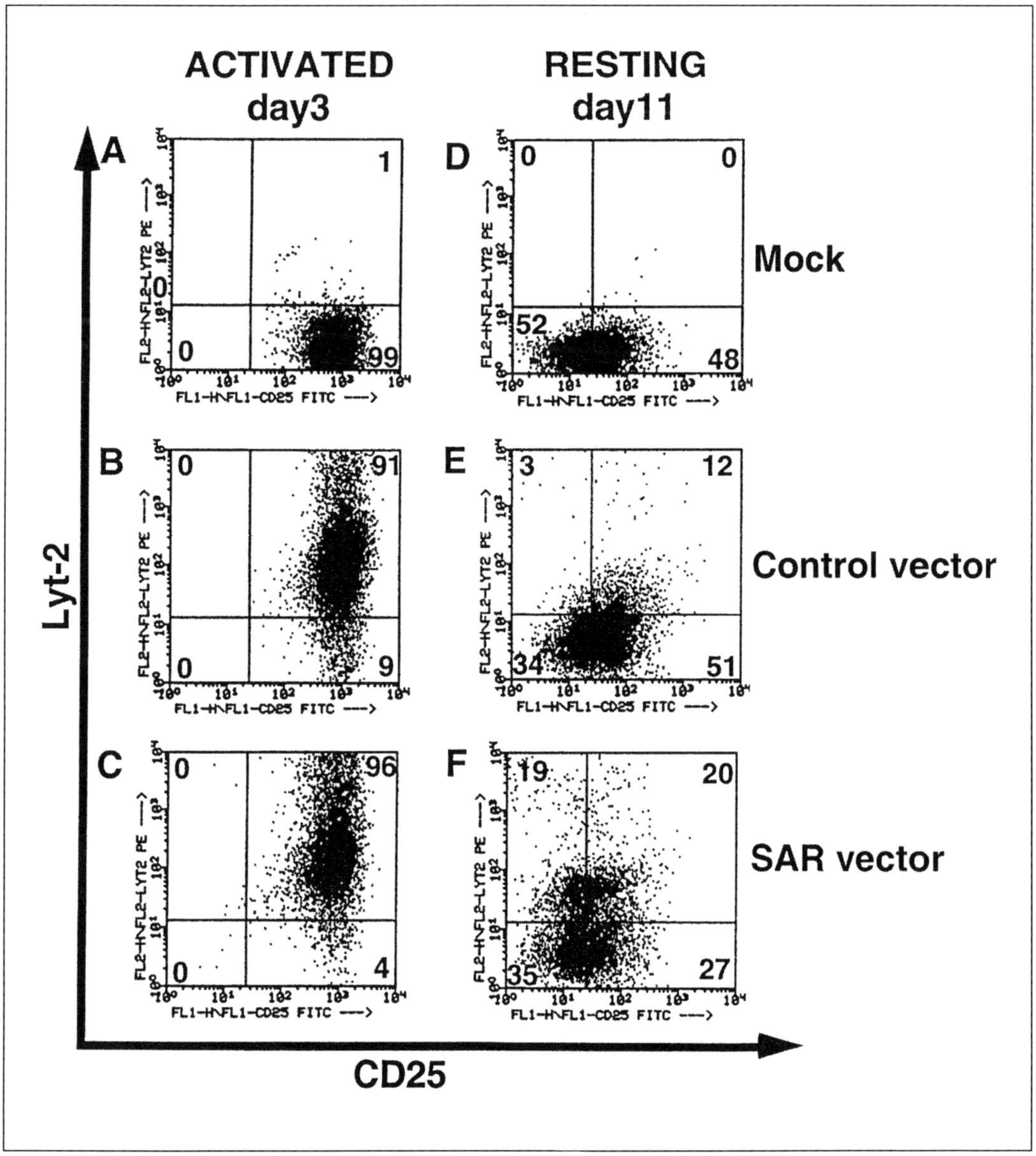

Figure 3. The SAR-containing LMiLy2S vector is efficiently expressed in resting T cells. Lyt-2-enriched LMiLy and LMiLy2S-transduced CD4$^+$ primary T cells were activated with PHA, IL-2, and irradiated allogenic feeder cells. On days 3 and 11 after stimulation, cell aliquots were stained with anti-CD25-FITC and anti-Lyt-2-PE antibodies and analyzed by FACScan (Becton Dickinson Immunocytometry Systems, San Jose, CA, USA). Numbers indicate the percentage of Lyt-2-positive cells in the respective quadrants. Gates for background fluorescence were set based on control isotype antibodies. Mock are untransduced control cells.

Table 1. Lyt-2 Surface Marker Expression in Activated and Resting Primary T Cells Transduced with the LMiLy and LMiLy2S Vectors

	Activated			Resting		
	Percent Lyt-2+ cells			Percent Lyt-2+ cells		
Tissue[a]	LMiLy	LMiLy2S	Ratio LMiLy2S/ LMiLy	LMiLy	LMiLy2S	Ratio LMiLy2S/ LMiLy
1	91	96	1.05	15	39	2.6
2	65	64	0.98	7	26	3.7
3	73	86	1.18	33	48	1.5
4	95	96	1.01	35	55	1.6
Average			1.06 ± 0.08			2.4 ± 0.9 (P<0.1)[b]

Samples were analyzed on day 3 (Activated) and day 10–12 (Resting) after stimulation with PHA, IL-2, and allogeneic feeder cells. Gates for the Lyt-2+ cells were set using untransduced T cells as control (see Figure 2, panels A and D).

[a]Tissues 1 and 4 are thymocytes, and 2 and 3 are peripheral blood lymphocytes.

[b]The significance of differences in mean values was determined using paired *t* test analysis.

exclude this possibility, we have inserted a random DNA sequence into the LMiLy vector or deleted the entire NCR [vectors LMiLy-800 and LMiLy-NCR(-) in Figure 4]. None of these modifications had any effect on vector expression, indicating that the observed increase in expression, was indeed due to the positive effect of the SAR element (Figure 4).

Optimization of SAR-Containing Retroviral Vectors

To further optimize the SAR-containing vectors, we have inserted the IFN-SAR into the LMiLy vector at various positions and in the reverse orientation. The SAR was able to enhance vector expression irrespective of the position in the vector. However, the best results were observed when the element was placed just upstream of the 3′ LTR (vector LMiLyS in Figure 4). To our surprise, the enhancing effect was observed only when the SAR sequence was present in the reverse orientation. When the SAR element was in the forward orientation, transgene expression was even lower than it was with the parental vector without SAR (compare LMiLy2S-F and LMiLy vectors in Figure 4).

SARs are defined by the specific binding to the isolated nuclear scaffold (9). We have cloned eight different SAR sequences into the LMiLy vector and analyzed the expression in resting T cells (Figure 5). The Chinese hamster dihydrofolate reductase (*DHFR*) SAR (38), three human chromosome 19 SARs (52), and three subfragments of the hematopoietic serine protease 5′ and the 3′ SARs (28) showed no activity (Figure 5). The only SAR that did enhance retroviral vector expression was the HSP-31 subfragment of the human hematopoietic serine protease 3′ SAR (28). However, the effect was less than with the IFN-β SAR (Figure 5). Thus, only certain SAR sequences have the ability to enhance retroviral vector expression.

Table 2. Expression of the Lyt-2 Surface Marker in the $CD25^-$ and $CD25^+$ Subpopulations of Resting Primary T Cells Transduced with the LMiLy and LMiLy2S Vectors

	Percent Lyt-2+ cells						Relative transgene expression (Lyt-2 mean fluorescence intensity)					
	CD25- fraction			CD25+ fraction			CD25-Lyt-2+ fraction			CD25+Lyt-2+ fraction		
Tissue*	LMiLy	LMiLy2S	Ratio LMiLy2S/ LMiLy	LMiLy	LMiLy2S	Ratio LMiLy2S/ LMiLy	LMiLy	LMiLy2S	Ratio LMiLy2S/ LMiLy	LMiLy	LMiLy2S	Ratio LMiLy2S/ LMiLy
1	3	19	6.3	12	20	1.7	25	41	1.6	32	48	1.5
2	1	11	11	6	15	2.5	23	35	1.5	26	39	1.5
3	7	20	2.9	26	28	1.1	27	46	1.7	44	76	1.7
4	9	22	2.4	26	33	1.3	23	41	1.8	32	54	1.7
Average			5.7 ± 3.4		1.7 ± 0.5				1.65 ± 0.1			1.6 ± 0.1

Samples were analyzed 10 to12 days after stimulation with PHA, IL-2, and allogeneic feeder cells. Gates for the $CD25^-$ fractions were set using control isotype antibodies. Quantitative analysis of the Lyt-2 staining was performed when approximately 50% of the total cells fell into the $CD25^-$ gate (see Figure 2, panel D).

*Tissues 1 and 4 are thymocytes, and 2 and 3 are peripheral blood lymphocytes.

Gene Transfer into HSPCs and the Effect of the IFN-β SAR on Retroviral Vector Expression in Differentiated Cells

For many gene therapy applications, HSPCs are the target cells of choice. Therefore, we were interested in the effect of the IFN-SAR sequence on vector expression in lymphoid and myeloid cells derived from retrovirally transduced HSPCs. We have employed the SCID-hu Thy/Liv mouse model to generate T cells from transduced HSPCs. Mobilized peripheral blood CD34+ HSPC from granulocyte-colony stimulating factor (G-CSF) mobilized healthy donors were cultured in vitro for two days in the presence of IL3, IL6, and stem cell factor, and then subjected to two rounds of transduction with either control LMiLy vector or the SAR-containing LMiLyS vector (Figure 4). Forty-eight hours after the final round of transduction, we detected 32% and 22% CD34+Lyt-2+ cells with the LMiLy and LMiLyS vectors (Figure 6). Cells were then sorted to 84% to 86% purity for CD34 and Lyt-2 (data not shown) and injected directly into the conjoint organ of irradiated SCID-hu Thy/Liv mice.

Eight weeks after HSPC transplantation, Thy/Liv grafts were removed and examined for donor cell content by staining for HLA marker MA2.1 and for vector expression by staining for the Lyt-2 antigen. Lyt-2 surface marker expression was detected in 3/7 mice reconstituted with the LMiLyS and in 2/5 mice reconstituted

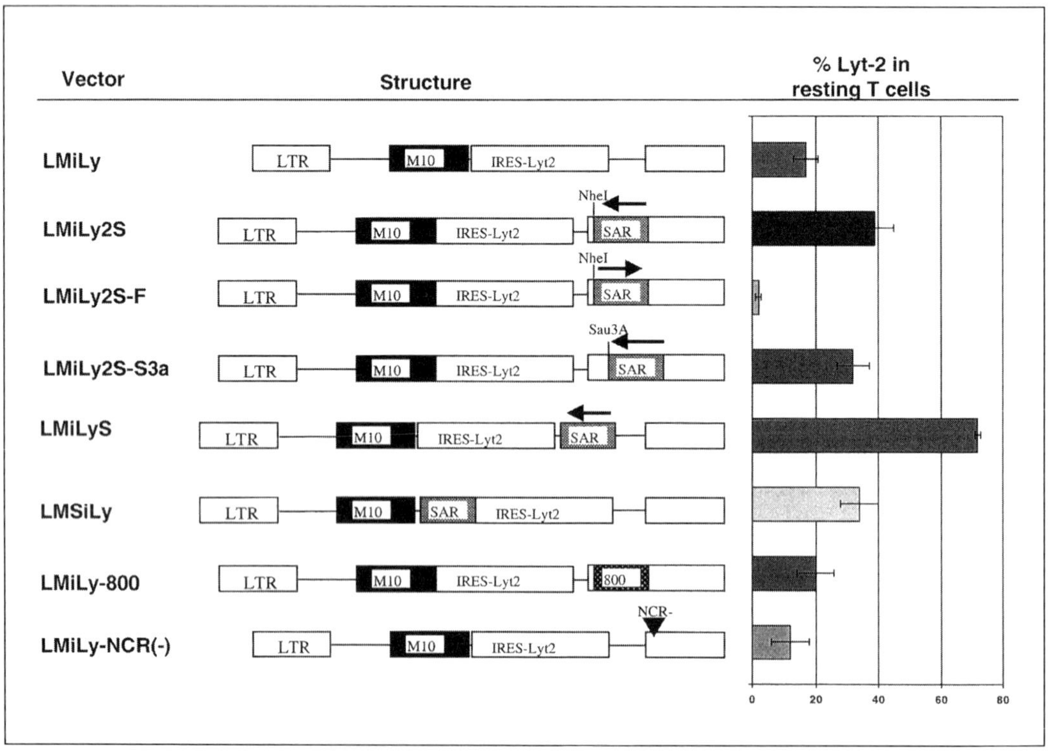

Figure 4. Optimization of SAR-containing retroviral vectors. Primary T cells were transduced with different retroviral vectors as described in the Materials and Methods section, and the Lyt-2 transgene expression was measured in resting cells. An average from two independent experiments is shown. Error bars indicate standard deviation. M10, *RevM10* gene; LTR, long terminal repeat; SAR, IFN-β scaffold attachment region; IRES-Lyt2, internal ribosomal entry site linked to the Lyt-2 gene; 800, 800-bp *Pvu*II-*Rsa*I fragment from the 3′ untranslated region of the human *NGFr* gene; NCR, negative control region. Arrow indicates orientation of the IFN-SAR sequence.

with the LMiLy-transduced HSPC (Figure 6). When corrected for the percentage of donor cells in the grafts, the levels of Lyt-2-positive cells were somewhat higher for the LMiLyS (15%, 0.73%, and 7.6%) than for the LMiLy (2% and 5.6%) vector (Figure 6). However, the most significant difference between the two vectors was in the relative expression level per cell, which was measured as Lyt-2 mean fluorescence intensity. Cells harboring SAR-containing LMiLyS vector showed distinct Lyt-2-positive populations that were 3 to 6 times brighter compared with the LMiLy-marked cells (Figure 6). The frequency of gene-marked cells was determined by depositing individual MA2.1-positive cells into 96-well-format polymerase chain reaction (PCR) plates and performing sensitive DNA PCR. Overall, 27% to 55% of thymocytes were transgene positive, and there was no marked difference between the LMiLyS and the LMiLy vectors (data not shown), indicating that irrespective of the vector used, only a fraction of marked cells expressed levels of Lyt-2 transgene high enough to be detected by FACS analysis. These results indicate that the SAR element can increase vector expression, but it does not seem to confer position-independent expression.

To generate transduced primary monocytes we have established an in vitro protocol in which transduced CD34+ cells were differentiated into monocytes and at the same time selected with trimetrexate. For these experiments, we have used the control LNiD and the SAR-containing LNiDS vectors that encode a surface marker, the human nerve growth factor receptor

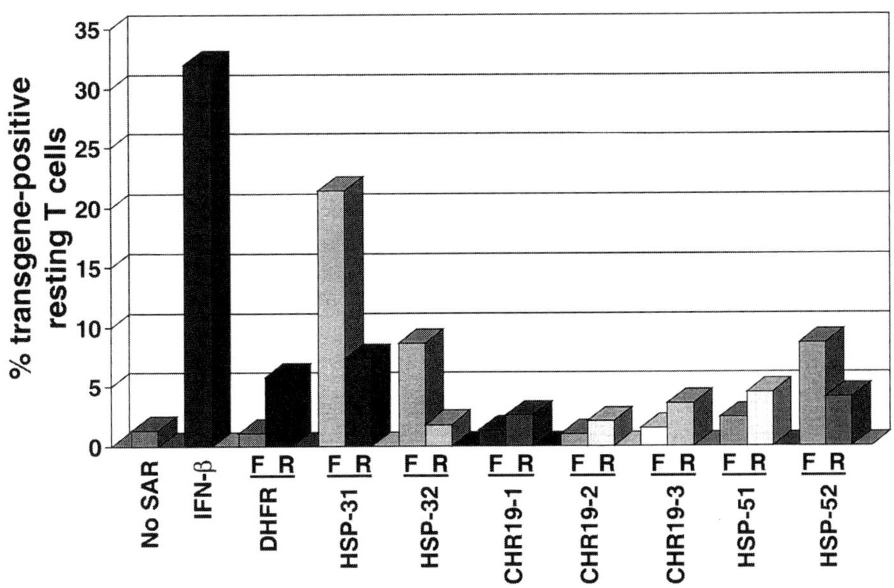

Figure 5. Only certain SARs can enhance retroviral vector expression. Eight different SAR sequences were cloned (in both orientations) into the LMiLy vector just upstream of the 3′ LTR, like in the LMiLyS vector (Figure 4). These are chinese hamster DHFR SAR (DHFR), three human chromosome 19 SARs (CHR19-1, CHR19-2, and CHR19-3), two subfragments of the hematopoietic serine protease 5′ SAR (HSP-51 and HSP-52), and the two subfragments of the 3′ SAR (HSP-31 and HSP-32). F and R indicate forward and reverse orientations of the SAR sequences. Primary T cells were transduced with the SAR-containing vectors as described in the text, and the Lyt-2 transgene expression was measured in resting cells. No SAR and the IFN-β indicate the LMiLy and the LMiLyS vectors, respectively.

(NGFr), and a resistance marker, the mutant *DHFR* gene (Figure 7) (construction of these vectors is described in detail in Materials and Methods). CD34+ HSPCs were transduced with the LNiD and LNiDS vectors and cultured on Sys-1 stromal cells in the presence of IL-3, stem cell factor, and granulocyte macrophage colony stimulating factor (GM-CSF) cytokines to promote myeloid differentiation (37). Trimetrexate was added into the medium to select for DHFR-expressing cells. Trimetrexate selection was continued until all cells in the mock-transduced samples died (usually two weeks). At that point, the trimetrexate was removed, the medium was supplemented with M-CSF, and the culture was continued for an additional three weeks. At the end of the fifth week, 50% to 80% of the cells were $CD33^+CD14^+CD4^+$ monocyte/macrophages (data not shown). The analysis of NGFr expression on the monocyte/macrophages at the end of the five-week culture is shown in Figure 7. The cells transduced with the SAR-containing LNiDS vector expressed three- to sixfold higher levels of NGFr marker compared with the control LNiD vector, both in terms of the percent of NGFr-positive cells (87% versus 18%) and mean NGFr fluorescence intensity (270 versus 81 arbitrary units). Thus, the IFN-SAR element can enhance retroviral vector expression levels in the T cells and monocyte/macrophages derived from transduced HSPC.

Application of SAR-Containing Vectors for HIV Gene Therapy

We have used HIV gene therapy as a model to demonstrate the usefulness of the SAR-containing retroviral vectors. $CD4^+$ T

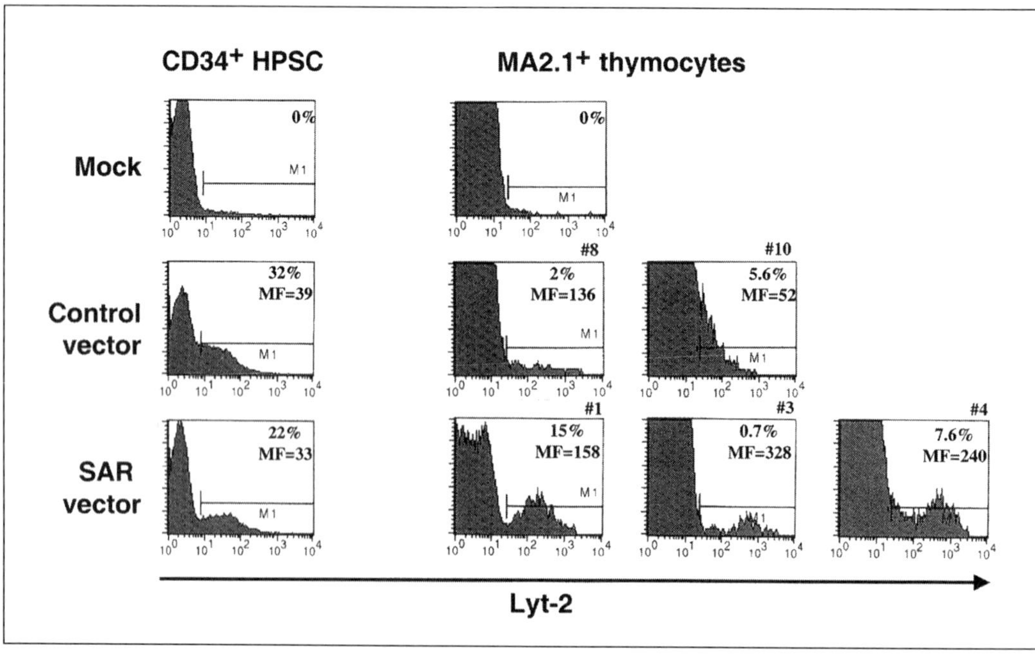

Figure 6. Analysis of Lyt-2 transgene expression in thymocytes from HSPC-reconstituted SCID-hu Thy/Liv grafts. (A) Mobilized peripheral blood $CD34^+$ HSPC were transduced with the LMiLy and LMiLyS vectors and analyzed for Lyt-2 expression two days after transduction. (B) At eight weeks after transplantation of SCID-hu Thy/Liv mice with Lyt-2-enriched HSPC, thymus/liver grafts were harvested and freshly isolated thymocytes examined for surface Lyt-2 (transgene) expression on donor HLA-positive cells ($MA2.1^+$ thymocytes in a MA2.1-negative host background). Percentages of $Lyt2^+$ cells were determined after subtracting background noise observed in mock-transduced reconstituted controls. Numbers (#) above FACS plots indicate individual SCID-hu animals. MF, relative Lyt-2 mean fluorescence intensity.

cells are primary targets for human immunodeficiency virus type 1 (HIV-1) infection. HIV-1 virus efficiently infects and replicates in fully activated (inoculated on day 2–3 after activation) and partially activated CD4+ T cells (inoculated on day 4–5 after activation) but does not replicate in resting cells (inoculated on day 8 or after activation) (57,72). The *RevM10* gene, a transdominant mutant of the HIV-1 Rev protein (45), can inhibit HIV replication in primary CD4+ T cells (57,69), but relatively high levels of RevM10 protein are required to achieve effective inhibition (57). Furthermore, the inhibitory effect of the LMiLy-encoded *RevM10* gene on HIV-1 replication is drastically reduced in partially activated T cells, probably because of a rapid decrease in vector expression during the infection experiment (57). To evaluate the effect of the SAR sequence on *RevM10*-mediated inhibition of HIV replication, we have used LMiLy and LMiLy2S vectors (Figure 4). CD4+ T cells were enriched from peripheral blood mononuclear cells from normal healthy

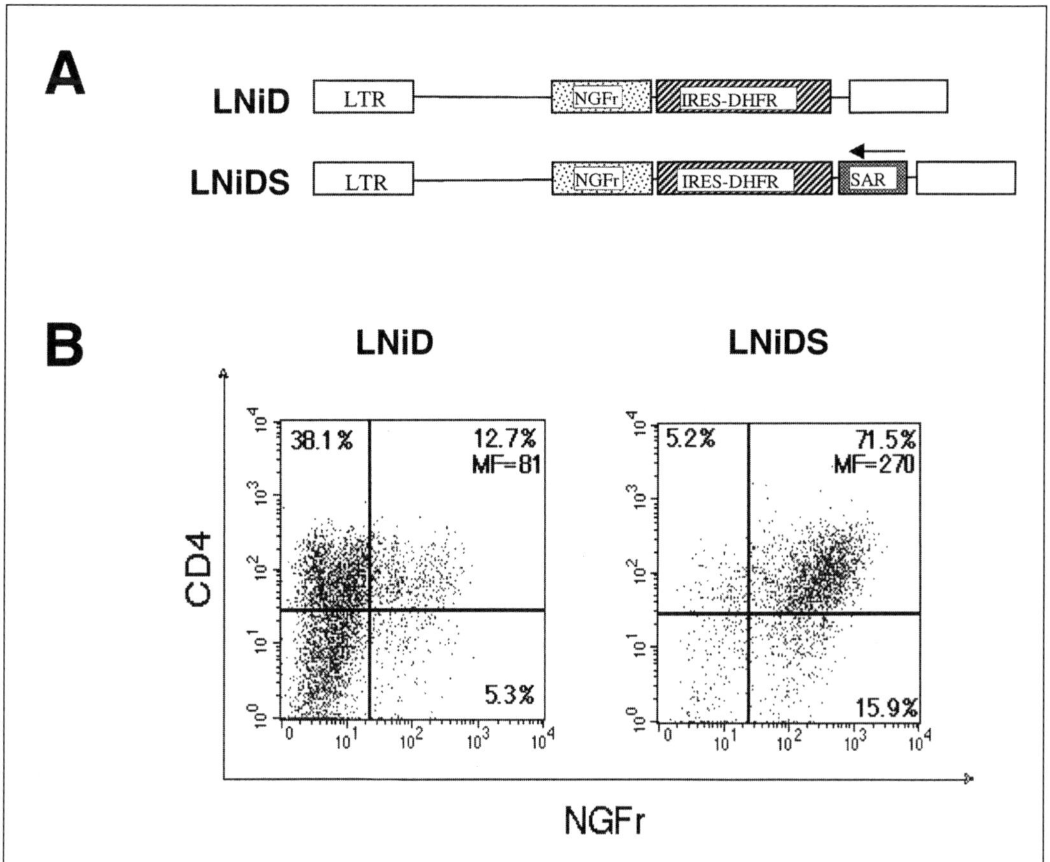

Figure 7. Effect of the IFN-SAR on vector expression in primary monocyte/macrophages. (A) The LNiD and the LNiDS retroviral vectors encode two genes, a surface marker, the truncated NGFr receptor (35,45), and a selection marker, the murine DHFR *L22Y* gene. Expression of the DHFR gene is mediated by the IRES element. (B) Mobilized peripheral blood CD34+ HSPCs were transduced with the LNiD and the LNiDS vectors and cultured on Sys-1 stromal cells for five weeks. During the first two weeks of culture, trimetrexate was included in the medium to select for vector-transduced cells. On week 5, nonadherent cells were harvested, stained with the anti-CD4-APC and anti-NGFr-FITC antibodies and analyzed by FACscan. A representative experiment is shown. Similar results were obtained in four independent experiments. Numbers indicate percentage of cells in the respective quadrants. MF, NGFr mean fluorescence intensity.

donors by depleting CD8+ cells using immunomagnetic beads yielding 90% enriched cell populations (57). Cells were transduced with the LMiLy and LMiLy2S vectors and inoculated with the HIV-1 JR-CSF strain on day 5 after stimulation with PHA, IL-2, and feeder cells. HIV replication was monitored over a period of nine days by measuring p24 antigen in cell supernatants. The LΔMiLy vector, which does not encode the RevM10 protein (22), was used as a negative control. In the representative experiment shown in Figure 8, the SAR-containing vector was one order of magnitude more potent in inhibiting HIV replication than the control vector. Thus, the IFN-SAR element can significantly enhance the antiviral efficacy of the RevM10 vectors and could certainly be useful for other gene therapy strategies that rely on persistent transgene expression levels in the T cell compartment.

CONCLUSIONS

We have described here the utility of the IFN-SAR sequence for enhancing retroviral vector expression in hematopoietic cells. The SAR element can prevent expression shut-off in resting T cells and enhances the level of transgene expression in T and myeloid cells derived from transduced HSPCs. However, it does not seem to confer full position-independent expression. Our preliminary data suggest that, at least in cell lines, the SAR element can reduce position effect but cannot abolish it completely

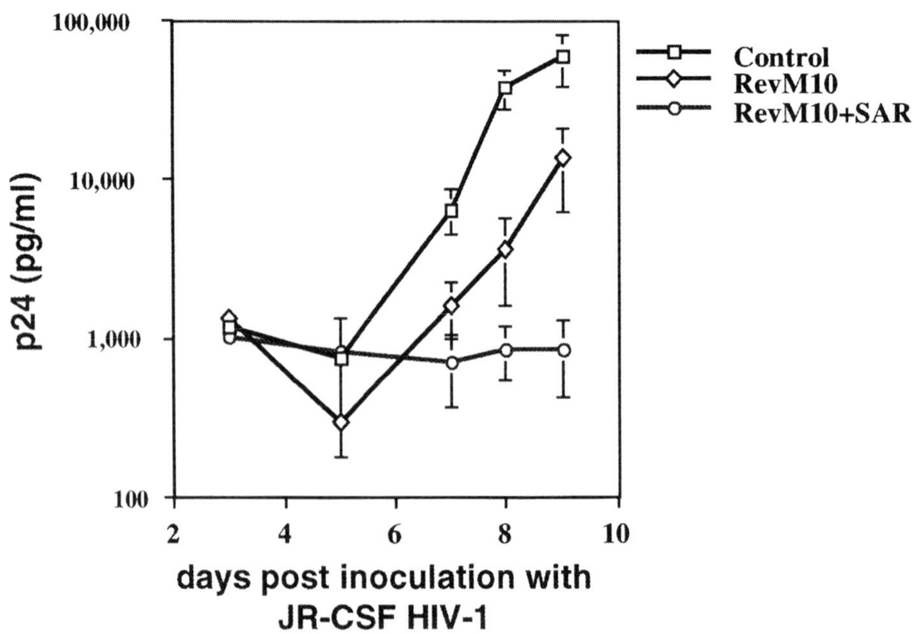

Figure. 8. Improved *RevM10*-mediated inhibition of HIV replication in primary T cells using SAR-containing retroviral vector. Primary CD4+ T cells transduced with the LMiLy, LMiLy2S, and control LΔMiLy vectors were harvested on day 5 after stimulation with PHA, IL-2, and feeder cells, and inoculated with the HIV-1 JR-CSF virus at 1:1 dilution. Viral replication was monitored over a period of nine days by measuring p24 antigen concentration in cell supernatants.

(I. Plavec and M. Agarwal, unpublished data). The particularly attractive feature of the IFN-SAR element is that it improved expression of all retroviral vectors we tested so far. These include MoMLV-, MPSV-, MESV-, SFFV-, and MSCV-derived vectors (I. Plavec and M. Agarwal, unpublished data). It could therefore be used with different vectors and for various applications.

Besides the practical utility of the SAR element, there are a number of interesting questions related to the SAR's mode of action that need to be addressed. Obviously there are features other than binding to the nuclear scaffold that are necessary for the SAR sequence to be able to enhance retroviral vector expression. Out of nine SARs tested, only two had a detectable effect on vector expression (IFN-SAR and HSP-31 SAR). Furthermore, the activity of the IFN-SAR element was orientation dependent, a feature that clearly distinguishes it from the classical enhancer elements. Better understanding of these issues will help us rationally design further improvements toward achieving sustained and high-level retroviral vector expression.

MATERIALS AND METHODS

Construction of Recombinant Retroviral Vectors and Retrovirus-Producing Cells

The retroviral vectors derived from the LMiLy vector (Figure 4) have been described previously (1,57). The vectors LNiD and LNiDS (Figure 7) are derived from the LxSN construct (50). These vectors contain the murine *DHFR* selectable marker gene (62), which carries a leucine to tyrosine substitution at amino acid position 22. This mutation renders the DHFR enzyme resistant to inhibition by pholate analog and trimetrexate (42). The *DHFR* gene is linked to the internal ribosomal entry site (IRES) from the encephalomyocarditis virus (EMCV) (34). The IRES-DHFR fusion was generated by PCR and inserted (as *Xho*I-*Bgl*II fragment) into LxSN in place of the SV40-*neo* gene (50). Subsequently, the PCR-generated, 899-bp *Bam*HI-*Hin*dIII-truncated *NGFr* gene fragment, (N-terminal 288 amino acids including the extracellular and transmembrane portion of the protein (36), was inserted in front of the IRES-DHFR into the *Hpa*II site of the vector generating the LNiD vector (Figure 7). The LNiDS vector was constructed by inserting the interferon β SAR sequence (800-bp *Hin*dIII-*Bam*HI fragment from the pCL plasmid (41) in reverse orientation into the *Nsi*I site in the 3′ untranslated region of the *DHFR* gene. Vesicular stomatitis virus glycoprotein (VSV-G)-pseudotyped retroviral particles were generated by co-transfecting retroviral vector plasmid DNA and a VSV-G expression plasmid (15) into MoMLV *gag-pol*-expressing gp47 cells as described (55,60). Forty-eight hours after transfection, culture supernatants were used to inoculate amphotropic ProPak-A packaging cells (60). Following transduction, transgene-expressing ProPak-A cells were enriched either using fluorescence-activated cell sorting (for the LMiLy-derived vectors) or by selection in 200 nM trimetrexate (US Bioscience, West Conshohocken, PA, USA) to generate polyclonal producer cell populations. Retroviral vector supernatants were prepared as described. All producer cells tested negative for replication-competent retrovirus (RCR) by S^+L^- assay on PG4 cells (27).

Transduction of Primary T Cells and Analysis of Transgene Expression in Activated and Resting Cells

Total nucleated cell fraction (buffy coat) was isolated from peripheral blood of healthy donors (peripheral blood lymphocytes) using Ficoll® gradient. The cells were cultured in a medium that supports T cell growth (TOC medium: RPMI supplemented with 1× minimal essential medium [MEM] vitamin solution [Life Technologies, Rockville, MD, USA], insulin-transferrin-sodium selenite supplement [Sigma, St. Louis, MO, USA], 10% fetal bovine serum [HyClone, Logan, UT, USA], 2 μg/mL PHA, 40 U/mL IL-2, and allogeneic JY feeder cells). Retroviral vector transduction was performed by centrifugation at 2000× *g* of 5×10^5 cells with 1 mL retroviral supernatant supplemented with 8 μg/mL polybrene for 3 hours at 34°C. This procedure was repeated on two consecutive days. Transduced cells were routinely enriched to greater than 90% surface-marker-positive cells by FACS using anti-Lyt-2 (R-PE) antibody (BD PharMingen, San Diego, CA, USA). For analysis of transgene expression, cells were stimulated with PHA plus feeder cells as described above. At various points in time afterstimulation, aliquots of cells were stained with anti-CD25 (fluorescein isothyocyanate) and anti-Lyt-2 (R-PE) antibodies and analyzed on a FACScan (Becton Dickinson, San Jose, CA, USA).

Analysis of T Cells Derived from Retrovirus-Transduced HSPCs in the SCID-hu Mouse

Mobilized peripheral blood (MPB) HSPC samples were obtained from healthy donors treated with GM-CSF (0.25 mg/m^2; Baxter Biotech Immunotherapy Division, Irvine, CA, USA) with informed consent from the donors. Leukophoresis samples were enriched (generally >90%) for CD34$^+$ cells using an immuno-magnetic bead selection device (Baxter Healthcare Corporation). Thirty-four selected MPB tissues were phenotyped for HLA MA2.1 (FITC-conjugated antibody was prepared at SyStemix from hybridomas obtained from ATCC, Rockville, MD, USA). Before transduction, cells were cultured for two days at 5 to 10×10^5 cells/mL in Whitlock/Witte (W/W) medium (50% Iscove's modified Dulbecco's medium [IMDM], 50% RPMI, 10% fetal calf serum [FCS], 4×10^{-5} mol/L 2-mercapto-ethanol, 5 mmol/L sodium pyruvate, 10 mmol/L HEPES, 100 U/mL penicillin, 100 mg/mL streptomycin and 4 mmol/L glutamine) supplemented with 20 ng/mL IL-3, 20 ng/mL IL-6, and 100 ng/mL SCF, and then subjected to two rounds of transduction as described above with either LMiLy or LMiLyS supernatants. At 48 hours after the second transduction, cells were stained for surface CD34 (SR) and Lyt2 (PE), and sorted for CD34$^+$Lyt2$^+$ on a FACStar cell sorter (Becton Dickinson) as described in detail previously (11). The SCID-hu mice were constructed by surgical transplantation of human fetal thymus/liver fragments into C. B-17 scid/scid mice as described (51). All thymus/liver tissues had been identified as negative for expression of the HLA MA2.1 marker. Three to five months after transplantation of thymus and liver fragments, SCID-hu mice, thymus/liver grafts were given 400 rads total body irradiation, followed by transduced HSPC reconstitution, as described previously in detail (11,20). 1×10^5 LMiLy- or LMiLyS-transduced and sorted CD34$^+$Lyt2$^+$ HSC were injected directly into thymus/liver grafts in SCID-hu mice. Control mice were injected with mock transduced unsorted HSPCs. Eight weeks after reconstitution, thymocytes were recovered from the grafts and analyzed for the level of donor reconstitution (MA2.1 FITC), and the expression of the LMiLy- or LMiLyS-encoded Lyt2 (Lyt2 PE) by flow cytometry (FACScan).

Transduction of MPB CD34$^+$ Cells and Differentiation into Monocyte/Macrophage Lineage

CD34$^+$ MPB HSPCs were cultured on Sys-1 feeders (6) at 5 to 10×10^5 cells/mL in W/W medium (50% IMDM, 50% RPMI, 10% dialyzed FCS, 4×10^{-5} mol/L

2-mercaptoethanol, 5 mmol/L sodium-pyruvate, 10 mmol/L HEPES, 100 U/mL penicillin, 100 mg/mL streptomycin, and 4 mmol/L glutamine) supplemented with 20 ng/mL rhuIL-3, 20 ng/mL rhuSCF, and 100 ng/mL rhuGM-CSF. To render Sys-1 cells TMTX resistant, the cells were transduced with the LNiD vector before co-culture with CD34 cells. For transduction, CD34 cells were first prestimulated in Sys-1 co-culture for 24 hours, and then on days 2 and 3 were harvested and transduced with the LNiD and LNiDS vectors by centrifugation as described above. Starting from day 6, 400 nmol/L trimetrexate was added daily to cultures. Selection continued until mock-transduced cells died (usually 9–12 days). Following selection, nonadherent cells were harvested, transferred to fresh Sys-1 cells, and cultured for an additional three weeks in the above medium supplemented with rhuM-CSF 100 ng/mL to promote monocytic differentiation. At four- to five-day intervals, cells were stained for surface CD33, CD14, CD4, and NGFr and analyzed on a FACScalibur™ (Becton Dickinson).

HIV Infection of Primary T Cells

On day 5 following stimulation with PHA and feeder cells, T cells were washed and resuspended in TOC medium containing IL-2 only; 2 to 3×10^4 cells (75 µL) were mixed with 75 µL undiluted JR-CSF HIV-1 virus stock (1×10^4 to 1×10^5 $TCID_{50}$/mL) and plated in triplicate in round-bottom, 96-well plates. Cells were cultured overnight and the following day, 125 µL medium was removed and replaced with 135 µL fresh TOC plus IL-2. Cell supernatants were harvested every two to three days over a period of 20 days. HIV-1 p24 antigen concentration in the culture supernatants was determined using an enzyme-linked immunosorbent assay (ELISA) kit (Dupont/NEN, Boston, MA, USA).

ACKNOWLEDGMENTS

The authors thank Jennifer Auten, Kathy Moss, Jingyi Chen, Michele Pineda, Franck Morel, and Tim Austin for technical assistance, and Lesley Murray for critical reading of the manuscript. CEMSS cells and JR-CSF HIV-1 virus were obtained through the AIDS Research and Reference Reagent Program, Division of AIDS, NIAID, NIH.

REFERENCES

1. **Agarwal, M., T.W. Austin, F. Morel, J. Chen, E. Böhnlein, and I. Plavec.** 1998. Scaffold attachment region-mediated enhancement of retroviral vector expression in primary T cells. J. Virol. *72*:3720-3728.
2. **Anderson, F.** 1998. Human gene therapy. Nature (Suppl.) *392*:25-30.
3. **Armentano, D., S.F. Yu, P.W. Kantoff, T. von Ruden, W.F. Anderson, and E. Gilboa.** 1987. Effect of internal viral sequences on the utility of retroviral vectors. J. Virol. *61*:1647-1650.
4. **Bauer, Jr., T.R., W.R.A. Osborne, W.W. Kwok, and D.D. Hickstein.** 1994. Expression from leukocyte integrin promoters in retroviral vectors. Hum. Gene Ther. *5*:709-716.
5. **Baum, C., S. Hegewisch-Becker, H.G. Eckert, C. Stocking, and W. Ostertag.** 1995. Novel retroviral vectors for efficient expression of the multidrug resistance (mdr-1) gene in early hematopoietic cells. J. Virol. *69*:7541-7547.
6. **Baum, C.M., I.L. Weissman, A.S. Tsukamoto, A.M. Buckle, and B. Peault.** 1992. Isolation of a candidate human hematopoietic stem-cell population. Proc. Natl. Acad. Sci. USA *89*:2804-2808.
7. **Beck-Engeser, G., C. Stocking, U. Just, L. Albritton, M. Dexter, E. Spooncer, and W. Ostertag.** 1991. Retroviral vectors related to the myeloproliferative sarcoma virus allow efficient expression in hematopoietic stem and progenitor cell lines, but retroviral infection is reduced in more primitive cells. Hum. Gene Ther. *2*:61-70.
8. **Bode, J., Y. Kohwi, L. Dickinson, T. Joh, D. Klehr, C. Mielke, and T. Kohwi-Shigematsu.** 1992. Biological significance of unwinding capability of nuclear matrix-associated DNAs. Science *255*:195-197.
9. **Bode, J., T. Schlake, M. Ríos-Ramírez, C. Mielke, M. Stengert, V. Kay, and D.W. Klehr.** 1995. Scaffold/matrix-attachment regions (S/MAR): Structural properties creating transcriptionally active loci. Academic Press, Orlando.
10. **Bonifer, C., N. Yannoutsos, G. Kruger, F. Grosveld, and A.E. Sippel.** 1994. Dissection of the locus control function located on the chicken lysozyme gene domain in transgenic mice. Nucleic Acids Res. *22*:4202-4210.

11.**Bonyhadi, M., K. Moss, A. Voytovitch, J. Auten, C. Kalfoglou, I. Plavec, S. Forestell, L. Su, E. Böhnlein, and H. Kaneshima.** 1997. RevM10-expressing T cells derived in vivo from transduced human hematopoietic stem-progenitor cells inhibit human immunodeficiency virus replication. J. Virol. *71*:4707-4716.

12.**Boulikas, T.** 1993. Nature of DNA sequences at the attachment regions of genes to the nuclear matrix. J. Cell. Biochem. *52*:14-22.

13.**Brenner, M.K.** 1995. Human somatic gene therapy: progress and problems. J. Int. Med. *237*:229-239.

14.**Bunnell, B.A., M. Metzger, E. Byrne, R.A. Morgan, and R.E. Donahue.** 1997. Efficient in vivo marking of primary CD4+ T lymphocytes in nonhuman primates using a gibbon ape leukemia virus-derived retroviral vector. Blood *89*:1987-1995.

15.**Burns, J.C., T. Friedmann, W. Driever, M. Burrascano, and J.K. Yee.** 1993. Vesicular stomatitis virus G glycoprotein pseudotyped retroviral vectors: concentration to very high titer and efficient gene transfer into mammalian and nonmammalian cells. Proc. Natl. Acad. Sci. USA *90*:8033-8037.

16.**Challita, P.M. and D.B. Kohn.** 1994. Lack of expression from a retroviral vector after transduction of murine hematopoietic stem cells is associated with methylation in vivo. Proc. Natl. Acad. Sci. USA *91*:2567-2571.

17.**Colicelli, J. and S.P. Goff.** 1986. Isolation of a recombinant murine leukemia virus utilizing a new primer tRNA. J. Virol. *57*:37-45.

18.**Dang, Q., D. Walker, S. Taylor, C. Allan, P. Chin, J. Fan, and J. Taylor.** 1995. Structure of the hepatic control region of the human apolipoprotein E/C-I gene locus. J. Biol. Chem. *270*:22577-22585.

19.**Diaz, P., D. Cado, and A. Winoto.** 1994. A locus control region in the T cell receptor α/δ locus. Immunity *1*:207-217.

20.**DiGiusto, D.L., R. Lee, J. Moon, K. Moss, T. O'Toole, A. Voytovich, D. Webster, and J.J. Mulé.** 1996. Hematopoietic potential of cryopreserved and ex-vivo manipulated cord blood progenitors evaluated in vitro and in vivo. Blood *87*:1261-1271.

21.**Dranoff, G., E. Jaffee, A. Lazenby, P. Golumbek, H. Levitsky, K. Brose, V. Jackson, H. Hamada, D. Pardoll, and R.C. Mulligan.** 1993. Vaccination with irradiated tumor cells engineered to secrete murine granulocyte-macrophage colony-stimulating factor stimulates potent, specific, and long-lasting antitumor immunity. Proc. Natl. Acad. Sci. USA *90*: 3539-3543.

22.**Escaich, S., C. Kalfoglou, I. Plavec, S. Kaushal, J.D. Mosca, and E. Böhnlein.** 1995. RevM10-mediated inhibition of HIV-1 replication in chronically infected T cells. Hum. Gene Ther. *6*:625-634.

23.**Flanagan, J.R., A.M. Krieg, E.E. Max, and A.S. Khan.** 1989. Negative control region at the 5′ end of murine leukemia virus long terminal repeats. Mol. Cell. Biol. *9*:739-746.

24.**Forestell, S.P., E. Böhnlein, and R.J. Rigg.** 1995. Retroviral end-point titer is not predictive of gene transfer efficiency: implications for vector production. Gene Ther. *2*:723-730.

25.**Greaves, D.R., F.D. Wilson, G. Lang, and D. Kioussis.** 1989. Human CD2 3′-flanking sequences confer high-level, T-cell specific, position independent gene expression in transgenic mice. Cell *56*:979-986.

26.**Grosveld, F., G. Bloom Van Assendelft, D.R. Greaves, and G. Kollias.** 1987. Position-independent, high level expression of the human β-globin gene in transgenic mice. Cell *51*:975-985.

27.**Haapala, D.K., S.D. Robay, S.D. Oroszlan, and W.P. Tsai.** 1985. Isolation from cats of an endogenous type C virus with a novel glycoprotein. J. Virol. *53*:827-833.

28.**Hanson, R.D. and T.J. Ley.** 1992. A-T-rich scaffold attachment regions flank the hematopoietic serine protease genes clustered on chromosome 14q11.2. Blood *79*:610-618.

29.**Hantzopoulos, P.A., B.A. Sullenger, G. Ungers, and E. Gilboa.** 1989. Improved gene expression upon transfer of the adenosine deaminase minigene outside the transcriptional unit of a retroviral vector. Proc. Natl. Acad. Sci. USA *86*:3519-3523.

30.**Hawley, R.G., L. Covarrubias, T. Hawley, and B. Mintz.** 1987. Handicapped retroviral vectors efficiently transduce foreign genes into hematopoietic stem cells. Proc. Natl. Acad. Sci. USA *84*:2406-2410.

31.**Hawley, R.G., A.Z.C. Fong, B.F. Burns, and T.S. Hawley.** 1992. Transplantable myeloproliferative disease induced in mice by an interleukin 6 retrovirus. J. Exper. Med. *176*:1149-1164.

32.**Hawley, R.G., F.H.L. Lieu, A.Z.C. Fong, and T.S. Hawley.** 1994. Versatile retroviral vectors for potential use in gene therapy. Gene Ther. *1*:136-138.

33.**Higgs, D.R., W.G. Wood, A.P. Jarman, J. Sharpe, J. Lida, I.-M. Pretorius, and H. Ayyub.** 1990. A major positive regulatory region located far upstream of the human α-globin gene locus. Genes Dev. *4*:1588-1601.

34.**Jang, S.K., M.V. Davies, R.J. Kaufman, and E. Wimmer.** 1989. Initiation of protein synthesis by internal entry of ribosomes into 5′ nontranslated region of encephalomyocarditis virus RNA in vivo. J. Virol. *63*:1651-1660.

35.**Jenuwein, T., W.C. Forrester, L.A. Fernández-Herrero, G. Laibe, M. Dull, and R. Grosschedl.** 1997. Extension of chromatin accessibility by nuclear matrix attachment regions. Nature *385*:269-272.

36.**Johnson, D., A. Lanahan, C.R. Buck, A. Sehgal, C. Morgan, E. Mercer, M. Bothwell, and M. Chao.** 1986. Expression and structure of the human NGF receptor. Cell *47*:545-554.

37.**Junker, U., C.S. Kalfoglou, J.J. Moon, M.K. Beck, E. Kaneshima, and E. Böhnlein.** 1998. Inhibition of human immunodeficiency virus type 1 replication in myelomonocytic cells derived from retroviral vector-transduced peripheral blood progenitor cells. Hum. Gene Ther. *9*:333-340.

38.**Kaes, E. and L.A. Chasin.** 1987. Anchorage of the chinese hamster dihydropholate reductase gene to the nuclear scaffold occurs in an intragenic region. J. Mol. Biol. *198*:677-692.

39.**Kalos, M. and R.E.K. Fournier.** 1995. Position-independent transgene expression mediated by boundary elements from the apolipoprotein B chromatin domain. Mol. Cell. Biol. *15*:198-207.

40.**Kioussis, D. and R. Festenstein.** 1997. Locus control regions: overcomming heterochromatin-induced gene inactivation in mammals. Curr. Opin. Genet. Dev. *7*:614-619.

41.**Klehr, D., K. Maass, and J. Bode.** 1991. Scaffold-attached regions from the human interferon β domain can be used to enhance stable expression of genes under the control of various promoters. Biochemistry *30*:1264-1270.

42.**Lewis, W.S., V. Cody, N. Galitsky, J.R. Luft, W. Pangborn, S.K. Chunduru, H.T. Spencer, J.R. Appleman, and R.L. Blakley.** 1995. Methotrexate-resistant variants of human dihydrofolate reductase with substitutions of leucine 22. J. Biol. Chem. *270*:5057-5064.

43.**Lim, B., D.A. Williams, and S.H. Orkin.** 1987. Retrovirus-mediated gene transfer of human adenosine deaminase: expression of functional enzyme in murine hematopoietic stem cells in vivo. Mol. Cell. Biol. *7*:3459-3465.

44.**Lu, M., N. Zhang, M. Maruyama, R.G. Hawley, and A.D. Ho.** 1996. Retrovirus-mediated gene expression in hematopoietic cells correlates inversely with growth factor stimulation. Hum. Gene Ther. *7*:2263.

45.**Malim, M.H., S. Böhnlein, J. Hauber, and B.R. Cullen.** 1989. Functional dissection of the HIV-1 Rev *trans*-activator—derivation of a *trans*-dominant repressor of Rev function. Cell *58*:205-214.

46.**Mavilio, F., G. Ferrari, S. Rossini, N. Nobili, C. Bonini, G. Casorati, C. Traversari, and C. Bordignon.** 1994. Peripheral blood lymphocytes as target cells of retroviral vector-mediated gene transfer. Blood *83*:1988-1997.

47.**McIvor, R.S., M.J. Johnson, A.D. Miller, S. Pitts, S.R. Williams, D. Valerio, D.W. Martin, Jr., and I.M. Verma.** 1987. Human purine nucleoside phosphorylase and adenosine deaminase: gene transfer into cultured cells and murine hematopoietic cells by using recombinant amphotropic retroviruses. Mol. Cell. Biol. *7*:838-846.

48.**McKnight, R.A., A. Shamay, L. Sankaran, R.J. Wall, and L. Hennighausen.** 1992. Matrix-attachment regions can impart position-independent regulation of a tissue-specific gene in transgenic mice. Proc. Natl. Acad. Sci. USA *89*:6943-6947.

49.**Miller, A.D.** 1992. Human gene therapy comes of age. Nature *357*:455-460.

50.**Miller, A.D. and G.J. Rosman.** 1989. Improved retroviral vectors for gene transfer and expression. BioTechniques *7*:980-990.

51.**Namikawa, R., K.N. Weilbaecher, H. Kaneshima, E.J. Yee, and J.M. McCune.** 1990. Long-term human hematopoiesis in the SCID-hu mouse. J. Exp. Med. *172*:1055-1063.

52.**Nikolaev, L.G., T. Tsevegiyn, S.B. Akopov, L.K. Ashworth, and E.D. Sverdlov.** 1996. Construction of a chromosome specific library of human MARs and mapping of matrix attachment regions on human chromosome 19. Nucleic Acids Res. *24*:1330-1336.

53.**Novak, U., E.A.S. Harris, W. Forrester, M. Groudine, and R. Gelinas.** 1990. High-level β-globin expression after retroviral transfer of locus activation region-containing human β-globin gene derivatives into murine erythroleukemia cells. Proc. Natl. Acad. Sci. USA *87*:3386-3390.

54.**Pawliuk, R., C.J. Eaves, and R.K. Humphries.** 1997. Sustained high-level reconstitution of the hematopoietic system by preselected hematopoietic cells expressing a transduced cell-surface antigen. Hum. Gene Ther. *8*:1595-1604.

55.**Pear, W.S., G.P. Nolan, M.L. Scott, and D. Baltimore.** 1993. Production of high-titer helper-free retroviruses by transient transfection. Proc. Natl. Acad. Sci. USA *90*:8392-8396.

56.**Petersen, R., G. Kempler, and E. Barklis.** 1991. A stem cell-specific silencer in the primer-binding site of retrovirus. Mol. Cell. Biol. *11*:1214-1227.

57.**Plavec, I., M. Agarwal, K.E. Ho, M. Pineda, J. Auten, J. Baker, H. Matsuzaki, S. Escaich, M. Bonyhadi, and E. Böhnlein.** 1997. High trans-dominant RevM10 protein levels are required to inhibit HIV-1 replication in cell lines and primary T cells: implication for gene therapy of AIDS. Gene Ther. *7*:128-139.

58.**Plavec, I., T. Papayannopoulou, C. Maury, and F. Meyer.** 1993. A human β-globin gene fused to the human β-globin locus control region is expressed at high levels in erythroid cells of mice engrafted with retrovirus-transduced hematopoietic stem cells. Blood *81*:1384-1392.

59.**Poljak, L., C. Seum, T. Mattioni, and U.K. Laemmli.** 1994. SARs stimulate but do not confer position independent gene expression. Nucleic Acids Res. *22*:4386-4394.

60.**Rigg, J.R., J. Chen, J.S. Dando, S.P. Forestell, I. Plavec, and E. Böhnlein.** 1996. A novel human amphotropic packaging cell line: high titer, complement resistance, and improved safety. Virology *218*:290-295.

61.**Rivière, I., K. Brose, and R.C. Mulligan.** 1995. Effect of retroviral vector design on expression of human adenosine deaminase in murine bone marrow transplant recipients engrafted with genetically modified cells. Proc. Natl. Acad. Sci. USA *92*:6733-6737.

62.**Robbins, P.B., D.C. Skelton, X.J. Yu, S. Halene, E.H. Leonard, and D.B. Kohn.** 1998. Consistent, persistent expression from modified retroviral vectors in murine hematopoietic stem cells. Proc. Natl. Acad. Sci. USA *95*:10182-10187.

63.**Sadelain, M., C.H. Wang, M. Antoniou, F. Grosveld, and R.C. Mulligan.** 1995. Generation of a high-titer retroviral vector capable of expressing high levels of the human β-globin gene. Proc. Natl. Acad. Sci. USA *92*:6728-6732.

64.**Simonsen, C.C. and A.D. Levinson.** 1983. Isolation and expression of an altered mouse dihydrofolate reductase cDNA. Proc. Natl. Acad. Sci. USA *80*:2495-2499.

65.**Spencer, H.T., S.E.H. Sleep, J.E. Rehg, R.L. Blakley, and B.P. Sorrentino.** 1996. A gene transfer strategy for making bone marrow cells resistant to trimetrexate. Blood *87*:2579-2587.

66.**Su, L., R. Lee, M. Bonyhadi, H. Matsuzaki, S. Forestell, S. Escaich, E. Böhnlein, and H. Kaneshima.** 1997. Hematopoietic stem cell-based gene therapy for acquired immunodeficiency syndrome: efficient transduction and expression of RevM10 in myeloid cells in vivo and in vitro. Blood *89*:2283-2290.

67. **Tagawa, M., L. Nakauchi, L.A. Herzenberg, and G.P. Nolan.** 1986. Formal proof that different-size Lyt-2 polypeptides arise from differential splicing and post-transcriptional regulation. Proc. Natl. Acad. Sci. USA *83*:3422-3426.

68. **Thompson, E.M., E. Christians, M.-G. Stinnakre, and J.-P. Renard.** 1994. Scaffold attachment region stimulate HSP70.1 expression in mouse preimplantation embryos but not in differentiated tissues. Mol. Cell. Biol. *14*:4694-4703.

69. **Vandendriesche, T., M.K.L. Chuah, L. Chiang, H.K. Chang, B. Ensoli, and R.A. Morgan.** 1995. Inhibition of clinical human immunodeficiency virus (HIV) type 1 isolates in primary CD4+ T lymphocytes by retroviral vectors expressing anti-HIV genes. J. Virol. *69*:4045-4052.

70. **Williams, D.A., S.H. Orkin, and R.C. Mulligan.** 1986. Retrovirus-mediated transfer of human adenosine deaminase gene sequences into cells in culture and into murine hematopoietic cells in vivo. Proc. Natl. Acad. Sci. USA *83*:2566-2570.

71. **Wilson, J.M., O. Danos, M. Grossman, D.H. Raulet, and R.C. Mulligan.** 1990. Expression of human adenosine deaminase in mice reconstituted with retrovirus-transduced hematopoietic stem cells. Proc. Natl. Acad. Sci. USA *87*:439-443.

72. **Zack, J.A., S.J. Arrigo, S.R. Weitsman, A.S. Go, A. Haislip, and I.S.Y. Chen.** 1990. HIV-1 entry into quiescent primary lymphocytes: molecular analysis reveals a labile, latent viral structure. Cell *61*:213-222.

26 Kinetics of Transduction Using Retroviral Vectors: Enhancement by Complex Formation with Cationic Liposomes

Colin Porter
Chester Beatty Laboratories, Institute of Cancer Research, London, England, UK

INTRODUCTION

Retroviral vectors have found widespread use both in the laboratory and, more recently, in the clinic. Three quarters of current clinical gene therapy protocols involve viral vectors, and approximately two thirds of these use vectors based on murine C-type retroviruses. Although vectors based on murine leukemia virus (MLV) are relatively well understood, and packaging cell systems are now highly developed to produce high titers of recombinant retroviruses free of replication-competent virus (11), many aspects of their use are still far from optimized. For instance, the infection process is intrinsically inefficient, and many of the details of the attachment and entry mechanism are unknown. This chapter will address what is known of the mechanism of transduction, the methods used to enhance transduction with recombinant retroviruses, and means to enhance transduction kinetics with the use of cationic liposomes.

THE MECHANISM OF RETROVIRAL TRANSDUCTION

Successful transduction of a target cell by a retrovirus entails a number of steps. The first stage is an encounter between the virus and the cell. A nonspecific interaction is required to stabilize the encounter and enable a specific virus envelope/cell surface receptor interaction to occur. The virus is internalized prior to, or as a result of, a membrane fusion event triggered by envelope/receptor binding. Fusion delivers the viral core to the cytoplasm. Reverse transcription of the viral RNA genome within the core and transport of the preintegration complex to the vicinity of the nucleus is required. Finally, when the nuclear membrane breaks down during mitosis, this complex gains entry to the nucleus, and the proviral cDNA is integrated into the replicated cellular DNA. Each of these steps can potentially regulate the overall efficiency of gene delivery, which will also depend on the biochemical processes of the

Viral Vectors: Basic Science and Gene Therapy
Edited by A. Cid-Arregui and A. García-Carrancá

target cell. For instance, mitosis must occur within a reasonable period after entry since the internalized virus has a finite half-life. Variations in the efficiency of negotiating each of these hurdles in different target cells lead to widely varying susceptibilities to infection. The overall process is of very low efficiency, with the physical-infectious particle ratio on the order of 100 to 1000:1 (4,56).

Adsorption

The specific adsorption of virus from the medium is inefficient and is governed by a number of physicochemical factors (39). Following exposure to an adherent target cell layer for 2 hours, there is little detectable depletion of ecotropic MLV (MLV-E) (52) or amphotropic MLV (MLV-A) (34) vectors. For infection of adherent cells by virus present in the overlying medium, a significant limitation has been attributed to the Brownian movement of the viral particles (9). The initial encounter of virus and cells is thus diffusion-limited, and only those virions close to the cell layer will be captured before the virus is removed or decays in the medium. Infectivity is thus dependent upon virus concentration and cell density (9,34,52). However, the association rate constant for virus binding to cells in suspension (determined by indirect immunofluorescent staining for virus envelope) is significantly lower than the calculated diffusion limitation, which suggests that binding may be rate-limiting rather than encounter (56). The data were consistent with a simple, bimolecular, noncooperative model for virus binding. Virus binding is likely to involve two steps: *(i)* an initial nonspecific interaction and then *(ii)* a specific interaction. Nonspecific interaction has been proposed to enable viral scanning of the cell surface to facilitate specific receptor recognition (19). This interaction is likely to be influenced by dipole-dipole attraction, electrostatic repulsion, hydration forces caused by water exclusion, and steric forces from large molecular weight molecules that prevent close contact. It is, therefore, unsurprising that various polymers, such as polybrene, are able to modulate this step and improve the mass transfer of virus to the cell surface (see below).

Receptor Binding

The specific interaction involves the viral envelope protein and its receptor on the cell surface (54). Different packaging cells enable production of vectors pseudotyped with a number of different envelope proteins utilizing distinct receptors (40). Receptor mRNA expression shows some tissue specificity that may influence transduction efficiency for different target cells (23), but is not in itself predictive of efficiency (51). The ecotropic receptor density for murine 3T3 fibroblasts was determined using free envelope protein binding to be 7.4 to 8.9×10^5 receptors per cell (15). A similar study using an isolated envelope receptor binding domain estimated the amphotropic receptor density at 7.7×10^4 receptors per cell for 3T3 cells and 1.4×10^5 per cell for human TE671 cells (8). These numbers would seem to be in excess of the requirement. However, there is evidence that receptor abundance is limiting for MLV-A transduction of rat fibroblasts, which naturally possess 1.8×10^4 receptors per cell, whereas a derivative with eightfold increased levels is correspondingly more susceptible (27). In contrast, MLV-E transduction of nonsusceptible cells following ecotropic receptor expression is not dependent upon receptor abundance above a low threshold (not quantified), and it has been proposed that a putative accessory factor may be more of a limitation (52). Support for the hypothesis that only a subset of binding sites are functional receptors comes from the observation that infection with amphotropic vectors is more sensitive to competition with the receptor-binding domain of the amphotropic envelope protein than is virus binding (8).

Multiple binding events are required both to appose and destabilize the two membranes. Binding of free envelope proteins is not cooperative (15,27), but it is not known if there is cooperativity in the multivalent binding for a single virion. The equilibrium dissociation constant for ecotropic MLV binding to 3T3 cells is on the order of 1000 times less than that for free envelopes (56). Binding of the surface (SU) envelope domain is thought to lead to conformational changes for envelope and receptor (53), allowing the hydrophobic transmembrane (TM) domain to trigger the membrane fusion process and so release the viral core to the cytoplasm. The capacity to mediate fusion requires the postassembly priming of TM by C-terminal cleavage by the virus-encoded protease (42). Additionally, receptor clustering on the surface of the infected cell and putative cellular factors may potentiate membrane fusion (24,45).

Internalization

In addition to adsorption, the postbinding efficiency of retroviral infection is poor. Indeed, only 1% to 2% of bound virions were infectious in one study (4). Entry of MLV-E into 3T3 and SC-1 fibroblasts was studied; the study compared physical ([^{3}H]-Leu radiolabeled) particles and infectious particles (determined by reverse transcriptase release after replication) throughout the binding, internalization, and degradation stages of infection. Both physical and infectious particles bound and were internalized with similar kinetics (half times of 2 to 3 hours), showing first-order dependence on virus concentration. Fifteen percent of bound particles were retained at the cell surface, and up to 75% of internalized particles were subsequently degraded in lysosomes. Degraded material appeared in the medium after a short lag but could be inhibited with lysosomotropic agents such as chloroquine. Use of such agents demonstrated that passage through an acidic compartment shortly after internalization is also necessary for infectivity. Entry into 3T3 cells has also been studied by tracking particle-associated reverse transcriptase (1). Virus adsorbed to the cell surface at 4°C appeared in the cytoplasm as intact particles after raising the temperature to 37°C. The level subsequently declined because of the uncoating of intracellular virions.

Electron microscopic immunolabeling studies of 3T3 cells infected with ecotropic MLV have similarly shown that most virus initially enters the cells as intact particles (43). This is followed by the release of some particles from vesicles into the cytoplasm with concomitant loss of envelope protein. Viral cores remain intact during intracytoplasmic transport and eventually disrupt at the nucleus with nuclear entry of nucleocapsid and integrase, but not matrix or capsid proteins. Those particles that fail to fuse following internalization are degraded, which is consistent with the observation that essentially all internalized envelope proteins are degraded, whereas a fraction (30% in SC-1 cells) of the core proteins is resistant (3). While reduced binding can occur at 4°C, fusion is temperature dependent. For MLV-E infection of most cell types, it is also pH dependent and requires acidification for the release of viral cores from endosomes, which is consistent with the above observations (31). However, acidification per se does not trigger envelope fusogenicity (38), and a requirement for SU cleavage by a pH-dependent protease (cathepsin) has been postulated. On the basis of these data, it seems that ecotropic MLV infection involves the internalization of intact particles by endocytosis, leading either to membrane fusion for cytoplasmic release of virion cores or to degradation of virions that fail to fuse. The efficiency of membrane fusion and core release versus destruction within the endosome is likely to be a significant controlling factor since many internalized virions are degraded. There is no such acidification requirement for amphotropic MLV infectivity, which suggests that entry may be caused by direct fusion at the cell surface. Infection of rat XC sarcoma cells is an exception to the pH dependence of MLV-E entry, suggesting that target cell characteristics may influence the route of internalization (31). Target cell cytoskeletal components are involved with the requirement for an intact actin network at an early stage (prior to internalization) for both pH-dependent and pH-independent MLV-E entry, and for intact microtubules at a stage postinternalization for pH-dependent entry only (24).

Reverse transcription takes place within the internalized viral core. The virus-specific enzyme is copackaged with the RNA template and tRNA primer but is dependent upon the target cell nucleotide pool. Depletion of this pool may be a block to transduction in some cases, such as primitive hematopoietic cells (2). Finally, successful transduction of the target cell depends on its status in the cell cycle (32). Breakdown of the nuclear membrane is necessary for integration (44) and must occur before the intracellular decay of the preintegration complex. The half-life of the latter for ecotropic MLV is on the order of 5.5 to 7.5 hours, which is likely to be restrictive for slowly dividing cells (5). Mathematical modeling shows that the intracellular half-life is the most significant limitation to the maximal infectivity that may be obtained (6). In this regard, vectors derived from the lentivirus family of retroviruses, for which infection is independent of cell cycle, offer particular advantages (37).

METHODS OF INCREASING EFFECTIVE TITER OR TRANSDUCTION EFFICIENCY

Many approaches have been adopted in an effort to increase the efficiency of transduction using retroviral vectors. Traditionally, efficiency is recorded in terms of functional titer, which is the concentration of colony-forming units per mL (cfu/mL) of virus preparation. In practice, this means an end-point titer that results under conditions of low multiplicity of infection (moi). The value obtained is not only dependent upon the target cell used but also on many of the details of how the experiment is conducted. Variables such as the number of target cells and the volume of medium in which the virus is applied are important in determining the titer calculated from the sample volume used because the number of transduction events depends on the number of cells and the final virus concentration. The time for which the cells are exposed will also affect the number of transduction events.

Although relative end-point titer is a convenient measure for some purposes, application of retroviral vectors, especially clinical application, will involve the use of undiluted virus, and the transduction efficiency achievable at high moi is a more significant measure. However, in a survey of vector supernatants, there was no correlation of end-point titer and transduction efficiency obtained with undiluted stocks (14). Moreover, specific treatments that increase end-point titer do not always enhance transduction efficiency; on the contrary, they may lead to loss of efficiency. Concentration by ultrafiltration (14) (unpublished data) can result in efficient increases in end-point titer but reduction of transduction efficiency, suggesting that an inhibitor of transduction is co-concentrated by this procedure. Transduction similar to that obtained prior to concentration may usually be obtained by dilution of the concentrate. Different target cells can be differentially susceptible to the inhibitor (unpublished data). The nature of the inhibitory effect likely differs for different producer cells and has been variably attributed to inactive particles, shed envelope proteins, or nonviral components such as proteoglycan (14,29,34).

Polybrene is a polyamine frequently included during the infection process to enhance transduction (50). Other cations, such as diethylaminoethyl (DEAE)-dextran, protamine sulphate, liposomes, or the basic protein histone IIA can similarly improve infectious titer (12,20,46,49). Such experiments typically involve exposure of target cells for extensive periods of time (4 hours or more), and these cationic agents appear to work by electrostatic neutralization to facilitate cell adsorption. The magnitude of the enhancement is variable, the nature of the target cell being an important determinant, but is typically on the order of tenfold. However, the precise effect in any given situation is likely to depend on additional parameters, and other physicochemical processes may contribute. For instance, the polybrene concentration for optimal transduction enhancement can depend on the serum content of the medium, and polybrene additionally reduces vector stability in solution (7). Extended exposure to cationic liposomes can also cause virus aggregation and reduce transduction efficiency.

A number of methods to increase transduction are based on physical means to facilitate the initial vector/cell interaction. These include calcium phosphate precipitation of the vector onto the target cells (36), or centrifugation of target cells in the presence of the vector (47) to increase exposure of cells to virus. In the former case, addition of calcium chloride to tissue culture medium that contains the retrovirus results in a calcium phosphate coprecipitate that sediments onto the target cells, thereby enhancing virus cell contact and subsequent adsorption. Centrifugation presumably also works by stabilizing an initial interaction, thus facilitating productive binding to the cell surface receptor. Flow-through transduction, based on the limitation of contact between an adherent target cell layer and viral particles subject to Brownian movement in the overlying medium, similarly aims to increase cellular exposure (9,10). In this system, fluid flow through the cell bed supported on a porous membrane results in directed motion of the viral particles towards the target cells, leading to greater exposure to virus than would be obtained in a static system and correspondingly greater transduction.

Fragments of fibronectin have been found to facilitate transduction if present during the infection process (35). Recombinant peptides that encompass the cell-binding domain (containing the RGDS amino acid motif of type III repeat number 10, which is recognized by the integrin VLA-5), the heparin-binding domain (type III repeats 12–14) and the VLA-4 binding site of the CS1 domain are most effective (18). Virus has been shown to bind to such peptides attached to tissue culture dishes and to increase the transduction of target cells capable of binding via VLA-4/VLA-5. Colocalization of viruses and target cells is necessary, with the fibronectin polypeptide acting as a bridge. Enhanced transduction results from more efficient interactions of viruses and target cells colocalized on the surface of the tissue culture dish rather than in three dimensions.

THE IMPORTANCE OF KINETIC CONSIDERATIONS

While many of these measures may be valuable for laboratory studies, they can find only limited applicability for clinical gene therapy with the exception of ex vivo strategies. In vivo application of recombinant retroviral vectors is subject to many additional constraints. The stability of the viral particles when exposed to body fluids is an important factor. Recent work has identified the mechanism behind the instability of vectors derived from the original murine packaging cell systems upon exposure to human serum (48), and novel cell lines avoiding this problem have been generated (13). The removal of vector from the site of administration due to blood flow, depletion of available vector due to nonspecific binding to inappropriate cells, and clearance by cells of the immune system are also factors that may militate against efficient transduction in vivo. To combat these, it will be necessary to achieve transduction of the intended target cells rapidly. Enhancement of infection during brief exposure to the vector is altogether a different problem from the enhancement of end-point titers following extended exposure, yet the analysis of the kinetics of retroviral transduction has been largely overlooked.

The equilibrium dissociation constant for specific virus binding, determined by indirect immunofluorescence assay, has been estimated for MLV-E binding to 3T3 cells in suspension (56). The value obtained suggests that virus binding is strongly favored at equilibrium. However, equilibrium conditions are relevant only to artificial experimental systems, and a high affinity constant does not necessarily imply a fast reaction rate (55). The critical parameter for application under more realistic conditions is the kinetic rate association constant. The same authors estimated the kinetic association and dissociation constants (the ratio of which should be the same as the equilibium constant), concluding that the association rate for specific virus binding is significantly less than the (theoretical) rate of encounter (56), consistent with the notion that an initial nonspecific but rapidly associating interaction may precede specific binding (19).

Studies that have looked at the effect of various agents or treatments on the efficiency of infection have tended to concentrate on end-point titers or transduction efficiencies following extended exposure, as discussed above. Although there may be some expectation that this reflects the kinetics of the infection process, significant differences in initial rates would not be obvious.

MODULATION OF RETROVIRUS TRANSDUCTION WITH CATIONIC LIPOSOMES

All of the approaches to enhance transduction discussed above aim to increase the formation of receptor-bound virions, either by improving target cell encounter or by facilitating a specific receptor interaction. However, from the preceding discussion, it is clear that there is plenty of scope to improve the efficiency of postbinding events. Enhancing membrane fusion to release a larger proportion of virus cores to the cytoplasm rather than their being routed for lysosomal degradation may also be achievable. One approach to this end is to modify the structure of the envelope protein (28) (see also Chapter 23) but an alternative would be to use agents that modulate membrane structure and potentiate fusion, such as cationic liposomes.

Many cationic lipids have been synthesized during the development of nonviral means of gene transfer using cationic liposomes. Such liposomes are unilamellar, typically on the order of 100 to 200 nm in diameter, and form a variety of different structures when complexed with DNA molecules. Endocytosis of the complex is the major route of entry, and membrane fusion or destabilization is often enhanced by the inclusion of a neutral helper lipid (dioeoyl phosphatidylethanolamine [DOPE]) although some cationic lipids are themselves fusogenic (17). Cell surface proteoglycans, particlularly heparan sulphate, are involved in transfection with some cationic liposomes (33).

Several cationic liposomes have also shown the ability to modulate infection with viruses, including retroviruses. Cationic liposomes composed of DOTMA/cholesterol have been shown to enhance the infectivity and replication of human immunodeficiency virus (HIV) (25,26). The optimal effect required the presence of the liposome during virus exposure, was similar to that obtained with polybrene or DEAE-dextran, and could be nullified in the presence of polyanions, indicating that the mechanism of enhancement was electrostatic (see above). Preincubation before exposure to target cells led to virus fusion with the liposomes and significant loss of efficacy. Various preparations of cationic liposomes can also enhance MLV infection (20,22,41,47,49). Information regarding liposome formulations discussed in this chapter is given in Table 1. A recent publication from this laboratory has shown that pretreatment of several pseudotypes of MLV with cationic liposomes can significantly enhance transduction rates (discussed in detail below) (41).

The studies with HIV and one study with MLV (20) showed that liposome-mediated infection was still dependent upon the appropriate viral receptor. However, receptor-independent transduction can also be achieved with some virus/liposome/cell combinations (21,41). Lipofectin® Reagent (DOTMA/DOPE)(Life Technologies, Rockville, MD, USA), despite failing to enhance transduction with MLV-E or MLV-A in this study, enabled infection of cells lacking receptors (21). DC-Chol/DOPE, LipofectAMINE™ (DOSPA/DOPE) (Life Technologies), and DOTAP, in addition to enhancing viral transduction, also mediated transduction with wrongly enveloped or nonenveloped virus particles (41). Nevertheless, the efficiency of transduction using a cognate interaction was considerably greater. From such observations, it is clear that cationic liposomes are more than simply cations acting via charge neutralization (such as polybrene), for which transduction maintains a strict requirement for envelope/receptor interaction. In addition to electrostatic effects, their fusogenic properties, manifest in the absence of envelope-mediated fusion, are likely to influence receptor-dependent virus entry.

TRANSDUCTION RATE ENHANCEMENT WITH DC-Chol/DOPE

An MLV-A vector for expressing β-galactosidase was harvested under serum-free conditions from a human-based packaging cell line (13) and used to infect human rhabdomyosarcoma TE671 target cells. Target cells were plated the previous day in 24-well plates at a density of 2.5×10^4 cells/well. The medium was removed, replaced with 250 µL optiMEM™ (Life Technologies), and 100 µL pretreated virus was added. Pretreatment involved mixing appropriately diluted virus harvests with an equal volume of agent for 30 minutes at room temperature. Unless otherwise stated, the agent used was DC-Chol/DOPE (16) at a concentration of 50 µg/mL in optiMEM, resulting in a final liposome concentration of 2.5 µg/350 µL per well. Two commercial cationic liposome preparations were similarly used: DOSPA/DOPE and DOTAP (Roche Molecular Biochemicals, Indianapolis, IN, USA). The cells were exposed for varying times, after which unbound virus was removed and replaced with growth medium. Two days later, the cells were stained histochemically to detect

Table 1. Cationic Liposome Formulations

Cationic Lipid[a]	Chemical Structure[b]	Co-lipid + Formulation[c]	Commercial Name/ Availability[d]	References[e]
DOTMA	2-chain (oleoyl) Monovalent	DOPE, 1:1 (w/w)	Lipofectin	20,21,47
DOSPA	2-chain (oleoyl) Multivalent	DOPE, 3:1 (w/w)	LipofectAMINE	20,41,47,49
DOTAP	2-chain (oleoyl) Monovalent	None	DOTAP	20,41
DC-Chol	Cholesterol-based Monovalent	DOPE, 3:2 (mol/mol)		41,47

[a]The full names of the cationic lipids are as follows:
DOTMA: N-[1-(2,3-dioleoyloxy)propyl]-N,N,N-trimethylammonium chloride
DOSPA: 2,3-dioleoyloxy-N-[2(spermine-carboxamido)ethyl]-N,N-dimethyl-1-propanaminium trifluoroacetate
DOTAP: N-[1-(2,3-dioleoyloxy)propyl]-N,N,N-trimethylammonium methylsulfate
DC-Chol: 3β[N-(N′,N′-dimethylaminoethane)-carbamoyl]cholesterol

[b]The hydrophobic anchor is either two aliphatic chains of monounsaturated C18 (oleoyl) or cholesterol. The chemical linkage to the headgroup is ether for DOTMA and DOSPA, ester for DOTAP, or carbamoyl for DC-Chol. The headgroup is monovalent quaternary amine for DOTMA and DOTAP, monovalent tertiary amine for DC-Chol, or polyamine for DOSPA.

[c]Where indicated, the neutral co-lipid DOPE (dioleoyl phosphatidylethanolamine) is included with the cationic lipid in the formulation of cationic liposomes. The formulation refers to the ratio of cationic lipid:DOPE.

[d]The DC-Chol/DOPE used in the studies presented here and in Reference 41 was a generous gift from Dr. L. Huang and Dr. F.L. Sorgi, University of Pittsburgh (16).

[e]Reference 17 may be consulted for a review of cationic liposomes as DNA transfection reagents. Other references given here refer to the use of such reagents with MLV, as discussed in the text.

β-galactosidase expression, and the proportion of positive (blue) cells was counted. The moi was determined from this figure according to the formula moi = -(logP)/0.4343, where P is the proportion of negative cells (41). As used here, moi defines the number of transduction events as a proportion of the number of target cells (i.e., the average transduction per cell).

The moi for virus without pretreatment increased linearly over 4 hours at a rate that was concentration dependent. Virus that was pretreated with cationic liposomes consisting of DC-Chol/DOPE was studied in parallel. Such pretreatment resulted in greatly enhanced kinetics over the first 40 minutes of exposure, followed by a reduced rate of transduction (Figure 1). A similar result was obtained for liposomes consisting of DOSPA/DOPE or DOTAP. The initial rate of transduction was dependent upon liposome concentration, but the dose-response was different for the different liposomes (Figure 2). This emphasizes the need for optimization of transduction conditions. Virus pretreated in the same way with polybrene also showed enhanced kinetics, less marked than that seen with DC-Chol/DOPE but at a sustained rate, such that equivalent transduction was seen after exposure for 4 hours (Figure 1).

Optimal rate enhancements were on the

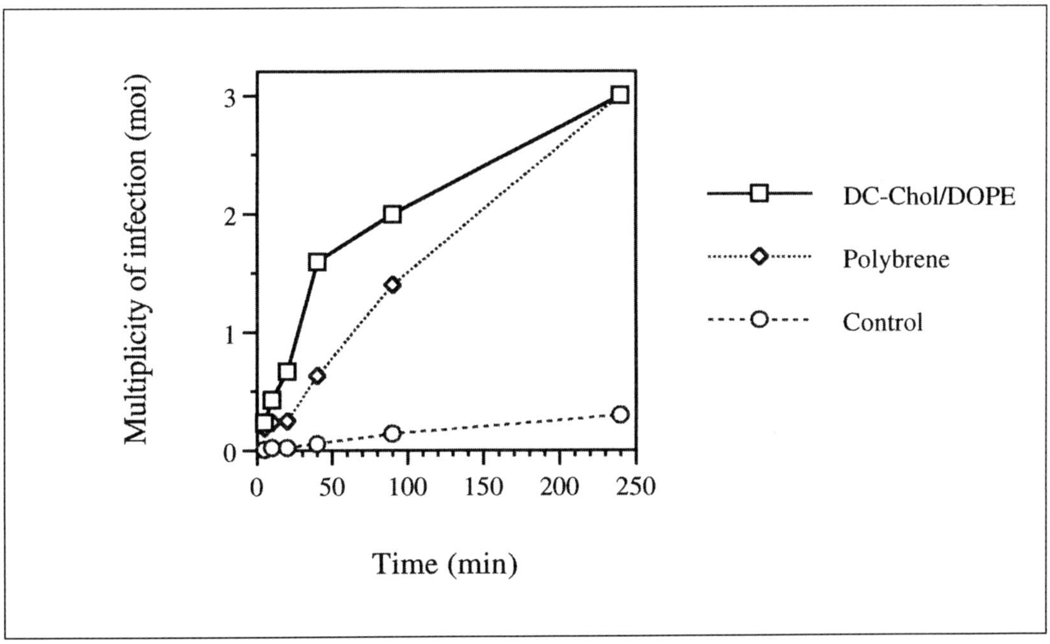

Figure 1. Transduction rate enhancement of a retroviral vector with cationic liposomes. An amphotropic retroviral vector was pre-incubated with DC-Chol/DOPE and added to TE671 target cells. After exposure for the time indicated, the reagents were replaced with growth medium. The proportion of histochemically positive cells was determined two days later, from which the moi was calculated. Polybrene was similarly used for a final concentration of 8 μg/mL. The virus stock (titer 2×10^6 cfu/mL when used untreated) was initially diluted 1:10 to prevent saturation of the assay.

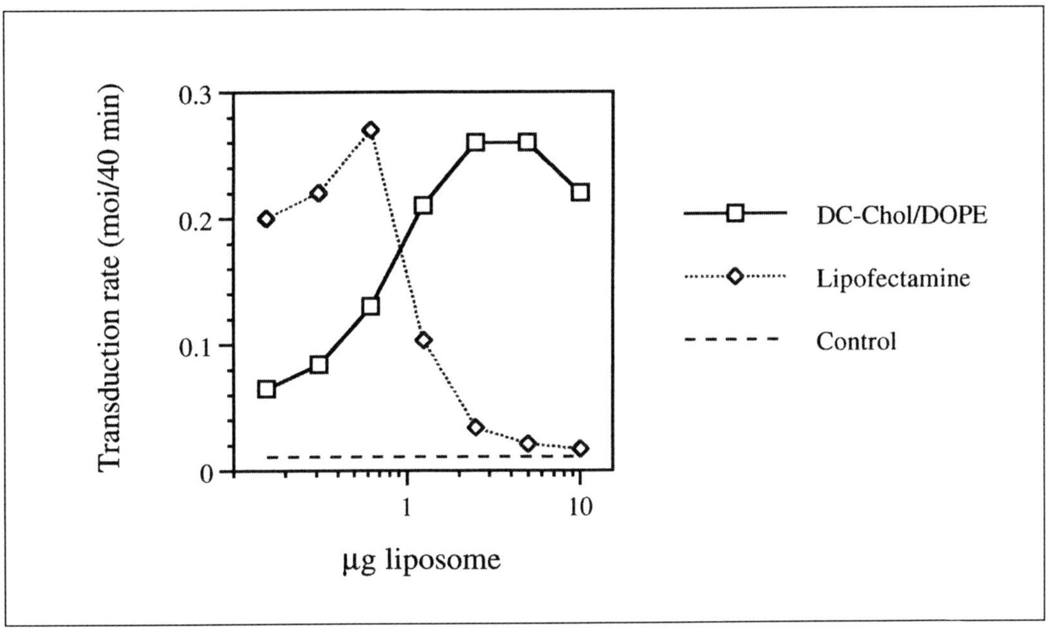

Figure 2. Cationic liposome dose-response curves. Virus was pre-incubated with serial dilutions of DC-Chol/DOPE or LipofectAMINE, and target cells were exposed for 40 min. The moi after exposure for this time is a measure of the initial transduction rate, which varied according to the concentration and identity of the liposome used. The data presented indicate the amount of liposome per 350-μL final volume. The dashed line (control) shows the rate obtained without pretreatment.

order of 10-fold and 20-fold for polybrene and DC-Chol/DOPE, respectively. Enzyme activity and provirus copy numbers increased proportionately, indicating that the effect was to increase the number of independent transduction events rather than to generate large aggregates. Consistent with this, the initial rate showed first-order dependence on virus concentration, and the activity of the liposome-treated virus was recoverable after 0.45-μm filtration. The virus was shown to form a stable complex with the liposome by demonstrating that the transduction rate showed first-order dependence on virus concentration after preincubation in contrast to the higher-order dependence obtained if virus and liposome were independently diluted before combining (Figure 3). Complex formation was rapid (within 5 minutes), and transduction was not significantly affected by serum present during preincubation or exposure of the target cells. However, extended periods of preincubation with DOTAP resulted in loss of enhancement and correlated with the appearance of aggregated material. The precise nature of the active complex is unclear, but fusion of virus membrane and liposome bilayers is a possibility given the rapid fluorescent dequenching (indicative of membrane fusion) observed upon incubation of octadecyl-rhodamine-labeled HIV with DOTMA-containing liposomes (25). Importantly, the data shown in Figure 3 indicate that pretreatment generates a stable liposome-modified virus that demonstrates significantly enhanced transduction efficiency during brief target cell exposure.

The optimal order of combining virus, liposomes, and target cells was studied. When DC-Chol/DOPE was preincubated with the cells and then removed before the addition of virus, the transduction rate enhancement was markedly reduced. The maximum effect was seen when the virus was pretreated in the absence of the target cells; this result is consistent with the for-

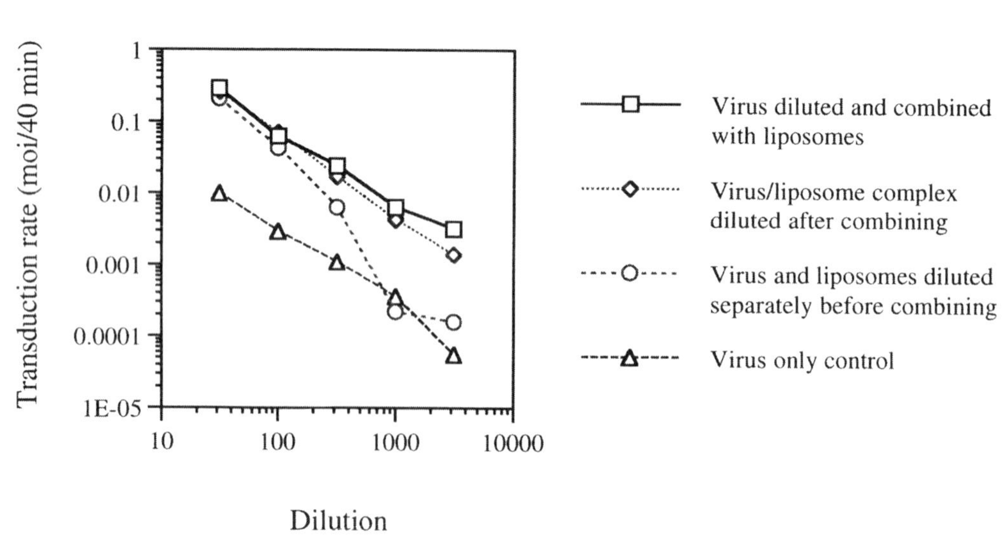

Figure 3. Formation of a stable virus/liposome complex. The transduction rate (moi after exposure for 40 min) showed first-order dependence on virus concentration for virus diluted before treatment (or untreated control) or for virus pretreated with DC-Chol/DOPE. This contrasts with the higher-order dependence seen if the virus and liposome were diluted separately, resulting in a loss of enhancement at high dilution, as would pertain if there was no stable association of the two.

mation of a stable complex. In contrast, polybrene pretreatment of the cells was at least as effective as pretreatment of the virus, consistent with its proposed role in membrane charge neutralization. The actions of both agents were not additive when virus was incubated with a combination of the two, nor was there a significant enhancement using liposome-treated virus for infection of polybrene-treated cells. This may imply some competition for virion association and/or overlapping activities.

There was no effect of DC-Chol/DOPE, and only a small effect of polybrene, when added in warm medium to cells with virus already bound but not internalized by incubation at 0°C. In contrast, there was substantial enhancement following the warming of cells prebound with virus that had been pretreated with either agent, which is consistent with improved binding. The time course of internalization, measured by treating the warmed cells with trypsin at various points in time to destroy surface-exposed virions and then replating them to allow completion of the infection process, was rapid in all cases. There was no difference in dissociation of prebound virus according to treatment during maintenance at 0°C before warming. The initial rate for binding and internalization was measured by exposure of the cells for 5 minutes followed by trypsin treatment. Approximately 90% of virus binding in 5 minutes could be inactivated when untreated or pretreated with polybrene, compared with 75% when treated with DC-Chol/DOPE.

The simplest interpretation of all these data is that both polybrene and DC-Chol/DOPE enhance virus binding and that DC-Chol/DOPE additionally enhances postbinding internalization/fusion. The possibility that DC-Chol/DOPE treatment is protective during exposure to trypsin cannot, however, be ruled out. Fusion may be direct at the cell surface, as is thought to be the case for MLV-A , or subsequent to endocytosis, the primary route of cationic liposome-mediated DNA entry.

The experiments described above were conducted using an MLV-A-pseudotyped vector for infection of TE671 cells; virus was harvested from TE671-based packaging cells expressing the amphotropic envelope (13). However, the effects were shown to be more general by using 3T3 target cells for MLV-A and also MLV-E-pseudotyped vector, and additionally by using RD114-pseudotyped vector for infection of TE671 cells. MLV-A, MLV-E, and RD114 use distinct cellular receptors. For generation of these different pseudotypes, producer cells that express the different envelope proteins were used (13,40). The magnitude of the effect varied for the different virus/target cell combinations, but was evident in all cases (Figure 4). Included in these experiments were combinations of viruses with nonpermissive cells, and also virus particles lacking envelope proteins. Cationic liposomes, but not polybrene, were able to mediate transduction independent of an envelope/receptor interaction, presumably because of their fusogenic properties. The efficiency of such envelope-independent transduction with DC-Chol/DOPE was significantly (at least 100 times) less than that dependent upon envelope, showing the superiority of fusion triggered by a cognate receptor interaction. Nevertheless, this observation is consistent with a mode of action for the cationic liposomes in which binding is enhanced, because of their cationic properties, and for which envelope-mediated fusion is also facilitated because of their membrane destabilization properties. DOTAP demonstrated the greatest envelope-independent transduction, with rates 10% to 20% of those obtained for untreated MLV-A, but was less effective at promoting envelope-dependent entry. Moreover, for MLV-E infection of 3T3 cells, DOTAP not only failed to enhance undiluted virus but also significantly reduced the activity of dilutions, suggesting competitive modes of entry. This may be related to the need for MLV-E to pass through a low-pH environment to trigger fusion, whereas DOTAP may mediate fusion before such a stage.

The rationale for embarking upon the above experiments was to enhance the transduction kinetics for improved application of retroviral vectors. To investigate the effect of cationic liposome treatment of virus on transduction in vivo, a xenograft model of intraperitoneal growth of human mesothelioma cells was used. The MLV-A vector used in the above in vitro experiments, alone or complexed with DC-Chol/DOPE, was administered to the peritoneal cavity two days following tumor establishment. On day 14, ascitic and solid

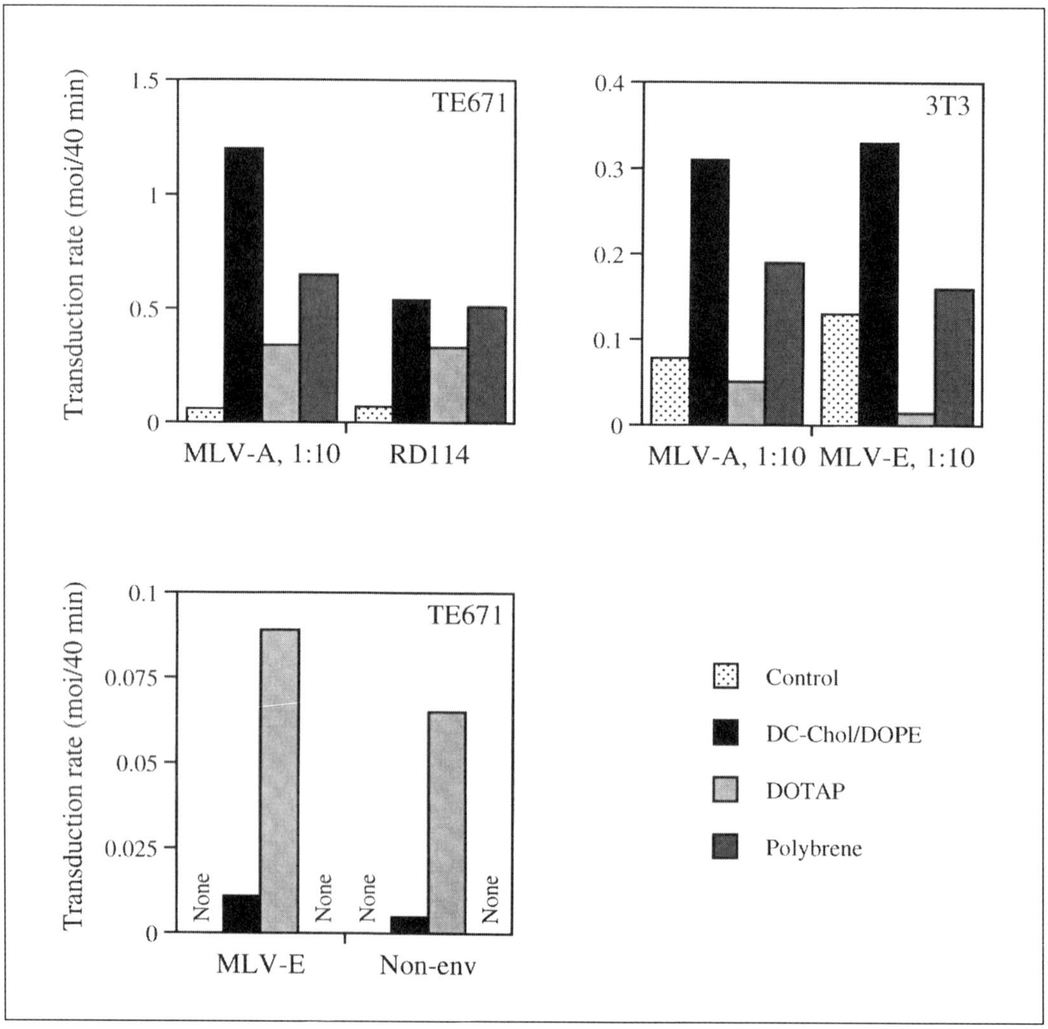

Figure 4. Effect of liposome pretreatment for different vector pseudotype/target cell combinations. Top panels: Virus, diluted where indicated, was pretreated with DC-Chol/DOPE, DOTAP, or polybrene, and the initial transduction rate was determined. Vectors pseudotyped with MLV-A, MLV-E, and RD114 viral envelopes were used for TE671 or 3T3 target cells. Bottom panel: Human TE671 target cells were exposed to ecotropic vector or to vector particles lacking any specific envelope proteins, pre-treated as indicated. No transduction was seen when untreated or treated with polybrene, consistent with the lack of an envelope/receptor interaction. However, pretreatment with liposome, in particular DOTAP, did facilitate transduction independent of such an interaction.

mesothelioma cells were isolated, and the transduction efficiency and levels of enzyme expression were determined. Both parameters were similarly enhanced four- to fivefold in tumors from animals that received the liposome-complexed virus (Figure 5). This enhancement was statistically highly significant (41). Moreover, similar data showing a tenfold enhancement have since been obtained using an MLV-E vector in combination with DC-Chol/DOPE for in vivo transduction of a murine intraperitoneal tumor (30). Clearly, these experiments do not reveal the kinetics of the process of infection in vivo. Nevertheless, the improvements observed are entirely consistent with the suggestion that enhanced transduction kinetics will lead to enhanced gene transfer in settings where other biological effects are likely to limit vector availability.

CONCLUSIONS

I have argued that the process of infection/transduction by retroviruses is intrinsically of low efficiency. Specific receptor binding and fusion to release viral cores to the cytoplasm are two stages that may be most amenable to modulation. I have also argued that, for relevance to any proposed application of retroviral vectors for gene therapy, the kinetics of the transduction process are of critical importance. Cationic liposomes form stable complexes with such vectors and significantly improve the initial rate of transduction. The mechanism for this enhancement likely involves both improved virus binding to the target cell and improved efficiencies of internalization/fusion. Such an association offers a means of enhancing the gene transfer achievable in a brief period of exposure of

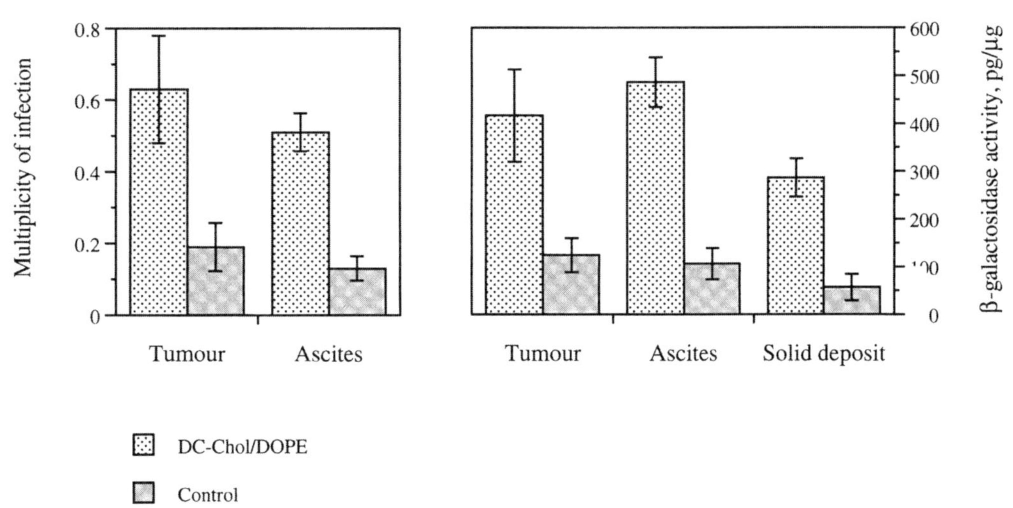

Figure 5. In vivo transduction of tumor xenograft. Human mesothelioma cells were grown in the peritoneal cavity of athymic (nu/nu) mice. Virus (500 μL) was delivered, with or without DC-Chol/DOPE pretreatment, two days after injection of 10^7 tumor cells. There were five animals per group. At day 14, ascites cells and disaggregated tumor deposits were reestablished in culture to determine the transduction efficiency and enzyme activity. Additionally, solid deposits were lysed for immediate enzyme assay. Values are SEM.

target cells, such as is likely to be the case for in vivo delivery. Indeed the efficacy of this approach has been demonstrated in two in vivo tumor transduction models. Given the clinical experience now available regarding the use of cationic liposomes, these agents are likely to find application in gene therapy protocols involving retroviral vectors.

REFERENCES

1. **Aboud, M., R. Shoor, and S. Salzberg.** 1979. Adsorption, penetration and uncoating of murine leukemia virus studied by using its reverse transcriptase. J. Virol. *30*:32-37.
2. **Agrawal, Y.P., R.S. Agrawal, A.M. Sinclair, D. Young, M. Maruyama, F. Levine, and A.D. Ho.** 1996. Cell-cycle kinetics and VSV-G pseudotyped retrovirus-mediated gene transfer in blood-derived CD34+ cells. Exp. Hematol. *24*:738-747.
3. **Andersen, K.B.** 1985. The fate of the surface protein gp70 during entry of retrovirus into mouse fibroblasts. Virology *142*:112-120.
4. **Andersen, K.B. and B.A. Nexo.** 1983. Entry of murine retrovirus into mouse fibroblasts. Virology *125*:85-98.
5. **Andreadis, S.T., D. Brott, A.O. Fuller, and B.O. Palsson.** 1997. Moloney murine leukemia virus-derived retroviral vectors decay intracellularly with a half-life in the range of 5.5 to 7.5 hours. J. Virol. *71*:7541-7548.
6. **Andreadis, S.T. and B.O. Palsson.** 1996. Kinetics of retrovirus mediated gene transfer: the importance of intracellular half-life of retroviruses. J. Theor. Biol. *182*:1-20.
7. **Andreadis, S. and B.O. Palsson.** 1997. Coupled effects of polybrene and calf serum on the efficiency of retroviral transduction and the stability of retroviral vectors. Hum. Gene Ther. *8*:285-291.
8. **Battini, J.-L., P. Rodrigues, R. Muller, O. Danos, and J.M. Heard.** 1996. Receptor-binding properties of a purified fragment of the 4070A amphotropic murine leukemia virus envelope glycoprotein. J. Virol. *70*: 4387-4393.
9. **Chuck, A.S., M.F. Clarke, and B.O. Palsson.** 1996. Retroviral infection is limited by Brownian motion. Hum. Gene Ther. *7*:1527-1534.
10. **Chuck, A.S. and B.O. Palsson.** 1996. Consistent and high rates of gene transfer can be obtained using flow-through transduction over a wide range of retroviral titers. Hum. Gene Ther. *7*:743-750.
11. **Collins, M.K.L. and C.D. Porter.** 1999. Retroviral vectors. *In* L.J. Fairbairn and N.G. Testa (Eds.), Blood Cell Biochem, Vol. 8: Hematopoiesis and Gene Therapy. Kluwer Academic/Plenum Publishers.
12. **Cornetta, K. and W.F. Anderson.** 1989. Protamine sulfate as an effective alternative to polybrene in retrovirus-mediated gene transfer. J. Virol. Methods *23*:187-194.
13. **Cosset, F.-L., Y. Takeuchi, J.L. Battini, R.A. Weiss, and M.K.L. Collins.** 1995. High titre packaging cells producing recombinant retroviruses resistant to human serum. J. Virol. *69*:7430-7436.
14. **Forestell, S.P., E. Bohnlein, and R.J. Rigg.** 1995. Retroviral end-point titer is not predictive of gene transfer efficiency: implications for vector production. Gene Ther. *2*:723-730.
15. **Ganguly, K., V.S. Kalyanaraman, and M.G. Sarngadharan.** 1983. Analysis of the interaction between Rauscher murine leukemia virus and murine cell membrane receptor by in vitro binding assay. Cancer Lett. *18*:79-86.
16. **Gao, X. and L. Huang.** 1991. A novel cationic liposome reagent for efficient transfection of mammalian cells. Biochem. Biophys. Res. Commun. *179*:280-285.
17. **Gao, X. and L. Huang.** 1995. Cationic liposome-mediated gene transfer. Gene Ther. *2*:710-722.
18. **Hanenberg, H., X.L. Xiao, D. Dilloo, K. Hashino, I. Kato, and D.A. Williams.** 1996. Colocalization of retrovirus and target cells on specific fibronectin fragments increases genetic transduction of mammalian cells. Nat. Med. *2*:876-882.
19. **Haywood, A.M.** 1994. Virus receptors: binding, adhesion strengthening, and changes in viral structure. J. Virol. *68*:1-5.
20. **Hodgson, C.P. and F. Solaiman.** 1996. Virosomes: cationic liposomes enhance retroviral transduction. Nat. Biotechnol. *14*:339-342.
21. **Innes, C.L., P.B. Smith, R. Langenbach, K.R. Tindall, and L.R. Boone.** 1990. Cationic liposomes (lipofectin) mediate retroviral infection in the absence of specific receptors. J. Virol. *64*:957-961.
22. **Kaneko, Y. and A. Tsukamoto.** 1996. Cationic liposomes enhance retrovirus-mediated mulitnucleated cell formation and retroviral transduction. Cancer Lett. *105*:39-44.
23. **Kavanaugh, M.P., D.G. Miller, W. Zhang, W. Law, S.L. Kozak, D. Kabat, and A.D. Miller.** 1994. Cell-surface receptors for gibbon ape leukemia virus and amphotropic murine retrovirus are inducible sodium-dependent phosphate symporters. Proc. Natl. Acad. Sci. USA *91*:7071-7075.
24. **Kizhatil, K. and L.M. Albritton.** 1997. Requirements for different components of the host cell cytoskeleton distinguish ecotropic murine leukemia virus entry via endocytosis from entry via surface fusion. J. Virol. *71*:7145-7156.
25. **Konopka, K., B.R. Davis, C.E. Larsen, D.R. Alford, R.J. Debs, and N. Duzgunes.** 1990. Liposomes modulate human immunodeficiency virus infectivity. J. Gen. Virol. *71*:2899-2907.
26. **Konopka, K., L. Stamatatos, C.E. Larsen, B.R. Davis, and N. Duzgunes.** 1991. Enhancement of human immunodeficiency virus type 1 infection by cationic liposomes: the role of CD4, serum and liposome-cell interactions. J. Gen. Virol. *72*:2685-2696.
27. **Kurre, P., H.-P. Kiem, J. Morris, S. Heyward, J.-L. Battini, and A.D. Miller.** 1999. Efficient transduction by an amphotropic retrovirus vector is dependent on high-level expression of the cell surface receptor. J Virol. *73*:495-500.

28. **Lavillette, D., M. Maurice, C. Roche, S.J. Russell, M. Sitbon, and F.-L. Cosset.** 1998. A proline-rich motif downstream of the receptor binding domain modulates conformation and fusogenicity of murine retroviral envelopes. J. Virol. *72*:9955-9965.
29. **LeDoux, J.M., J.R. Morgan, R.G. Snow, and M.L. Yarmush.** 1996. Proteoglycans secreted by packaging cell lines inhibit retrovirus infection. J. Virol. *70*:6468-6473.
30. **Lukacs, K.V., C.D. Porter, O.E. Pardo, R.E. Oakley, R.M. Steel, D.V. Judd, J.E. Browning, D.M. Geddes, and E.W.F.W. Alton.** 1999. In vivo transfer of bacterial marker genes results in differing levels of gene expression and tumour progression in immunocompetent and immunodeficient mice. Hum. Gene Ther. *10*:2373-2379.
31. **McClure, M.O., M.A. Sommerfelt, M. Marsh, and R.A. Weiss.** 1990. The pH independence of mammalian retrovirus infection. J. Gen. Virol. *71*:767-773.
32. **Miller, D.G., M.A. Adam, and A.D. Miller.** 1990. Gene transfer by retrovirus vectors occurs only in cells that are actively replicating at the time of infection. Mol. Cell. Biol. *10*:4239-4242.
33. **Mislick, K.A. and J.D. Baldeschwieler.** 1996. Evidence for the role of proteoglycans in cation-mediated gene transfer. Proc. Natl. Acad. Sci. USA *93*:12349-12354.
34. **Morgan, J.R., J.M. LeDoux, R.G. Snow, R.G. Tompkins, and M.L. Yarmush.** 1995. Retrovirus infection: effect of time and target cell number. J. Virol. *69*:6994-7000.
35. **Moritz, T., V.P. Patel, and D.A. Williams.** 1994. Bone marrow extracellular matrix molecules improve gene transfer into human hematopoietic cells via retroviral vectors. J. Clin. Invest. *93*:1451-1457.
36. **Morling, F.J. and S.J. Russell.** 1995. Enhanced transduction efficiency of retroviral vectors coprecipitated with calcium phosphate. Gene Ther. *2*:504-508.
37. **Naldini, L., U. Blomer, P. Gallay, D. Ory, R. Mulligan, F.H. Gage, I.M. Verma, and D. Trono.** 1996. In vivo gene delivery and stable transduction of nondividing cells by a lentiviral vector. Science *272*:263-267.
38. **Nussbaum, O., A. Roop, and W.F. Anderson.** 1993. Sequences determining the pH dependence of viral entry are distinct from the host range-determining region of the murine ecotropic and amphotropic retrovirus envelope proteins. J. Virol. *67*:7402-7405.
39. **Palsson, B. and S. Andreadis.** 1997. The physicochemical factors that govern retrovirus-mediated gene transfer. Exp. Hematol. *25*:94-102.
40. **Porter, C.D., M.K.L. Collins, C.S. Tailor, M.H. Parkar, F.-L. Cosset, R.A. Weiss, and Y. Takeuchi.** 1996. Comparison of efficiency of infection of human gene therapy target cells via four different retroviral receptors. Hum. Gene Ther. *7*:913-919.
41. **Porter, C.D., K.V. Lukacs, G. Box, Y. Takeuchi, and M.K.L. Collins.** 1998. Cationic liposomes enhance the rate of transduction by a recombinant retroviral vector in vitro and in vivo. J. Virol. *72*:4832-4840.
42. **Rein, A., J. Mirro, J.G. Haynes, S.M. Ernst, and K. Nagashima.** 1994. Function of the cytoplasmic domain of a retroviral transmembrane protein: p15E-p2E cleavage activates the membrane fusion capability of the murine leukemia virus env protein. J. Virol. *68*:1773-1781.
43. **Risco, C., L. Menendez-Arias, T.D. Copeland, P.P.d. Silva, and S. Oroszlan.** 1995. Intracellular transport of the murine leukemia virus during acute infection of NIH3T3 cells: nuclear import of nucleocapsid protein and integrase. J. Cell. Sci. *108*:3039-3050.
44. **Roe, T., T.C. Reynolds, G. Yu, and P.O. Brown.** 1993. Integration of murine leukemia virus DNA depends on mitosis. EMBO J. *12*:2099-2108.
45. **Siess, D.C., S.L. Kozak, and D. Kabat.** 1996. Exceptional fusogenicity of Chinese hamster ovary cells with murine retroviruses suggests roles for cellular factor(s) and receptor clusters in the membrane fusion process. J. Virol. *70*:3432-3439.
46. **Singh, D. and P.W.J. Rigby.** 1996. The use of histone as a facilitator to improve the efficiency of retroviral gene transfer. Nucleic Acids Res. *24*:3113-3114.
47. **Swaney, W.P., F.L. Sorgi, A.B. Bahnson, and J.A. Barranger.** 1997. The effect of cationic liposome pretreatment and centrifugation on retrovirus-mediated gene transfer. Gene Ther. *4*:1379-1386.
48. **Takeuchi, Y., C.D. Porter, K.M. Strahan, A.F. Preece, K. Gustafsson, F.-L. Cosset, R.A. Weiss, and M.K. Collins.** 1996. Sensitization of cells and retroviruses to human serum by (alpha 1-3) galactosyltransferase. Nature *379*:85-88.
49. **Themis, M., S.J. Forbes, L. Chan, R.G. Cooper, C.J. Etheridge, A.D. Miller, H.J.F. Hodgson, and C. Coutelle.** 1998. Enhanced in vitro and in vivo gene delivery using cationic agent complexed retrovirus vectors. Gene Ther. *5*:1180-1186.
50. **Toyoshima, K. and P.K. Vogt.** 1969. Enhancement and inhibition of avian sarcoma viruses by polycations and polyanions. Virology *38*:414-426.
51. **Uckert, W., G. Willimsky, F.S. Pedersen, T. Blankenstein, and L. Pedersen.** 1998. RNA levels of human retrovirus receptors Pit1 and Pit2 do not correlate with infectability by three retroviral pseudotypes. Hum. Gene Ther. *9*:2619-2627.
52. **Wang, H., R. Paul, R.E. Burgeson, D.R. Keene, and D. Kabat.** 1991. Plasma membrane receptors for ecotropic murine retroviruses require a limiting accessory factor. J. Virol. *65*:6468-6477.
53. **Weiss, R.A.** 1993. Cellular receptors and viral glycoproteins involved in retroviral entry, p. 1-108. *In* J. Levy (Ed.), The Retroviridae, Vol. 2. Plenum Press, New York.
54. **Weiss, R.A. and C.S. Tailor.** 1995. Retrovirus receptors. Cell *82*:531-533.
55. **Williams, A.F.** 1991. Out of equilibrium. Nature *352*:473-474.
56. **Yu, H., N. Soong, and W.F. Anderson.** 1995. Binding kinetics of ecotropic (Moloney) murine leukemia retrovirus with NIH3T3 cells. J. Virol. *69*:6557-6562.

27 Retrovirally Mediated Genetic Marking of Murine and Human Hematopoietic Repopulating Cells

Antonio Bernad[1,2], Florencio Varas[1,3], Manuel Ramírez[1,3], José C. Segovia[1,3], Guillermo Güenechea[1,3], Jordi Barquinero[4], and Juan A. Bueren[1,3]

[1]CNB-CIEMAT Associated Unit, Madrid; [2]Department of Immunology and Oncology, National Biotechnology Center, Autonomous University, Madrid; [3]Unit of Molecular and Cellular Biology, CIEMAT, Madrid; [4]Institute of Oncologic Research, Department of Criobiology and Cellular Therapy, Barcelona, Spain

INTRODUCTION

Progressive refinement of the techniques for gene transfer and expression of exogenous proteins in eukaryotic cells has allowed considerable progress in studies on gene regulation, genetic marking, and gene therapy, particularly in those concerning the immune system (1,7,13,58). Based on its extensive experience in transplantation and ex vivo manipulation, the lymphohematopoietic system has been the focus of the first studies aimed at the phenotypic modification of somatic cells in experimental animals and also in human therapy.

A number of methods have been proposed for the genetic modification of hematopoietic cells, of which retroviral-mediated gene transfer has become one of the most efficient tools in experimental and clinical protocols. This procedure offers a number of advantages over other gene transfer methods, including *(i)* high stability of the transduced sequences upon integration, *(ii)* transfer of a reduced number of gene constructs (usually one) per target genome, and *(iii)* safe procedures for retroviral vector packaging.

Because of the properties described above and given that provirus integration occurs essentially at random, retroviral vectors can serve as genetic markers and can be consequently used to retrieve the transduced population and trace the progeny of individual clones after transplantation (23,42). Before retroviral-mediated marking became available, chromosomal markers induced by sublethal irradiation (53)

Viral Vectors: Basic Science and Gene Therapy
Edited by A. Cid-Arregui and A. García-Carrancá

and radioisotopic marking of test populations with ^{111}In (26,28,46) were routine techniques. Both methods yielded interesting results, but the low efficiency of the former approach and the progressive reduction in radioactive tracer in the latter limited their spectrum of applications. Retroviral marking has permitted confirmation and implementation of previous conclusions and allowed the development of new strategies based on the stable integration of transgenes into the host genome.

Earlier studies in the mouse hematopoietic system showed that the primitive hematopoietic stem cells (HSC) can give rise to all hematopoietic cell lineages. The clonal composition and dynamics of the hematopoietic system were not clarified, however, until the application of retroviral gene-marking strategies. These approaches demonstrated the maintenance of retrovirally marked HSC clones for long periods of time in transplanted recipients (14,15, 23,34,35) and evidenced the monoclonal or oligoclonal composition of the reconstituted system (42,67). Other studies showed, however, that hematopoiesis is maintained by a relatively large number of simultaneously functioning, short-life (1–3 months) clones that usually grow locally with little or no dispersion among different regions of the hematopoietic system (24). Retrovirally mediated gene transfer has been used to compare the efficiency of retroviral transduction of different sources of HSC using defined ex vivo transduction protocols (6) and to investigate the clonal repertoire of hematopoietic precursors that are mobilized to peripheral blood following the administration of hematopoietic growth factors (69).

During the last few years, several protocols for testing transplantation of transduced gene-marked hematopoietic grafts have been approved, and clinical trials have been completed in some instances. Gene marking has thus been used to determine the effectiveness of ex vivo purging methods applied to patients affected with neuroblastoma, acute leukemia, or chronic myelogenous leukemia (8,13). In several instances, patients transplanted with gene-marked autografts showed the presence of the transgene in relapsed malignant cells, which indicated a contribution of the autograft in the relapse (10–13,20). Further, the susceptibility to retroviral infection and repopulating ability of two distinct HSC sources from a patient have been evaluated using gene-marking studies (10,13).

Until recently, most gene transfer methodology was optimized using colony-forming unit cells (CFU-C) or long-term culture-initiating cells (LTC-IC) as target populations. Relatively high transduction efficiencies of these progenitors have been widely reported (22,31,49). However, the use of similar vectors and gene transfer procedures has generated only modest results in transducing the precursors responsible for long-term reconstitution in large experimental animals or human patients (11,12,17,20,25,37). This indicates that the capacity of CFU-C and LTC-IC to be transduced is not a predictive parameter of the transducibility of the long-term repopulating cells in these species.

New models to characterize the functionality of human repopulating cells have been developed and used to test the efficacy of gene transfer protocols on early repopulating cells. Immunodeficient beige/nude/xid (bnx) and nonobese diabetic severe combined immunodeficient (NOD/SCID) mice have been engrafted with human repopulating cells (18,19,52). Bnx recipients can be engrafted with human myeloid and T-cell progeny, but not with B cells. The repopulation of NOD/SCID mice with human hematopoietic cells is generally higher when compared with that observed in bnx mice and is most marked in the myeloid and B-cell lineages, but not in the T-cell differentiation lineage. Whereas the human precursors that repopulate NOD/SCID recipients are more refractory to retroviral infection than the CFU-C and LTC-IC precursors (29,41), recent studies have shown the feasibility of transducing the NOD/SCID repopulating cells with retroviral vectors that encode the nerve growth factor receptor or the green fluorescent protein (GFP) as marker (60,66). In view of the new ex vivo gene transfer methods and of new generations of vectors, it is generally believed that evolving clinical protocols using transduction of human repopulating cells will result in successful alternatives for the treatment of congenital and acquired diseases of the hematopoietic system.

RETROVIRAL GENE MARKING OF MURINE HEMATOPOIETIC REPOPULATING CELLS: APPLICATIONS TO STUDIES ON STEM CELL DYNAMICS IN TRANSPLANTATION MODELS

Retrovirally mediated gene transfer to HSC usually represents the method of choice to enable persistent and stable modification of the lymphohematopoietic system. Retroviral vectors are replication-deficient but infection-competent pseudoviruses; they are most frequently derived from the Moloney murine leukemia virus (MoMLV). The integrative step of these vectors occurs efficiently only when the target cell is proliferating during the infection process (48) and specifically when the cell passes through the mitotic phase (59) which enables access of the preintegrative complex to the nuclear compartment. HSCs not only constitute a very scarce population, but most of these cells persist in a quiescent state under stationary conditions (43). Several experimental approaches have been pursued to facilitate the activation and enrichment of this population in the bone marrow (BM) or the peripheral blood of HSC donors. In experimental animals, particularly in the mouse, 5-fluorouracil is commonly administered two to five days before BM harvest to kill cycling cells and thus, through homeostatic/feedback regulatory mechanisms, promote proliferative activation of the HSC population (30).

Based on previously reported investigations and by assaying the optimal combination of various parameters involved in the transduction and survival of hematopoietic precursors, we developed an efficient protocol, infection/expansion/selection (IES), for HSC transduction of murine BM (3). Briefly, BM cells harvested four to five days after intravenous 5-fluorouracil injection were infected with pXT1 (9) retroviral particles. Target cells were subsequently cultured ex vivo with interleukin-3 and stem cell factor for a maximum of three days. Under these conditions, the content of committed and pluripotent hematopoietic progenitors in the graft was expanded (51,70), which improved the hematopoietic recovery of transplanted recipients (50,62) and preserving stem cell viability (33,70). Finally, transduced cells were positively selected in vitro based on the resistance to G418, a neomycin analog, conferred by the neomycin phosphotransferase gene provided by the retroviral vector. The transplantation of IES grafts into mice subjected to optimized irradiation conditioning (68) resulted in an essentially complete lymphohematopoietic reconstitution of recipients by transduced cells (3,69). Using this murine model, we have studied the relationship between graft size and number of repopulating clones in the recipient. Previous reports indicated oligoclonality of the hematopoietic system in mice transplanted with transduced grafts (22,32,34,42). These results have frequently been explained as a consequence of the limiting number of transduced HSCs provided by the transplanted graft (47,63,64). To determine if the oligoclonality of the reconstituted hematopoiesis is a consequence of limited graft size, recipient mice were transplanted with increasing numbers of IES-transduced cells, ranging from 1×10^5 to 2.3×10^6 cells. At eight months after transplantion, recipients were sacrificed, and the number of repopulating clones in their lymphohematopoietic organs was determined by Southern blot analysis. To facilitate the proliferative activation of potentially silent clones, BM cell suspensions from primary recipients were subjected to two further serial transplants by transplanting 1×10^7 BM cells from each primary recipient into a secondary recipient. Twelve days later, 5×10^6 BM cells from each secondary recipient were transplanted into tertiary recipients, which were killed 12 days later. The self-renewal capacity of repopulating clones of primary recipients and the dynamics of clonal activation were investigated by comparing the transduced splenic clones in primary, secondary, and tertiary recipients.

The analysis of repopulating clones in lymphohematopoietic organs of primary recipients is presented in Figure 1. In accordance with other studies (23,32,34, 42), our results support the hematopoetic oligoclonality of recipients transplanted with transduced BM grafts. Although the graft size in recipients receiving large samples (2.3×10^6 cells) was 23-fold higher than those receiving small grafts (1×10^5 BM cells), no correlation was observed with the number of repopulating clones. Only one clone was detected in the mouse transplanted with 1×10^5 cells, and the increase in the number of clones in mice transplanted with 5×10^5 to 2.3×10^6 cells was clearly not dose dependent (5–7 transduced clones). No clonal activation of silent clones was evidenced after two serial BM transplantations (Figure 2). On the contrary, a simplification of the clonal repertoire was generally observed in secondary and tertiary recipients with respect to primary recipients. These results suggest that a reduced number of transplanted precursors with a repopulating advantage over other repopulating clones, either intrinsic or conditioned by the hematopoietic niche, predominantly reconstitute the recipient, regardless of the total content of clones infused into the recipient. Our data suggest this is the case, provided a threshold number of stem cells is transplanted.

RETROVIRALLY MEDIATED GENE TRANSFER OF HUMAN HEMATOPOIETIC PRECURSORS

In contrast to the relatively high efficiency of retroviral vectors in transducing murine hematopoietic repopulating cells, earlier studies have evidenced more difficulties in transducing HSCs in large-animal models (29,41). In addition, important problems related to transgene expression in the hematopoietic progeny have been observed in vivo. Nonetheless, though the road toward achieving this goal may be long and beset with difficulties, it appears to be a promising course.

Transduction of Human Hematopoietic Precursors with GFP-Expressing Retroviral Vectors

GFP has emerged as one of the most promising marker molecules in biomedical research. Cloned in 1992 from the jellyfish *Aequorea victoria*, its cDNA encodes a 238 amino acid protein with a molecular weight of 27 kDa (57). A number of mutants and codon-optimized forms of the molecule have been developed that confer different, useful light absorption and emission properties. Enhanced GFP (EGFP) is a red-shifted variant containing two mutations in the chromophore region and more than 190 silent base changes corresponding to human codon-usage preferences. Together, these changes result in an approximately 35-fold increase in fluorescence when expressed in mammalian cells. Since EGFP can be excited by blue light (peak at 490 nm), cells expressing EGFP can be analyzed in a conventional argon ion laser cytometer.

A variety of retroviral vectors containing GFP or related molecules has been developed, and stable packaging cell lines generating high-titer retroviral vectors have become available for HSC gene transfer (4,45,55). Among other advantages for gene transfer studies, no cofactor is needed for EGFP detection, and it allows direct analy-

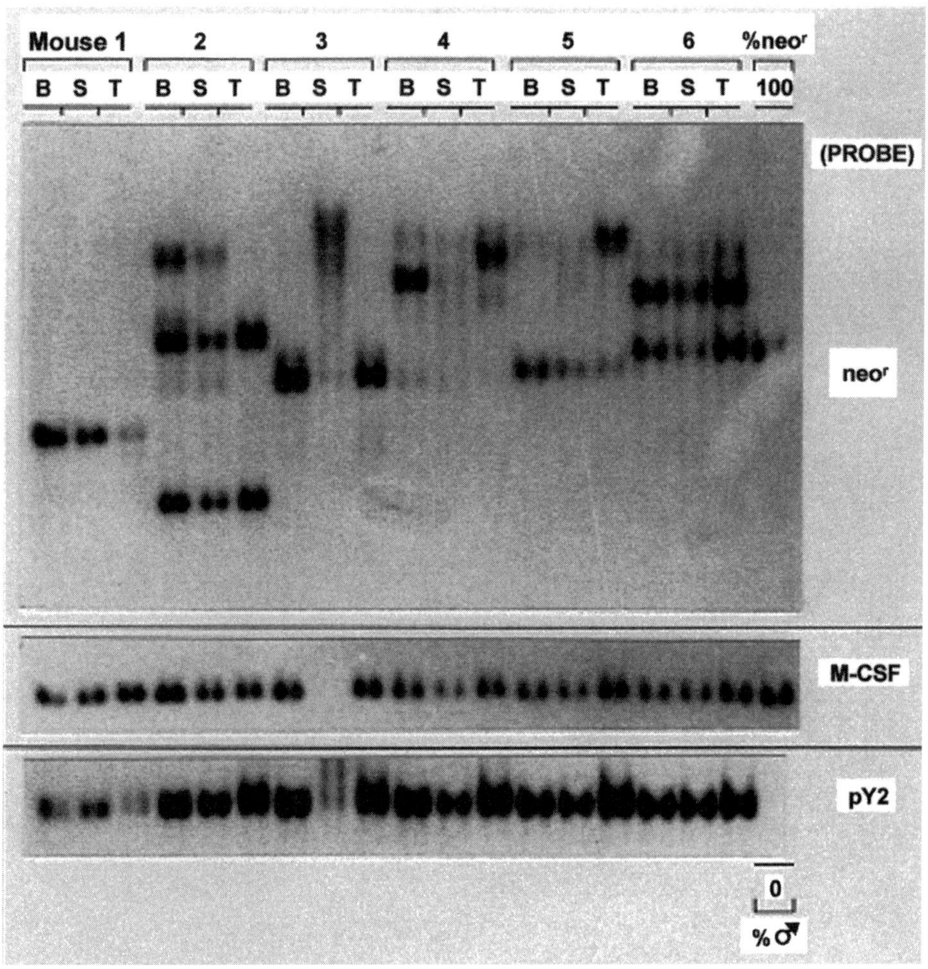

Figure 1. Analysis of transduced repopulating clones in the bone marrow (B), spleen (S), and thymus (T) of (C57Bl/6J × BALB/c)F_1 female mice killed eight months after transplantation with 1×10^5 (mouse 1), 5×10^5 (mice 2–4), 1×10^6 (mouse 5), and 2.3×10^6 (mouse 6) syngenic male cells transduced by the IES protocol. The clonal makeup of transduced repopulating clones was evidenced by hybridization with a *neo*^{R} gene probe. The extent of exogenous repopulation (male cells) and DNA loading variations were determined, respectively, by hybridizations with probes from the pY2 plasmid (39) and the M-CSF single-copy gene (38). A control lane of 100% *neo*^{R}/0% male DNA was included for comparison and obtained from the 3T3-I cell line (3), which bears one single copy of the pXT1 vector and does not hybridize with the male pY2 probe.

sis of gene transfer efficiency in real time by flow cytometry or by direct visual monitoring of cultures under an inverted fluorescence microscope. In addition, in HSC gene transfer studies, transduction efficiencies can be evaluated accurately in hematopoietic colonies using methods that include CFU-C (Figure 3), cobblestone area-forming cells (CAFC), or specific phenotypically defined cell populations such as CD34$^+$ or CD34$^+$/CD38^{low} cells (Figure 4).

In HSC gene therapy, the transduction efficiency of true HSC is an important issue. The preservation of the engraftment ability of these long-term repopulating cells and the long-term expression of the transgene in the recipient are nonetheless critical steps for the successful development of gene therapy protocols in hematopoiesis. The goal is to obtain maximal transduction

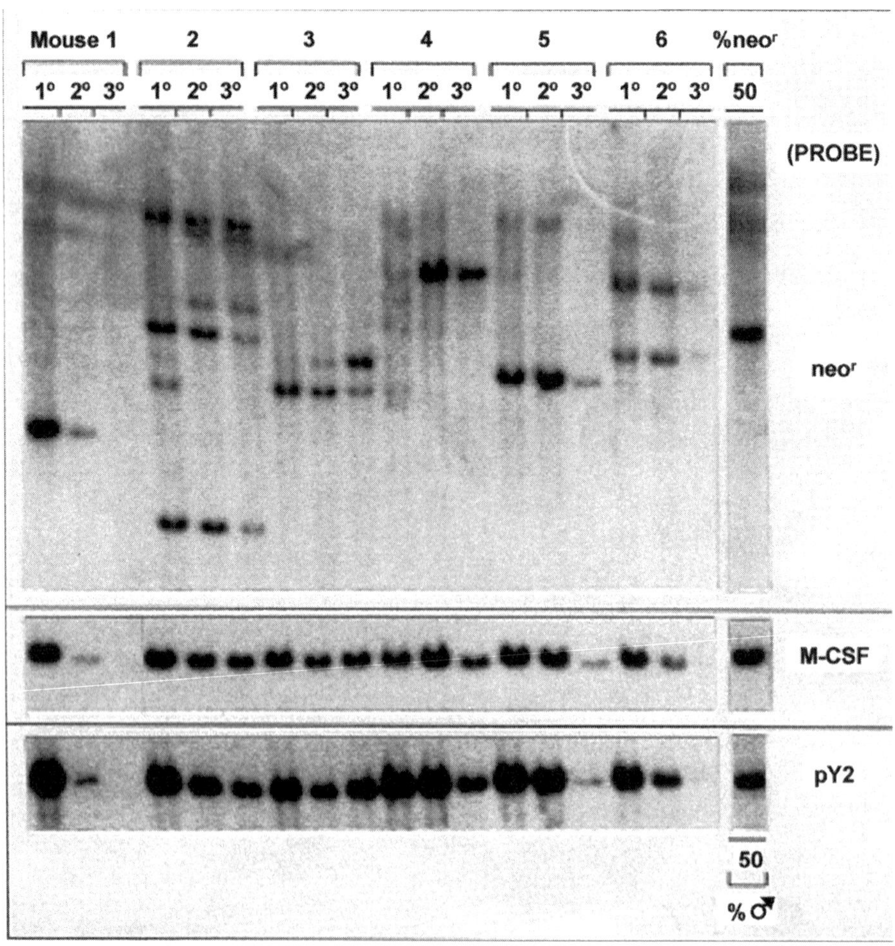

Figure 2. Analysis of transduced repopulating clones contributing to spleen populations in (C57BL/6J × BALB/c)F$_1$ female recipients transplanted with syngeneic male IES cell grafts, as in Figure 1. Marrow grafts from primary recipients were subjected to two serial transplants. Secondary and tertiary recipients were killed 12 days after transplantation, their spleens removed and processed for DNA extraction. The transduced repopulating clones in the spleens of secondary and tertiary recipients are presented in lanes 2 and 3, respectively. Hybridization was carried out with the probes described in Figure 1. A control lane (50% *neo^R*/50% male DNA) composed of the same DNA ratio from male spleen and the 3T3-I cell line (see Figure 1) was included for comparison.

efficiency of HSCs, while preserving their in vivo repopulating ability.

The main limiting factors of transduction efficiency include vector titer, expression of the appropriate viral receptor, cycling rate of the target cell, and finally, the method used to promote ex vivo interaction of the vector with the target cell. Among the different procedures to favor interaction between the retroviral vectors and target cells, the use of fibronectin-coated plates has recently been shown to be one of the most efficient, and, in addition, it probably induces cells to divide (72). Viral titers depend largely on the clone of vector-producer cell line selected. Supernatants of retroviral vector-producing cells typically contain up to 1×10^6 to 1×10^7 infectious units (iu)/mL. Expression of the appropriate viral receptor is also necessary for retroviral transduction. It has been suggested that immature hematopoietic cells express low levels of the amphotropic retroviral receptor (Pit-2), which could partially explain why these cells are relatively resistant to transduction by these vectors. Pseudotyped vectors containing other viral envelope proteins, such as the gibbon ape leukemia virus (GALV) envelope or the vesicular stomatitis virus G protein (VSV-G; see Chapter 24), could be more efficient than amphotropic vectors for HSC gene transfer (36,54).

Efficient delivery of genes into HSC is necessary but not sufficient for permanent correction of genetic diseases. Transduced cells or their progeny must repopulate the recipient and express the transgene in the long term, for which promoters play a crucial role. Silencing of transgene expression occurs both in vitro and in vivo and is associated, at least in some cases, with methylation in the promoter region (16). Since most current retroviral vectors only integrate

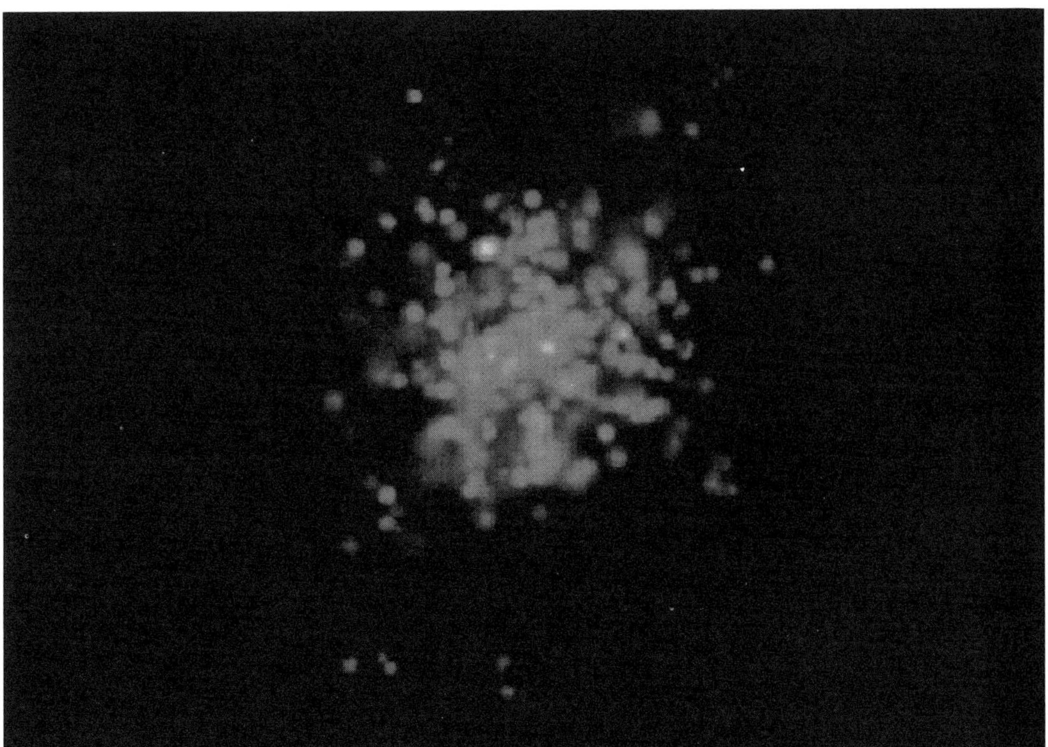

Figure 3. Cord blood-derived CFU-GM expressing EGFP (400× optical magnification). (See color plate A7.)

during cell division, and given that HSCs are predominantly quiescent, mitosis must be induced in these target cells for efficient gene transfer. Most retroviral transduction protocols thus include the use of a variety of growth factors or growth factor mixtures, including stem cell factor (SCF), megakaryocyte growth and development factor (MGDF) or thrombopoietin (TPO), interleukin 3 (IL-3), interleukin 6 (IL-6), granulocyte colony-stimulating factor (G-CSF), granulocyte and macrophage colony-stimulating factor (GM-CSF), and flt-3 ligand (FL). The use of these growth factors should be carefully considered in hematopoietic grafts, given the evidence that factors or

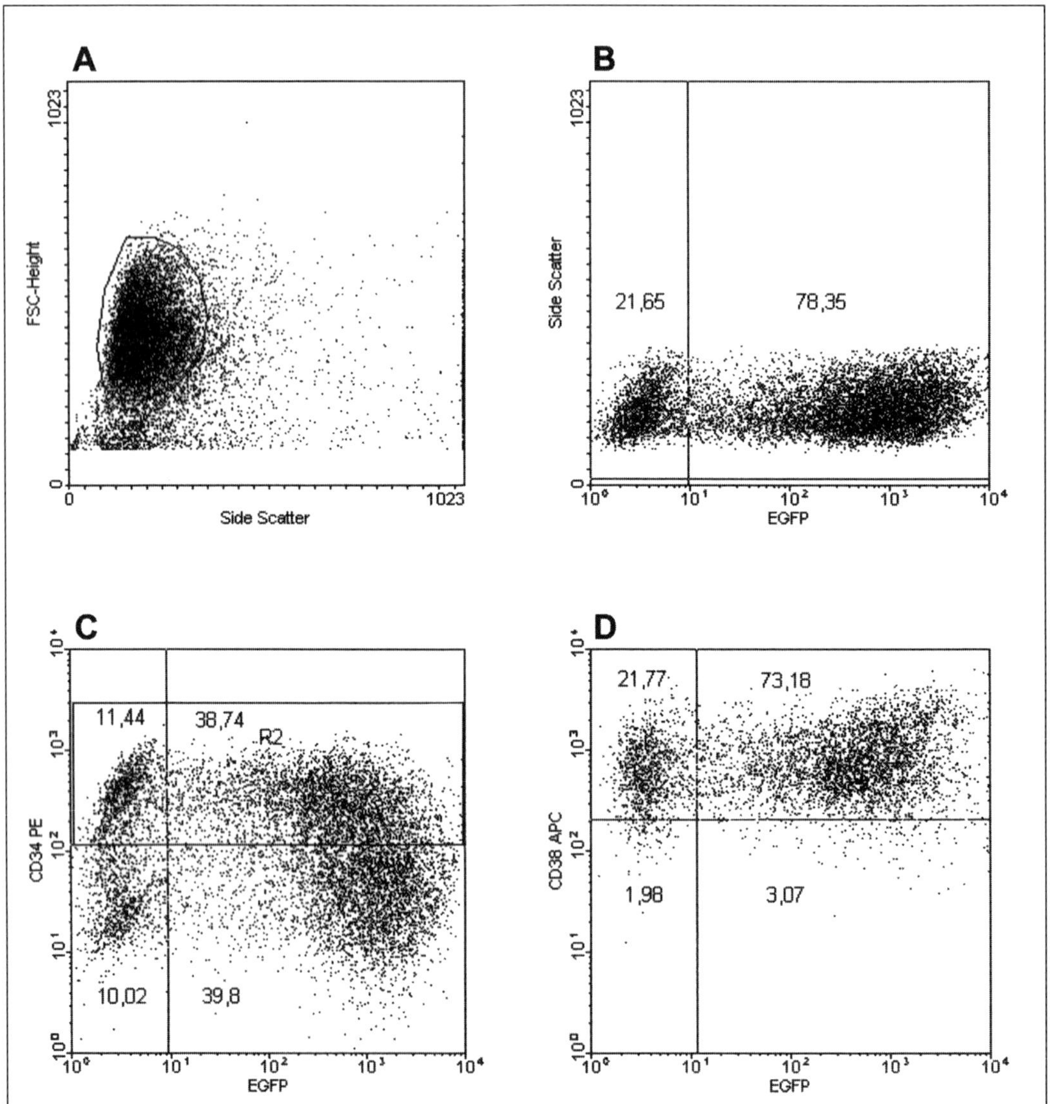

Figure 4. FACS analysis of CD34+ cells from cord blood after triple labeling (CD34-PE, CD38-APC, and EGFP). Percentages of events analyzed are shown for each cell population. (A) Dot plot represents forward and side scatter parameters. (B) EGFP expression in cells gated in dot plot A. (C) CD34-PE vs. EGFP expression. (D) CD38-APC vs. EGFP expression in CD34+ (gate R2 in C)/CD38low (lower percentile 5 of APC fluorescence). Note that approximately two-thirds of cells with this phenotype express EGFP.

culture conditions that promote retroviral transduction can hamper the engraftment ability and the repopulating potential of the ex vivo cultured cells (27). Some investigators avoid the use of IL-3 because it may induce cell differentiation and depletion of the stem cell pool, although the biology of cytokine effects on stem cells is not yet sufficiently understood to ensure the existence of an optimal growth factor combination for this purpose. Efficient retroviral gene transfer into human hematopoietic cells can currently be achieved using short-term protocols that include 24- or 48-hour preincubation with SCF, MGDF, and FL, and subsequent feeding of liquid cultures with vector-containing supernatant for two consecutive days on recombinant fibronectin fragment-coated surfaces.

Using many of the tools mentioned above to facilitate retrovirus-mediated gene transfer (infection on recombinant fibronectin fragment-coated surfaces, in the presence of SCF, MGDF, and FL), we observed that between 30% and 92% (median 61%, $n = 22$) of cord blood-derived $CD34^+$ cells express the *EGFP* transgene. Under these conditions, a median of 40% of $CD34^+/CD38^{low}$ cells expressed EGFP fluorescence, as assessed by flow cytometry (Figure 4). These results indicate that efficient transduction of early human hematopoietic precursors took place under these conditions.

Repopulating Immunodeficient Mice with Retrovirally Transduced Human Hematopoietic Cells

HSCs are defined by their capacity for long-term repopulation of the entire hematopoietic system of a myeloablated recipient. Data from the follow-up of patients subjected to hematopoietic transplantation consequently provide the most reliable information on the functional properties of human HSCs. Considerable effort has nonetheless been dedicated to defining in vitro and in vivo assays that could be used as surrogate assays to reflect the HSC function of a human hematopoietic sample (2,5,56). Among the assays developed to date, the reconstitution of sublethally irradiated immunodeficient mice with human cells is considered the most restrictive assay to define the human HSC function of a graft (21,41).

The immunodeficient NOD/LtSz-SCID/SCID mice combine the T and B deficiencies associated with the SCID mutation, together with the characteristic defects of the NOD background: alterations in serum complement, macrophage, and natural killer (NK) cell activities (61). NOD/SCID mice are better hosts for human hematopoietic cells than SCID mice and are currently one of the preferred in vivo systems for the study of human hematopoiesis (40). The transplantation of sublethally irradiated NOD/SCID mice with human BM, cord blood, or mobilized peripheral blood allows the engraftment of human repopulating cells (SCID repopulating cells [SRC]), considered the putative HSC in humans. This in vivo model has been used in a variety of studies related to human hematopoiesis [i.e., analysis of engraftment potential (44) or frequency (71) of human HSCs in different grafts, studies of ex vivo expansion (27), mobilization of HSC from BM to peripheral blood (65), and finally, studies of gene transfer into HSC (60)].

To study the effects of newly improved procedures in the transduction of human HSCs, we analyzed the expression of an easily detectable transgene reporter, *EGFP*, in transduced human cells that had been transplanted into NOD/SCID recipients. Transduced $CD34^+$ cord blood cells were transplanted into sublethally irradiated NOD/SCID mice. One month after transplantation, a BM aspirate was drawn from each animal, and the proportion of human cells (estimated as CD45+ cells) and primitive human CD34+ cells expressing the *EGFP* transgene were determined by flow cytometry. As shown in a representative mouse (Figure 5), the reporter transgene was well expressed, both in the bulk of CD45+ and the more primitive CD34+ cells harvested from BM of recipient mice. This indicates that human cord blood CD34+ cells can be cultured and transduced in vitro with optimized retroviral vectors under defined culture conditions. Long-term posttransplant analyses have indicated that this procedure is efficient for transducing the self-renewing human HSC. In addition, preliminary gene expression studies suggest that the transgene seems to be expressed in the human hematopoietic cells in vivo for extended periods of time. (J. Barquinero et al., unpublished data).

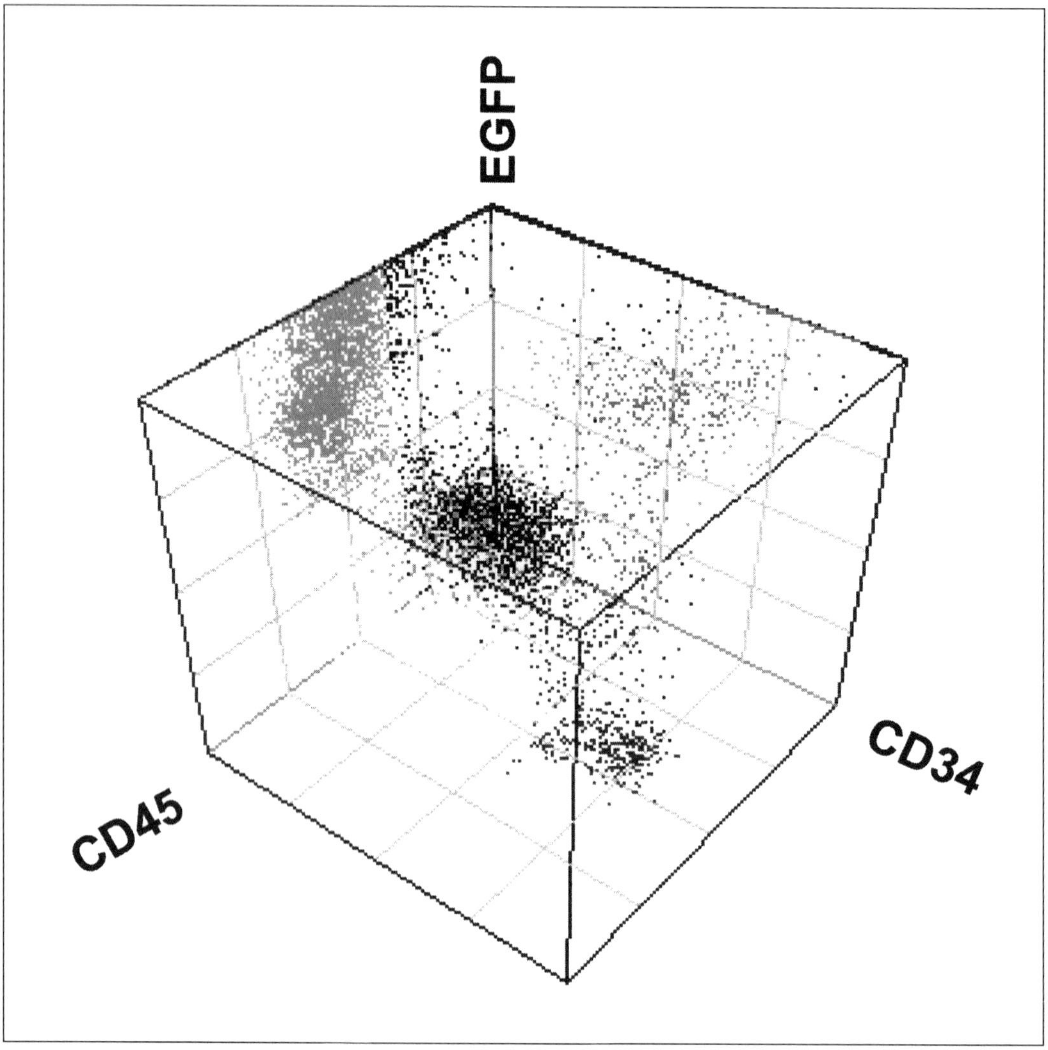

Figure 5. EGFP expression in human CD45+ and CD45+CD34+ cells obtained from a BM biopsy of NOD-SCID mice at 30 days after transplantation. BM cells were labeled with specific antobodies for the CD45 and CD34 human antigens and analyzed by flow cytometry. Yellow: CD45+/CD34- cells negative to EGFP; blue: CD45+/CD34+ cells negative to EGFP; light green: CD45+/CD34- cells positive to EGFP; dark green: human CD34+ cells expressing the EGFP protein. See text for information on percentage values. **(See color plate A8.)**

PROTOCOLS

Transduction of Murine BM HSCs Using the IES Protocol

We have developed an IES protocol (3) for efficient transduction of murine HSCs by means of pXT1-derived retroviral particles packaged in the ecotropic Ψ2 cell line. The pXT1 retroviral vector (9) encodes neomycin phosphotransferase (neo^R) from the MoMLV long terminal repeat (LTR). The use of either different selectable markers or packaging cell lines may require the corresponding modifications to optimize results.

IES Protocol

Materials

- 5-fluorouracil (5-FU) (Roche Molecular Biochemicals, Indianapolis, IN, USA). Supplied as a stock liquid solution at 50 mg/mL. This drug is intravenously injected into mice via a lateral tail vein at a dosage of 150 mg/kg body weight in a final volume of 200 µL. For this purpose, the stock solution may be diluted with Iscove's modified Dulbecco's medium (IMDM; Life Technologies, Rockville, MD, USA).
- SCF (Genetics Institute, Cambridge, MA, USA). Provided as recombinant soluble murine SCF. One unit of activity is the half the maximal activity as measured by the MO-7 proliferation assay.
- Virus. A batch of high-titer infective supernatant (100–200 mL, ≥1–2 × 10^6 cfu neo^R/mL) tested for the absence of helper virus.
- WEHI-3b conditioned medium (WCM). Conditioned medium from WEHI-3b cell line is used as a source of IL-3.

Procedure

1. Four to five days after administering a single dose of 5-FU to 8- to 12-week-old donor male mice, their hind limbs (tibiae and femora) are perfused at room temperature under sterile conditions with IMDM. Cellular aggregates are dispersed by repeated passages through a 25G needle, and the resulting cell suspension is not subjected to hemolysis.
2. Donor BM is mixed with pXT1-derived infective supernatants (1–3 × 10^6 cfu neo^R/mL) in culture flasks to a final concentration of 1–2 × 10^5 cells/mL and supplemented with 20% fetal bovine serum (FBS), 20% WCM, 0.2 U/mL SCF, and 5 µg/mL polybrene (Aldrich Chemical, St. Louis, MO, USA). The resulting multiplicity of infection (moi) should be above 5.
3. After a 6-hour infection period at 37°C in a fully humidified atmosphere with 5% CO_2 in air, culture flasks are scraped, and cells are collected after centrifugation at 400× *g* for 5 minutes.
4. The cell pellet is resuspended (2 × 10^6 cells/mL) in fresh IMDM, supplemented as previously described, but increasing SCF to a final concentration of 4 U/mL and eliminating polybrene.
5. Forty hours later, the neomycin analogue G418 is added to the culture at a final concentration of 1 mg/mL. This selection step is extended for 24 hours.
6. Cells are collected as described in step 3, washed once, and resuspended in fresh IMDM.
7. The infection efficiency is estimated by seeding different cell numbers into

clonogenic assays both in vitro (CFU-GM assay) and in vivo (CFU-S assay). Transduced cells are also injected into lethally irradiated female mice to test the ability of the transplanted cells to reconstitute their lymphohematopoietic system. All these assays are described below.

Production of Infective Supernatant for the Transduction of Murine BM Populations

According to the proposed IES protocol, it is necessary to produce a batch of high-titer infective supernatant (≥1–2 × 10^6 cfu neo^R/mL).

Materials

- Gelatin stock solution: 10% gelatin (Bio-Rad Laboratories, Hercules, CA, USA) solution in phosphate-buffered saline (PBS) is prepared, autoclaved, and stored at 4°C.
- Virus. High-titer retroviral producer cells are routinely cultured in Dulbecco's modified Eagle's medium (DMEM; Life Technologies) supplemented with 75 iu/mL penicillin (Sigma, St. Louis, MO, USA), 75 mg/mL streptomycin sulfate (Sigma), and 10% heat-inactivated FBS (Life Technologies) at 37°C in a fully humidified atmosphere with 5% CO_2 in air. When indicated, cells are grown with IMDM supplemented with 5% FBS, maintaining the other culture conditions as above.

Procedure

1. The selected Ψ2-pXT1 clone is grown and expanded in 75-cm^2 flasks pretreated with 1% gelatin to avoid massive detachment during supernatant harvest.
2. When the cell monolayers reach 70% to 80% confluency, the medium is replaced by 20 mL fresh IMDM supplemented with 5% FBS. Supernatants are harvested 24 hours later, clarified by low-speed centrifugation, filtered, and stored frozen at -80°C. The titer of frozen supernatant batches are stable for long periods.
3. The supernatants are titered on NIH 3T3 cells by conventional assays (3). Before being used for the IES protocol, the corresponding supernatant must be tested for the absence of competent helper virus using a genome rescue assay (3).

Long-Term Analysis of Lymphohematopoietic-Reconstituted Mice

At least four months after transplantation of a transduced graft, surviving animals are sacrificed, and their hematopoietic organs (BM, spleen, and thymus) are removed and processed for DNA extraction. Recipient BM can be also assayed to determine its content in committed CFU-GM progenitors, as well as transplanted into secondary recipients to quantify the extent of transduction into the pluripotent CFU-S population and the preservation of the long-term repopulating ability.

Materials

- Bacto-agar (Becton Dickinson, Franklin Lakes, NJ, USA). Stock 3% (wt/vol) is prepared in hot, distilled water, previously sterilized by autoclaving. Solidified stock is dissolved by incubation in a hot water bath before use.
- Horse serum (Life Technologies). Heat-inactivated at 56°C for 30 minutes.

- Irradiation equipment (Philips MG 324 X-ray unit [Hamburg, Germany] or equivalent).
- Tellesniczky's solution. Fixing solution composed of 44% ethanol, 31% acetic acid, and 2.3% formaldehyde. After a few minutes immersion in this solution, spleen colonies are easily visualized because of their white color over the brown splenic background.
- WCM. Conditioned medium from WEHI-3b cell line is used as the source of IL-3.

Procedures

CFU-GM Assay

1. Cells are resuspended at different concentrations in IMDM supplemented with 25% horse serum, 10% WCM, and mixed with Bacto-agar to a 0.3% final concentration. This suspension should be brought to a final volume of 3.5 mL.
2. Aliquots of 1 mL are seeded into each of three 35-mm ϕ plastic culture dishes. After seeding, the plates are incubated for 5 minutes at 4°C to facilitate solidification.
3. Cells are cultured at 37°C in a fully humidified atmosphere with 5% CO_2 in air.
4. Seven days later, colonies containing 50 or more cells are scored under a dissecting microscope.

The number of CFU-GM progenitors resistant to G418 (0.5 mg/mL) is expressed as relative to the number of colonies growing in absence of G418.

CFU-S Assay

1. Female mice (12–24 weeks old) are conditioned by total body irradiation. (In our hands, a 5 Gy + 4h + 5 Gy X-ray irradiation protocol with a dose rate of 1.03 Gy/minute works properly.)
2. Within the first 4 hours from completion of irradiation conditioning, groups of 8 to 10 recipient mice are independently injected via a lateral vein of the tail with a 200–500 μL different dilutions of the cell suspension to be assayed.
3. Transplanted mice are housed individually and given sterilized food and acidified water (pH 2.8) ad libitum.
4. Surviving animals are sacrified 12 days later, and their spleens are removed and fixed in Tellesniczky's solution to facilitate the quantification of the splenic colonies under a dissecting microscope. If molecular analysis is required, splenic colonies are individually dissected without fixation and processed for DNA extraction.

Long-Term Repopulating Ability Assay

1. Female mice (12 weeks old) are conditioned by total body irradiation. In our hands, a 6 Gy + 4 hours + 6 Gy X-ray irradiation protocol with a dose rate of 1.03 Gy/min works properly.
2. Within the first 4 hours from completion of irradiation conditioning, groups of recipient mice are independently injected via a lateral vein of the tail with a 200 to 500 μL volume with appropriated cell numbers from the suspension to be assayed.
3. Transplanted mice are housed individually, especially during the first month after irradiation, and given sterilized food and acidified water (pH 2.8) ad libitum.
4. Animals surviving at least four months after transplantation are sacrificed, and their hematopoietic

organs are removed and processed for DNA extraction or clonogenic assays.

Exogenous and Clonal Contribution to the Lymphohematopoietic System from Long Term Repopulated Mice

1. Genomic DNA samples (10 μg) are digested overnight with a restriction enzyme (e.g., *Bam*HI) cutting outside the proviral genome of the pXT1-retroviral vector.
2. Digested samples are electrophoresed on a 0.8% agarose gel in TBE buffer 1× at low voltage (1–2 V/cm).
3. DNA fragments are transferred from agarose gels to nylon membranes using the alkaline method.
4. The membranes are prehybridized for 3 hours at 42°C in 10 mmol/L Tris 7.5, 50% formamide, 5× standard saline citrate (SSC), 5× Denhardt's reagent, and 0.5% sodium dodecyl sulfate (SDS).
5. Membranes are next hybridized overnight with the proper probe (see below) under the same conditions used for prehybridization, but supplemented with 10% dextran sulfate and 50 μg/mL denatured carrier DNA.

   To detect integrated pXT1 retroviral sequences, hybridization with a probe for the neo^R gene is performed. A specific probe for mouse Y-chromosome sequences is used to test the exogenous contribution in the female recipients. To verify the amount of DNA load, membranes are stripped and rehybridized with a probe for a single-copy gene (e.g., macrophage colony-stimulating factor [M-CSF]).
6. Wash membranes three times for 1 hour each at 56°C in 0.1 × SSC and 0.5% SDS, followed by one wash in 2× SSC.
7. Exposed wet to X-ray films.

Evaluation of Retroviral Transduction of Human CD34+ Hematopoietic Cells Using the NOD-SCID Mice Model

Transduction of Human CD34+ Hematopoietic Cells Isolated from Cord Blood Samples

Materials

- Antibodies: Direct-labeled antibodies were purchased from Becton Dickinson and Coulter (Hialeah, FL, USA) and maintained at 4°C. They were allophycocyanin (APC)-conjugated anti-CD38 antibody (Clone HIT2 [Caltag, Burlingame, CA, USA]), phycoerytrin (PE)-conjugated anti-CD34 antibody (Clone 8G12 [Becton Dickinson]), QBEND-10 antibody (mouse antihuman CD34 [Miltenyi Biotec, Auburn, CA, USA]), and MACS microbeads (Miltenyi Biotech).
- CD34 purification: VarioMACS CD34 progenitor cell isolation kit (Miltenyi Biotec).
- CFU-GM culture: Enriched semisolid culture media from StemCells Technologies (Vancouver, BC, Canada).
- Cytokines: Recombinant human interleukin 6, (rhIL-6 [Immunex, Seattle, WA, USA]), human stem cell factor (rhSCF [Immunex]), human FLT-3 ligand (rhFL [Immunex]), and human megacaryocyte-derived growth factor (rhMGDF [Amgen]) were resuspended at 20 ng/mL in PBS- + 0.01% bovine serum albumin (BSA), filtered, and stored at -20°C until use.
- EPICS Elite ESP cytometer (Coulter).
- Fibronectin recombinant fragment CH-269 (Takara Shuzo, Otsu, Japan).
- Retrovirus-producer cell line. The PG13/EGFP7 clone used here was obtained by transduction of PG13 packaging cells (provided by D. Miller, Fred

Hutchinson Cancer Research Center, Seattle, WA, USA) with supernatants from PA317/EGFP1 cells, which contain the pSF-EGFP plasmid (45). After extensive analysis, the clone termed PG13/EGFP7, which exhibited the highest titer, was selected for further experiments. This clone has yielded viral titers in the range of 1×10^6 infectious particles/mL for more than a year.

- Solutions: PBE (PBS containing 0.5% BSA [Fraction V; Sigma] and 5 mmol/L EDTA).

Procedures

- Cord Blood Sample Preparation and $CD34^+$ Cell Selection

Cells were obtained from umbilical cord blood after normal full-term deliveries, according to the protocol approved by the Ethical Committee of our institution. $CD34^+$ cells were isolated from cord blood samples by an indirect immunomagnetic method, as indicated below.

1. Mononuclear cells are obtained by layering the blood sample onto Ficoll-Hypaque (1.077 g/mL [Amersham Pharmacia Biotech, Uppsala, Sweden]), and centrifugation at 400× *g* for 30 minutes. Cells from the interface layer are collected, washed three times by adding at least three volumes of PBE buffer, sedimented at 400× *g* for 15 minutes, and finally, resuspended in 2 mL PBE buffer.
2. To purify the $CD34^+$ fraction, mononuclear cells are subjected to immunomagnetic separation, following the manufacturer's recommendations. Briefly, mononuclear cells are washed and resuspended in PBE. Cells are incubated first with antihuman CD34 antibody (QBEND-10) in the presence of human IgG as a blocking reagent, and then with an antimouse antibody coupled with MACS microbeads (Miltenyi Biotech). Labeled cells are filtered through a 30-μm nylon mesh before loading onto a column installed in the magnetic field. Trapped cells are eluted after the column is removed from the magnet and subjected to a second MACS column passage for further depletion of contaminant $CD34^-$ cells. The purity of the $CD34^+$ population ranged from 95% to 98%, as evaluated by flow cytometry using an EPICS Elite ESP analyzer.
3. $CD34^+$-enriched cells are used directly or cryopreserved in liquid nitrogen, with 10% and 20% dimethyl sulfoxide fetal bovine serum (FBS).

- Retroviral Transduction of $CD34^+$ Cells with PG13/EGFP7 Viral Supernatants

1. Culture flasks are coated with recombinant fibronectin at 4 μg/cm² for at least 2 hours.
2. $CD34^+$ cells are stimulated for 24–48 hours with rhIL-6, rhSCF, rhFL, and rhMGDF—all of them used at 100 ng/mL. Cells are incubated at a cell density of $0.5–1 \times 10^5$/mL in 25-cm² canted-neck tissue culture flasks and maintained at 37°C in 5% CO_2.
3. After the stimulation period, 90% of the medium is replaced daily with 0.45-μm filtered supernatant containing fresh vector, growth factors (see step 2), and 4 μg/mL protamine sulfate for two consecutive days. In each experiment, a mock transduction is done in parallel using the same medium, growth factors, and protamine sulfate, but without addition of viral vector.

- Fluorescence-Activated Cell Sorting (FACS) Analysis of $CD34^+/CD38^{low}$ Cells

The transduction efficiency of hematopoietic cells can be assessed directly by FACS analysis. A minimum of 1×10^5 cells are washed with PBE and stained with PE-conjugated anti-CD34 antibody and/or APC-conjugated anti-CD38 antibody for 20 minutes at room temperature. Cells are washed and analyzed using an EPICS Elite ESP cytometer equipped with an argon ion laser tuned at 488 nm and a He-Ne laser tuned at 620 nm. PE- and APC-conjugated mouse isotypic antibodies serve as controls. A first gate including only the living cells can be used to estimate the percentage of $CD34^+/EGFP^+$ cells. A second gate on $CD34^+$ cells is used to analyze EGFP expression within the $CD34^+/CD38^{low}$ cell population.

- Evaluation of Retroviral Transduction of CFU-GM Progenitors

CFU-GM colonies are grown in enriched semisolid culture media from StemCells Technologies. Appropriate amounts of cells are seeded and cultured for 14 days at 37°C in a humidified atmosphere at 5% CO_2 in air. Colonies are scored under an inverted microscope equipped with epifluorescent illumination. The percent of green colonies versus total colonies was used as an estimation of the transduction efficiency of the infection procedure.

Transplantation of Transduced Human $CD34^+$ Hematopoietic Cells from Cord Blood Cells into NOD-SCID Mice

Materials

- Animals: We used NOD/SCID mice (a strain deficient in Fc receptors, complement function, natural killer, B- and T-cell function) as recipients of human hematopoietic cells. Mice were purchased from the Jackson Laboratory (Bar Harbor, ME, USA). All animals were handled under sterile conditions and maintained under microisolators.
- Antibodies: Antihuman-CD45-PECy5 (Clone J33 [Immunotech, Marseille, France]). Antihuman-CD34-PE (Anti-HPCA-2 [Becton Dickinson]).
- Irradiation equipment (Philips MG 324 X-ray unit or equivalent).
- Lysis solution: 0.155 mol/L NH_4Cl + 0.01 mol/L $KHCO_3$ + 1×10^{-4} mol/L EDTA.
- Washing solution: PBA (PBS solution with 0.1% BSA and 0.02% sodium azide).

Procedures

- Transplantation Procedure

1. Before transplantation, 6- to 8-week-old mice are total-body irradiated with 2.5 to 3.0 Gy of X-rays (300 kV, 10 mA).
2. Within the 4-hour postirradiation period, cell suspensions are injected through the lateral tail vain in a final volume of 0.2 to 0.5 μL. Animals are prewarmed under a heating lamp to visualize the veins.

- Analysis of Human Hematopoiesis in NOD-SCID Mice Transplanted with Retrovirally Transduced Human Grafts

1. BM samples are obtained under anesthesia by aspiration through the knee joint.
2. Aliquots of 1–5×10^5 cells/tube were stained for 25 minutes at 4°C with antihuman-CD45-PECy5 in combination with antihuman-CD34-PE. Thereafter, red blood cells are lysed by adding 2.5 mL lysis solution and incubated at room temperature for 10 minutes. Cells are then washed in PBA, resuspended in PBA and 2 μg/mL propidium iodide (PI), and analyzed by flow cytometry.

3. For each analysis a total of 10 000 viable (PI$^-$) cells with appropriate forward and side scatter (FS/SSC) were collected. As controls of nonspecific staining, cells can be labeled with conjugated-nonspecific antibodies of the appropiate isotype.

CONCLUSIONS AND PERSPECTIVES

Based on earlier experimental work, several clinical applications of retroviral gene marking are coming into practice. Further work needs to be done on experimental transplantation models, however, to expand the potential of gene marking in the analysis of HSC dynamics under steady-state conditions and in post-BM transplant situations. The progressive improvement of gene-marking strategies will clarify the contribution of hematopoietic autografts in the relapse of transplanted cancer patients. The work presented here supports the observation that the oligoclonal reconstitution pattern generally observed in transplanted recipients is not dependent on the size of the inoculated graft; this conclusion may have important implications for gene therapy approaches requiring stable and predictable expression of the therapeutic gene.

A second aspect we have introduced is the current difficulties for retroviral transduction of early human hematopoietic precursors. Here we show that, under certain defined ex vivo conditions and using *EGFP* as the reporter gene, it is possible to transduce human hematopoietic precursors capable of in vivo repopulating NOD/SCID mice. The transduced cells remain in the transplanted animal and contribute significantly to the human hematopoiesis that takes place in these animals. They express the reporter gene throughout the period of analysis. Further investigation to demonstrate long-term persistence and expression of the transgene in the human hematopoietic progeny is in progress.

The optimization of retroviral vectors, the development of new ex vivo manipulation procedures for hematopoietic grafts, and the generation of improved experimental in vivo models of human hematopoiesis have given rise to expectations for converting HSC gene therapy from a long-standing promise into a reality. Although further studies in large-animal models are required before routine procedures of gene transfer will be applicable to humans, recent advances in the field show genuine promise. The incorporation of new, optimized vectors (i.e., lentivirus-derived vectors) will doubtless bring even nearer the promising future of gene transfer in medicine.

REFERENCES

1. **Anderson, W.F.** 1998. Human gene therapy. Nature *392*:25-30.
2. **Audet, J., P.W. Zandstra, C.J. Eaves, and J.M. Piret.** 1998. Advances in hematopoietic stem cell culture. Curr. Opin. Biotechnol. *9*:146-151.
3. **Bernad, A., F. Varas, J.M. Gallego, J.M. Almendral, and J.A. Bueren.** 1994. Ex vivo expansion and selection of retrovirally transduced bone marrow: an efficient methodology for gene-transfer to murine lympho-haemopoietic stem cells. Br. J. Haematol. *87*:6-17.
4. **Bierhuizen, M.F., Y. Westerman, T.P. Visser, W. Dimjati, A.W. Wognum, and G. Wagemaker.** 1997. Enhanced green fluorescent protein as selectable marker of retroviral-mediated gene transfer in immature hematopoietic bone marrow cells. Blood *90*:3304-3315.
5. **Bock, T.A.** 1997. Assay systems for hematopoietic stem and progenitor cells. Stem Cells (Suppl. 1) *15*:185-195.
6. **Bodine, D.M., N.E. Seidel, M.S. Gale, A.W. Nienhuis, and D. Orlic.** 1994. Efficient retrovirus transduction of mouse pluripotent hematopoieitc stem cells mobilized into peripheral blood by treatment with granulocyte colony-stimulating factor and stem cell factor. Blood *84*:1482-1491.
7. **Bordignon, C.** 1998. A new chance for SCID gene therapy. Nat. Med. *4*:19-20.
8. **Bordignon, C., L.D. Notarangelo, N. Nobili, G. Ferrari, G. Casaroti, P. Panina, E. Mazzolari, D. Maggioni, C. Rossi, P. Servida, et al.** 1995. Gene therapy in peripheral blood lymphocytes and bone marrow for ADA-immunodeficient patients. Science *270*:470-475.

9. **Boulter, C.A. and E.F Wagner.** 1987. A universal retroviral vector for efficient constitutive expression of exogenous genes. Nucleic Acids Res. *15*:7194.

10. **Brenner, M.K.** 1994. Genetic marking and manipulation of hematopoieitic progenitor cells using retroviral vectors. Immunomethods *5*:204-210.

11. **Brenner, M.K., D.R. Rill, M.S. Holladay, H.E. Heslop, R.C. Moen, M. Buschle, R.A. Krance, V.M. Santana, W.F. Anderson, and J.N. Ihle.** 1993. Gene marking to determine whether autologous marrow infusion restores long-term haemopoiesis in cancer patients. Lancet *342*:1134-1137.

12. **Brenner, M.K., D.R. Rill, R.C. Moen, R.A. Krance, J. Mirro, Jr., W.F. Anderson, and J.N. Ihle.** 1993. Gene-marking to trace origin of relapse after autologous bone-marrow transplantation. Lancet *341*:85-86.

13. **Brenner, M.K., D.R. Rill, R.C. Moen, R.A. Krance, H.E. Heslop, J. Mirro, W.F. Anderson, and J.N. Ihle.** 1994. Gene Marking and autologous bone marrow transplantation. Ann. N.Y. Acad. Sci. *716*:204-217.

14. **Capel, B., R. Hawley, L. Covarrubias, T. Hawley, and B. Mintz.** 1989. Clonal contributions of small numbers of retrovirally marked hematopoieitc stem cells engrafted in unirradiated neonatal W/WV. Proc. Natl. Acad. Sci. USA *86*:4564 -4568.

15. **Capel, B., R.G. Hawley, and B. Mintz.** 1990. Long- and short-lived murine hematopoieitic stem cell clones individually identified with retroviral integration markers. Blood *75*:2267-2270.

16. **Challita, P.M. and D.B. Kohn.** 1994. Lack of expression from a retroviral vector after transduction of murine hematopoietic stem cells is associated with methylation in vivo. Proc. Natl. Acad. Sci. USA *91*:2567-2571.

17. **Cornetta, K., E.F. Srour, A. Moore, A. Davidson, E.R. Broun, R. Hromas, R.C. Moen, R.A. Morgan, L. Rubin, W.F. Anderson, et al.** 1996. Retroviral gene transfer in autologous bone marrow transplantation for adult acute leukemia. Hum. Gene Ther. *7*:1323-1329.

18. **Dao, M.A., C.H. Hannum, D.B. Kohn, and J.A. Nolta.** 1997. FLT3 ligand preserves the ability of human CD34$^+$ progenitors to sustain long-term hematopoiesis in immune-deficient mice after ex vivo retroviral-mediated transduction. Blood *89*:446-456.

19. **Dao, M.A., A.J. Shah, G.M. Crooks, and J.A. Nolta.** 1998. Engraftment and retroviral marking of CD34$^+$ and CD34$^+$CD38$^-$ human hematopoietic progenitors assessed in immune-deficient mice. Blood *91*:1243-1255.

20. **Deisseroth, A.B., Z. Zu, D. Claxton, E.G. Hanania, S. Fu, D. Ellerson, L. Goldberg, M. Thomas, K. Janicek, W.F. Anderson, et al.** 1994. Genetic marking shows that Ph$^+$ cells present in autologous transplants of chronic myelogenous leukemia (CML) contribute to relapse after autologous bone marrow in CML. Blood *83*:3068-3076.

21. **Dick, J.E., M. Bhatia, O. Gan, U. Kapp, and J.C. Wang.** 1997. Assay of human stem cells by repopulation of NOD/SCID mice. Stem Cells (Suppl. 1) *15*:199-203.

22. **Dick, J.E., S. Kamel-Reid, B. Murdoch, and M. Doedens.** 1991. Gene transfer into normal human hematopoietic cells using in vitro and in vivo assays. Blood *78*:624-634.

23. **Dick, J.E., M.C. Magli, D. Huszar, R.A. Phillips, and A. Bernstein.** 1985. Introduction of a selectable gene into primitive stem cells capable of long-term reconstitution of the hemopoietic system of W/Wv mice. Cell *42*:71-79.

24. **Drize, N.J., J.R. Keller, and J.L. Chertkov.** 1996. Local clonal analysis of the hematopoieitc system shows that multiple small short-lived clones maintain life-long hematopoiesis in reconstituted mice. Blood *88*:2927-2938.

25. **Dunbar, C.E., M. Cottler-Fox, J.A. O'Shaughnessy, S. Doren, C. Carter, R. Berenson, S. Brown, R.C. Moen, J. Greenblatt, F.M. Stewart, et al.** 1995. Retrovirally marked CD34-enriched peripheral blood and bone marrow cells contribute to long-term engraftment after autologous transplantation. Blood *85*:3048-3057.

26. **Fisher, B., B.S. Packard, E.J. Read, J.A. Carrasquillo, C.S. Carter, S.L. Topalian, Y.C. Yang, P. Yolles, S.M. Larson, and S.A. Rosenberg.** 1989. Tumor localization of adoptively transferred indium-111 labeled tumor infiltrating lymphocytes in patients with metastatic melanoma. J. Clin. Oncol. *7*:250-261.

27. **Güenechea, G., J.C. Segovia, B. Albella, M. Lamana, M. Ramirez, C. Regidor, M.N. Fernandez, and J.A. Bueren.** 1999. Delayed engraftment of nonobese diabetic/severe combined immunodeficient mice transplanted with ex vivo-expanded human CD34(+) cord blood cells. Blood *93*:1097-1105.

28. **Griffith, K.D., E.J. Read, J.A. Carrasquillo, C.S. Carter, Y.C. Yang, B. Fisher, P. Aebersold, B.S. Packard, M.Y. Yu, and S.A. Rosenberg.** 1989. In vivo distribution of adoptively transferred indium-111-labeled tumor infiltrating lymphocytes and peripheral blood lymphocytes in patients with metastatic melanoma. J. Natl. Cancer. Inst. *81*:1709-1717.

29. **Hao, Q.L., F.T. Thiemann, D. Petersen, E.M. Smogorzewska, and G.M. Crooks.** 1996. Extended long-term culture reveals a highly quiescent and primitive human hematopoietic progenitor population. Blood *88*:3306-3313.

30. **Harrison, D.E. and C.P. Lerner.** 1991. Most primitive hematopoietic stem cells are stimulated to cycle rapidly after treatment with 5-fluorouracil. Blood *78*:1237-1240.

31. **Hughes, P.F., C.J. Eaves, D.E. Hogge, and R.K. Humphries.** 1989. High-efficiency gene transfer to human hematopoietic cells maintained in long-term marrow culture. Blood *74*:1915-1922.

32. **Jordan, C.T. and I.R. Lemischka.** 1990. Clonal and systemic analysis of long-term hematopoiesis in the mouse. Genes Dev. *4*:220-232.

33. **Katayama, N., S.C. Clark, and M. Ogawa.** 1993. Growth factor requirement for survival in cell-cycle dormancy of primitive murine lymphohematopoietic progenitors. Blood *81*:610-616.

34. **Keller, G. and R. Snodgrass.** 1990. Life span of multipotential hematopoieitic stem cells in vivo. J. Exp. Med. *171*:1407-1418.

35. **Kiem, H.-P., B. Darovsky, C. von Kalle, S. Goehle,**

D. Stewart, T. Graham, R. Hackman, F.R. Appelbaum, H.J. Deeg, and A.D. Miller. 1994. Retrovirus-mediated gene transduction into canine peripheral blood repopulating cells. Blood *83*:1467-1473.

36. **Kiem, H.P., S. Heyward, A. Winkler, J. Potter, J.M. Allen, A.D. Miller, and R.G. Andrews.** 1997. Gene transfer into marrow repopulating cells: comparison between amphotropic and gibbon ape leukemia virus pseudotyped retroviral vectors in a competitive repopulation assay in baboons. Blood *90*:4638-4645.

37. **Kohn, D.B., K.I. Weinberg, J.A. Nolta, L.N. Heiss, C. Lenarsky, G.M. Crooks, M.E. Hanley, G. Annett, J.S. Brooks, A. el-Khoureiy, et al.** 1995. Engraftment of gene-modified umbilical cord blood cells in neonates with adenosine deaminase deficiency. Nat. Med. *1*:1017-1023.

38. **Ladner, M.B., G.A. Martin, J.A. Noble, V.P. Wittman, M.K. Warren, M. McGrogan, and E.R. Stanley.** 1988. cDNA cloning and expression of murine macrophage colony-stimulating factor from L929 cells. Proc. Natl. Acad Sci. USA *85*:6706-6710.

39. **Lamar, E.E. and E. Palmer.** 1984. Y-encoded, species-specific DNA in mice: evidence that the Y chromosome exists in two polymorphic forms in inbred strains. Cell *37*:171-177.

40. **Lapidot, T., Y. Fajerman, and O. Kollet.** 1997. Immune-deficient SCID and NOD/SCID mice models as functional assays for studying normal and malignant human hematopoiesis. J. Mol. Med. *75*:664-673.

41. **Larochelle, A., J. Vormoor, H. Hanenberg, J.C. Wang, M. Bhatia, T. Lapidot, T. Moritz, B. Murdoch, X.L. Xiao, I. Kato, et al.** 1996. Identification of primitive human hematopoietic cells capable of repopulating NOD/SCID mouse bone marrow: implications for gene therapy. Nat. Med. *2*:1329-1337.

42. **Lemischka, I., D.H. Raulet, and R.C. Mulligan.** 1986. Developmental potential and dynamic behaviour of hematopoieitc stem cells. Cell *45*:917-927.

43. **Lerner, C. and D.E. Harrison.** 1990. 5-fluorouracil spares hemopoietic stem cells responsible for long-term repopulation. Exp. Hematol. *18*:114-118.

44. **Leung, W., M. Ramirez, E.M. Novelli, and C.I. Civin.** 1998. In vivo engraftment potential of clinical hematopoietic grafts. J. Invest. Med. *46*:303-311.

45. **Limon, A., J. Briones, T. Puig, M. Carmona, O. Fornas, J.A. Cancelas, M. Nadal, J. Garcia, F. Rueda, and J. Barquinero.** 1997. High-titer retroviral vectors containing enhanced green fluorescent protein (EGFP) for efficient expression in hematopoietic cells. Blood *90*:3316-3321.

46. **Lotze, M.T., B.R. Line, D.J. Mathisen, and S.A. Rosenberg.** 1980. The in vivo distribution of autologous human and murine lymphoid cells grown in T cell growth factor (TCGF): implications for the adoptive immunotherapy of tumor. J. Immunol. *125*:1487-1493.

47. **Micklem, H.S., J.E. Lennon, J.D. Answell, and R.A. Gray.** 1987. Numbers and dispersion of repopulating hematopoietic cell clones in radiation chimeras as functions od injected cell dose. Exp. Hematol. *15*:251-257.

48. **Miller, D.G., A.A. Mohammed, and A.D Miller.** 1990. Gene transfer by retrovirus vectors occurs only in cells that are actively replicating at the time of infection. Mol. Cell. Biol. *10*:4239-4242.

49. **Moritz, T., V.P. Patel, and D.A. Williams.** 1994. Bone marrow extracellular matrix molecules improve gene transfer into human hematopoietic cells via retroviral vectors. J. Clin. Invest. *93*:1451-1457.

50. **Muench, M.O. and M.A.S. Moore.** 1992. Accelerated recovery of peripheral blood cell counts in mice transplanted with in vitro cytokine-expanded hematopoietic progenitors. Exp. Hematol. *20*:611-618.

51. **Muench, M.O., J.G. Schneider, and M.A.S. Moore.** 1992. Interactions among colony-stimulating factors, IL-1, IL-6 and Kit-ligand in the regulation of primitive murine hematopoietic cells. Exp. Hematol. *20*:339-349.

52. **Nolta, J.A., M.A. Dao, S. Wells, E.M. Smogorzewska, and D.B. Kohn.** 1996. Transduction of pluripotent human hematopoietic stem cells demonstrated by clonal analysis after engraftment in immune-deficient mice. Proc. Natl. Acad. Sci. USA *93*:2414-2419.

53. **Nowell, P.C., B.E. Hirsch, D.H. Fox, and D.B. Wilson.** 1970. Evidence for the existence of multipotential lympho-hematopoietic stem cells in adult rat. Cell. Physiol. *75*:151-158 .

54. **Orlic, D., L.J. Girard, C.T. Jordan, S.M. Anderson, A.P. Cline, and D.M. Bodine.** 1996. The level of mRNA encoding the amphotropic retrovirus receptor in mouse and human hematopoietic stem cells is low and correlates with the efficiency of retrovirus transduction. Proc. Natl. Acad. Sci. USA *93*:11097-11102.

55. **Persons, D.A., J.A. Allay, E.R. Allay, R.J. Smeyne, R.A. Ashmun, B.P. Sorrentino, and A.W. Nienhuis.** 1997. Retroviral-mediated transfer of the green fluorescent protein gene into murine hematopoietic cells facilitates scoring and selection of transduced progenitors in vitro and identification of genetically modified cells in vivo. Blood *90*:1777-1786.

56. **Ploemacher, R.E.** 1997. Stem cells: characterization and measurement. Baillieres Clin. Haematol. *10*:429-444.

57. **Prasher, D.C., V.K. Eckenrode, W.W. Ward, F.G. Prender-Gast, and M.J. Cormier.** 1992. Primary structure of the Aequorea victoria green fluorescent protein. Gene *111*:229-233.

58. **Prockop, D.J.** 1997. Marrow stromal cells as stem cells for nonhematopoietic tissues. Science *276*:71-74.

59. **Roe, T., T.C. Reynolds, G. Yu, and P.O. Brown.** 1993. Integration of murine leukemia virus DNA depends on mitosis. EMBO J. *12*:2099-2108.

60. **Schilz, A.J., G. Brouns, H. Knoss, O.G. Ottmann, D. Hoelzer, A.A. Fauser, A.J. Thrasher, and M. Grez.** 1998. High efficiency gene transfer to human hematopoietic SCID-repopulating cells under serum-free conditions. Blood *92*:3163-3171.

61. **Shultz, L.D., P.A. Schweitzer, S.W. Christianson, B. Gott, I.B. Schweitzer, B. Tennent, S. McKenna, L. Mobraaten,T.V. Rajan, D.L. Greiner, and E.H. Leiter.** 1995. Multiple defects in innate and adaptive immunologic function in NOD/LtSz-scid mice. J. Immunol. *154*:180-191.

62.**Serrano, F., F. Varas, A. Bernad, and J.A. Bueren.** 1994. Accelerated and long-term hematopoietic engraftment in mice transplanted with ex vivo expanded bone marrow. Bone Marrow Transplant. *4*:855-862.

63.**Smith, L.G., I.L. Weissman, and S. Heimfeld.** 1991. Clonal analysis of hematopoietic stem-cell differentiation in vivo. Proc. Natl. Acad. Sci. USA *88*:2788-2792.

64.**Spangrude, G.J., L. Smith, N. Uchida, K. Ikuta, S. Heimfeld, J. Friedman, and I.L. Weissman.** 1991. Mouse hematopoietic stem cells. Proc. Natl. Acad. Sci. USA *87*:7433-7437.

65.**van der Loo, J.C., H. Hanenberg, R.J. Cooper, F.Y. Luo, E.N. Lazaridis, and D.A. Williams.** 1998. Nonobese diabetic/severe combined immunodeficiency (NOD/SCID) mouse as a model system to study the engraftment and mobilization of human peripheral blood stem cells. Blood *92*:2556-2570.

66.**van Hennik, P.B., M.M. Verstegen, M.F. Bierhuizen, A. Limon, A.W. Wognum, J.A. Cancelas, J. Barquinero, R.E. Ploemacher, and G. Wagemaker.** 1998. Highly efficient transduction of the green fluorescent protein gene in human umbilical cord blood stem cells capable of cobblestone formation in long-term cultures and multilineage engraftment of immunodeficient mice. Blood *92*:4013-4022.

67.**Van Zant, G., J.J. Chen, and K. Scott-Micus.** 1991. Developmental potential of hematopoieitc stem cells determined using retrovirally marked allophenic marow. Blood *77*:756-763.

68.**Varas, F., A. Bernad, J.M. Almendral, and J.A. Bueren.** 1996. Relevance of myeloablative conditioning in the engraftment of limiting numbers of normal and genetically marked lympho-hematopoietic stem cells. Bone Marrow Transplant. *18*:981-989.

69.**Varas, F., A. Bernad, and J.A. Bueren.** 1996. Granulocyte colony-stimulating factor mobilizes into peripheral blood the complete clonal repertoire of hematopoietic precursors residing in the bone marrow of mice. Blood *88*:2495-2501.

70.**Varas, F., A. Bernad, and J.A. Bueren.** 1998. Restrictions in the stem cell function of murine bone marrow grafts after ex vivo expansion of short-term repopulating progenitors. Exp. Hematol. *26*:100-109.

71.**Wang, J.C., M. Doedens, and J.E. Dick.** 1997. Primitive human hematopoietic cells are enriched in cord blood compared with adult bone marrow or mobilized peripheral blood as measured by the quantitative in vivo SCID-repopulating cell assay. Blood *89*:3919-3924.

72.**Yokota, T., K. Oritani, H. Mitsui, K. Aoyama, J. Ishikawa, H. Sugahara, I. Matsumura, S. Tsai, Y. Tomiyama, Y. Kanakura, and Y. Matsuzawa.** 1998. Growth-supporting activities of fibronectin on hematopoietic stem/progenitor cells in vitro and in vivo: structural requirement for fibronectin activities of CS1 and cell-binding domains. Blood *91*:3263-3272.

28

Towards the Genetic Manipulation of the Hematopoietic System: Tumor-Specific Gene Therapy Strategies

Cesar Cobaleda and Isidro Sánchez-García
Department of Cellular Differentiation and Proliferation, Institute of Biochemical Microbiology, CSIC–University of Salamanca, Salamanca, Spain

INTRODUCTION

The main problem facing the effective treatment of patients with cancer is how to distinguish between tumor and normal cells. The chimeric molecules created by cancer-associated chromosomal abnormalities are ideal targets for therapy because they are unique to the transformed cells. In this chapter, we review some advances made to selectively destroy tumor cells through specific inhibition of products resulting from chromosomal translocations.

Chromosomal Translocations and Tumor Development

Cancer is a complex phenomenon that results from acquired genetic changes accumulating in a cell that proliferates to form a clone with increasingly aggressive behavior (23). The genetic changes occuring in cancer cells involve protooncogenes and/or tumor-supression genes. Activation of protooncogenes and subsequent expression of oncoproteins can occur as a consequence of different events (Figure 1): *(i)* point mutations, *(ii)* gene amplification, and *(iii)* chromosomal abnormalities (e.g., deletions, inversions, or translocations). Among these mechanisms, the importance of chromosomal translocations has been clear since early cytogenetic studies showed recurrent chromosomal changes in specific types of tumors (47). More recently, molecular genetic analyses of consistent chromosomal rearrangements have identified either previously known or new protooncogenes spanning the chromosomal breakpoints (13,61,64), thus confirming the pivotal role of chromosomal translocations in tumor development. Virtually all solid tumors and most hematopoietic neoplasias show abnormalities in the kariotype; translocations are more frequently found in hematological malignancies than in solid tumors. Although balanced chromosomal translocations are well known in mesenchymal tumors, they have rarely been described in those of epithelial origin.

There are two main consequences of chromosomal rearrangements (Figure 2). The first is illustrated by the case of a T-cell receptor (TCR) gene's or an immunoglob-

Viral Vectors: Basic Science and Gene Therapy
Edited by A. Cid-Arregui and A. García-Carrancá

ulin gene's being translocated near to a protooncogene because of a mistake in the process of V(D)J recombination and joining, which in normal conditions should give rise to a mature B or T cell with specific TCR or immunoglobulins. Such wrong recombination often leads to activation of the protooncogene by deregulating its expression and affecting its normal function: *(i)* in the control of cell proliferation, as in the case of the c-*MYC* gene in Burkitt's lymphoma (61); *(ii)* in the process of regulated cell death, as *Bcl-2* gene in follicular lymphoma (12); or *(iii)* in the regulation of differentiation processes, as Tal 1, LMO 1, LMO 2, or HOX-11 (27,35,41, 61,68) in T-acute lymphoblastic leukemias.

Another possible outcome of a chromosomal rearrangement is gene fusion (i.e., when the break occurs within a gene on each chromosome involved), thus leading to the juxtaposition of part of each gene and the creation of a composite gene. As breakpoints disrupt each of the genes involved within an intron, coding sequences of each gene are maintained in the same reading frame, and the resulting chimeric transcripts usually encode hybrid proteins. Gene fusions are, in practice, the most common repercussion of chromosomal rearrangements in human cancers (leukemias and solid tumors). The first well-documented example of a chimeric protein came from the cloning of the Philadelphia chromosome breakpoint. In Ph^{1+}-leukemias, the translocation t(9;22)(q34;q11) rearranges the *BCR* gene and the c-ABL protein tyrosine kinase gene. Depending on the precise breakpoint within the *BCR* gene, alternative BCR-ABLp^{210} and BCR-ABLp^{190} chimeric proteins are produced, which are characteristic of chronic myelogenus leukemia and acute lymphoblastic leukemia, respectively (18). Many more fusion proteins have been described since then, most of them in hematopoietic cancers, but also in solid tumors (13,61). Molecular cloning of the breakpoints of chromosomal translocations in humar cancers has led to the identification of different classes of putative oncogenes thought to contribute to tumor phenotype. The expanding list includes growth factors, growth factor receptors, signal transducers, protein kinases, and transcriptional activators. Most genes activated or generated by chromosomal translocations in human cancers share a common structural composi-

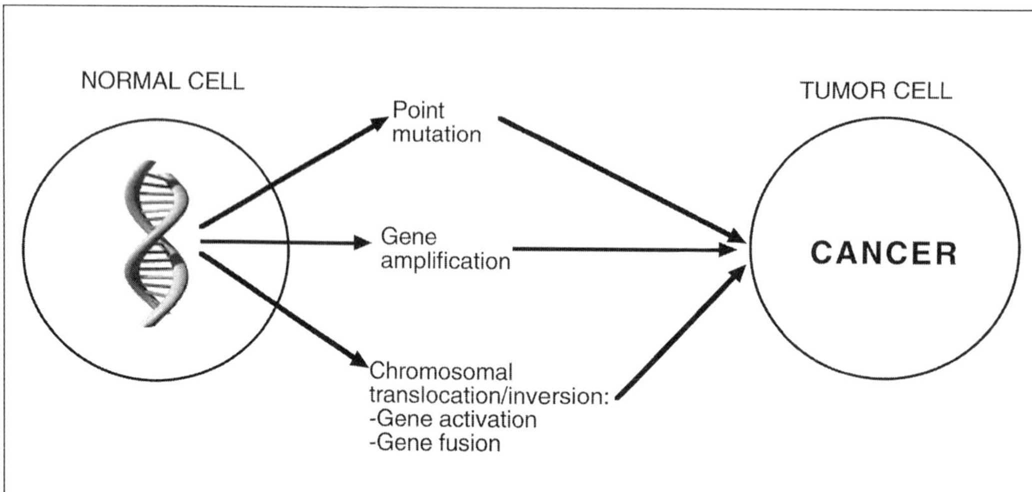

Figure 1. Main genetic events involved in the transformation of a normal cell into a tumor one.

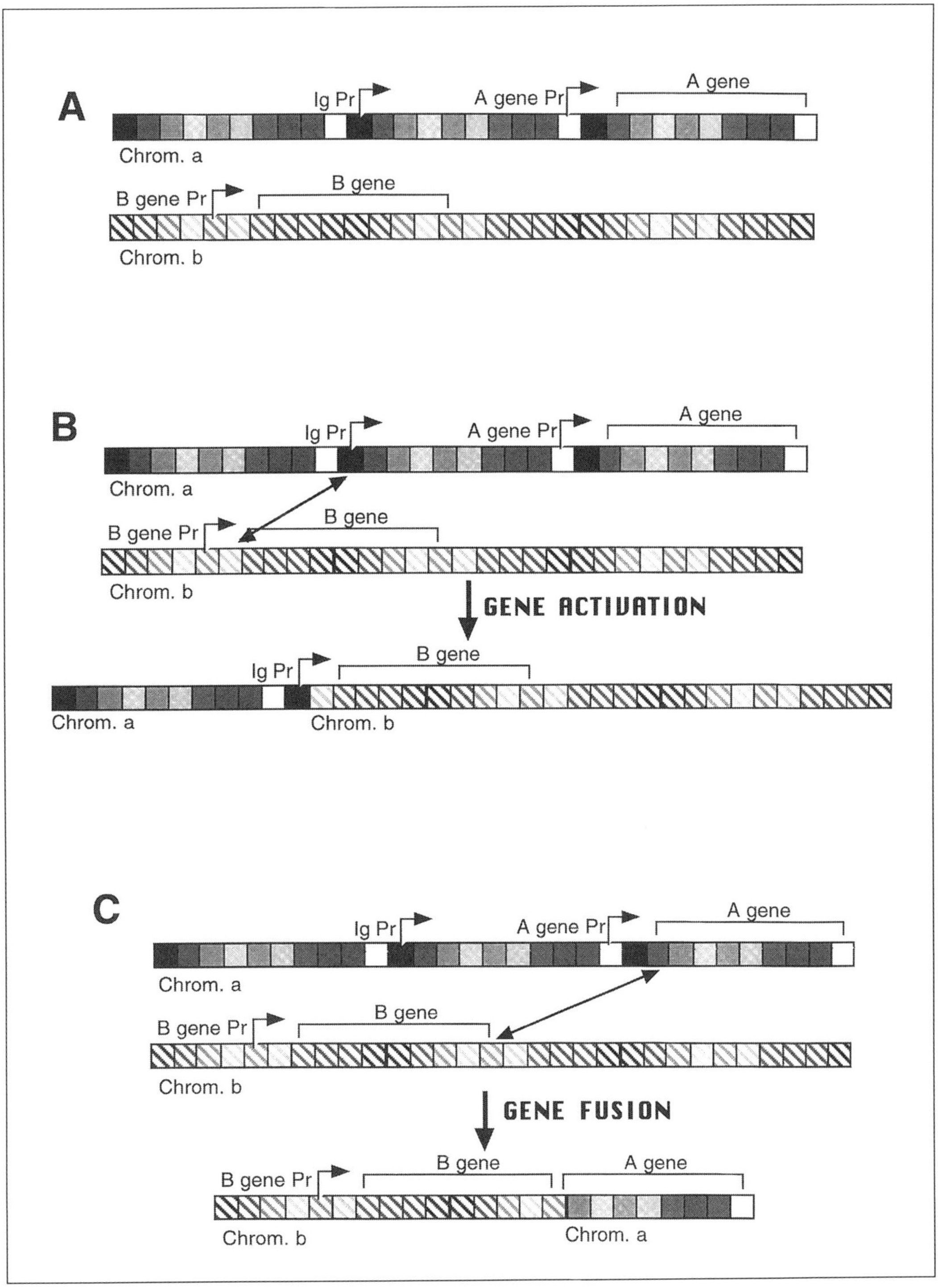

Figure 2. **The two main consequences of chromosomal translocations.** (A) Normal chromosomes "a" and "b" with their respective genes "A" and "B" and their corresponding promoters. In chromosome "a", an immunoglobulin promoter (Ig Pr) also appears. (B) Activation of "B" gene as consequence of a translocation (double-headed arrow) that leaves this gene under the control of the Ig promoter in chromosome "a". (C) Creation of a fusion gene "B-A" under the control of the "B" gene promoter as consequence of the translocation (double-headed arrow).

tion; they are mainly transcription factors with an activation domain linked to a DNA-binding domain or to a domain that can interact with another protein that binds DNA (65). Probably, such common features reflect not only a similar structure but also a similar mechanism of action that is at the basis of their oncogenic potential.

The evidence for a direct relationship between chromosomal rearrangements and tumor development comes not only from cytogenetic and molecular analyses of translocation products but also from *(i)* immunodeficient mice implanted with human cancer cells or cells transduced with retroviruses expressing chimeric proteins and *(ii)* transgenic mouse models in which DNA constructs resembling the structure of chromosomal breakpoints have been introduced into the germ line. Although different phenotypes can arise from the same activated/fused gene, due to differences in genetic background, promoter utilized (tissue-specific or housekeeping), or cell type used for infection, these mouse models corroborate that there is a direct relationship between a concrete activated/fused gene and a tumor type. Transgenic mice that overexpress genes known to be activated as a result of chromosomal translocations develop tumors. For instance, mice expressing a c-*myc* gene under the control of the immunoglobulin Eμ promoter/enhancer develop leukemias and lymphomas (1,38,69). Also, Eμ-directed *Bcl-2* gene expression causes follicular B-lymphocyte accumulation (43,44) in normal mice and promotes B-lymphoid development in SCID mice (78). Lck-promoter-driven expression of Tal-1 induces T-acute lymphoblastic leukemia in mice models (14,34), and *HOX11* gene expression controlled by an immunoglobulin promoter produces B-lineage lymphomas (30). Similar conclusions could be drawn from studies in which fusion genes resulting from chromosomal rearrangement were used to generate transgenic or knock-in mice. Knock-in mice for the product of the t(9;11)(p22;q23) translocation, which encodes the myeloid lymphoid leukemia-AF9 protein, developed acute myelogenous leukemia (17). Knock-in mice expressing the BCR-ABLp^{190} gene product suffered from acute lymphoblastic leukemia (8). In addition, transgenic mice expressing the product of translocation t(1;19)(q23;p13), the fusion oncogene *E2A-PBX 1*, under the control of an Eμ-promoter, made T-lymphoblastic lymphomas (19). Further, 30% of transgenic mice that express a promyelocytic leukemia-retinoic acid receptor A (PML-RARA) fusion oncogene under the control of the human cathepsin G promoter developed acute myelogenous leukemia after a long latency period (26).

Altogether, these results show beyond any doubt that those activation/fusion products resulting from chromosomal translocations are all that is necessary for the tumor phenotype to appear in an otherwise normal cellular context. However, the precise evolution and differentiation grade of the tumor cell obviously depend on factors other than the translocation itself (e.g., those derived from the cell type undergoing the translocation, environmental circumstances, etc.). In addition, transgenic mice provide an attractive system to test gene therapy approaches before starting clinical trials (e.g., by assessing the efficiency of gene transduction protocols into hematopoietic stem cells).

Targeted Gene Therapy of Chromosomal Translocations

As already explained, in many cases, a single gene fusion created by a chromosomal translocation can be all that is needed to transform a cell from a normal state to a tumoral phenotype (67). Chromosomal translocation implies severe DNA damage and as such should conduce the cell undergoing it to rapid programmed death through a DNA-damage checkpoint pathway. The only way that the affected cell has to evade its apoptotic end relies precisely on what makes it different (i.e., the product of the chromosomal translocation), which for a tumor to be developed should provide by itself a double function: *(i)* rescue the cell from apoptosis, and *(ii)* promote and maintain a proliferative state in the cell. This double effect has been demonstrated for the BCR-ABLp^{190} and BCR-ABLp^{210} proteins generated in Ph^{1+}-leukemias, which induce proliferation and inhibit the apoptotic pathway mediated by c-*MYC*. The block is produced via a *Ras*-dependent upregulation of the antiapoptotic gene *Bcl2* (66,67). Therefore, it can be concluded that these chimeric proteins are necessary for tumor growth and are thus ideal targets for gene therapy.

A key aspect in the effective treatment of cancer is how to distinguish between tumor and normal cells. Chimeric messengers and proteins resulting from chromosomal translocations are ideal therapeutic targets since they are unique to tumor cells and are essential for the maintenance of the tumor clone. Thus, inhibition of expression of a chimeric gene by antitumor agents (Figure 3) will eliminate malignant cells without affecting normal ones. Gene therapy strategies can be directed against the different elements involved in malignancy that results from chromosomal translocation, from the breakpoint DNA to the synthesis of a functional chimeric protein. For instance, expressing zinc finger proteins designed to bind specifically to the chromosomal breakpoint can block transcription (10,11). Such zinc fingers seem to recognize the oncogenic DNA without affecting the normal DNA. In fact, DNA-binding peptides containing three zinc finger motifs bind specifically to a unique 9-bp sequence of a *BCR-ABL* fusion oncogene in preference to the parent genomic sequences and can significantly repress transcription of the oncogene in cultured cells (11). However, zinc fingers need to be designed against the specific fusion sequence, which would imply a patient-specific design of peptides, because of sequence variability at the fusion point. Other therapeutic agents, perhaps more promising, are those directed against chimeric mRNAs (i.e., antisense RNA, nucleotides, and ribozymes). Antisense RNAs directed against the b3a2 *BCR/ABL* breakpoint reduced BCR-ABLp^{210} mRNA and protein levels more than 90% in transduced cells with high specificity, thus restoring normal function in cells transfected with a *BCR-ABL* cDNA (5,90). Szczylik et al. (81) showed in vitro that antisense oligodeoxynucleotides added to culture media selectively inhibited proliferation of *BCR-ABL*$^{+}$ leukemia cells. In vivo experiments in SCID mice transplanted with Ph^{+} leukemia cells showed that systemic treatment of leukemic mice with a *BCR-ABL*-specific antisense oligodeoxynucleotide (26 bases) increased their survival rate (71).

Ribozymes are likely to become useful tools in gene therapy for cancer. Because of their catalytic activity, they may greatly accelerate the rate of degradation of specific oncogene messengers. They should be designed to find a compromise between specificity and activity (29) in such a way that they can recognize oncogenic fusion messengers and leave unaffected the normal cellular messengers. One of the substrates more frequently used to test ribozymes as tools in the therapy of cancer has been the *BCR-ABL* oncogene. Leukemia cell lines treated with ribozymes directed against *BCR-ABL* (31) have a slower proliferation rate (39) and show reduced mRNA and protein levels (40,72) and decreased BCR-ABL kinase activity (70). Recently, an allosterically controllable ribozyme has been constructed that has the ability to induce apoptosis in Ph^{1+}

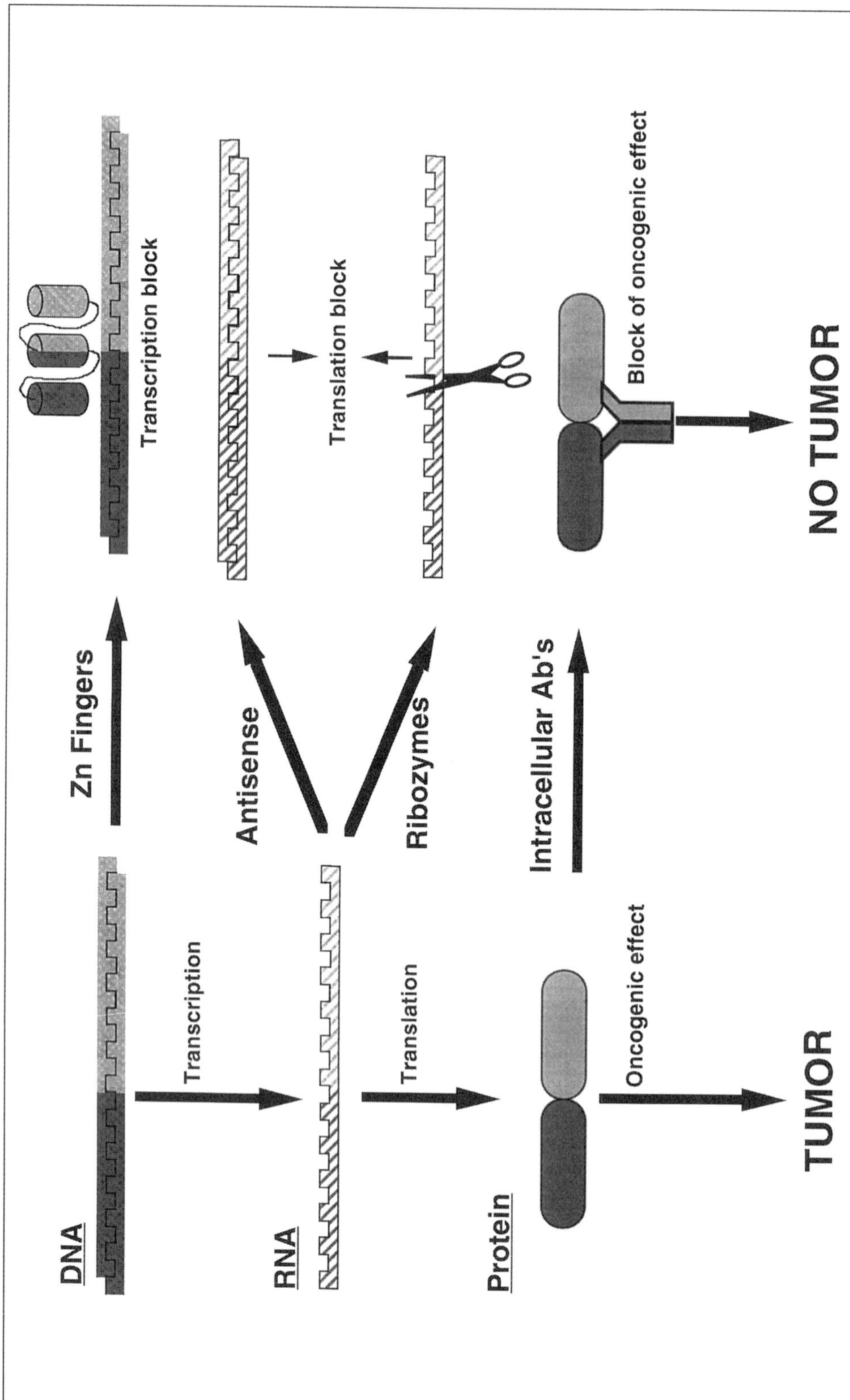

Figure 3. Antitumor action of gene therapy agents directed against the fusion product of chromosomal translocation. The antitumor agent can act at three levels: *(i)* zinc fingers block transcription interacting at the DNA level; *(ii)* ribozymes and antisense RNAs block translation by degrading or interacting with the fusion mRNA; and *(iii)* intracellular antibodies inhibit the oncogenic activity at the fusion protein level.

leukemia cells without affecting, in all lines tested, the messengers of endogenous BCR and ABL products (36).

The strategies described above and some others (e.g., triple helix formation, decay strategies, and intracellular antibodies) are being explored and soon will surely be the basis for a gene therapy for cancer, providing the specificity, selectivity, and efficiency that is required for destruction of chromosomal translocation products and hence the death of the tumor cells.

USE OF RETROVIRAL VECTORS TO DELIVER SPECIFIC ANTITUMOR AGENTS

There have been remarkable advances in the understanding of the molecular biology of cancer; these advances provide new mechanisms for selective tumor destruction. As we have shown above, the molecular characterization of tumor-specific chromosomal translocations has revealed that fusion proteins are behind the majority of cancers. Chimeric genes generated by the chromosomal rearrangements encode these fusion proteins. Because they are unique to the disease (i.e., only exist in tumor cells), such chimeric molecules represent ideal therapeutic targets. Indeed, inhibition of chimeric gene expression by antitumor agents kills the leukemic cells without affecting the normal cells (13,61). A major challenge, however, is to deliver specific antitumor agents into every tumor cell of a patient because, if this is not achieved, untreated malignant cells will emerge as resistant clones.

Recombinant retroviruses provide an attractive vehicle for gene transfer based on their capacity for highly efficient infection and nontoxic integration into the genome of a wide range of cell types. Although the potential for retrovirus-mediated gene transfer efficiency approaches 100%, this may not be realized because of low viral titers or the failure to stimulate cells to divide, which is required for successful integration (7,83). Moreover, expression of the transferred gene(s) may not reach the desired levels or be sustained (54). Such problems are evident in current efforts to apply retroviral gene transfer to gene therapy of cancer (3,57,62,63,84).

Recently, we have tested the utility of coexpressing a specific antitumor agent and a cDNA that encodes a truncated human cell surface antigen CD5 for the immediate postinfection selection of the hematopoietic cells transduced with the retrovirus that expresses the specific antitumor product (25) (bicistronic vectors; Figure 4). Fluorescence-activated cell sorting (FACS) analysis in combination with functional studies showed that, under the conditions used, all cells selected expressed CD5 within 48 hours of termination of the infection procedure. Moreover, the fraction of CD5-positive cells did not decrease with extended propagation of selected cells, thereby indicating that the surface marker lacks deleterious effects on cell growth effect and viability. The use of a truncated CD5 cell surface antigen as a selectable marker for gene transfer offers significant advantages over those conferring resistance to toxic drugs. These include the rapid and quantitative detection of transferred gene expression in the desired target cell population by flow cytometry and the efficient and nontoxic selection of transduced target cells by FACS (57). Moreover, these results showed that expression of the transferred genes was not sustained since a significant proportion of hematopoietic cells did not express surface marker in nine-month-old recipient mice. This fact has to be taken into account for treatments that depend on the generation of drug-resistant hematopoietic cells and in strategies in which the inhibition of the leukemic phenotype relies on continuous expression of the transferred gene.

Bicistronic Vectors and Selectable Markers

Selectable marker genes that have been or could be incorporated into bicistronic vectors for selection of transduced cells after infection fall into two main categories (Figure 4): *(i)* those conferring resistance to toxic drugs, whose use implies a treatment of cells with the appropriate drug for positive selection of transduced cells, and *(ii)* those driving expression of a cell surface marker that can be used for FACS selection of immunostained transduced cells, thus avoiding drug toxicity (58). Many strategies have been devised that confer resistance to toxic compounds such as neomycin (20,22,33,37), hygromycin (89), chloramphenicol (88), mycophenolic acid (79), trimetrexate (2,74), or methotrexate (16,46,75,86). For instance, a retroviral vector containing antisense sequences against the fusion junction of b3a2 type BCR-ABLp^{210} and a murine-mutated dihidrofolate reductase (DHFR) has been shown to render transfected BCR-ABLp^{210}-leukemic hematopoietic progenitors methotrexate-resistant and nontumorigenic (90). Another example of a selectable marker conferring resistance to cell-killing drugs is the multidrug resistance gene (MDR1 or *ABCB1*), which provides resistance to colchicine, anthracyclins, vinca alkaloids, etoposides, and taxol (59,73). In addition, MDR1 is a plasma membrane protein and can be used for FACS selection of transduced cells since normal bone marrow cells have unusually low levels of MDR1 (21,85). However, the use of these markers for selection protocols carries disadvantages that include nonspecific drug toxicity and difficul-

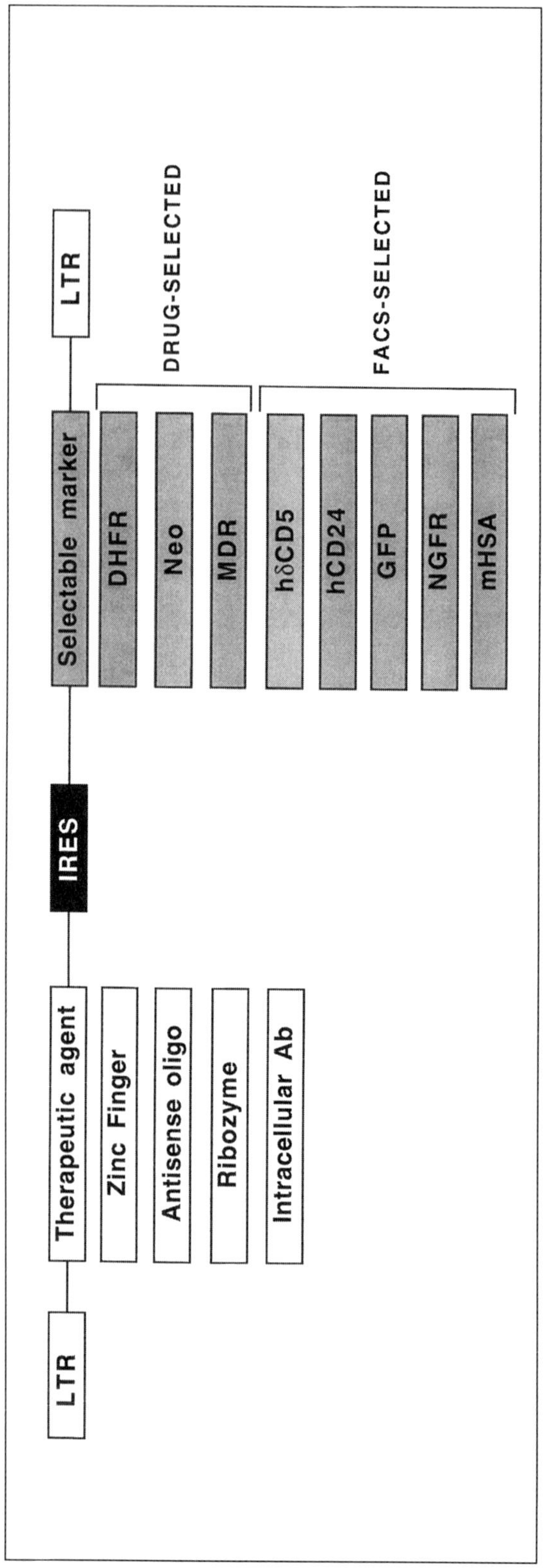

Figure 4. Scheme of bicistronic retroviral vectors encoding an antitumor-specific gene therapy agent and a marker that allows either drug or FACS selection of transduced cells. See text for details.

ties for quantitating expression. The bacterial β-galactosidase (LacZ) and the human placental alkaline phosphatase genes have also been used to select transduced cells in vitro (53,76,77), and as reporter molecules both in vitro (28,76,77,87) and in vivo (24,50,51,60). These markers are also problematic because of endogenous expression in some cell types and the requirement for transport of fluorogenic substrates across the cellular membranes.

Therefore, the easiest way to select transduced cells may be the expression of a specific cell surface marker that can be immunodetected and sorted. As mentioned above, we applied this type of strategy by using a truncated human CD5 antigen to achieve in mice a complete reconstitution of the hematopoietic system with FACS-selected, retrovirally transduced bone marrow cells (25). Other membrane markers that have been used for FACS selection of infected cells are the low-affinity human nerve growth factor receptor (NGFR) (9,42), the murine CD4 (4), the human hematopoietic cell surface antigen CD24 (55), and its murine homologue, the heat-stable antigen (HSA) (15). With the HSA marker and a bicistronic vector carrying the glucocerebrosidase gene, it has been possible to correct the enzyme deficiency in cells from Gaucher patients (45). All these markers imply the use of antibody staining, which is not feasible for some assays, such as the analysis of gene transfer efficiency into clonogenic hematopoietic colonies in methylcellulose. This is the reason for the use of the *Aequorea victoria* green fluorescent protein (GFP), whose natural fluorescence requires no substrates or cofactors and can be directly scanned by flow cytometry, thereby allowing selection of cells in vitro and in vivo (57) without the need for antibody staining or extensive cell manipulation. In this context, high levels of gene transfer and specific selection of transduced cells by sorting have been achieved using a retroviral bicistronic vector carrying GFP and a human *DHFR* gene (56).

At the end of this chapter, we include a protocol guideline on strategies designed to use transgenic mouse models carrying gene fusions like those created by chromosomal translocations.

CONCLUSIONS AND PERSPECTIVES

A recurring theme in the molecular pathogenesis of human cancer is the activation of cellular protooncogenes by acquired chromosomal aberrations, the most frequent of which are chromosomal translocations. Characterization of these chromosomal lesions has led to the identification of a growing number of candidate fusion genes implicated in the causation of human cancers. The resulting chimeric molecules represent ideal therapeutic targets since they are unique to the cancer cell. Inhibition of expression of the chimeric gene by antitumor products kills the leukemic cells while leaving intact the normal ones (13,61). A major challenge is to deliver these specific antitumor products into every tumor cell and avoid recurrence caused by intact malignant clones. The possibility of generating a functional hematopoietic system using only retrovirally transduced cells should greatly enhance the power of studies aimed at addressing questions such as the generation of cancer-negative hematopoiesis (25). Data discussed here suggest an approach to in vitro gene therapy in hematopoietic cancer. Cancer-free hematopoiesis might be restored by autografting patients with bone marrow cells infected with virus that expresses a specific antitumor product against the tumor-associated fusion junctions and a selectable marker, followed by selection to ensure that all reinfused stem cells have integrated functional retroviral sequences. The selected (cured) stem cells should reconstitute a normal hematopoietic system, and the specific antitumor agent should eradicate all tumor stem cells. However, the first problem to be solved is the efficient transduction/selection of 100% of cells. Many efforts are currently being addressed in the search of a more efficient vector for transduction of nondividing cells. Lentiviruses (such as human immunodeficiency virus 1 [HIV-1]) are able to infect nondividing cells, but vectors based on these viruses raise concerns about safety for obvious reasons. Attempts to transfer into murine or feline retroviral vectors the specific signals from HIV that allow transduction of nondividing cells have not been successful (23). Recently, it has been possible to use just 1/5 of the HIV genome (which excludes all genes causing pathology) in a retroviral vector (32), and chances of recombination have been further reduced by the use of a non-HIV envelope protein that pseudotypes the vector with vesicular stomatitis virus (VSV) G protein, which also makes the vector able to transduce a broad range of tissues and cell types (32) in a safe manner. In these lentiviral vectors, the transgene is enclosed between the long terminal repeats and a packaging sequence (52) and can be produced at a concentration of greater than 1×10^9 virus particles/mL. When these vectors were used to infect rodent brain, liver, muscles, eye, or pancreatic islet cells, they provided sustained expression that did not seem to be subjected to inactivation (6,49). In hematopoietic stem cells, some reports (80,82) showed that HIV-1-based vectors could efficiently transduce G_0/G_1 human hematopoietic stem cells, although sustained expression was not achieved. Nevertheless, more recent data have shown that human $CD34^+$ cells can be transduced with high efficiency with HIV-based vectors and that reporter (GFP) gene expression in transduced cells transplanted to NOD/SCID mice is not inactivated, at least for 22 weeks (duration of the experiment) (48).

In the near future, the study of the mechanism by which the cell recognizes and inactivates foreign promoters (methylation being the main candidate for this second function) will allow scientists to design more efficient vectors that will be able to maintain the expression of the therapeutic gene for all the life span of the cell (and its descendants), thus providing a complete remission of the disease by the regeneration of a cancer-free hematopoietic system.

PROTOCOLS

We describe here a protocol to explore the possible use of cell surface markers for tagging and selection of functional retrovirally transduced bone marrow cells. We will focus on strategies designed for studies using cancer mouse models.

Construction of Bicistronic Retroviral Vectors

Bicistronic retroviral vectors are based on a simplified retroviral vector that does not contain a dominant selectable marker gene. The precise elements subcloned into this retroviral vector are sequences corresponding to the therapeutic agent, the picornaviral internal ribosome-entry site (IRES) sequence, and the portion of the cDNA that encompasses either selectable marker of those listed in Figure 4. In the case of FACS-selected markers, the cDNA inserted should only contain the complete extracytoplasmic coding region. Usually, the specific DNA fragments to be inserted into the retroviral vector are generated by reverse transcription polymerase chain reaction (RT-PCR) with primers that included initiation and termination codons. The authenticity of this construct must be confirmed by DNA sequencing in order to reveal no PCR errors or cloning artifacts. The proviral structure of this type of vector is shown in Figure 4.

Cell Lines

The ecotropic retrovirus packaging cell line ψ-cre is used to generate helper-free recombinant retrovirus. This cell line is maintained in Dulbecco's modified Eagle medium (DMEM) supplemented with 10% calf serum. Viral packaging cell lines are maintained in the same medium supplemented with 1 mg/mL neomycin analog G418 (Life Technologies, Rockville, MD, USA). The mouse hematopoietic Ba/F3 cell line is maintained in DMEM with 10% fetal calf serum (FCS) and 5% of WEHI-3B-conditioned medium as a source of interleukin-3 (IL-3). This cell line allows researchers to test the specificity of any strategy designed against the fusion genes created by chromosomal translocations and to maintain live cells by adding IL-3 to the medium. All cells are cultured at 37°C in a humidified atmosphere of 5% CO_2 in air.

Retrovirus Production

ψ-cre cells are first transfected with either 20 μg bicistronic retroviral vector or 1 μg pMC1neo vector using the calcium phosphate precipitation method. After G418 selection (1 mg/mL), the best clones are selected on the basis of their surface marker expression. Usually, they have a titer of 0.5 to 1 (copy number) as assessed by titering in NIH 3T3 cells. To test the specificity and efficacy of any of the strategies outlined in Figure 4, Ba/F3 cells are infected by a two-day coculture with the bicistronic retroviral producer cell line in DMEM supplemented with 10% calf serum, polybrene (8 μg/mL), and IL-3. After coculture, Ba/F3 cells infected by the virus are monitored for expression of surface marker.

Transduction of Murine Bone Marrow Cells

Donor bone marrow cells are harvested from the limbs of C57 BL/6J male mice (cancer mouse models) five days after injection with 5-fluorouracil (5-FU) (150 mg/kg body weight). Bone marrow cells are pre-stimulated in the presence of growth factors and cocultivated for two days over vector-producing fibroblasts in the presence of polybrene (8 μg/mL). The growth factors used for pre-stimulation are 200 U murine IL-3/mL (Amgen, Thousand Oaks, CA, USA), 200 U human IL-6/mL (Amgen), 200 U human IL-1α/mL (Immunex, Seattle, WA, USA) and 100 ng mast cell growth factor/mL (Immunex). Transduced bone marrow cells (approximately 20%) are monitored by surface marker expression.

Bone Marrow Transplant and Sample Collection

Recipient female C57 BL/6J mice (8–12 weeks old) (cancer mouse models) are irradiated with two split doses of 600 cGy 2 hours apart. Sorted transduced bone marrow cells are injected into the tail vein of the irradiated mice at 2–4 × 10^6 cells per mouse for long-term reconstitution. Animals are sacrified, and hematopoietc tissues are collected for FACS analysis and tumorigenic assays.

Labeling of Cells

To sort surface marker-positive cells, bicistronic retroviral-infected cells are incubated on ice for 40 minutes with the FITC-directed conjugated antihuman surface marker monoclonal antibody (BD Pharmingen, San Diego, CA, USA), washed twice with PBS + 1% FCS and resuspended in PBS + 1% FCS. Propidium iodide (Sigma, St. Louis, MO, USA) is included in the last wash to distinguish dead cells before analysis by flow cytometry using a FACScan™ cell analyzer (Becton Dickinson, San Jose, CA, USA). Cells are sorted on a FACStar™ Plus (Becton Dickinson) equipped with a 5-W argon and a 30-mW helium neon laser. Cells are collected in sterile Eppendorf® vials in medium with 50% FCS. Surface marker expression among specific hematopoietic cell types is analyzed by staining hematopoietic tissues (bone marrow, spleen, and thymus) with antihuman surface marker-phycoerythrin in combination with FITC-labeled Sca-1 to identify hematopoietic stem cells, Gr-1 for granulocytes, CD11b (Mac-1) for macrophages, TER-119 for erythroid cells, CD3-e, CD4 (L3T4) and CD8 (Ly-2) for T cells, and CD45R (B220) for B cells.

ACKNOWLEDGMENTS

We apologize to all those whose work has not been directly referenced due to space constraints. We thank Dr. Angel Cid-Arregui for critical reading and helpful comments on the manuscript. Research in the authors' laboratory is supported by European Comission (BMH4-CT96-0375), DGCYT (UE96-0041, PB96-0816, and 1FD97-0360), Fundación Científica of the AECC, FIS (99/0935), and NIH Grant 1 R01 CA79955-01. C.C. is supported by Fundación Ferrer Investigación.

REFERENCES

1. **Adams, J.M., A.W. Harris, C.A. Pinkert, L.M. Corcoran, W.S. Alexander, S. Cory, R.D. Palminter, and R.L. Brinster.** 1985. The c-*myc* oncogene driven by immunoglobulin enhancers induces lymphoid malignancy in transgenic mice. Nature *318*:533-538.
2. **Allay, J.A., D.A. Persons, J. Galipeau, J.M. Riberdy, R.A. Ashmun, R.L. Blakley, and B.P. Sorrentino.** 1998. In vivo selection of retrovirally transduced hematopoietic stem cells. Nat. Med. *4*:1136-1143.
3. **Anderson, W.F.** 1998. Human gene therapy. Nature *392*:25-30.
4. **Bauer, T.R., Jr. and D.D. Hickstein.** 1997. Transduction of human hematopoietic cells and cell lines using a retroviral vector containing a modified murine CD4 reporter gene. Hum. Gene Ther. *8*:243-252.
5. **Bhatia, R. and C.M. Verfaillie.** 1998. Inhibition of BCR-ABL expression with antisense oligodeoxinucleotides restores β1 integrin-mediated adhesion and proliferation inhibition in chronic myelogenous leukemia hematopoietic progenitors. Blood *91*:3414-3422.
6. **Blomer, U., L. Naldini, T. Kafri, D. Trono, I.M. Verma, and F.H. Gage.** 1997. Highly efficient and sustained gene transfer in adult neurons with a lentivirus vector. J. Virol. *71*:6641-6649.
7. **Brenner, M.K.** 1996. Gene transfer into hematopoietic cells. N. Engl. J. Med. *335*:337-339.
8. **Castellanos, A., B. Pintado, E. Weruaga, R. Arevalo, A. Lopez, A. Orfao, and I. Sanchez-Garcia.** 1997. A BCR-ABLp190 fusion gene made by homologous recombination causes B-cell acute lymphoblastic leukemias in chimeric mice with independence of the endogenous bcr product. Blood *90*:2168-2174.
9. **Cheng, L., C. Du, C. Lavau, S. Chen, J. Tong, B.P. Chen, R. Scollay, R.G. Hawley, and B. Hill.** 1998. Sustained gene expression in retrovirally transduced, engrafting human hematopoietic stem cells and their lympho-myeloid progeny. Blood *92*:83-92.
10. **Choo, Y., A. Castellanos, B. García-Hernández, I. Sánchez-García, and A. Klug.** 1997. Promoter specific activation of gene expression directed by bacteriophage-selected zinc-fingers. J. Mol. Biol. *273*:525-532.
11. **Choo, Y., I. Sánchez-García, and A. Klug.** 1994. In vivo repression by a site-specific DNA-binding protein designed against an oncogenic sequence. Nature *372*:642-645.
12. **Cleary, M.L. and J. Sklar.** 1985. Nucleotide sequence of a t14,18) chromosomal breakpoint in follicular lymphoma and demonstration of a breakpoint-cluster region near a transcriptionally active locus on chromosome 18. Proc. Natl. Acad. Sci. USA 82:7439-7443.
13. **Cobaleda, C., J. Pérez-Losada, and I. Sánchez-García.** 1998. Chromosomal abnormalities and tumor development: from genes to therapeutic mechanism. Bioessays *20*:922-930.
14. **Condorelli, G.L., F. Facchiano, M. Valtieri, E. Proietti, L. Vitelli, V. Lulli, K. Huebner, C. Peschle, and C.M. Croce.** 1996. T-cell-directed TAL-1 expression induces T-cell malignancies in transgenic mice. Cancer Res. *56*:5113-5119.
15. **Conneally, E., P. Bardy, C.J. Eaves, T. Thomas, S. Chappel, E.J. Shpall, and R.K. Humphries.** 1996. Rapid and efficient selection of human hematopoietic cells expressing murine heat-stable antigen as indicator of retroviral-mediated gene transfer. Blood *87*:456-464.
16. **Corey, C.A., A.D. DeSilva, C.A. Holland, and D.A. Williams.** 1990. Serial transplantation of methotrexate-resistant bone marrow: protection of murine recipients from drug toxicity by progeny of trans-

duced stem cells. Blood *75*:337-343.

17. **Corral, J., I. Lavenir, H.I. Impey, A.J. Warren, A. Forster, T.A. Larson, S. Bell, A.N.J. McKenzie, G. King, and T.H. Rabbits.** 1996. An Mll-AF9 fusion gene made by homologous recombination causes acule leukemia in chimeric mice: a method to create fusion oncogenes. Cell *85*:853-861.
18. **De Klein, A., A.G. van Kessel, G. Grosveld, C.R. Bartram, and A. Hagemeijer.** 1982. A cellular oncogene is translocated to the Philadelphia chromosome in chronic myelocytic leukemia. Nature *300*:765-767.
19. **Dedera, D.A., E.K. Waller, D.P. LeBrun, A. Sen-Majumdar, M.E. Stevens, G.S. Barsh, and M.L. Cleary.** 1993. Chimeric homeobox gene *E2A-PBX1* induces proliferation, apoptosis and malignant lymphomas in transgenic mice. Cell *74*:833-843.
20. **Dick, J.E., M.C. Magli, D. Huszar, R.A. Phillips, and A. Bernstein.** 1985. Introduction of a selectable gene into primitive stem cells capable of long-term reconstitution of hematopoietic system W/Wv mice. Cell *42*:71-79.
21. **Eckert, H.-G., M. Stockschländer, U. Just, S. Hegewisch-Becker, M. Grez, A. Uhde, A. Zander, W. Ostertag, and C. Baum.** 1996. High-dose multidrug resistance in primary human hematopoietic progenitor cells transduced with optimized retroviral vectors. Blood *88*:3407-3415.
22. **Eglitis, M.A., P. Kantoff, E. Gliboa, and W.F. Anderson.** 1985. Gene expression in mice after high efficiency retroviral mediated gene transfer. Science *230*:1395-1398.
23. **Fearon, E.R.** 1997. Human cancer syndromes: clues to the origin and nature of cancer. Science *278*:1043-1050.
24. **Ferry, N., O. Duplessio, D. Houssin, O. Danos, and J.-M. Heard.** 1991. Retroviral mediated gene transfer into hepatocytes in vivo. Proc. Natl. Acad. Sci. USA *88*:8377-8381.
25. **García-Hernández, B., A. Castellanos, A. López, A. Orfao, and I. Sánchez-García.** 1997. Murine hematopoietic reconstitution after tagging and selection of retrovirally transduced bone marrow cells. Proc. Natl. Acad. Sci. USA *94*:13239-13244.
26. **Grisolano, J.L., R.L. Wesselschmidt, P.G. Pelicci, and T.J. Ley.** 1997. Altered myeloid development and acute leukemia in transgenic mice expressing PML-RARα under control of cathepsin G regulatory sequences. Blood *89*:376-387.
27. **Hatano, M., C.W.M. Roberts, M. Minden, W.M. Crist, and S.J. Korsmeyer.** 1991. Deregulation of a homeobox, HOX11, by the t(10,14) in T cell leukemia. Science *253*:79-82.
28. **Henthorn, P., P. Zervos, M. Raducha, H. Harris, and T. Kadesch.** 1988. Expression of a human placental alkaline phosphatase gene in transfected cells: use as a reporter for studies of gene expression. Proc. Natl. Acad. Sci. USA *85*:6342-6346.
29. **Herschlag, D.** 1991. Implications of ribozyme kinetics for targeting the cleavage of specific RNA molecules in vivo: more isn't always better. Proc. Natl. Acad. Sci. USA *88*:6291-6295.
30. **Hough, MR., M.D. Reis, R. Singaraja, D.M. Bryce, S. Kamel-Reid, I. Dardick, M.L. Breitman, and I.D. Dube.** 1998. A model for spontaneous B-lineage lymphomas in IgHα-Hox11 transgenic mice. Proc. Natl. Acad. Sci. USA *95*:13853-13858.
31. **James, H.A. and I. Gibson.** 1998. The therapeutic potential of ribozymes. Blood *91*:371-382.
32. **Kafri, T., U. Blömer, D.A. Peterson, F.H. Gage, and I.M. Verma.** 1997. Sustained expression of genes delivered direcly into liver and muscle by lentiviral vectors. Nat. Genet. *17*:314-317.
33. **Keller, G., C. Paige, E. Gliboa, and E.F. Wagner.** 1985. Expression of a foreign gene in myeloid and lymphoid cells derived from multipotent hematopoietic precursors. Nature *318*:149-154.
34. **Kelliher, M.A., D.C. Seldin, and P. Leder.** 1996. Tal-1 induces T cell acute lymphoblastic leukemia accelerated by casein kinase IIα. EMBO J. *15*:5160-5166.
35. **Kennedy, M.A., R. González-Sarmiento, U.R. Kees, F. Lampert, N. Dear, T. Boehm, and T.J. Rabbitts.** 1991. HOX11, a homeobox-containing T-cell ongogen on human chromosome 10q24. Proc. Natl. Acad. Sci. USA *88*:8900-8904.
36. **Kuwabara, T., M. Warashina, T. Tanabe, K. Tani, S. Asano, and K. Taira.** 1998. A novel allosteric trans-activated ribozyme, the maxizyme, with exceptional specificity in vitro and in vivo. Mol. Cell *2*:617-627.
37. **Laneuville, P., W. Chang, S. Kamel-Reid, A.A. Fauser, and J.E. Dick.** 1988. High-efficiency gene transfer and expression in normal human hematopoietic cells with retrovirus vectors. Blood *71*:811-814.
38. **Langdon, W.Y., A.W. Harris, S. Cory, and J.M. Adams.** 1986. The c-*myc* oncogene perturbs B lymphocyte development in Em-*myc* transgenic mice. Cell *47*:11-18.
39. **Lange, W., E.M. Cantin, J. Finke, and G. Dolken.** 1993. In vitro and in vivo effects of synthetic ribozymes targeted against BCR/ABL mRNA. Leukemia *7*:1786-1794.
40. **Leopold, L.H., S.K. Shore, T.A. Newkirk R.M.V. Reddy, and E.P. Reddy.** 1995. Multi-unit ribozyme-mediated cleavage of *bcr-abl* mRNA in a myeloid leukemias. Blood *85*:2162-2170.
41. **Lu, M., Z. Gong, W. She, and A.D. Ho.** 1991. The tcl-3 proto-oncogene altered by chromosomal translocation in T-cell leukemia codes for a homeobox protein. EMBO J. *10*:2905-2910.
42. **Mavilio, F., G. Ferrari, S. Rossini, N. Nobili, C. Bonini, G. Casorati, C. Traversari, and C. Bordignon.** 1994. Peripheral blood lymphocytes as target cells of retroviral vector-mediated gene transfer. Blood *83*:1988-1997.
43. **McDonnell, T.J., N. Deane, F.M. Platt, G. Nunez, U. Jaeger, J.P. McKearn, and S.J. Korsmeyer.** 1989. *bcl*-2-Immunoglobulin transgenic mice demonstrate extended B cell survival and follicular lymphoproliferation. Cell *57*:79-88.
44. **McDonnell, T.J., G. Nunez, F.M. Platt, D. Hockenberry, L. London, J.P. McKearn, and S.J. Korsmeyer.** 1990. Deregulated Bcl-2-Immunoglobulin transgene expands a resting but responsive immunoglobulin M and D-expressing B-cell population. Mol. Cell. Biol. *10*:1901-1907.
45. **Medin, J.A., M. Migita, R. Pawliuk, S. Jacobson, M. Amiri, S. Kluepfel-Stahl, R.O. Brady, R.K. Humph-**

ries, and S. Karlsson. 1996. A bicistronic therapeutic retroviral vector enhables sorting of transduced CD34+ cells and corrects the enzyme deficiency in cells from Gaucher patients. Blood *87*:1754-1762.

46. **Miller, A.D., M.F. Law, and I.M. Verma.** 1985. Generation of helper-free amphotropic retroviruses that transduce a dominant-acting methotrexate-resistant dihydrofolate reductase gene. Mol. Cell. Biol. *5*:431-437.

47. **Mitelman, F., F. Mertens, and B.A. Johansson.** 1997. A breakpoint map of recurrent chromosomal rearrangements in human neoplasia. Nat. Genet. *15*:417-474.

48. **Miyoshi, H., K.A. Smith, D.E. Mosier, I.M. Verma, and B.E. Torbett.** 1999. Transduction of human CD34+ cells that mediate long-term engrafment of NOD/SCID mice by HIV vectors. Science *283*:682-686.

49. **Miyoshi, H., M. Takahashi, F.H. Gage, and I.M. Verma.** 1997. Stable and efficient gene transfer into the retina using an HIV-based lentiviral vector. Proc. Natl. Acad. Sci. USA *94*:10319-10323.

50. **Nabel, E., G. Plautz, F.M. Boyce, C. Stanley, and G.J. Nabel.** 1989. Recombinant gene expression in vivo within endothelial cells of the arterial wall. Science *244*:1342-1344.

51. **Nabel, E., G. Plautz, and G.J. Nabel.** 1990. Site-specific gene expression in vivo by direct gene transfer into the arterial wall. Science *249*:1285-1288.

52. **Naldini, L., U. Blömer, P. Gallay, D. Ory, R. Mulligan, F.H. Gage, I.M. Verma, and D. Trono.** 1996. In vivo gene delivery and stable transduction of nondividing cells by a lentiviral vector. Science *272*:263-267.

53. **Nolan, G.P., S. Fiering, J.-F. Nicolas, and L.A. Herzenberg.** 1988. Fluorescence-activated cell analysis and sorting of viable mammalian cells based on B-D-galactosidase activity after transduction of *Escherichia coli* lacZ. Proc. Natl. Acad. Sci. USA *85*:2603-2607.

54. **Palmer, T.D., R.A. Hock. W.R.A. Osborne, and A.D. Miller.** 1987. Efficient retrovirus mediated transfer and expression of a human adenosine deaminase gene in diploid skin fibroblasts from an adenosine deaminase deficient human. Proc. Natl. Acad. Sci. USA *84*:1055-1059.

55. **Pawliuk, R., R. Kay, P. Lansdorp, and R.K. Humphries.** 1994. Selection of retrovirally transduced hematopoietic cells using CD24 as a marker of gene transfer. Blood *84*:2868-2877.

56. **Persons, D., J.A. Allay, E.A. Allay, R.J. Smeyne, R.A. Ashmun, B.P. Sorrentino, and A.W. Nienhuis.** 1997. Retroviral-mediated transfer of the green fluorescent protein gene into murine hematopoietic cells facilitates scoring and selection of transduced progenitors in vitro and identification of genetically modified cells in vivo. Blood *90*:1777-1786.

57. **Persons, D., J.A. Allay, J.M. Riberdy, R.P. Wersto, R.E. Donahue, B.P. Sorrentino, and A.W. Nienhuis.** 1998. Use of the green fluorescent protein as a marker to identify and track genetically modified hematopoietic cells. Nat. Med. *4*:1201-1205.

58. **Phillips, K., T. Gentry, G. McCowage, E. Gilboa, and C. Smith.** 1996. Cell-surface markers for assesing gene transfer into human hematopoietic cells. Nat. Med. *2*:1154-1156.

59. **Podda, S., M. Ward, A. Himelstein, C. Richardson, E. De La Flor-Weiss, L. Smith, M. Gottesman, I. Pastan, and A. Bank.** 1992. Transfer and expression of the human multiple drug resistance gene into live mice. Proc. Natl. Acad. Sci. USA *89*:9676-9680.

60. **Price, J., D. Turner, and C. Cepko.** 1987. Lineage analysis in vertebrate nervous system by retrovirus-mediated gene transfer. Proc. Natl. Acad. Sci. USA *84*:156.

61. **Rabbitts, T.H.** 1994. Chromosomal translocations in human cancer. Nature *372*:143-149.

62. **Reynolds, P.N., M. Feng, and D.T. Curiel.** 1999. Chimeric viral vectors—the best of both worlds? Mol. Med. Today (January 1999) *5*:25-31.

63. **Robbins, P.D., H. Tahara, and S.S. Ghivizzani.** 1998. Viral vectors for gene therapy. Trends Biotechnol. *16*:35-40.

64. **Sánchez-García, I.** 1997. Consequences of chromosomal abnormalities in tumor development. Ann. Rev. Genet. *31*:429-453.

65. **Sánchez-García, I., H. Axelson, and T.H. Rabbitts.** 1995. Functional diversity of LIM proteins: amino-terminal activation domains in the oncogenic proteins RBTN1 and RBTN2. Oncogene *10*:1301-1306.

66. **Sánchez-García, I. and G. Grütz.** 1995. Tumorigenic activity of the *BCR-ABL* oncogenes is mediated by *BCL2*. Proc. Natl. Acad. Sci. USA *92*:5287-5291.

67. **Sánchez-García, I. and D. Martín-Zanca.** 1997. Regulation of Bcl-2 gene expression by BCR-ABL is mediated by Ras. J. Mol. Biol. *267*:225-228.

68. **Sánchez-García, I. and T.H. Rabbitts.** 1993. LIM-domain proteins in leukemia and development. Semin. Cancer Biol. *4*:349-358.

69. **Schmidt, E.V., P.K. Pattengale, L. Leir, and P. Leder.** 1988. Transgenic mice bearing the human c-*myc* gene acticated by an immunoglobulin enhancer: a pre-B cell lymphoma model. Proc. Natl. Acad. Sci. USA *85*:6047-6051.

70. **Shore, S.K., P.M. Nabissa, and E.P. Reddy.** 1993. Ribozyme-mediated cleavage of BCR-ABL oncogene transcript: in vitro cleavage of RNA and in vivo loss of p210 protein-kinase activity. Oncogene *8*:3183-3188.

71. **Skorski, T., M. Nieborowska-Skorska, N.C. Nicolaides, C. Szczylik, P. Iversen, R.V. Iozzo, G. Zon, and B. Calabretta.** 1994. Supression of Philadelphia[1] leukemia cell growth in mice by BCR-ABL antisense oligodeoxinucleotide. Proc. Natl. Acad. Sci. USA *91*:4504-4508.

72. **Snyder, D.S., Y. Wu, J.L. Wang, J.J. Rossi, P. Swiderski, B.E. Kaplan, and S.J. Forman.** 1993. Ribozyme-metiated inhibition of *bcr-abl* gene expression in a Philadelphia chromosome positive cell line. Blood *82*:600-605.

73. **Sorrentino, B.P., S.J. Brandt, D. Bodine, M. Gottesman, I. Pastan, A. Cline, and A.W. Nienhuis.** 1992. Selection of drug-resistant bone marrow cells in vivo after retroviral transfer of human *MDR1*. Science *257*:99-103.

74. **Spencer, H.T., S.E.H. Sleep, J.E. Rehg, R.L. Blakley, and B.P. Sorrentino.** 1996. A gene transfer strategy for making bone marrow cells resistant to trimetrex-

ate. Blood *87*:2579-2587.

75. **Stead, R.B, W.W. Kwok, R. Storb, and A.D. Miller.** 1988. Canine model for gene therapy: inefficient gene expression in dogs reconstituted with autologous marrow infected retroviral vectors. Blood *71*:742.

76. **Strair, R.K., M. Towle, P.W. Heald, and B.R. Smith.** 1990. Retroviral mediated gene transfer and expression of exogenous genes in primary lymphoid cells: assaying for a viral transactivatior activity in normal and malignant cells. Blood *76*:1201-1208.

77. **Strair, R.K., M. Towle, and B.R. Smith.** 1990. Retroviral mediated gene transfer into bone marrow progenitor cells: use of beta-galactosidase as a selectable marker. Nucleic Acids Res. *18*:4759-4762.

78. **Strasser, A., A.W. Harris, L.M. Corcoran, and S. Cory.** 1994. Bcl-2 expression promotes B- but not T-lymphoid development in *scid* mice. Nature *368*:457-460.

79. **Stuhlmann, H., R. Cone, R.C. Mulligan, and R. Jaenisch.** 1984. Introduction of a selectable gene into different animal tissue by a retrovirus recombinant vector. Proc. Natl. Acad. Sci. USA *81*:7151-7155.

80. **Sutton, R.E., H.T.M. Wu, R. Rigg, E. Böhnlein, and P.O. Brown.** 1998. Human immunodeficiency virus type 1 vectors efficiently transduce human hematopoietic stem cells. J. Virol. *72*:5781-5788.

81. **Szczylik, C., T. Skorski, N.C. Nicolaides, L. Manzella, L. Malaguarnera, D. Venturelli, A.M. Gewirtz, and B. Calabretta.** 1991. Selective inhibition of leukemia cell proliferation by BCR-ABL antisense oligodeoxinucleotides. Science *253*:562-565.

82. **Uchida, N., R.E. Sutton, A. Friera, D. He, M.J. Reitsma, W.C. Chang, G. Veres, R. Scollay, and I.L. Weissman.** 1998. HIV, but not murine leukemia virus, vectors mediate high efficiency gene transfer into freshly isolated G_0/G_1 human hematopoietic stem cells. Proc. Natl. Acad. Sci. USA *95*:11939-11944.

83. **Van Beusechem, V.W. and D. Valerio.** 1996. Gene transfer into hematopoietic stem cells of nonhuman primates. Hum. Gene. Ther. *7*:1649-1668.

84. **Verma, I.M. and N. Somia.** 1997. Gene therapy—promises, problems and prospects. Nature *389*:239-242.

85. **Ward, M., C. Richardson, P. Pioli, L. Smith, S. Podda, S. Goff, C. Hesdorffer, and A. Bank.** 1994. Transfer and expression of the human multiple drug resistance gene in human $CD34^+$ cells. Blood *84*:1408-1414.

86. **Williams, D.A., S.H. Orkin, and R.C. Mulligan.** 1986. Retrovirus-mediated transfer of adenosine deaminase gene sequences into cells in culture and into murine hematopoietic cells in vivo. Proc. Natl. Acad. Sci. USA *83*:2566-2570.

87. **Wilson, J.M., D.M. Jefferson, J.R. Chowdhury, P.M. Novikoff, D.E. Johnston, and R.C. Mulligan.** 1988. Retrovirus-mediated transduction of adult hepatocytes. Proc. Natl. Acad. Sci. USA *85*:3014-3018.

88. **Wood, P.A., G.E. Herman, C.Y. Chao, W.E. O'Brien, and A.L. Beaudet.** 1986. Retrovirus-mediated gene transfer of a arginosuccinate synthetase into culture rodent cells and human citrulinemic fibroblasts. Cold Spring Harb. Symp. Quant. Biol. *51*:1027-1032.

89. **Yang, Z., A.J. Korman, J. Cooper, D. Pious, R.S. Acolla, R.C. Mulligan, and J.L. Strominger.** 1987. Expression of HLA-DR antigen in human class II mutant B-cell lines by double infection with retrovirus vectors. Mol. Cell. Biol. *7*:3923-3928.

90. **Zhao, R.C.H., R.S. McIvor, J.D. Griffin, and C.M. Verfaillie.** 1997. Gene therapy for chronic myelogenous leukemia (CML): a retroviral vector that renders hematopoietic progenitors methotrexate-resistant and CML progenitors functionally normal and nontumorigenic in vivo. Blood *90*:4687-4698.

FAMILY: Retroviridae
GENUS: Lentivirinae (HIV)

29 Cellular Transduction by Lentiviral Vectors

Richard E. Sutton
Department of Molecular Virology and Microbiology, Department of Medicine, and the Center for Cell and Gene Therapy, Baylor College of Medicine, Houston, TX, USA

INTRODUCTION

Human immunodeficiency virus type 1 (HIV-1) is the etiologic agent of acquired immunodeficiency syndrome (AIDS). HIV, by interacting with both the cellular transmembrane protein CD4 and any of a number of serpentine chemokine receptors (e.g., CXCR4, Bob, Bonzo, or CXCR5), infects $CD4^+$ helper T cells and macrophages (65). Following reverse transcription, viral cDNA randomly integrates into the host genomic DNA, much akin to all other retroviruses. After integration, long terminal repeat (LTR)-driven transcription occurs, such that viral proteins are expressed and progeny virions are assembled, budded, and released to infect other receptor-bearing cells.

HIV is a lentivirus, so named because of its long latency period, morphology, tropism for macrophages, antigenic variability, and the presence of additional regulatory and accessory genes not present in other retroviruses. In addition, the lentiviruses have the common pathological features of long (but variable) clinically silent periods, persistent viral replication, central nervous system involvement, and destruction of a specific type of peripheral blood mononuclear cell. In the absence of therapy against HIV, there is an inexorable decline of $CD4^+$ helper T cells, but whether that is due to the direct effects of viral replication, the immune (cytolytic) response directed towards the infected lymphocyte, or some type of indirect immunomodulatory reaction is still subject to investigation and debate.

The complex genome of HIV (Figure 1) encodes the three structural/enzymatic genes *gag*, *pol*, and *env*; the two regulatory genes *tat* and *rev*; and the four accessory proteins *vif*, *vpu*, *vpr*, and *nef* (18). The Gag polyprotein is cleaved by viral protease to form matrix, capsid, and nucleocapsid; the Pol precursor is cleaved to form reverse transcriptase, protease, and integrase; and the glycoprotein envelope gp160 is cleaved (likely by furin, a host cell protease) to gp120 and gp41. These structured and enzymatic proteins have the standard functions similar to the oncoretroviruses. The other encoded proteins are unique to lentiviruses. Tat, in conjunction with the

Viral Vectors: Basic Science and Gene Therapy
Edited by A. Cid-Arregui and A. García-Carrancá

cellular proteins cyclin T1 and CDK9, hyperphosphorylates the carboxyterminal domain of RNA polymerase II, thereby increasing its processivity (25,68–70). Typically, the presence of Tat transactivates the LTR 100- to 500-fold. Rev, which shares an intron with Tat, multimerizes along the Rev-response-element (RRE), a structured RNA element within the same intron (12). Cellular factors also bind to Rev, and these complexes are exported from the nucleus. This nuclear transport pathway allows unspliced or singly spliced messenger RNAs to be translated in the cytoplasm of infected cells. Thus, the structural/enzymatic mRNAs can be translated, along with some of the accessory proteins (e.g., Vif, Vpr, and Vpu).

Nef is produced from a fully spliced mRNA. Its known functions are to downregulate CD4 and class I major histocompatibility complexes (MHCs) (12). It can augment viral infectivity and bind Src family members. The true function of Vif is unknown, although it appears to counteract an HIV inhibitory factor naturally found in primary T cells and some T-cell lines (40,60). Vpu acts as an adapter protein to fully degrade endocytosed CD4 (41), and Vpr causes cell-cycle arrest in G2/M (27,31,54). The latter may allow increased expression of viral proteins and prevent the infected cell from undergoing apoptosis (24,35).

Lentiviruses share another important property, which is the ability to infect mitotically inactive cells. Oncoretroviruses, such as Rous sarcoma virus and murine leukemia virus (MLV), require cell division and nuclear membrane disassembly to gain access to the chromatin of the host cell (39,57). Thus, actively cycling cells are the

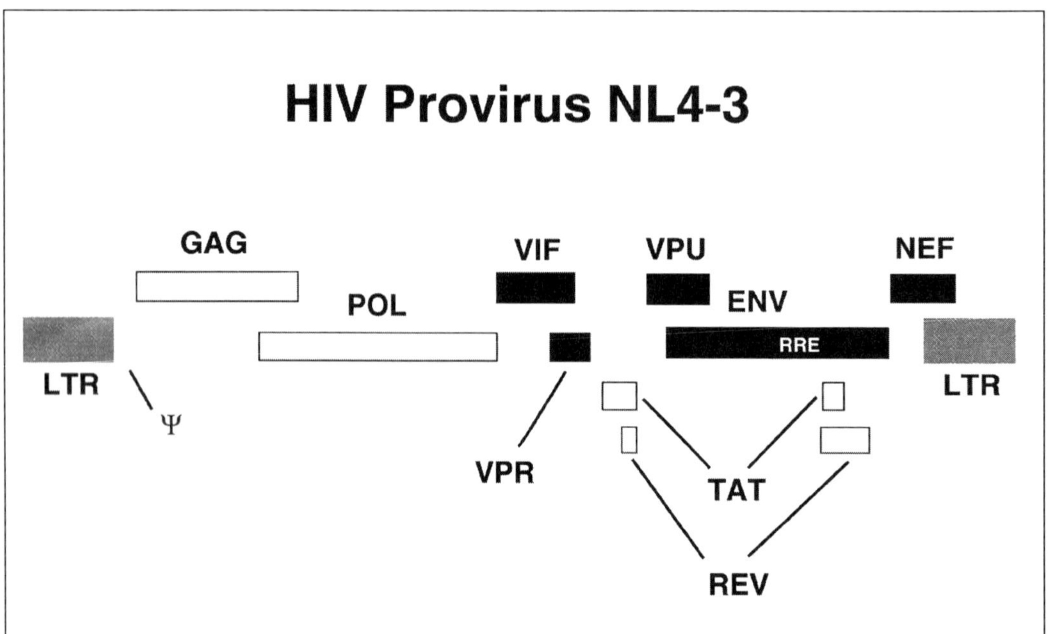

Figure 1. Structure of HIV provirus NL4-3, the T-tropic HIV isolate on which most HIV vectors are based. LTRs are indicated in light grey, essential genes that must be provided in the producer line are indicated by white boxes, and nonessential genes are indicated by black boxes. The nonessential genes (with the exception of the RRE) can be deleted from the vector with little or no loss of titer. Although not indicated, the regulatory genes *tat* and *rev* can be provided on separate plasmids for added safety. In addition, the 5′ U3 can be replaced with an alternative promoter, and an internal portion of the 3′ U3 can be deleted with no change in viral titer. Ψ=packaging signal, RRE=Rev response element. Viral gene products are not precisely to scale.

targets for these retroviruses. In contrast, HIV and the other lentiviruses can infect cycling and noncycling (nondividing) cells. This means that HIV can infect quiescent helper T cells, central nervous system microglia, and tissue macrophages.

The viral proteins that allow infection of mitotically resting cells are still subject to some debate. Initially and fortuitously, matrix was identified as being crucial in the ability to infect resting cells in the setting of a virus lacking Vpr (8). Matrix was demonstrated to have a canonical and functional nuclear localization signal (NLS) consisting of a series of basic amino acid residues (67). Subsequently, a small fraction of matrix (1%) in the infected cell was shown to be phosphorylated on the ultimate tyrosine residue, and this phosphorylation was critical to the nuclear import function of matrix (21,22). Others, however, have not observed this phosphorylation pattern (7,19), and one group has been unable to demonstrate any ability of matrix to colocalize to the nucleus (17).

The second viral protein implicated in the nuclear import of the viral core is Vpr (28). The ability of Vpr to do so is separate from its ability to cause cell-cycle arrest. Although Vpr has no canonical NLS, it has been observed to colocalize to the inner nuclear envelope (66). It is unknown how Vpr gets imported into the nucleus, and, in many cases, the presence of Vpr is unnecessary in terms of HIV's ability to infect nondividing cells (see below).

The third viral protein that may have a role in nuclear import of the viral core is integrase, the enzyme responsible for integrating the viral cDNA into the host genome (20). Integrase appears to have a bipartite, canonical NLS, and whether this is separable from its ability to bind DNA (to carry out its enzymatic activity) is unknown. Why HIV would potentially have three proteins involved in the nuclear import of the viral core in resting cells is a matter of speculation. It is clear that HIV replicates poorly in truly quiescent cells, and HIV replication is markedly increased once helper T cells are activated by ligation of the T-cell receptor (37,72,73). However, the cells that HIV initially infects are likely to be terminally differentiated tissue macrophages found in and near mucosal (rectal and vaginal) epithelial surfaces. Thus, successful replication in these cells mandates an ability of the viral nucleoprotein complex to traverse an intact nuclear pore.

INITIAL HIV VECTORS

The initial HIV vectors were designed to quickly assay the effects of different HIV envelope mutations on the association of gp41 and gp120 subunits, CD4 binding, and syncytium formation (membrane fusion). These vectors were constructed before it was clear that HIV could infect nondividing cells. In the initial report, a small *Bgl*II fragment (located within *env*) was removed, and a chloramphenicol acetyltransferase (CAT) reporter gene was inserted in the *nef* coding sequence of the T-tropic HXBIII provirus to render it replication defective (29). This was co-transfected with a gp160 expression construct in either the monkey cell line COS-1 or human T-cell line Jurkat. The co-transfection procedure allowed for the production of a filterable infectious viral particle that had a defective HIV core pseudotyped with an HIV envelope. These particles could only undergo a single cycle of replication since the target cells would only have the defective viral core, and not the envelope, integrated into their genomes. Measurement of CAT activity of the target cells quantitated the level of infection. The investigators showed that infection was sensitive to the reverse transcriptase inhibitor azidothymidine and the anti-CD4 antibody OKT4a. The effects of various mutations in gp120 on viral infectivity were also demonstrated.

CAT activity, although quantitative, unfortunately provides no information with regard to viral titer. In addition, Helseth et al. provided no evidence that the defective viral particles had similar physicochemical properties to wild-type HIV (29). Page et al. constructed an HIV vector in which a larger deletion in envelope was replaced by a simian virus 40 (SV40)-gpt (*Escherichia coli* hypoxanthine-guanine phosphoribosyl-transferase) cassette (48). They were able to show that the ultrastructural features of these pseudotyped particles produced in COS cells were similar to that of replication-competent HIV. Bald (i.e., not containing envelope) and pseudotyped particles had identical densities compared to wild-type HIV, as measured on a sucrose gradient. Because they had inserted the gpt marker, they were able to titer the cell-free HIV(HXB2) virus, which varied between 825 to 3000 infectious units (iu)/mL on HeLa T4 cells. They also pseudotyped the HIV core with the heterologous Moloney MLV envelope 4070A (amphotropic, A-MoMLV) on both human and mouse cells, a process that was reasonably efficient. They went on to pseudotype HIV cores with the human T-cell leukemia virus type I (HTLV-I) envelope both in a cell-free and cocultivation assay, and obtained titers close to 1000 on human cells (38).

Poznansky et al. further refined their HIV vector by constructing a transfer vector that only had both viral LTRs and neomycin phosphotransferase driven by an internal promoter (53). This vector was *trans*-complemented by co-transfection with either a full-length provirus with a small deletion in the packaging signal or with both an HIV envelope expression construct and a provirus that had a deletion in the packaging signal and envelope. In either case, the titer on Jurkat targets was reported to be between 1×10^4 and 1×10^5 iu/mL, as determined by serially diluting virus stocks, although in some cases replication-competent virus was present. This work demonstrated that a relatively simple HIV vector could be fully *trans*-complemented by a defective provirus.

Delwart et al. constructed a defective, *env*-deleted vector encoding hygromycin B phosphotransferase (hyg^r); the highest cell-free titers they were able to obtain with gp120 pseudotypes were approximately 500 iu/mL (13). They, along with Page et al., were able to confirm the essential role of the V3 loop of gp160 for viral infectivity using this pseudotype system (13,49). Buchschacher and Panganiban were also able to construct a Tat-inducible HIV vector that had both LTRs, the primer binding site, the major splice donor, and a portion of Gag with either hyg^r or a portion of the hematagglutinin protein of influenza A virus (6). They in fact suggested that such a vector could be used for HIV gene therapy in a transduced target cell population, provided that a suitable anti-HIV gene could be expressed that would induce cell death, interfere with the normal viral life cycle, or provoke a cytotoxic T-cell response after being infected with wild-type HIV.

GENERATION OF HIGH-TITER HIV VECTORS

A major breakthrough occurred with the demonstration that high-titer, pseudotyped HIV particles could be generated using vesicular stomatitis virus (VSV) G glycoprotein as an envelope, and that these particles could infect nondividing cells. Naldini et al. (46) used a vector system very similar to Poznansky et al. with the following modifications: *(i)* they deleted a portion of *vpu* and *env* from the packaging construct, which was driven by a cytomegalovirus (CMV) promoter and terminated by a cellular polyadenylation signal; *(ii)* they employed either MLV 4070A or VSV G as a heterologous envelope; *(iii)* their transfer vector had the reporter (luciferase or LacZ) driven by an internal promoter; and *(iv)* they used the very efficient expression system of 293T cells to transiently produce virus.

With this three-component system, titers of 1×10^5 iu/mL were routinely obtained, as assayed on adherent human cells, and VSV G-pseudotyped viral supernatants could be concentrated up to 1000-fold (but more typically 100-fold) by ultracentrifugation without more than 20% loss of iu. G0-, G1-, and G2-arrested human epithelial cells (HeLa) and rat fibroblasts (208F) were shown to be 10- to 100-fold more transducable with HIV compared with MLV-based vectors. The HIV vectors were shown to transduce adult rat neurons, oligodendrocytes, and astrocytes in vivo after stereotactic injection with concentrated viral stocks, and expression was persistent for at least 30 days. Human monocyte-derived macrophages were also transduced using the luciferase reporter HIV vector, and transduction was markedly inhibited by using a vector with a truncated *vpr* and a nonfunctional matrix NLS. Adequate transduction was also dependent on a functional integrase. No helper virus was detected by two different assays (activation of an HIV LTR-LacZ in HeLa targets or detection of capsid in targets after serial passage). This report was the first demonstration that HIV vectors could have true clinical utility for gene transfer into nondividing cells.

At approximately the same time, Akkina et al. reported high-efficiency gene transfer into human $CD34^+$ hematopoietic stem cells using HIV (VSV G) particles (2). The HIV vector they employed had a small deletion in *env*, and the luciferase reporter was placed within *nef*. True transduction efficiency was not measured since the primary readout was relative light units, but semiquantitative polymerase chain reaction (PCR) suggested on average there were 2.8 proviral copies per cell. However, because the starting cell population was heterogeneous, it remained possible that certain cells were transduced very efficiently and others not at all (see below). Transduced cells were placed into methylcellulose, and PCR, performed on granulocyte-macrophage colonies, detected vector sequences in all seven of them. This result may be confounded by contaminating DNA from the original transient transfection of COS monkey cells, which is extremely hard to eliminate even with extensive DNase treatment. Even so, this report suggests that these hematopoietic progenitors, which represent the ultimate target in somatic cell gene therapy, are transducable with HIV-based vectors.

Since then, many reports have shown that a myriad of different types of nondividing (or slowly dividing) cells (ranging from cardiac myocytes to pancreatic islet cells) can be efficiently transduced with lentiviral vectors (Table 1), suggesting that if they are ever approved for human use, these vectors would have their own niche for therapeutic gene transfer into nondividing cells.

FURTHER IMPROVEMENTS IN LENTIVIRAL VECTOR DESIGN

Since those initial reports, a number of improvements have been made in the first-generation lentiviral vectors. Many investigators have deleted additional genes in the packaging vector to demonstrate that the accessory gene products (e.g., Nef, Vif, Vpr, and Vpu) are unnecessary for production of high-titer virus and transduction of growth-arrested cells and macrophages in vitro and neurons in vivo (36,44,64,75). In one system, the packaging vector had *gag-pol* driven by a CMV immediate early (IE) promoter along with the RRE, and the

Table 1. Transduction of Primary Cells by HIV Vectors

Primary Cell Type	Reporter Gene	Transduction Efficiency (%)	Persistence of Expression	Reference
Rat retina	GFP	80	12 weeks	43
adult rat neurons	GFP/LacZ	unknown	6 months	5
human CD34+	Heat stable antigen (CD24)	83	not tested	56
human CD34+Lin-Thy+	Alkaline phosphatase	50-60	4-6 weeks	64
human endothelial cells	GFP/LacZ/Alkaline phosphatase	100	not tested	R.E.S., unpublished
human airway epithelium	LacZ	60	not tested	26
human cardiac myocytes	HIV antigens	50	not tested	55
human pancreatic islet cells	GFP/LacZ	62	5 days	32
mouse pancreatic islet cells	LacZ	5	30 days	23
rat hepatocytes	GFP	90	22 weeks	33
rat skeletal myocytes	GFP	49	8 weeks	33
human vascular smooth muscle cells	GFP/LacZ	100	2 weeks	R.E.S., unpublished
human corneal cells	LacZ	90–100	5 days	R.E.S., unpublished

minimal transfer vector had intact LTRs, Ψ, RRE, *rev*, and the CMV IE-driving LacZ (36). Titers of VSV G pseudotypes were 3 $\times$ 10^5 iu/mL. Substitution of the constitutive transport element (CTE) for the RRE reduced titers 99%.

A third-generation, split genome packaging system was developed with some additional safety features against the production of replication-competent recombinant virus (15). The 5′ LTRs of both the packaging and transfer vector were made chimeric such that transcription was independent of Tat. Titers remained within a factor of 5 (lowest being 4.6×10^6 iu/mL), and the vectors still adequately transduced human T lymphocytes (14%). These Tat-independent vectors efficiently transduced adult rat neurons in vivo. Rev was also removed from the packaging vector. When it was supplied *in trans* in the 293T producer cells, end point titers were still as high as 1×10^7 iu/mL. The elimination of Rev from the packaging vector provides an additional safety measure (not present in MLV) since nuclear export of unspliced, proviral HIV RNA is absolutely Rev- and RRE-dependent.

Two groups have developed self-inactivating vectors (42,74). These vectors have deletions in the 3′ U3 such that during the process of transduction the deletion is duplicated to the 5′ LTR to render the LTR nonfunctional. Thus, transgene transcription in the transduced targets must occur from an internal promoter. This type of vector may be safer since insertional activation of a cellular oncogene by the crippled LTR would be less likely to occur. In the case of MLV, historically, these self-inactivating vectors have been plagued by low titers (30,47,71). The HIV-based self-inactivating vectors of Miyoshi et al. had the 5′ U3 replaced by the CMV IE promoter and a deletion from -454 to -13 of the 3′ U3 (numbers are relative to the transcription start site at the U3/R border) (42). These vectors had essentially the same titer as the parent transfer vector and had the expected structure after genomic integration. Importantly, their ability to be rescued was reduced 99.99%. They maintained their capacity to transduce retinal and neuronal cells in vivo, and transcriptional interference was actually reduced, thereby resulting in higher levels of transgene expression. Zufferey et al. made comparable HIV-based self-inactivating vectors and obtained similar results (74).

HIV vectors have also demonstrated additional versatility by their ability to be pseudotyped with many different heterologous envelopes. Early on, HIV cores were pseudotyped with amphotropic 4070A MLV envelope and HTLV-1 and HTLV-2 envelopes (38,62), and most investigators now use VSV G because titers are higher, the envelope is more stable, and the receptor for VSV G, phosphotidylserine, is omnipresent (59). Unfortunately, expression of VSV G is cytotoxic, presumably because of uncontrolled intracellular membrane fusion. Mochizuki et al. demonstrated that other RNA virus envelopes such as Mokola-G and Rabies-G gave titers that were approximately 10% of those of VSV and pseudotyped particles on a variety of targets (44). The envelope of the retrovirus human foamy virus poorly pseudotyped HIV cores, but the Ebola (Zaire) pseudotype titers were approximately 10^5 iu/mL on human cells (unpublished data). These results suggest that assembly of HIV in terms of core association with envelope may be quite flexible, and it might well be worthwhile to systematically test viral envelopes from all taxa to find one that affords high titer, low cytopathicity, and potentially altered host range.

Some effort has also been directed towards producing stable HIV packaging cell lines. Initial investigations focused on using the HIV envelope itself, gp160. Srinivasakumar et al. produced an African green monkey kidney cell line that expressed SV40 large T antigen, HIV *gag-pol*, *env*, and *rev* under the control of the CMV IE promoter (61). The transfer vector, which was transiently transfected, had a deletion extending from *gag* to *env*, with a hygromycin resistance marker. Titers were strictly dependent on the presence of CD4 on the surface of the targets and were as high as 9×10^3 iu/mL. To make the transfer vector Rev independent, the CTE was inserted, and titers were roughly 75% of the controls and not dependent on Rev. Nef expression also increased titer roughly 10- to 15-fold, an effect that is specific to the HIV envelope (1).

Corbeau et al. developed a similar packaging system based in HeLa cells (10). The stably integrated packaging vector was essentially an intact HIV provirus with a deletion in Ψ (between the major splice donor and the start site of Gag). The transfer vector in this case was also stably integrated and consisted of intact LTRs, a portion of Gag, the RRE, and the marker xanthine-guanine phosphoribosyltransferase (*E. coli* gpt). Titers on HeLa T4 cells were as high as 1.4×10^5 iu/mL, and no replication-competent virus was detected. In a further modification of this stable packaging system, the HIV-1 transfer vector was replaced with an HIV-2 vector containing a CTE to confer Rev independence (11). Titers were as high as 1×10^4 iu/mL on $CD4^+$ cells.

Despite the utility of this type of semi-homologous packaging system, it has its limitations in terms of titer, safety, and host range of the transducing particles. Recently, Kafri et al. adapted the inducible VSV G system to produce higher-titer HIV-pseudotyped particles in a 293T-based stable packaging cell line (34). In this case, all of the HIV genes with the exception of envelope were transcribed from a single expression cassette, and VSV G and green fluorescent protein (GFP) were transcribed from a bidirectional tetracycline-regulated promoter. Pseudotyped particles of titer of at least 1×10^6 iu/mL were produced for three to four days following induction in the absence of doxycycline and the presence of sodium butyrate, and these particles efficiently transduced adult rat neurons. No replication-competent virus was detected. Unfortunately, titer was not maintained because of the cytopathic effect of VSV G. It is possible that even higher titers can be obtained once producer clones are individually analyzed.

Because HIV-1 is a known human pathogen, some effort has been directed towards developing non-HIV-1 vectors, such as HIV-2 (a less pathogenic human lentivirus) (4,58), caprine arthritis-encephalitis virus (CAEV) (45), and feline immunodeficiency virus (FIV) (52). The results of these studies are summarized in Table 2. The greatest progress has been made with FIV-based vectors. A chimeric 5′ LTR was constructed to allow transcription of the packaging vector in human cells. The transfer vectors also had chimeric LTRs, a portion of Gag, the RRE, and an internal promoter driving LacZ or GFP. Titers were as high as 5×10^5 iu/mL on cat and human cells, and transduction occurred in growth-arrested cells. Human neurons and macrophages were efficiently transduced.

One argument that has been advanced publicly is that FIV might be better than HIV as a vector because it is not known to be pathogenic to man, based on the absence of seroconversion of people who live with FIV-infected cats. However, since FIV is not transmitted casually, the same statement applies equally well to HIV. The second argument is that FIV replicates poorly in human cells because of low levels of transcription from the LTR, thus the possibility of making a replication-competent virus is diminished. However, the viral vector has been specifically engineered to be transcriptionally active in human cells.

Table 2. Non-HIV-1 Lentiviral Vectors

Lentivirus	Special Features[b]	Titer (iu/mL)[a]	Comments	Reference
CAEV	carries both neo^R and LacZ or fusion of $bleo^R$ and LacZ	<200 (goat)	replication-competent CAEV used as helper virus; poor titer due to defective packaging	45
FIV	CMV-FIV chimeric 5′ LTR, 3-component system, CMV-LacZ or CMV-GFP marker (opp. orientation)	0.9×10^3-5.1×10^5 (human) 2.7×10^4 to 4.9×10^5 (cat)	95% transduction efficiency of human neurons, titer of 9.2×10^4 iu/mL on human macrophages	52
HIV-2	multiple deletions in 5′ UTR, multiple configurations of transgenes	not determined	not yet a true vector system: really a study of packaging and Tat-dependent expression	4, 58
HIV-2	HIV-1 packaging cell line expressing gp120 HIV-2 transfer vector with CTE (Rev independent)	10^4 (human macrophages)	transduced $CD4^+$ cells	11
HIV-2	packaging vector has deletion in Ψ and *env*; transfer vector has SV40-*LacZ* or CMV-GFP.	8.6×10^5 (HeLa)	efficiently transduced neurons and macrophages; $CD34^+$ transduced by PCR, but no expression seen	51

[a]Titer (infectious or transduction units/mL) is typically determined by end-point dilution on a highly susceptible cell target. This is not equivalent to transduction efficiency. A titer of at least 10^6 iu/mL is usually required for >95% transduction efficiency.

[b]neo^R=neomycin phosphotranferase, LacZ=β-galactosidase, $bleo^R$=bleomycin resistance, UTR=untranslated region, CTE=constitutive transport element of Mason-Pfizer monkey virus, Ψ=packaging signal.

Clinical-grade vectors based on HIV might have several advantages compared with the other lentiviruses. The pathogenesis of replication-competent HIV is well known (unlike engineered FIV in humans), and tests to detect replication-competent lentivirus (RCL) are quite advanced (although not yet validated). In addition, HIV is a human virus, and titers on human cells are about 1000% higher than that of FIV. Titer is the single most important factor in determining transduction efficiency of a given cell type (9), and a log reduction in titer may result in very poor transduction efficiency for certain target cell populations (63). Much more is known about the actions of HIV gene products and the host-viral relationship compared with most, if not all, other known viruses.

EXTENDED UTILITY OF LENTIVIRAL VECTORS

The initial reports used HIV-based vectors to examine the host range of wild-type HIV and pseudotyped particles and the effects of gp120 mutations on HIV envelope function. The most obvious use of these vectors today is the transduction of nondividing cells. However, these vectors have other, often overlooked, uses. For example, an HIV vector was used to identify several novel chemokine receptors that function in both HIV and simian immunodeficiency virus (SIV) entry (14). This cloning was accomplished by transducing NIH 3T3 cells that express human CD4 with an MLV cDNA expression library. The transductants were then infected with HIV-puro (HIV vector that carries puromycin acetyltransferase gene) pseudotyped with an M (macrophage)-tropic HIV envelope. Genomic DNA was prepared from the individual puromycin-resistant colonies; the cDNA inserts were recovered by PCR and were shown to confer susceptibility to M-tropic HIV. Similarly, HIV vectors pseudotyped with HTLV-I envelope are now being used to identify the HTLV-I receptor, which is more omnipresent (62). Since these vectors can be engineered not to cause cell-cycle arrest or apoptosis in the targets, individual HIV accessory proteins can thus be studied. This approach showed that a deletion in both *vif* and *vpr* was necessary for persistent expression in hematopoietic stem cells (64).

Along the same lines, these vectors can also be used to study the pathophysiology of HIV infection. Near-synchronous, 100% infection of both primary and established T cells can be accomplished using high-titer, VSV G-pseudotyped viral stocks. After varying lengths of time, messenger RNA is prepared from infected cells, reverse transcribed in the presence of deoxynucleotides coupled to CY dyes, and used to probe cDNA microarrays, which contain up to 16 000 cloned human genes on a single glass slide. The expression profile, as analyzed by a confocal laser scanner, is then compared to that of mock-infected cells. This protocol allows the identification of cloned genes whose expression is either increased or decreased, and the results of these analyses are in progress (R. Pillai, et al., unpublished data). It is hoped that this methodology will reveal novel surrogate markers of HIV infection and potential therapeutic targets.

HIV-based vectors can also be used to screen for inhibitors of HIV infection at both the level of entry and postentry. This can accomplished in a high-throughput manner by using vectors that have GFP and yellow fluorescent protein (YFP) reporters and using flow cytometry in a 96-well format (3). If one is searching for inhibitors of entry, then a dual infection of appropriate target cells (in the presence of inhibitor) can be performed using HIV-YFP (VSV G) and HIV-GFP (ADA), where ADA is an M-tropic HIV envelope.

A specific decrease in the GFP signal would suggest that the inhibitor is acting at the level of viral entry. To uncover postentry inhibitors, a dual infection of targets can be performed, using MLV-GFP (VSV G) and HIV-YFP (VSV G). A specific decrease in the YFP signal would be consistent with a postentry block. Further testing of the inhibitor should elucidate the block to HIV replication. Of course, this screen would only uncover inhibitors up to the point of transcription; viral assembly and budding inhibitors (such as the HIV protease inhibitors currently in clinical use) cannot be identified. Nevertheless, this type of screen could easily be accomplished on libraries of 100 000 compounds.

Finally, HIV vectors may be used as an AIDS candidate vaccine, akin to live, attenuated virus. In this case, the vector would have most, if not all, of the viral genes present with the notable exception of having a transgene expression cassette inserted into a large deletion in *env*. Viral stocks can be pretested in vitro to be certain no RCL is present. As an example, high-titer, replication-defective viral stocks, which have been pseudotyped with a heterologous envelope (VSV G or amphotropic 4070A) and concentrated, are used to immunize transgenic mice that express human cyclin T1 (one of the species-specific Tat cofactors). Mice can be infected parenterally (i.e., intravenously or intramuscularly) or mucosally (i.e., rectally or vaginally) and boosted by the same route. Both the T-cell and B-cell responses to the HIV antigens can be quantitated by proliferative/cytotoxic assays and isotype-specific enzyme-linked immunosorbent assays, respectively. In addition, persistence and localization of infected cells can be determined by immunocytochemistry and in situ hybridization. Although spreading infection will not occur, enough cells should be initially transduced and continue to express viral genes to obtain an adequate immune response. Unfortunately, no HIV challenge can yet be performed in the mouse, but the success of these preliminary studies may provide the impetus for testing this vaccine strategy by using pseudotyped SIV in macaques.

FUTURE DIRECTIONS

Substantial progress has been made in the last few years in the field of lentiviral vectors. I expect the further improvements on vector design will include cell-targeted vectors, episomal, nonintegrating vectors, cell type-specific and inducible vectors, and embedded vectors (i.e., an adenovirus vector that produces all the necessary gene products to *trans*-complement the lentiviral transfer vector that is also present in the adenovirus). This latter approach has been sucessfully used in the case of MLV (16).

It is evident that lentiviral vectors are headed for clinical use, paralleling the track already established by the type C oncoretroviruses. In vitro efficacy has clearly been established for a number of primary human cell types. There is no theoretical or experimental barrier to making lentiviral vectors just as safe, if not safer, than the current MLV vectors undergoing clinical testing. It is tacitly understood, however, that these vectors will be used in patients who have a terminal illness (e.g., metastatic malignancy) or in AIDS patients who have failed conventional treatments (e.g., highly active antiretroviral therapy). In the latter case, typical serum viral RNA loads are between 1×10^5 and 1×10^6/mL, and concerns have been voiced that recombination will occur between the HIV vector and replicating wild-type HIV. If that happens and a replication-competent virus is produced, then the transgene can spread from cell to cell in the infected patient and conceivably be passed from patient to patient.

These fears would certainly be justified

if the transgene was oncogenic or deleterious to the host. However, for AIDS patients, the transgenes under consideration are, for the most part, directed against HIV itself. These include RRE and Tat responsive RNA (TAR) decoys, Rev and Tat dominant-negative proteins, and antisense viral mRNAs. Thus, viral spread of these genes may in fact inhibit HIV replication. The worst-case scenario would be viral spread of an anti-HIV gene that would then result in HIV strains that are resistant to that specific gene. This would likely occur eventually, even in the absence of a replication-competent HIV carrying that gene, because cells transduced with the protective gene would then be applying a selective pressure for viral escape mutants. It is interesting to note that despite extensive study and in vitro use of these anti-HIV genes, no viral escape mutants have been definitively identified. Given the collective experience with antiretroviral agents, coupled with the rapid replication rate of the virus and error-prone reverse transcriptase (0.33 bp/genome), viral escape from even the most efficacious anti-HIV genetic therapy is a near certainty.

I anticipate that after safety and efficacy have been established in the patients described above, the clinical use of HIV vectors will likely expand to conditions that affect other terminally differentiated, nondividing cell types. Fifty years from now, I expect that lentiviral vectors, like any of a number of other vectors in preclinical and clinical testing, will have found their own small niche in the therapeutic armamentarium of the 21st century.

REFERENCES

1. **Aiken, C.** 1997. Pseudotyping human immunodeficiency virus type 1 (HIV-1) by the glycoprotein of vesicular stomatitis virus targets HIV-1 entry to an endocytic pathway and suppresses both the requirement for Nef and the sensitivity to cyclosporin A. J. Virol. *71*:5871-5877.
2. **Akkina, R.K., R.M. Walton, M.L. Chen, Q.X. Li, V. Planelles, and I.S. Chen.** 1996. High-efficiency gene transfer into CD34+ cells with a human immunodeficiency virus type 1-based retroviral vector pseudotyped with vesicular stomatitis virus envelope glycoprotein G. J. Virol. *70*:2581-2585.
3. **Anderson, M.T., I.M. Tjioe, M.C. Lorincz, D.R. Parks, L.A. Herzenberg, and G.P. Nolan.** 1996. Simultaneous fluorescence-activated cell sorter analysis of two distinct transcriptional elements within a single cell using engineered green fluorescent proteins. Proc. Natl. Acad. Sci. USA *93*:8508-8511.
4. **Arya, S.K., M. Zamani, and P. Kundra.** 1998. Human immunodeficiency virus type 2 lentivirus vectors for gene transfer: expression and potential for helper virus-free packaging. Hum. Gene Ther. *9*:1371-1380.
5. **Blomer, U., L. Naldini, T. Kafri, D. Trono, I.M. Verma, and F.H. Gage.** 1997. Highly efficient and sustained gene transfer in adult neurons with a lentivirus vector. J. Virol. *71*:6641-6649.
6. **Buchschacher, G.L. and A.T. Panganiban.** 1992. Human immunodefiency virus vectors for inducible expression of foreign genes. J. Virol. *66*:2731-2739.
7. **Bukrinskaya, A.G., A. Ghorpade, N.K. Heinzinger, T.E. Smithgall, R.E. Lewis, and M. Stevenson.** 1996. Phosphorylation-dependent human immunodeficiency virus type 1 infection and nuclear targeting of viral DNA. Proc. Natl. Acad. Sci. USA *93*:367-371.
8. **Bukrinsky, M.I., S. Haggerty, M.P. Dempsey, N. Sharova, A. Adzhubel, L. Spitz, P. Lewis, D. Goldfarb, M. Emerman, and M. Stevenson.** 1993. A nuclear localization signal within HIV-1 matrix protein that governs infection of non-dividing cells. Nature *365*:666-669.
9. **Miller, A.D.** 1997. Development and applications of retroviral vectors, p. 437-473. *In* J.M. Coffin, S.H. Hughes, and H.E. Varmus (Eds.), Retroviruses. CSH Laboratory Press, Cold Spring Harbor, NY.
10. **Corbeau, P., G. Kraus, and F. Wong-Staal.** 1996. Efficient gene transfer by a human immunodeficiency virus type 1 (HIV-1)-derived vector utilizing a stable HIV packaging cell line. Proc. Natl. Acad. Sci. USA *93*:14070-14075.
11. **Corbeau, P., G. Kraus, and F. Wong-Staal.** 1998. Transduction of human macrophages using a stable HIV-1/HIV-2-derived gene delivery system. Gene Ther. *5*:99-104.
12. **Cullen, B.R.** 1998. HIV-1 auxiliary proteins: making connections in a dying cell. Cell *93*:685-692.
13. **Delwart, E.L., G.L. Buchschacher, Jr., E.O. Freed, and A.T. Panganiban.** 1992. Analysis of HIV-1 envelope mutants and pseudotyping of replication-defective HIV-1 vectors by genetic complementation. AIDS Res. Hum. Retroviruses *8*:1669-1677.
14. **Deng, H.K., D. Unutmaz, V.N. KewalRamani, and D.R. Littman.** 1997. Expression cloning of new receptors used by simian and human immunodeficiency viruses. Nature *388*:296-300.
15. **Dull, T., R. Zufferey, M. Kelly, R.J. Mandel, M. Nguyen, D. Trono, and L. Naldini.** 1998. A third-generation lentivirus vector with a conditional packaging system. J. Virol. *72*:8463-8471.
16. **Feng, M., W.H. Jackson, Jr., C.K. Goldman, C. Ran-**

court, M. Wang, S.K. Dusing, G. Siegal, and D.T. Curiel. 1997. Stable in vivo gene transduction via a novel adenoviral/retroviral chimeric vector. Nat. Biotechnol. *15*:866-870.

17. **Fouchier, R.A., B.E. Meyer, J.H. Simon, U. Fischer, and M.H. Malim.** 1997. HIV-1 infection of non-dividing cells: evidence that the amino-terminal basic region of the viral matrix protein is important for Gag processing but not for post-entry nuclear import. EMBO J. *16*:4531-4539.
18. **Frankel, A.D. and J.A. Young.** 1998. HIV-1: fifteen proteins and an RNA. Annu. Rev. Biochem. *67*:1-25.
19. **Freed, E.O., G. Englund, F. Maldarelli, and M.A. Martin.** 1997. Phosphorylation of residue 131 of HIV-1 matrix is not required for macrophage infection. Cell *88*:171-174.
20. **Gallay, P., T. Hope, D. Chin, and D. Trono.** 1997. HIV-1 infection of nondividing cells through the recognition of integrase by the importin/karyopherin pathway. Proc. Natl. Acad. Sci. USA *94*:9825-9830.
21. **Gallay, P., S. Swingler, C. Aiken, and D. Trono.** 1995. HIV-1 infection of non-dividing cells: C-terminal tyrosine phophorylation of the viral matrix protein is a key regulator. Cell *80*:379-388.
22. **Gallay, P., S. Swingler, J. Song, F. Bushman, and D. Trono.** 1995. HIV nuclear import is governed by the phosphotyrosine-mediated binding of matrix to the core domain of integrase. Cell *83*:569-575.
23. **Gallichan, W.S., T. Kafri, T. Krahl, I.M. Verma, and N. Sarvetnick.** 1998. Lentivirus-mediated transduction of islet grafts with interleukin 4 results in sustained gene expression and protection from insulitis. Hum. Gene Ther. *9*:2717-2726.
24. **Goh, W.C., M.E. Rogel, C.M. Kinsey, S.F. Michael, P.N. Fultz, M.A. Nowak, B.H. Hahn, and M. Emerman.** 1998. HIV-1 Vpr increases viral expression by manipulation of the cell cycle: a mechanism for selection of Vpr in vivo. Nat. Med. *4*:65-71.
25. **Gold, M.O., X. Yang, C.H. Herrmann, and A.P. Rice.** 1998. PITALRE, the catalytic subunit of TAK, is required for human immunodeficiency virus Tat transactivation in vivo. J. Virol. *72*:4448-4453.
26. **Goldman, M.J., P.S. Lee, J.S. Yang, and J.M. Wilson.** 1997. Lentiviral vectors for gene therapy of cystic fibrosis. Hum. Gene. Ther. *8*:2261-2268.
27. **He, J., S. Choe, R. Walker, P. Marzio, D.O. Morgan, and N.R. Landau.** 1995. Human immunodeficiency virus type 1 viral protein R (Vpr) arrests cells in the G_2 phase of the cell cycle by inhibiting $p34^{cdc2}$ activity. J. Virol. *69*:6705-6711.
28. **Heinzinger, N.K., M.I. Bukinsky, S.A. Haggerty, A.M. Ragland, V. Kewalramani, M.A. Lee, H.E. Gendelman, L. Ratner, M. Stevenson, and M. Emerman.** 1994. The Vpr protein of human immunodeficiency virus type 1 influences nuclear localization of viral nucleic acids in nondividing host cells. Proc. Natl. Acad. Sci. USA *91*:7311-7315.
29. **Helseth, E., M. Kowalski, D. Gabuzda, U. Olshevsky, W. Haseltine, and J. Sodroski.** 1990. Rapid complementation assays measuring replicative potential of human immunodeficiency virus type 1 envelope glycoprotein mutants. J. Virol. *64*:2416-2420.
30. **Hwang, J.J., L. Li, and W.F. Anderson.** 1997. A conditional self-inactivating retrovirus vector that uses a tetracycline-responsive expression system. J. Virol. *71*:7128-7131.
31. **Jowett, J.B.M., V. Planelles, B. Poon, N.P. Shah, M.-L. Chen, and I.S.Y. Chen.** 1995. The human immunodeficiency virus type 1 vpr gene arrests infected T cells in the G_2+ M phase of the cell cycle. J. Virol. *69*:6304-6313.
32. **Ju, Q., D. Edelstein, M.D. Brendel, D. Brandhorst, H. Brandhorst, R.G. Bretzel, and M. Brownlee.** 1998. Transduction of non-dividing adult human pancreatic beta cells by an integrating lentiviral vector. Diabetologia *41*:736-739.
33. **Kafri, T., U. Blomer, D.A. Peterson, F.H. Gage, and I.M. Verma.** 1997. Sustained expression of genes delivered directly into liver and muscle by lentiviral vectors. Nat. Genet. *17*:314-317.
34. **Kafri, T., H. van Praag, L. Ouyang, F.H. Gage, and I.M. Verma.** 1999. A packaging cell line for lentivirus vectors. J. Virol. *73*:576-584.
35. **Kewalramani, V.N., C.S. Park, P.A. Gallombardo, and M. Emerman.** 1996. Protein stability influences human immunodeficiency virus type 2 Vpr virion incorporation and cell cycle effect. Virology *218*:326-334.
36. **Kim, V.N., K. Mitrophanous, S.M. Kingsman, and A.J. Kingsman.** 1998. Minimal requirement for a lentivirus vector based on human immunodeficiency virus type 1. J. Virol. *72*:811-816.
37. **Korin, Y.D. and J.A. Zack.** 1998. Progression to the G1b phase of the cell cycle is required for completion of human immunodeficiency virus type 1 reverse transcription. J. Virol. *72*:3161-3168.
38. **Landau, N.R., K.A. Page, and D.R. Littman.** 1991. Pseudotyping with human T-cell leukemia virus type I broadens the human immunodeficiency virus host range. J. Virol. *65*:162-169.
39. **Lewis, P.F. and M. Emerman.** 1994. Passage through mitosis is required for oncoretroviruses but not for the human immunodeficiency virus. J. Virol. *68*:510-516.
40. **Madani, N. and D. Kabat.** 1998. An endogenous inhibitor of human immunodeficiency virus in human lymphocytes is overcome by the viral Vif protein. J. Virol. *72*:10251-10255.
41. **Margottin, F., S.P. Bour, H. Durand, L. Selig, S. Benichou, V. Richard, D. Thomas, K. Strebel, and R. Benarous.** 1998. A novel human WD protein, h-beta TrCp, that interacts with HIV-1 Vpu connects CD4 to the ER degradation pathway through an F-box motif. Mol. Cell. *1*:565-574.
42. **Miyoshi, H., U. Blomer, M. Takahashi, F.H. Gage, and I.M. Verma.** 1998. Development of a self-inactivating lentivirus vector. J. Virol. *72*:8150-8157.
43. **Miyoshi, H., M. Takahashi, F.H. Gage, and I.M. Verma.** 1997. Stable and efficient gene transfer into the retina using an HIV-based lentiviral vector. Proc. Natl. Acad. Sci. USA *94*:10319-10323.
44. **Mochizuki, H., J.P. Schwartz, K. Tanaka, R.O. Brady, and J. Reiser.** 1998. High-titer human immunodeficiency virus type 1-based vector systems for gene delivery into nondividing cells. J. Virol. *72*:8873-8883.

45.**Mselli-Lakhal, L., C. Favier, M.F. Da Silva Teixeira, K. Chettab, C. Legras, C. Ronfort, G. Verdier, J.F. Mornex, and Y. Chebloune.** 1998. Defective RNA packaging is responsible for low transduction efficiency of CAEV-based vectors. Arch. Virol. *143*:681-695.

46.**Naldini, L., U. Blomer, P. Gallay, D. Ory, R. Mulligan, F.H. Gage, I.M. Verma, and D. Trono.** 1996. In vivo gene delivery and stable transduction of nondividing cells by a lentiviral vector. Science *272*:263-267.

47.**Olson, P., S. Nelson, and R. Dornburg.** 1994. Improved self-inactivating retroviral vectors derived from spleen necrosis virus. J. Virol. *68*:7060-7066.

48.**Page, K.A., N.R. Landau, and D.R. Littman.** 1990. Construction and use of a human immunodeficiency virus vector for analysis of virus infectivity. J. Virol. *64*:5270-5276.

49.**Page, K.A., S.M. Stearns, and D.R. Littman.** 1992. Analysis of mutations in the V3 domain of gp160 that affect fusion and infectivity. J. Virol. *66*:524-533.

50.**Poeschla, E., P. Corbeau, and F. Wong-Staal.** 1996. Development of HIV vectors for anti-HIV gene therapy. Proc. Natl. Acad. Sci. USA *93*:11395-11399.

51.**Poeschla, E., J. Gilbert, X. Li, S. Huang, A. Ho, and F. Wong-Staal.** 1998. Identification of a human immunodeficiency virus type 2 (HIV-2) encapsidation determinant and transduction of nondividing human cells by HIV-2-based lentivirus vectors. J. Virol. *72*:6527-6536.

52.**Poeschla, E.M., F. Wong-Staal, and D.J. Looney.** 1998. Efficient transduction of nondividing human cells by feline immunodeficiency virus lentiviral vectors. Nat. Med. *4*:354-357.

53.**Poznansky, M., A. Lever, L. Bergeron, W. Haseltine, and J. Sodroski.** 1991. Gene transfer into human lymphocytes by a defective human immunodeficiency virus type 1 vector. J. Virol. *65*:532-536.

54.**Re, F., D. Braaten, E.K. Franke, and J. Luban.** 1995. Human immunodeficiency virus type 1 Vpr arrests the cell cycle in G2 by inhibiting the activation of $p34^{cdc2}$-cyclin B. J. Virol. *69*:6859-6864.

55.**Rebolledo, M.A., P. Krogstad, F. Chen, K.M. Shannon, and T.S. Klitzner.** 1998. Infection of human fetal cardiac myocytes by a human immunodeficiency virus-1-derived vector. Circ. Res. *83*:738-742.

56.**Reiser, J., G. Harmison, S. Kluepfel-Stahl, R.O. Brady, S. Karlsson, and M. Schubert.** 1996. Transduction of nondividing cells using pseudotyped defective high-titer HIV type 1 particles. Proc. Natl. Acad. Sci. USA *93*:15266-15271.

57.**Roe, T., T.C. Reynolds, G. Yu, and P.O. Brown.** 1993. Integration of murine leukemia virus DNA depends on mitosis. EMBO J. *12*:2099-2108.

58.**Sadaie, M.R., M. Zamani, S. Whang, N. Sistron, and S.K. Arya.** 1998. Towards developing HIV-2 lentivirus-based retroviral vectors for gene therapy: dual gene expression in the context of HIV-2 LTR and Tat. J. Med. Virol. *54*:118-128.

59.**Schlegel, R., T.S. Tralka, M.C. Willingham, and I. Pastan.** 1983. Inhibition of VSV binding and infectivity by phosphotidylserine: is phosphotidylserine a VSV-binding site? Cell *32*:639-646.

60.**Simon, J.H., N.C. Gaddis, R.A. Fouchier, and M.H. Malim.** 1998. Evidence for a newly discovered cellular anti-HIV-1 phenotype. Nat. Med. *4*:1397-1400.

61.**Srinivasakumar, N., N. Chazal, C. Helga-Maria, S. Prasad, M.L. Hammarskjold, and D. Rekosh.** 1997. The effect of viral regulatory protein expression on gene delivery by human immunodeficiency virus type 1 vectors produced in stable packaging cell lines. J. Virol. *71*:5841-5848.

62.**Sutton, R.E. and D.R. Littman.** 1996. Broad host range of human T-cell leukemia virus type 1 demonstrated with an improved pseudotyping system. J. Virol. *70*:7322-7326.

63.**Sutton, R.E., M.J. Reitsma, N. Uchida, and P.O. Brown.** Transduction of human progenitor hematopoietic stem cells by human immunodeficiency virus type 1 vectors is cell-cycle dependent. J. Virol. *73*:3649-3660.

64.**Sutton, R.E., H.T. Wu, R. Rigg, E. Bohnlein, and P.O. Brown.** 1998. Human immunodeficiency virus type 1 vectors efficiently transduce human hematopoietic stem cells. J. Virol. *72*:5781-5788.

65.**Unutmaz, D., V.N. KewalRamani, and D.R. Littman.** 1998. G protein-coupled receptors in HIV and SIV entry: new perspectives on lentivirus-host interactions and on the utility of animal models. Semin. Immunol. *10*:225-236.

66.**Vodicka, M.A., D.M. Koepp, P.A. Silver, and M. Emerman.** 1998. HIV-1 Vpr interacts with the nuclear transport pathway to promote macrophage infection. Genes Dev. *12*:175-185.

67.**von Schwedler, U., R.S. Kornbluth, and D. Trono.** 1994. The nuclear localization signal of the matrix protein of human immunodeficiency virus type 1 allows the establishment of infection in macrophages and quiescent T lymphocytes. Proc. Natl. Acad. Sci. USA *91*:6992-6996.

68.**Wei, P., M.E. Garber, S.M. Fang, W.H. Fischer, and K.A. Jones.** 1998. A novel CDK9-associated C-type cyclin interacts directly with HIV-1 Tat and mediates its high-affinity, loop-specific binding to TAR RNA. Cell *92*:451-462.

69.**Yang, X., M.O. Gold, D.N. Tang, D.E. Lewis, E. Aguilar-Cordova, A.P. Rice, and C.H. Herrmann.** 1997. TAK, an HIV Tat-associated kinase, is a member of the cyclin-dependent family of protein kinases and is induced by activation of peripheral blood lymphocytes and differentiation of promonocytic cell lines. Proc. Natl. Acad. Sci. USA *94*:12331-12336.

70.**Yang, X., C.H. Herrmann, and A.P. Rice.** 1996. The human immunodeficiency virus Tat proteins specifically associate with TAK in vivo and require the carboxyl-terminal domain of RNA polymerase II for function. J. Virol. *70*:4576-4584.

71.**Yu, S.F., T. von Ruden, P.W. Kantoff, C. Garber, M. Seiberg, U. Ruther, W.F. Anderson, E.F. Wagner, and E. Gilboa.** 1986. Self-inactivating retroviral vectors designed for transfer of whole genes into mammalian cells. Proc. Natl. Acad. Sci. USA *83*:3194-3198.

72.**Zack, J.A., S.J. Arrigo, S.R. Weitsman, A.S. Go, A. Haislip, and I.S.Y. Chen.** 1990. HIV-1 entry into quiescent primary lymphocytes: molecular analysis reveals a labile, latent viral structure. Cell *61*:213-222.

73.**Zack, J.A., A.M. Haislip, P. Krogstad, and I.S.Y.**

Chen. 1992. Incompletely reverse-transcribed human immunodeficiency virus type 1 genomes in quiescent cells can function as intermediates in the retroviral life cycle. J. Virol. *66*:1717-1725.

74. **Zufferey, R., T. Dull, R.J. Mandel, A. Bukovsky, D. Quiroz, L. Naldini, and D. Trono.** 1998. Self-inactivating lentivirus vector for safe and efficient in vivo gene delivery. J. Virol. *72*:9873-9880.

75. **Zufferey, R., D. Nagy, R.J. Mandel, L. Naldini, and D. Trono.** 1997. Multiply attenuated lentiviral vector achieves efficient gene delivery in vivo. Nat. Biotechnol. *15*:871-875.

FAMILY: Retroviridae
GENUS: Spumavirinae (Foamy Viruses)

30 Foamy Viruses: Molecular Features and Therapeutic Potential

Claire L. Hill and Myra O. McClure
Department of G.U. Medicine and Communicable Diseases, Jefferiss Research Laboratories, Imperial College School of Medicine at St. Mary's, London, England, UK

INTRODUCTION

The Spumaviridae (more commonly known as foamy viruses [FVs]) are one of the seven genera of retroviruses. FVs are highly cytopathic in cell culture with infection of cultured cells by the virus giving rise to vacuolization of cells and multinucleated syncytia. When viewed by light microscopy, infected cells have a foamy appearance, hence their name (Figure 1). One interesting characteristic of FVs is that although they demonstrate a strong cytopathic effect in cell culture, they appear to be benign in their infected hosts. They were first discovered as cell culture contaminants in 1954 (18) but were not isolated until several years later from tissue of a Kenyan nasopharyngeal carcinoma (NPC) patient (2). At the time, this constituted the first human foamy virus (HFV) isolation and, indeed, the first retrovirus to be isolated from human tissue. However, this claim has not stood the test of time.

Until recently, FVs had received little research attention primarily because of their lack of associated pathology in infected hosts. However, in recent years, there has been greater interest in their molecular biology because of their potential use as vectors for the delivery and expression of therapeutic genes. In this review, we consider the molecular properties of FVs and their pathogenesis before describing the basic FV vectors that have been published to date.

Molecular Biology

FV particles are 100 to 140 nm in diameter and have a distinct morphology with prominent surface spikes (5–15 nm in length) and a central uncondensed core, similar to the B- and D-type retroviruses (Figure 2). In comparison with other retroviruses that are assembled at the plasma membrane, FV capsid assembly occurs mainly in the cytoplasm before budding. The FV genome is 13 kb in length and belongs to the complex group of retroviruses (17). This means that in addition to the *gag*, *pol*, and *env* genes that encode the structural proteins found in all retroviruses, the FV genome encodes several accessory open reading frames (ORFs) at the 3′ end

Viral Vectors: Basic Science and Gene Therapy
Edited by A. Cid-Arregui and A. García-Carrancá

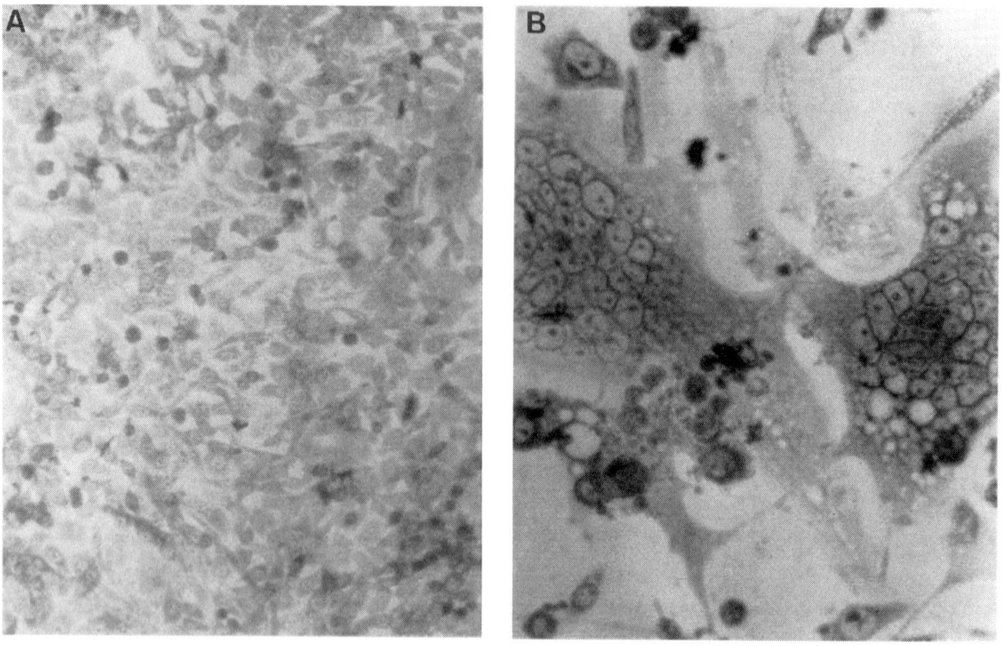

Figure 1. Infection of cell cultures by FV. Light microscopy image demonstrating the formation of syncytia in FV-infected cells. (A) Uninfected BHK-21 cells; (B) infected with HFV.

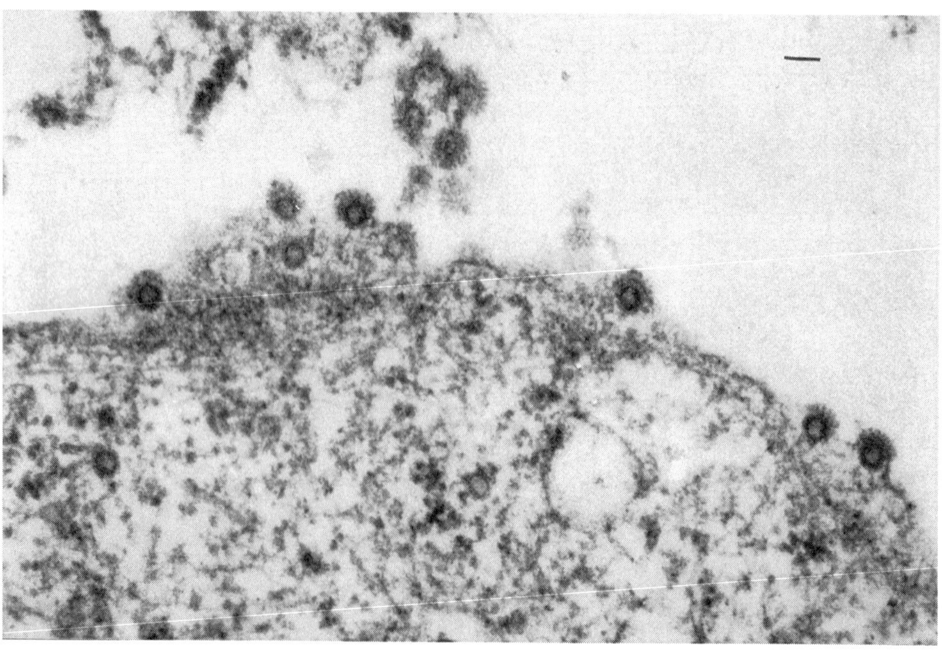

Figure 2. Photograph of FV particles. Thin-section electron micrograph of HFV particles budding from Sup-T1 cells. The long spikes characteristic of FVs are evident. The bar represents 100 nm.

of the genome known as *bel* genes (between *env* and long terminal repeat [LTR]) (Figure 3) (24). Of the additional genes only *tas* (previously known as *bel-1*) is essential for virus replication in vitro (5,42,83). *Tas* encodes a 36-kDa transcriptional transactivator that induces transcription from both the viral LTR and an internal promoter located upstream of the accessory ORFs in the *env* gene (37,39,60). The functions of the *bel-2* and *bel-3* ORFs are unknown, but they are dispensable for viral replication in vitro (5,83). Bet, a protein resulting from mRNA spliced from the *tas* ORF into the *bel-2* ORF, is a 56-kDa protein expressed to high levels in HFV-infected cells. Its function remains unknown.

Some aspects of the FV replication cycle (Figure 4) vary from that of most other retroviruses. For example, the Pol protein of FVs is expressed independently of Gag, from a spliced subgenomic mRNA, unlike other retroviruses where Gag and Pol are produced as a fusion protein (19,46). In addition, FV particle release requires the coexpression of Gag and Env (22), whereas other retroviruses only require Gag. Following particle formation, not all FV virions bud spontaneously from cellular membranes (22) as Env contains an endoplasmic reticulum retrieval signal (25) which leads to many of the virions remaining in the cytoplasm (Figure 4). (Further properties of the FV Env protein and its role in virus host range will be discussed in more detail in Chapter 31.) Finally, one of the most unusual features of FVs is the presence of double-stranded DNA, of a similar length to that of the provirus, found in the virus particle (82). This suggests that reverse transcriptase has already occurred in the virion before infection and may take place late in the replication cycle (54); this characteristic is in contrast to other retroviruses in which reverse transcription occurs in the early phase of replication following virus entry into the cell and uncoating. In this respect, FVs replication is analogous to the replication cycle of hepadnaviruses (82).

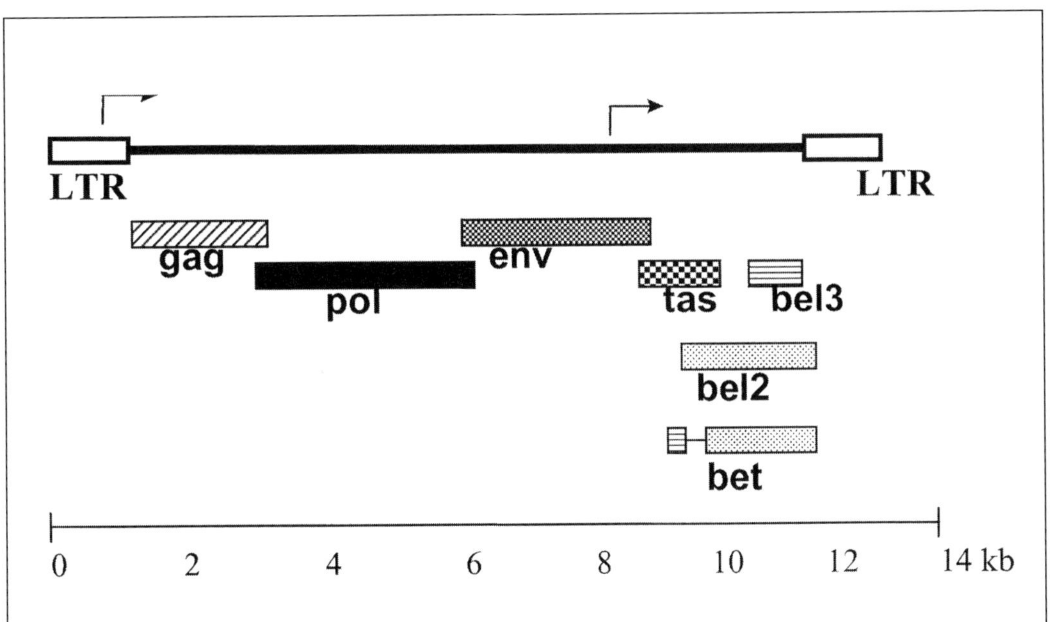

Figure 3. Organization of the HFV genome. A schematic of the 13-kb HFV provirus genome indicating the positions of the LTRs *gag, pol, env,* and the accessory ORFs. Arrows indicate the transcriptional promoters.

FOAMY VIRUS ISOLATES, SEROEPIDEMIOLOGY, AND DISEASE ASSOCIATION

FVs infect diverse mammalian species, for example, cats (61), sea lions (40), cattle (36), chimpanzees (29,67), and other primates such as African green monkeys (67,72) and macaques (67,72). In all these animals, the virus induces persistent infections and high-titer neutralizing antibodies, but no apparent pathology (76). Some of these isolates have been cloned and sequenced including the bovine (32), feline (78), and chimpanzee isolates (29).

The first HFV was isolated from cultures of an NPC taken from a Kenyan patient (2,20), and further FV isolations from human tissues have been subsequently reported on a number of occasions. For example, a virus related to the macaque virus, SFV-1, was isolated from peripheral blood mononuclear cells (PBMCs) cultured from a leukemia patient (81). Isolates from patients with de Quervain's subacute granulomatous thyroiditis are well documented (69,70), as is the HFV cultured from the brain of a British patient with encephalopathy (15). However, only the prototype remains available to the scientific community for investigation. Hence, it is from this strain that an infectious molecular clone has been derived (59), its genomic sequence analyzed (24,50), and its molecular biology profiled (23,44).

As early as 1978, Nemo et al. (56) contested the observation by Epstein and coworkers that HFV was serotypically unrelated to other known FVs (20). This was based on the demonstration that by indirect immunofluorescence microscopy and neutralization assays, HFV was closely

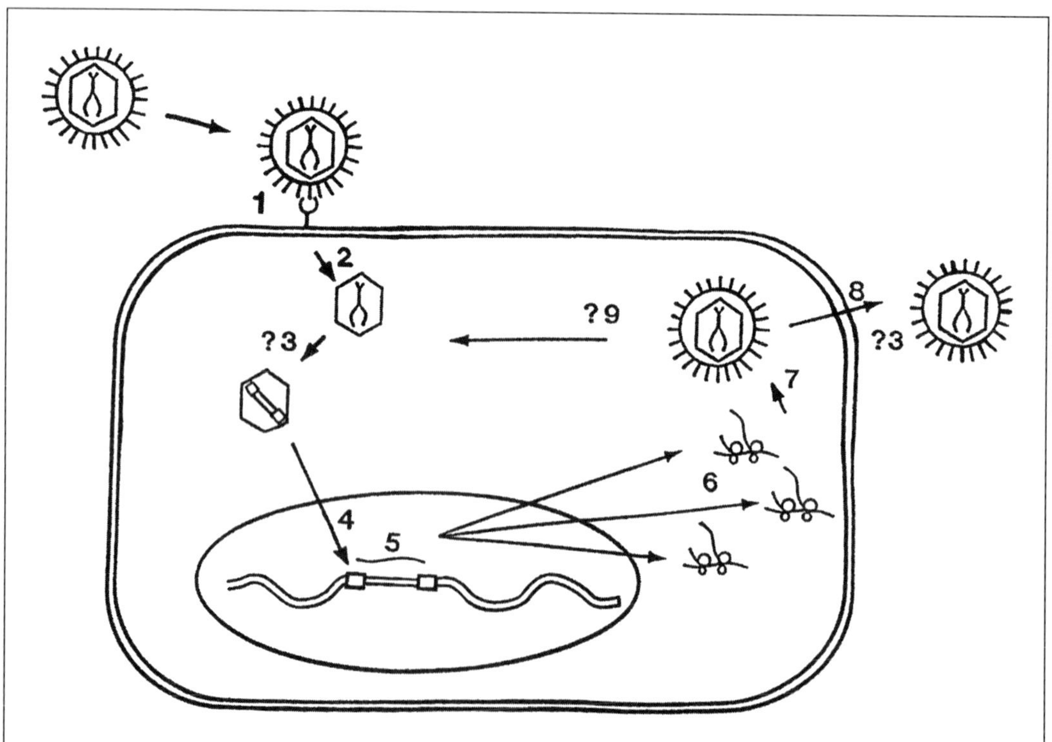

Figure 4. A schematic of the FV replication pathway. (1) Adsorption to the cell surface; (2) virus uncoating; (3) reverse transcription; (4) integration; (5) transcription; (6) translation; (7) virus assembly; (8) viral budding; and (9) viral retention and recycling.

related to SFV-6. They further deduced that since HFV, SFV-6, and SFV-7 were indistinguishable by complement fixation tests, and since the human isolate was not neutralized by human sera taken from nasopharyngeal carcinoma or Burkitt's lymphoma patients, it represented a variant of SFV-6—there being some immunological cross-reactivity between the two. Indeed, they went further to suggest that the human isolates represented either chance human infection with a simian FV or laboratory contamination. However, these reservations were largely ignored until in recent years interest in the natural seroprevalence of FV infection worldwide has been stimulated by curiosity in the potential for foamy viruses to be used as gene vectors (62).

To address this question and to screen large numbers of sera for the presence of HFV antibodies, Ali et al. (3) developed a sensitive and specific enzyme-linked immunosorbent assay (ELISA) using recombinant HFV Gag and Env proteins and assayed over 5000 sera collected from people in diverse geographic regions. A high percentage of the individuals screened were immunosuppressed as a result of HIV infection and, therefore, it might be argued that they would be more likely to harbor novel viral infections, as for example was the case when HHV-6 was discovered (63) and when the isolation of a D-type retrovirus from an AIDS patient with a B-cell lymphoma was reported (11). The significance of Ali's study was not so much the larger sample size in contrast to previous seroepidemiological surveys of HFV infection, but that ELISA-positive samples were subjected to additional serological and, where PBMCs were available, molecular assays.

No serum sample from Europe, the United States, or South America tested positive in this ELISA. A strong ELISA reactivity was found in a significant proportion of the population of the Pacific Islands, in Central Africa (Malawi), and in Uganda—areas previously cited as having a high level of HFV infection (1,47,49,55). Of particular significance was that the survey failed to detect any seropositivity among 419 Kenyan samples taken from HIV-infected individuals, some of whom had malignancies such as lymphoma and Kaposi's sarcoma. Because of a persistently high background caused by nonspecific serological reactivity produced in all assays using some sera (e.g., those from the Pacific Islands and parts of Africa), it proved critical that those test sera that were reactive in ELISA were confirmed as positive by Western blot, neutralization, and polymerase chain reaction (PCR). Some of those assays, like PCR, were unavailable at the time of Achong and Epstein's earlier studies (1), were carried out on very few samples (four sera tested by immunofluorescence microscopy in the Loh study [47] were further assayed for neutralizing antibody at an insufficient dilution to minimize background staining) or were simply unconvincing (Western blot analysis by Mahnke [49]).

These findings raised questions about the provenance of HFV. Recent sequence and functional analyses carried out by several groups endorse the phylogenetic and antigenic closeness of SFV-6 and SFV-7 to HFV (8,14,29). This, taken together with the modern serological epidemiological evidence, strongly suggests that HFV is not a virus naturally found in the human population. In the absence of any HFV isolate other than the prototype being available for scrutiny, and in view of the fact that the human population does not appear to be a natural reservoir for HFV infection, it follows that reports associating HFV with a variety of disease states should be viewed with caution.

FVs have never been clearly associated with any disease in any naturally or experimentally infected animal, although transient immunodeficiency in rabbits has been reported following experimental infection with SFV-7 (33). While inbred strains of mice like all infected animals appear to establish an asymptomatic but persistent infection, even in the presence of a high-titer, neutralizing antibody response (13,35,64), transgenic mice expressing all or part of the HFV genome developed a severe neurodegenerative disorder (12). The most recent foamy virus serotype to be discovered (SFV-11) was isolated from an orangutan with encephalopathy and myopathy (51), but since the syndrome appears to be restricted to this isolated case, a causal link between the retroviral infection and the disease state of the animal is not justified.

In humans, FV infection has been claimed to be associated with several illnesses including chronic fatigue syndrome (26,49), Graves disease (41,49,77), and multiple sclerosis (48,49,71). However, these reports need to be viewed with caution since the HFV association with Graves disease could not be independently confirmed (68), and contamination and over-interpretation of single test results could not be ruled out (67). Historical reports of the isolation of FV from patients who suffered from tumors (81), encephalopathy (15), and de Quervain thyroiditis (69) are undermined by a loss of the isolates in question and a lack of supporting data. On rare occasions where accidental human infection with HFV or SFV has occurred, individuals concerned have remained completely healthy (28,58,66). In summary, there remains no evidence of an association between FV infection and any disease in animals or humans.

POTENTIAL FOR GENE THERAPY

Ever since the first attempt to cure a gene defect, initiated in 1990 for adenosine deaminase deficiency (10), there has been considerable interest in the use of gene therapy for the treatment of genetic defects, the prevention and therapy of infectious diseases (e.g., AIDS), and tumor therapy. One of the major challanges surrounding gene therapy is finding ways to deliver genetic material efficiently to the cell and maintain long-term expression. Vectors can be either nonviral (e.g., liposome/DNA complexes) or viral (e.g., retroviruses, herpesviruses, adenoviruses [Ads], and adeno-associated viruses [AAVs]) (4). Retroviruses have the advantage of being able to integrate into the genetic material of the host cell genome, thereby ensuring stable expression. Hence, 80% of current research uses retroviral vectors, of which murine leukemia virus (MLV) dominates by far.

FVs have the potential to be utilized as an alternative retroviral delivery system for several reasons. First, they have an amphotropic host range. To date, no cell type resistant to FV infection has been found. HFV infects every mammalian cell type that has been tested in our laboratory in addition to epithelial, neuronal, myeloid, and fibroblastoid cells, as well as cell lines of reptilian and avian origin (30,34,35,58, 62,82). This broad host range makes FVs ideal vector candidates for DNA delivery in gene therapy, particularly for the treatment of genetic diseases in which all cell types need to be targeted. FVs have some of the longest genomes of all characterized retroviruses—13 kb compared with, for example, 8 kb for MLV—thereby enabling the virus to package more foreign DNA. Finally, FVs are not inactivated by complement/human serum, although modifications in culture conditions can circumvent this problem with MLV (73).

The regulation of FV gene expression is tightly controlled. The Tas protein of HFV is an essential transcriptional transactivator, and viruses are not capable of virus replication in its absence. If Tas were to be deleted from vector constructs, and the U3 sequences in the 5′ LTR replaced by a constitutive promoter (e.g., the cytomegalovirus promoter), then the requirement of Tas for vector synthesis would be eliminated. This would constitute a gene transfer system in which an essential key gene for virus replication is excluded from the system. Of potential relevance to vector safety is that the FV Gag, Pol, and Env structural proteins are synthesized independently of each other, which allows for their expression on three different plasmids in packaging cell lines. The elimination of Tas and independent expression of the structural proteins would together reduce the likelihood of producing replication-competent retrovirus (RCR) as a result of recombination to a negligable risk.

An integral part of the retroviral replication cycle is integration of the viral DNA into the host chromosomal DNA. In the case of oncovirus (e.g., MLV) replication host cell proliferation is a prerequisite, while lentiviruses (e.g., HIV) are able to infect productively metabolically active nondividing cells (43). There is some suggestion that FVs may infect nondividing cells better than MLV (see below).

VECTOR DESIGN

Studies on the use of FV as vectors have recently emerged from several laboratories (7,31,57,62,65,74). Early FV vector design has involved deletion of some or all of the ancillary ORFs (*bel* genes) from HFV molecular clones (59) known to be dispensable for viral replication in vitro. For example, some vectors are deleted in the *bel-2* and *bel-3* region (65), some deleted in *tas* and *env* (62), or, alternatively, deleted in *tas* and *bel-2* or all three *bel* genes (7,74). In each of these constructs, a reporter gene such as CAT (65), neomycin phosphotransferase, alkaline phosphatase (62,74), puromycin-resistance, or β-glucuronidase (7) were used to assess transduction. Due to the absence of the viral transactivator Tas in some of these vectors, they required either a BHK-21-derived cell line engineered to express Tas Bel-1 (known as BHK/Bel-1 cells) (7), or replacement of the U3 region of the 5′ LTR with a constitutive promoter (74) to allow transcription to occur.

Replication-competent vectors produced a typical FV cytopathic effect (CPE) when transfected into BHK-21 cells (65). Replication-incompetent vectors were able to transduce a variety of primate cell lines (7,57,62,74) including hematopoietic progenitor cells (31). Vector titers in excess of 1×10^5 transducing units per mL have been generated (7,65,74). In addition, helper-free FV vectors have been developed in which the essential *tas* gene is eliminated from the system (74); this development demonstrates a potentially safer gene transfer system in which wild-type virus can not be regenerated.

One aspect of FV vectors for gene delivery that merits further investigation is their cell-cycle dependence. A study from our laboratory showed that although FVs were not able to infect growth-arrested RD-CD4 cells, a productive infection was observed when these cells were allowed to pass through mitosis (9). Other data using FV vectors demonstrated that stationary-phase cultures can be successfully transduced, although at a much lower rate than in proliferating cells, and that effective gene expression occurs after cells are stimulated to divide (62). Thus, although FVs do not infect or transduce nondividing cells to the extent that lentiviruses do, they are slightly less dependent on cell division than MLV (62). Further, FVs appear to survive in nondividing cells and retain their ability for gene expression in cells when they are stimulated to divide later on (9,62). FVs have also been shown to transfer genes to hematopoietic stem/progenitor cells (31). Hence, vectors based on FVs have an advantage over MLV vectors, which demonstrate poor transduction rates in hematopoietic stem cells, most likely because of the lack of cell division (53).

PACKAGING SEQUENCES

Initial experiments using first-generation FV vectors summarized here have proved to be very promising. However, these vectors also transfer coding sequences for viral proteins; this action limits the space available for the incorporation of foreign genes of interest. Identification of the minimal elements of the genome necessary for transfer of the nucleic acid was required to maximize packaging capacity. Genomic RNA is packaged in a highly specific process involving interactions between the viral structural proteins and *cis*-acting sequences on the viral RNA called psi (Ψ) or encapsidation signals (45). The packaging sequences for a number of retroviruses have been identified and are commonly found in close proximity to the 5′ end in the untranslated leader region of the genomic RNA and, in some cases, can extend into the *gag* ORF (6,45). For HIV-1, additional sequences in the *env* ORF have been shown to enhance transfer efficiency (38).

Sequences necessary for transfer of FV RNA into virions have been determined by several groups (21,27,79). FVs require that two distinct regions of the viral genome are present simultaneously for efficient transfer. In HFV, as with previously described retroviral Ψ signals, the first sequence is located at the 5′ end of the viral RNA and encompasses the untranslated leader region and the beginning of the *gag* ORF, which includes the dimer linkage site (21,27). In addition, a second element that appears to be unique to FVs lies at the 3′ end of the *pol* gene (21,27). Similar data for the packaging signals of SFV-1 have been described (79), which suggests that two sequences are necessary for transfer for all members of this genus and that the second polypurine tract appears to be dispensible.

These data identify the minimal sequence requirement for HFV vector transfer and will facilitate the development of FV vectors lacking sequences encoding viral proteins. Of the constructs designed and used by each of these groups, those that demonstrated most efficient transduction contained just 3.5 kb of the FV proviral sequences (21,27). Therefore, this allows for the insertion of at least 10 kb of foreign DNA without exceeding the viral packaging limit.

A FOAMY VIRUS PACKAGING CELL LINE

Retroviral vector design has moved from a single vector carrying the viral structural genes, reporter gene, and gene of interest to a system consisting of two elements: *(i)* a packaging cell line and *(ii)* a vector that encodes the therapeutic gene of interest. This system is favored since it eliminates many of the potential hazards in generating replication-competent viruses, minimizes the potential for target cell damage, and, in turn, increases their acceptance as safe genetic tools in medicine. To develop a packaging cell line, a prerequisite is the identification of the viral genome sequence(s) necessary for packaging or transfer of the viral RNA in virions. As described, FVs require two *cis*-acting elements to achieve efficient vector transfer by helper viruses (21,27,79).

Currently, attempts are being made to produce a FV cell line capable of packaging vectors that contain the minimum and most efficient packaging sequences. All retroviral packaging cell lines developed to date contain two plasmids—one encoding *gag* and *pol* and a second encoding *env* (16). FVs have the advantage that the *gag* and *pol* genes can be separated, and therefore, FV packaging cell lines can contain up to three plasmids encoding the Gag, Pol, and Env proteins. The use of three separate plasmids has significant safety value, as at least three recombination events would have to occur to generate replication-competent wild-type virus (52). Such a cell line, lacking the packaging signals necessary for encapsidation of the DNA transcripts, would produce empty virions. We and others are currently designing first-generation BHK-21 and 293 packaging cell lines that constitutively and functionally express the HFV structural proteins without inducing CPE. The expression vectors used contain the Kozak sequence upstream of the start codons for optimal translation and are driven by the CMV promoter, removing the need for Tas. A packaging cell line derived from SFV-1 has already been produced (80).

The FV vectors used in conjunction with FV packaging cell lines will lack all viral sequences not required for efficient packaging, reverse transcription, integration, and gene expression. Thus, the minimal sequence requirement for HFV vector transfer will be 3.5 kb. Initially, vectors will encode a reporter gene (e.g., β-gal) before being engineered to express proteins of therapeutic interest. Following successful production of high-titer FV vectors encoding a particular gene of interest, they will be analyzed in established cell lines, primary macrophages, lymphocytes, and CD34-positive stem cells. Further development of FV vectors could involve modulation of the viral host range by creating pseudotyped vector systems (see Chapter 31). In retroviruses, host range is determined by the Env protein (75). It is thought that the host cell range of the more commonly studied retroviral MLV-derived vectors can be extended using the foamy virus envelope. Alternatively, the packaging limit of MLV vectors could be increased from 9 kb to 13 kb by using FV capsids (see Chapter 31).

ACKNOWLEDGMENTS

We are grateful to D. Cain and Dr. I. Chrystie (St Thomas's Medical School, London) for Figures 1 and 2, respectively. The authors' research is supported by the European Commission and the Wellcome Trust.

REFERENCES

1. **Achong, B.G. and M.A. Epstein.** 1978. Preliminary seroepidemiological studies on the human syncytial virus. J. Gen. Virol. *40*:175-181.
2. **Achong, B.G., P.W. Mansell, M.A. Epstein, and P. Clifford.** 1971. An unusual virus in culture from a nasopharyngeal carcinoma. J. Natl. Cancer Inst. *46*:299-307.
3. **Ali, M., G.P. Taylor, R.J. Pitman, D. Parker, A. Rethwilm, R. Cheingsong-Popov, J.N. Weber, P.D. Bieniasz, J. Bradley, and M.O. McClure.** 1996. No evidence of antibody to human foamy virus in widespread human populations. AIDS Res. Hum. Retroviruses *12*:1473-1483.
4. **Anderson, W.F.** 1998. Human gene therapy. Nature *392*:25-30.
5. **Baunach, G., B. Maurer, H. Hahn, M. Kranz, and A. Rethwilm.** 1993. Functional analysis of human foamy virus accessory reading frames. J. Virol. *67*:5411-5418.
6. **Berkowitz, R., J. Fischer, and S.P. Goff.** 1996. RNA packaging. Curr. Top. Microbiol. Immunol. *214*:177-218.
7. **Bieniasz, P.D., O. Erlwein, A. Aguzzi, A. Rethwilm, and M.O. McClure.** 1997. Gene transfer using replication-defective human foamy virus vectors. Virology *235*:65-72.
8. **Bieniasz, P.D., A. Rethwilm, R. Pitman, M.D. Daniel, I. Chrystie, and M.O. McClure.** 1995. A comparative study of higher primate foamy viruses, including a new virus from a gorilla. Virology *207*:217-228.
9. **Bieniasz, P.D., R.A. Weiss, and M.O. McClure.** 1995. Cell cycle dependence of foamy retrovirus infection. J. Virol. *69*:7295-7299.
10. **Blaese, R.M., K.W. Culver, A.D. Miller, C.S. Miller, C.S. Carter, T. Fleisher, M. Clerici, G. Shearer, L. Chang, Y. Chiang, et al.** 1995. T lymphocyte-directed gene therapy for ADA$^-$ SCID: initial trial results after 4 years. Science *270*:475-480.
11. **Bohannan, R.C., L.A. Donehower, and R.J. Ford.** 1991. Isolation of a type D retrovirus from B-cell lymphomas of a patient with AIDS. J. Virol. *65*:5663-5672.
12. **Bothe, K., A. Aguzzi, H. Lassmann, A. Rethwilm, and I. Horak.** 1991. Progressive encephalopathy and myopathy in transgenic mice expressing human foamy virus genes. Science *253*:555-557.
13. **Brown, P., M.C. Moreau-Dubois, and D.C. Gajdusek.** 1982. Persistant asymptomatic infection of the

laboratory mouse by simian foamy virus type 6: a new model of retrovirus latency. Arch. Virol. *71*:229-234.
14.**Brown, P.W., G.J. Nemo, and D.C. Gajdusek.** 1978. Human foamy virus: further characterisation, seroepidemiology and relationship to chimpanzee foamy viruses. J. Infect. Dis. *137*:421-427.
15.**Cameron, K.R., S.M. Birchall, and M.A. Moses.** 1978. Isolation of foamy virus from patient with dialysis encephalopathy. Lancet *2*:796.
16.**Cosset, F.L., C. Legras, Y. Chebloune, P. Savatier, P. Thoraval, J.L. Thomas, J. Samarut, V.M. Nigon, and G. Verdier.** 1990. A new avian leukosis virus-based packaging cell line that uses two separate transcomplementing helper genomes. J. Virol. *64*:1070-1078.
17.**Cullen, B.R.** 1991. Human immunodeficiency virus is a prototypic complex retrovirus. J. Virol. *65*:1053-1056.
18.**Enders, J. and T. Peebles.** 1954. Propagation in tissue cultures of cytopathogenic agents from patients with measles. Proc. Soc. Biol. Med. *86*:277-287.
19.**Enssle, J., I. Jordan, B. Maurer, and A. Rethwilm.** 1996. Foamy virus reverse transcriptase is expressed independently from the Gag protein. Proc. Natl. Acad. Sci. USA *93*:4137-4141.
20.**Epstein, M.A., B.G. Achong, and G. Ball.** 1974. Further observations on a human syncytial virus from a nasopharyngeal carcinoma. J. Natl. Cancer Inst. *53*:681-688.
21.**Erlwein, O., P.D. Bieniasz, and M.O. McClure.** 1998. Sequences in pol are required for transfer of human foamy virus based vectors. J. Virol. *72*:5510-5516.
22.**Fischer, N., M. Heinkelein, D. Lindemann, J. Enssle, C. Baum, E. Werder, H. Zentgraf, J.G. Müller, and A. Rethwilm.** 1998. Foamy virus particle formation. J. Virol. *72*:1610-1615.
23.**Flugel, R.M.** 1991. Spumaviruses: a group of complex retroviruses. J. Acquir. Immune Defic. Syndr. *4*:739-750.
24.**Flugel, R.M., A. Rethwilm, B. Maurer, and G. Darai.** 1987. Nucleotide sequence analysis of the *env* gene and its flanking regions of the human spumaretrovirus reveals two novel genes. EMBO J. *6*:2077-2084.
25.**Goepfert, P.A., K.L. Shaw, G.D. Ritter, Jr., and M.J. Mulligan.** 1997. A sorting motif localises the foamy virus glycoprotein to the endoplasmic reticulum. J. Virol. *71*:778-784.
26.**Gow, J.W., K. Simpson, A. Schliephake, W.M. Behan, L.J. Morrison, H. Cavanagh, A. Rethwilm, and P.O. Behan.** 1992. Search for retrovirus in the chronic fatigue syndrome. J. Clin. Pathol. *45*:1058-1061.
27.**Heinkelein, M., M. Schmidt, N. Fischer, A. Moebes, D. Lindemann, J. Enssle, and A. Rethwilm.** 1998. Characterization of a cis-acting sequence in the pol region required to transfer human foamy virus vectors. J. Virol. *72*:6307-6314.
28.**Heneine, W., W.M. Switzer, P. Sandstrom, J. Brown, S. Vedapuri, C.A. Schable, A.S. Khan, N.W. Lerche, M. Schweizer, D. Neumann-Haefelin, et al.** 1998. Identification of a human population infected with simian foamy viruses. Nat. Med. *4*:403-407.
29.**Herchenroder, O., R. Renne, D. Loncar, E.K. Cobb, K.K. Murthy, J. Schneider, A. Mergia, and P.A. Luciw.** 1994. Isolation, cloning and sequencing of simian foamy viruses from chimpanzees (SFVcpz): high homology to human foamy virus (HFV). Virology *201*:187-199.
30.**Hill, C.L., P.D. Bieniasz, and M.O. McClure.** 1999. Properties of human foamy virus relevant to its potential use a as a vector for gene therapy. J. Gen. Virol. *80*:2003-2009.
31.**Hirata, R.K., A.D. Miller, R.G. Andrews, and D.W. Russell.** 1996. Transduction of hematopoietic cells by foamy virus vectors. Blood *88*:3654-3661.
32.**Holzschu, D.L., M.A. Delaney, R.W. Renshaw, and J.W. Casey.** 1998. The nucleotide sequence and spliced *pol* mRNA levels of the nonprimate spumavirus bovine foamy virus. J. Virol. *72*:2177-2182.
33.**Hooks, J.J. and B. Dentrick-Hooks.** 1979. Simian foamy virus-induced immunosupression in rabbits. J. Gen. Virol. *44*:383-390.
34.**Hooks, J.J. and B. Dentrick-Hooks.** 1981. Spumavirinae foamy virus group infections: comparative aspects and diagnosis, p. 559-618. *In* E. Kurstak and C. Kurstak (Eds.), Comparative Diagnosis of Viral Disease. Academic Press, New York.
35.**Hooks, J.J. and C.J. Gibbs.** 1975. The foamy viruses. Bacteriol. Rev. *39*:169-185.
36.**Johnson, R.H., J. de la Rosa, I. Abher, I.G. Kertayadnya, K.W. Entwistle, G. Fordyce, and R.G. Holroyd.** 1988. Epidemiological studies of bovine spumavirus. Vet. Microbiol. *16*:25-33.
37.**Kang, Y., W.S. Blair, and B.R. Cullen.** 1998. Identification and functional characterization of a high-affinity Bel-1 DNA binding site located in the human foamy virus internal promoter. J. Virol. *72*:504-511.
38.**Kaye, J.F., J.H. Richardson, and A.M. Lever.** 1995. cis-acting sequences involved in human immunodeficiency virus type 1 RNA packaging. J. Virol. *69*:6588-6592.
39.**Keller, A., K.M. Partin, M. Löchelt, H. Bannert, R.M. Flügel, and B.R. Cullen.** 1991. Characterization of the transcriptional trans activator of human foamy retrovirus. J. Virol. *65*:2589-2594.
40.**Kennedy-Stoskopf, S., M.K. Stoskopf, M.A. Eckhaus, and J.D. Strandberg.** 1986. Isolation of a retrovirus and a herpesvirus from a captive California sea lion. J. Wildl. Dis. *22*:156-164.
41.**Lagaye, S., P. Vexiau, V. Morozov, V. Guenebaut-Claudet, J. Tobaly-Tapiero, M. Canivet, G. Cathelineau, J. Peries, and R. Emanoil-Ravier.** 1992. Human spumaretrovirus-related sequences in the DNA of leukocytes from patients with Graves disease. Proc. Natl. Acad. Sci. USA *89*:10070-10074.
42.**Lee, A.H., H.Y. Lee, and Y.C. Sung.** 1994. The gene expression of human foamy virus does not require a post-transcriptional transactivator. Virology *204*:409-413.
43.**Lewis, P., M. Hensel, and M. Emerman.** 1992. Human immunodeficiency virus infection of cells arrested in the cell cycle. EMBO J. *11*:3053-3058.
44.**Linial, M.L.** 1999. Foamy viruses are nonconventional retroviruses. J. Virol. *73*:1747-1755.
45.**Linial, M.L. and A.D. Miller.** 1990. Retroviral RNA

packaging: sequence requirements and implications. Curr. Top. Microbiol. Immunol. *157*:125-152.
46.**Lochelt, M. and R.M. Flugel.** 1996. Human foamy virus pol gene is expressed as a pro-pol polyprotein and not as a gag-pol fusion protein. J. Virol. *62*:1120-1124.
47.**Loh, P.C., F. Mutsuuru, and C. Mizumoto.** 1980. Seroepidemiology of human syncitial virus: antibody prevalence in the Pacific. Intervirology *13*:87-90.
48.**Lycke, J., B. Svennerholm, A. Svenningsson, W. Muranyi, R.M. Flugel, and O. Andersen.** 1994. Human spumaretrovirus antibody reactivity in multiple sclerosis. J. Neurol. *241*:204-209.
49.**Mahnke, C., P. Kashaiya, J. Rossler, H. Bannert, A. Levin, W.A. Blattner, M. Dietrich, J. Luande, M. Lochelt, and A.E. Friedman-Kien.** 1992. Human spumavirus antibodies in sera from African patients. Arch. Virol. *123*:243-253.
50.**Maurer, B., H. Bannert, G. Darai, and R.M. Flugel.** 1988. Analysis of the primary structure of the long terminal repear and the *gag* and *pol* genes of the human spumaretrovirus. J. Virol. *62*:1590-1597.
51.**McClure, M.O., P.D. Bieniasz, T.F. Schulz, I.L. Chrystie, G. Simpson, A. Aguzzi, J.G. Hoad, A. Cunningham, J. Kirkwood, and R.A. Weiss.** 1994. Isolation of a new foamy retrovirus from orangutans. J. Virol. *68*:7124-7130.
52.**Miller, A.D.** 1992. Retroviral vectors. Curr. Top. Microbiol. Immunol. *158*:1-24.
53.**Miller, D.G., M.A. Adam, and A.D. Miller.** 1990. Gene transfer by retrovirus vectors occurs only in cells that are actively replicating at the time of infection. Mol. Cell. Biol. *10*:4239-4242.
54.**Moebes, A., J. Enssle, P.D. Bieniasz, M. Heinkelein, D. Lindemann, M. Bock, M.O. McClure, and A. Rethwilm.** 1997. Human foamy virus reverse transcription that occurs late in the viral replication cycle. J. Virol. *71*:7305-7311.
55.**Muller, H.K., G. Ball, M.A. Epstein, B.G. Achong, G. Lenoir, and A. Levin.** 1980. The prevalence of naturally occurring antibodies to human syncytial virus in East African populations. J. Gen. Virol. *47*:399-406.
56.**Nemo, G.J., P.W. Brown, C.J. Gibbs, and D.C. Gajdusek.** 1978. Antigenic relationship of human foamy virus to the simian foamy viruses. Infect. Immun. *20*:69-72.
57.**Nestler, U., M. Heinkelein, M. Lucke, J. Meizensberger, W. Scheurlenm, A. Kretschmer, and A. Rethwilm.** 1997. Foamy virus vectors for suicide gene therapy. Gen. Ther. *4*:1270-1277.
58.**Neumann-Haefelin, D., A. Rethwilm, G. Bauer, F. Gudat, and H.H. zur Hausen.** 1983. Characterization of a foamy virus isolated from Ceropithecus aethiops lymphoblastoid cells. Med. Microbiol. Immunol. *172*:75-86.
59.**Rethwilm, A., G. Baunach, K. Netzer, B. Maurer, B. Borisch, and V. Meulen.** 1990. Infectious DNA of the human spumaretrovirus. Nucleic Acids Res. *18*:733-738.
60.**Rethwilm, A., O. Erlwein, G. Baunach, B. Maurer, and V. Meulen.** 1991. The transcriptional transactivator of human foamy virus maps to the bel1 genomic region. Proc. Natl. Acad. Sci. USA *88*:941-945.
61.**Riggs, J.L., L.S. Oshiro, D.O.N. Taylor, and E.H. Lennette.** 1969. Syncytium-forming agent isolated from domestic cats. Nature *222*:1190-1191.
62.**Russell, D.W. and A.D. Miller.** 1996. Foamy virus vectors. J. Virol. *70*:217-222.
63.**Salahuddin, S.Z., V.A. Dharam, P.D. Markham, S.F. Josephs, M.K. Sturzenegger, G. Halligan, P. Biberfield, F. Wong-Staal, B. Kramarsky, and R.C. Gallo.** 1986. Isolation of a new virus, HBLV, in patients with lymphoproliferative disorders. Science *234*:596-601.
64.**Schmidt, M., S. Niewiesk, J. Heeny, A. Aguzzi, and A. Rethwilm.** 1997. Mouse model to study the replication of primate foamy viruses. J. Gen. Virol. *78*:1929-1933.
65.**Schmidt, M. and A. Rethwilm.** 1995. Replicating foamy virus based vectors directing high level expression of foreign genes. Virology *210*:167-178.
66.**Schweizer, M., V. Falcone, J. Gänge, R. Turek, and D. Neumann-Haefelin.** 1997. Simian foamy virus isolated from an accidentally infected human individual. J. Virol. *71*:4821-4824.
67.**Schweizer, M., R. Turek, H. Hahn, A. Schliephake, K. Netzer, G. Eder, M. Reinhardt, A. Rethwilm, and D. Neumann-Haefelin.** 1995. Markers of foamy virus infections in monkeys, apes and accidentally infected humans: appropriate testing fails to confirm suspected foamy virus prevalence in humans. AIDS Res. Hum. Retroviruses *11*:161-170.
68.**Schweizer, M., R. Turek, M. Reinhardt, and D. Neumann-Haefelin.** 1994. Absence of foamy virus DNA in Graves' disease. AIDS Res. Hum. Retroviruses *10*:601-605.
69.**Stancek, D. and M. Gressnerova.** 1974. A viral agent isolated from a patient with subacute thyroiditis de Quervain type. Acta. Virol. *18*:365.
70.**Stancek, D., M. Stancekova, P. Janotka-Hnilica, and D. Oravec.** 1975. Isolation and some serological and epidemiological data on the viruses recovered from patients with subacute thyroiditis de Quervain. Med. Microbiol. Immunol. (Berl.) *161*:133-144.
71.**Svenningsson, A., J. Lycke, B. Svennerholm, S. Gronowitz, and O. Andersen.** 1992. No evidence for spumavirus or oncovirus infection in relapsing-remitting multiple sclerosis. Ann. Neurol. *32*:711-714.
72.**Swack, N.S., R.A. Schoentag, and G.D. Hsiung.** 1970. Foamy virus infection of rhesus and green monkeys in captivity. Am. J. Epidemiol. *92*:79-83.
73.**Takeuchi, Y., C.D. Porter, K.M. Strahan, A.F. Preece, K. Gustafsson, F.L. Cosset, R.A. Weiss, and M.K.L. Collins.** 1996. Sensitization of cells and retroviruses to human sera by (alpha-1-3) galactosyltransferase. Nature *379*:85-88.
74.**Trobridge, G.D. and D.W. Russell.** 1998. Helper-free foamy virus vectors. Hum. Gene Ther. *9*:2517-2525.
75.**Weiss, R.A.** 1984. Experimental biology and assay of RNA tumour viruses, p. 209-260. *In* R.A. Weiss, N. Teich, H. Varmus, and J. Coffin (Eds.), RNA Tumour Viruses, 2nd ed. CSH Laboratory Press, Cold Spring Harbor, NY.
76.**Weiss, R.A.** 1988. A virus in search of a disease. Nature *333*:497-498.
77.**Wick, G., K. Trieb, A. Aguzzi, H. Recheis, H.**

Anderl, and B. Grubeck-Lobenstein. 1993. Possible role of human foamy virus in Graves' disease. Intervirology *35*:101-107.

78. **Winkler, I., J. Bodem, L. Haas, M. Zemba, H. Delius, R. Flower, R.M. Flugel, and M. Lochelt.** 1997. Characterization of the genome of feline foamy virus and its proteins shows distinct features different from those of primate spumaviruses. J. Virol. *71*:6727-6741.

79. **Wu, M., S. Chari, T. Yanchis, and A. Mergia.** 1998. cis-acting sequences required for simian foamy virus type 1 vectors. J. Virol. *72*:3451-3454.

80. **Wu, M. and A. Mergia.** 1999. Packaging cell lines for simian foamy virus type 1 vectors. J. Virol. *73*:4498-4501.

81. **Young, D., J. Samuels, and J.K. Clarke.** 1973. A foamy virus of possible human origin isolated in BHK-21 cells. Arch. Gesamte Virusforsch. *42*:228-234.

82. **Yu, S.F., D.N. Baldwin, S.R. Gwynn, S. Yendapalli, and M.L. Linial.** 1996. The human foamy virus replication pathway is distinct from that of retroviruses and hepadnaviruses. Science *268*:1579-1582.

83. **Yu, S.F. and M. Linial.** 1993. Analysis of the role of the bel and bet open reading frames of human foamy virus by using a quantitative assay. J. Virol. *67*:6618-6624.

31 Aspects of Foamy Virus Vectors: Host Range and Replication-Competent Vectors

Dirk Lindemann[1], Martin Heinkelein[1], and Axel Rethwilm[2]
[1]Institute of Virology and Immunology, University of Würzburg, Würzburg; [2]Institute of Virology, Carl Gustav Carus Faculty of Medicine, Technical University of Dresden, Dresden, Germany

INTRODUCTION

Foamy viruses (FVs) or spumaviruses are a separate subclass of retroviruses. Because of special properties and unique features of their replication strategy, they have received increased interest as potential transfer vectors for gene therapeutic applications in recent years. Many of these unique features (e.g., the bipartite packaging sequence, the independent expression of the Pol protein from a separately spliced mRNA, and the basic biology of FVs) are discussed in detail in Chapter 30. In this article, we will concentrate on two special properties of FVs and their consequences for the development of FV vector systems. First, the FV envelope (Env) protein, which plays an important role in the FV particle release process and is the origin of the broad host range of FVs, and second, the benign nature of the FV infections in their natural hosts, mainly nonhuman primates, which gave rise to the development of replication-competent FV vectors and their potential use for cancer therapy.

THE FV ENVELOPE PROTEIN AND ITS ROLE IN FV PARTICLE RELEASE

Retroviral Env proteins are generally synthesized as precursor proteins that are cleaved by a cellular protease into surface (SU) and transmembrane (TM) subunits during their transport to the cell surface. These different Env subunits serve two major functions. Initially, the SU subunit is responsible for docking of the viral particle to the host cell by interacting with the cellular receptor. Subsequently, fusion of the viral membrane with the cellular membrane is mediated by a hydrophobic stretch of amino acids at or near the N terminus of the TM subunit, called the fusion peptide, allowing the viral capsid to be released into the cytoplasm of the host cell (10,19).

Two distinct forms of the human isolate of the chimpanzee foamy virus (CFV/hu) Env protein have been described to date (12,13,22,27).

The minor form is a 170-kDa Env-Bet fusion protein that is generated by an alternative splicing mechanism using an evolu-

Viral Vectors: Basic Science and Gene Therapy
Edited by A. Cid-Arregui and A. García-Carrancá

tionarily conserved splice donor/acceptor pair localized within the Env open reading frame (ORF) (12,22). Its biological role is currently unknown. However, Env-Bet is not essential for particle release and in vitro replication of CFV/hu (22) and will therefore not be discussed in further detail. The predominant form of the CFV/hu Env protein is expressed as a 130-kDa precursor glycoprotein that is cleaved by an unknown cellular protease into an 80- to 90-kDa SU and a 47-kDa TM subunit during its transport to the cell surface and incorporation into the viral particle (14). Unique structural features that distinguish gp130 from most other retroviral Env proteins are the extremely long putative membrane spanning domain (MSD) and the very short cytoplasmic domain (CyD) of the TM subunit (Figure 1). Furthermore, because of a retrieval signal present in the cytoplasmic domain, most of the 130-kDa CFV/hu Env protein is retained in the endoplasmic reticulum (ER) in the absence of expression of other CFV/hu structural genes or by inactivation of the ER retrieval signal (14,15). FVs' capsid assembly strategy resembles that of the B/D-type retroviruses, which preform capsids (A-type particles) in the cytoplasm of the infected cells before the viral particle buds across cellular membranes (8,38). Therefore, it is thought that ER retrieval of the gp130 results in the efficient incorporation of FV Env into FV particles, which bud predominantly into intracellular compartments, presumably the ER. However, budding of FV is also observed at the plasma membrane (3,11). Most retroviruses, such as the murine leukemia viruses (MLVs), the lentiviruses, or the B/D-type retroviruses, require only the cellular expression of the capsid protein (Gag) for the assembly of viral core structures, their enclosure by lipid membranes of the host cell, and the release of virus-like particles from the expressing cell into the supernatant (8,38). However, unlike all other retroviruses, FVs require the coexpression of the Env protein for viral particle egress (3,11). In cells transfected with Env-deficient FV proviral clones or vectors no virus particle release could be detected. Instead, an accumulation of naked capsids in the cytoplasm of the cell was observed (11). This may point to a more specific interaction of the FV Env protein with the FV capsid as observed for other retroviruses. Recently, we have identified the MSD of FV gp130 as an important determinant for FV particle release (28). Furthermore, we could show that neither the MLV Env protein nor the vesicular stomatitis virus (VSV) glycoprotein G, which efficiently pseudotypes other retroviral capsids, is capable of supporting FV particle release and results in pseudotyped FV particles (28). Surprisingly, replacement of the MSD and/or CyD of these Env proteins with the corresponding domains of FV Env was not able to overcome the block in

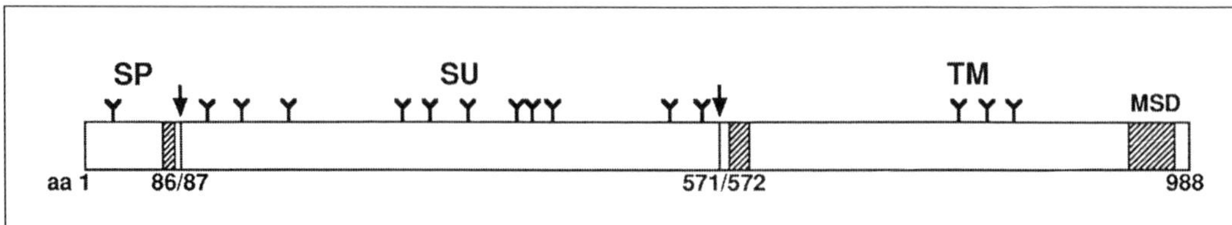

Figure 1. Schematic illustration of the CFV/hu envelope structural organization. The CFV/hu Env protein is predicted to be posttranslationally cleaved by cellular proteases first after amino acid 86 to release the signal peptide (SP) and second after amino acid 571 to generate the SU and TM subunits. Hydrophobic stretches in the CFV/hu sequence, at the C terminus of the SP, near the N terminus (fusion peptide) and close to the C terminus (MSD) of the TM subunit, are indicated as hatched boxes. Potential N-glycosylation sites are indicated as Y. The human isolate of the CFV/hu was previously named HFV.

FV particle release, thereby indicating that additional FV Env domains are involved in FV particle egress. Therefore, a more detailed analysis of the role of gp130 in the FV particle release is necessary to be able to generate pseudotyped FV vectors with altered host range.

We have shown previously (21) that the FV Env protein can be used to pseudotype the currently most commonly used retroviral vectors based on MLV to utilize the unique features and advantages of the FV Env protein (see below). The original FV Env protein was readily incorporated into MLV retroviral vectors, albeit with low efficiency. By creating a chimeric Env protein (having part of the FV CyD replaced by that of the MLV Env protein), we were able to achieve a more efficient particle incorporation of the Env protein that resulted in higher titers of the recombinant vector particles generated (21).

Broad Host Range and Other Features Mediated by the FV Env Protein

FVs are characterized by a very broad host range. To date, we and others (17,18, 21,25,33,34,39,41, and unpublished data) have not been able to find a cell line or cell type that clearly could not be infected by FVs or FV-based vectors. A comprehensive list of various target cells examined in different reports is given in Table 1. The receptor of FV is currently unknown. Because of the broad host range, however, it seems to be a ubiquitiously expressed cell surface molecule that is conserved among different species.

Besides its capacity to infect cells of various origins, there are additional advantages of the FV Env protein as compared with other viral Env proteins used with retroviral vectors. These advantages may make it especially well suited as an in vivo gene delivery system. First, retroviral vectors carrying the FV Env protein do not require the addition of polycations such as polybrene or protamine for efficient transduction of target cells (21,33). Second, such vectors are not inactivated by the human complement.

REPLICATION-COMPETENT FV VECTORS

A replication-competent vector derived from a nonpathogenic virus could be an interesting agent to use in different approaches that require efficient gene transfer into eukaryotic cells in vitro and in vivo. FVs are nonpathogenic retroid viruses whose replication strategy, although sharing some similarities, differs from retroviruses and hepadnaviruses (24,32). There are some features of FVs that make them attractive candidates for the development of gene transfer vectors: *(i)* their broad host cell range (see above), *(ii)* an extended packaging ability as compared with commonly used retroviral vectors, and *(iii)* stable expression of the inserted foreign gene in the target cell (5,33,34). Furthermore, replication-competent FV vectors that carry a reporter gene should provide a sensitive tool for the study of various aspects of virus biology.

For these reasons, we decided to develop a replication-competent FV vector system (pFOV) based on the human isolate of CFV/hu, previously named human foamy virus (HFV).

Construction and Molecular Biology of Replication-Competent FV Vectors

Besides the *gag*, *pol*, and *env* ORFs, FVs bear at least two additional ORFs located between the *env* gene and the 3′ long terminal repeat (LTR) (Figure 2). The first one of the accessory ORFs, *bel-1*, encodes a potent 36-kDa *trans*-acting factor that stimulates transcription driven by FV pro-

Table 1. Host Range of FV and Cell Types Susceptible to CFV/hu Infection

Species	Cell line	Cell type	References
human	HT1080	fibrosarcoma	21,39
	primary	fibroblasts	18,33
	MRC-5	fibroblasts	34
	293	kidney epithelial	25 and unpublished data
	primary CD34+	bone marrow	18
	primary	bone marrow stroma cells	18
	primary	peripheral blood lymphocytes	unpublished data
	EBV transf.	B-cell blasts	unpublished data
	CEM	T cell	25
	HUT 78	T cell	25
	Jurkat	T cell	41 and unpublished data
	A.301	T cell	unpublished data
	H92.1.7	erythroblastoid	25,41
	U937	myeloid-monocytic	41
	THP-1	myeloid	25
	KG-1	myeloid	25 and unpublished data
	K-562	myeloid	18 and unpublished data
	U87	glioblastoma	unpublished data
	Hep3b	liver epithelial	unpublished data
monkey	COS-7	kidney fibroblast	33
	Vero	kidney epithelial	33
	primary CD34+	bone marrow	18
mouse	NIH3T3	fibroblast	21
	L929	fibroblast	21
	MEL	erythroleukemia	18
	primary CD34+	bone marrow	18
rat	NRK	kidney epithelial	33
	208F		33
hamster	BHK-21	kidney fibroblast	18,21,25,33,41
	CHO-K1	ovary epithelial	33
mink	Mv.1.Lu	lung epithelial	21
quail	QT-6	fibrosarcoma	21
chicken	primary	embryo, fibroblasts	33
dog	Cf2Th	thymus fibroblast	33
cat	CCC-81		33
	AH 927		33
bovine	MDBK	kidney epithelial	33
	primary	choroid plexus	33

moters (32). The second ORF, *bel-2*, is preferentially expressed from a multiply spliced mRNA that joins a *bel-1* exon to the *bel-2* ORF, giving rise to the 60-kDa Bet protein (4,24). Interestingly, only the *bel-1* regulatory gene was found to be essential for virus replication. CFV/hu clones with mutations in the *bel-2* ORF were shown to be replication competent in a variety of tissue culture cells. In addition, they replicate to yield virus titers that are comparable to or only slightly lower than those of the wild-type virus (4,20,40). Therefore, we reasoned that CFV/hu mutants deleted in the *bel-2* ORF could be used as a vector to express foreign DNA. To this end, a vector system was designed based on a plasmid (pFOV) that carries the cDNA sequence corresponding to an infectious clone of CFV/hu. This plasmid contains the coding sequences required for the production of replication-competent FV vectors when transfected into fibroblast cell lines. The packaging capacity of these vectors is in the range of 2 kb.

For the generation of the pFOV plas-

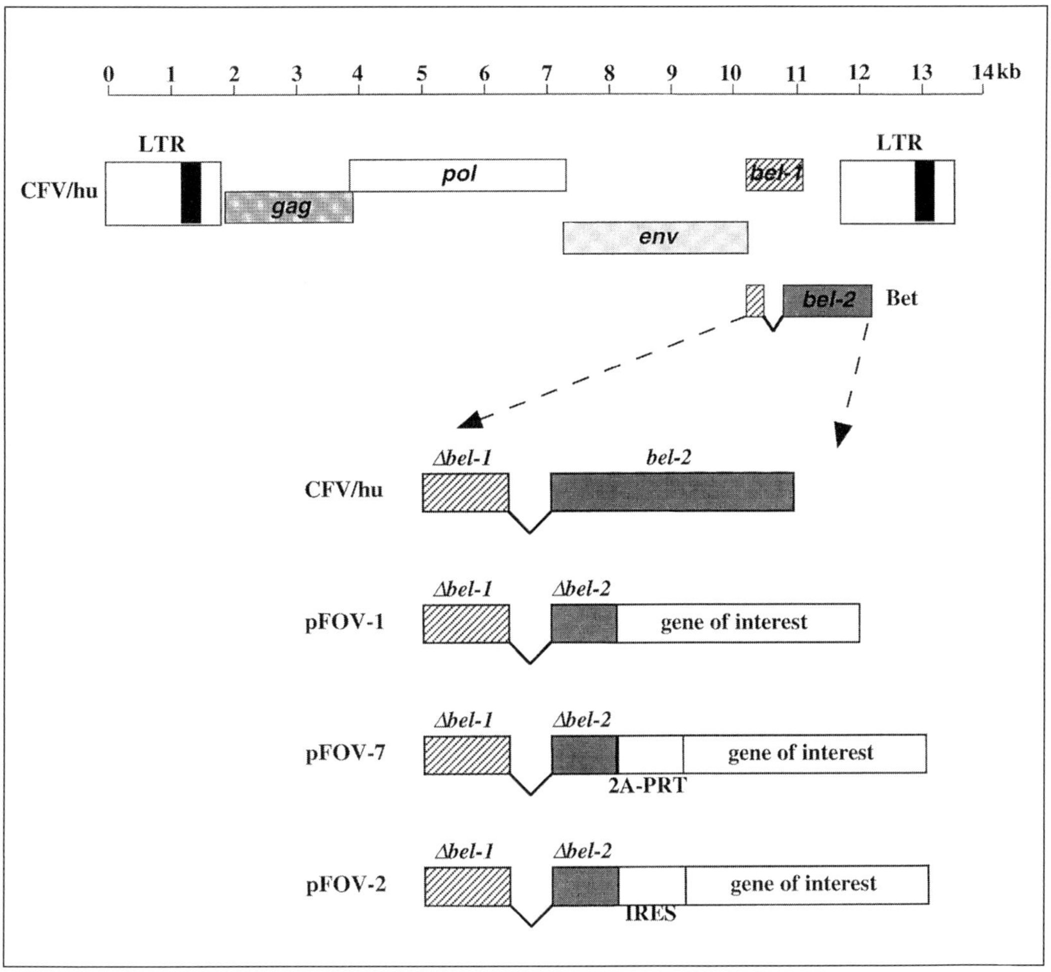

Figure 2. Genome organization of wild-type FV and the replication-competent FV vector sytem pFOV. In this vector system, the *bet* reading frame is deleted, and foreign genes are initially expressed as C-terminal fusion proteins to the truncated Bet protein (pFOV-1). The self-cleaving 2A protease of FMDV (2A-PRT) was introduced to facilitate the cleavage of the fusion partners (pFOV-7). Alternatively, an IRES was inserted, which allows cap site-independent translation and expression of a foreign gene (pFOV-2).

mid, the *bel-2* genomic region was deleted from immediately downstream of the *bel-1* ORF to the polypurine tract just before the 3′ LTR. A small polylinker was introduced in place of the deleted sequence to facilitate insertion of foreign genes, which are expressed as fusion proteins to the C-terminally truncated Bet protein (pFOV-1; Figure 2). To further improve this vector system, an additional sequence was introduced upstream of the polylinker sequence coding for the self-cleaving foot-and-mouth disease virus (FMDV) 2A protease (pFOV-7; Figure 2). In this vector, the foreign gene is initially expressed as a Bet fusion protein similar to the pFOV-1 vector. However, the presence of the FMDV 2A protease between the two fusion partners results in their separation via autocatalytic cleavage. Alternatively, an internal ribosomal entry site (IRES) was inserted, allowing for cap site-independent translation of the foreign sequence (pFOV-2; Figure 2).

Several reporter genes including the chloramphenicol acetyl transferase (*CAT*), the luciferase, and the green fluorescent protein (*GFP*) genes were inserted into the different vectors described above. Transfection of the vector plasmids into BHK-21 cells resulted in the generation of replication-competent virus vectors. When different cell lines were infected with these vectors, their replication abilities were only slightly reduced as compared to the wild-type virus. Vector titers ranging between 1×10^4 and 1×10^7 infectious units (iu)/mL could be easily produced. Cells infected by these vectors were analyzed by Western blot analysis and were assayed for reporter activities showing good expression of functional recombinant proteins (26,34, and unpublished data). These data indicate that the FOV vector system provides a useful tool to transduce eukaryotic cells by a nonpathogenic replication-competent FV vector (34).

Replication-Competent FV Vectors for Suicide Gene Therapy of Cancer

Viral vector-mediated genetic delivery of enzymes that convert prodrugs into toxic, active drugs in tumors has gained increasing interest among cancer therapists (9,37). In most experimental and clinical protocols, retrovirus vectors are used for gene delivery, and trials in human glioblastoma patients have been initiated using an MLV-derived packaging cell line and an MLV-derived vector expressing the herpes simplex virus type 1 (HSV-1) thymidine kinase (*tk*) (2). After injection of vector-producing packaging cells into tumors, the viral vector is able to perform one round of reverse transcription and integration. Since MLV-derived vectors only integrate into proliferating cells, glioblastoma cells are much more susceptible to infection by the virus vector than the normal tissue surrounding the tumor. However, recent studies using this approach have shown only limited success in humans and led the authors to conclude that improvement of present gene delivery systems is required (29).

In an attempt to improve the strategy of virus-mediated destruction of tumor cells, we have initiated studies using the HSV-1 *tk* and other suicide genes inserted into our replication-competent FV vectors. Ideally, the viral vectors to be used for cancer therapy should: *(i)* be nonpathogenic, *(ii)* be of primate origin to ensure efficient infection of human cells, *(iii)* preferentially integrate into proliferating cells, *(iv)* allow the stable insertion and expression of a functional suicide gene, and *(v)* have no antibodies present in the human population to avoid problems arising from neutralization of the vector. Primate FVs fulfill criteria *ii* (6,16,30), *iii* (7), and *v* (1,35). With respect to the first criterion, no pathogenicity has so far been convincingly associated with FV infection (30,31,36). Therefore, we tested the expression of different suicide genes using the vector system described above and have tested their stability and functionality in the context of virus replication (criterion *iv*) in vitro.

Recombinant viruses were constructed by inserting either the HSV-1 *tk*, the *Escherichia coli* cytosine deaminase (*cd*), or purine nucleoside phosphorylase (*pnp*) into the FV-derived, replication-competent vector (pFOV-7; Figure 2). Expression and stability of the inserted foreign genes were analyzed by immunoblot and polymerase chain reaction. The functionality of the suicide genes was determined either by a metabolic assay using virus vector-infected cells or by treatment with the respective prodrug in vitro. In terms of vector stability and effectiveness of specific cell killing, a virus transducing the *pnp* gene (FOV-7/*pnp*) was superior over vectors using the other two suicide genes (27).

Preliminary experiments indicate that this suicide vector is able to replicate in tumors induced experimentally in vivo, and that suicide gene expression is detectable in the tumor cells by Western blot analysis (M. Heinkelein et al., unpublished data). The antitumor potential of the vector is currently under investigation.

These studies demonstrate that replication-competent FV vectors can be used for the delivery of suicide genes into tumor cells in vitro and in vivo. Further studies are required, however, to show whether this approach will be successful to eliminate experimental tumors in vivo.

REFERENCES

1. **Ali, M., G.P. Taylor, R.J. Pitman, D. Parker, A. Rethwilm, R. Cheingsong Popov, J.N. Weber, P.D. Bieniasz, J. Bradley, and M.O. McClure.** 1996. No evidence of antibody to human foamy virus in widespread human populations. AIDS Res. Hum. Retroviruses *12*:1473-1483.
2. **Anderson, W.F.** 1998. Human gene therapy. Nature *392*:25-30.
3. **Baldwin, D.N. and M.L. Linial.** 1998. The roles of Pol and Env in the assembly pathway of human foamy virus. J. Virol. *72*:3658-3665.
4. **Baunach, G., B. Maurer, H. Hahn, M. Kranz, and A. Rethwilm.** 1993. Functional analysis of human foamy virus accessory reading frames. J. Virol. *67*:5411-5418.
5. **Bieniasz, P.D., O. Erlwein, A. Aguzzi, A. Rethwilm, and M.O. McClure.** 1997. Gene transfer using replication-defective human foamy virus vectors. Virology *235*:65-72.
6. **Bieniasz, P.D., A. Rethwilm, R. Pitman, M.D. Daniel, I. Chrystie, and M.O. McClure.** 1995. A comparative study of higher primate foamy viruses, including a new virus from a gorilla. Virology *207*:217-228.
7. **Bieniasz, P.D., R.A. Weiss, and M.O. McClure.** 1995. Cell cycle dependence of foamy retrovirus infection. J. Virol. *69*:7295-7299.
8. **Boulanger, P. and I. Jones.** 1996. Use of heterologous expression systems to study retroviral morphogenesis. Curr. Top. Microbiol. Immunol. *214*:237-260.
9. **Deonarain, M.P., R.A. Spooner, and A.A. Epenetos.** 1995. Genetic delivery of enzymes for cancer therapy. Gene Ther. *2*:235-244.
10. **Einfeld, D.** 1996. Maturation and assembly of retroviral glycoproteins. Curr. Top. Microbiol. Immunol. *214*:133-176.
11. **Fischer, N., M. Heinkelein, D. Lindemann, J. Enssle, C. Baum, E. Werder, H. Zentgraf, J.G. Muller, and A. Rethwilm.** 1998. Foamy virus particle formation. J. Virol. *72*:1610-1615.
12. **Giron, M.L., H. de The, and A. Saib.** 1998. An evolutionarily conserved splice generates a secreted env-Bet fusion protein during human foamy virus infection. J. Virol. *72*:4906-4910.
13. **Giron, M.L., F. Rozain, M.C. Debons Guillemin, M. Canivet, J. Peries, and R. Emanoil Ravier.** 1993. Human foamy virus polypeptides: identification of env and bel gene products. J. Virol. *67*:3596-3600.
14. **Goepfert, P.A., K.L. Shaw, G.D. Ritter, Jr., and M.J. Mulligan.** 1997. A sorting motif localizes the foamy virus glycoprotein to the endoplasmic reticulum. J. Virol. *71*:778-784.
15. **Goepfert, P.A., G. Wang, and M.J. Mulligan.** 1995. Identification of an ER retrieval signal in a retroviral glycoprotein. Cell *82*:543-544.
16. **Herchenroder, O., R. Renne, D. Loncar, E.K. Cobb, K.K. Murthy, J. Schneider, A. Mergia, and P.A. Luciw.** 1994. Isolation, cloning, and sequencing of simian foamy viruses from chimpanzees (SFVcpz): high homology to human foamy virus (HFV). Virology *201*:187-199.
17. **Hill, C.L., P.D. Bieniasz, and M.O. McClure.** 1999. Properties of human foamy virus relevant to its potential as a vector for gene therapy. J. Gen. Virol. *80*:2003-2009.
18. **Hirata, R.K., A.D. Miller, R.G. Andrews, and D.W. Russell.** 1996. Transduction of hematopoietic cells by foamy virus vectors. Blood *88*:3654-3661.
19. **Hunter, E.** 1997. Viral entry and receptors, p. 71-119. *In* J.M. Coffin, S.H. Hughes, and H.E. Varmus (Eds.), Retroviruses. CSH Laboratory Press, Cold Spring Harbor, NY.
20. **Lee, C.W., J. Chang, K.J. Lee, and Y.C. Sung.** 1994.

The Bel1 protein of human foamy virus contains one positive and two negative control regions which regulate a distinct activation domain of 30 amino acids. J. Virol. *68*:2708-2719.

21. **Lindemann, D., M. Bock, M. Schweizer, and A. Rethwilm.** 1997. Efficient pseudotyping of murine leukemia virus particles with chimeric human foamy virus envelope proteins. J. Virol. *71*:4815-4820.

22. **Lindemann, D. and A. Rethwilm.** 1998. Characterization of a human foamy virus 170-kilodalton Env-Bet fusion protein generated by alternative splicing. J. Virol. *72*:4088-4094.

23. **Linial, M.L.** 1999. Foamy viruses are unconventional retroviruses. J. Virol. *73*:1747-1755.

24. **Löchelt, M., R.M. Flugel, and M. Aboud.** 1994. The human foamy virus internal promoter directs the expression of the functional Bel 1 transactivator and Bet protein early after infection. J. Virol. *68*:638-645.

25. **Mikovits, J.A., P.M. Hoffman, A. Rethwilm, and F.W. Ruscetti.** 1996. In vitro infection of primary and retrovirus-infected human leukocytes by human foamy virus. J. Virol. *70*:2774-2780.

26. **Nestler, U., M. Heinkelein, M. Lucke, J. Meixensberger, W. Scheurlen, A. Kretschmer, and A. Rethwilm.** 1997. Foamy virus vectors for suicide gene therapy. Gene Ther. *4*:1270-1277.

27. **Netzer, K.O., A. Rethwilm, B. Maurer, and V. ter Meulen.** 1990. Identification of the major immunogenic structural proteins of human foamy virus. J. Gen. Virol. *71*:1237-1241.

28. **Pietschmann, T., M. Heinkelein, M. Heldmann, H. Zentgraf, A. Rethwilm, and D. Lindemann.** 1999. Foamy virus capsids require the cognate envelope protein for particle export. J. Virol. *73*:2613-2621.

29. **Ram, Z., K.W. Culver, E.M. Oshiro, J.J. Viola, H.L. DeVroom, E. Otto, Z. Long, Y. Chiang, G.J. McGarrity, L.M. Muul, et al.** 1997. Therapy of malignant brain tumors by intratumoral implantation of retroviral vector-producing cells [see comments]. Nat. Med. *3*:1354-1361.

30. **Rethwilm, A.** 1995. Regulation of foamy virus gene expression. Curr. Top. Microbiol. Immunol. *193*:1-24.

31. **Rethwilm, A.** 1996. Unexpected replication pathways of foamy viruses. J. Acquir. Immune Defic. Syndr. Hum. Retrovirol. 13 Suppl *1*:S248-253.

32. **Rethwilm, A., O. Erlwein, G. Baunach, B. Maurer, and V. ter Meulen.** 1991. The transcriptional transactivator of human foamy virus maps to the bel 1 genomic region. Proc. Natl. Acad. Sci. USA *88*:941-945.

33. **Russel, D.W. and A.D. Miller.** 1996. Foamy virus vectors. J. Virol. *70*:217-222.

34. **Schmidt, M. and A. Rethwilm.** 1995. Replicating foamy virus-based vectors directing high level expression of foreign genes. Virology *210*:167-178.

35. **Schweizer, M., R. Turek, H. Hahn, A. Schliephake, K.O. Netzer, G. Eder, M. Reinhardt, A. Rethwilm, and D. Neumann Haefelin.** 1995. Markers of foamy virus infections in monkeys, apes, and accidentally infected humans: appropriate testing fails to confirm suspected foamy virus prevalence in humans. AIDS Res. Hum. Retroviruses *11*:161-170.

36. **Schweizer, M., R. Turek, M. Reinhardt, and D. Neumann Haefelin.** 1994. Absence of foamy virus DNA in Graves' disease. AIDS Res. Hum. Retroviruses *10*:601-605.

37. **Sikora, K.** 1994. Genetic approaches to cancer therapy [editorial]. Gene Ther. *1*:149-151.

38. **Swanstrom, R. and J.W. Wills.** 1997. Synthesis, assembly, and processing of viral proteins, p. 263-334. *In* J.M. Coffin, S.H. Hughes, and H.E. Varmus (Eds.), Retroviruses. CSH Laboratory Press, Cold Spring Harbor, NY.

39. **Trobridge, G.D. and D.W. Russell.** 1998. Helper-free foamy virus vectors. Gene Ther. *9*:2517-2525.

40. **Yu, S.F. and M.L. Linial.** 1993. Analysis of the role of the bel and bet open reading frames of human foamy virus by using a new quantitative assay. J. Virol. *67*:6618-6624.

41. **Yu, S.F., J. Stone, and M.L. Linial.** 1996. Productive persistent infection of hematopoietic cells by human foamy virus. J. Virol. *70*:1250-1254.

FAMILY: Orthomyxoviridae
GENUS: Influenza Virus

32 Transfectant Influenza Viruses as Vaccine Vectors

Adolfo García-Sastre
Department of Microbiology, Mount Sinai School of Medicine, New York, NY, USA

MOLECULAR BIOLOGY OF INFLUENZA VIRUSES

Influenza viruses are segmented negative-strand RNA viruses belonging to the group of the orthomyxoviruses (17). Human influenza viruses belong to three major types: A, B, and C. Influenza A viruses are the most important type from the point of view of human and animal pathogenicity. Influenza A viruses comprise different virus subtypes, according to the antigenicity of their external glycoproteins, hemagglutinin (HA) and neuraminidase (NA). Figure 1 shows a schematic representation of an influenza A virus particle. The outside of the virus consists of a lipid envelope that contains three different viral transmembrane proteins: the HA, NA, and M2 proteins. Underneath the envelope there is a protein layer of matrix (M1) protein. The genomic RNA is located in the inside of the virus particle. There are eight different viral RNA segments that are tightly complexed into ribonucleoproteins (RNP) by association with the nucleoprotein (NP) protein. A few molecules of the viral RNA-dependent RNA polymerase are also associated with the RNPs. This polymerase is a complex of three viral proteins: PB1, PB2, and PA (P proteins). In addition, virions contain a few molecules of the nuclear export protein (NEP or NS2) protein, which was previously thought to be nonstructural. Influenza A viruses also encode a nonstructural protein (NS1) that is only found in virus-infected cells.

The replication cycle of influenza A viruses has been schematically depicted in Figure 2. Infection of a permissive cell by an influenza A virus is initiated by binding of a viral particle to sialic acid-containing receptors through the HA protein. The virus becomes internalized into an endosome, where the HA undergoes a conformational change caused by the acidification of the endosomal milieu. The new conformation of the HA is responsible for the fusion between the viral and endosomal membrane, thereby resulting in the release of the RNPs into the cytoplasm of the infected cell. Uncoating (i.e., M1 release from the RNP complexes) is mediated by proton pumping into the inside of the virus particle

Viral Vectors: Basic Science and Gene Therapy
Edited by A. Cid-Arregui and A. García-Carrancá

while in the endosome. The M2 is responsible for this ion channel activity, and M2 inhibitors (e.g., amantadine) are being used as antiviral agents against influenza A viruses. After uncoating, the RNPs are transported into the nucleus of the infected cells, where RNA replication and transcription take place. Viral RNAs (vRNA) are transcribed into single mRNA species by the viral polymerase. Therefore, every RNA segment encodes only one viral protein, except for the M and NS RNA segments, whose mRNAs undergo alternative splicing, and there are translated into the M1 and M2, and NS1 and NEP (NS2) proteins, respectively. In addition to mRNA transcription from vRNA, the viral polymerase catalyzes the synthesis of full-length complementary RNAs (cRNA) using vRNA templates. The cRNAs are used as replicative intermediates

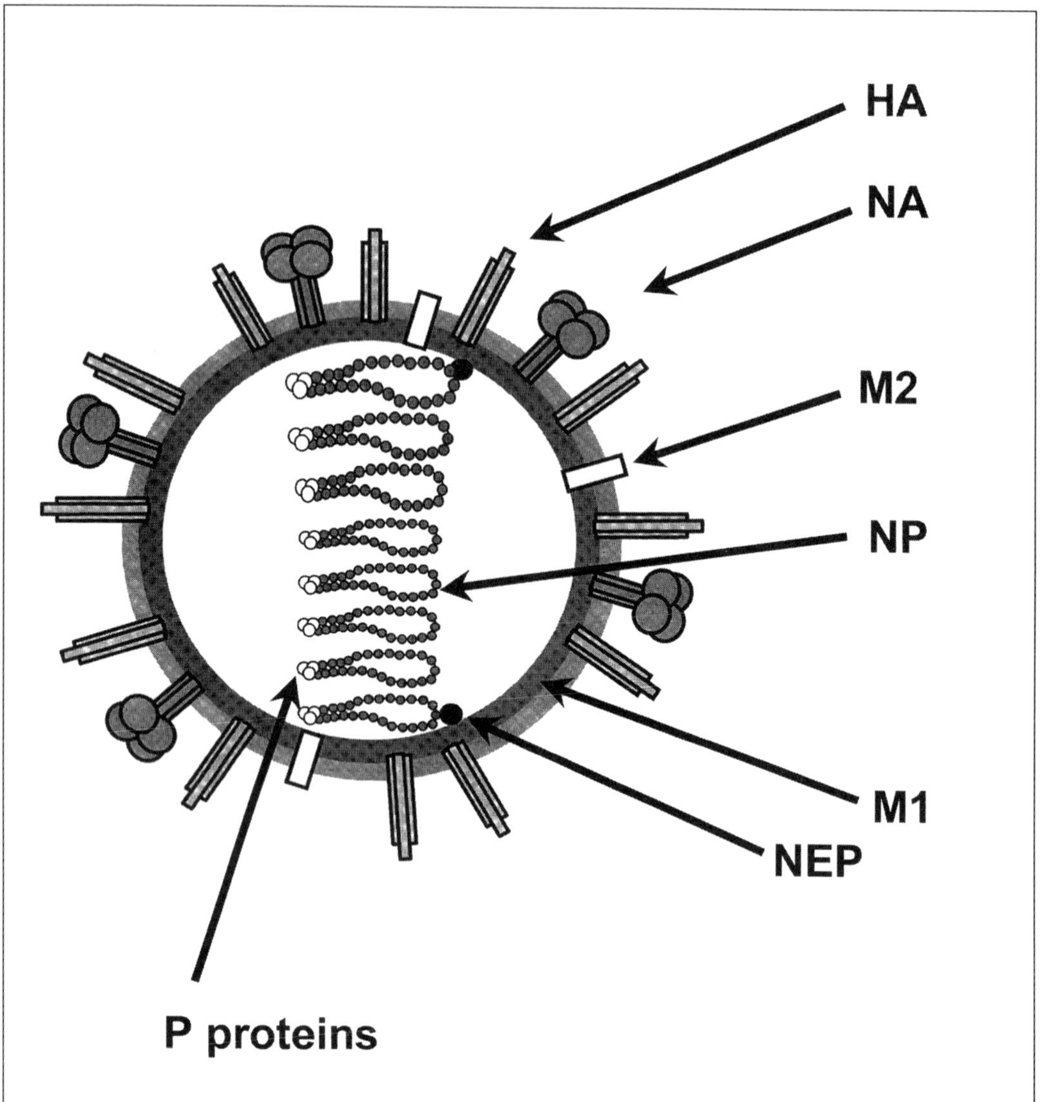

Figure 1. Schematic representation of an influenza A virus particle. Structural protein components are indicated. Represented structures are not to scale.

for the synthesis of new vRNA molecules, which are encapsidated into RNPs. Newly synthesized M1 and NEP proteins seem to be responsible for the export of the new RNPs into the cytoplasm (25,33). Finally, newly formed viruses bud from the cellular plasma membrane, where the HA, NA, and M2 proteins have been incorporated. The NA protein participates in virus release by removing sialic acids from the viral glycoproteins, thereby preventing the aggregation of the virions through HA-sialic acid interactions. Of the ten proteins encoded by influenza A viruses, only one protein has been described to be nonstructural, the NS1 protein. This protein is not required for viral replication in several systems. The NS1 seems to be involved in interacting with different host cell factors, thereby resulting in the evasion of the type I interferon-mediated antiviral responses of the host (13).

GENETIC MANIPULATION OF THE INFLUENZA VIRUS GENOME

The genomes of DNA and of positive-strand RNA viruses are in most cases able to initiate infection and virus particle formation when transfected into permissive cells. However, the genomes of negative-strand RNA viruses are not infectious, and they need to be transcribed into mRNA by the viral RNA-dependent RNA polymerase to initiate viral protein synthesis and replication. In the case of these viruses, the ini-

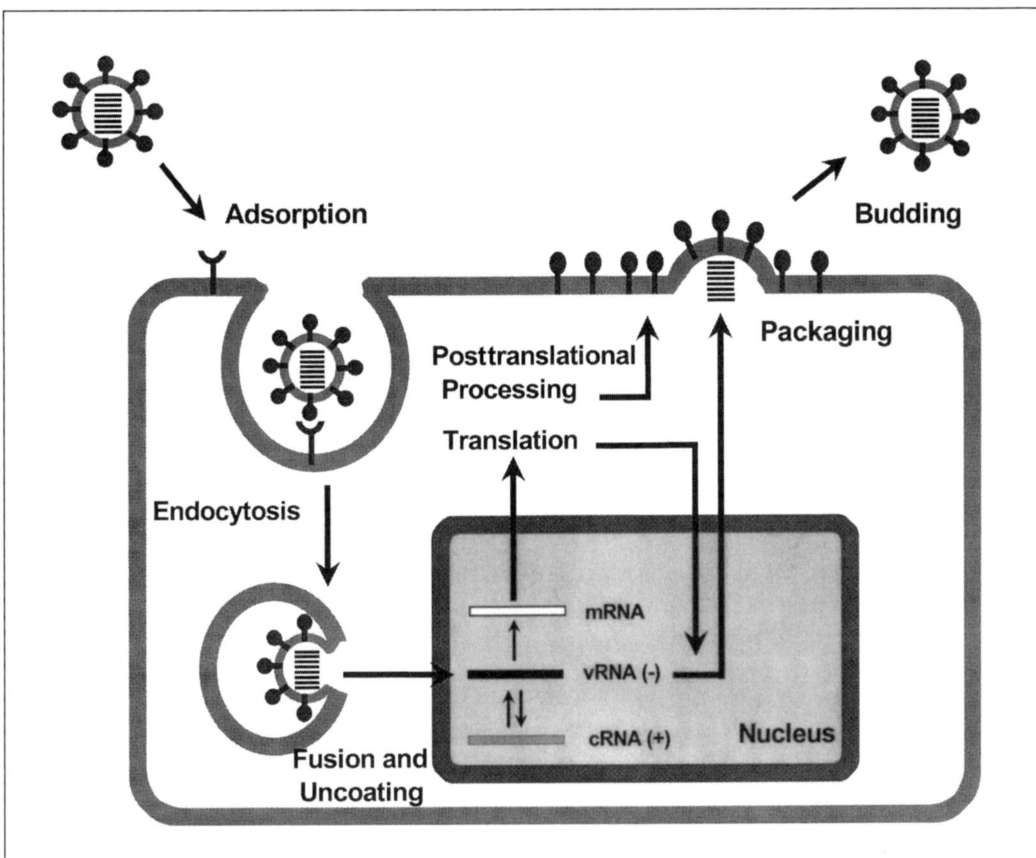

Figure 2. Schematic representation of the replication cycle of influenza A virus. Represented components are not to scale.

tiation of infection requires at least the presence in the host cell of a functional viral RNA template (i.e., encapsidated genomic or antigenomic RNA in the form of RNP) and of the viral RNA polymerase. A reverse genetics method to genetically manipulate the genome of influenza A virus was first described by Enami et al. (8) (Figure 3A). This method is based on the reconstitution in vitro of biologically active viral RNPs and their transfection into cells infected with a helper influenza virus. The transfected RNPs correspond to one of the eight RNA segments of influenza virus. RNP reconstitution is achieved by incubating a plasmid-derived viral RNA with purified NP and P proteins. The helper influenza virus provides all the viral functions required for the transcription, replication, and packaging into new viruses of the transfected RNPs. The transfected/infected cells generate new helper viruses as well as recombinant (transfectant) viruses; the latter contains the recombinant RNA segment. At this point, a selection method is used to separate transfectant from helper viruses. RNP transfection-based reverse genetics methods to rescue genetically modified PB2, HA, NP, NA, M, and NS segments of influenza A virus, as well as the HA and NA segments of influenza B virus, have been successful (1,5,9,20,41,46,50). In each particular case, the transfected RNA segment confers a selectable property to the transfectant virus, such as a wider host tropism or the ability to replicate under different selective pressures, such as high temperature, antiviral drugs, or neutralizing antibodies against the helper viruses. For example, an HA segment of the H1 subtype can be RNP-transfected into cells infected with a helper virus of the H3 subtype. Helper viruses present in the viral progeny can then be neutralized using specific antisera against H3 HAs. Transfectant H1 viruses will not be neutralized by the H3 antisera.

Other methods to generate transfectant influenza viruses have recently been described. For instance, in vitro RNP reconstitution can be substituted by in vivo RNP reconstitution by co-transfecting expression plasmids for a viral RNA segment and for the NP and P proteins (Figure 3B) (38). This method eliminates the laborious need for purifying the NP protein required for in vitro RNP reconstitution. The precise noncoding ends of the expressed viral RNA segment, which are required for efficient replication, transcription, and packaging of the RNA, are achieved by using a truncated *PolI* promoter and a ribozyme. These two elements flank the cDNA corresponding to the sequences of the RNA segment in the expression plasmid.

The use of a helper virus can also be eliminated by transfecting a full complement of eight RNP complexes (Figure 3C) (7). In this method, seven of the eight RNPs are derived from purified viruses in which one particular RNA segment has been depleted by incubating purified RNPs with the corresponding cDNA and digesting the generated RNA-DNA hybrid molecules with RNaseH. The resulting RNP preparation can be complemented in vitro with plasmid-derived RNA corresponding to the missing RNP segment. The complementing RNP restores the function of the digested RNP. This technique can be applied to any of the eight viral RNA segments, thereby eliminating the requirement of a selection method against the helper virus (see also note added in proof).

FOREIGN ANTIGEN EXPRESSION BY TRANSFECTANT INFLUENZA VIRUS VECTORS

The development of reverse genetics methods for the genetic manipulation of influenza viruses allowed the construction of transfectant influenza viruses that express foreign antigens. This was first achieved by Luytjes et al. (24). A synthetic RNA segment was engineered to contain the bacterial chloramphenicol acetyl transferase (CAT) open reading frame (ORF) (in negative polarity) flanked by the 5′ and 3′ noncoding regions of the nonstructural gene of the influenza A virus. This RNA was RNP-transfected into helper virus-infected cells. The transfected RNPs were amplified and packaged into new infectious viruses. Infection of new cells with the viral progeny resulted in expression of the CAT protein (Figure 4A). However, the CAT RNA segment was rapidly lost after a couple of virus passages, most likely because this RNA is not required for virus infectivity. The use of a stronger viral promoter allows the maintenance of the foreign extra gene in the virus progeny for several passages (53). This was achieved by mutating positions 3 and 8 at the 3′ end of the wild-type (wt) viral RNA promoter, thereby resulting in higher promoter activity. However, the levels of the ninth extra segment vary from preparation to preparation, depending on the passage number and on the multiplicity of infection (moi) used in each passage. This makes the standardization of different viral vector preparations difficult. Nevertheless, these vectors may have future applications as vaccine delivery agents.

Stable maintenance of a foreign coding sequence by influenza virus vectors has been achieved by linking the foreign sequence to the coding sequence of an essential viral product in the same influenza virus gene. One approach was based on the insertion of the foreign sequence into the ORF of an influenza virus protein (Figure 4B) (21). Recombinant influenza viruses that express short amino acid sequences (epitopes) derived from a variety of different pathogens, including *Plasmodium* (19), human immunodeficiency virus type 1 (HIV-1) (18), lymphocytic choriomeningitis virus (4), *Pseudomonas aeruginosa* (15), and herpes simplex virus type 2 (2), have been generated using this strategy. The for-

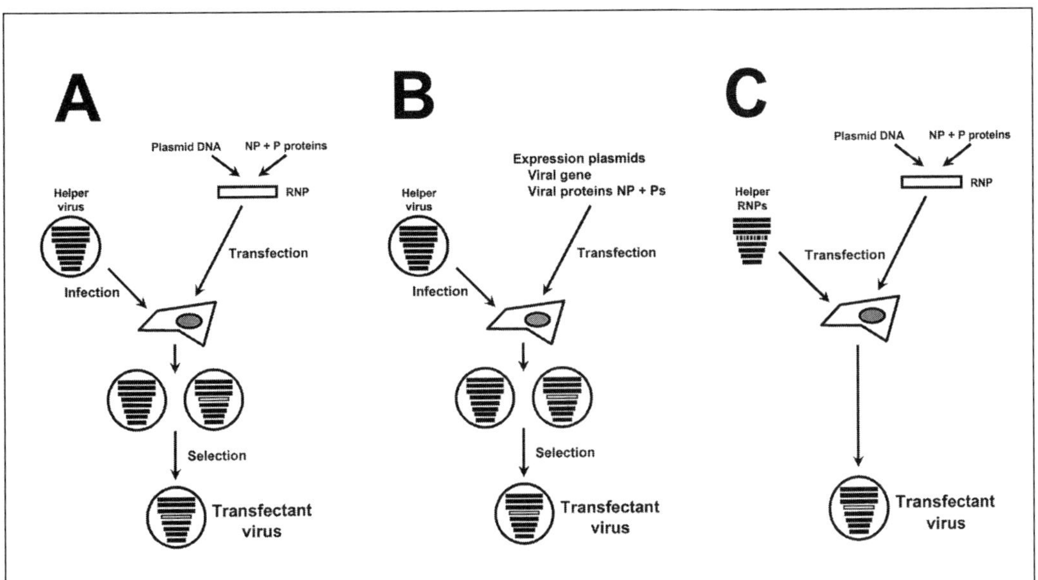

Figure 3. Reverse genetics methods to manipulate the genome of influenza viruses. Helper viral genes are represented as black rectangles. Transfected viral genes are represented as white rectangles. (A) Classical RNP transfection. (B) Plasmid transfection. (C) RNP knock-out.

eign epitopes were inserted into the amino acid sequence of the HA or NA proteins of the recombinant viruses. The success of this approach is limited to the insertion of sequences resulting in functional HA or NA chimeric proteins (i.e., resulting in functional [infectious] viruses). A second approach consisted of the use of genetically engineered viral segments that encode two independent (foreign and viral) proteins (Figure 4C). This was achieved by the insertion of a ribosomal entry site between the two coding sequences (Figure 5A) (14).

Alternatively, a viral RNA segment can be designed to encode a fusion protein that undergoes self-cleavage (Figure 5B) (36). This strategy was used to generate a transfectant influenza A virus that expresses CAT. Interestingly, the bacterial CAT sequences were maintained upon passaging in the majority of the infectious viruses of the viral progeny (36). The use of bicistronic viral segments to generate influenza virus vectors that encode foreign polypeptides seems to be limited to sequences of less than 220 amino acid residues in length. However, this length limitation might be overcome in the near future (see below).

Using the techniques described above, we have constructed transfectant influenza virus vectors expressing a variety of different antigens, including cytoplasmic, transmem-

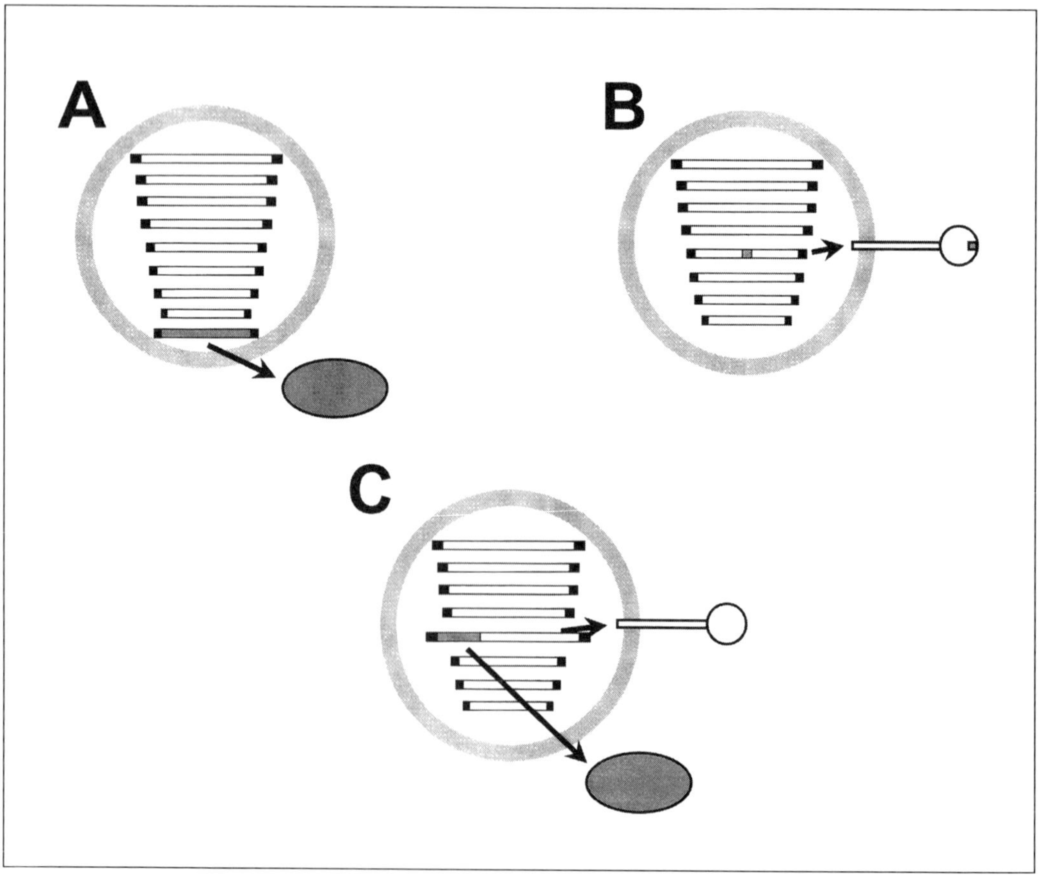

Figure 4. Influenza virus vectors. Noncoding regions at the end of the influenza virus genes are represented as black boxes. Additional (foreign) sequences are represented as grey boxes. (A) Expression of a foreign protein from an additional RNA segment. (B) Expression of a foreign epitope as part of the open reading frame of a chimeric viral protein. (C) Expression of a foreign protein from a bicistronic influenza virus gene that encodes a viral protein (represented in white) as well as the foreign protein (represented in grey).

brane, and secreted polypeptides. Transmembrane proteins expressed by these vectors can be targeted for incorporation into viral envelopes if they are designed to contain the transmembrane and cytoplasmic domains of the viral HA protein (14,53).

INDUCTION OF ANTIBODY RESPONSES BY INFLUENZA VIRUS VECTORS

Most of the antibody response that is induced in the host upon infection with an

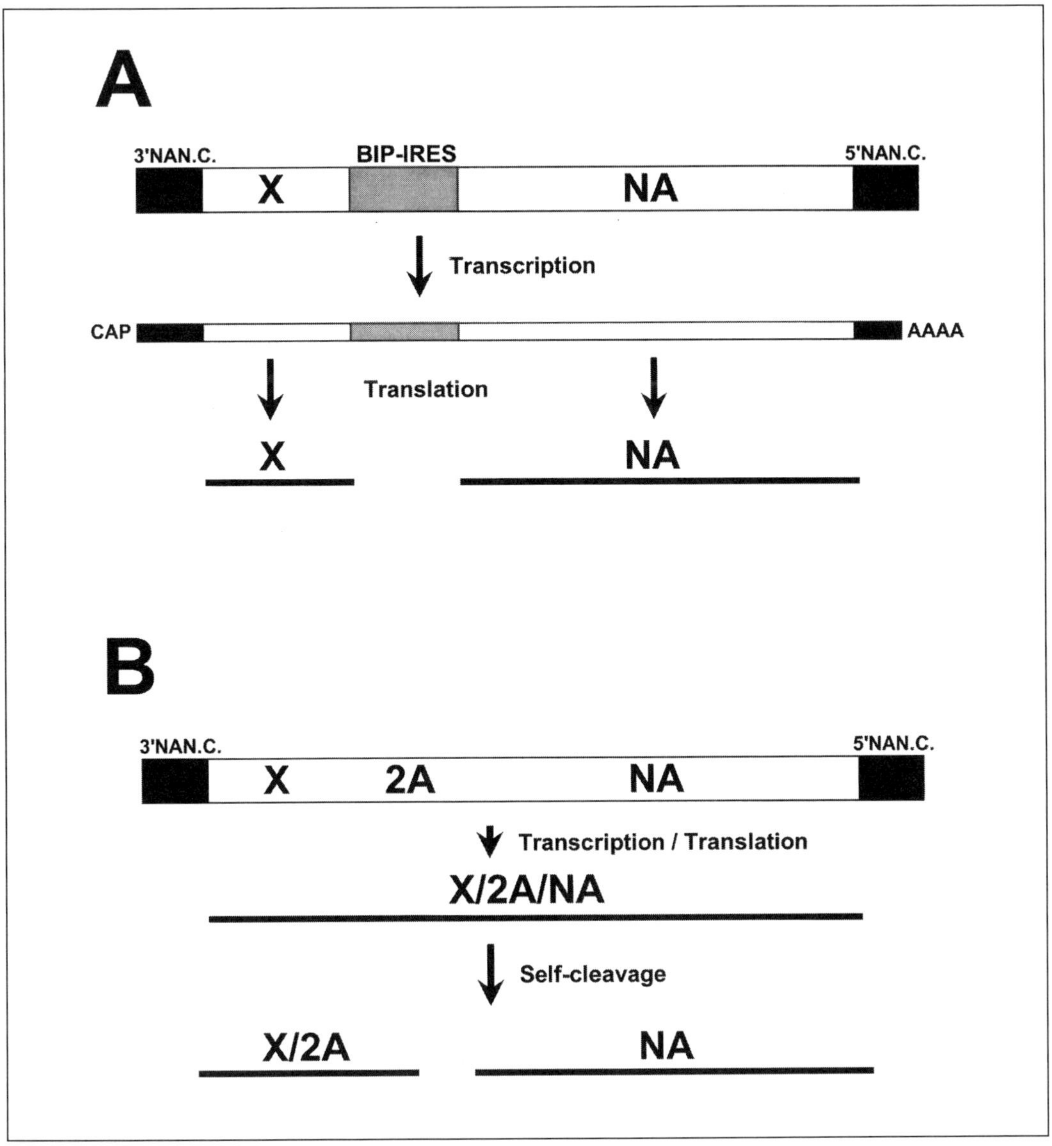

Figure 5. Schematic representation of influenza virus RNAs, which allow expression of a viral (NA) protein and of a foreign (X) protein. 3′ and 5′ noncoding sequences are indicated (3′NAN.C., 5′NAN.C.) (A) The RNA gene contains two open reading frames (ORFs) that are separated by an RNA sequence that encodes the IRES of the human immunoglobulin heavy-chain binding protein (BIP). The first ORF is translated into a foreign protein via the usual cap-dependent mechanism of initiation of translation. The second ORF is translated into the NA via internal ribosomal initiation of translation. (B) The RNA gene encodes a polyprotein (X/2A/NA) that is proteolytically self-cleaved into two polypeptides, X/2A and NA. The foreign polypeptide contains 16 amino acids fused to its carboxy terminal that are derived from the 2A self-protease motif of foot-and-mouth-disease virus.

influenza virus is directed against the HA glycoprotein, specifically against the antigenic sites A, B, C, D, and E of this protein. Therefore, it was predicted that recombinant influenza viruses that contain a foreign epitope inserted into one of these antigenic sites would be efficient inducers of antibody responses against this epitope. Li et al. (21) generated a transfectant influenza virus in which the six-amino-acid loop of the antigenic site B of its HA (H1 subtype) was replaced by the corresponding loop of an H3 subtype HA. This chimeric virus was able to induce a neutralizing antibody response in mice against H1 and H3 subtypes of the influenza A virus.

An influenza virus vector with the ability to induce antibodies against HIV-1 in infected mice was first engineered by Li et al. (18). This viral vector contained 12 amino acids derived from the V3 loop of gp160 of HIV-1 inserted into the antigenic site B of the HA. A second influenza virus vector that expresses an HIV-1-derived sequence was constructed by Muster et al. (30). In this case, the amino acid sequence ELDKWAS was inserted into the site B of the HA. This gp41-derived sequence was previously found to be recognized by a human monoclonal antibody that was able to neutralize a variety of HIV-1 strains, including primary isolates (31). Mice immunized with this virus developed a long-lasting neutralizing antibody response against HIV-1 at both systemic and mucosal levels; this antibody response indicates that influenza virus vectors are efficient inducers of mucosal immunity (10,29). Mucosal immunity might play a key role in preventing disease by pathogens such as HIV-1 that enter the body through mucosal surfaces.

We have also generated influenza virus vectors that express foreign epitopes derived from bacterial and parasitic pathogens in the antigenic site B of their HAs. A transfectant influenza virus that expresses an epitope from the outer membrane protein F of *P. aeruginosa* afforded antibody-mediated protection against this bacterial pathogen in a mouse model system (15, 45). Induction of protective antibody responses in mice against malaria sporozoites was achieved using a transfectant influenza virus expressing an epitope derived from the circumsporozoite protein of *Plasmodium* (27,40). All of these results suggest that influenza virus vectors expressing antigenic epitopes in their HA molecules are potent inducers of antibody responses against the expressed epitope.

INDUCTION OF CELLULAR IMMUNE RESPONSES BY INFLUENZA VIRUS VECTORS

It is believed that an efficient vaccine against pathogens such as HIV-1 and *Plasmodium* will require the induction of both humoral and cellular responses. Antibodies might not be able to completely neutralize infection, and once the pathogen becomes intracellular, the infected cells are best eliminated by cytotoxic T cells (CTLs) that recognize specific peptides (CTL epitopes) from the pathogen, which are associated with major histocompatibility complex (MHC) molecules at the surface of the infected cell. The presence of a specific T-cell memory might mediate a fast and potent T-cell response when the host becomes infected by the pathogen, thereby contributing to disease prevention. Therefore, it is thought that an ideal AIDS or malaria vaccine should also induce T-cell memory against HIV-1 or *Plasmodium*, respectively.

Since influenza A viruses are potent inducers of cellular immune responses, the use of influenza virus vectors that express CTL epitopes from different pathogens as inducers of protective cellular immune responses against these pathogens has been investigated. A transfectant influenza virus that expresses a murine CTL epitope derived from the circumsporozoite protein of *Plasmodium yoelii*, a murine malaria parasite, was generated (19). This epitope was inserted into the ORF of the HA protein. The transfectant virus was able to induce a CTL response in mice that was specific for the plasmodial peptide. Interestingly, the response was efficiently boosted by a second viral vector (e.g., recombinant vaccinia virus) that expresses the same epitope (19). It appears that priming with the first viral vector also results in an antibody response against the vector that prevents reinfection with the same vector. Therefore, a second administration of the same vector is not able to efficiently restimulate cellular immunity, most likely because of the lack of infection and MHC class I presentation. Efficient boosting is only achieved by using a second viral vector that is not neutralized by the antibody response against the first vector (28). The use of a combined immunization protocol involving influenza and vaccinia virus recombinants expressing malarial epitopes resulted in high levels of protection in mice against murine malaria (40). Transfectant influenza viruses that express epitopes from human malarial parasites have also been successfully generated and tested in mice (27).

CTL epitopes derived from HIV-1 proteins have also been included into transfectant influenza virus vectors. These viruses were able to induce HIV-1-specific cellular immune responses in mice (18). As in the case with the malaria vectors, combined administrations of influenza and vaccinia virus vectors expressing the same HIV-1 CTL epitope induced high levels of HIV-1-specific CTLs (16). Other examples of transfectant influenza virus that express CTL epitopes from different viruses (e.g., lymphocytic choriomeningitis virus and herpes simplex virus type 2) have been described (2,4). Immunizations of mice with these transfectant viruses induced protection against a challenge with the corresponding virus. In addition, a transfectant influenza virus that expresses a model tumor-associated antigen recognized by $CD8^+$ T cells was able to mediate tumor regression in vaccinated mice (39). Finally, the ability of influenza virus vectors to present foreign epitopes in professional antigen-presenting cells to $CD4^+$ T cells has also been demonstrated (47). All of these results demonstrate that, at least in the mouse system, transfectant influenza viruses expressing antigens recognized by specific T cells are good inducers of cellular immune responses against the expressed antigens.

TRANSFECTANT INFLUENZA VIRUSES WITH ATTENUATED PHENOTYPES

The reverse genetics methodology has also been applied to generate attenuated transfectant viruses. Stably attenuated influenza viruses might be used as live vaccines against influenza, if they are more immunogenic than the available inactivated vaccines. In addition, the introduction of attenuating mutations in influenza virus vectors that express foreign antigens reduces the risk of potential side effects of the vectors, such as the induction of influenza-like symptoms in the vaccinees.

The insertion of coding sequences from foreign antigens into the genome of influenza viruses resulted in transfectant viruses with attenuated phenotypes (4). However, it is difficult to predict the attenuation levels that different antigens might confer to the chimeric transfectant viruses. We have investigated three different methods to rationally attenuate influenza viruses by genetic techniques. These methods are based on the introduction of mutations that target the replication, the transcription, and the protein products of the influenza virus genes. Luo et al. (23) described a transfectant influenza virus whose NA gene replicated to lower levels than those in wt influenza viruses. This virus was constructed (32) by mutating the 3′ and 5′ noncoding ends of the gene, which contain the *cis*-acting signals required for RNA replication (22,37,43, 49). Specifically, the noncoding ends of the NA gene of influenza A virus were changed by the corresponding ends of the NS gene of influenza B virus. The transfectant virus generated a high proportion of noninfectious viruses, which lacked the NA gene, and was attenuated in mice (23). Immunizations performed in mice with this attenuated virus resulted in protective immune responses against wt influenza virus. Other specific deletions or substitutions in the noncoding ends of the influenza virus genes also resulted in attenuated transfectant viruses with a replication-impaired RNA segment (52).

Recently, Fodor et al. (11) introduced base pair substitutions at a dsRNA structure of the NA gene of the influenza virus. This structure is an important component of the viral promoter (12), and it is formed by association between the 3′ and 5′ noncoding ends of the influenza virus genes. The sequence of the three first base pairs is conserved among different influenza A virus strains and RNA segments. Mutations were made that changed the sequence, but not the complementarity, of these three base pairs. All virus mutants showed wt levels of replication of their NA genes. However, mutants in which the base pair CG was changed to AU at positions 10 and 11 of the 3′ and 5′ ends, respectively, showed reduced levels of transcription of their NA genes, thereby resulting in attenuation of the viruses. Reversion to wt sequences in these mutant viruses would require two nucleotide changes. In addition, single changes are predicted to result in even more attenuated viruses because the complementarity (i.e., the structure of the promoter) will be affected. Therefore, it is likely that the attenuating mutation will be stable.

A third way to attenuate influenza A viruses is based on the deletion of genes that encode auxiliary viral proteins. However, it was believed until recently that all ten influenza virus proteins were essential for virus viability. Surprisingly, we were able to show that the only nonstructural protein of the virus, the NS1 protein, was dispensable for virus replication. Specifically, a transfectant NS1 knockout influenza A virus grew efficiently in type I interferon-deficient mice but was highly attenuated in wt mice (13). These results suggest that the NS1 protein is a virally encoded inhibitor of the type I interferon responses. Deletions of the carboxy terminal of the NS1 protein were well tolerated in recombinant influenza viruses (6). Transfectant influenza A viruses containing deletions in the NS1 gene might represent ideal attenuated strains for human use. Other approaches to attenuating influenza viruses by reverse genetics techniques include the insertion of multiple temperature-sensitive mutations in their genes (34,47).

FUTURE AND PERSPECTIVE

The present experimental evidence indicates that influenza virus vectors are efficient inducers of cellular and humoral protective immune responses in mice. Safety and efficacy studies in humans will evaluate their potential use as live vaccines in humans against different pathogens and/or tumor malignancies. The absence of a DNA replicative phase in the replication cycle of influenza viruses eliminates the risk of chromosomal integration. Additional safety measures might include the use of genetically attenuated transfectant viruses. The availability of different influenza virus serotypes could circumvent the problems of preexisting immunity against the vectors in humans. Furthermore, packaging size limitations may be overcome by eliminating accessory proteins from the viral genome, such as the NS1 protein, and by designing new reverse genetics methods (see also note added in proof). In this respect, it is noteworthy that the recently developed helper virus-free methods to generate infectious nonsegmented negative-strand RNA viruses have been used to obtain rabies, VSV, respiratory syncytial, and measles virus vectors among others, with the ability to express long foreign sequences (3,26,42,44). Influenza virus vectors as well as other negative-strand RNA virus vectors may be used in the near future as key components in vaccine formulations against different infectious agents such as HIV-1.

MATERIALS AND METHODS

Generation of Stocks of Influenza A Viruses

Influenza A virus stocks are usually prepared by growing the virus in tissue culture or in embryonated chicken eggs. Madin-Darby canine kidney (MDCK) cells are often used to grow the virus in tissue culture. These cells can be maintained in minimal essential medium (MEM) containing 10% heat-inactivated fetal calf serum (FCS). The following protocol refers to infection of confluent monolayers of MDCK cells in 35-mm tissue culture dishes (containing approximately 1×10^6 cells) and might be upscaled to bigger dishes. Tissue culture medium is removed from the dishes, and the cell monolayers are washed with phosphate-buffered saline (PBS). After removing the PBS, 100 µL of a preparation containing between 1×10^2 and 1×10^3 plaque-forming units (pfu) of influenza virus in PBS are added to the cells. To prevent drying of the cells during the incubation period, additional PBS (approximately 200 µL) might be added together with the virus inoculum. Cells are incubated at room temperature for one hour, and the virus inoculi are removed and substituted by 2 mL MEM without FCS containing 0.2% bovine serum albumin (BSA) and 1 µg/mL trypsin. Infected cell monolayers are then incubated at 37°C in a CO_2 incubator. Typically, cell monolayers show almost 100% cytopathic effect (CPE) at day 2 postinfection. At this moment, supernatants are harvested, clarified by centrifugation at 3000 rpm (1000× *g*) for 30 minutes and aliquoted. Aliquots are quickly frozen using dry ice and can be stored at -80°C for more than one year. A few strains of influenza A virus, such as influenza A/WSN/33 (WSN) virus, do not require trypsin for growth in MDCK cells. In addition, WSN virus can be grown in Madin-Darby bovine kidney (MDBK) cells in reinforced Eagle's medium (REM) containing 0.2% BSA.

Ten- to eleven-day-old chicken eggs can also be used to grow stocks of influenza A viruses. Typically, between 1×10^2 and 1×10^3 pfu influenza virus in 100 µL are inoculated with the help of a 1-mL syringe into the allantois chamber of the eggs. After inoculation, the injection points are sealed using hot wax, and the eggs are incubated for two days at 37°C and overnight at 4°C. Infected eggs are then carefully opened, and the allantoic fluid is harvested and clarified by centrifugation at 3000 rpm for 30 minutes. These virus preparations can be stored as described above for the tissue culture preparations.

Influenza A Virus Titrations

Hemagglutination Assay

Fifty microliters of 0.5% chicken red blood cells in PBS are mixed in a 96-well plate with 50 µL serial twofold dilutions in

PBS of a preparation of influenza virus. The plate is incubated at 4°C for one hour. The hemagglutination titer corresponds to the higher dilution showing hemagglutinating activity (51).

Plaque Assay

MDCK cell monolayers in 35-mm dishes are infected as described above with serial tenfold dilutions in PBS of a sample that contains influenza viruses. During the one-hour incubation time, agar overlay medium is prepared by first mixing 50 mL 2× MEM medium containing 0.4% BSA and 2 μg/mL trypsin with 17 mL distilled water, 1 mL 1% DEAE-dextran, and 2 mL 5% $NaHCO_3$. This medium is warmed to 37°C. In addition, a solution of 2% oxoid agar is melted and kept at 42°C. When the virus inoculi are removed, 30 mL agar solution are added to the warm medium, and 2 mL agar-containing medium are added per dish. After 10 minutes at room temperature, the agar solidifies, and the dishes are incubated upside down for two days at 37°C in a CO_2 incubator. Virus plaques can be directly visualized or stained using crystal violet.

Purification of Viral NP for RNP Transfection

Viral NP to be used in RNP transfection is purified from influenza virions essentially as described by Parvin et al. (35). Viruses present in allantois fluid preparations (at 1×10^8 pfu/mL) are pelleted by centrifugation at 25 000 rpm (90 000× *g*) for two hours in a SW25 rotor. A virus pellet corresponding to a preparation of approximately 150 eggs is resuspended in 1 mL 1.5% Triton® N-101, 10 mg/mL lysolecithin, 100 mmol/L Tris HCl (pH 8.1), 100 mmol/L KCl_2, 5 mmol/L $MgCl_2$, 5% glycerol, and 1.5 mmol/L dithiothreitol (DTT). Viruses are disrupted by incubation in this solution for 25 minutes at 31°C. The sample is fractionated by centrifugation in a 33% to 70% glycerol (weight/volume) step gradient in the presence of 150 mmol/L NaCl and 50 mmol/L Tris HCl (pH 7.8). This gradient has steps of 33% (1.5 mL), 40% (0.375 mL), 50% (0.75 mL), and 70% (1 mL) glycerol. The disrupted viruses are centrifuged through the gradient at 250 000× *g* in a SW50.1 rotor for four hours at 4°C, and fractions of 300 μL are collected. Fractions enriched in RNPs are identified by sodium dodecyl sulfate-polyacrylamide gel electrophoresis (SDS-PAGE) of 1 μL from each fraction and silver staining. The RNPs band close to the bottom of the gradient. These RNP-enriched fractions are collected and subjected to a second-gradient ultracentrifugation with steps of 1.5 mol/L CsCl/30% glycerol (0.8 mL), 2.0 mol/L CsCl/35% glycerol (1.15 mL), 2.5 mol/L CsCl/40% glycerol (1.65 mL), and 3 mmol/L CsCl/45% glycerol (0.5 mL) in 100 mmol/L NaCl, and 50 mmol/L Tris HCl (pH 7.6). This gradient is centrifuged at 250 000× *g* in a SW50.1 rotor for 25 hours at 4°C. NP-containing fractions, usually close to the top of the gradient, are again identified by SDS-PAGE. The fractions are pooled and dialyzed against 50 mmol/L Tris HCl (pH 7.5), 100 mmol/L NaCl, 10 mmol/L $MgCl_2$, 1 mmol/L DTT, and 10% glycerol, and then against 50 mmol/L Tris HCl (pH 7.5), 100 mmol/L NaCl, 10 mmol/L $MgCl_2$, 1 mmol/L DTT, and 50% glycerol. The NP preparations can be aliquoted, quickly frozen, and stored at -80°C.

Generation of Transfectant Influenza Viruses by RNP Transfection

Plasmid containing the cDNA of one of the influenza virus genes in negative sense under the control of a truncated T7 or T3 promoter is linearized using an appropriate restriction enzyme (8). Runoff RNA transcripts are generated from 0.5 μg template in 50 μL volume using T3 or T7 RNA polymerase. Ten microliters of purified NP are added to the in vitro transcription reaction. Reactions are carried for 15 minutes at 37°C in transcription buffer containing 10 mmol/L DTT, 2 mmol/L NTPs, 100 U RNAsin, and 100 U phage polymerase. The resulting RNPs are DEAE-transfected into cells previously infected with a helper influenza virus at an moi of 2 (9). For this purpose, cell monolayers in 35-mm dishes are infected with 100 μL helper virus preparation at 2×10^7 pfu/mL as described above. Virus inoculi are then removed, and cells are treated for 30 minutes at room temperature with 1 mL PBS containing 0.1 mg/mL gelatin, 0.033% DEAE-dextran, and 0.5% dimethyl sulfoxide. This solution is then removed, and the in vitro reconstituted RNPs (50 μL) are mixed with 100 μL PBS and 0.1 mg/mL gelatin, and then the solution is added to the cells. After one hour of incubation at room temperature, the cells are washed and infection medium (1.5 mL) is added. Several cell lines, including MDBK and Vero 2-2 cells, are suitable substrates for RNP transfection. After 18 hours of incubation at 37°C, supernatants are passaged under selective conditions, thereby allowing growth of the transfectant virus, but preventing growth of the helper virus. Transfectant viruses are plaque-purified three times before their analysis.

Generation of Transfectant Influenza Viruses by Plasmid Transfection

Approximately 1×10^6 293 or Vero 2-2 cells are transfected using a lipofection reagent with five different plasmids. These plasmids include mammalian expression vectors for the NP (2 μg), PB2 (1 μg), PB1 (1 μg), and PA (1 μg) proteins, as well as an expression vector (4 μg) for an influenza virus RNA segment under the transcriptional control of a truncated human polymerase I promoter (38). A typical transfection involves 15 minutes of incubation at room temperature of the cDNAs with 30 μL *N*-[1-(2,3-dioloyloxy)propyl]-*N*,*N*,*N*-trimethylammonium methylsulfate (DOTAP) in a total volume of 150 μL 20 mmol/L HEPES (pH 7.5), followed by the addition of the DNA-DOTAP complexes to the culture medium. After transfection, cells are incubated for 36 hours at 37°C and then infected with a helper influenza A virus at an moi of 2. Eighteen hours after infection, supernatants are passaged under selective conditions.

***Note added in proof*:** Recently, helper virus-free rescue systems for the generation of transfectant influenza A viruses entirely from plasmid DNA have been developed (Neuman, G., W. Watanabe, H. Ito, S. Watanabe, H. Goto, P. Gao, M. Hughes, D.R. Perez, et al. 1999. Generation of influenza A viruses entirely from cloned cDNAs. Proc. Natl. Acad. Sci. USA *96*: 9345-9350; Fodor, E., L. Devenish, O.G. Engelhardt, P. Palese, G.G. Brownlee, and A. García-Sastre. 1999. Rescue of influenza A virus from recombinant DNA. J. Virol. *73*:9679-9682; Hoffmann, E., G. Neumann, Y. Kawaoka, G. Hobom, and R.G. Webster. 2000. A DNA transfection system for generation of influenza A virus from eight plasmids. Proc. Natl. Acad. Sci. USA *97*:6108-6113). In these systems, the eight negative-sense genomic viral RNAs are transcribed from *PolI* expression plasmids. The three influenza virus P proteins and the NP are expressed from *PolII* expression plasmids. As a result, transfected cells produce recombinant influenza viruses derived from the transfected plasmids.

ACKNOWLEDGMENTS

This work was partially supported by an NIH grant to A. García-Sastre.

REFERENCES

1. **Barclay, W.S. and P. Palese.** 1995. Influenza B viruses with site-specific mutations introduced into the HA gene. J. Virol. *69*:1275-1279.
2. **Blaney, J.E., Jr., E. Nobusawa, M.A. Brehm, R.H. Bonneau, L.M. Mylin, T.M. Fu, Y. Kawaoka, and S.S. Tevethia.** 1998. Immunization with a single major histocompatibility complex class I-restricted cytotoxic T-lymphocyte recognition epitope of herpes simplex virus type 2 confers protective immunity. J. Virol. *72*:9567-9574.
3. **Bukreyev, A., E. Camargo, and P.L. Collins.** 1996. Recovery of infectious respiratory syncytial virus expressing an additional, foreign gene. J. Virol. *70*:6634-6641.
4. **Castrucci, M.R., S. Hou, P.C. Doherty, and Y. Kawaoka.** 1994. Protection against lethal lymphocytic choriomeningitis virus (LCMV) infection by immunization of mice with an influenza virus containing an LCMV epitope recognized by cytotoxic T lymphocytes. J. Virol. *68*:3486-3490.
5. **Castrucci, M.R. and Y. Kawaoka.** 1995. Reverse genetics system for generation of an influenza A virus mutant containing a deletion of the carboxyl-terminal residue of M2 protein. J. Virol. *69*:2725-2728.
6. **Egorov, A., S. Brandt, S. Sereinig, J. Romanova, B. Ferko, D. Katinger, A. Grassauer, G. Alexandrova, H. Katinger, and T. Muster.** 1998. Transfectant influenza A viruses with long deletions in the NS1 protein grow efficiently in Vero cells. J. Virol. *72*:6437-6441.
7. **Enami, M.** 1997. Improved technique to genetically manipulate influenza virus, p. 19. Frontiers of RNA Virus Research. The Oji International Seminar in Natural Science, Kyoto, Japan.
8. **Enami, M., W. Luytjes, M. Krystal, and P. Palese.** 1990. Introduction of site-specific mutations into the genome of influenza virus. Proc. Natl. Acad. Sci. USA *87*:3802-3805.
9. **Enami, M. and P. Palese.** 1991. High-efficiency formation of influenza virus transfectants. J. Virol. *65*:2711-2713.
10. **Ferko, B., D. Katinger, A. Grassauer, A. Egorov, J. Romanova, B. Niebler, H. Katinger, and T. Muster.** 1998. Chimeric influenza virus replicating predominantly in the murine upper respiratory tract induces local immune responses against human immunodeficiency virus type 1 in the genital tract. J. Infect. Dis. *178*:1359-1368.
11. **Fodor, E., P. Palese, G.G. Brownlee, and A. García-Sastre.** 1998. Attenuation of influenza A virus mRNA levels by promoter mutations. J. Virol. *72*:6283-6290.
12. **Fodor, E., D.C. Pritlove, and G.G. Brownlee.** 1995. Characterization of the RNA-fork model of virion RNA in the initiation of transcription in influenza A virus. J. Virol. *69*:4012-4019.
13. **García-Sastre, A., A. Egorov, D. Matassov, S. Brandt, D.E. Levy, J.E. Durbin, P. Palese, and T. Muster.** 1998. Influenza A virus lacking the NS1 gene replicates in interferon-deficient systems. Virology *252*:324-330.
14. **García-Sastre, A., T. Muster, W.S. Barclay, N. Percy, and P. Palese.** 1994. Use of a mammalian internal ribosomal entry site element for expression of a foreign protein by a transfectant influenza virus. J. Virol. *68*:6254-6261.
15. **Gilleland, H.E., Jr., L.B. Gilleland, J. Staczek, R.N. Harty, A. García-Sastre, O.G. Engelhardt, and P. Palese.** 1997. Chimeric influenza viruses incorporating epitopes of outer membrane protein F as a vaccine against pulmonary infection with *Pseudomonas aeruginosa*. Behring Inst. Mitt. *98*:291-301.
16. **Gonzalo, R.M., D. Rodriguez, A. García-Sastre, J.R. Rodriguez, P. Palese, and M. Esteban.** 1999. Enhanced CD8+ T cell response to HIV-1 env by combined immunization with influenza and vaccinia virus recombinants. Vaccine *17*:887-892.
17. **Lamb, R.A. and R.M. Krug.** 1996. *Orthomyxoviridae*: the viruses and their replication, p. 1353-1395. *In* B.N. Fields, D.M. Knipe, P.M. Howley, R.M. Chanock, J.L. Melnick, T.P. Monath, B. Roizman, and S.E. Straus (Eds.), Fields Virology, 2nd ed. Lippincott-Raven, Philadelphia.
18. **Li, S., V. Polonis, H. Isobe, H. Zaghouani, R. Guinea, T. Moran, C. Bona, and P. Palese.** 1994. Chimeric influenza virus induces neutralizing antibodies and cytotoxic T cells against human immunodeficiency virus type 1. J. Virol. *67*:6659-6666.
19. **Li, S., M. Rodrigues, D. Rodriguez, J.R. Rodriguez, M. Esteban, P. Palese, R.S. Nussenzweig, and F. Zavala.** 1993. Priming with recombinant influenza virus followed by administration of recombinant vaccinia virus induces CD8+ T-cell-mediated protective immunity against malaria. Proc. Natl. Acad. Sci. USA *90*:5214-5218.
20. **Li, S., M. Xu, and K. Coelingh.** 1995. Electroporation of influenza virus ribonucleoprotein complexes for rescue of the nucleoprotein and matrix genes. Virus Res. *37*:153-161.
21. **Li, S.Q., J.L. Schulman, T. Moran, C. Bona, and P. Palese.** 1992. Influenza A virus transfectants with chimeric hemagglutinins containing epitopes from different subtypes. J. Virol. *66*:399-404.
22. **Li, X. and P. Palese.** 1992. Mutational analysis of the promoter required for influenza virus virion RNA synthesis. J. Virol. *66*:4331-4338.
23. **Luo, G., M. Bergmann, A. García-Sastre, and P. Palese.** 1992. Mechanism of attenuation of a chimeric influenza A/B transfectant virus. J. Virol. *66*:4679-4685.
24. **Luytjes, W., M. Krystal, M. Enami, J.D. Parvin, and P. Palese.** 1989. Amplification, expression, and packaging of a foreign gene by influenza virus. Cell *59*:1107-1113.
25. **Martin, K. and A. Helenius.** 1991. Nuclear transport of influenza virus ribonucleoproteins: the viral matrix protein (M1) promotes export and inhibits import. Cell *67*:117-130.
26. **Mebatsion, T., M.J. Schnell, J.H. Cox, S. Finke, and K.-K. Conzelmann.** 1996. Highly stable expression of a foreign gene from rabies virus vectors. Proc. Natl. Acad. Sci. USA *93*:7310-7314.
27. **Miyahira, Y., A. García-Sastre, D. Rodriguez, J.R. Rodriguez, K. Murata, M. Tsuji, P. Palese, M. Esteban, F. Zavala, and R.S. Nussenzweig.** 1998. Recom-

binant viruses expressing a human malaria antigen can elicit potentially protective immune CD8+ responses in mice. Proc. Natl. Acad. Sci. USA *31*:3954-3959.

28.**Murata, K., A. García-Sastre, M. Tsuji, M. Rodrigues, D. Rodriguez, J.R. Rodriguez, R.S. Nussenzweig, P. Palese, M. Esteban, and F. Zavala.** 1996. Characterization of in vivo primary and secondary CD8+ T cell responses induced by recombinant influenza and vaccinia viruses. Cell. Immunol. *173*:96-107.

29.**Muster, T., B. Ferko, A. Klima, M. Purtscher, A. Trkola, P. Schulz, A. Grassauer, O.G. Engelhardt, A. García-Sastre, P. Palese, and H. Katinger.** 1995. Mucosal model of immunization against human immunodeficiency virus type 1 with a chimeric influenza virus. J. Virol. *69*:6678-6686.

30.**Muster, T., R. Guinea, A. Trkola, M. Purtscher, A. Klima, F. Steindl, P. Palese, and H. Katinger.** 1994. Cross-neutralizing activity against divergent human immunodeficiency virus type 1 isolates induced by the gp41 sequence ELDKWAS. J. Virol. *68*:4031-4034.

31.**Muster, T., F. Steindl, M. Purtscher, A. Trkola, A. Klima, G. Himmler, F. Ruker, and H. Katinger.** 1993. A conserved neutralizing epitope on gp41 of human immunodeficiency virus type 1. J. Virol. *67*:6642-6647.

32.**Muster, T., E.K. Subbarao, M. Enami, B.R. Murphy, and P. Palese.** 1991. An influenza A virus containing influenza B virus 5′ and 3′ noncoding regions on the neuraminidase gene is attenuated in mice. Proc. Natl. Acad. Sci. USA *88*:5177-5181.

33.**O'Neill, R.E., J. Talon, and P. Palese.** 1998. The influenza virus NEP (NS2 protein) mediates the nuclear export of viral ribonucleoproteins. EMBO J. *17*:288-296.

34.**Parkin, N.T., P. Chiu, and K. Coelingh.** 1997. Genetically engineered live attenuated influenza A virus vaccine candidates. J. Virol. *71*:2772-2778.

35.**Parvin, J.D., P. Palese, A. Honda, A. Ishihama, and M. Krystal.** 1989. Promoter analysis of influenza virus RNA polymerase. J. Virol. *63*:5142-5152.

36.**Percy, N., W.S. Barclay, A. García-Sastre, and P. Palese.** 1994. Expression of a foreign protein by influenza A virus. J. Virol. *68*:4486-4492.

37.**Piccone, M.E., A. Fernandez-Sesma, and P. Palese.** 1993. Mutational analysis of the influenza virus vRNA promoter. Virus Res. *28*:99-112.

38.**Pleschka, S., R. Jaskunas, O.G. Engelhardt, T. Zürcher, P. Palese, and A. García-Sastre.** 1996. A plasmid-based reverse genetics system for influenza A virus. J. Virol. *70*:4188-4192.

39.**Restifo, N.P., D.R. Surman, H. Zheng, P. Palese, S.A. Rosenberg, and A. García-Sastre.** 1998. Transfectant influenza A viruses are effective recombinant immunogens in the treatment of experimental cancer. Virology *249*:89-97.

40.**Rodrigues, M., S. Li, K. Murata, D. Rodriguez, J.R. Rodriguez, I. Bacik, J.R. Bennink, J.W. Yewdell, A. García-Sastre, R.S. Nussenzweig, et al.** 1994. Influenza and vaccinia viruses expressing malaria CD8+ T and B cell epitopes. Comparison of their immunogenicity and capacity to induce protective immunity. J. Immunol. *153*:4636-4648.

41.**Rowley, K.V., R. Harvey, and W.S. Barclay.** 1999. Isolation and characterization of a transfectant influenza B virus altered in RNA segment 6. J. Gen. Virol. *80*:2353-2359.

42.**Schnell, M.J., L. Buonocore, M.A. Whitt, and J.K. Rose.** 1996. The minimal conserved transcription stop-start signal promotes stable expression of a foreign gene in vesicular stomatitis virus. J. Virol. *70*:2318-2323.

43.**Seong, B.L. and G.G. Brownlee.** 1992. A new method for reconstituting influenza polymerase and RNA in vitro: a study of the promoter elements for cRNA and vRNA synthesis in vitro and viral rescue in vivo. Virology *186*:247-260.

44.**Singh, M. and M.A. Billeter.** 1999. A recombinant measles virus expressing biologically active human interleukin-12. J. Gen. Virol. *80*:101-106.

45.**Staczek, J., H.E. Gilleland Jr., L.B. Gilleland, R.N. Harty, A. García-Sastre, O.G. Engelhardt, and P. Palese.** 1998. A chimeric influenza virus expressing an epitope of outer membrane protein F of *Pseudomonas aeruginosa* affords protection against challenge with *P. aeruginosa* in a murine model of chronic pulmonary infection. Infect. Immun. *66*:3990-3994.

46.**Subbarao, E.K., Y. Kawaoka, and B.R. Murphy.** 1993. Rescue of an influenza A virus wild-type PB2 gene and a mutant derivative bearing a site-specific temperature-sensitive and attenuating mutation. J. Virol. *67*:7223-7228.

47.**Subbarao, E.K., E.J. Park, C.M. Lawson, A.Y. Chen, and B.R. Murphy.** 1995. Sequential addition of temperature-sensitive missense mutations into the PB2 gene of influenza A transfectant viruses can effect an increase in temperature sensitivity and attenuation and permits the rational design of a genetically engineered live influenza A virus vaccine. J. Virol. *69*:5969-5977.

48.**Walker, W.S., M.R. Castrucci, M.Y. Sangster, R.T. Carson, and Y. Kawaoka.** 1997. HEL-Flu: an influenza virus containing the hen egg lysozyme epitope recognized by CD4+ T cells from mice transgenic for an alphabeta TCR. J. Immunol. *159*:2563-2566.

49.**Yamanaka, K., N. Ogasawara, H. Yoshikawa, A. Ishihama, and K. Nagata.** 1991. In vivo analysis of the promoter structure of the influenza virus RNA genome using a transfection system with an engineered RNA. Proc. Natl. Acad. Sci. USA *88*:5369-5373.

50.**Yasuda, J., D.J. Bucher, and A. Ishihama.** 1994. Growth control of influenza A virus by M1 protein: analysis of transfectant viruses carrying the chimeric M gene. J. Virol. *68*:8141-8146.

51.**Zambon, M.** 1998. Laboratory diagnosis of influenza, p. 291-313. *In* K.G. Nicholson, R.G. Webster, and A.J. Hay (Eds.), Textbook of Influenza. Blackwell Science, London.

52.**Zheng, H., P. Palese, and A. García-Sastre.** 1996. Nonconserved nucleotides at the 3′ and 5′ ends of an influenza A virus RNA play an important role in viral RNA replication. Virology *217*:242-251.

53.**Zhou, Y., M. Konig, G. Hobom, and E. Neumeier.** 1998. Membrane-anchored incorporation of a foreign protein in recombinant influenza virions. Virology *246*:83-94.

FAMILY: Picornaviridae
GENUS: Enterovirus (Poliovirus)

33 Picornaviruses as Tools for Antigen Delivery

Steffen Mueller and Eckard Wimmer
Department of Molecular Genetics and Microbiology, School of Medicine, State University of New York at Stony Brook, Stony Brook, NY, USA

INTRODUCTION

General Remarks

It is an appealing idea to employ picornaviruses as vectors for the delivery of immunogens. Most picornaviruses are rapidly replicating, and their genomes are not integrated into the host cell's genome because of their exclusively cytoplasmic life cycle. Infection with a picornavirus leads to strong expression of viral antigens over a limited time span, followed by the virus' clearance through the host's immune system. Their fast, lytic replication cycle makes it unnecessary for the virus to modulate the immune defenses of the infected host. Picornaviruses generally infect their host through the gastrointestinal or respiratory surfaces, thereby eliciting a strong mucosal immune response. All these features are highly desirable in a vaccine vector. Hence, a great deal of effort has been made to develop picornaviruses, most notably poliovirus (PV), as antigenic vectors. However, the genetic properties of picornaviruses, in particular their genomic plasticity and high rate of RNA recombination, pose some obstacles for the development of such expression vectors. In this chapter, we will critically review several strategies to express exogenous sequences in a picornavirus background.

Poliovirus—The Picornavirus Paradigm

PV, the causative agent of poliomyelitis, belongs to the genus *Enterovirus* of the *Picornaviridae*, a family of small (Latin, *pico*), nonenveloped, positive-stranded RNA viruses. Coxsackieviruses and echoviruses, which compose the majority of the *Enterovirus* genus, have a wide pathological spectrum. They cause diseases such as meningitis, encephalitis, myocarditis, paralysis, and diarrhetic and respiratory diseases (78). Other members of the picornavirus family include the genera of *Rhinovirus* (common cold virus) with over 100 serotypes, *Cardiovirus* (encephalomyocarditis virus [EMCV] Theiler's mouse encephalomyelitis virus, or Mengo virus), and *Apthovirus* (foot-and-mouth disease virus [FMDV]).

Viral Vectors: Basic Science and Gene Therapy
Edited by A. Cid-Arregui and A. García-Carrancá

Although no longer a major public health hazard in the developed world, several factors have made PV the best understood and the most important virus in *Enterovirus* research. The virus possesses a very stable capsid, it grows to high titers in sturdy tissue culture cells (e.g., HeLa), and is easy to purify. These factors together with a low biosafety level make PV an ideal compound to study and cultivate. At the height of the poliomyelitis epidemics around the middle of the twentieth century, huge research efforts that parallel the human immunodeficiency virus (HIV) research of the present day were undertaken to contain this infectious disease. Aside from a wealth of clinical findings, many of the basic virological techniques were established during this period, including virus purification protocols, crystallization of the virion, an efficient tissue culture system, and plaque assay analysis (13,36,38,112,118). Never before had a virus spurred so much public interest as the "crippler" did. The March of Dimes raised millions of dollars for PV research, an effort that eventually culminated in the introduction of two PV vaccines developed independently by Jonas Salk (killed virus) and by Albert Sabin (live, attenuated, oral vaccine) (116,117). Soon poliomyelitis epidemics vanished in the Western World. Remarkably, PV is now targeted by the World Health Organization (WHO) to be eradicated worldwide during the first half of this decade.

After a temporary decline in PV research that nevertheless led to the elucidation of the chemical structure of PV (61), a second wave of activity ensued with the dawn of molecular cloning. Although it had been known for decades that the naked RNA genome of picornaviruses is infectious (3,24), it was not until 1981 that Racaniello and Baltimore cloned a full-length genomic cDNA clone of PV and found that plasmids harboring this cDNA are infectious when transfected into mammalian cells (106). This opened seemingly unlimited possibilities of genetically manipulating the PV genome with DNA-based techniques that had already been established. The infectivity of the original cDNA clones by direct DNA transfection was very low (10 plaque-forming units [pfu]/μg DNA) but could be increased considerably (1000 pfu/μg DNA) by including simian virus 40 (SV40) transcription and replication signals on the plasmid (119). Cloning the infectious cDNA downstream of a bacteriophage T7 promoter facilitated in vitro RNA transcription with phage T7 RNA polymerase and RNA transfection of tissue culture cells—the method of choice for obtaining infectious recombinant virus today (126). Using this method, the specific infectivity could be increased by another two magnitudes, which facilitated even the recovery of viruses with debilitating mutations that were previously assumed nonviable. With the advent of new reverse transription polymerase chain reaction (RT-PCR) techniques, infectious cDNA clones contained on a single plasmid can now be generated in a matter of weeks (123). In vitro translation in cell-free HeLa extracts of genomic T7 transcript RNAs allows PV to be studied at the level of the initiation of cap-independent translation, viral protein synthesis, and autoproteolytic processing of the viral polyprotein (18,32,84). It represents an important checkpoint in the successful cloning of new recombinant virus constructs and provides useful information about the effect of a particular genetic manipulation on these early steps in the viral life cycle. Under optimized conditions, PV RNA, purified from virions or in the form of T7 transcripts, can initiate the complete replicative cycle of translational initiation, polyprotein synthesis, proteolytic processing, RNA replication, and genome encapsidation, thereby leading to de novo synthesis of infectious virus (84).

PV is the first and only virus that could be replicated outside a living cell thus far. This exciting technology opened new avenues for the study of the PV life cycle in the absence of a cell membrane barrier.

Humans are the only natural hosts to be infected by PV, and all nonhuman primates can be infected experimentally with the virus. However, transgenic mice have been generated that express the human PV receptor CD155 (formerly called PVR). These transgenic mice are highly susceptible to wild-type PV infection by intracerebral, intravenous, or intraperitoneal inoculation (but not by the oral route, the port of infection in humans), and they develop clinical symptoms of poliomyelitis, including flaccid paralysis and respiratory arrest (63,109). This inexpensive animal model presents an invaluable asset for testing recombinant PVs as well as PV-based expression or vaccine vectors with respect to their pathogenicity, immune response, and immune protection under real-life conditions in a mammalian host.

GENETICS OF PV—A PRIMER

Genome Organization

The PV genome consists of a positive-stranded RNA molecule of about 7500 nucleotides (61). A relatively long 5′ nontranslated region (NTR) of 742 nucleotides is followed by a single open reading

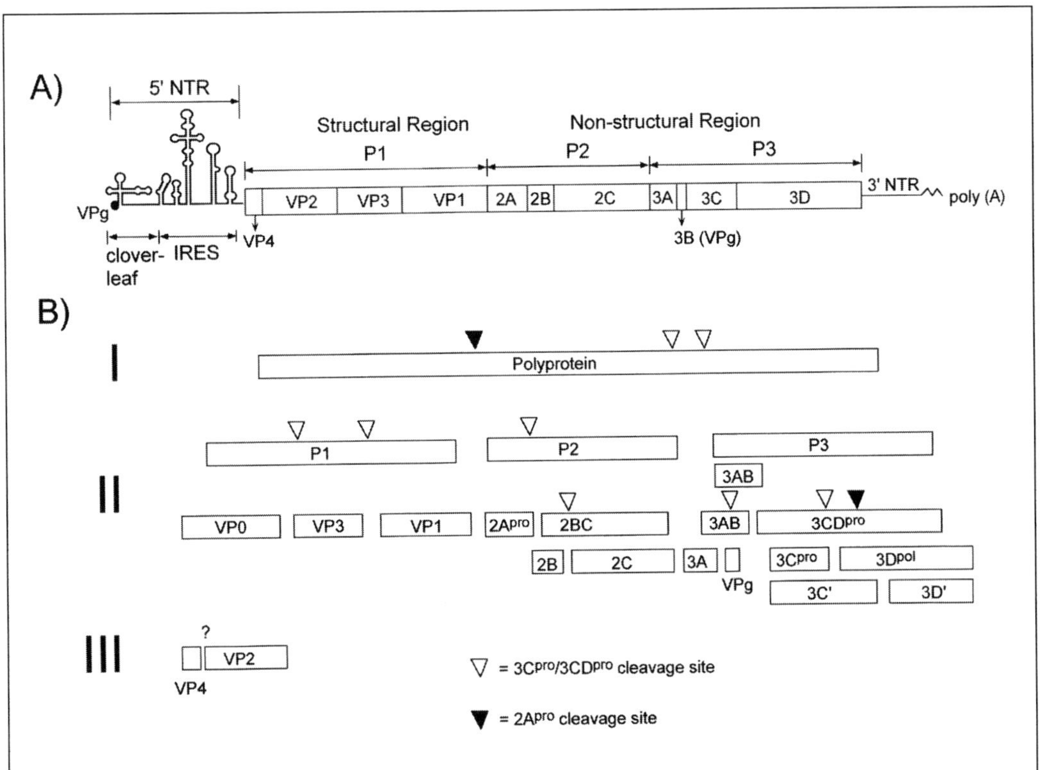

Figure 1. The poliovirus genome and polyprotein processing. (A) Genomic organization of poliovirus. The secondary structure of the 5′ NTR shown here is a consensus model of various proposals (68,97,99,119). (B) The three stages of polyprotein processing: I, *cis*-cleavages by $2A^{pro}$ and $3CD^{pro}$; II, *trans*-cleavage events generating the majority of poliovirus proteins; III, maturation cleavage. The question mark indicates that the mechanism of the cleavage is obscure. Open triangles indicate cleavage by $3C^{pro}$/$3CD^{pro}$ and closed triangles cleavage by $2A^{pro}$.

frame (ORF) that codes for a polyprotein of 2209 amino acids, and a 3′ NTR of 70 nucleotides (31,61,107). The very 3′ end of the viral RNA is composed of a stretch of poly(A) (131) varying around 60 adenylate residues in length (2) (Figure 1A). Upon viral infection, the vRNA is released into the cytoplasm of the host cell where it serves as a messenger RNA and is directly translated into the polyprotein. However, unlike most eukaryotic mRNAs, the PV genomic RNA lacks the m^7G cap structure essential for translational initiation of normal eukaryotic mRNA (92). Instead, a small viral protein, VPg, is covalently linked to the 5′ terminus (69,70,91,127).

The 5′ NTR harbors two known *cis*-acting elements whose structures are generally conserved throughout the *Enterovirus* and *Rhinovirus* genera (111). Three stem loops within the 5′ 88 nucleotides form a clover leaf-like secondary structure (9,111) that has been implicated in the formation of the replication complex during positive-strand RNA synthesis (9). However, the largest portion of the 5′ NTR is taken up by a second *cis*-acting element, a genetic element first identified in encephalomyocarditis virus (54,55), which facilitates the assembly of the translational initiation complex and the binding of the ribosomes to the RNA (94,95). Since the events leading to initiation of polioviral translation are cap independent, and the formation of the initiation complex does not occur at the 5′ terminus of the PV genomic RNA, this complex structural and functional element has been named internal ribosomal entry site (IRES) (54,55). Following the discovery of the IRES in the encephalomyocarditis virus (54,55), similar elements were identified in all other picornaviruses (14,17). IRES-dependent translation is now well accepted as a novel mechanism of eukaryotic cap-independent protein synthesis. Aside from picornaviral genomic RNAs, several cellular mRNAs have been proposed to also utilize IRES-mediated translation (1,53,110,125,130).

Polyprotein Processing

Translation of the single ORF of PV results in an unstable polyprotein that immediately undergoes autocatalytic processing. A cascade of concerted proteolytic cleavage events, catalyzed by the virus-encoded proteinase entities $2A^{pro}$, $3CD^{pro}$, and $3C^{pro}$, yields the set of functional viral proteins necessary for virus proliferation (Figure 1B). Processing of the polyprotein can be divided into three stages. First $2A^{pro}$ carries out a *cis*-cleavage of the Tyr-Gly bond at its own N terminus (50). This *cis*-cleavage leads to the release of the capsid precursor P1. Once translation has been completed, proteinase $3CD^{pro}$, which maps to the C terminus, performs another *cis*-cleavage, similar to $2A^{pro}$, to sever itself from the C terminus of the polyprotein. In a highly ordered succession of *trans*-cleavage events by $3CD^{pro}$, the nonstructural proteins 2A, 2BC, 3AB, 2B, 2C, 3A, VPg (3B), $3C^{pro}$, and $3D^{pol}$, as well as the capsid proteins VP0, VP1, and VP3, are then released from their precursors (47,49,64). $3C^{pro}$ and $3CD^{pro}$ exclusively cleave Gln-Gly bonds. The amino acid sequence of the immediate surroundings and the location of the Gln-Gly pair within the protein seem to determine the efficiency of the cleavage. Out of 13 Gln-Gly pairs within the polyprotein, 8 represent actual substrates for $3CD^{pro}$, whereas only 2 out of 10 Tyr-Gly pairs are cleaved by $2A^{pro}$ (49). An alanine residue in the P4 position (Ala-X-X-Gln-Gly) has been linked to fast cleavage (16,20), as observed between 2C and 3A, 2A and 2B, or VP3 and VP1. In the last stage of protein processing, after assembly of the virus particle, VP0 is cleaved by an unknown mechanism to yield VP4 and VP2. This cleavage is required for infectivity of the virus and is therefore referred to as maturation cleavage (46).

Coding of the genomic information in the form of a single polyprotein excludes the possibility of any regulation in the expression of individual PV proteins at the transcriptional or translational level (61). However, in infected cells, PV proteins are not present at an equimolar ratio. All $3C^{pro}/3CD^{pro}$ cleavages occur at a Gln*Gly pair, that of $2A^{pro}$ at a Tyr*Gly pair. The individual cleavage recognition signals for $3C^{pro}/3CD^{pro}$, however, differ slightly from each other in the sequences upstream and downstream of the scissile bond so that the sites are not cleaved with the same efficiency. This leads to a disproportional accumulation of the viral proteins and their precursors. Polymerase $3D^{pol}$, for instance, is apparently only needed in small amounts to synthesize viral positive- and negative-strand RNA templates. Facilitated by a slow cleavage between $3C^{pro}$ and $3D^{pol}$ (48), most of the 3D polypeptide is, therefore, present in its precursor form, $3CD^{pro}$, the major proteinase employed in polyprotein processing. Interestingly, $3CD^{pro}$ does not possess any polymerase activity (48), while, on the other hand, it is capable of performing almost all the functions of $3C^{pro}$ (83). This ability to encode several functions in the same primary amino acid sequence increases the genomic information while allowing the virus to keep a small genome (128).

Proteolytic processing does not depend on any cellular proteinases (128). Moreover, the polioviral proteinases $2A^{pro}$ and $3C^{pro}$ are actively involved in inhibition of host cell functions. $2A^{pro}$ is responsible for the cleavage of eIF4G, a component of the cap-recognizing complex eIF-4F, thereby shutting off cap-dependent host cell translation (65,122). $3C^{pro}$ has been shown to inactivate transcription factor TFIIIC (23) and to cleave the TATA box binding protein (27), which leads to an inhibition of cellular transcription.

Genetic Recombination in RNA Viruses

Genetic recombination occurs with high frequency in positive-strand RNA virus-infected cells (57,59,128). RNA recombination has been slowly accepted as an important factor for genetic variation in RNA viruses (128). The mechanism of recombination is poorly understood, but for picornaviruses, it occurs by copy choice at homologous regions as opposed to strand breakage and reunion (59,60, 128). An estimated 10% to 20% of all PV genomes undergo recombination in a single growth cycle (59), and recombination between sibling strands occurs at an astoundingly high frequency (20). In cells infected with a single-genotype virus, recombination (isogenic) mainly serves to compensate for the high error rate of the viral RNA polymerase (29,30,33,52,108) during RNA replication and to maintain the wild-type genotype. Furthermore, recombination can give rise to variants more viable under certain conditions, and it may accelerate evolution (128).

In the copy choice model of picornaviruses (34,56,57,60,67,114,124,128), nascent RNA strands switch templates, presumably during negative-strand synthesis (60). Two mechanisms have been proposed to explain how this template switching occurs. The first one depends on self-complementary sequences (inverted repeats), which can lead to the local formation of heteroduplexes of two RNA strands of the same polarity. Initiating transcription on the 3′ end of the first strand, the RNA polymerase may stall when it reaches the region of the intergenomic base pairing. Instead of unwinding the heteroduplex and proceeding to replicate the first template, the replication complex may just dissociate from the template and reassociate with the homologous sequence (i.e., the complementary sequence to the 3′ end of the nascent RNA molecule) of the second template (115,128). This model is favored by two observations. First, genetic recombination of PV is precise. That is, no extra nucleotides are inserted at the crossover site nor are any nucleotides lost. Second, the site of crossover changes when replication occurs at different temperatures (35). The second mechanism favors pausing of replication caused by nucleotide content at the site of recombination (59) followed by dissociation of the nascent RNA strand and reannealing at a complementary region of the same or another template molecule (85,99). Illegitimate recombination can occur with high frequency when extra sequences have been inserted into the viral genome. These extra sequences could either be of the viral genome itself (20,21) or of foreign origin (72,85). The net result of the illegitimate recombination event is either the complete removal of the extra sequence, partial deletion, or genetic rearrangements. The model of pausing caused by nucleotide content is more suitable to explain illegitimate recombination between unrelated sequences because only short direct repeats seem to be needed as parting and anchoring sites to promote template switching (85,99).

EXPRESSION OF FOREIGN SEQUENCES IN A PV BACKGROUND

General Considerations

As one of the fastest replicating picornaviruses, PV expresses its cognate proteins in relatively large quantities. However, if a foreign gene is inserted into the PV genome, PV is now only a marginal expression vector in terms of recombinant protein yield. This is because most insertions of foreign sequences interfere with efficient replication. A simple calculation will serve to illustrate the expression level of PV. Under optimal conditions, a tissue culture cell infected with PV1 Mahoney produces 1000 pfu of virus, which correspond to about 100000 virus particles. Every particle contains 60 copies of each capsid protein. Assuming that not all capsid proteins synthesized by the cell will form particles, the cell must have produced up to 1×10^7 capsid proteins and, consequently, the same amount of all polypeptides contained in the polyprotein. 1×10^9 cells could, therefore, express 1 to 2×10^{16} copies of a foreign protein as part of the polyprotein, or 1.7 to 3.3×10^{-8} mol, corresponding to 330 to 660 μg of a 20-kDa protein. In addition, the loss of the foreign gene by recombination reduces the yield of foreign protein (see below). Therefore, PV is not known to be a high-level expression vector. Nevertheless, a number of favorable properties of the live-attenuated PV vaccines (Sabin strains) (115,128) would make the virus an attractive vector for the development of a new generation of vaccines, provided that it was possible to stably express foreign antigenic sequences in the PV genome. In over 40 years of vaccination, the Sabin vaccines have proven to be safe and efficient. The live virus has the advantage of a self-replicating system that delivers more antigen over a longer period of time at the site of natural infection when compared to a killed virus or a protein vaccine. This is likely to be one of the reasons for the strong mucosal and humoral immunity observed with the Sabin vaccines (78). Furthermore, mucosal immunity is a desirable feature of any vaccine, if the first encounter with a pathogen occurs at the mucosal surfaces of the gastrointestinal tract or the respiratory system. Finally, the ease of administering live PV-based vaccines via the oral route is of great practical advantage over delivery by needle injection.

Another reason for expressing exogenous sequences in PV relates to studies of the life cycle of the virus and its pathogenesis. Data concerning the primary site of virus proliferation in the course of human infection are still fragmentary. In particular, it is not clear why invasion of the virus into the central nervous system (CNS) occurs only at a frequency of 1% to 2% of naïve infected individuals, regardless of the genetics or age of the infected person. Available evidence suggests that the virus can spread to the CNS via crossing the blood-brain barrier (129) or by retrograde axonal transport (43). To what extent either of these pathways operate during a normal infection is not clear, although muscle injury clearly favors retrograde axonal transport (43). To answer these important questions of PV pathogenesis, suitable marker genes expressed within the polio genome could be of great advantage to detect replicating PV in situ in infected CD155-transgenic mice.

PV Replicons

Picornaviruses are genetic minimalists. Their genomes are streamlined to such an extent that none of the gene products can be spared. Except for some small deletion within the 5′ NTR, removal of any coding sequences is generally incompatible with viral proliferation, with one notable exception. In laboratory strains of PV that are passaged at high multiplicities of infection (moi), mutant genomes containing large deletions within the capsid coding region can be isolated. These genomes can replicate, and they can be encapsidated *in trans* into capsids provided by the parental virus that is present in the same cell. All such naturally occuring deletion variants, termed defective interfering (DI) particles (128), maintain the correct reading frame following the deletion, an observation which suggests that at least one of the replication proteins must be provided *in cis* by the DI genome to be replicated, rather than by the helper virus (66,74). Furthermore, the existence of DI particles showed that neither the capsid proteins nor their respective coding sequence are necessary for RNA replication, and that the elusive encapsidation signal cannot be located in this region of the PV genome. Consequently, it was found to be possible to replace P1 coding sequences partly or completely with a foreign gene, while retaining the approximate size of the genome (22,96). The latter is an important feature, since packaging constraints limit the maximum size of a recombinant genome to about 120% of the wild-type (4). Upon RNA transfection, such vectors, referred to as replicons, are capable of cap-independent translation, RNA replication, and proteolytic processing. *cis*-cleavage by $2A^{pro}$ proteinase at its natural cleavage site (VP1-2A) resulted in the release of the foreign gene from the polyprotein. For optimal expression and genome replication, the heterologous gene is expressed as a fusion with all or a fraction of the VP4 capsid protein sequence at the N terminus, and parts of VP1 at the C terminus (Figure 2A). An additional 2A cleavage site can be introduced between VP4 and the foreign gene to produce an authentic N terminus (104). However, all of the P1 coding region can be deleted, leaving only a minimal 2A cleavage site at the C terminus of the inserted gene (Figure 2A). This allows the expression of native proteins with as little as four extra amino acids at the C terminus (104). In this case, if no P1 stuffer sequences are present, it is important that the inserted foreign gene be large enough to ensure a genome above the minimal length required for encapsidation. Representatives of proteins expressed in PV replicon vectors include HIV-1 Gag, Pol, Env (22,102), mouse interleukin-2 (15), chloramphenicol acetyltransferase (CAT) (96), firefly luciferase (8,26,103), and the green fluorescent protein (GFP) (unpublished results).

If cells are preinfected with a recombinant vaccinia virus, termed VV-P1, that

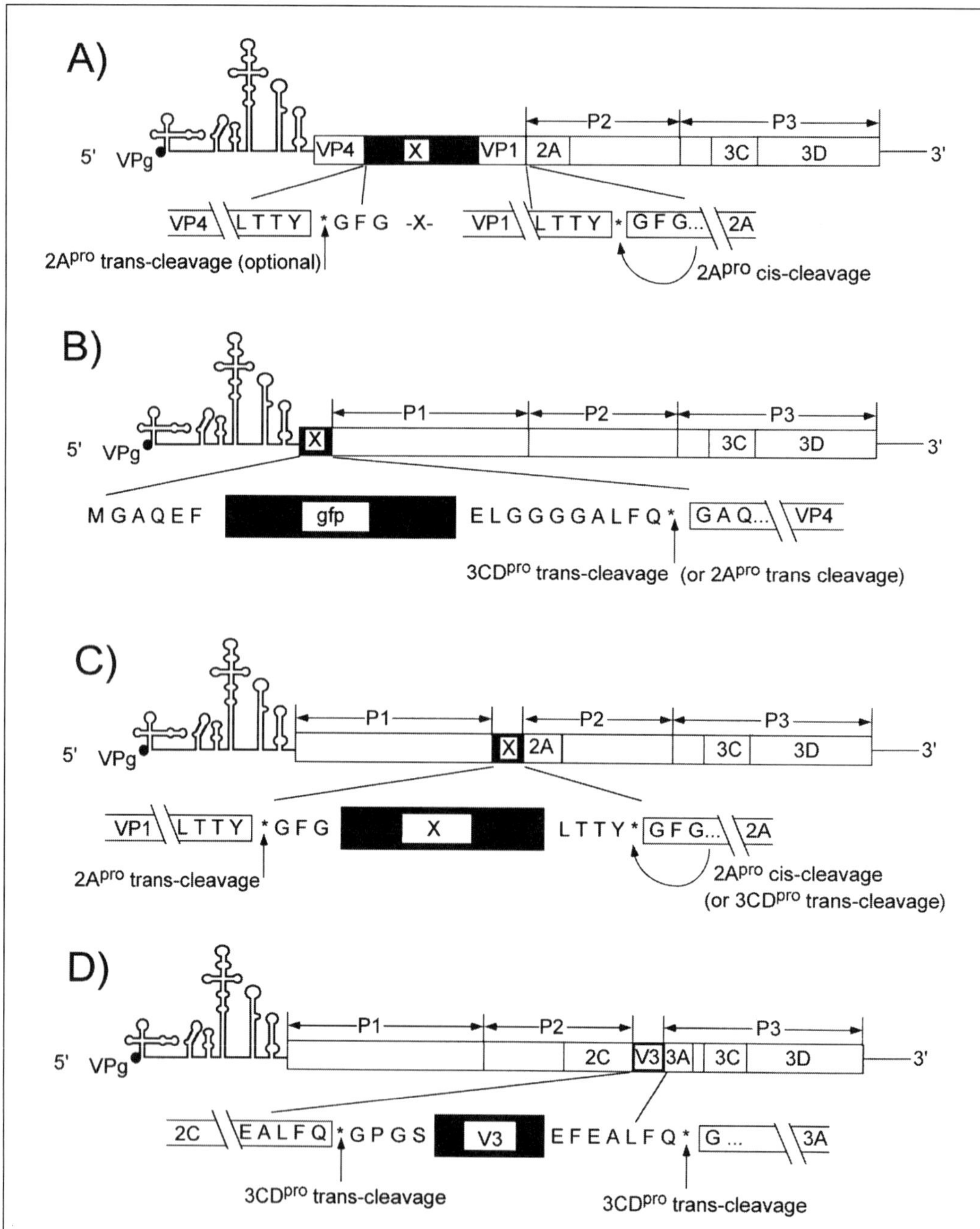

Figure 2. Expression of exogenous sequences in poliovirus. (A) Poliovirus replicon. The foreign gene X replaces a large portion of the P1 region. Upon RNA transfection the X-ΔP1 fusion protein is released by 2A^{pro} *cis*-cleavage at the authentic cleavage site. To generate infectious particles, cells are preinfected with recombinant vaccinia virus, providing the P1 proteins in *trans.* (B) Polyprotein fusion vector. Heterologous sequences are fused to 5′ end of the PV ORF and expressed as a fusion with P1. An artificial 3C^{pro}/3CD^{pro} or 2A^{pro} cleavage site ensures that the foreign protein is cleaved off post-translationally to leave an intact, unaltered VP4. (C) Polyprotein fusion vector. The exogenous sequence is inserted at the P1-P2 junction, with a 2A^{pro} and a 3C^{pro}/3CD^{pro} cleavage sites at its N′ terminus and C′ terminus, respectively. Both enzymes cleave in *trans* and no changes to the coding sequences of PV proteins are made. Alternatively, the cleavage at the C-terminus can be performed by a 2A^{pro} *cis*-cleavage. (D) Coding sequences for the V3 loop of HIV gp120 were inserted between 2C and 3A. V3 peptide is released from the polyprotein by two 3C^{pro}/3CD^{pro} *trans*-cleavage events.

provides the PV capsid precursor protein P1 *in trans*, these replicons are encapsidated and yield infectious particles (12,101). The encapsidated replicons can now be used to infect PV-susceptible cells and be passaged in the presence of the vaccinia virus expression vector VV-P1 providing the PV capsids. Alternatively, replicon RNA can be packaged into capsid provided by coinfection with a proliferation-competent PV. Supplying the capsid protein by means of a recombinant vaccinia virus presents a clear advantage over coinfection with a polio helper virus. The encapsidated replicons can easily be separated from the vaccinia packaging virus and be highly purified, thereby removing any viable virulent virus.

An obvious benefit of encapsidated replicons as immunogenic agents would be their safety because the lack of capsid-coding genes eliminates the chance of uncontrolled proliferation. Any proliferation in a vector recipient can lead to a virulent virus by genetic variation. On the other hand, this inability of proliferation prevents the replicon from spreading in the immunized host. Indeed, the expression of the foreign gene by the vector is restricted to a single cell, that of primary infection. Consequently, an immunogenic replicon will be readily cleared, in stark contrast to a live attenuated virus vector. As a result, higher doses and more booster inoculations would have to be given to obtain an immune response comparable to that of a live vaccine. It is also unlikely to obtain humoral immunity by way of oral administation, so that both oral and intramuscular inoculations are indicated to receive mucosal and humoral immunity. Nonetheless, work by C.D. Morrow and his colleagues has demonstrated that CD155-transgenic mice immunized with encapsidated replicons show an antibody response against a foreign gene after two booster immunizations (11,81). Immunization of mice with a replicon containing the C-fragment of tetanus toxin led to protective immunity against lethal doses of tetanus toxin (104).

Large-scale production of encapsidated replicon, in light of using them as antigenic vectors, is labor-intensive and costly since it may take up to 50 passages to obtain a sufficient amount of virions (103). Because of the truncations of P1, the size of the replicon genomes does not differ significantly from wild-type, which may contribute to their apparent genetic stability. Porter et al. (101) reported no apparent genetic rearrangements of such a vector for at least 14 passages. Since the replication is restricted to a single growth cycle upon infection of a cell, selection pressure is greatly limited. Better-replicating variants (as a result of deleting the futile foreign sequence) cannot unfold their full potential of outcompeting the parental expression vector. There might even be a selective advantage to keeping the foreign sequence because its deletion could result in generation of a genome too short for encapsidation. The largest deletions that have been identified in naturally occuring DI particles compose only 16% of the genome (66). With all its capsid genes as well as the inserted foreign gene deleted, a replicon RNA (now 35% smaller than wild-type) would most likely not be encapsidated.

Antigenic Hybrid Vectors

PV consists of three serotypes, PV1, PV2, and PV3, which differ immunologically with respect to four major neutralization antigenic sites (N-Ags) on the surface of the virus capsid. Once the location of the N-Ags was mapped (37,79,86) and the crystal structure of PV was solved, it became evident that the regions that contain the N-Ags form loops protruding from the surface of the virion, which renders them easily accessible to the host's immune system (51). This favorable presentation on the surface taken together with the structural flexibility that such loops may offer made the N-Ags prime candidates for the insertion of foreign epitopes. However, the structural properties of N-Ags restrict their genetic alteration. For example, N-AgIIIa and N-AgIIIb are discontinous antigenic sites, being composed of residues not only from different portions of the same capsid protein but also from different capsid proteins (for a structural presentation of the N-Ags on different capsid proteins, see Reference 97). N-AgI in the BC-loop of capsid protein VP1, however, consists of a continuous sequence of amino acids, and it is thus suitable for genetic manipulation. The BC-loop of VP1 consists of some 10 amino acids, all of which contribute to the formation of N-AgI. Early work demonstrated that N-AgI of PV type 1 (PV1) can be replaced by that of PV type 3 (PV3), resulting in an antigenic hybrid virus that elicited neutralizing antibodies against both PV1 and PV3 (19,89). Interestingly, the replacement of the VP1 BC-loop in PV1 with that of PV2 (strain Lansing) yielded a virus with the host range of PV2 (Lansing). That is, it was able to kill mice (90). The replacement of the BC-loop with heterologous sequences was later extended to the insertion of antigenic sites of other viruses into the BC-loop of VP1. The list includes neutralization immunogenic site IA of human rhinovirus type 14 (7), peptides of hepatitis A virus capsid (71), epitopes of HIV-1 gp160/gp120 (28,39) and of human papillomavirus 16 (58).

Similar results were achieved with epitopes of nonviral immunogens. Antigenic hybrids that express epitopes of the major outer membrane protein (MOMP) of *Chlamydia trachomatis*, a causative agent of sexually transmitted diseases, induced a strong immune response in rabbits against the MOMP protein (87,88). Antisera of these rabbits were able to neutralize chlamydial infectivity in tissue culture and by passive immunization in the conjunctival epithelia of cynomolgus monkeys (87,88).

Whereas all other PV-based immunogenic vectors require active expression of an antigen within the host, antigenic hybrids, by virtue of displaying the antigen on the virion surface, can also be delivered as a killed virus vaccine or as a live vaccine in unsusceptible hosts—a feature that would greatly increase its range of application in live stock. For example, Kitson and colleagues replaced PV1 (Sabin) N-AgI with antigenic site 1 of FMDV and showed that one such chimera could protect immunized guinea pigs from a challenge with FMDV (62).

Although the BC-loop of VP1 seems to possess an inherent flexibility to accommodate foreign sequences, there is no consensus as to what epitopes can be inserted without interfering with proteolytic processing of the polyprotein (due to change in folding) and with capsid structure. Ideally, the number of foreign amino acids inserted should be equal or only slightly larger than the number of amino acids removed from the BC-loop (25). In one case, nine extra amino acids could be inserted into the BC-loop without negatively affecting the virus (80). Since the heterologous sequence is expressed as a permanent fusion with capsid proteins, larger inserts are more likely to interfere with protein folding of the capsid, proteolytic processing of the capsid precursors, interaction with the cellular receptor, or possibly with uncoating of the virus upon entry into the cell. Indeed, antigenic hybrid vectors are often found to display inferior proliferation and/or poor antigen presentation. Finding a suitable epitope to insert into PV antigenic sites that does not interfere with viral proliferation and infectivity, but at the same time is highly immunogenic therefore constitutes a process of trial and error. Smith et al. have tried to shorten the process of finding suitable BC-loop constructs by sophisticated peptide library display methods for the selection of picornavirus capsid chimeras (121). Nevertheless, no virus surface chimera suitable for human trials has yet been found.

Polyprotein Fusion Vectors

In the previous two strategies, either a part of the genome was removed to accommodate insertion of a foreign gene, thereby sacrificing the ability to form virions (generation of replicons), or permanent changes were made to the virus capsid (antigenic hybrids). A more recent method that avoids both these limitatons is based on the unique genetic organization of cardioviruses. Here, a short leader protein, L, precedes the capsid precursor within the polyprotein. The function of of the L protein in cardioviruses is unknown (45). Immediately following translation, the L protein is severed from the polyprotein by the cardiovirus-encoded proteinases $3C^{pro}$. Altmeyer et al. made in-frame insertions of foreign proteins into the L protein of Mengo virus (5,6). The recombinant fusion protein is released from the polyprotein by using the indigenous $3C^{pro}$ cleavage site between L and VP4. By expressing a nine amino acid long cytotoxic T lymphocyte (CTL) epitope of lymphocytic coriomeningitis virus (LCMV), the authors were able to immunize mice against a lethal challenge with LCMV (6). In another experiment, the same group inserted 147 amino acids of HIV-1 gp120 and reported that the resulting recombinant virus was genetically stable for four passages (5), but eventually deleted the foreign sequence (R. Altmeyer, personal communication) Points to consider in favor of Mengo virus for its development as a recombinant virus vaccine vector are its broad host range and the existence of attenuated subtypes.

The idea behind the Mengo vector was applied to PV in a very similar fashion. PV does not naturally encode a leader protein. Nonetheless, insertion of several coding sequences of foreign genes immediately upstream of the P1 region (just like a cardiovirus leader) generated a viral genome that is longer than the wild-type (Figure 2B) and yielded viable virus (10,77,85). An artificial $3C^{pro}/3CD^{pro}$ cleavage site between the exogenous sequence and the PV polyprotein facilitated *trans*-cleavage by $3C^{pro}/3CD^{pro}$ to release the foreign protein and yield an authentic P1 region that was subsequently processed into functional capsid proteins. Some of the proteins expressed by this method include rotavirus VP4 and VP7, hepatitis B surface antigen, herpes simplex virus type 2 glycoprotein D (76,77), HIV-1 gag, cholera toxin subunit B (10), and GFP (85). Similarly, coding sequences could be inserted at the VP1-2A junction (76,77) (Figure 2C). However, at either insertion site, the genetic stabilty of this type of polyprotein fusion construct greatly decreases with increasing size of the inserted sequence and may also depend on the nature of the foreign gene (76,77). Mattion et al. (76,77), using this type of vector, were unable to isolate stable virus from constructs that express exogenous sequences larger than 303 nucleotides (at either insertion site, N terminal or at the P1-2A junction). With larger inserts, for example, p17-p24 of HIV-1 gag (1161 nucleotides), we found that deletions of part or all of the insert occurred during the first passage, presumably during the early rounds of genome replication immediately after RNA transfection (85). Although Andino et al. (10) initially reported that the same recombinant virus, mo-hgag, was genetically stable and elicited humoral and mucosal immune response in transgenic mice and cynomolgus monkeys, respectively (10), it was later shown that the construct developed by Andino et al. was also highly unstable (85). That is, the viral RNAs containing the insert were quasi-infectious (40), and only deletion variants emerged, the deletions occuring by illegitimate recombination (85). Interestingly, not all virus variants isolated after RNA transfection with the HIV-1 Gag fusion construct reverted back to the wild-type genotype. The majority retained between 50 and 150 nucleotides of the heterologous sequences, which could now be stably passaged for at least six passages (85). Sequence analysis revealed that many of these variants still contained at least one of the antigenic sites of the inserted Gag antigen, located between amino acids 11 and 25 of p17 (85,113). Based on our observations (85), it is certain that the very weak immune response reported in test animals (10) was induced by the truncated variants containing coding sequences for only small fragments of the foreign proteins, rather than by the parental virus vectors. Judging from the available data, PV polyprotein fusion vectors are therefore impractical as delivery vehicles for the expression of large antigens, a severe limitation on the usefulness of this strategy of vaccine construction. Generating an array of vectors by inserting individual antigenic peptides of a certain antigen, rather than the whole gene in one vector, might present a viable alternative to stably retain the immunogenic information of the particular antigen. However, we predict that long inserts like the *gag* gene of HIV, conferring no advantage to the wild-type genomes with respect to proliferation, will be too much of a genetic burden for PV to survive more than a few rounds of genome replication.

In another approach, Mandl and colleagues inserted a 600-nucleotide fragment of the chicken ovalbumin gene at the P1-2A junction (Figure 2C) (75). Despite the fact that the resulting virus was highly unstable and started deleting the insert in the second passage on HeLa cells, they were able to show that mice immunized with this vector showed a CTL response. Furthermore, the immunized animals resisted a tumor challenge with melanoma cells that express chicken ovalbumin as a model tumor antigen (75). The animals were immunized with 1×10^6 to 1×10^7 pfu of virus followed by three booster inoculations with 1×10^6 pfu of virus each (75). This corresponds to a titer of 5×10^{10} pfu if humans were ever to receive such a treatment. The first immunization would require a nearly 10000-fold higher amount of virus than used in PV vaccinations with Sabin strains. Under these circumstances, a vaccine vector that already is highly susceptible to genetic rearrangement may create a high risk of vaccine-related disease. It should also be mentioned that the vectors described by Mandl et al. (75) are constructed in a wild-type PV1 (Mahoney) background. For any application in humans, these vectors will have to be constructed in a viral background with a proven attenuation phenotype (Sabin strains). This in turn would debilitate the virus vector further, and even higher titers may have to be inoculated. Moreover, the vector cannot be passaged to produce large amounts of virus because of the rampant loss of its foreign insert. Therefore, only virus directly isolated from RNA transfection could be used for inoculation. Together with the necessary high titers, this would also make the use of such a vector an extremely expensive and labor-intensive undertaking. Finally, the nearly 100% protective immunity against PV of the population within the United States would pose problems of whether the vaccine would "take."

An interesting variation of a polyprotein fusion vector was a PV genome in which the coding sequence for the V3 loop of HIV gp120 (42 amino acids) was inserted between the coding sequences of 2C and 3A (see Figure 2D) (93). Remarkably, the resulting virus grew with wild-type kinetics, and its genome was stable for at least six passages (93). This might present yet another strategy for the development of novel vaccines or vehicles for gene delivery.

Figure 3 shows HeLa cells that were infected with a PV polyprotein fusion vector, PVM-gfp (Figure 2B), expressing the GFP of *Aequorea victoria* (85). Picornaviruses expressing reporter genes as this one does could be great assets for the study of viral pathogenesis in vivo, if they could be improved to stably retain the foreign gene.

A critical requirement for the good growth characteristics in this vector system is the fast and efficient cleavage at the artificially introduced proteinase cleavage site. A $2A^{pro}$ cleavage site was found to be superior to a $3C^{pro}/3CD^{pro}$ at both insertion sites, N terminal or at the P1-$2A^{pro}$ junction (75,86,132).

Dicistronic Vectors

The discovery of the IRES warranted the development of dicistronic PV constructs (82). This made it possible to express foreign sequences in a separate ORF, leaving the PV polyprotein unchanged (4). In W1-DICAT (Figure 4C), translation of the *CAT* gene is driven by the PV IRES, whereas translation of the PV polyprotein is reinitiated by the IRES of EMCV inserted upstream of the PV ORF (4). In another construct, the EMCV IRES was placed downstream of the P1 region, thus separating translation of P1 from the translation of the nonstructural proteins (Figure 4A) (82). Heterologous sequences that were fused to the N terminus of 2A were cotranslationally cleaved off by $2A^{pro}$ *cis*-cleavage (Figure 4B). Both type of constructs, with genomes up to 17% larger than wild-type, yielded viable viruses that expressed the foreign gene product at least in the initial passages. However, upon further passaging, the virus deleted parts or all of the exogenous gene together with the EMCV IRES (4,72). One further disadvantage of dicistronic

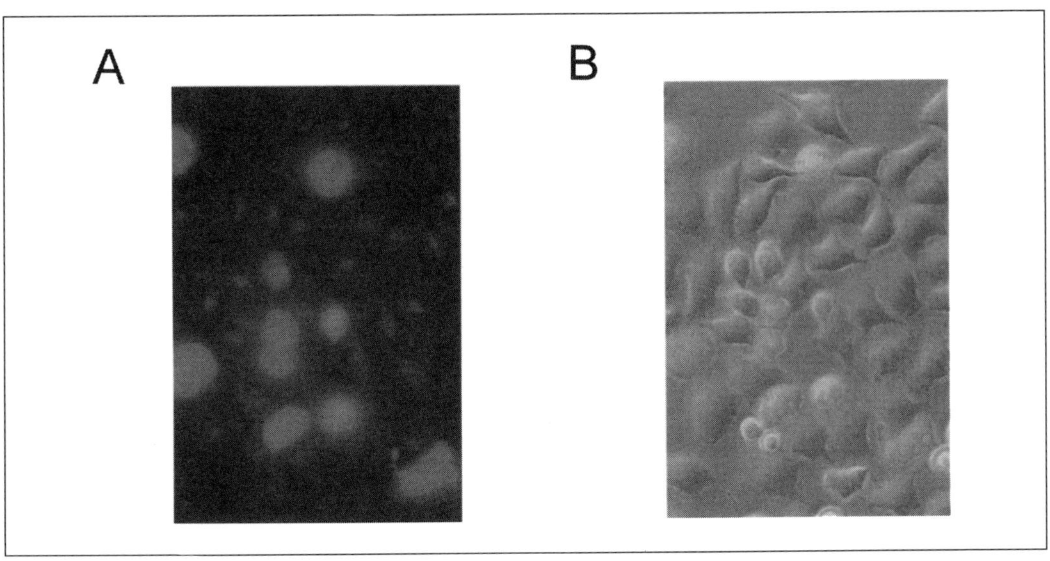

Figure 3. HeLa cells infected with poliovirus polyprotein fusion vector expressing GFP (85). (A) Fluorescence microscopy. Cells were seeded on coverslips, infected, and mounted after 7 hours in PBS (unfixed) and photographed immediately. (B) Double exposure photography. Fluorescence microscopy was performed and the first exposure was made. Then, without rewinding the film and moving the object, a second exposure of the brightfield image was taken. This method allows fluorescing and nonfluorescing cells to be visualized on the same photograph. *Note:* The image as it appears is not actually observed under the microscope. **(See color plate A8.)**

expression vectors is that about 50% of the heterologous sequence that can be inserted is already taken up by the second IRES, thereby considerably limiting the feasible size of the gene to be expressed. Insertion of the EMCV IRES at several other sites within the PV genome, including the 2B-2C, 2C-3A, 3A-3B, 3B-3C, or 3C-3D junctions, abolished viral replication, an observation indicating that certain PV protein precursors are functionally required at some stage during viral replication (93).

Overall, the dicistronic viruses were not any more or less stable than the polypro-

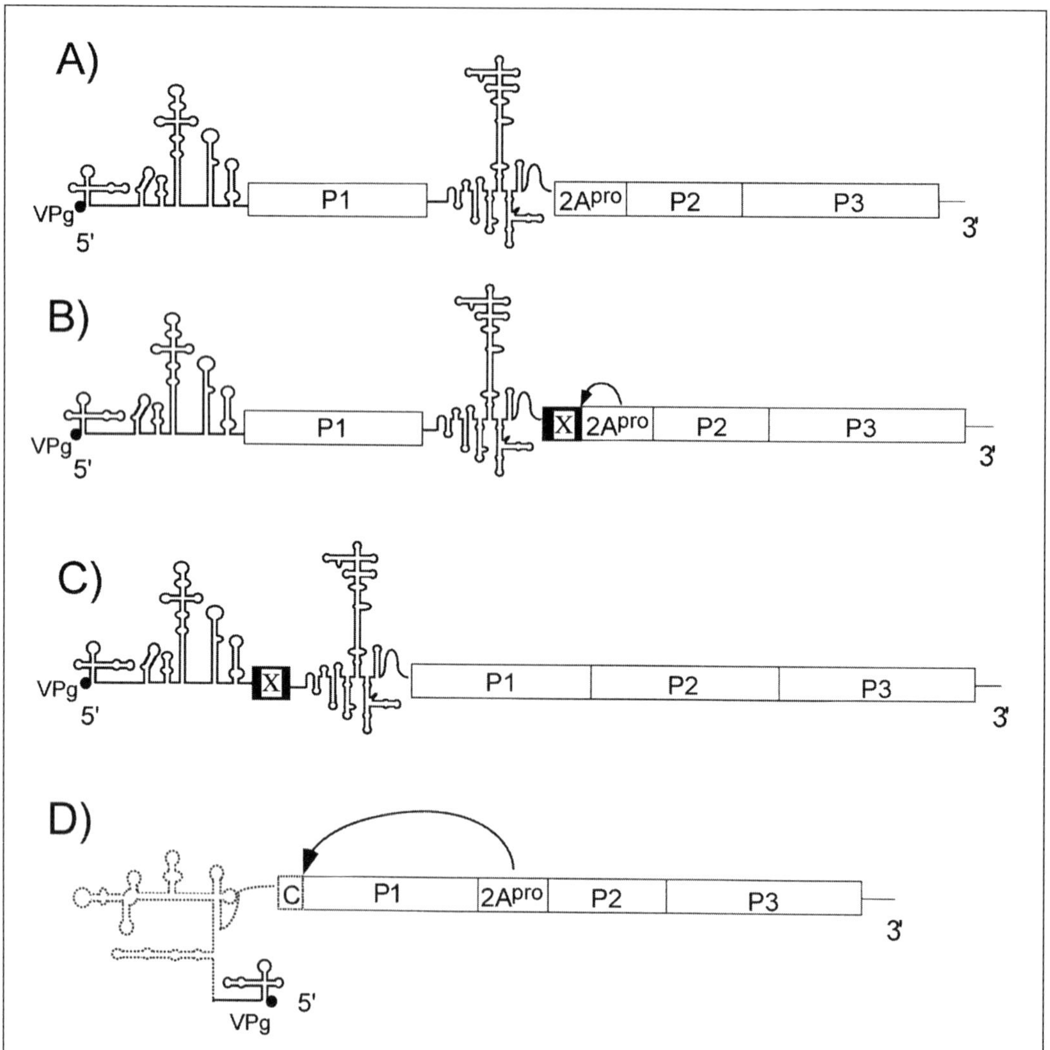

Figure 4. Polioviruses carrying heterologous IRES elements. (A) Insertion of an EMCV IRES at the P1-P2 junction disrupts the reading frame of the polyprotein and generates a dicistronic virus. As a result, two polypeptides, P1 and P2-P3, are synthesized. (B) Dicistronic vector with an EMCV IRES placed in between P1 and P2 to drive expression of a heterologous protein fused to the N′ terminus of 2Apro. Subsequently the protein is released by 2Apro *cis*-cleavage of the Tyr-Gly bond at the N′ terminus of 2A. (C) Dicistronic vector, in which an EMCV IRES is inserted immediately upstream from the PV ORF to initiate translation of the polyprotein. A foreign sequence between the two IRESs is translated from the PV IRES in a separate ORF, therefore no artificial cleavage event is necessary. (D) HCV IRES chimera. The poliovirus IRES is replaced by the HCV IRES plus part of the HCV core protein (C). The core protein is post-translationally cleaved off the polyprotein by 2Apro.

tein fusion constructs containing inserts larger than approximately 100 nucleotides.

Expression of Viral RNA Elements in PV

Since heterologous IRES are functional in dicistronic PVs, the question arose whether the PV IRES could be replaced entirely by an IRES of another virus. Indeed, changing the polio IRES for those of EMCV, human rhinovirus (HRV) type 2 and 14, and Coxsackievirus B4 and A9 yielded viable virus with excellent growth characteristics in HeLa cells (4,41,44). The promiscuity of PV to utilize a foreign IRES is even more surprising in a chimera that utilizes the IRES element of hepatitis C virus (HCV) (Figure 4D), a flavivirus only very distantly related to PV (73,132). This system allows the study of heterologous IRES in a well-replicating PV, which is especially useful for the HCV-IRES chimera since there is no tissue culture system available for the study of HCV. This HCV chimera is also interesting in that it stably expresses the N-terminal 123 amino acids of the HCV core protein. The core protein RNA sequence (but not its translation product) is absolutely necessary for the proper function of the HCV IRES, and it is therefore stably retained by this virus (73,132). The core polypeptide that is translated as an in-frame fusion with the P1 precursor is very efficiently cleaved off the polyprotein through an artificial $2A^{pro}$ (Figure 4D) just as in the case of the polyprotein fusion vector described above.

In another chimeric virus, the insertion of a HRV2 IRES instead of the PV IRES resulted in a virus, PV1 (RIPO), that grows with wild-type characteristics in a standard HeLa tissue culture system, but is attenuated in human neuroblastoma cell line. PV1 (RIPO) is furthermore highly neuroattenuated in the transgenic mouse model and in experimental monkeys (41,42). This model constitutes an excellent tool to address questions of IRES-dependent tissue and host tropism and may eventually lead us to an understanding of the attenuation of the Sabin vaccine strains.

CONCLUDING COMMENTS

If PV is to be used as an expression vector for antigens of heterologous human pathogens, any construct of a recombinant virus will have to be carefully analyzed first for the emergence of deletion variants and scrutinized continuously for its long-term genetic stability. To the authors' knowledge, none of the vectors used to immunize test animals were ever recovered from the animals and subjected to thorough analysis. Most of the recombinant PV vectors are severely impaired in replication, and they undergo genetic rearrangements in tissue culture. In some cases, the vectors do not survive a single round of replication since the deleted variants outcompete the parental virus because of vast differences in fitness. It can be assumed that this genetic phenomenon also operates in mammals sensitive to PV infection—humans included. Moreover, in infected animals with an intact immune system, selection pressure for optimal virus growth will be much higher as compared with a tissue culture system.

Irrespective of the strategy used to express foreign antigens in PV, there are several problems common to all of them. The maximum length of foreign sequences is restricted by the maximum length of the RNA genome that can still be encapsidated. Furthermore, the inserted RNA sequence may interfere with correct genome folding, thereby disrupting the function of *cis*-acting elements of the RNA genome. If expressed as a fusion to the polyprotein, the foreign protein has to be cleaved off efficiently from the polyprotein to release the functional PV proteins and not to accumulate nonfunctional precursors. Finally, the exogenous amino acid sequence can lead to misfolding of the polyprotein or its cleavage products. This may present a particular obstacle if the heterologous sequence is to be expressed as a permanent part of a PV polypeptide, as described above for the expression of (rather small) heterologous sequences within the BC-loop of VP1. In our view, none of the PV-based expression systems described in this chapter presents as yet a serious contender for consideration as an immunogenic vector. The genetic plasticity of the PV genome is just too great to warrant a stable expression of the antigen. Finally, we have to keep in mind the targeted erradication of wild-type PV, which will further complicate any PV-based vectors' expressing a foreign gene.

REFERENCES

1. **Abramson, R.D., T.E. Dever, and W.C. Merrick.** 1988. Biochemical evidence supporting a mechanism for cap-independent and internal initiation of eukaryotic mRNA. J. Biol. Chem. *263*:6016-6019.
2. **Ahlquist, P. and P. Kaesberg.** 1979. Determination of the length distribution of poly (A) at the 3′ terminus of the virion RNAs of EMC virus, poliovirus, rhinovirus, RAV-61 and CPMV and of mouse globin nRNA. Nucleic Acids Res. *7*:1195-1204.
3. **Alexander, H.E., G. Koch, I.M. Mountain, K. Sprunt, and O.V. Damme.** 1958. Infectivity of ribonucleic acid of poliovirus on HeLa cell monolayers. Virology *5*:172.
4. **Alexander, L., H.H. Lu, and E. Wimmer.** 1994. Polioviruses containing picornavirus type 1 and/or type 2 internal ribosomal entry site elements: genetic hybrids and the expression of a foreign gene. Proc. Natl. Acad. Sci. USA *91*:1406-1410.
5. **Altmeyer, R., N. Escriou, M. Girad, A. Palmenberg, and S. van der Werf.** 1994. Attenuated Mengo virus as a vector for immunogenic human immunodeficiency virus type 1 glycoprotein 120. Proc. Natl. Acad. Sci. USA *91*:9775-9779.
6. **Altmeyer, R., M. Girad, S. van der Werf, V. Mimic, L. Seigneur, and M.-F. Saron.** 1995. Attenuated Mengo virus: a new vector for live recombinant vaccines. J. Virol. *69*:3193-3196.
7. **Altmeyer, R., A.D. Murdin, J.J. Harber, and E. Wimmer.** 1991. Construction and characterization of a poliovirus/rhinovirus antigenic hybrid. Virology *184*:636-644.
8. **Andino, R., G.E. Rieckhof, P.L. Achacoso, and D. Baltimore.** 1993. Poliovirus RNA synthesis utilizes an RNP complex formed around the 5′-end of viral RNA. EMBO J. *12*:3587-3598.
9. **Andino, R., G.E. Rieckhof, and D. Baltimore.** 1990. A functional ribonucleoprotein complex forms around the 5′ end of poliovirus RNA. Cell *63*:369-380.
10. **Andino, R., D. Silvera, S.D. Suggett, P.L. Achacoso, C.J. Miller, D. Baltimore, and M.B. Feinberg.** 1994. Engineering poliovirus as a vaccine vector for the expression of diverse antigens. Science *265*:1448-1451.
11. **Ansardi, D.C., Z. Moldoveanu, D.C. Porter, D.E. Walker, R.M. Conry, A.F. LoBuglio, S. McPherson, and C.D. Morrow.** 1994. Characterization of poliovirus replicons encoding carcinoembryonic antigen. Cancer Res. *54*:6359-6364.
12. **Ansardi, D.C., D.C. Porter, and C.D. Morrow.** 1991. Coinfection with recombinant vaccinia viruses expressing poliovirus P1 and P3 proteins results in polyprotein processing and formation of empty capsid structures. J. Virol. *65*:2088-2092.
13. **Bachrach, H. and C. Schwerdt.** 1952. Purification studies on Lansing poliomyelitis virus, pH stability, CNS extraction and butanol. Purification experiments. I. J. Immunol. *69*:551-560.
14. **Bandyopadhyay, P.K., C. Wang, and H.L. Lipton.** 1992. Cap-independent translation and replication by the 5′ untranslated region of Theiler's Murine Encephalomyelitis virus. J. Virol. *66*:6249-6256.
15. **Basak, S., S. McPherson, S. Kang, J.F. Collawn, and C.D. Morrow.** 1998. Construction and characterization of encapsidated poliovirus replicons that express biologically active murine interleukin-2. J. Interferon Cytokine Res. *18*:305-313.
16. **Blair, W.S. and B.L. Semler.** 1991. Role for the P4 amino acid residue in substrate utilization by the poliovirus 3CD proteinase. J. Virol. *65*:6111-6123.
17. **Borman, A. and R.J. Jackson.** 1992. Initiation of translation of human Rhinovirus RNA: mapping the internal Ribosome entry site. Virology *188*:685-696.
18. **Brown, B. and E. Ehrenfeld.** 1979. Translation of poliovirus RNA in vitro: changes in cleavage pattern and initiation sites by ribosomal salt wash. Virology *97*:396-405.
19. **Burke, K.L., G. Dunn, M. Ferguson, P.D. Minor, and J.W. Almond.** 1988. Antigen chimaeras of poliovirus as potential new vaccines. Nature *332*:81-82.
20. **Cao, X. and E. Wimmer.** 1996. Genetic variation of the poliovirus genome with two VPg coding units. EMBO J. *15*:23-33.
21. **Cao, X. and E. Wimmer.** 1995. Intragenomic complementation of a 3AB mutant in dicistronic polioviruses. Virology *209*:315-326.
22. **Choi, W.S., R.S. Pal-Gosh, and C.D. Morrow.** 1991. Expression of human immunodeficiency virus type 1(HIV-1) gag, pol, and env proteins from chimeric HIV-1 poliovirus minireplicons. J. Virol. *65*:2875-2883.
23. **Clark, M.E., T. Hämmerle, E. Wimmer, and A. Dasgupta.** 1991. Poliovirus proteinase 3C converts an active form of transcription factor IIIC to an inactive form: a mechanism for inhibition of host cell polymerase III transcription by poliovirus. EMBO J.

10:2941-2947.

24.**Colter, J.S., H.H. Bird, A.W. Moyer, and R.A. Brown.** 1957. Infectivity of ribonucleic acid isolated from virus infected tissues. J. Virol. *4*:522.

25.**Crabbe, M.J., D.J. Evans, and J.W. Almond.** 1990. Modelling of poliovirus. HIV-1 antigen chimaeras. FEBS Lett. *271*:194-198.

26.**Cuconati, A., W. Xiang, F. Lahser, T. Pfister, and E. Wimmer.** 1998. A protein linkage map of the P2 nonstructural proteins of poliovirus. J. Virol. *72*:1297-1307.

27.**Das, S. and A. Dasgupta.** 1993. Identification of the cleavage site and determinants required for poliovirus 3C^{pro}-catalyzed cleavage of human TATA-binding transcription factor TBP. J. Virol. *67*:3326-3331.

28.**Dedieu, J.F., J. Ronco, S. van der Werf, J.M. Hogle, Y. Henin, and M. Girard.** 1992. Poliovirus chimeras expressing sequences from the principal neutralization domain of human immunodeficiency virus type 1. J. Virol. *66*:3161-3167.

29.**Domingo, E.** 1992. Genetic variation and quasi-species. Curr. Opin. Genet. Dev. *2*:61-63.

30.**Domingo, E. and J.J. Holland.** 1988. High error rates, population equilibrium, and evolution of RNA replication systems, p. 3-36. *In* E. Domingo, J.J. Holland, and P. Ahlquist (Eds.), RNA Genetics, Vol. 3. CRC Press, Boca Raton.

31.**Dorner, A.J., L.F. Dorner, G.R. Larsen, E. Wimmer, and C.W. Anderson.** 1982. Identification of the initiation site of poliovirus polyprotein synthesis. J. Virol. *42*:1017-1028.

32.**Dorner, A.J., B.L. Semler, R.J. Jackson, R. Hanecak, E. Duprey, and E. Wimmer.** 1984. In vitro translation of poliovirus RNA: utilization of internal initiation sites in reticulocyte lysate. J. Virol. *50*:507-514.

33.**Drake, J.W.** 1993. Rates of spontaneous mutation among RNA viruses. Proc. Natl. Acad. Sci. USA *90*:4171-4175.

34.**Duggal, R., A. Cuconati, M. Gromeier, and E. Wimmer.** 1997. Genetic recombination of poliovirus in a cell-free system. Proc. Natl. Acad. Sci. USA *94*: 13786-13791.

35.**Duggal, R. and E. Wimmer.** 1999. Temperature-dependent alteration of cross-over sites in poliovirus recombination. Virology *258*:30-41.

36.**Dulbecco, R. and M. Vogt.** 1954. Plaque formation and isolation of pure lines with poliomyelitis viruses. J. Exp. Med. *99*:167-182.

37.**Emini, E.A., B.A. Jameson, A.J. Lewis, G.R. Larsen, and E. Wimmer.** 1982. Poliovirus neutralization epitopes: analysis and localization with neutralizing monoclonal antibodies. J. Virol. *43*:997-1005.

38.**Enders, J.F., T.H. Weller, and F.C. Robbins.** 1949. Cultivation of the Lansing strain of poliomyelitis virus in cultures of various human embryonic tissues. Science *109*:85-87.

39.**Evans, D.J., J. McKeating, J.M. Meredith, K.L. Burke, K. Katrak, A. John, M. Ferguson, P.D. Minor, R.A. Weiss, and J.W. Almond.** 1989. An engineered poliovirus chimaera elicits broadly reactive HIV-1 neutralizing antibodies. Nature *339*:385-388.

40.**Gmyl, A.P., E.V. Pilipenko, S.V. Maslova, G.A. Belov, and V.I. Agol.** 1993. Functional and genetic plasticities of the poliovirus genome: quasi- infectious RNAs modified in the 5′-untranslated region yield a variety of pseudorevertants. J. Virol. *67*:6309-6316.

41.**Gromeier, M., L. Alexander, and E. Wimmer.** 1996. Internal ribosomal entry site substitution eliminates neurovirulence in intergeneric poliovirus recombinants. Proc. Natl. Acad. Sci. USA *93*:2370-2375.

42.**Gromeier, M., B. Bossert, M. Arita, A. Nomoto, and E. Wimmer.** 1999. Dual stem loops within the poliovirus internal ribosomal entry site control neurovirulence. J. Virol. *73*:958-964.

43.**Gromeier, M. and E. Wimmer.** 1998. Mechanism of injury-provoked poliomyelitis. J. Virol. *72*:5056-5060.

44.**Gromeier, M., S. Mueller, D. Solecki, B. Bossert, G. Bernhard, and E. Wimmer.** 1997. Determinants of poliovirus neurovirulence. J. Neurovirol. (Suppl. 1) *3*:S35-S38.

45.**Gromeier, M., E. Wimmer, and A.E. Gorbalenya.** 1999. Genetics, pathogenesis and evolution of picornaviruses, p. 287-343. *In* E. Domingo, R.G. Webster, and J.J. Holland (Eds.), Origin and Evolution of Viruses. Academic Press, London.

46.**Harber, J.J., J. Bradley, C.W. Anderson, and E. Wimmer.** 1991. Catalysis of poliovirus VP0 maturation cleavage is not mediated by serine 10 of VP2. J. Virol. *65*:326-334.

47.**Harris, K.S., C.U.T. Hellen, and E. Wimmer.** 1990. Proteolytic processing in the replication of picornaviruses. Semin. Virol. *1*:323-333.

48.**Harris, K.S., S.R. Reddigari, M.J. Nicklin, T. Hammerle, and E. Wimmer.** 1992. Purification and characterization of poliovirus polypeptide 3CD, a proteinase and a precursor for RNA polymerase. J. Virol. *66*:7481-7489.

49.**Hellen, C.U., H.G. Krausslich, and E. Wimmer.** 1989. Proteolytic processing of polyproteins in the replication of RNA viruses. Biochemistry *28*:9881-9890.

50.**Hellen, C.U., C.K. Lee, and E. Wimmer.** 1992. Determinants of substrate recognition by poliovirus 2A proteinase. J. Virol. *66*:3330-3338.

51.**Hogle, J.M., M. Chow, and D.J. Filman.** 1985. Three-dimensional structure of poliovirus at 2.9 A resolution. Science *229*:1358-1365.

52.**Holland, J.J., K. Spindler, E. Horodyski, S. Grabau, S. Nichol, and S. VandePol.** 1982. Rapid evolution of RNA genomes. Science *215*:1577-1585.

53.**Iizuka, N., C. Chen, Q. Yang, G. Johannes, and P. Sarnow.** 1995. Cap-independent translation and internal initiation of translation in eukaryotic cellular mRNA molecules. Curr. Top. Microbiol. Immunol. *203*:155-177.

54.**Jang, S.K., M.V. Davies, R.J. Kaufman, and E. Wimmer.** 1989. Initiation of protein synthesis by internal entry of ribosomes into the 5′ nontranslated region of encephalomyocarditis virus RNA in vivo. J. Virol. *63*:1651-1660.

55.**Jang, S.K., H.G. Kräusslich, M.J. Nicklin, G.M. Duke, A.C. Palmenberg, and E. Wimmer.** 1988. A segment of the 5′ nontranslated region of encephalomyocarditis virus RNA directs internal entry of ribosomes during in vitro translation. J. Virol. *62*:2636-

2643.
56.**Jarvis, T.C. and K. Kirkegaard.** 1992. Poliovirus RNA recombination: mechanistic studies in the absence of selection. EMBO J. *11*:3135-3145.
57.**Jarvis, T.C. and K. Kirkegaard.** 1991. The polymerase in its labyrinth: mechanisms and implications of RNA recombination. Trends Genet. *7*:186-191.
58.**Jenkins, O., J. Cason, K.L. Burke, D. Lunney, A. Gillen, D. Patel, D.J. McCance, and J.W. Almond.** 1990. An antigen chimera of poliovirus induces antibodies against human papillomavirus type 16. J. Virol. *64*:1201-1206.
59.**King, A.M.** 1988. Preferred sites of recombination in poliovirus RNA: an analysis of 40 intertypic crossover sequences. Nucleic Acids Res. *16*:11705-11723.
60.**Kirkegaard, K. and D. Baltimore.** 1986. The mechanism of RNA recombination in poliovirus. Cell *47*:433-443.
61.**Kitamura, N., B.L. Semler, P.G. Rothberg, G.R. Larsen, C.J. Adler, A.J. Dorner, E.A. Emini, R. Hanecak, J.J. Lee, S. van der Werf, C.W. Anderson, and E. Wimmer.** 1981. Primary structure, gene organization and polypeptide expression of poliovirus RNA. Nature *291*:547-553.
62.**Kitson, J.D., K.L. Burke, L.A. Pullen, G.J. Belsham, and J.W. Almond.** 1991. Chimeric polioviruses that include sequences derived from two independent antigenic sites of foot-and-mouth disease virus (FMDV) induce neutralizing antibodies against FMDV in guinea pigs. J. Virol. *65*:3068-3075.
63.**Koike, S., C. Taya, T. Kurata, S. Abe, I. Ise, H. Yonekawa, and A. Nomoto.** 1991. Transgenic mice susceptible to poliovirus. Proc. Natl. Acad. Sci. USA *88*:951-955.
64.**Kräusslich, H.G. and E. Wimmer.** 1988. Viral proteinases. Ann. Rev. Biochem. *57*:701-754.
65.**Krausslich, H.G., M.J. Nicklin, H. Toyoda, D. Etchison, and E. Wimmer.** 1987. Poliovirus proteinase 2A induces cleavage of eucaryotic initiation factor 4F polypeptide p220. J. Virol. *61*:2711-2718.
66.**Kuge, S., L. Saito, and A. Nomoto.** 1986. Primary structure of poliovirus defective interfering particle genomes and possible generation mechanism of the particles. J. Mol. Biol. *192*:473-487.
67.**Lai, M.M.C.** 1992. RNA recombination in animal and plant viruses. Annu. Rev. Microbiol. *56*:61-79.
68.**Le, S.Y., J.H. Chen, N. Sonenberg, and J.V. Maizel.** 1992. Conserved tertiary structure elements in the 5′ untranslated region of human Enteroviruses and Rhinoviruses. Virology *191*:858-866.
69.**Lee, Y.F., A. Nomoto, B.M. Detjen, and E. Wimmer.** 1977. A protein covalently linked to poliovirus genome RNA. Proc. Natl. Acad. Sci. USA *74*:59-63.
70.**Lee, Y.F., A. Nomoto, and E. Wimmer.** 1976. The genome of poliovirus is an exceptional eukaryotic mRNA. Prog. Nucleic Acid Res. Mol. Biol. *19*:89-96.
71.**Lemon, S.M., W. Barclay, M. Ferguson, P. Murphy, L. Jing, K. Burke, D. Wood, K. Katrak, D. Sangar, P.D. Minor, et al.** 1992. Immunogenicity and antigenicity of chimeric picornaviruses which express hepatitis A virus (HAV) peptide sequences: evidence for a neutralization domain near the amino terminus of VP1 of HAV. Virology *188*:285-295.
72.**Lu, H.H., L. Alexander, and E. Wimmer.** 1995. Construction and genetic analysis of dicistronic polioviruses containing open reading frames for epitopes of human immunodeficiency virus type 1 gp120. J. Virol. *69*:4797-4806.
73.**Lu, H.H. and E. Wimmer.** 1996. Poliovirus chimeras replicating under the translational control of genetic elements of hepatitis C virus reveal unusual properties of the internal ribosomal entry site of hepatitis C virus. Proc. Natl. Acad. Sci. USA *93*:1412-1417.
74.**Lundquist, R.E., M. Sullivan, and V.J. Maizel, Jr.** 1979. Characterization of a new isolate of poliovirus defective interfering particles. Cell *18*:759-769.
75.**Mandl, S., L.J. Sigal, K.L. Rock, and R. Andino.** 1998. Poliovirus vaccine vectors elicit antigen-specific cytotoxic T cells and protect mice against lethal challenge with malignant melanoma cells expressing a model antigen. Proc. Natl. Acad. Sci. USA *95*:8216-8221.
76.**Mattion, N.M., P.A. Reilly, E. Camposano, S.L. Wu, S.J. DiMichele, S.T. Ishizaka, S.E. Fantini, J.C. Crowley, and C. Weeks-Levy.** 1995. Characterization of recombinant polioviruses expressing regions of rotavirus VP4, hepatitis B surface antigen, and herpes simplex virus type 2 glycoprotein D. J. Virol. *69*:5132-5137.
77.**Mattion, N.M., P.A. Reilly, S.J. DiMichele, J.C. Crowley, and C. Weeks-Levy.** 1994. Attenuated poliovirus strain as a live vector: expression of regions of rotavirus outer capsid protein VP7 by using recombinant Sabin 3 viruses. J. Virol. *68*:3925-3933.
78.**Melnick, J.L.** 1996. Enteroviruses: Polioviruses, Coxsackieviruses, Ecoviruses, and newer Enteroviruses, p. 655-712. *In* B.N. Fields, D.M. Knipe and P.M. Howley (Eds.), Fields Virology, 3rd ed., Vol. 1. Raven-Lippincott Press, Philadelphia.
79.**Minor, P.D., G.C. Schild, J. Bootman, D.M. Evans, M. Ferguson, P. Reeve, M. Spitz, G. Stanway, A.J. Cann, R. Hauptmann, et al.** 1983. Location and primary structure of a major antigenic site for poliovirus neutralization. Nature *301*:674-679.
80.**Mirzayan, C., R. Ingraham, and E. Wimmer.** 1991. Specificity of the polioviral proteinase 3C towards genetically engineered cleavage sites in the viral capsid. J. Gen. Virol. *72*:1159-1163.
81.**Moldoveanu, Z., D.C. Porter, A. Lu, S. McPherson, and C.D. Morrow.** 1995. Immune responses induced by administration of encapsidated poliovirus replicons which express HIV-1 gag and envelope proteins. Vaccine *13*:1013-1022.
82.**Molla, A., S.K. Jang, A.V. Paul, Q. Reuer, and E. Wimmer.** 1992. Cardioviral internal ribosomal entry site is functional in a genetically engineered dicistronic poliovirus. Nature *356*:255-257.
83.**Molla, A., A.V. Paul, M. Schmid, S.K. Jang, and E. Wimmer.** 1993. Studies on dicistronic polioviruses implicate viral proteinase 2Apro in RNA replication. Virology *196*:739-747.
84.**Molla, A., A.V. Paul, and E. Wimmer.** 1991. Cell-free, de novo synthesis of poliovirus. Science *254*:1647-1651.
85.**Mueller, S. and E. Wimmer.** 1998. Expression of foreign proteins by poliovirus polyprotein fusion: analy-

sis of genetic stability reveals rapid deletions and formation of cardioviruslike open reading frames. J. Virol. *72*:20-31.

86.**Murdin, A.D., H.H. Lu, M.G. Murray, and E. Wimmer.** 1992. Poliovirus antigenic hybrids simultaneously expressing antigenic determinants from all three serotypes. J. Gen. Virol. *73*:607-611.

87.**Murdin, A.D., H. Su, M.H. Klein, and H.D. Caldwell.** 1995. Poliovirus hybrids expressing neutralization epitopes from variable domains I and IV of the major outer membrane protein of Chlamydia trachomatis elicit broadly cross-reactive C. trachomatis-neutralizing antibodies. Infect. Immun. *63*:1116-1121.

88.**Murdin, A.D., H. Su, D.S. Manning, M.H. Klein, M.J. Parnell, and H.D. Caldwell.** 1993. A poliovirus hybrid expressing a neutralization epitope from the major outer membrane protein of Chlamydia trachomatis is highly immunogenic. Infect. Immun. *61*:4406-4414.

89.**Murray, M.G., R.J. Kuhn, M. Arita, N. Kawamura, A. Nomoto, and E. Wimmer.** 1988. Poliovirus type 1/type 3 antigenic hybrid virus constructed in vitro elicits type 1 and type 3 neutralizing antibodies in rabbits and monkeys. Proc. Natl. Acad. Sci. USA *85*:3203-3207.

90.**Murray, M.G., J. Bradley, X.F. Yang, E. Wimmer, E.G. Moss, and V.R. Racaniello.** 1988. Poliovirus host range is determined by a short amino acid sequence in neutralization antigenic site I. Science *241*:213-215.

91.**Nomoto, A., B. Detjen, R. Pozzatti, and E. Wimmer.** 1977. The location of the polio genome protein in viral RNAs and its implication for RNA synthesis. Nature *268*:208-213.

92.**Nomoto, A., Y.F. Lee, and E. Wimmer.** 1976. The 5′ end of poliovirus mRNA is not capped with m7G(5′)ppp(5′)Np. Proc. Natl. Acad. Sci. USA *73*:375-380.

93.**Paul, A.V., J.A. Mugavero, A. Molla, and E. Wimmer.** 1998. Internal ribosomal entry site scanning of the poliovirus polyprotein: implications for proteolytic processing. Virology *250*:241-253.

94.**Pelletier, J., G. Kaplan, V.R. Racaniello, and N. Sonenberg.** 1988. Cap-independent translation of poliovirus mRNA is conferred by sequence elements within the 5′-noncoding region. Mol. Cell. Biol. *8*:1103-1112.

95.**Pelletier, J. and N. Sonenberg.** 1988. Internal initiation of translation of eukaryotic mRNA directed by a sequence derived from poliovirus RNA. Nature *334*:320-325.

96.**Percy, N., W.S. Barclay, M. Sullivan, and J.W. Almond.** 1992. A poliovirus replicon containing the chloramphenicol acetyltransferase gene can be used to study the replication and encapsidation of poliovirus RNA. J. Virol. *66*:5040-5046.

97.**Pfister, T., C. Mirzayan, and E. Wimmer.** 1999. Poliovirus: molecular biology, p. 1330-1348. *In* R.G. Webster and A. Granoff (Eds.), The Encyclopedia of Virology, 2nd ed. Academic Press, San Diego.

98.**Pilipenko, E.V., V.M. Blinov, L.I. Romanova, A.N. Sinyakov, S.V. Maslova, and V.I. Agol.** 1989. Conserved structural domains in the 5′-untranslated region of picornaviral genomes: an analysis of the segment controlling translation and neurovirulence. Virology *168*:201-209.

99.**Pilipenko, E.V., A.P. Gmyl, and V.I. Agol.** 1995. A model for rearrangements in RNA genomes. Nucleic Acids Res. *23*:1870-1875.

100.**Poeyry, T., L. Kinnunen, and T. Hovi.** 1992. Genetic variation in vivo and proposed functional domains of the 5′ noncoding region of poliovirus RNA. J. Virol. *66*:5313-5319.

101.**Porter, D.C., D.C. Ansardi, W.S. Choi, and C.D. Morrow.** 1993. Encapsidation of genetically engineered poliovirus minireplicons which express human immunodeficiency virus type 1 Gag and Pol proteins upon infection. J. Virol. *67*:3712-3719.

102.**Porter, D.C., D.C. Ansardi, and C.D. Morrow.** 1995. Encapsidation of poliovirus replicons encoding the complete human immunodeficiency virus type 1 gag gene by using a complementation system which provides the P1 capsid protein in trans. J. Virol. *69*:1548-1555.

103.**Porter, D.C., D.C. Ansardi, J. Wang, S. McPherson, Z. Moldoveanu, and C.D. Morrow.** 1998. Demonstration of the specificity of poliovirus encapsidation using a novel replicon which encodes enzymatically active firefly luciferase. Virology *243*:1-11.

104.**Porter, D.C., L.R. Melsen, R.W. Compans, and C.D. Morrow.** 1996. Release of virus-like particles from cells infected with poliovirus replicons which express human immunodeficiency virus type 1 Gag. J. Virol. *70*:2643-2649.

105.**Porter, D.C., J. Wang, Z. Moldoveanu, S. McPherson, and C.D. Morrow.** 1997. Immunization of mice with poliovirus replicons expressing the C-fragment of tetanus toxin protects against lethal challenge with tetanus toxin. Vaccine *15*:257-264.

106.**Racaniello, V.R. and D. Baltimore.** 1981. Cloned poliovirus complementary DNA is infectious in mammalian cells. Science *214*:916-919.

107.**Racaniello, V.R. and D. Baltimore.** 1981. Molecular cloning of poliovirus cDNA and determination of the complete nucleotide sequence of the viral genome. Proc. Natl. Acad. Sci. USA *78*:4887-4891.

108.**Reanney, D.C.** 1984. The molecular evolution of RNA viruses. Symp. Soc. Gen. Microbiol. *35*:175-196.

109.**Ren, R., F. Costantini, E.J. Gorgacz, J.J. Lee, and V.R. Racaniello.** 1990. Transgenic mice expressing a human poliovirus receptor: a new model for poliomyelitis. Cell *63*:353-362.

110.**Rhoads, R.E. and B.J. Lamphear.** 1995. Cap-independent translation of heat shock messenger RNAs. Curr. Top. Microbiol. Immunol. *203*:131-153.

111.**Rivera, V.M., J.D. Welsh, and J.V. Maizel.** 1988. Comparative sequence analysis of the 5′ noncoding region of the enteroviruses and rhinoviruses. Virology *165*:42-50.

112.**Robbins, F. and J. Enders.** 1950. Tissue culture techniques in study of animal viruses. Am. J. Med. Soc. *220*:316-338.

113.**Robert-Hebmann, V., S. Emiliani, M. Resnicoff, F. Jean, and C. Devaux.** 1992. Subtyping of human immunodeficiency virus isolates with a panel of monoclonal antibodies: identification of conserved and divergent epitopes on p17 and p25 core proteins. Mol. Immunol. *29*:1175-1183.

114. **Romanova, L.I., V.M. Blonov, E.A. Tolskaya, E.G. Viktorova, and M.S. Kolenikova.** 1986. The primary structure of intertypic poliovirus recombinants: a model of recombination btween RNA genomes. Virology *155*:202-213.

115. **Sabin, A.** 1956. Pathogenesis of poliomyelitis. Science *123*:1151-1157.

116. **Sabin, A.B.** 1957. Properties of attenuated polioviruses and their behavior in human beings. Spec. Pub. of the N.Y. Acad. of Sci. *5*:113-127.

117. **Salk, J.E., U. Krech, J.S. Younger, B.L. Bennett, L.J. Lewis, and P.L. Bazeley.** 1954. Formaldehyde treatment and safety testing of experimental poliomyelitis vaccines. Am. J. Pub. Health *44*:563.

118. **Schaffer, F.L. and C.E. Schwerdt.** 1955. Crystallization of purified MEF-I poliomyelitis virus particles. Proc. Natl. Acad. Sci. USA *41*:1020-1023.

119. **Semler, B.L., A.J. Dorner, and E. Wimmer.** 1984. Production of infectious poliovirus from cloned cDNA is dramatically increased by SV40 transcription and replication signals. Nucleic Acids Res. *12*:5123-5141.

120. **Skinner, M.A., V.R. Racaniello, G. Dunn, J. Cooper, P.D. Minor, and J.W. Almond.** 1989. New model for the secondary structure of the 5′ non-coding RNA of poliovirus is supported by biochemical and genetic data that also shows that RNA secondary structure is important in neurovirulence. J. Mol. Biol. *207*:379-392.

121. **Smith, A.D., S.C. Geisler, A.A. Chen, D.A. Resnick, B.M. Roy, P.J. Lewi, E. Arnold, and G.F. Arnold.** 1998. Human rhinovirus type 14:human immunodeficiency virus type 1 (HIV-1) V3 loop chimeras from a combinatorial library induce potent neutralizing antibody responses against HIV-1. J. Virol. *72*:651-659.

122. **Sonenberg, N.** 1987. Regulation of translation by poliovirus. Adv. Virus Res. *33*:175-204.

123. **Tellier, R., J. Bukh, S. Emerson, and R. Purcell.** 1996. Amplification of the full-length hepatitis A virus genome by long reverse transcription-PCR and transcription of infectious RNA directly from the amplicon. Proc. Nat. Acad. Sci. USA *93*:4370-4373.

124. **Tolskaya, E., L. Romanova, M. Kolesnikova, and V. Agol.** 1983. Intertypic recombination in poliovirus: genetic and biochemical studies. Virology *124*:121-132.

125. **Vagner, S., M.C. Gensac, A. Maret, F. Bayard, F. Amalric, H. Prats, and A.C. Prats.** 1995. Alternative translation of human fibroblast growth factor 2 mRNA occurs by internal entry of ribosomes. Mol. Cell. Biol. *15*:35-44.

126. **van der Werf, S., J. Bradley, E. Wimmer, F.W. Studier, and J.J. Dunn.** 1986. Synthesis of infectious poliovirus RNA by purified T7 RNA polymerase. Proc. Natl. Acad. Sci. USA *83*:2330-2334.

127. **Wimmer, E.** 1982. Genome-linked proteins of viruses. Cell *28*:199-201.

128. **Wimmer, E., C.U. Hellen, and X. Cao.** 1993. Genetics of poliovirus. Annu. Rev. Genet. *27*:353-436.

129. **Yang, W.X., T. Terasaki, K. Shiroki, S. Ohka, J. Aoki, S. Tanabe, T. Nomura, E. Terada, Y. Sugiyama, and A. Nomoto.** 1997. Efficient delivery of circulating poliovirus to the central nervous system independently of poliovirus receptor. Virology *229*:421-428.

130. **Ye, X., P. Fong, N. Iizuka, D. Choate, and D.R. Cavener.** 1997. Ultrabithorax and Antennapedia 5′ untranslated regions promote developmentally regulated internal translation initiation. Mol. Cell. Biol. *17*:1714-1721.

131. **Yogo, Y. and E. Wimmer.** 1972. Polyadenylic acid at the 3′-terminus of poliovirus RNA. Proc. Natl. Acad. Sci. USA *69*:1877-1882.

132. **Zhao, W.D., E. Wimmer, and F.C. Lahser.** 1999. Poliovirus/hepatitis C virus (internal ribosomal entry site-core) chimeric viruses: improved growth properties through modification of a proteolytic cleavage site and requirement for core RNA sequences but not for core-related polypeptides. J. Virol. *73*:1546-1554.

FAMILY: Togaviridae

GENUS: Alphavirus (Semliki Forest Virus, Sindbis Virus)

34 Strategies and Applications of Alphavirus Vectors in Gene Therapy

Stephen M. Smith[1] and Kuan-Teh Jeang[2]

[1] *St. Michael's Medical Center and the New Jersey Medical School–UMDNJ, Newark, NJ;* [2] *Molecular Virology Section, National Institute of Allergy and Infectious Diseases, National Institutes of Health, Bethesda, MD, USA*

INTRODUCTION

Most vector strategies for gene therapy or antisense delivery deploy vectors that permanently transduce the target cell (8). However, in many instances, stable transduction is neither necessary nor desired. Instead, vectors that act transiently and have no effect on the host's genome may be preferred over DNA-integrating vectors. For example, treatment of an infectious disease with an antisense vector would only require short-term expression of the antisense molecule. To this end, we and others have explored the utility of alphavirus vectors.

Alphaviruses are positive-strand RNA viruses. Alphavirus replication occurs purely in the cytoplasm and does not utilize a DNA intermediate (25). Upon entering a cell, the viral genome is translated directly into proteins. The early products include the four nonstructural proteins (NSP1–4). These proteins interact and serve to produce a negative-sense genomic RNA. From this negative-strand genomic RNA, many full-length genomic RNA copies are transcribed per cell. In addition to the full-length RNAs, the alphavirus replicon can also produce up to 100 000 copies of smaller subgenomic RNA per cell. The subgenomic RNAs encode for and are translated into the structural proteins. These structural proteins interact with the genomic RNA at the plasma membrane to assemble new viral particles.

The alphavirus family includes Semliki Forest virus (SFV), Sindbis virus (SIN), and the human pathogens of eastern and western equine encephalitis. SFV and SIN have been studied extensively and safely in laboratory settings (14,18,23,31). These studies have revealed that the nonstructural and structural components of an alphavirus can be segregated and can complement each other *in trans*. Hence, structural gene products, when provided *in trans* for an incomplete genome containing the nonstructural components of the alphavirus, produce single-cycle infectious viral particles. Implicit in this approach is the ability to replace in the viral genome, downstream of the subgenomic promoter, the structural genes with heterologous genes/sequences of interest. The resulting

Viral Vectors: Basic Science and Gene Therapy
Edited by A. Cid-Arregui and A. García-Carrancá

chimeric genome is then complemented by another vector that supplies the structural gene products *in trans* to produce particles that are infectious for a single cycle. These single-cycle particles can transduce susceptible cells to produce large amounts of the heterologous gene/sequence of interest. Because up to 25% of the total proteins in an infected cell is produced from the transducing alphavirus (14), many investigators have used these vectors for protein expression in vivo and ex vivo. The latter use has been predominantly in vaccine research (2,26), where short-term expression of large amounts of an antigen may be ideal. Additionally, alphaviruses can be effectively pseudotyped with a heterologous envelope and retain high titers of infectivity (15,22). Others have exploited these properties for delivery of effector RNAs or for purposes of targeting a specific cell population.

Helper-Dependent and Helper-Independent Alphavirus Vectors

The alphavirus genome can be engineered in various ways to achieve different effects for gene therapy. As shown in Figure 1, alphavirus structural genes can be removed from the viral genome and placed onto a separate plasmid. This structural gene-containing plasmid (helper) is used to complement the nonstructural genome (alphavirus vector) in generating single-cycle infectious particles. A heterologous gene of interest (*trans* gene) can then be inserted downstream of the subgenomic promoter in the alphavirus vector. To produce infectious particles, both alphavirus vector and helper plasmids are linearized with a restriction enzyme and separately transcribed into RNAs using SP6 polymerase. These RNAs are cointroduced into

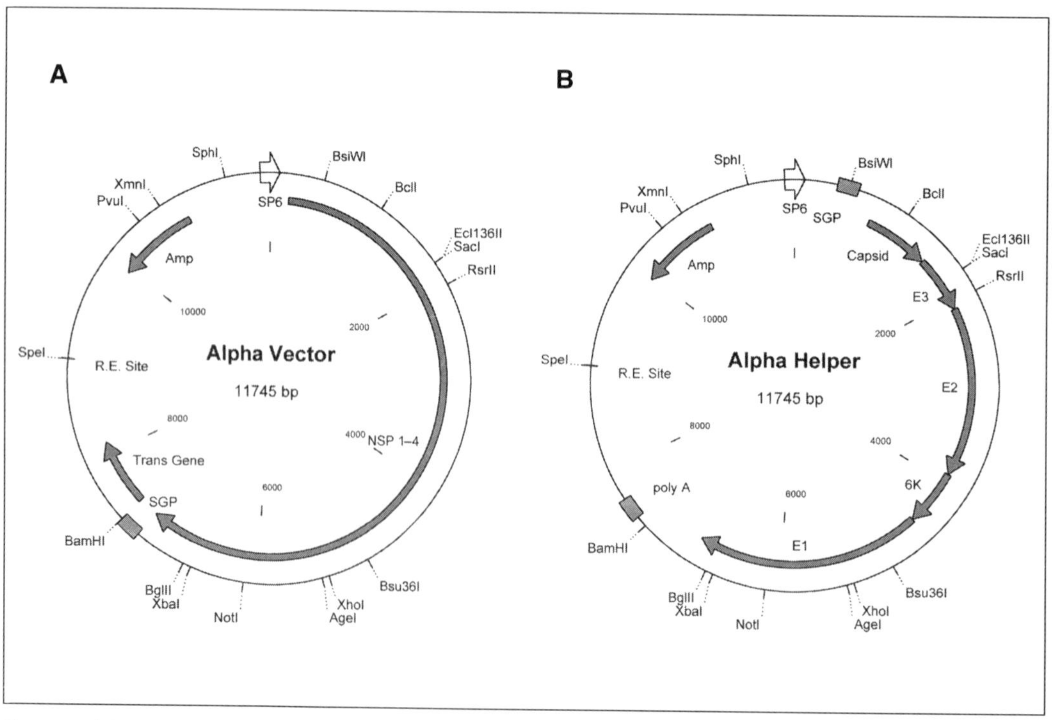

Figure 1. Alphavirus helper constructs. (A) The alphavirus vector (Alpha Vector) and (B) the helper vector (Alpha Helper). SP6, SP6 RNA polymerase promoter; NSP1–4, nonstructural proteins 1–4; SGP, subgenomic promoter; R.E. site, restriction enzyme site; Amp, gene for ampicillin resistance protein.

cells (e.g., BHK). Inside cells, NSP1–4, encoded by the alphavirus vector RNA, are translated. NSP1–4 function to produce full-length, negative-sense RNAs, from which large amounts of sense genomic and subgenomic RNA are transcribed. Because the helper RNA also has a subgenomic promoter, large amounts of helper sense RNA are transcribed. From these sense RNAs, the structural proteins are produced in high levels that serve for viral particle formation. Only the genomic RNA, which has the packaging signal and contains the heterologous gene of interest, is packaged into particles. After 24 hours, high titers of these single-cycle infectious particles are secreted into the culture supernatant and can be used to infect mammalian and insect cells.

In an alternative approach, the structural genes may be left intact within the alphavirus vector plasmid, and an additional subgenomic promoter can be positioned downstream of the structural genes (Figure 2). Heterologous gene(s)/sequence(s) can then be placed downstream of this second subgenomic promoter. This helper-free construct can then be used to produce infectious virions that can then spread the heterologous sequence via multiple rounds of infection and reinfection of susceptible cells. A third approach leaves the genomic structure undisturbed and modifies the envelope to restrict entry into cells that express specific receptors (Figure 3). Below, examples of these three strategies are presented and discussed.

Alphavirus that Delivers a Functional Ribozyme Targeted to HIV

Ribozymes are considered potentially useful candidate effectors to specifically downregulate expression of genes (3,5,7, 9,10,20,21). However, successful applications of ribozymes face several barriers. First, ribozyme concentration must exceed substrate target concentration intracellularly by several fold in order to mediate a significant effect (28). Second, most RNA polymerase II promoters are not sufficiently active to produce high enough levels of ribozymes (9,11). Third, specific and efficient delivery of therapeutic genes has proven to be difficult (1,12,30). Consequently, the use of an alphavirus vector to deliver and express an antiviral ribozyme (24) seems to merit consideration. Previously, we have described a hammerhead ribozyme that specifically cleaves a sequence in the U5 region of human immunodeficiency virus type 1 (HIV-1) RNA (4). In using alphavirus, we placed this ribozyme under the control of a subgenomic promoter in an SFV vector that lacks structural genes. The resulting chimeric vector was termed SFVRz (24).

Initially, we tested the ability of the 53-nucleotide U5 ribozyme to cleave its substrate when embedded within an 8.3-kb RNA molecule (24). We isolated RNA from a BHK long terminal repeat (LTR) CAT cell line, which expresses the HIV U5 target RNA as part of a CAT mRNA. An in vitro-transcribed ribozyme (60 nucleotides) and an in vitro-transcribed genomic length SFVRz (8314 nucleotides) both effectively cleaved the target RNA in an expected fashion. The cleavage was analyzed by ribonuclease protection assay (RPA) with a probe that asymmetrically overlapped the target region. The RPA showed the predicted size fragments after cleavage of the target RNA; uncut target RNA produced the expected protection of a full-length probe (Figure 4). The ribozyme, despite being embedded within the SFV genome, functioned efficiently to cleave its substrate RNA.

We also used RPA to demonstrate the time course and expression level of SFV in cells following transduction with viral particles. In this case, the probe encompassed the sequences around the SFV subgenomic promoter. Hence, the protected product from a

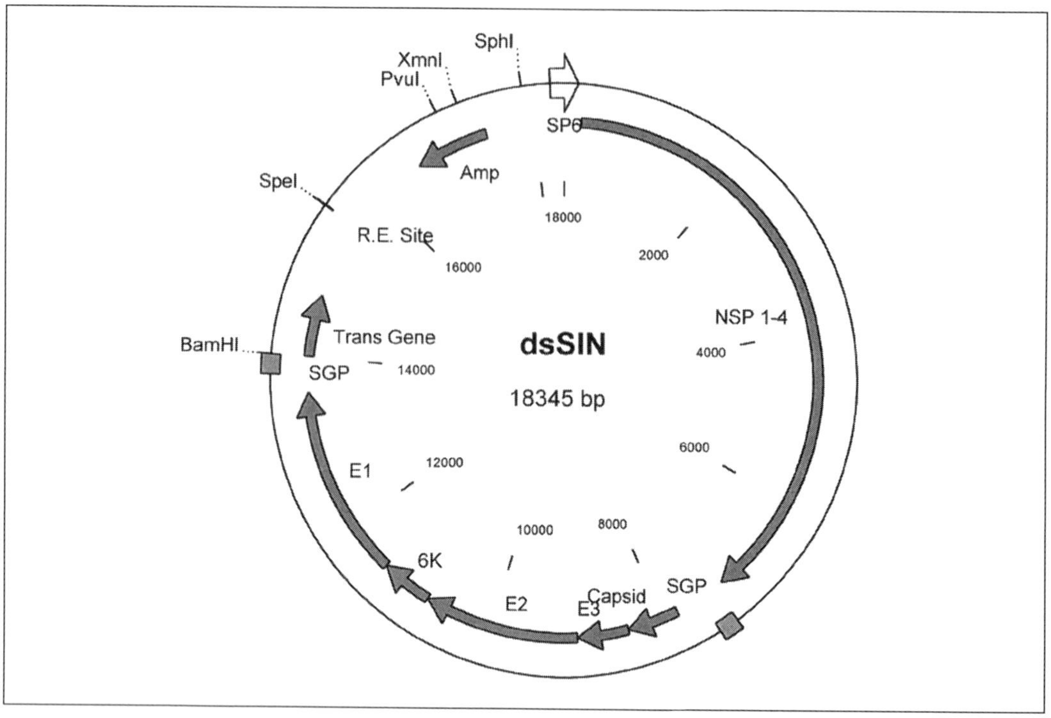

Figure 2. The double subgenomic promoter Sindbis vector (dsSIN).

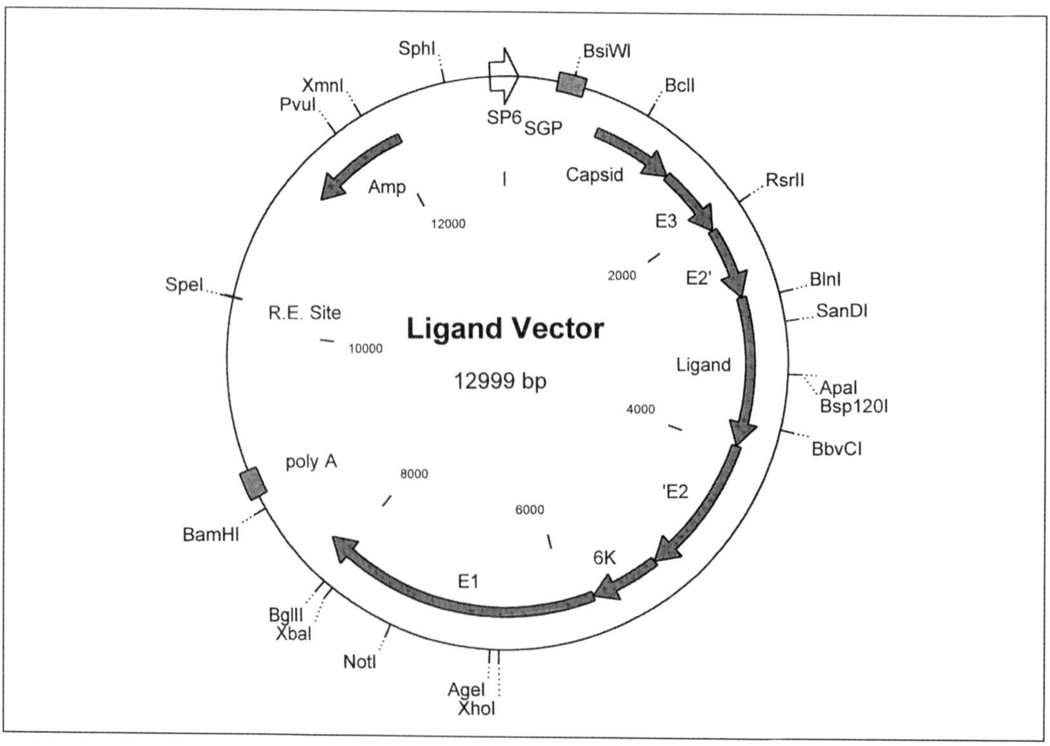

Figure 3. The ligand vector. The ligand sequence interrupts the sequence of the structural gene, E2.

probe that hybridized to the genomic RNA was 18 nucleotides longer than the probe that hybridized with subgenomic RNA. These analyses showed that both subgenomic and genomic RNAs are transcribed at high levels by four hours postinfection. Interestingly, the level of ribozyme expression from the SFV vector was at a level that exceeded by several fold even that of the abundant housekeeping actin mRNA.

SFVRz was introduced into a BHK cell line expressing HIV LTR CAT mRNA. Levels of target RNA were efficiently detected by RPA. The target RNA was effectively reduced after transduction of this cell line via electroporation or infection with SFV particles (Figure 5). The amount of CAT protein was also effectively reduced by this ribozyme. Thus, SFV can be used to deliver high levels of

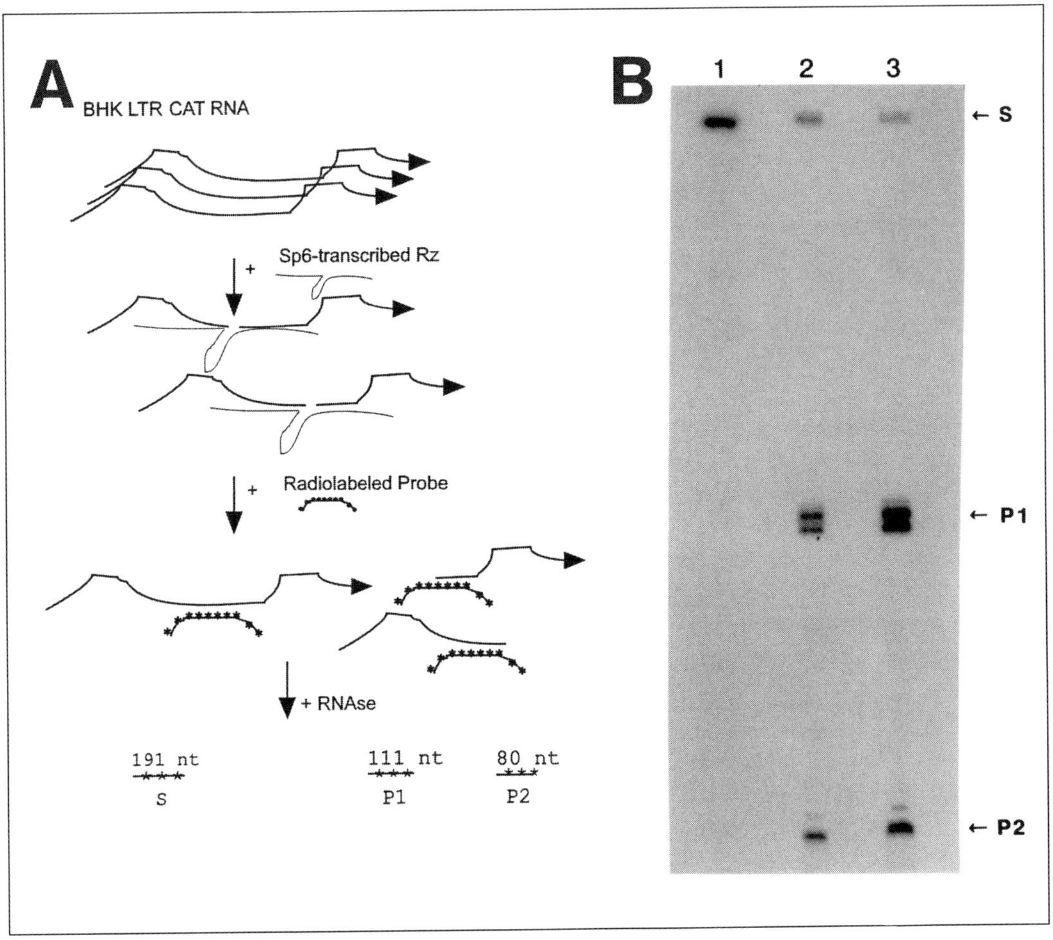

Figure 4. An 8.3-kb SFVRz chimeric RNA maintains the same catalytic activity as a small U5 ribozyme. (A) Schematic illustration of an RPA that discriminates between probe-protected uncleaved CAT mRNA substrate (S; 191 nucleotides [nt]) and ribozyme-generated cleavage products (P1 and P2; 111 and 80 nucleotides, respectively). (B) RPA analysis demonstrating ribozyme activity in SFVRz chimeric RNA. A 60-nucleotide control U5 ribozyme (lane 3) or an 8.3-kb SFVRz RNA (lane 2) was transcribed in vitro. BHK LTR CAT total-cell RNA was isolated and used as a substrate. The BHK LTR CAT total-cell RNA was mixed with buffer (lane 1), SFVRZ RNA (lane 2), or the 60-nucleotide U5 ribozyme. Cleavage of the U5-CAT mRNA present in BHK LTR CAT total-cell RNA was monitored by RPA followed by denaturing gel electrophoresis and autoradiography. Both SFVRz RNA (lane 2) and control U5 ribozyme (lane 3) produced the expected cleavage products, P1 (111 nucleotides) and P2 (80 nucleotides). Buffer alone produced no cleavage of substrate. [From Smith, Maldarelli, and Jeang (24), with permission.]

ribozyme, which can effectively and specifically downregulate a targeted RNA.

Expression of an Antisense Sequence Using Alphavirus Vector

Others have exploited the broad host range of SIN to target dengue virus (6,16,17). To obtain a vector that spreads via multicycle infection, the structural genes were maintained intact. To express a heterologous sequence, a second subgenomic promoter was placed distal to the structural genes; this double subgenomic (dsSIN) was used for delivery of an antisense RNA (Figure 2) (17). In one example, under the control of a second subgenomic promoter, a 567-bp RNA from the prM coding region of the dengue virus type 2 was placed in an antisense (D2prMa) orientation (17). Infectious virus was made from this construction via electroporation of the in vitro-transcribed RNA genome. Infection of cells with the antisense-expressing virus, D2prMa, protected mosquito cells from infection with dengue virus; in parallel, infection with the control virus conferred no protection. Through multiple cycles of replication, the novel chimeric genome maintained its ability to express the antisense RNA from the additional subgenomic promoter.

In vivo protection with alphavirus expressing the same antisense was also

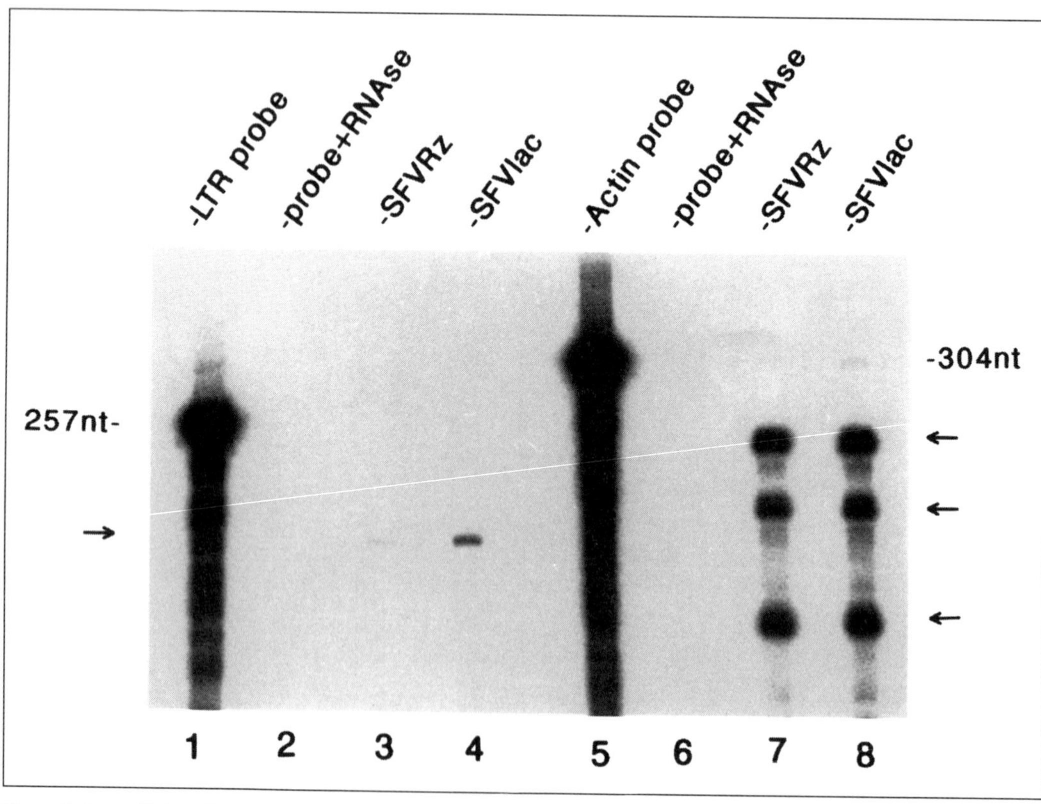

Figure 5. Intracellular efficacy of virally transduced SFVRz. BHK LTR CAT cells were transduced with SFVRz or SFVlac viral particles. Total RNAs were harvested 16 hours later and analyzed with LTR-specific (lanes 3 and 4) or actin-specific (lanes 7 and 8) probes. An RPA with the LTR-specific probes shows significantly reduced amounts of LTR-CAT mRNA in SFVRz-transduced (lane 3) versus SFVlac-transduced (lane 4) cells. Control RPA with a murine actin probe shows equivalent actin mRNA signals in the SFVRz (lane 7) and the SFVlac (lane 8) samples. Lane 1 and 5 contain input LTR and actin probes, respectively. Lanes 2 and 6 contain the same probes digested with RNase. [From Smith, Maldarelli, and Jeang (24), with permission.]

demonstrated (16). In this regard, *Aedes aegypti* mosquitoes were intrathoracically coinjected with D2prMa and dengue virus type 2. A control SIN that does not express an antisense RNA was tested in parallel. RNAs from the alphaviruses, including the antisense gene, were found distributed in various tissues 14 days after infection. The D2prMa-infected mosquitoes had no dengue virus infection of their salivary glands, while the control counterparts had substantial dengue virus replication in the glands. This protection correlated with the presence of D2prMa RNA in these salivary glands. Interestingly, the D2prMa-infected mosquitos did have dengue virus infection of their midguts. This result was explained by the fact that dengue and SIN infect different cells in the mosquito midgut. However, infection with D2prMa effectively abrogated dengue virus transmission. Saliva samples from animals coinfected with dengue virus and D2prMa or the control virus were inoculated into the thoracic cavities of naïve mosquitoes. Only one mosquito of 26 tested from the D2prMa saliva samples became infected with dengue virus. The control group developed dengue infection in 20 of 26 mosquitoes. In summary, this alphavirus antisense vector specifically and potently blocked the ability of mosquitoes to transmit dengue infection.

Modified Envelope Proteins

Most viral vectors developed for gene therapy use envelope-dependent entry of cells, which can broadly affect many cell types. However, in most therapeutic situations, transduction of the majority of host cell types is not desired. Rather, a small minority of cells is generally the desired target population. Consequently, gene therapy strategies that use viruses with broad tropism would needlessly transduce an excess of cells. Additionally, in certain situations, such as in malignancy or in an infection, the goal of therapy is to specifically remove small subpopulations of cells. One approach in these situations is to use an envelope, which confers specific entry to targeted cells. Therefore, various explorations of cell-specific gene therapy with modified viral envelopes have been performed (29). However, one shortcoming is that many viruses show reduced transduction efficiency after manipulation or mutation of their envelope proteins.

Interestingly, alphaviruses seem to tolerate envelope manipulations relatively well. Thus, alphaviruses with altered envelopes that target certain cell populations have been developed (15,22). Ohno and colleagues have effected two strategies to achieve cell type-specific delivery of alphaviruses. In the first strategy, the investigators placed the Fc-binding portion of the staphylococcal protein A within the SIN envelope protein, E2, between codons 71 and 74 (15). This alteration permitted viral assembly but blocked virus binding to its normal receptor, laminin (27). Through the chimerical envelope protein, E2, and protein A subunit, viral particles were able to bind IgG molecules via the Fc-binding portion. Additionally, when a monoclonal antibody specific for CD4 was incubated with the virus particles, the viruses could specifically infect CD4-expressing cells. The investigators showed that the antibody and its target could be changed and the specificity of viral entry would change accordingly. The same group used a slightly different approach to target human choriocarcinoma cell lines (22). The authors designed a similar SIN, which expressed the human chorionic gonadotropin (hCG) glycoprotein. The ligand (hCG) interacts with the luteinizing hormone (LH)/chorionic gonadotropin (CG) receptor. Since hCG is formed by two, noncovalently linked subunits, the coding sequences for the alpha and beta subunits were placed sequentially within the E2 protein reading frame. This chimerical envelope again permitted viral assembly, but again blocked normal viral entry. Instead, the hCG-expressing alphavirus could only infect cells expressing the LH/CG receptor, which is expressed at high levels on choriocarcinoma cells. Using the hCG-expressing vector, the authors showed specific infection and killing of choriocarcinoma cell lines in vitro. The hCG-expressing alphavirus did not infect or kill cells that did not express the LH/CG receptor. Alphaviruses cause apoptotic cell death after infection (13). Consequently, these viruses could potentially be targeted to attack cancer. Many malignancies express specific surface molecules, which are found at low or undetectable levels in other tissues. Potentially, these viruses could be applied therapeutically to infect and kill these cancer cells selectively.

In summary, alphaviruses have many in vitro and in vivo features that make them attractive candidates for RNA and gene therapy. This family of viruses can be used to deliver high levels of antisense or ribozyme RNAs that function appropriately. Additionally, the alphavirus envelope tolerates manipulations that permit targeting of the viral vector to specific cell populations. Future studies will likely lead to effective applications of alphavirus vectors as biological therapeutics.

REFERENCES

1. **Anderson, W.F.** 1998. Human gene therapy. Nature *392*:25-30.
2. **Atkins, G.J., B.J. Sheahan, and P. Liljestrom.** 1996. Manipulation of the Semliki Forest virus genome and its potential for vaccine construction. Mol. Biotechnol. *5*:33-38.
3. **Cech, T.R.** 1993. The efficiency and versatility of catalytic RNA: implications for an RNA world. Gene *135*:33-36.
4. **Dropulic, B., N.H. Lin, M.A. Martin, and K.T. Jeang.** 1992. Functional characterization of a U5 ribozyme: intracellular suppression of human immunodeficiency virus type 1 expression. J. Virol. *66*:1432-1441.
5. **Dropulic, B., S.M. Smith, and K.T. Jeang.** 1994. Activation and inactivation of gene expression using RNA sequences. Adv. Pharmacol. *30*:247-270.
6. **Gaines, P.J., K.E. Olson, S. Higgs, A.M. Powers, B.J. Beaty, and C.D. Blair.** 1996. Pathogen-derived resistance to dengue type 2 virus in mosquito cells by expression of the premembrane coding region of the viral genome. J. Virol. *70*:2132-2137.
7. **Homann, M., S. Tzortzakaki, K. Rittner, G. Sczakiel, and M. Tabler.** 1993. Incorporation of the catalytic domain of a hammerhead ribozyme into antisense RNA enhances its inhibitory effect on the replication of human immunodeficiency virus type 1. Nucleic Acids Res. *21*:2809-2814.
8. **Israel, M.A.** 1993. Molecular approaches to cancer therapy. Adv. Cancer Res. *61*:57-85.
9. **Kashani-Sabet, M., T. Funato, V.A. Florenes, O. Fodstad, and K.J. Scanlon.** 1994. Suppression of the neoplastic phenotype in vivo by an anti-ras ribozyme. Cancer Res. *54*:900-902.
10. **Kashani-Sabet, M., T. Funato, T. Tone, L. Jiao, W. Wang, E. Yoshida, B.I. Kashfinn, T. Shitara, A.M. Wu, J.G. Moreno, et al.** 1992. Reversal of the malignant phenotype by an anti-ras ribozyme. Antisense Res. Dev. *2*:3-15.
11. **Kashani-Sabet, M. and K.J. Scanlon.** 1995. Application of ribozymes to cancer gene therapy. Cancer Gene Ther. *2*:213-223.
12. **Knoell, D.L. and I.M. Yiu.** 1998. Human gene therapy for hereditary diseases: a review of trials. Am. J. Health Syst. Pharm. *55*:899-904.
13. **Levine, B., Q. Huang, J.T. Isaacs, J.C. Reed, D.E. Griffin, and J.M. Hardwick.** 1993. Conversion of lytic to persistent alphavirus infection by the bcl-2 cellular oncogene. Nature *361*:739-742.
14. **Liljestrom, P. and H. Garoff.** 1991. A new generation of animal cell expression vectors based on the Semliki Forest virus replicon. Biotechnology *9*:1356-1361.
15. **Ohno, K., K. Sawai, Y. Iijima, B. Levin, and D. Meruelo.** 1997. Cell-specific targeting of Sindbis virus vectors displaying IgG-binding domains of protein A. Nat. Biotechnol. *15*:763-767.
16. **Olson, K.E., S. Higgs, P.J. Gaines, A.M. Powers, B.S. Davis, K.I. Kamrud, J.O. Carlson, C.D. Blair, and B.J. Beaty.** 1996. Genetically engineered resistance to dengue-2 virus transmission in mosquitoes [see comments]. Science *272*:884-886.
17. **Powers, A.M., K.E. Olson, S. Higgs, J.O. Carlson, and B.J. Beaty.** 1994. Intracellular immunization of mosquito cells to LaCrosse virus using a recombinant Sindbis virus vector. Virus Res. *32*:57-67.
18. **Rice, C.M.** 1992. Examples of expression systems based on animal RNA viruses: alphaviruses and influenza virus. Curr. Opin. Biotechnol. *3*:523-532.
19. **Rossi, J.J.** 1995. Controlled, targeted, intracellular expression of ribozymes: progress and problems. Trends Biotechnol. *13*:301-306.
20. **Rossi, J.J., D. Elkins, N. Taylor, J. Zaia, S. Sullivan, and J.O. Deshler.** 1991. Exploring the use of antisense, enzymatic RNA molecules (ribozymes) as therapeutic agents. Antisense Res. Dev. *1*:285-288.
21. **Sarver, N., E.M. Cantin, P.S. Chang, J.A. Zaia, P.A. Ladne, D.A. Stephens, and J.J. Rossi.** 1990.

Ribozymes as potential anti-HIV-1 therapeutic agents. Science *247*:1222-1225.

22. **Sawai, K. and D. Meruelo.** 1998. Cell-specific transfection of choriocarcinoma cells by using Sindbis virus hCG expressing chimeric vector. Biochem. Biophys. Res. Commun. *248*:315-323.

23. **Schlesinger, S.** 1993. Alphaviruses—vectors for the expression of heterologous genes. Trends Biotechnol. *11*:18-22.

24. **Smith, S.M., F. Maldarelli, and K.T. Jeang.** 1997. Efficient expression by an alphavirus replicon of a functional ribozyme targeted to human immunodeficiency virus type 1. J. Virol. *71*:9713-9721.

25. **Strauss, J.H. and E.G. Strauss.** 1994. The alphaviruses: gene expression, replication, and evolution [published erratum appears in Microbiol. Rev. 58(4):806, 1994]. Microbiol. Rev. *58*:491-562.

26. **Tubulekas, I., P. Berglund, M. Fleeton, and P. Liljestrom.** 1997. Alphavirus expression vectors and their use as recombinant vaccines: a minireview. Gene *190*:191-195.

27. **Wang, K.S., R.J. Kuhn, E.G. Strauss, S. Ou, and J.H. Strauss.** 1992. High-affinity laminin receptor is a receptor for Sindbis virus in mammalian cells. J. Virol. *66*:4992-5001.

28. **Wang, S. and B.J. Dolnick.** 1993. Quantitative evaluation of intracellular sense: antisense RNA hybrid duplexes. Nucleic Acids Res. *21*:4383-4391.

29. **Wickham, T.** 1997. Short-order Sindbis vector targeting [news]. Nat. Biotechnol. *15*:717.

30. **Wivel, N.A. and J.M. Wilson.** 1998. Methods of gene delivery. Hematol. Oncol. Clin. North Am. *12*:483-501.

31. **Xiong, C., R. Levis, P. Shen, S. Schlesinger, C.M. Rice, and H. V. Huang.** 1989. Sindbis virus: an efficient, broad host range vector for gene expression in animal cells. Science *243*:1188-1191.

35 Efficient Expression of Proteins with Recombinant Alphaviruses

Angel Cid-Arregui[1,2], Nestor Morales-Peza[3], Victoria Juárez[4], Maria José García-Iglesias[5], and Alejandro García-Carrancá[3,6]

[1]Institute of Applied Tumor Virology, German Cancer Research Center, Heidelberg, Germany; [2]Max Planck Institute of Molecular Cell Biology and Genetics, Dresden, Germany; [3]Department of Molecular Biology, Institute of Biomedical Research, UNAM, Mexico City, Mexico; [4]Laboratory of Animal Resources, European Molecular Biology Laboratory, Heidelberg, Germany; [5]Department of Animal Pathology, Faculty of Veterinary Science, University of León, León, Spain; [6]National Institute of Cancerology (INCAN), Mexico City, Mexico

INTRODUCTION

Sindbis virus (SIN) and Semliki Forest virus (SFV) are the most intensively characterized members of the alphavirus group. They have proven useful for investigating fundamental questions related to the synthesis, posttranslational modification, and sorting of integral membrane proteins (58). Recently, expression vectors have been derived from these viruses that are proving convenient for applications that require high levels of expression of recombinant protein in a short time (52,58). In this chapter, we describe the current types of alphavirus vectors, protocols for the production of viral vector stocks, and some illustrative results obtained in vitro and in vivo with recombinant vector expressing a reporter protein.

Viral Replication

Alphaviruses are among the simplest enveloped viruses. Their genome consists of a unique single-stranded RNA molecule of positive polarity that functions directly as mRNA since it is 5′ capped and 3′ polyadenylated (63,64). The length of the alphavirus genomes without the poly(A) tail is about 12 000 nucleotides (63). The first two thirds of the viral genome from its 5′ end encode the four nonstructural proteins, while the 3′ third, contained in the subgenomic 26S RNA, codes for the capsid and envelope proteins. Table 1 summarizes the proteins encoded by SFV and their known functions.

The life cycle of alphaviruses, which takes place in the cytoplasm of the infected cell, comprises the following major events (Figure 1):

Viral Vectors: Basic Science and Gene Therapy
Edited by A. Cid-Arregui and A. García-Carrancá

Virus Entry and Uncoating

The E2 and E3 components of the spike bind to a cellular receptor. Several putative receptors have been identified for SIN and SFV in different cell types (30,48,68,71). The virion is internalized by receptor-mediated endocytosis. The low pH of the endosome activates the fusogenic capacity of E1, which acts to facilitate fusion of the viral envelope with the endosomal membrane (46,70). After release into the cytoplasm, the nucleocapsid disassembles by a mechanism that likely involves interaction of the capsid (C) protein with ribosomes (59,69,77).

Translation and Replication

Once released from the capsid, the viral genome is immediately translated to produce the nonstructural proteins (NSPs) (7). A precursor polyprotein is synthesized first, which is then cleaved into the four

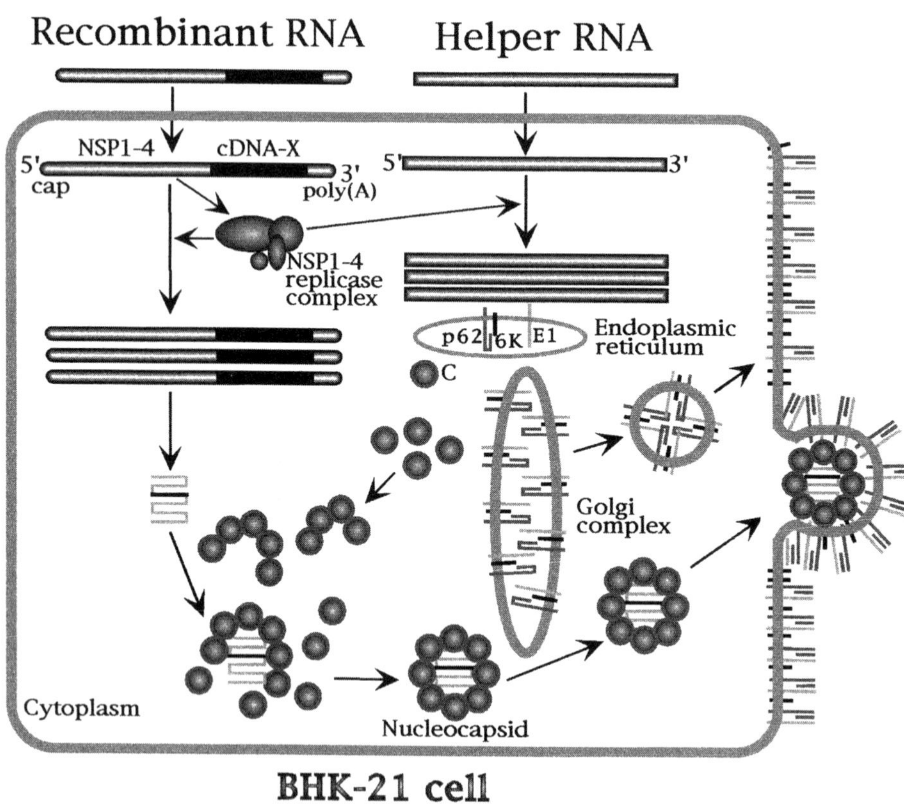

Figure 1. Cellular events during packaging of SFV vectors. Recombinant and helper mRNAs enter the cell upon transfection. The viral replicase complex, produced from the recombinant mRNA, starts to replicate both types of mRNA. The capsid (C) protein is synthesized and released into the cytoplasm, while the envelope proteins travel through the Golgi complex, where they undergo glycosylation and other posttranslational modifications, to be constitutively transported to the plasma membrane. The capsid protein forms nucleocapsids only with the recombinant RNA molecules since the packaging sequence has been deleted in the helper RNA. Finally, the nucleocapsids reach the plasma membrane where the C protein binds the cytoplasmic tail of the envelope glycoproteins and budding occurs. In polarized cells, budding takes place preferentially in a defined plasma membrane region (e.g., the somatodendritic domain of neurons and the basolateral surface of epithelial cells) (54). In insect cells, budding occurs intracellularly, and the virions are transported in vesicles to the surface.

Table 1. Sizes and Functions of SFV Proteins

	Amino Acids	kDa	Description/Function
Nonstructural Proteins			
NSP1	537		Methyl-transferase (5′ capping activity)
NSP2	798		Autoprotease (helicase)
NSP3	482		Phosphoprotein; unknown function
NSP4	614		Polymerase activity
Structural Proteins			
Capsid	267	33	Nucleocapsid; autoprotease
p62	482	62	Precursor to E2 and E3
E2	418	52	Spike component (major antigen for neutralizing antibodies)
E3	64	10	N-terminal portion of p62; unknown function
6K	60	6	Membrane-associated; contains signal sequence for E1
E1	438	49	Spike component; fusogen at pH <6

NSP1–4 by a protease activity that has been mapped to the C-terminal domain of NSP2 (15,62). When a sufficient amount of NSPs has been produced, replication of the viral genome takes place. The original plus strand of SFV serves as template for the synthesis of full-length minus strands, which are transcribed into plus-strand mRNAs (7,34). In addition, a subgenomic RNA species of about 4000 nucleotides (26S RNA) is rapidly transcribed from the full-length minus strands, which corresponds to the 3′ third of the viral genome (19,35,36). The 26S RNA is translated into a polyprotein that undergoes self-cleavage to release the five structural proteins (Table 1) (37).

Virion Assembly and Release

The structural proteins are synthesized as a precursor polypeptide in the sequence C-E3E2(p62)-6K-E1 (20,58). The C protein has a protease activity that cleaves itself from the precursor polypeptide chain while synthesis proceeds (29,44). This uncovers a signal sequence that targets the growing polypeptide chain to the endoplasmic reticulum (22), where it is cotranslationally translocated and cleaved by host cell signalase into the three viral spike proteins p62 (E3 and E2), 6K, and E1 (37,44). These proteins undergo glycosylation and other posttranslational modifications as they progress through the Golgi stacks. Finally, the complex is constitutively transported to the plasma membrane. In the *trans*-Golgi vesicles, p62 is cleaved to yield E2 and E3 (12,72). In the cytoplasm, the C protein assembles with the 42S RNA genome to form nucleocapsids (18,45,76), which migrate to the plasma membrane where they interact with the 6K protein of the spike (39). Finally, the viral particles bud from the plasma membrane, and enveloped virions are released.

Wild-type SFV is among the viruses that most actively replicate (58). A single infected cell can produce about 1×10^5 viral particles before exhausting the cell machinery. This implies approximately 1×10^7 to 1×10^8 molecules of viral protein (63).

Alphavirus Vectors

Alphaviruses present three main advantageous features that facilitate their use as expression vectors (21,52,58,64): *(i)* high replication capacity, *(ii)* efficient translation of the viral RNA, and *(iii)* infection of a wide range of vertebrate hosts and cell types. The alphavirus genome is a self-replicative mRNA molecule. Therefore, transfection of cells with naked viral RNAs transcribed in vitro from the cDNAs results in productive infection (11,53). SIN virus and SFV cDNAs have been engineered to incorporate heterologous RNA sequences in substitution of those of the viral genome that specify the structural proteins (38,78). The resulting recombinant mRNAs are self-replicative molecules (replicons). In addition, they retain the packaging signal and, therefore, can be incorporated into infective virions in cells cotransfected with helper mRNA that encode the structural proteins of the virus.

There are two types of alphavirus vectors:

1. **Replication-incompetent vectors.** The heterologous sequence is inserted into the viral genome downstream of the sequences encoding the NSPs in place of the sequence coding for the structural proteins. This RNA is self-replicative and contains a packaging signal, but to be packaged into virions, it requires a complementing viral mRNA that specifies the structural proteins. Complementation can be provided by a defective interfering (DI, deletion mutants unable to replicate themselves in the absence of the parental virus) helper RNA, which cannot be packaged itself (5,38).

2. **Replication-competent vectors.** Some defective SIN viruses used as helpers retain the packaging signal and, therefore, can be packaged together with the replicon, which in turn can be amplified through multiple passages (5,23). A second type of vector uses an RNA molecule containing two subgenomic mRNA promoters—one that specifies the synthesis of the viral structural proteins and the other the heterologous protein (28,51).

In addition to using alphaviruses as expression vectors, naked recombinant viral genomes can also be used as efficient virus-free expression systems by simply transfecting cells with subgenomic alphavirus mRNAs that express the NSPs together with a recombinant heterologous protein. Alternatively, cDNAs of such subgenomic mRNAs can be expressed under the control of a strong eukaryotic promoter (32).

RECOMBINANT SFVs

The SFV replicon consists of a subgenomic mRNA coding for the viral NSPs (the replicase complex) and the heterologous protein of choice, but not for the capsid and envelope proteins (Figure 1). The viral genome is manipulated as a cDNA copy inserted into a pGEM-based plasmid downstream of an SP6 promoter (38) (Figure 2). First, the heterologous sequence is subcloned, thereby replacing the structural genes, under the control of the 26S subgenomic RNA promoter (25,36). Then, a convenient amount of recombinant mRNA is synthesized in vitro by the SP6 polymerase and used to transfect a producer cell line (e.g., BHK-21 cells), which leads to the synthesis of high amounts of NSPs and heterologous protein. In addition, the replicase complex synthesizes high numbers of replicons (Figure 1). In a variant form of this replicon, the heterologous coding sequence is inserted downstream of the capsid gene, whose sequence contains a translational enhancer, which drives expression levels one order of magnitude higher than the RNA that lacks capsid sequences (60).

Packaging of replicon RNA (Figures 1 and 3) is achieved by cotransfecting the cells with in vitro-transcribed mRNA (helper mRNA) that codes for the capsid and envelope proteins, also from the 26S RNA promoter (Figure 2). The helper RNA cannot be packaged, however, because it lacks the packaging signal, which is located at the end of the NSP1 open reading frame (ORF) (18,73). In an attempt to make the system safer, Berglund et al. (2) developed a helper that carries three point mutations in the sequence coding for the p62 precursor glycoprotein, which results in noninfectious particles that need to be activated by mild treatment with protease (chymotrypsin) to become infectious. Rather than employing a unique helper mRNA, two helper mRNAs, one coding for the capsid protein and the other for the spike proteins, can also be used to minimize the risk of wild-type

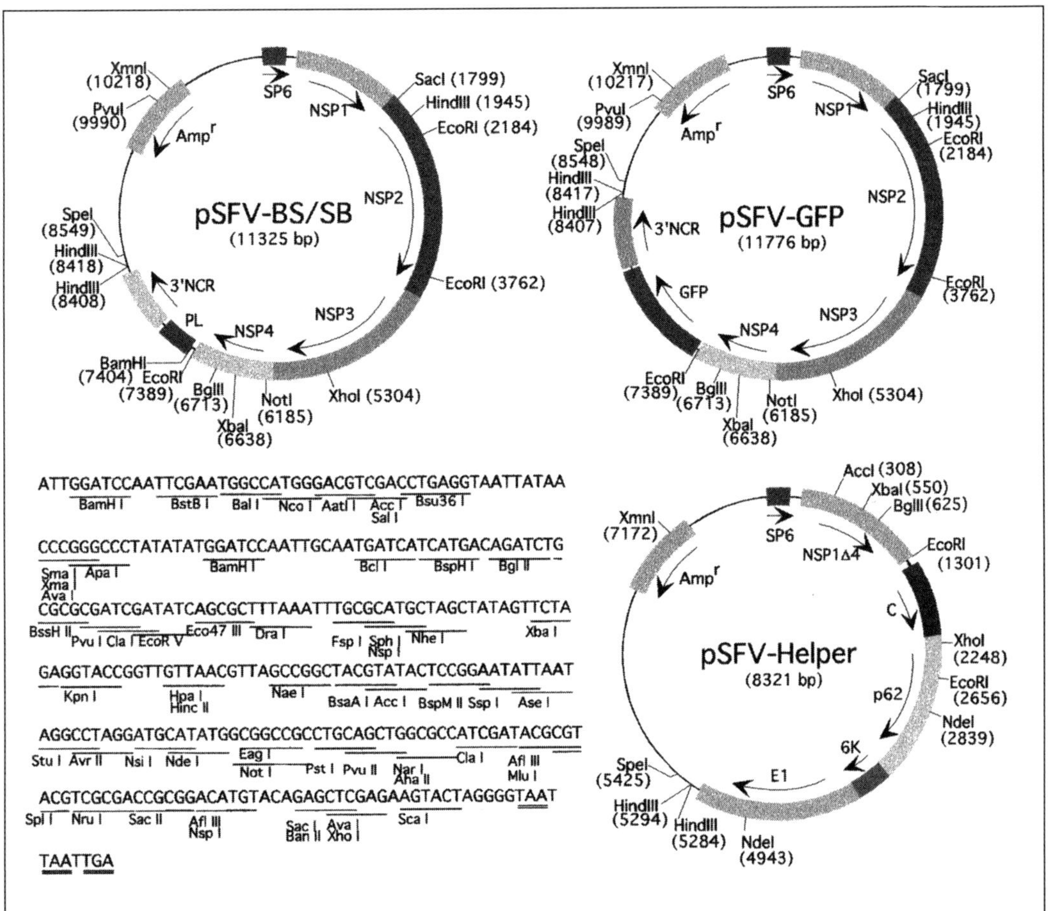

Figure 2. Plasmids used for production of recombinant SFVs. Plasmids pSFV-BS/SB are derivatives of pSFV-1 (38) and contain a polylinker with 49 restriction sites in either orientation downstream the 26S subgenomic promoter. The complete sequence and restriction map of the polylinker is shown at the bottom. Plasmid pSFV-GFP is a pSFV derivative in which the sequence coding for EGFP has been inserted into the polylinker. Also shown is the pSFV-helper plasmid used for in vitro transcription of mRNA coding for the structural proteins of the virus (C, p62, 6K and E1). *Amp^r*, ampicillin resistance gene; NCR, noncoding region; NSP1-4, nonstructural proteins; PL, polylinker.

virus production through RNA-RNA recombination (61).

Packaging and Expression Efficiencies

Approximately 1×10^3 to 1×10^5 viral particles can be released from a single cell infected by wild-type SFV (63). Transfection of 1×10^6 BHK-21 cells with recombinant viral RNA and incubation for 24 to 36 hours yields virus suspensions of about 1×10^8 to 1×10^9 infectious units (iu) (38).

Both the high number of copies of the viral genome produced by the replicase complex (75) and the low degree of secondary structure of the subgenomic 26S mRNA in its 5′ region (1), which facilitates ribosome binding, result in high rates of transcription of the heterologous protein in a very short time. Thus, infection of 5×10^6 BHK-21 cells with recombinant virus at a multiplicity of infection (moi, the virus to cell ratio) of 1, and incubation for 48 to 72 hours yields approximately 200 to 400 µg recombinant protein. Apparently, there are no major differences in the yields when using an moi ranging from 1 to 100 (38).

Biohazard and Biosafety Aspects

SFV and SIN virus were isolated some 50 years ago, and ever since they have been the subject of study in many laboratories, without any accidental infection in humans reported to date.

SFV is not a serious pathogen for humans (55,56). It does not pose a serious threat to lower vertebrates either, unless it is inoculated directly into the brain where it produces focal necrosis. In rare, naturally occurring infections of humans (in central Africa), after an incubation period of 3 to 12 days, SFV produced a mild disease of abrupt onset, with a rapid rise in temperature to 38.5°C to 40.5°C (102°F–105°F) associated with arthralgia, myalgia, and, in 60% to 80% of cases, rash (42,55).

Wild-type SFV particles can appear in cells cotransfected with recombinant and helper RNAs when packaging recombinant virus. This is likely due to strand switching by the replicase resulting in the production of replicative forms of the SFV genome, which has been estimated to occur at a frequency of 1×10^{-6} (23,74). Therefore, it is advised that recombinant virus be handled under biosafety level 2 conditions.

PROTOCOLS FOR PACKAGING AND TITRATION OF RECOMBINANT SFVs

Once the recombinant DNA has been subcloned into the SFV plasmid (see "Plasmids" below), the recombinant virus can be produced in high amounts within two days (Figure 3).

Materials

Plasmids

PSFV-BS/SB (Figure 2)

This plasmid was obtained by inserting a polylinker into pSFV1 (38), which contains a cDNA clone of SFV after deletion of the 26S coding region (ORFs C-p62-6K-E1), except for the sequence that encodes the last 49 amino acids of E1 and the complete 3′ noncoding region. This sequence, together with a short stretch 5′ in the SFV genome (nucleotides 1–247) were shown to be essential for SFV RNA replication (33,35). The viral cDNA is placed downstream of an SP6 promoter in a pGEM plasmid. The polylinker, which can be used for the insertion of heterologous cDNAs, is located 31 bp downstream of the 26S transcription initiation site and is followed by translational stop codons in all three reading frames (Figure 2). No translation initiation motive is present in the SFV sequence. This plasmid is used for in vitro production of transcripts that encode the NSPs of the virus (NSP1–4, replicase complex), after linearization with *Spe*I, which is a unique restriction site located close to the 3′ end of the viral sequence.

pSFV-Helper (Figure 2)

This plasmid contains a subgenomic SFV cDNA, in which most of the nonstructural coding region has been deleted, downstream of an SP6 promoter. It is used for producing helper mRNA that encodes the structural proteins of the virus (C, p62, 6K, and E1) required for assembling recombinant SFV virions. For in vitro transcription, the plasmid is linearized with *Spe*I for which there is a unique site. The SFV mRNA transcribed from this plasmid contains the 5′ and 3′ sequences essential

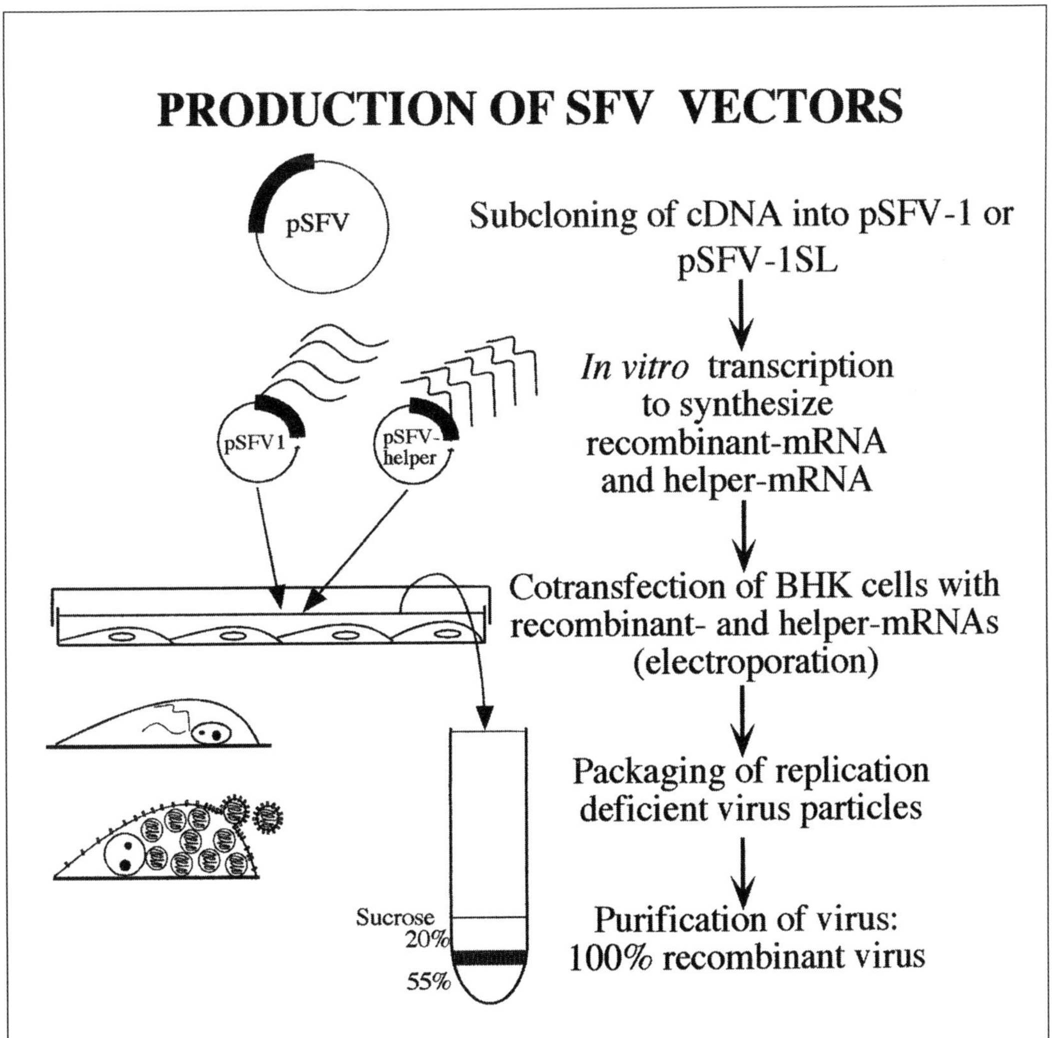

Figure 3. Steps for construction and purification of SFV vectors. The plasmids that contain recombinant SFV carrying a cDNA X sequence (pSFV-X) and the helper sequences (pSFV-helper) are linearized with *Spe*I and transcribed in vitro with the SP6 polymerase. The recombinant and helper mRNA species are mixed 1:1 and used to transfect an 80% to 90% confluent monolayer of BHK-21 cells. After 40 to 48 hours of incubation, the virus is purified from the supernatant.

for RNA replication (35,47,66), yet it lacks the sequence needed for packaging into virions. The helper mRNA cannot replicate when transfected into cells unless it is cotransfected together with mRNA derived from the pSFV plasmid, which provides the replicase complex.

pSFV-GFP

It is a derivative of pSFV1 that contains a cDNA that encodes the 238 amino acids of the green fluorescent protein (GFP). GFP was isolated from the bioluminescent jellyfish *Aequorea victoria* (50) and was shown to retain its fluorescence capacity when expressed in eukaryotic cells (6). GFP is a monomer that emits green fluorescent light upon excitation with ultraviolet or blue light. Its visualization, however, is limited by the low amount of light that it emits (less than one photon per molecule), as compared with most fluorescent compounds (several thousands of photons per molecule). Therefore, several GFP mutants have been developed to improve the fluorescent emission of the protein. GFPmut1 (9) (EGFP; Clontech Laboratories, Palo Alto, CA, USA) is one such mutant. It contains two amino acid substitutions in the chromophore that make it fluoresce 4- to 35-fold brighter than the wild-type protein when excited with blue light. In addition, EGFP contains more than 190 silent mutations that transform the gene into a sequence of almost entirely preferred human codons (27).

Cells

BHK-21 Cells (ATCC-CL13; American Type Culture Collection, Manassas, VA, USA):

Hamster syrian kidney cells are fibroblasts with a karyotype of 2n=44, pseudodiploid that are subclones of parent line derived from five 1-day-old unsexed hamster kidneys. They are cultured in BHK medium (see below) with 5% CO_2 at 37°C. The subculture routine consists of splitting of confluent cultures 1:3 to 1:6 (i.e., seeding at 2–4 × 104 cells/cm^2 using trypsin/EDTA). This cell line is well suited for virus replication studies (e.g., poliovirus, rabies, foot-and-mouth disease, vesicular stomatitis virus, herpes simplex virus, Ad25, alphaviruses, and other arboviruses).

Hippocampal Neurons

Primary cultures of rat hippocampal neurons were prepared from E18 embryos as described elsewhere (24). The cells were plated onto poly-L-lysine-coated coverslips (15-mm ø) and maintained in serum-free medium supplemented with N-2 supplement (Life Technologies, Rockville, MD, USA) (3) and fibroblast growth factor (Roche Molecular Biochemicals, Indianapolis, IN, USA) to a final concentration of 1 ng/mL.

Equipment

- Electrotransfection device (Bio-Rad Laboratories [Hercules, CA, USA] or equivalent) and cuvettes
- Fluorescence microscope

Reagents, Solutions, and Media

- Antibodies: Anti-EGFP, rabbit polyclonal (Clontech Laboratories, Palo Alto, CA, USA); peroxidase-conjugated anti-mouse IgG+IgM, rabbit polyclonal (Jackson ImmunoResearch Laboratories, West Grove, PA, USA)
- BHK culture medium: Glasgow minimum essential medium (GMEM), 10% tryptose phosphate, 2 mmol/L glutamine, 5% fetal calf serum (FCS), 0.1 U/mL penicillin, and 0.1 µg/mL streptomycin
- Enhanced chemiluminescence kit (ECL® RNP2106, Amersham Pharma-

cia Biotech, Piscataway, NJ, USA)

- DAPI (Molecular Probes, Eugene, OR, USA). Stock 1 mg/mL in water, stored at -20°C. Working dilution: 1 μg/mL.
- Fixative: 2% paraformaldehyde (PFA)/ 2% sucrose in phosphate-buffered saline (PBS)
- Hank's balanced salt solution (HBSS): 8.0 g NaCl, 0.4 g KCl, 0.14 g $CaCl_2$, 0.1 g $MgSO_4$, 0.1 g $MgCl_2$, 0.06 g Na_2HPO_4, 0.06 g KH_2PO_4, 1 g glucose, and 0.35 g $NaHCO_3$, made up to 1 L with deionized water
- Infection medium (for BHK cells): GMEM supplemented with 0.2% bovine serum albumin (BSA); 2 mmol/L glutamine; 20 mmol/L HEPES, pH 7.0; 0.1 U/mL penicillin; and 0.1 μg/mL streptomycin. For infection of other cell types, substitute GMEM by the appropriate medium.
- Mowiol (Calbiochem-Novabiochem, San Diego, CA, USA). See Chapter 16 for a protocol to prepare the working solution.
- Dulbecco's PBS: 8 g NaCl, 0.2 g KCl, 1.44 g Na_2HPO_4, and 0.24 g KH_2PO_4. Adjust pH to 7.0–7.2 with HCl and make up to 1 L with deionized water.
- STE buffer (100 mmol/L NaCl; 50 mmol/L Tris-HCl, pH 7.4; and 0.5 mmol/L EDTA); autoclave.
- Sucrose: solutions 20% and 55% in STE buffer; autoclave.
- Transcription kit: mMESSAGE mMACHINE™ SP6 (Ambion, Austin, TX, USA)

Purification of Recombinant SFVs

See Figure 3 for a schematic flow chart of the protocol.

Preparation of cells

BHK-21 cells are used to produce large amounts of recombinant virus. The cells should have reached the late logarithmic phase by the time of transfection. Two days before the transfection day (see below), split a confluent dish (10-cm ø) 1:3 into 10-cm ø dishes.

Preparation of Helper and Recombinant mRNAs

Linearization of Plasmids:

1. Digest approximately 2 mg pSFV-helper and approximately 2 mg pSFV-X (where X is the cDNA sequence specifying a protein of interest; e.g., EGFP) with *Spe*I, in final reaction volumes of 100 mL.
2. Check digestion in a 0.7% agarose gel.
3. When the digestion has been completed, raise volume to 200 mL with water.
4. Extract consecutively with 1 volume of first phenol-chloroform and with chloroform/isoamylalcohol (24:1, vol/vol).
5. Precipitate DNA with 3.5 volumes of 100% ethanol/7.5 mol/L ammonium acetate (6:1, vol/vol), wash pellet with cold 70% ethanol, and resuspend in 10 mL diethylpyrocarbonate (DEPC)-water.

In Vitro Transcription Reaction

Most mRNAs have a 5′ 7-methyl guanosine residue (cap), which serves both in the translation initiation process and to protect the mRNA from degradation by intracellular nucleases. Capped in vitro transcripts are synthesized by including cap analog to the transcription reaction. The ratio of cap analog to guanosine-5′-triphosphate (GTP) to be used in the reaction is an important factor to consider since *(i)* the cap analog is an expensive reagent, *(ii)* high levels of cap analog significantly reduce the yield of the reaction, and *(iii)* unincorporated cap analog is a potent inhibitor of transcription initiation. We advise the use of a cap analog to GTP ratio of 2 to 4:1.

Prepare the following reaction mixture using the mMESSAGE mMACHINE SP6 in vitro transcription kit:

- 2 µL 10× reaction buffer
- 10 µL rNTPs (including $m^7G(5')ppp(5')G$ cap analog)
- Up to 6 µL template plasmid (1 µg) linearized with *Spe*I
- 2 µL enzyme mixture

If necessary, complete volume to 20 µL with nuclease-free water. Incubate reaction mixture at 37°C for 1 to 2 hours (depending on the size of the cDNA subcloned in the pSFV plasmid), preferably in a cabinet incubator or a heated room.

Notes

Pipet the reaction components at room temperature to avoid precipitation of DNA from spermidine present in reaction buffer. Check the quality and efficiency of transcription by running a 1-µL aliquot of reaction mixture on a normal (nondenaturing) 0.7% agarose gel. The RNA band should be much more intense than the band of DNA template. Keep the reaction mix at -80°C in aliquots of 10 µL until needed for transfection. There is no need to digest the template DNA with DNase I since the molar ratio of DNA to RNA should be very low by the end of the reaction. Purification of the transcribed RNA is not required for electrotransfection of cells. However, for other transfection methods, such as lipofection, the RNA should be purified (e.g., by using glass milk [Qiagen, Valencia, CA, USA] or by lithium chloride precipitation).

Transfection of BHK-21 Cells with In Vitro-Transcribed SFV mRNAs

BHK-21 cells can be easily and efficiently transfected by electrotransfection (electroporation) as described below. However, other transfection methods can be used such as lipofection with FuGENE™ (Roche Molecular Biochemicals), GenePORTER™, or LIPOFECTAMINE™ 2000 (Life Technologies, Rockville, MD, USA).

1. Prepare cell suspension by growing BHK-21 cells to late logarithmic phase (80%–90% confluence) in BHK culture medium in a 10-cm ø dish.

- Wash cells twice with PBS.
- Add 1 mL trypsin-EDTA. Incubate at room temperature for 1 to 2 minutes until the cells detach.
- Disaggregate cell clumps to get a single cell suspension by pipetting gently back and forth with a Pasteur pipet.
- Transfer the cells to a 15-mL sterile plastic tube containing 5 mL BHK medium. Pipet back and forth 5 to 10 times. Then, add 5 mL BHK medium and pipet again 5 to 10 times to disaggregate any clumps that may have formed. If necessary, check under the microscope that no large clumps of cells remain in the suspension. At this point, the cells can be counted using a hemocytometer. Keep the tube on ice if harvesting will not be performed immediately.
- Harvest cells by centrifugation in a tabletop centrifuge at 200× *g* (approximately 1000 rpm) for 10 minutes at 4°C.
- Aspirate supernatant and resuspend the cells in 5 mL PBS by gently pipetting back and forth 5 to 10 times. Centrifuge as above.
- Resuspend the cell pellet in 0.7 mL PBS by pipetting gently (5–10 times), using a Pasteur pipet, until a single cell suspension is obtained. The concentration of cells in the suspension should be about 1×10^7 cells/mL.
- Using a Pasteur pipet, transfer the cells to a chilled electroporation cuvette (1 mL; 4 mm between electrodes). Keep on ice until performing electrotransfection. Avoid any foaming when transferring the cells to the electroporation cuvette.

2. Preparation of RNA mixture:

- Thaw aliquots of helper and recombinant SFV RNAs transcribed previously in vitro.
- Mix helper and recombinant RNAs (1:1) in a microcentrifuge tube by pipetting back and forth.
- Spin down for a few seconds in the microcentrifuge.
- Chill the tube on ice until electroporation.

3. Electrotransfection:

- Just before electroporation, add the helper plus recombinant mRNA mixture to the cell suspension in the electroporation cuvette, cap tightly, and mix by inverting the cuvette a few times.
- Place the cuvette into the electroporation device.
- Set at 0.85 kV and 25 µF. Set pulse controller to "Off", which means that the electrical field to be applied to the fluid (cell suspension) will be exactly the voltage set on the control panel.
- Pulse twice. After each pulse, the display should show a time constant of 0.4. The composition of the PBS and its pH are critical parameters of the efficiency of electrotransfection. Before the electroporation of cells, it is advisable to pulse a cuvette filled with only PBS to check that the time constant is 0.4.
- Immediately after the electroporation, transfer the cell suspension with a Pasteur pipet to a 10-cm ø dish containing 12 mL BHK medium. Pipet gently back and forth several times until a homogeneous cell suspension is obtained.
- Incubate the cells for 24 to 36 hours at 37°C with 5% CO_2.

Harvesting and Purification of Virus

Visualize the cells under the microscope to check that no microbiological contamination occurred during electroporation and that cytopathic effects caused by the virus are evident. (Most cells should appear rounded and almost detached from the dish surface.)

1. Transfer the culture supernatant to a 15-mL plastic tube.
2. Centrifuge for 15 minutes at low speed to remove cells in suspension (approximately 2 000× *g* or 3500 rpm in a tabletop centrifuge).
3. Aspirate 12 mL supernatant. Discard remaining medium together with the pellet. Make aliquots (500 µL) and store at -80°C or proceed as follows for sucrose gradient purification.
4. Carefully pipet 6 mL supernatant onto a discontinuous sucrose cushion of 2 mL 55% sucrose (in STE) and 5 mL 20% sucrose (in STE) in an SW40 polycarbonate tube. Proceed the same way with the remaining 6 mL medium.
5. Centrifuge at approximately 150 000× *g* (approximately 35 000 rpm in an SW40 rotor) for 2 hours.
6. Carefully remove the tubes from the ultracentrifuge and place them on ice until you begin to recover the virus.
7. Remove 6 mL medium from each tube and discard.
8. Carefully remove 4.5 mL 20% sucrose and discard.
9. The viral band is at the 20% to 55% sucrose interphase. However, the band is hardly visible, if at all; therefore, for the first time, it is advisable to keep a reasonable part of the gradient above and below this interphase to ensure that the virus is recovered. Aspirate with a pipet 10 aliquots of 0.1 mL and transfer them to microcentrifuge tubes on ice (pool virus from both ultracentrifuge tubes).
10. Place aliquots of 20 µL into cryotubes, freeze them quickly in ethanol/dry ice, and store them at -80°C until needed. Just before use, transfer a tube on ice,

and wait until the virus suspension thaws, mix gently, and dilute appropriately in infection medium.

Titration of Recombinant Virus Stocks

Since recombinant SFVs are not able to replicate, the titer of the vectors cannot be determined by plaque assay. Instead, the number of recombinant iu/mL of stock is estimated by immunofluorescence using antibodies against the heterologous protein that they express.

Titration of vector stocks can be performed as follows:

1. Two days before infecting for titration, split a 90% confluent dish of BHK-21 cells 1:2 into a 6-cm ø dish containing 12 coverslips (10-mm ø).
2. The day of infection, thaw an aliquot of virus stock and prepare a series of dilutions (e.g., 1:100, 1:200, 1:500, 1:1 000) in infection medium (see Materials section).
3. Wash the coverslips three times in PBS.
4. Using sterile forceps, transfer coverslips right side up onto a piece of Parafilm™ in a 10-cm ø bacteriological dish.
5. Pipet 50 µL of each dilution of virus stock onto separate coverslips, taking care that their entire surface is covered by the fluid.
6. Incubate the coverslips for 1 hour at 37°C, 5% CO_2, to allow adsorption of the virus to the cell surface.
7. Remove virus-containing medium and wash coverslips twice with PBS.
8. Transfer coverslips to a 24-multiwell plate containing 0.5 mL of BHK medium per well.
9. Incubate overnight at 37°C/5% CO_2.
10. Remove medium and wash coverslips twice with PBS.
11. Fix the cells by incubating the coverslips in 2% PFA/4% sucrose in PBS for 15 minutes at room temperature. Quench with 50 mmol/L ammonium chloride. For GFP visualization, drain and mount coverslips on a drop of Mowiol. For immunofluorescence, permeabilize cells in 0.2% Triton® X-100 in PBS for 5 minutes at room temperature, wash with PBS, and proceed with incubations in primary and secondary antibody solutions.

The titer of the virus can be determined by counting the total number of cells infected in the entire coverslip and dividing this number by the volume of virus suspension used. Typical titer values are in the range of 1×10^7 to 1×10^8 iu/mL. A total of 1×10^8 to 1×10^9 viral particles can be obtained from a subconfluent 10-cm ø dish ($3–4 \times 10^6$ cells).

Notes

The above protocol describes optimal conditions for production of SFV using BHK-21 cells, which are highly efficient packaging cells that can be transfected efficiently by electroporation. For best results, use BHK-21 cells of low passages (below 20). If other cell types are to be used, optimal transfection conditions should be established first. Best packaging efficiencies are obtained when the cells are transfected with a mixture containing helper and recombinant RNA at a molar ratio of 1:1 (38).

It is advisable to perform all infections of cells in culture in a standard way (i.e., as described for titration of viral stocks). This ensures the comparability of the data and allows more precise calculations of the effective moi used in the experiment. If necessary, the volume of medium and the amount of virus can be scaled up in order to infect cells growing on plastic dishes instead of coverslips.

Infections should be performed in serum-free medium buffered with HEPES at pH

7.0. Viral infectivity drops dramatically if the cells are incubated with the virus in alkaline medium (e.g., pH 7.5). Undiluted stock of purified virus should not be used to infect cells or for animal immunization because of its high sucrose content. Dilutions of stocks of at least 100-fold are advisable. The efficiency of the infection seems to be higher at temperatures of 30°C to 35°C.

For many purposes, there is no need to purify the virus. Instead, the medium can be clarified by centrifugation (optionally, it can be filtered trough 0.45-µm filters), aliquoted (e.g., 100 µL), and snap frozen in liquid nitrogen. The SFV and SIN virions are stable through repeated freezing and thawing and in storage at 4°C (56). However, they lose infectious activity rapidly at 56°C or when exposed to detergents or organic solvents.

When using virus expressing GFP reporters, the green fluorescence can be observed in living and in fixed cells. Wild-type GFP and its derivatives are seen throughout the cytoplasm and the nucleus. In neurons, even the finest branches are filled with reporter protein after 12 to 24 hours in culture in amounts that allow their visualization (Figure 4). GFP fluorescence persists after PFA fixation; however, fixation with methanol should be avoided.

For visualization of GFP fluorescence in living cells, it is advisable to supplement the culture medium with N-acetyl-L-cysteine (NAC), which acts as a scavenger for free radicals and protects the cells from damage caused by UV light (43). After having exposed the cells to UV light, rinse the coverslip briefly in PBS, transfer to a dish containing the appropriate culture medium (without NAC), and return to the incubator. A comprehensive description of time-lapsed video microscopy methods can be found elsewhere (4).

Complementary protocols on immunocytochemistry, metabolic labeling, surface biotinylation, and immunoprecipitation of cell lines and primary hippocampal neurons infected with SFV have been described previously (13).

APPLICATIONS OF ALPHAVIRUS VECTORS

The simple composition of alphaviruses, their powerful replication machinery, and the efficient translation of the viral mRNAs make these viruses and their derived vectors suitable for studies requiring high levels of protein expression in a short time (21,31,57). Indeed, they are particularly advantageous when the target cells are highly differentiated and hence difficult to transfect by other means, as in the case of neurons (8,16,17,26,49). However, the activity of the viral replicase cannot be modulated. Therefore, infection by the recombinant vectors ends up with the destruction of the cell within a few days—the reason they have been called "suicide vectors" (57). In spite of this, the efficient expression driven by alphavirus vectors and their broad cell-type specificity make them appropriate for short-term studies in vitro, which may benefit from the expression of high levels of recombinant protein (e.g., for biochemical analyses). Moreover, these vectors are becoming interesting tools for genetic immunization and vaccine development purposes because, in addition to an initial presentation of recombinant antigen by the major histocompatibility complex (MHC) class I (28), the subsequent lysis of the infected cells should lead to a release of high amounts of antigen. Examples of the application of SFV vectors for these two types of studies are given below.

Gene Transfer into Cultured Neurons and Neuroepithelial Cells Using SFV Vectors

Wild-type SFV has proven to be a useful

tool when applied to the study of some cellular functions, such as the sorting of surface proteins in polarized epithelial cells and neurons (8,16,54). Likewise, SFV expression vectors have been used recently to express heterologous plasma membrane proteins in cultured neurons to analyze their intracellular processing and sorting signals (14,65). These studies have revealed the suitability of SFV vectors for the expression of recombinant protein in neurons, which are quite difficult to transfect by other means. Within the first hours after infection (12–20 hours), the infected neurons remain morphologically and functionally intact (8,17,49). However, these vectors are not appropriate for studies requiring prolonged expression since the viral replicase causes progressive and irreversible damage to the cell by exhausting its metabolic pools while producing recombinant protein and viral replicase.

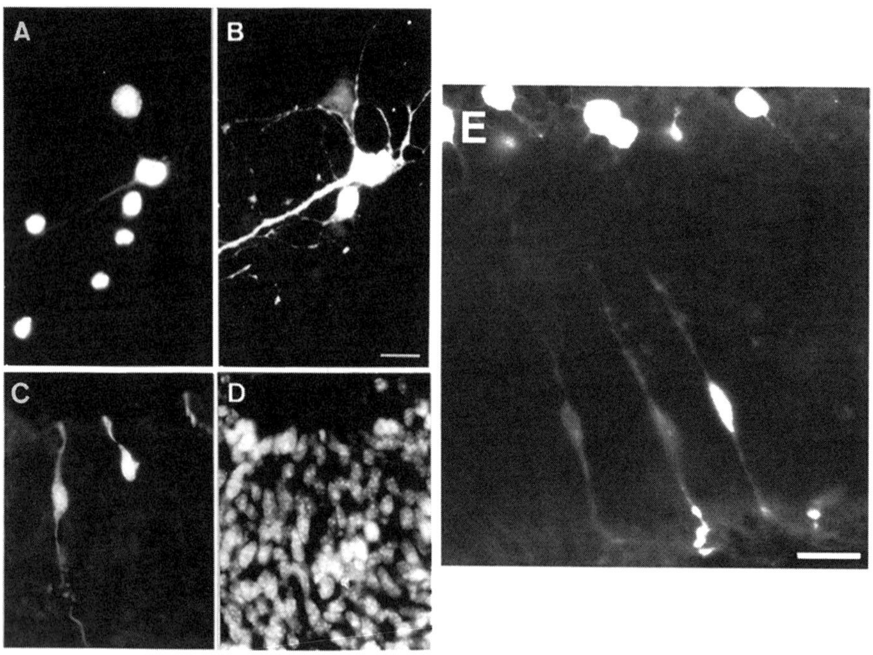

Figure 4. Expression of EGFP in cultured neurons and neuroepithelial cells infected with SFV-EGFP. (A and B) Rat hippocampal neurons (after 4 days in culture) growing on glass coverslips were infected for 1 hour with recombinant SFV-EGFP (at an moi of approximately 10), washed twice with HBSS, and cultured in conditioned medium (24). After 6 hours, the cells were fixed with 2% paraformaldehyde, quenched for 10 minutes in 50 mmol/L ammonium chloride, counterstained for 15 seconds with DAPI, and mounted with a drop of Mowiol. (A) For DAPI staining, the fluorescence in some of the neurites is due to cross talk between the blue and green channels. (B) EGFP fluorescence of two infected cells; note the intense fluorescence in the bigger neurites and that the protein is already filling the finest branches that lie on the neighboring cells. (C–E) Slices (200 µm) from the hemispheres of day 15 mouse embryos were incubated for 1 hour with the virus (1×10^5 iu/mL in neurobasal medium; see Chapter 16), washed once, and incubated in fresh medium at 36°C, 5% CO_2. After 20 hours, the slices were washed in PBS, fixed in 2% paraformaldehyde, embedded in sucrose, and snap-frozen in liquid nitrogen for cryosectioning (C–D). Sections (10 µm) were permeabilized with 0.5% Triton X-100 for 5 minutes, immunostained with polyclonal anti-EGFP and FITC-conjugated anti-rabbit antibodies (C), and counterstained with DAPI (D) before mounting with Mowiol. Alternatively, fixed slices were permeabilized with 0.5% Triton X-100 for 10 minutes, immunostained as in panel C, and whole-mounted with Mowiol (E). The EGFP-expressing neuroepithelial cells (C and E) exhibit a typical bipolar morphology with one process toward the luminal side (top side in C–E) and the other extending into deeper layers of the neuroepithelium to reach the basal lamina, which is not visible here. Panel C and D micrographs were made in a conventional fluorescence microscope (Axiovert; Carl Zeiss, Thornwood, NY, USA). The image in panel E was obtained in a confocal microscope (TCS; Carl Zeiss) and resulted from the integration of ten consecutive images. Bar, (A and B) 10 µm, (C and D) 20 µm, and (E) 25 µm.

SFV Infection of Hippocampal Neurons in Culture

To establish optimal conditions for the expression of recombinant protein in cultured neurons with SFV vectors, we have used the SFV-EGFP reporter virus to infect primary cultures of neurons established from the hippocampi of day 18 rat embryos (24). Time-course analyses of neurons infected with recombinant SFV-EGFP at an moi of 50 to 100 showed visible levels of reporter in the cell bodies after 1 hour of incubation (following 1 hour of virus adsorption; see protocol for infection of BHK cells above). At 3 hours, the protein was seen in all neuronal processes, and after 6 hours, even the finest and more distant processes showed intense fluorescence (Figure 4B). Infections at lower moi (0.1–1) showed somewhat delayed expression of the reporter, suggesting that multiple simultaneous infection of a single cell provides higher levels of viral and heterologous proteins in a shorter time. However, in agreement with previous reports (38), the maximal levels of protein reached at long incubation times (48 hours) were comparable at any moi, at least in fluorescence intensity of EGFP. Some differences in infectivity of the recombinant viruses were also noticed according to the degree of maturity (and hence polarization and branching) of the neurons. Thus, at the same moi, lower numbers of young neurons (one day in culture) were infected with SFV-EGFP as compared to older neurons (four days in culture) (Reference 8, and A. Cid-Arregui, unpublished data). A plausible explanation is the larger surface of the latter, which makes them more accessible to the virus. However, differences in the expression of the receptor for the virus during neuronal differentiation cannot be ruled out.

Concerning the viability of the neurons infected with SFV vectors, there is some contradictory data. While hippocampal neurons in culture have been reported to be irreversibly damaged by recombinant virus in less than 24 hours (49), neurons infected in hippocampal slice cultures appeared healthy five days after infection (17). The discrepancy could be explained either by differences in the moi used in those studies or by a higher robustness of the neurons when they are integrated in a tissue architecture as in the slice cultures. It has also been suggested that neurotoxic products released by the BHK-21 cells might be present in the viral stocks (17). In our studies, we have used only sucrose gradient-purified virus and have observed that infected neurons survive longer when the infections are performed at lower moi (less than 1). In addition, neurons from older cultures (e.g., two weeks), with long neurites that run together with those of neighboring cells, were more resistant to infection and showed normal morphology and expression of neuronal markers (e.g., MAP2) three to four days after infection (data not shown).

SFV Infection of Neuroepithelial Cells

We have also used SFV-EGFP to infect neuroepithelial cells in slices obtained from the brains of mouse embryos as described in Chapter 16. Although recombinant alphaviruses have been shown recently to infect differentiated neurons in hippocampal slices from postnatal rats (17,41), these vectors have never been reported to infect progenitor neuroepithelial cells in the developing nervous system. Figure 4 shows neuroepithelial cells infected with SFV-EGFP in in vitro-maintained brain slices from day 15 mouse embryos. The neuroepithelial cells were infected from their luminal pole using purified virus stocks diluted to 1×10^5 iu/mL. The EGFP was visible in the fine processes distant from the lumen already 5 hours after infection (the earliest time point analyzed). The infected neuroepithelial cells remained morphologically normal for at least 24 hours (the longest time analyzed), that is, as bipolar cells with a process toward the lumen and a cytoplasmic extension toward the basal lamina.

Application of SFV Vectors for Antibody Production and Vaccine Development

Alphavirus vectors can be engineered as genetic vaccines. Cellular and humoral responses have been shown against recombinant proteins expressed in mice after infection with recombinant alphaviruses (10,28,67,79). These studies showed that a single boost in mice with recombinant virus expressing heterologous antigen from a pathogen (e.g., influenza virus) elicits cellular and humoral immune responses that protect against a challenge with a virulent strain of that pathogen (67,79). In a different approach, London et al. (40) fused a heterologous epitope from Rift Valley fever virus (RVFV) to the structural proteins of SIN virus. Mice immunized with this recombinant virus became resistant to RVFV. These data illustrate the potential utility of alphaviruses as tools for the development of recombinant vaccines, although safety issues still need to be addressed.

We are currently testing the efficiency of antibody production in mice inoculated with recombinant SFV vectors that express heterologous proteins (e.g., EGFP). BALB/c mice were inoculated by various routes (e.g., intramuscular, intraperitoneal, subcutaneous, and oral) with different amounts of recombinant virus in the range of 1×10^4 to 1×10^5 iu. Infection did not cause any evident pathology to any of the mice tested, which were observed up to six months after the first challenge, although some animals were reboosted twice. Since high antibody titers were expected, the sera were analyzed by Western blot hybridization using protein extracts from transfected cells that express the proteins encoded by the recombinant vectors inoculated into the animals. In general, a single boost did not suffice to induce a humoral response that could be detected by this method. Figure 5 shows the presence of antibodies against EGFP in the serum of a mouse immunized twice with recombinant SFV-GFP. Our preliminary results indicate that SFV vectors can be useful for the quick

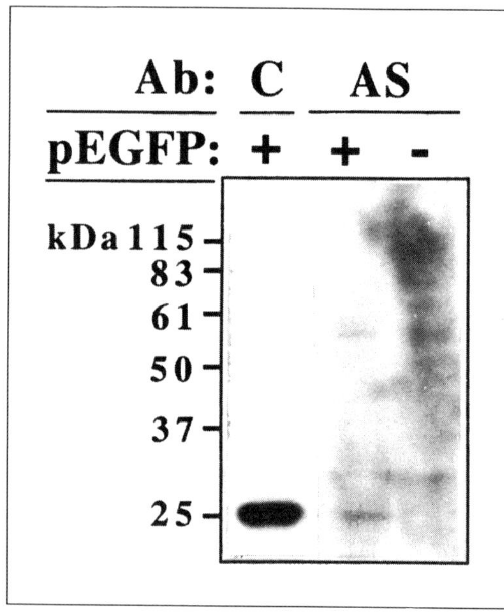

Figure 5. Anti-EGFP antibodies detected in the serum of a mouse immunized with SFV-GFP. The mice (BALB/c) were injected twice intraperitoneally with 1×10^6 iu of SFV-GFP with an interval of 10 days. One week after the last immunization, a blood sample (30 μL) was taken, and the serum was used for Western blot hybridization against total extracts from Vero 2-2 cells transfected with a plasmid expressing EGFP (lanes C+ and AS+) and from nontransfected cells (lane AS-). To reduce background caused by the secondary antibody, the serum was diluted 1:50 in PBS/0.2% Tween® 20/5% milk powder and incubated before hybridization with peroxidase-conjugated goat anti-mouse antibodies (diluted 1:50) for 30 min at 37°C, followed by 1 hour at room temperature. This premixture was diluted 1:10 in PBS/0.2% Tween 20/5% milk powder for hybridization (lanes AS+ and AS-). Lane C+ was hybridized with rabbit polyclonal anti-EGFP antibodies, detected subsequently by peroxidase-conjugated anti-rabbit antibodies. Finally, the peroxidase activity was demonstrated by chemiluminiscence. The 26-kDa band in the control C+ corresponds to EGFP and is visible also in AS+, but not in the negative control (AS-) of nontransfected cells. On the left, molecular weights of the prestained marker proteins (Bio-Rad Laboratories).

production of antibodies in mice. In addition, immunization tests in mice with recombinant alphavirus vectors that express either polypeptides or entire proteins could be helpful in determining their antigenic capacity before using them for vaccine development.

ACKNOWLEDGMENTS

We wish to thank H. Garoff for kindly providing the SFV plasmids and C.G. Dotti for helpful discussions and advice on the culture of hippocampal neurons. This work was supported by the Mexican Comisión Nacional de Ciencia y Tecnología and the German Bundesministerium für Bildung und Forschung. A. Cid-Arregui thanks W.B. Huttner and H. zur Hausen for support.

REFERENCES

1. **Berben-Bloemheuvel, G., M.A. Kasperaitis, H. van Heugten, A.A. Thomas, H. van Steeg, and H.O. Voorma.** 1992. Interaction of initiation factors with the cap structure of chimaeric mRNA containing the 5′-untranslated regions of Semliki Forest virus RNA is related to translational efficiency. Eur. J. Biochem. *208*:581-587.
2. **Berglund, P., M. Sjöberg, H. Garoff, G.J. Atkins, B.J. Sheahan, and P. Liljeström.** 1993. Semliki Forest virus expression system: production of conditionally infectious recombinant particles. Biotechnology *11*:916-920.
3. **Bottenstein, J.E. and G.H. Sato.** 1979. Growth of a neuroblastoma cell line in serum-free supplemented medium. Proc. Natl. Acad. Sci. USA *76*:719-731.
4. **Bradke, F. and G.C. Dotti.** 1998. Videomicroscopy of living microinjected hippocampal neurons in culture, p. 81-94. *In* A. Cid-Arregui and A. Garcia-Carranca (Eds.), Microinjection and Transgenesis: Strategies and Protocols. Springer-Verlag, Berlin.
5. **Bredenbeek, P.J., I. Frolov, C.M. Rice, and S. Schlesinger.** 1993. Sindbis virus expression vectors: packaging of RNA replicons by using defective helper RNAs. J. Virol. *67*:6439-6446.
6. **Chalfie, M., Y. Tu, E.G Euskirchen, W.W. Ward, and D.C. Prasher.** 1994. Green fluorescent protein as a marker for gene expression. Science *263*:802-805.
7. **Chambers, T.J., C.S. Hahn, R. Galler, and C.M. Rice.** 1990. Flavivirus genome organization, expression, and replication. Annu. Rev. Microbiol. *44*:649-688.
8. **Cid-Arregui, A., R.G. Parton, K. Simons, and C.G. Dotti.** 1995. Nocodazole-dependent transport, and brefeldin A—sensitive processing and sorting, of newly synthesized membrane proteins in cultured neurons. J. Neurosci. *15*:4259-4269.
9. **Cormack, B.P., R. Valdivia, and S. Falkow.** 1996. FACS-optimized mutants of the green fluorescent protein (GFP). Gene *173*:33-38.
10. **Davis, N.L., K.W. Brown, and R.E. Johnston.** 1996. A viral vaccine vector that express foreign genes in lymph nodes and protects against mucosal challenge. J. Virol. *70*:3781-3787.
11. **Davis, N.L., L.V. Willis, J.F. Smith, and R.E. Johnston.** 1989. In vitro synthesis of infectious venezuelan equine encephalitis virus RNA from a cDNA clone: analysis of a viable deletion mutant. Virology *171*:189-204.
12. **de Curtis, I. and K. Simons.** 1988. Dissection of Semliki forest virus glycoprotein delivery from the trans-Golgi network to the cell surface in permeabilized BHK cells. Proc. Natl. Acad. Sci. USA *85*:8052-8056.
13. **de Hoop, M.J. and A. Cid-Arregui.** 1996. Use of fowl plague, vesicular stomatitis and Semliki forest viruses to study the mechanisms of neuronal membrane polarity. *In* P.R. Lowenstein and L.W. Enquist (Eds.), Protocols for Gene Transfer in Neuroscience: Towards Gene Therapy of Neurological Disorders. John Wiley & Sons Ltd., Chichester, UK.
14. **De Strooper, B., W. Annaert, P. Cupers, P. Saftig, K. Craessaerts, J.S. Mumm, E.H. Schroeter, V. Schrijvers, M.S. Wolfe, W.J. Ray, et al.** 1999. A presenilin-1-dependent γ-secretase-like protease mediates release of notch intracellular domain. Nature *398*:518-522.
15. **Ding, M. and M.J. Schlesinger.** 1989. Evidence that Sindbis virus nsP2 is an autoprotease which processes the virus nonstructural polyprotein. Virology *171*:280-284.
16. **Dotti, C.G. and K. Simmons.** 1990. Polarized sorting of viral glycoproteins to the axon and dendrites of hippocampal neurons in culture. Cell *62*:63-72.
17. **Ehrengruber, M.U., K. Lundstrom, C. Schweitzer, C. Heuss, S. Schlessinger, and B.H. Gähwiler.** 1999. Recombinant Semliki Forest virus and Sindbis virus efficiently infect neurons in hippocampal slice cultures. Proc. Natl. Acad. Sci. USA *96*:7041-7046.
18. **Frolova, E., I. Frolov, and S. Schlesinger.** 1997. Packaging signals in alphaviruses. J. Virol. *71*:248-258.
19. **Garoff, H., A.M. Frischauf, K. Simons, H. Lehrach, and H. Delius.** 1980. Nucleotide sequence of cDNA coding for Semliki Forest virus membrane glycoproteins. Nature *288*:236-241.
20. **Garoff, H., C. Kondor-Koch, and H. Riedel.** 1982. Structure and assembly of alphaviruses. Curr. Top. Microbiol. Immunol. *99*:1-50.
21. **Garoff, H. and K.J. Li.** 1998. Recent advances in gene expression using alphavirus vectors. Curr. Opin. Biotechnol. *9*:464-469.
22. **Garoff, H., K. Simons, and B. Dobberstein.** 1978. Assembly of Semliki forest virus membrane glycoproteins in the membrane of the endoplasmic reticulum in vitro. J. Mol. Biol. *124*:587-600.
23. **Geigenmüller-Gnirke, U., B. Weiss, R. Wright, and**

S. Schlesinger. 1991. Complementation between Sindbis viral RNAs produces infectious particles with a bipartite genome. Proc. Natl. Acad. Sci. USA *88*:3253-3257.

24.**Goslin, K. and G. Banker.** 1991. Rat hippocampal neurons in low density cultures, p. 251-281. *In* G. Banker and K. Goslin (Eds.), Culturing Nerve Cells. MIT Press, Cambridge, MA.

25.**Grakoui, A., R. Levis, R. Raju, H.V. Huang, and C.M. Rice.** 1989. A cis-acting mutation in the Sindbis virus junction region which affects subgenomic RNA synthesis. J. Virol. *63*:5216-5227.

26.**Gwag, B.J., E.Y. Kim, B.R. Ryu, S.J. Won, H.W. Ko, Y.J. Oh, Y.G. Cho, S.J. Ha, and Y.C. Sung.** 1998. A neuron-specific gene transfer by a recombinant defective Sindbis virus. Brain Res. Mol. Brain Res. *63*:53-61.

27.**Haas, J., E.-C. Park, and B. Seed.** 1996. Codon usage limitation in the expression of HIV-1 envelope glycoprotein. Curr. Biol. *6*:315-324.

28.**Hahn, C.S., Y.S. Hahn, T.J. Braciale, and C.M. Rice.** 1992. Infectious Sindbis virus transient expression vectors for studying antigen processing and presentation. Proc. Natl. Acad. Sci. USA *89*:2679-2683.

29.**Hahn, C.S. and J.H. Strauss.** 1990. Site-directed mutagenesis of the proposed catalytic amino acids of the Sindbis virus capsid protein autoprotease. J. Virol. *64*:3069-3073.

30.**Helenius, A., B. Morein, E. Fries, K. Simons, P. Robinson, V. Schirrmacher, C. Terhorst, and J.L. Strominger.** 1978. Human (HLA-A and HLA-B) and murine (H-2K and H-2D) histocompatibility antigens are cell surface receptors for Semliki Forest virus. Proc. Natl. Acad. Sci. USA *75*:3846-3850.

31.**Huang, H.V., C.M. Rice, C. Xiong, and S. Schlesinger.** 1989. RNA viruses as gene expression vectors. Virus Genes *3*:85-91.

32.**Kohno, A., N. Emi, M. Kasai, M. Tanimoto, and H. Saito.** 1998. Semliki Forest virus-based DNA expression vector: transient protein production followed by cell death. Gene Ther. *5*:415-418.

33.**Kuhn, R.J., Z. Hong, and J.H. Strauss.** 1990. Mutagenesis of the 3′ nontranslated region of Sindbis virus RNA. J. Virol. *64*:1465-1476.

34.**Lemm, J.A., T. Rumenapf, E.G. Strauss, J.H. Strauss, and C.M. Rice.** 1994. Polypeptide requirements for assembly of functional Sindbis virus replication complexes: a model for the temporal regulation of minus and plus strands. EMBO J. *13*:2925-2934.

35.**Levis, R., S. Schlesinger, and H.V. Huang.** 1986. Deletion mapping of Sindbis virus DI RNAs derived from cDNAs defines the sequences essential for replication and packaging. Cell *44*:137–145.

36.**Levis, R., S. Schlesinger, and H.V. Huang.** 1990. Promoter for Sindbis virus RNA-dependent subgenomic RNA transcription. J. Virol. *64*:1726-1733.

37.**Liljeström, P. and H. Garoff.** 1991. Internally located cleavable signal sequences direct the formation of Semliki Forest virus membrane proteins from a polyprotein precursor. J. Virol. *65*:147–154.

38.**Liljeström, P. and H. Garoff.** 1991. A new generation of animal cell expression vectors based on the Semliki Forest virus replicon. Biotechnology *9*:1356-1361.

39.**Liljeström, P., S. Lusa, D. Huylebroeck, and H. Garoff.** 1991. In vitro mutagenesis of a full-length cDNA clone of Semliki Forest virus: the small 6,000-molecular-weight membrane protein modulates virus release. J. Virol. *65*:4107-4113.

40.**London, S.D., A.L. Schmaljohn, J.M. Dalrymple, and C.M. Rice.** 1992. Infectious enveloped RNA virus antigenic chimeras. Proc. Natl. Acad. Sci. USA *89*:207-211.

41.**Maletic-Savatic, M., R. Malinow, and K. Svoboda.** 1999. Rapid dendritic morphogenesis in CA1 hippocampal dendrites induced by synaptic activity [see comments]. Science *283*:1923-1927.

42.**Mathiot, C.C., G. Grimaud, P. Garry, J.C. Bouquety, A. Mada, A.M. Daguisy, and A.J. Georges.** 1990. An outbreak of human Semliki Forest virus infections in Central African Republic. Am. J. Trop. Med. Hyg. *42*:386-393.

43.**Mayer, M. and M. Noble. 1994.** N-acetyl-cystein is a pluripotent protector against cell dead and enhancer of tropic factor-mediated cell survival in vitro. Proc. Natl. Acad. Sci. USA *91*:7496-7500.

44.**Melançon, P. and H. Garoff.** 1986. Reinitiation of translocation in the Semliki forest virus structural polyprotein: identification of the signal for the E1 glycoprotein. EMBO J. *5*:1551-1560.

45.**Metsikko, K. and H. Garoff.** 1990. Oligomers of the cytoplasmic domain of the p62/E2 membrane protein of Semliki Forest virus bind to the nucleocapsid in vitro. J. Virol. *64*:4678-4683.

46.**Meyer, W.J., S. Gidwitz, V.K. Ayers, R.J. Schoep, and R.E. Johnston.** 1992. Conformational alteration of Sindbis virion glycoproteins induced by heat, reducing agents or low pH. J. Virol. *66*:3504-3513.

47.**Niesters, H.G.M. and J.H. Strauss.** 1990. Mutagenesis of the conserved 51-nucleotide region of Sindbis virus. J. Virol. *64*:1639-1647.

48.**Oldstone, M.B., A. Tishon, F.J. Dutko, S.I. Kennedy, J.J. Holland, and P.W. Lampert.** 1980. Does the major histocompatibility complex serve as a specific receptor for Semliki Forest virus? J. Virol. *34*:256-265.

49.**Olkkonen, V.M., P. Liljeström, H. Garoff, K. Simons, and C.G. Dotti.** 1993. Expression of heterologous proteins in cultured rat hippocampal neurons using the Semliki Forest virus vector. J. Neurosci. Res. *35*:445-451.

50.**Prasher, D.C., V.K. Eckenrode, W.W. Ward, F.G. Prendgast, and M.J. Cormier.** 1992. Primary structure of the *Aequorea victoria* green-fluorescent protein. Gene *111*:229-233.

51.**Raju, R. and H.V. Huang.** 1991. Analysis of Sindbis virus promoter recognition in vivo using novel vectors with two subgenomic mRNA promoters. J. Virol. *65*:2531-2540.

52.**Rice, C.M.** 1992. Examples of expression systems based on animal RNA viruses: alphaviruses and influenza virus. Curr. Opin. Biotechnol. *3*:522-532.

53.**Rice, C.M., R. Levis, J.H. Strauss, and H.V. Huang.** 1987. Production of infectious RNA transcripts from Sindbis virus cDNA clones: mapping of lethal mutations, rescue of a temperature-sensitive marker, and in vitro mutagenesis to generate defined mutants. J.

Virol. *61*:3809-3819.

54. **Rodriguez-Boulan, E.J. and D.D. Sabatini.** 1978. Asymmetric budding of viruses in epithelial monolayers: a model system for the study of epithelial polarity. Proc. Natl. Acad. Sci. USA *75*:5071-5075.

55. **Sanford, J.P.** 1998. Arbovirus Infections, 14th ed. McGraw-Hill, Hamburg.

56. **Schlesinger, M.J.** 1994. Sindbis and Semliki Forest Viruses, p. 1330-1333. *In* R.G. Webster and A. Granoff (Eds.), Encyclopedia of Virology. Academic Press, London.

57. **Schlesinger, S.** 1993. Alphaviruses—vectors for the expression of heterologous genes. Trends Biotechnol. *11*:18-22.

58. **Schlesinger, S. and M.J. Schlesinger.** 1996. Togaviridae: the viruses and their replication, p. 523-539. *In* B.N. Fields, D.M. Knipe, P.M. Howley, et al. (Eds.), Fundamental Virology, 3rd. ed. Lippincott-Raven Publishers, New York.

59. **Singh, I. and A. Helenius.** 1992. Role of ribosomes in Semliki forest virus nucleocapsid uncoating. J. Virol. *66*:7049-7058.

60. **Sjöberg, E.M., M. Suomalainen, and H. Garoff.** 1994. A significantly improved Semliki forest virus expression system based on translation enhancer segments from the viral capsid gene. Biotechnology *12*:1127-1131.

61. **Smerdou, C. and P. Liljeström.** 1999. Two-helper RNA system for production of recombinant Semliki forest virus particles. J. Virol. *73*:1092-1098.

62. **Strauss, E.G., R.J. De Groot, R. Levinson, and J.H. Strauss.** 1992. Identification of the active site residues in the NSP2 proteinases of Sindbis virus. Virology *191*:932-940.

63. **Strauss, E.G. and J.H. Strauss.** 1986. Structure and replication of the alphavirus genome, p. 35-82. *In* S. Schlesinger and M.J. Schlesinger (Eds.), The Togaviridae and Flaviviridae. Plenum Press, New York.

64. **Strauss, J.H. and E.G. Strauss.** 1988. Evolution of RNA viruses. Ann. Rev. Microbiol. *42*:657-683.

65. **Tienari, P.J., B. De Strooper, E. Ikonen, M. Simons, A. Weidemann, C. Czech, T. Hartmann, N. Ida, G. Multhaup, C.L. Masters, F. Van Leuven, K. Beyreuther, and C.G. Dotti.** 1996. The beta-amyloid domain is essential for axonal sorting of amyloid precursor protein. EMBO J. *15*:5218-5229.

66. **Tsiang, M., B.G. Weiss, and S. Schlesinger.** 1988. Effects of 5′-terminal modifications on the biological activity of defective interfering RNAs of Sindbis virus. J. Virol. *62*:47-53.

67. **Tsuji, M., C.C. Bergmann, Y. Takita-Sonoda, K. Murata, E.G. Rodrigues, R.S. Nussenzweig, and F. Zavala.** 1998. Recombinant Sindbis viruses expressing a cytotoxic T-lymphocyte epitope of a malaria parasite or of influenza virus elicit protection against the corresponding pathogen in mice. J. Virol. *72*:6907-6910.

68. **Ubol, S. and D. Griffin.** 1991. identification of a putative alphavirus receptor on mouse neural cells. J. Virol. *62*:47-53.

69. **Ulmanen, I., H. Söderlund, and L. Kääriäinen.** 1976. Semliki Forest virus capsid protein associates with the 60S ribosomal subunit in infected cells. J. Virol. *20*:203-210.

70. **Wahlberg, J.M. and H. Garoff.** 1992. Membrane fusion process of Semliki Forest virus I: low pH-induced rearrangements in spike protein quaternary structure preceeds virus penetration into cells. J. Cell Biol. *116*:339-357.

71. **Wang, K.-S., R.J. Kuhn, and E.G. Strauss.** 1992. High-affinity laminin receptor is a receptor for Sindbis virus in mammalian cells. J. Virol. *66*:4992-5001.

72. **Watson, D.G., J.M. Moehring, and T.J. Moehring.** 1991. A mutant CHO-K1 strain with resistance to Pseudomonas exotoxin A and alphaviruses fails to cleave Sindbis virus glycoprotein PE2. J. Virol. *65*:2332-2339.

73. **Weiss, B., U. Geigenmüller-Gnirke, and S. Schlesinger.** 1994. Interactions between Sindbis virus RNAs and a 68 amino acid derivative of the viral capsid protein further defines the capsid binding site. Nucleic Acids Res. *22*:780-786.

74. **Weiss, B.G. and S. Schlesinger.** 1991. Recombination between Sindbis virus RNAs. J. Virol. *65*:4017-4025.

75. **Wengler, G.** 1980. Effects of alphaviruses on host cell macromolecular synthesis, p. 459-472. *In* R.W. Schlesinger (Ed.), The Togaviruses. Academic Press, New York.

76. **Wengler, G., U. Boehge, G. Wengler, H. Bischoff, and K. Wahn.** 1982. The core protein of the alphavirus Sindbis virus assembles into core-like nucleoproteins with the viral genome RNA and with other single-stranded nucleic acids in vitro. Virology *118*: 401-410.

77. **Wengler, G., D. Würkner, and G. Wengler.** 1992. Identification of a sequence element in the alphavirus core protein which mediates interaction of cores with ribosomes and the disassembly of cores. Virology *191*:880-888.

78. **Xiong, C., R. Levis, P. Shen, S. Schlesinger, C.M. Rice, and H.V. Huang.** 1989. Sindbis virus: an efficient, broad host range vector for gene expression in animal cells. Science *243*:1188-1191.

79. **Zhou, X., P. Berglund, H. Zhao, P. Liljeström, and M. Jondal.** 1995. Generation of cytotoxic and humoral immune responses by nonreplicative recombinant Semliki Forest virus. Proc. Natl. Acad. Sci. USA *92*:3009-3013.

Part III: General Issues

36 Gene Therapy Using Viral Vectors: Strategy and Design Issues

Pedro A. Lazo and Mónica Yunta
Cancer Research Center, CSIC–University of Salamanca, and Carlos III Institute of Health, Majadahonda, Madrid, Spain

INTRODUCTION

Gene therapy (GT) can be considered an attempt to treat diseases at the molecular level by transducing appropriate genes into selected target cells. Thus, the aim is to provide functions that are either missing because of a genetic defect or required to modify the pathogenic mechanisms of a disease. Ultimately, GT pursues a genetic correction of cellular defects in the context of highly complex physiological conditions like those of the human organism. This implies important, yet unresolved problems that have led to the present apparent stagnation of GT in spite of major progress made lately in some fields (40), such as gene transfer technology or the transduction of cells and tissues ex vivo and in animal models. As a consequence, the application of GT has evolved towards more realistic goals than those initially intended. Thus, there has been a slowdown in studies on germ line GT, namely hereditary monogenic disorders, and a shift of focus towards somatic GT of high-prevalence diseases, such as cancer, cardiovascular pathologies, and diabetes.

GT studies have to take into account many different issues (see Table 1), ranging from the medical aspects of a specific disease to the basic laboratory work concerning the appropriate gene constructs and delivery systems. Among these, gene transfer technologies play a central role in the application of GT strategies, and viruses are getting more and more attention. Viral infections represent a natural mechanism by which foreign genetic information (the viral genes) is expressed in human cells. Therefore, viruses are the obvious choice for expression of therapeutic gene constructs (29). The viruses most commonly used in GT protocols are outlined in Table 2. All of them have limitations and, therefore, the present attempts to manipulate their properties to make the viruses more suitable in order to achieve a therapeutic goal are logical (25,28,43). However, it is a very different thing to transfer genes to carry out a viral life cycle than to transfer a foreign gene to correct a genetic defect or give a new property to the host cell, either

Viral Vectors: Basic Science and Gene Therapy
Edited by A. Cid-Arregui and A. García-Carrancá

Table 1. Factors Affecting the Selection of Vectors for Gene Therapy Design

Type of Genetic Modification
- Correction of genetic defect
- Supply of new function

Selection
- Tracer gene (e.g., *GFP*)
- New function (e.g., *neo*^{R})

Gene Construct: Structure and Expression
- Size of gene
  - Small: cDNA
  - Large: Genomic: complete or partial gene
- Regulation of Gene Expression
  - Level and persistence of expression
  - Regulation: Intracellular, paracrine, endocrine

Gene Transfer Method
- Viral: Several types
- Nonviral: liposomes, mammalian artificial chromosomes, gene gun, etc.

Cellular Target
- Ex vivo: modification in tissue culture, cell therapy
- In vivo: Animal models for specific diseases; humans

in vivo or ex vivo (cell culture). The introduction of a therapeutic gene in all cells of a tissue in vivo is not a realistic approach. However, in some cases, this problem can be overcome by ex vivo gene transfer.

THE THERAPEUTIC GENE

Mutation and Disease Phenotype: Correction of Monogenic Hereditary Diseases

Single-gene defects are the underlying cause of human hereditary monogenic diseases, which constitute a large and very heterogeneous group of potential targets for GT protocols. In the design of a GT protocol, the characteristics of the specific disease selected for treatment must be known at the genetic, cellular, and organism levels. The first thing to determine is whether the phenotype of the defective gene is displayed in a unique cell or several cell types, and whether the clinical manifestations have systemic consequences. In some cases, such as in muscular dystrophies, both a mutation and its disease phenotype coincide and are restricted to a unique cell type. Another important case, because of its high frequency, is exemplified by cystic fibrosis (CF) (4). The CF phenotype is mainly displayed in the respiratory airways, although there is a high heterogeneity in the cells displaying the CF phenotype (15). These cells are the primary target for GT, and not the epithelial cells of other organs, which also contain the mutation, but do not display any phenotype.

A very different situation is illustrated by monogenic diseases whose defect in expression occurs in a single organ where it presents no problem, but its clinical manifestation occurs at the systemic level. This is the case in deficiencies of secreted proteins. In hemophilia, the lack of production of normal clotting factors is restricted to liver cells, but because these proteins are secreted, the clinical phenotype is systemic. In hemophilia, the treatment can be performed from a different cell type in so far as secretion and plasma levels of the desired protein are obtained (19).

There is an additional level of complexity in those cases when we have to consider the type of gene regulation and its physiological consequences. The damage may be restricted to a single cell type, but the gene regulation might have consequences at the organism level, and other organs suffer the consequences of its alteration. This complex situation is represented by diseases such as insulin-dependent diabetes mellitus. In diabetes, the lack of insulin production by pancreatic β cells has consequences

Table 2. Characteristics of Viral Vectors Most Commonly Used for Gene Therapy

Viral Vector	Advantages	Disadvantages
Retrovirus	Well-understood biology Stable integration in host cell Efficient cell entry No viral genes in vector	Low titer Infection restricted to dividing cells Gene expression difficult to control Expensive and complex to prepare Limited insert capacity
Pseudotyped retrovirus	Higher efficiency of infection Broader host cell range than retroviruses	Similar to retrovirus
Adenovirus	High titers Efficient entry in most cells High level of expression Infection of nondividing cells Large insert capacity	Vector with many viral genes Strong host T- and B-cell immune response Generation of viruses that are replication competent Cell tropism not well understood
Adeno-associated virus	DNA integration at specific, non-pathogenic, chromosome sites Targets nonreplicating cells Nonpathogenic to humans Nonimmunogenic	Requires adenovirus replication to grow No helper cell line Limited insert size
Herpesvirus	High titers Neurotropic (cell specific)	Complex construction No packaging cell line Neurological problems
Vaccinia	Highly efficient gene transfer	Induces local and systemic immune responses
Fowlpox	High efficiency of transfer High level of expression High titers Large capacity	Highly immunogenic Transient gene expression

in the organism because of the fluctuations in the plasma levels of glucose and the inability to respond to them. In this disease, the therapy should attempt to achieve a regulated expression and secretion of insulin that should mimic the physiological response needed to maintain the glucose concentration. However, production of recombinant insulin does not necessarily need to be directed to the damaged β cells, which can be substituted by other cell types, such as liver cells, manipulated to produce insulin (9).

Structure and Expression of the Therapeutic Gene

The size of human genes is frequently very large and, therefore, not suitable for cloning in most current gene transfer systems. This size problem is an issue when using viral vectors because their capacity is limited by their need to package the DNA construct within the viral particle. Vector capacity can range from a few kilobases (4–8 kb) in adeno-associated viruses (AAVs) and retroviruses to very large constructs (20 kb Ads and vaccinia vectors). To overcome this problem, cDNAs or fragments of a gene are most often used in GT protocols. A drawback of this size restriction is the potential loss of regulatory sequences that are located within introns. But the relevance of this problem will be determined by the nature of the disease and the requirements for its particular solution.

In some cases, a gene fragment can be used, instead of the cDNA, provided that it can correct the genetic defect. This is the case in CF. The CF gene is very large, 230 kb with 27 exons, and is not suitable for cloning into viral vectors. However, the use of the conductance transmembrane region (CFTR [cystic fibrosis transmembrane conductance regulator]) of the gene can correct the defect of ion transport in epithelial cells. Consequently, most GT protocols for CF use the CFTR fragment (7,15).

In addition, the gene regulatory sequences of the virus might also interfere with the regulation of the therapeutic gene. Therefore, those constructs that only aim to achieve a level of expression high enough to correct a phenotype are more successful than those requiring a very fine regulation.

Physiological Aspects of Gene Therapy

Cells produce proteins either at constant or variable rates. Therefore, in any GT protocol, the physiological regulation and level of expression of the therapeutic gene is a major issue. The GT construct must reproduce the changes in the level of gene expression as a response to physiological demands; this approach is known as physiologically regulated gene therapy (PRGT). In this process of regulated gene expression, there are four general aspects that have to be the goal of any GT protocol to be effective and that have to be engineered in the design of the recombinant viral DNA construct (39). First, the correction has to take place in the specific cell type where the genetic defects display its phenotype, and not in other organs. Second, the gene expression of the construct has to respond to the specific stimulus in the correct target cell type. Third, the magnitude of the response should reach a level of expression similar to that occurring in the normal cell regulation; otherwise, if it is too low, it might have no effect, or if overexpressed, might be dangerous, as is the case in an overproduction of insulin. Fourth, the timing of the response has to follow a temporal pattern similar to that occurring in normal physiological conditions. The basal GT gene expression should be subphysiological to avoid undesirable effects, and the therapeutic levels should only be achieved once a suitable expression is produced as a specific response to a physiological demand. All of these parameters can be affected by the number of transfected cells, blood supply, rate of clearance, or catabolism of the recombinant protein. Furthermore, PRGT shares with all other expression systems the requirement that the protein must be processed and posttranlationally modified correctly in transfected cells.

To fulfill all the characteristics of PRGT protocols, the type of promoter/enhancer system used should combine several features. It must contain a response element to the specific stimulus; the strength of the promoter should result in a level of expression similar to the normal situation; in those cases where the targeted cell is different from the normal one, the construct also has to be adapted to function in the new cell. For example, in liver production of insulin, the construct must have elements that respond to glucose levels, and it should not be affected by the regulatory elements that will make the gene functional in liver cells (9,23). Therefore, the complexity of the construction ranges from simple gene promoter/enhancer regions to complex transactivator-dependent systems, which may be partly synthetic. All the principles required for the development of a PRGT approach have been developed using the inflammation-inducible promoter of the C3 complement gene as the experimental model (38).

VIRAL VECTORS IN GENE THERAPY

Viral Tissue Specificity and Cell Targeting

The cell-type specificity of viruses is a major limitation in their use as therapeutic agents, unless there is a concordance between the cell that manifests the pathology and the viral target. There are three factors that determine which cells are targeted by a given viral vector: *(i)* the distribution of specific cellular receptors for the virus, *(ii)* the viral dependence on cellular replication, and *(iii)* the cell specificity of viral promoters.

The cellular membrane proteins determine the interaction with viral particles and function as their receptors. Among them are the human fibroblast growth factor receptor 1 (GFR1), identified as a co-receptor for AAV (27), and certain integrins that also participate in viral entry as co-receptors for adenoviruses (Ads) (20) and AAV (36). In the case of Ad, the modification of its integrin-binding region can expand its infectivity for nonpermissive cells (14). Further knowledge of cellular co-receptors and the mechanism of viral entry is needed and will be of great help in the design of smart viruses to be used as therapeutical agents. Early approaches were directed to modify the cell-type specificity of retroviruses by replacing their envelope protein with envelope proteins of other viruses, such as vesicular stomatitis virus (VSV) (5) or the human immunodeficiency virus type 1 (HIV-1) (11). Pseudotyped retroviruses (1,2,11) have a different target cell repertory because their receptors are different.

The permissibility of a cell to carry out the viral life cycle also influences the specificity of vector targeting to some cells. Retroviruses depend on cellular replication for their life cycle, thus limiting their target cells to those actively replicating. Nondividing cells can be infected by viruses that have viral replication machinery, such as those of the herpesvirus family (e.g., HSV-1, cytomegalovirus, and Epstein-Barr virus), and are therefore suitable for infecting quiescent cell types. The target specificity of these latter viruses is further affected by other regulatory sequences that determine their suitability for one type of nonreplicating cell, but not others. This is exemplified by HSV-1, which infects neural, but rarely epithelial cells.

The third cellular element that conditions viral specificity is the use of viral regulatory sequences that are cell-type specific. In practice, the three elements are often associated. This situation is best illustrated by retroviruses, which only infect replicating cells and use cellular receptors present on lymphoid cells. In contrast, Ads can replicate in the absence of cellular replication and target epithelial cells. Therefore, these two viruses are used in GT protocols to target different organs.

Future development of new vectors implies the generation of chimeric viral vectors that combine elements of different viruses, which will make them better suited to each therapeutic strategy (28). A last and very important issue in the development of GT protocols is the availability of suitable animal models in which the GT treatment can be tested in an in vivo context. The importance of such models has been recently demonstrated by the correction of canine hemophilia by gene transfer of factor IX using AAV-based vectors (13,34).

Problems with Using Viruses for Gene Therapy

The use of genetically modified viruses for GT might present several problems, but their relative importance will vary depending on the type of virus used and the disease being treated. There are three major common problems related to the use of viral vectors for GT.

1. Integration of vector DNA into the host cell genome may occur. Viral DNA integration forms part of the viral life cycle of retroviruses and AAV. It might also occur with other viruses as a result of illegitimate DNA recombination. DNA integration in the host genome is a potentially oncogenic mutation, and for that reason it is not considered a suitable event in GT protocols when the aim of GT is to achieve a long-lasting effect in the treated individual. In the case of retroviruses, DNA integration is part of the life cycle; there is little control over where DNA integrates, and it has been clearly linked to oncogene activation (37). Therefore, in protocols using retroviruses, replication-incompetent viruses have to be engineered so that there is neither de novo retrovirus production nor the possibility of complementation that may facilitate viral dissemination. Viral DNA integration also occurs with AAV, but in this case integration occurs at a specific location on chromosome 19 where there appears to be no pathology associated with it (32).

2. The host humoral and cellular immune responses will cause the inactivation of the viral vector if several cycles of treatment are necessary (16). This problem is particularly important with large viruses, such as Ad and vaccinia vectors because they contain many viral immunogenic proteins. Other vectors such as herpes simplex virus (HSV) can trigger a local inflammatory reaction that might be of consequence if the inflammation occurs in the nervous system (8). To overcome this problem, different genetic manipulations of the viral DNA are performed to minimize their antigenic potential. Small viral vectors, such as those based on retroviruses, lentiviruses, or AAV, lack immunogenic viral genes and are unlikely to trigger a significant immune response.

3. The expression of the therapeutic gene from the viral vector may be transitory or long lasting. The duration of gene expression is an important factor in GT protocols since a construct with an initial good level of expression might become transcriptionally silent with time, and the correction might be only temporary. In many pathologies, there is a need for correction of the defect in the long run, as is the case of hereditary monogenic diseases.

Again, chimeric vectors, combining different properties of various types of virus, will help minimize all these potential problems (28).

MANIPULATION OF THE VIRAL LIFE CYCLE FOR GT STRATEGIES

In the last few years, there has been a shift in the strategies used to treat cancer and other diseases by GT. The common underlying principle of these new protocols is the manipulation of the viral life cycle properties, particularly the cytolytic effect, and the use of these properties to eliminate tumor or HIV-infected cells. These new approaches are a consequence of the advances in the understanding of the molecular bases of virus-host cell interactions. This type of novel GT protocol has been developed using reoviruses, Ads, HSVs, and rhabdoviruses.

HSV as Therapeutic Agent in Brain Tumors

Neurotropic viruses such as HSV-1 can

theoretically be of use to selectively target tumors located within the nervous system. However, these viruses can also cause inflammatory reactions with severe neurological complications. Recently a modified HSV-1, G207, has been developed that contains mutations in two viral genes (22,44). The first engineered mutation in HSV affects both of its γ34.5 alleles, a gene needed for viral replication in the brain, which causes encephalitis. The second mutation was made within the viral ICP6 gene by inserting the *E. coli LacZ* gene. The *ICP6* gene is required for viral growth in nondividing cells and has no effect in dividing cells. Therefore, the double HSV mutant, G207, is capable of replicating only in dividing cells, such as tumor cells, and does not cause infection or damage to normal nondividing neural cells. G207 has been shown to be avirulent upon intracranial inoculation in mice and HSV sensitive in nonhuman primates (22). Because of these results, the effect of the G207 virus has been tested in human malignant glioma (22) and meningioma cells (44) that were implanted in nude mice. In both experiments, G207 reduced tumor growth in a dose-dependent manner, without causing any noticeable neurological dysfunction or pathological change in the surrounding normal neural tissue (44). These observations suggest that G207 virus might be of potential use for the treatment of malignant meningiomas that are resistant to current therapies.

Reovirus Infection to Eliminate *ras*-Activated Cells

Reovirus is a ubiquitous, nonenveloped, double-stranded RNA virus that affects mainly the upper respiratory airway and the gastrointestinal tract and produces only mild symptoms. To lyse a cell, human reovirus requires an activated *ras* signaling pathway (6), necessary for the activity of the viral kinase that regulates viral replication (35). Fibroblasts, which are resistant to infection by reovirus become sensitive to infection when transfected with growth factor receptors such as EGFR or ERBB2. Also, fibroblasts with mutations in *sos* or *ras* genes have an enhanced reovirus infectivity. Ras proteins are regulators of mitogenic signals that are altered by mutations or alterations in receptors upstream in the signaling pathway (18).

The dependence of the reovirus life cycle on the function of a signaling pathway that is active in many tumor cells makes the idea of using this virus as an oncolytic agent very attractive. To test this possibility, Coffey et al. transformed fibroblasts with v-*erbB2* to activate the *ras* pathway and injected them to SCID mice (6). Infection of these animals with reovirus reduced tumor formation by more than 80% as compared to noninfected animals (6). The same result was obtained with a glioblastoma cell line overexpressing platelet-derived growth factor receptor (PDGFR). In both cases the elimination of tumor cells was caused by viral infection. However, the SCID mouse model has the disadvantage that there is no immune response against the virus, and, therefore, the results might be biased. Similar types of experiments were performed in immunocompetent mice in which tumors were transplanted as allografts. In the transplanted animals, reovirus infection resolved 70% of the tumors, and formation of an antibody response against the reovirus was similar to control animals, thereby suggesting that the effect of the virus occurred despite the immune reaction of the host. In addition, these animals showed no tumor recurrence at least four months after reovirus infection. Previous exposure of the animals to the virus does not appear to restrict its ability to destroy the tumor cells.

Many human tumors have activated the *ras* signaling pathway by *ras* point mutations, such as in 90% of pancreatic cancer cases, 50% of sporadic colon and lung cancers, and 30% of other tumor types (41). In those tumors that do not have *ras* mutations, the tumors have this signaling pathway activated by alteration of receptor tyrosine kinase receptors, such as ERBB2, PDGF, or EGFR (6,35). These tumors are potential targets for a reovirus-based therapy. Of 25 human tumor cell lines tested, 20 were very susceptible to reovirus cytolytic infection, which correlated with an activated mitogen-activated protein kinase (MAPK) pathway (6).

Ad as a Selective Cytolytic Agent for Tumor Cells

Great effort has been invested in the use of Ad as a viral vector for GT (16,43). However, in the last few years, a new strategy has been developed that relies on the use of the Ad oncolytic properties as a therapy by itself. Some modified Ads preferentially display their cytolytic activity in tumor cells, apparently without damaging normal cells, as exemplified by Onyx-015 (3). The molecular basis for this effect appears to be the result of a deletion within the E1B viral gene, which codes for its 55-kDa protein. The normal 55-kDa protein interacts with the normal p53 protein and blocks p53-dependent transcription; this interaction is necessary for efficient viral replication. Replication of the mutant Ad only takes place in those cells having an altered p53, among which are many tumor cell types (41). The cytolytic effect of Onyx-015 is mediated by interfering with the p53-dependent cell death/apoptosis pathway, which is required for productive infection (10,31). This observation suggests that this recombinant mutant virus might target tumor cells much more efficiently than normal cells because tumor cells with p53 mutations compensate for the alteration introduced within the 55-kDa viral protein, thus becoming permissible and allowing the viral life cycle to progress. Normal cells, such as endothelial cells, fibroblasts, and airway and mammary epithelial cells, have been shown to be highly resistant to lytic infection with Onyx-015 when compared to tumor cells of different origins. Cytopathic effects resulting from viral replication have been demonstrated in many human tumor cells of different histological types, and its efficacy was further increased when Onyx-015 was combined with chemotherapeutic drugs such as cisplatin or 5-fluorouracil (12).

Onyx-015 is now being tested in clinical trials for advanced head and neck tumors, and initial results appear to be promising when used in combination with chemotherapy (42).

Prostate-Specific Ad: Viral Targeting by Cell-Specific Promoters and Enhancers

Prostate cancer is one of the most common malignancies in men and is characterized by the expression of high levels of prostate-specific antigen (PSA). Ad-derived vectors are the most commonly used to target epithelial cells in GT protocols (43). The type of regulatory sequences present in the viral promoter-enhancers used is a major determinant of the cell-specific expression of a viral vector (17). Therefore, one could reason that viral gene expression could be targeted to a specific type of cell by using an appropriate combination of promoter-enhancer elements to express the exogenous gene from the viral vector. Such an approach has been developed recently by linking the expression of Ad *E1A* gene to a minimum promoter-enhancer derived from the *PSA* gene. Thus, the cytolytic life cycle of the resulting Ad (CN706) would be selectively restricted to cells expressing PSA and consequently could be used to treat prostate cancer (30). The viability of this prostate-specific Ad was tested by checking tumor formation in nude mice from xenografts of human tumor cell lines, which are positive or negative for PSA expression. The use of CN706 destroyed the tumors derived from PSA-positive cells, and expression of PSA antigen was abolished (30).

Experimental approaches based on the use of human cell-specific promoter-enhancers represent a novel strategy that could be of great interest in the design of tumor-specific vectors for GT since there are several antigens that are preferentially expressed in tumor cells, such as CEA (carcinoembryonic antigen) in colon cancer or MAGE (melanoma antigen) in melanoma.

Rhabdoviruses to Chase of HIV-1-Infected Cells

AIDS is one of the few diseases for which GT strategies are being developed. Knowledge of the biology of HIV-1-infected cells has led to the development of two promising approaches that use two viruses as therapeutic agents. One uses a modified form of rabies virus, and the other is based on the use of VSV. The recombinant forms of these two rhabdoviruses are capable of preferential replication and destruction of HIV-1-infected cells (26).

The mechanism for HIV-1 entry into a cell is determined by a cellular receptor that binds to viral gp120. After binding, another viral protein, gp41, mediates cell fusion. The cellular receptor is composed of two elements, CD4 and another co-receptor, which determines the specificity of the viral HIV-1 strain. This co-receptor is a chemokine receptor. The chemokine co-receptor CXCR4 is used in T cells, CCR5 in macrophages, and CCR2b or CCR3 in other cells susceptible to HIV-1 infection. Based on this knowledge, a strategy has been developed to interfere with receptor binding. As a consequence, HIV-1 cannot evade the block by mutating because this will also mean the loss of its ability to infect cells.

The first approach consists of a virus that expresses CD4 and the chemokine receptor CXCR4 as envelope proteins. Thus, the therapeutic virus will infect cells that contain HIV-1 and will eliminate them during its lytic cycle. To develop a virus with these characteristics, Mebatsion et al. (21) infected cells producing CD4 and CXCR4 with a mutant rabies virus that lacks its envelope G protein. During viral replication of this virus in the manipulated cells, the new virus incorporated in its envelope the cellular proteins CD4 and CXCR4 (21). This virus was capable of infecting cells expressing the HIV-1 gp120 and gp41 on its surface and carry out their lytic cycle. A major limitation of this vector is that it can only infect cells in a single round, thus HIV-1 infection might persist if a single infected cell survives.

In a further development of this initial approach, Schnell et al. used a vector in which the G envelope protein was replaced by a cassette that expresses the coreceptor pair CD4 and CXCR4 (33). This vector was capable of replicating and propagating in infected cells and of spreading out to infect more HIV-1-producing cells. The spread of the recombinant therapeutic virus should stop when no more HIV-1-infected cells are available. Therefore, it cannot propagate.

Although these strategies have not been applied yet to AIDS patients, they have opened up a new way to treat HIV-1 infection by using viral vectors designed to follow it wherever it is driving productive infection. As our knowledge of the nature and characteristics of HIV-1 co-receptors improves, more methods for targeting and eliminating the infected cells could be developed.

PERSPECTIVES

Future progress in the development of viral vectors for GT will depend on three factors. Recombinant viral GT vectors will combine elements derived from different viruses to tailor a higher selectivity in cell targeting. These viral vectors will contain complex promoter-enhancer regulatory elements to achieve the correct level of expression and physiological regulation of the therapeutic gene. Finally, progress in the isolation and manipulation of stem cells, which could reconstitute entire cell lineages, will open up a new era for GT protocols and strategies. It is reasonable to expect that by ex vivo manipulation of stem cells, many different hereditary disorders could be corrected. Such a possibility has been demonstrated by the correction of the immunodeficiency in NOD/SCID mice by the use of human $CD34^+$ cells tagged by transduction with an HIV-based vector expressing a fluorescent protein (24).

ACKNOWLEDGMENTS

Funded by grants from Fondo de Investigación Sanitaria (FIS98/0313), Comunidad de Madrid (8.1/0004/98), and CICYT (SAF 00/169).

REFERENCES

1. **Arai, T., K. Matsumoto, K. Saitoh, M. Ui, T. Ito, M. Murakami, Y. Kanegae, I. Saito, F.L. Cosset, Y. Takeuchi, and H. Iba.** 1998. A new system for stringent, high-titer vesicular stomatitis virus G protein-pseudotyped retrovirus vector induction by introduction of Cre recombinase into stable prepackaging cell lines. J. Virol. *72*:1115-1121.
2. **Balliet, J.W. and P. Bates.** 1998. Efficient infection mediated by viral receptors incorporated into retroviral particles. J. Virol. *72*:671-676.
3. **Bischoff, J.R., D.H. Kirn, A. Williams, C. Heise, S. Horn, M. Muna, L. Ng, J.A. Nye, A. Sampson Johannes, and A. Fattey.** 1996. An adenovirus mutant that replicates selectively in p53 deficient human tumor cells. Science *274*:373-376.
4. **Boucher, R.C.** 1999. Status of gene therapy for cystic fibrosis lung disease. J. Clin. Invest. *103*:441-445.
5. **Burns, J.C., T. Friedmann, W. Driever, M. Burrascano, and J. Yee.** 1993. Vesicular stomatitis virus G glycoprotein pseudotyped retroviral vectors: concentration to very high titer and efficient gene transfer into mammalian and non-mammalian cells. Proc. Natl. Acad. Sci. USA *90*:8033-8037.
6. **Coffey, M.C., J.E. Strong, P.A. Forsyth, and P.W.K. Lee.** 1998. Reovirus therapy of tumors with activated ras pathway. Science *282*:1332-1334.
7. **Davies, J.C., D.M. Geddes, and E.W.F.W. Alton.** 1998. Prospects for gene therapy for cystic fibrosis. Mol. Med. Today *4*:292-299.
8. **During, M.J. and L.A. Ashenden.** 1998. Towards gene therapy for the central nervous system. Mol. Med. Today *4*:485-493.
9. **Gros, L., L. Montoliu, E. Riu, L. Lebrigand, and F. Bosch.** 1997. Regulated production of mature insulin by non-B-cells. Hum. Gene Ther. *8*:2249-2259.
10. **Hall, A.R., B.R. Dix, S.J. O'Carroll, and A.W. Braithwaite.** 1998. p53-dependent cell death/apoptosis is required for a productive adenovirus infection. Nat. Med. *4*:1068-1072.
11. **Hansen, J.E., G.J. Gram, S.D. Nielsen, A. Sorensen, P.B. Jensen, M. Sehested, J.O. Nielsen, and M. Rorth.** 1997. Transduction potential of human retroviruses in highly proliferating small-cell lung cancer cells as well as non-proliferating hematopoietic stem cells. APMIS *105*:723-729.
12. **Heise, C., A. Sampson-Johannes, A. Williams, F. McCormick, D.D. Von Hoff, and D.H. Kirn.** 1997. ONYX-015, an E1B gene-attenuated adenovirus, causes tumor-specific cytolysis and antitumoral efficacy that can be augmented by standard chemotherapeutic agents. Nat. Med. *3*:639-645.
13. **Herzog, R.W., E.Y. Yang, L.B. Couto, J.N. Hagstrom, D. Elwell, P.A. Fields, M. Burton, D.A. Bellinger, M.S. Read, K.M. Brinkhous, et al.** 1999. Long-term correction of canine hemophilia B by gene transfer of blood coagulation factor IX mediated by adeno-associated viral vector. Nat. Med. *5*:56-63.
14. **Hidaka, C., E. Milano, P.L. Leopold, J.M. Bergelson, N.R. Hackett, R.W. Finberg, T.J. Wickham, I. Kovesdi, P. Roelvink, and R.G. Crystal.** 1999. CAR-dependent and CAR-independent pathways of adenovirus vector-mediated gene transfer and expression in human fibroblasts. J. Clin. Invest. *103*:579-587.
15. **Jiang, Q. and J.F. Engelhardt.** 1998. Cellular heterogeneity of CFTR expression and function in the lung: implications for gene therapy of cystic fibrosis. Eur. J. Hum. Genet. *6*:12-31.
16. **Kovesdi, I., D.E. Brough, J.T. Bruder, and T.J. Wickham.** 1997. Adenoviral vectors for gene transfer. Curr. Opin. Biotechnol. *8*:583-589.
17. **Lazo, P.A.** 1985. Pathogenesis of viral infections. N. Engl. J. Med. *312*:1574.
18. **Levitzki, A.** 1997. Targeting signal transduction for disease therapy. Med. Oncol. *14*:83-89.
19. **Linden, R.M. and S.L.C. Woo.** 1999. AAvant-garde gene therapy. Nat. Med. *5*:21-22.
20. **Mathias, P., M. Galleno, and G.R. Nemerow.** 1998. Interactions of soluble recombinat integrin alphaV beta5 with human adenoviruses. J. Virol. *72*:8669-8675.
21. **Mebatsion, T., S. Finke, F. Weiland, and K. Conzelmann.** 1997. A CXCR4/CD4 pseudotype rhabdovirus that selectively infects HIV-1 envelope protein-expressing cells. Cell *90*:841-847.

22.**Mineta, T., S.D. Rabkin, T. Yazaki, W.D. Hunter, and R.L. Martuza.** 1995. Attenuated multi-mutated herpes simplex virus-1 for the treatment of malignant gliomas. Nat. Med. *1*:938-943.

23.**Mitanchez, D., B. Doiron, R. Chen, and A. Kahn.** 1997. Glucose-stimulated genes and prospects of gene therapy for type I diabetes. Endocrinol. Rev. *18*:520-540.

24.**Miyoshi, H., K.A. Smith, D.E. Mosier, I.M. Verma, and B.E. Torbett.** 1999. Transduction of human CD34+ cells that mediate long-term engraftment of NOD/SCID mice by HIV vectors. Science *283*:682-686.

25.**Morsy, M.A. and C.T. Caskey.** 1999. Expanded-capacity adenoviral vectors—the helper dependent vectors. Mol. Med. Today *5*:18-24.

26.**Nolan, G.P.** 1997. Harnessing viral devices as pharmaceuticals: fighting HIV-1's fire with fire. Cell *90*:821-824.

27.**Qing, K., C. Mah, J. Hansen, S. Zhou, V. Dwarki, and A. Srivastava.** 1999. Human fibroblast growth factor receptor 1 is a co-receptor for infection by adeno-associated virus 2. Nat. Med. *5*:71-77.

28.**Reynolds, P.N., M. Feng, and D.T. Curiel.** 1999. Chimeric viral vectors—the best of both worlds. Mol. Med. Today *5*:25-31.

29.**Robbins, P.D. and S.C. Ghivizzani.** 1998. Viral vectors for gene therapy. Pharmacol. Ther. *80*:35-47.

30.**Rodriguez, R., E.R. Schuur, H.Y. Lim, G.A. Henderson, J.W. Simons, and D.R. Henderson.** 1997. Prostate attenuated replication competent adenovirus (ARCA) CN706: a selective cytotoxic for prostate-specific antigen-positive prostate cancer cells. Cancer Res. *57*:2559-2563.

31.**Rothmann, T., A. Hengstermann, N.J. Whitaker, M. Scheffner, and H. zur Hausen.** 1998. Replication of ONYX-015, a potential anticancer adenovirus, is independent of p53 status in tumor cells. J. Virol. *72*:9470-9478.

32.**Samulski, R.J.** 1993. Adeno-associated virus: integration at specific chromosomal locus. Curr. Opin. Genet. Dev. *3*:74-80.

33.**Schnell, M.J., J.E. Johnson, L. Buonocore, and J.K. Rose.** 1997. Construction of a novel virus that targets HIV-1-infected cells and controls HIV-1 infection. Cell *90*:849-857.

34.**Snyder, R.O., C. Miao, L. Meuse, J. Tubb, B.A. Donahue, H.-F. Lin, D.W. Stafford, S. Patel, A.R. Thompson, T. Nichols, et al.** 1999. Correction of hemophilia B in canine and murine models using recombinat adeno-associated viral vectors. Nat. Med. *5*:64-70.

35.**Strong, J.E., M.C. Coffey, D. Tang, P. Sabinin, and P.W.K. Lee.** 1998. The molecular basis of viral oncolysis: usurpation of the ras signaling pathway by reovirus. EMBO J. *17*:3351-3362.

36.**Summerfeld, C., J.S. Bartlett, and R.J. Samulski.** 1999. AlphaVbeta5 integrin: a co-receptor for adeno-associated virus type 2 infection. Nat. Med. *5*:78-82.

37.**Tsichlis, P.N. and P.A. Lazo.** 1991. Virus-host interactions and the pathogenesis of murine and human oncogenic retroviruses. Curr. Top. Microbiol. Immunol. *171*:95-170.

38.**Varley, A.W., S.M. Geiszler, R.B. Gaynor, and R.S. Munford.** 1997. A two-component expression system that responds to inflammatory stimuli in vivo. Nat. Biotechnol. *15*:1002-1006.

39.**Varley, A.W. and R.S. Munford.** 1998. Physiologically responsive gene therapy. Mol. Med. Today *4*:445-451.

40.**Verma, I.M. and N. Somla.** 1997. Gene therapy: promises, problems and prospects. Nature *389*:239-242.

41.**Vogelstein, B. and K.W. Kinzler.** 1998. The Genetic Basis of Human Cancer, 1st ed. McGraw-Hill, New York.

42.**Von Hoff, D.D.** 1998. Promising new agents for treatment of patients with colorectal cancer. Semin. Oncol. (Suppl. 11) *25*:47-52.

43.**Wilson, J.M.** 1996. Adenovirus as gene-delivery system. N. Eng. J. Med. *338*:1185-1187.

44.**Yazaki, T., H.J. Manz, S.D. Rabkin, and R.L. Martuza.** 1995. Treatment of human malignant meningiomas by G207, a replication- competent multimutated herpes simplex virus 1. Cancer Res. *55*:4752-4756.

37 Biosafety Recommendations for Working with Viral Vectors

Angel Cid-Arregui[1,2] and Victoria Juárez[3]
[1]Institute of Applied Tumor Virology, German Cancer Research Center, Heidelberg; [2]Max Planck Institute of Molecular Cell Biology and Genetics, Dresden; [3]European Molecular Biology Laboratory, Heidelberg, Germany

INTRODUCTION

Experience has demonstrated that experimental work performed in biology laboratories does not represent a serious risk for personnel, the environment, or the community. Most laboratory work can be carried out safely with a minimum of prudence, albeit with strict control of dangerous materials and adequate disposal of contaminated waste (5,8). Routine precautions, however, would not suffice when highly infectious agents or potentially dangerous biologic materials are to be handled. Genetically modified organisms, together with pathogens and infected laboratory animals, tissues, and cultures are major biohazard sources. Therefore, since the early 1970s, guidelines have been published for research using recombinant DNA (3,4,7) and for microbiological and biomedical practices. These include the concepts of containment and biosafety levels (BSL) (2), which have been defined to establish recommendations that should help prevent accidents causing biological hazard, which might cause infection of personnel or release of dangerous biological materials to the environment.

Work with viral vectors should involve consideration of risks associated with both recombinant DNA and infectious materials. This chapter describes biosafety recommendations for laboratory work using viral vectors in vitro and in laboratory animals, which follow those in the publication *Biosafety in Microbiological and Biomedical Laboratories* (2).

BIOSAFETY AND BSL

The term *biosafety* covers those special regulations, facilities, equipment, materials, and practices directed to prevent accidents that may threaten the security of laboratory personnel and the community and/or may lead to release of contaminated material to the environment. The concept of containment has been introduced to define facilities, equipment, and safe meth-

Viral Vectors: Basic Science and Gene Therapy
Edited by A. Cid-Arregui and A. García-Carrancá

ods for managing infectious agents in the laboratory (2). Two levels of containment are considered: *(i)* primary containment, which is intended to protect personnel and the laboratory environment, is provided by an appropriate practice and the use of adequate safety equipment, and *(ii)* secondary containment, required to protect the external environment and hence the community, is provided by special laboratory practices and facility construction.

This section describes recommendations for BSL-1–4 following those of the *Centers for Disease Control and Prevention and the National Institutes of Health* (2). Since routine work with the viral vectors described in this book involves BSL-1 and BSL-2, only these are described in detail here. In general, experiments in which DNA from pathogenic agents is transferred into viral vectors may be performed under BSL-2 containment. However, experiments in which DNA from dangerous pathogens is transferred into viral vectors should be performed under BSL-3 containment unless irreversible inactivation of pathogenicity can be demonstrated. In addition, for some studies, experiments may be designed that could either enhance pathogenicity (e.g., by insertion of oncogenes into their genome) or alter the host range of viral vectors (e.g., by modifying envelope proteins to make virions interact with ubiquitous receptors). In such cases, serious consideration should be given to using BSL-3 or BSL-4, and the applicable recommendations should be consulted elsewhere (1,2).

BSL-1

Applicability

This level is recommended for work with well-known agents that do not cause disease in healthy adult humans and that are of minimal potential hazard to the environment.

General Recommendations

1. The laboratory does not need to be located in an area of restricted traffic within the building. Access to the laboratory is restricted at the discretion of the laboratory director.
2. A biosafety manual should be prepared or adopted.
3. Laboratory personnel should be trained in the operation of equipment and in the techniques and procedures used in the laboratory and should be supervised by experienced scientists.
4. Eating, drinking, smoking, handling contact lenses, and applying topical treatments or cosmetics should not be allowed in the laboratory. Food and drinks should never be stored in or transported through the laboratory.
5. An insect and rodent control program should be active.

Facilities

6. The laboratory should be designed and maintained so that it can be easily cleaned. Refrigerators, tabletop centrifuges, and any other laboratory equipment should be placed so that they do not hinder regular cleaning; they should be maintained in clean condition.
7. Bench tops and furniture should be impervious to water and resistant to acids, alkalis, organic solvents, and moderate heat. Bench and shelf surfaces should be cleaned and disinfected whenever necessary and cleaned periodically with an alcohol-based disinfectant.
8. If the room has windows that open, they should be fitted with fly screens.

9. The laboratory should contain a sink and materials for hand washing and disinfecting, as well as an eye wash facility.

Safety Equipment

10. Special containment equipment is not required.
11. Laboratory coats or uniforms and clogs should be worn to prevent contamination of street clothes.
12. Gloves and protective eyewear should be worn to prevent contamination through splashes of microorganisms or other hazardous materials.

Laboratory Practices

13. Personnel should conduct procedures carefully to avoid creation of splashes and/or aerosols, and should wash their hands after handling materials and animals, after removing gloves, and before leaving the laboratory. Mouth pipetting should not be allowed; mechanical devices should be used instead.
14. Contaminated solid wastes should be collected in plastic bags and transported in leakproof metal containers to the autoclave. Contaminated fluids should be collected in leakproof autoclavable plastic flasks. Materials to be decontaminated offsite should be packaged and transported according to local, state, and federal regulations.

BSL-2

Applicability

BSL-2 is recommended for work involving agents of moderately potential hazard to personnel and the environment.

General Recommendations

1. Access to BSL-2 areas is restricted to authorized personnel who have specific training in handling pathogenic agents and in the techniques in use in the laboratory.
2. Access to the laboratory is restricted at the discretion of the laboratory director, who should establish policies that determine the requirements for authorization to enter the BSL-2 facilities. Immunodeficient persons should in no case be allowed to enter the BSL-2 area.
3. A biosafety manual should be prepared or adopted, and personnel should be trained in potential hazards, exposure evaluation procedures, and exposure prevention measurements, according to the risks and with the necessary updates.
4. Personnel should be tested for the agents handled in the laboratory and, when appropriate, should be vaccinated. Serum samples of at-risk personnel may be taken and stored before starting in the laboratory, and additional serum samples may be taken periodically.
5. Eating, drinking, smoking, chewing gum, handling contact lenses, and applying topical treatments or cosmetics should not be allowed in the laboratory. Food and drinks should never be stored in or transported through working areas.
6. Extreme precaution should be taken with contaminated wastes, materials, and in particular contaminated sharp items.
7. Physical containment equipment should be used.
8. An insect and rodent control program should be active.

Facilities

9. A hazard warning sign, labeled with the universal biohazard symbol, should be posted on the access door(s) to the laboratory. The warning should identify the infectious agent(s) in use, the responsible persons with their telephone numbers, and the special requirements for entering the room.
10. The laboratory should be constructed and maintained to facilitate cleaning. Materials and equipment should be placed so that they do not hinder regular cleaning. In addition, equipment should be cleaned and disinfected periodically and whenever necessary.
11. The laboratory should not contain desks or offices, which should be placed in a different, non-BSL-2 room.
12. Bench tops and furniture should be impervious to water and resistant to acids, alkalis, organic solvents, and moderate heat.
13. Preferably, the room should not have windows that open. If it does, they should be fitted with fly screens.
14. The laboratory should contain a sink and materials for hand washing and disinfecting, as well as an eye wash facility.

Safety Equipment

15. Biological safety cabinets (BSCs), preferably Class II (6), should be used for handling of concentrated stocks of infectious agents and for procedures that may generate infectious aerosols or splashes.
16. The facility should contain a small autoclave used exclusively for decontaminating wastes and materials.
17. Personnel should wear laboratory coats or uniforms and clogs, which should be used exclusively in the BSL-2 facility. Protective clothing should be laundered by the institution and in no case should be taken home by personnel.
18. Gloves and face protection should be worn to prevent contamination through splashes and aerosols, in particular when the microorganisms have to be manipulated outside the BSCs. It is advisable to use two pairs of gloves. Gloves should always be considered contaminated material, and, therefore, they should be disposed of as such and never be reused or worn outside the laboratory.

Laboratory Practices

19. Personnel should conduct procedures carefully to avoid creating splashes and/or aerosols. They should wash their hands after handling materials and animals, after removing gloves, and before leaving the laboratory. Mouth pipetting should not be allowed; mechanical devices should be used instead.
20. Preferably, disposable plastic pipettes should be used for cell culture. Only disposable cell scrapers for removing adherent cells should be used. Scraping for collection of infected cells should be carefully performed in a BSC Class II. Tissue culture dishes and flasks should be placed into a plastic box inside the incubator to avoid any leakage contaminating nonremovable parts. Tissue culture hoods and incubators should be maintained clean and free of contamination.
21. Work surfaces and laboratory equipment should be decontaminated on

a regular basis; after work with infectious materials; and immediately after contamination by infectious materials through splashes, spills, or other accident. Equipment that needs to be repaired should be decontaminated according to institutional, local, state, and federal regulations before it can be removed from the facility and sent away.

22. Contaminated solid wastes should be collected in double plastic bags and transported into leakproof metal containers to the autoclave. Contaminated media and buffers should be collected in leakproof autoclavable plastic flasks.
23. Contaminated sharp materials should be handled with extreme caution.
    - *(i)* The use of needles, scalpels, and glass items (e.g., slides, coverslips, pipettes, and glassware) should be restricted to those applications where there is no alternative.
    - *(ii)* Only disposable syringe-needle units, as well as disposable scalpels and razor blades should be used; once contaminated they should not be manipulated or reused.
    - *(iii)* If infectious agents have to be microinjected (e.g., into the cytoplasm of cells in culture or into the amniotic cavity of postimplantation embryos maintained in vitro) extreme care should be taken with the microinjection capillaries.
        - (a) When an automatic microinjector is used, the output pressure must be set to zero before a loaded capillary is connected to the adapter;
        - (b) Loaded capillaries should be fitted very carefully to the adapter wearing two pairs of gloves;
        - (c) Before removing the capillary, the output pressure of the device should be set to zero;
        - (d) A plastic screen adapted to the microscope binocular should protect the user from inoculation in case the capillary is accidentally shot by the microinjecting device.
    - *(iv)* Sharp items should be carefully discarded into hard-walled containers and decontaminated by autoclaving.
    - *(v)* Broken glassware should not be removed by hand, but by mechanical means.
24. Contaminated wastes should be decontaminated as soon as possible before disposal according to local, state, and federal regulations. Nondisposable contaminated materials should be decontaminated in the autoclave of the laboratory or transported to a decontamination area in hard-walled autoclavable containers. If materials are to be decontaminated off-site, they should be packaged and transported according to local, state, and federal regulations.
25. Animal carcasses should be introduced into leakproof plastic bags and stored in a freezer until they are transported to be incinerated in leakproof containers.
26. Accidents resulting in exposure to infectious agents should be immediately reported to the laboratory director, who should ensure that medical advice and treatment is provided and that appropriate measures for decontamination are taken. Written records should be kept.

BSL-3

This level is applicable to facilities in which work is done with infectious agents transmissible by inhalation that may cause severe, potentially lethal disease. Safety recommendations comprise all those of BSL-2 and additional ones concerning location, design and access to the laboratory, training of laboratory personnel, as well as safety clothes and equipment.

BSL-4

This level is required for work with extremely hazardous infectious agents that convey high risk of aerosol transmission and cause disease associated with elevated mortality rates. Safety recommendations comprise all those of BSL-3 and additional ones concerning location, design and access to the laboratory, training of laboratory personnel, as well as personal protective clothes and equipment.

ANIMAL BIOSAFETY LEVELS (ABSLs)

In general, experiments involving infections of commonly used laboratory animals (vertebrates) with viruses or viral vectors, require biosafety levels comparable to those recommended for in vitro experiments with the same agents (2). However, infected animals represent new hazards since they may spread infections to the rest of the colony and may be contagious to the animal handlers by producing aerosols and by scratching or biting. Ideally, animals used for studies on infectious agents should be maintained in ventilated cabinets, isolators, or individually ventilated cage racks.

ABSL-1

General Recommendations

1. Access to the animal facility should be restricted to authorized personnel that have been trained in handling infected animals and contaminated materials, and that are being updated on a regular basis and whenever considered necessary by the laboratory and animal facility directors or by themselves.
2. A biosafety manual should be prepared or edited.
3. Personnel should be trained in the operation of equipment and in the techniques and procedures used with the animals; they should be supervised by experienced scientists.
4. Eating, drinking, smoking, chewing gum, handling contact lenses, and applying topical treatments or cosmetics should not be allowed in animal rooms. Food and drinks should never be stored in or transported through working areas.
5. An insect and rodent control program should be active.

Facilities

6. The animal facility should be constructed and maintained to facilitate cleaning. Materials and equipment should be placed so that they do not hinder regular cleaning.
7. Working surfaces should be maintained clean and disinfected whenever necessary and periodically with an alcohol-based disinfectant.
8. The rooms should not have windows that open.
9. The facility should contain a sink and materials for hand washing and disinfecting, as well as an eye wash facility.

Safety Equipment

10. Special containment equipment is not required.
11. Laboratory coats or uniforms, clogs or overshoes, caps, and masks should be worn to prevent contamination of street clothes and minimize risk of coming in contact with infectious agents. Laboratory clothing should not be worn outside the animal facility.
12. Gloves and protective eyewear should be worn to prevent contamination through splashes of microorganisms or other hazardous materials.

Animal Laboratory Practices

13. Personnel should conduct procedures carefully to avoid creating splashes and/or aerosols. They should wash their hands before and after handling animals, after removing gloves, and before leaving the facility. Mouth pipetting should not be allowed; mechanical devices should be used instead.
14. Food and water dispensers, bedding materials, and any other elements of the animal cages should be removed carefully to minimize production of aerosols. Wastes from the cages should be dispensed according to institutional, local, state, and federal regulations. Cages should be decontaminated and washed manually or in a cage washer.
15. Contaminated solid wastes should be collected in plastic bags and transported in leakproof metal containers to the autoclave. Contaminated fluids should be collected in leakproof autoclavable plastic flasks. Materials to be decontaminated off-site should be packaged and transported according to local, state, and federal regulations.

ABSL-2

General Recommendations

1. Access to the animal facility is restricted at the discretion of the animal facility director in agreement with the laboratory director, who should establish policies that determine the requirements for authorization to enter the ABSL-2 facilities. Access should be restricted to authorized personnel who have been trained in handling pathogenic agents, infected animals, and contaminated materials.
2. A biosafety manual should be prepared or adopted. Personnel should be trained on potential hazards, exposure evaluation procedures, and exposure prevention measurements, according to the risks and with updates on a regular basis and whenever considered necessary by the laboratory and animal facility directors or by themselves.
3. Personnel should be tested for the agents handled in the laboratory and when appropriate should be vaccinated. Serum samples of at-risk personnel may be taken and stored before starting in the animal facility, and additional serum samples may be taken periodically. Immunodeficient persons should in no case be allowed to enter the ABSL-2 area.
4. Personnel should be trained in the operation of equipment and in the techniques and procedures used with the animals, and should be supervised by experienced scientists. Extreme precaution should be taken with contaminated wastes, materi-

als, and in particular contaminated sharp items.

5. Eating, drinking, smoking, chewing gum, handling contact lenses, and applying topical treatments or cosmetics should not be allowed in animal rooms. Food and drinks should never be stored in or transported through working areas.
6. An insect and rodent control program should be active.

Facilities

7. A hazard warning sign, labeled with the universal biohazard symbol, should be posted on the access door(s) to the laboratory. The warning should identify the infectious agent(s) in use, the responsible persons with their telephone numbers, and the special requirements for entering the room.
8. The animal facility should be constructed and maintained to facilitate cleaning. Materials and equipment should be placed so that they do not hinder regular cleaning.
9. The room should not have windows that open.
10. The animal facility should not contain desks or offices, which should be placed in a different, non-ABSL-2 room.
11. Working surfaces should be maintained clean and should be decontaminated after use and after any spill of contaminated tissue or fluid.
12. The facility should contain a sink and materials for hand washing and disinfecting, as well as an eye wash facility.

Safety Equipment

13. Special containment equipment is required. BSCs, preferably Class II, should be used for handling of concentrated stocks of infectious agents and for procedures that may generate infectious aerosols or splashes (e.g., inoculation of animals, dissection and necropsy of infected animals, collection of fluids, and tissue biopsies).
14. An autoclave for decontaminating infectious waste should be available in the animal facility.
15. Personnel should wear laboratory coats or uniforms and clogs, which should be of exclusive use for the ABSL-2 facility. Protective clothing should be laundered by the institution and in no case should be taken home by personnel.
16. Gloves and face protection (e.g., face shields and respirators) should be worn to prevent contamination through splashes and aerosols, in particular when the microorganisms have to be manipulated outside the BSCs. It is advisable to use two pairs of gloves. Gloves should always be considered contaminated material; therefore, they should be disposed as such and never be reused or worn outside the laboratory.

Animal Laboratory Practices

17. Personnel should conduct procedures carefully to avoid creating splashes and/or aerosols. They should wash their hands before and after handling animals and materials, after removing gloves, and before leaving the facility. Mouth pipetting should not be allowed; mechanical devices should be used instead.
18. Contaminated sharp materials should be handled with extreme caution.

(i) The use of needles, scalpels, sharp forceps, and glass items (e.g., slides, coverslips, pipettes, and glassware) should be restricted to those applications where there is no alternative.

(ii) Only disposable syringe-needle units, and disposable scalpels and razor blades should be used; once contaminated they should not be manipulated or reused.

(iii) Sharp items should be carefully discarded into hard-walled containers and decontaminated by autoclaving.

(iv) Broken glassware should never be removed by hand but rather by mechanical means.

19. Contaminated solid wastes, cultures, tissues, and specimens of body fluids should be collected in plastic bags and transported in leakproof metal containers to the autoclave. Contaminated media and buffers should be collected in leakproof autoclavable plastic flasks. Materials to be decontaminated off-site should be packaged and transported according to local, state, and federal regulations.
20. Contaminated wastes should be decontaminated as soon as possible before disposal according to local, state, and federal regulations. Nondisposable contaminated materials should be decontaminated in the autoclave of the laboratory or transported to a decontamination area in hard-walled autoclavable containers.
21. Cages should be decontaminated by autoclaving before they are washed.
22. Work surfaces and equipment should be decontaminated on a regular basis, after work with infectious materials; and immediately after contamination by infectious materials through splashes, spills, or other accident.
23. Contaminated equipment must be decontaminated before sending it for repair. Shipment should meet institutional, local, state, and federal regulations.
24. Accidents resulting in exposure to infectious agents should be immediately reported to the laboratory director, who should ensure that medical advice and treatment is provided and that appropriate measures for decontamination are taken. Written records should be kept.

REFERENCES

1. **Center for Disease Control, Office of Biosafety.** 1974. Classification of Etiologic Agents on the Basis of Hazard, 4th ed. US Department of Health, Education and Welfare, Public Health Service.
2. **Centers for Disease Control and Prevention and National Institutes of Health.** 1993. Biosafety in Microbiological and Biomedical Laboratories, 3rd ed. US Government Printing Office, Washington, DC.
3. **Federal Register.** 1976. Recombinant DNA Research Guidelines. *41*:27902-27943.
4. **Federal Register.** 1986. Guidelines for Research Involving Recombinant DNA Molecules. *51*:16958-16968.
5. **National Research Council.** 1989. Biosafety in the Laboratory: Prudent Practices for the Handling and Disposal of Infectious Materials. National Academy Press, Washington, DC.
6. **National Sanitation Foundation Standard 49.** 1992. Class II (Laminar Flow) Biohazard Cabinetry. Ann Arbor, MI.
7. **Office of Research Safety, National Cancer Institute, and the Special Committee of Safety and Health Experts.** 1978. Laboratory Safety Monograph: A Supplement to the NIH Guidelines for Recombinant DNA Research. National Institutes of Health, Bethesda, MD.
8. **US Environmental Protection Agency.** 1986. EPA guide for infectious waste management. US Environmental Protection Agency, Washington, DC. Publication EPA/530-5W-86-014.

RECOMMENDED READING

National Committee for Clinical Laboratory Standards (NCCLS). 1991. Protection of Laboratory workers from infectious disease transmitted by blood, body fluids, and tissue. Tentative Guideline. M29-T2, Vol. 11, No. 14.

Rutala, W.A. 1990. APIC guideline for selection and use of disinfectants. Am. J. Infect. Control *18*:99-117.

INFORMATIONAL WEB SITES

Massachussetts Institute of Technology
http://web.mit.edu/bso/www/bmbl/

Biosafety in Microbiological and Biomedical Laboratories
Health and Human Services Publication (CDC) 93-8395
http://siri.uvm.edu/library/cdc/html

Office of Health and Safety, Centers for Disease Control and Prevention Biosafety Homepage
http://www.cdc.gov/od/ohs/biosfty/biosfty.htm

American Biological Safety Association
http://www.absa.org/

Virtual library biotechnology catalog
http://www.cato.com/biotech/

Health Canada Laboratory Biosafety Guidelines
http://www.hc-sc.gc.ca/hpb/lcdc/biosafty/docs/index.html

University of Toronto Biosafety Program
http://www.toronto.edu/safety/bioshome.htm

Genetic Manipulation Advisory Committee, Australia
http://www.health.gov.au/tga/gene/gmac/gmac.htm

UK Advisory Committee on Releases to the Environment
http://www.shef.ac.uk/~doe/

Japanese Science and Technology Agency
http://www.sta.go.jp/index-e.html

European Biosafety Association
http://www.ebsa.be/

Database on Biosafety and Risk Assessment in Biotechnology
http://www.icgeb.trieste.it/biosafety/bsfdata1.htm

INDEX

Note: Page numbers with an "f" or a "t" indicate references to figures and tables, respectively.

A

B

C

D

E

G

Q

R

S

U

V